Studies in Economic Theory

Editors

Titles in the Series

M. A. Khan and *N. C. Yannelis* (Eds.)
Equilibrium Theory
in Infinite Dimensional Spaces

C. D. Aliprantis, K. C. Border
and *W. A. J. Luxemburg* (Eds.)
Positive Operators, Riesz Spaces, and Economics

D. G. Saari
Geometry of Voting

C. D. Aliprantis and *K. C. Border*
Infinite Dimensional Analysis

J.-P. Aubin
Dynamic Economic Theory

M. Kurz (Ed.)
Endogenous Economic Fluctuations

J.-F. Laslier
Tournament Solutions and Majority Voting

A. Alkan, C. D. Aliprantis and *N. C. Yannelis*
(Eds.)
Theory and Applications

J. C. Moore
Mathematical Methods for Economic Theory 1

J. C. Moore
Mathematical Methods for Economic Theory 2

M. Majumdar, T. Mitra and *K. Nishimura*
Optimization and Chaos

K. K. Sieberg
Criminal Dilemmas

M. Florenzano and *C. Le Van*
Finite Dimensional Convexity and Optimization

K. Vind
Independence, Additivity, Uncertainty

T. Cason and *C. Noussair* (Eds.)
Advances in Experimental Markets

F. Aleskerov and *B. Monjardet*
Utility Maximization. Choice and Preference

N. Schofield
Mathematical Methods in Economics
and Social Choice

C. D. Aliprantis, K. J. Arrow, P. Hammond,
F. Kubler, H.-M. Wu and *N. C. Yannelis* (Eds.)
Assets, Beliefs, and Equilibria in Economic Dynamics

Dionysius Glycopantis
Nicholas C. Yannelis
Editors

Differential Information Economies

Springer

Professor Dionysius Glycopantis
City University
Department of Economics
Northhampton Square
London EC1V OHB
Great Britain
Email: dyglcopantis@yahoo.co.uk

Professor Nicholas C. Yannelis
University of Illinois
Department of Economics
330 Commerce West Building
Champaign, IL 61820, USA
Email: nyanelli@uiuc.edu

Parts of the papers of this volume have been published in the journal *Economic Theory*.

Cataloging-in-Publication Data applied for

Library of Congress Control Number: 2004114699

A catalog record for this book is available from the Library of Congress.

Bibliographic information published by Die Deutsche Bibliothek
Die Deutsche Bibliothek lists this publication in the Deutsche Nationalbibliografie;
detailed bibliographic data available in the internet at *http://dnb.ddb.de*

ISBN 3-540-21424-0 Springer Berlin Heidelberg New York

Springer is a part of Springer Science+Business Media
springeronline.com

Printed in Germany

Cover design: Erich Kirchner, Heidelberg
Production: Helmut Petri
Typesetting: Steingraeber
Printing: betz-druck

SPIN 10984734 Printed on acid-free paper – 42/3130 – 5 4 3 2 1 0

Preface

It is not an exaggeration to say that all economic activity or all contracts among individuals in a society are made under conditions of uncertainty or incomplete information. Indeed, the need to introduce uncertainty in the classical Walrasian equilibrium model, was felt by Arrow and Debreu (1954) and it is explicitly modeled in Chapter 7 of the Theory of Value of Debreu (1959).

In Chapter 7 of the well-known treatise of Debreu (1959), it is suggested that once preferences and/or initial endowments are state dependent, i.e. they depend on a finite number of states of nature of the world, and agents, who are completely informed, maximize ex ante expected utility, then all the results on the existence and optimality of the Walrasian equilibrium continue to hold.

Radner (1968) went a step further, by introducing asymmetric or differential information into the Arrow–Debreu model. In particular, he assigned to each agent, in addition to his/her random initial endowment and random utility function, a private information set, which is a measurable partition of the exogenously given probability measure space (which describes the states of nature of the world).

Radner (1968) noticed that if all net trades reflect the private information of each agent, (i.e. they are measurable with respect to σ-algebra that his/her partition generates), then, again the standard existence and optimality results of the Walrasian equilibrium concept continue to hold, although some prices might be negative. This *Walrasian expectations equilibrium* notion (or *Radner equilibrium*) is an ex ante notion, and captures the idea of contracts under asymmetric information.

A related concept, called "*rational expectations equilibrium*" (REE), was also studied, e.g., Kreps (1977), Radner (1979), Allen (1981), among others. The REE is an interim notion and in this set up agents maximize (interim) expected utility functions, conditioned on their own private information as well as the information that the equilibrium prices generate.

Given the fact that there is asymmetric information, a variety of equilibrium concepts could be put forward. We find the condition of the measurability of an allocation for REE as the best way to proceed. Otherwise the agents calculate their demands without being certain about their initial endowments realization.

Unlike, the Radner equilibrium, the REE need not exist in well behaved economies, as shown by Kreps (1977), and may not be Pareto optimal, unless utility functions are state independent, and also it may not be Bayesian incentive compatible and may not be implementable as a perfect Bayesian equilibrium[1] of an extensive form game, (Glycopantis–Muir–Yannelis (2003)).

[1] Definitions of all concepts in this preface are given in the introductory chapter in this volume.

The above two notions, i.e., Radner equilibrium and REE are non-cooperative solution concepts, i.e., agents maximize ex ante or interim utility functions subject to their own budget constraint and their own, initial or eventual, private information constraint, independently of each other, and without sharing their own private information.

In seminal papers Wilson (1978) and Myerson (1984) introduced differential information in the core and the Shapley value respectively. Notice, that once cooperation is allowed, then a basic problem which arises is how the private information will be shared among the agents in a coalition. For example, *pooling* of information may not be the best alternative for an agent who is well informed and is expected to cooperate with a non-well informed agent. Also, using *common knowledge* information within a coalition may not be such a great idea for a well informed agent who cannot take advantage of his/her fine private information.

To put it differently the issue of incentive compatibility of the information asymmetries becomes a real problem that needs to be addressed. After all, agents do not want to be cheated, and at the same time they would like to write efficient contracts. This of course poses the following question: *Is it possible for agents to write incentive compatible and Pareto optimal contracts?* Let us answer this question by considering a simple two agents example.

Example 0.1 There are two Agents, 1 and 2, and three equally probable states of nature denoted by a, b, c and one good per state denoted by x. The utility functions, initial endowments and private information sets are given as follows:

$$
\begin{aligned}
u_1(w, x_1) &= \sqrt{x_1}, && \text{for } w = a, b, c \\
u_2(w, x_2) &= \sqrt{x_2}, && \text{for } w = a, b, c \\
e_1(a, b, c) &= (10, 10, 0), && \mathcal{F}_1 = \{\{a, b\}, \{c\}\} \\
e_2(a, b, c) &= (10, 0, 10), && \mathcal{F}_2 = \{\{a, c\}, \{b\}\}.
\end{aligned}
$$

Notice that a “fully”, pooled information, Pareto optimal, (i.e. a weak fine core outcome) is

$$
\begin{aligned}
x_1(a, b, c) &= (10, 5, 5) \\
x_2(a, b, c) &= (10, 5, 5).
\end{aligned} \tag{1}
$$

However, this outcome is *not* incentive compatible because if the realized state of nature is a, then Agent 1 has an incentive to report that it is state c, (notice that Agent 2 cannot distinguish state a from state c) and become better off. In particular, Agent 1 will keep her initial endowment in the event $\{a, b\}$ which is 10 units and receive another 5 units from Agent 2, in state c, (i.e., $u_1(e_1(a) + x_1(c) - e_1(c)) = u_1(15) > u_1(x(a)) = 10$) and becomes better off. Obviously Agent 2 is worse off. Similarly, Agent 2 has an incentive to report b when he observes $\{a, c\}$

This example demonstrates that *“full or ex post Pareto optimality” is not necessarily compatible with incentive compatibility.*

Most importantly, as it is known from Krasa–Yannelis (1994), the individual measurability of allocations in the one good case characterizes incentive compatibility. Thus, the only candidate in the above example for an incentive compatible allocation is the initial endowment which is dominated by the allocation in (1). Consequently, full Pareto optimal and incentive compatible allocations *need not exist* as the above example demonstrates.[2]

Thus, if we were to produce positive existence results for cooperative solution concepts which guarantee incentive compatibility we should not insist on full Pareto optimality but some "constrained informational" Pareto optimality. Indeed, by defining cores and values in differential information economies imposing measurability constraints, one is able to prove existence and incentive compatibility of cooperative solution concepts, (e.g. Yannelis (1991) and Krasa–Yannelis (1994)).

The following example will illustrate the role of the *private information measurability of an allocation.*

Example 0.2 There are two Agents, 1 and 2, two goods denoted by x and y and two equally probable states denoted by $\{a, b\}$. The agents' characteristics are:

$$\begin{aligned} &u_1(w, x_1, y_1) = \sqrt{x_1 y_1}, && \text{for } w = a, b \\ &u_2(w, x_2, y_2) = \sqrt{x_2 y_2}, && \text{for } w = a, b \\ &e_1(a, b) = ((10, 0), (10, 0)), && \mathcal{F}_1 = \{a, b\} \\ &e_2(a, b) = ((10, 8), (0, 10)), && \mathcal{F}_2 = \{\{a\}, \{b\}\}. \end{aligned}$$

The feasible allocation below is Pareto optimal (interim, ex post and ex ante).

$$\begin{aligned} ((x_1(a), y_1(a)), (x_1(b), y_1(b))) &= ((5, 4), (5, 5)) \\ ((x_2(a), y_2(a)), (x_2(b), y_2(b))) &= ((15, 4), (5, 5)). \end{aligned} \tag{2}$$

However, the allocation in (2) above is not incentive compatible because if b is the realized state of nature Agent 2 can report state a and become better off, i.e.,

$$\begin{aligned} u_2(e_2(b) + (x_2(a), y_2(a)) - e_2(a)) &= u_2((0, 10) + (15, 4) - (10, 8)) \\ &= u_2(5, 6) > u_2(x_2(b), y_2(b)) = u_2(5, 5). \end{aligned}$$

Notice that the allocation in (2) is *not* $\mathcal{F}_1$-measurable (i.e., measurable with respect to the private information of Agent 1). Hence, *an individually rational, efficient (interim, ex ante, ex post) without the $\mathcal{F}_i$-measurability ($i = 1, 2$) condition need not be incentive compatible.*

Observe that one can restore the incentive compatibility simply by making the allocation in (2) above $\mathcal{F}_i$-measurable for each i, ($i = 1, 2$). In particular, the $\mathcal{F}_i$-measurable allocation below is incentive compatible, and private information ($\mathcal{F}_i$-measurable) Pareto optimal.

$$\begin{aligned} (x_1(a), y_1(a)), (x_1(b), y_1(b)) &= ((5, 5), (5, 5)) \\ (x_2(a), y_2(a)), (x_2(b), y_2(b)) &= ((15, 3), (5, 5)). \end{aligned}$$

The importance of the measurability condition in restoring incentive compatibility and of course guaranteeing the existence of an optimal contract is obvious in the above example and this approach was introduced by Yannelis (1991).

[2] This example is taken from Koutsougeras–Yannelis (1993). It should be noted that Prescott-Townsend (1984) observed that the set of Pareto optimal and incentive compatible allocations may not be convex and therefore need not exist. See also Allen (2003).

It is worth pointing out that two important, new features that distinguish the "partition approach" of modeling differential information (e.g., Radner (1968), Wilson (1978), among others) and the mechanism design or Harsanyi-type modeling approach (e.g., Myerson (1984), among others).

First, as the examples above indicated, initial endowments are random and therefore the definition of incentive compatibility is different than the one found in the mechanism design literature where initial endowments, if they are explicitly stated, are typically assumed to be constant.

Second, as the reader will observe, in several of the papers in this volume the incentive compatibility is coalitional rather than individual. It is not difficult to see by means of examples that contracts that are *individual incentive compatible may not be coalitional incentive compatible* and therefore may not be viable. We believe that the coalitional incentive compatibility is more appropriate for multilateral contracts. The following demonstrates this.

Example 0.3 Consider a three person differential information economy, with Agents 1, 2, 3, two goods denoted by x, y, and the three equal probable states are denoted by a, b, c. The agents' utility functions, random initial endowments and private information sets are as follows:

$$\begin{aligned} u_i(x_i, y_i) &= \sqrt{x_i y_i}, & i &= 1,2,3, \\ e_1(a,b,c) &= ((20,0),(20,0),(20,0)), & \mathcal{F}_1 &= \{a,b,c\} \\ e_2(a,b,c) &= ((0,10),(0,10),(0,5)), & \mathcal{F}_2 &= \{\{a,b\},\{c\}\} \\ e_3(a,b,c) &= ((10,10),(10,10),(20,30)), & \mathcal{F}_2 &= \{\{a\},\{b\},\{c\}\}. \end{aligned}$$

The allocation below is individual incentive compatible but *not* coalitional.

$$\begin{aligned} ((x_1(a),y_1(a)),(x_1(b),y_1(b)),(x_1(c),y_1(c))) &= ((10,5),(10,5),(12.5,7.5)) \\ ((x_2(a),y_2(a)),(x_2(b),y_2(b)),(x_2(c),y_2(c))) &= ((10,5),(10,5),(2.5,2.5)) \\ ((x_3(a),y_3(a)),(x_3(b),y_3(b)),(x_3(c),y_3(c))) &= ((10,10),(10,10),(25,25)). \end{aligned} \tag{3}$$

Notice that only Agent 3 can cheat Agents 2 and 3 in state a or b, by announcing b and a respectively, but has no incentive to do so. Hence, allocation (3) is individual incentive compatible. However, Agents 2 and 3 can form a coalition and when state c occurs they report to Agent 1 state b. Thus, Agent 1 gets $(10, 5)$ instead of $(12.5, 7.5)$ and Agents 2 and 3 distribute among themselves 2.5 units of each good, and clearly are better off.

The reader may wonder if the new cooperative solution concepts in a differential information economy provide any new insights that cannot be captured by the REE or Walrasian expectations equilibrium. The following example demonstrates this.

Example 0.4 Consider a three person economy, with Agents 1, 2, 3, one good denoted by x, and three equally probable states denoted by a, b, c. The agents' utility function, initial endowments, and private information sets are as follows:

$$\begin{aligned} u_i &= \sqrt{x_i}, & i &= 1,2,3 \\ e_1(a,b,c) &= (5,5,0)), & \mathcal{F}_1 &= \{\{a,b\},\{c\}\} \end{aligned}$$

$$e_2(a,b,c) = (5,0,5), \qquad \mathcal{F}_2 = \{\{a,c\},\{b\}\}$$
$$e_3(a,b,c) = (0,0,0), \qquad \mathcal{F}_2 = \{\{a\},\{b\},\{c\}\}.$$

The allocation below is $\mathcal{F}_i$-measurable $(i = 1,2,3)$ and cannot be improved upon by any $\mathcal{F}_i$-measurable, and feasible redistributions of the initial endowments of any coalition (this is the *private core*, Yannelis (1991)):

$$x_1(a,b,c) = (4,4,1)$$
$$x_2(a,b,c) = (4,1,4)$$
$$x_3(a,b,c) = (2,0,0). \tag{4}$$

Notice that the allocation in (4) is incentive compatible in the sense that Agent 3 is the only one who can cheat Agents 1 and 2 if the realized state of nature is a. However, Agent 3 has no incentive to misreport state a since this is the only state she gets positive consumption, and in any case one of Agents 1 or 2 will be able to tell the lie. Neither is it possible, as it can be easily seen, to form a coalition, profitable to both members, and misreport the state they have observed. Finally, notice that if Agent 3 had "bad" information, i.e., $\mathcal{F}_3' = \{a,b,c\}$, then, in a private core allocation, she gets zero consumption in each state. Thus, *advantageous information* is taken into account.

Contrary to the private core allocation in (4) above, neither the REE nor the Walrasian expectations equilibrium (or Radner equilibrium) can capture this phenomenon. Both concepts ignore Agent 3 no matter how fine her private information is, contrary to the private core which rewarded Agent 3 who used her superior information to make a Pareto improvement for the whole economy.

The above example suggests that the REE may not be the appropriate concept to capture contracts under asymmetric information and a new *price expectations equilibrium* might be needed. Indeed, work in this direction is recently being done by Tourky–Yannelis (2003) who are introducing a personalized price expectations equilibrium notion. This notion exists in situations that the REE and Radner equilibrium fail to exist and it can characterize the private core.

Before we close, we would like to remark that despite the fact that this book has discussed some successful alternative equilibrium concepts, other than the REE and Radner equilibrium, the issue of modeling continuum economies (perfectly competitive economies in the sense of Aumann (1964)) is still open. The problem seems to center in defining precisely the idea of each agent's private information as being negligible.

Acknowledgements. We are deeply indepted to our coauthor Allan Muir for his thougthful comments and suggestions. We also wish to thank Roko Aliprantis who encouraged us to put this book in the series *Studies in Economic Theory*. Werner Müller the Editor of Economics in Springer (now bought by Kluwer) and his staff were extremely helpful and patient during the preparation of this volume, and we sincerely appreciate their kind assistance. Needless to say without the contribution of all the authors this volume would never have been produced.

References

Allen, B.: Generic existence of completely revealing equilibria with uncertainty, when prices convey information. Econometrica **49**, 1173–1199 (1981)

Allen, B.: Incentives and market games with asymmetric information. Economic Theory **21**, 527–544 (2003)

Arrow, K.J., Debreu, G.: Existence of an equilibrium in a competitive economy. Econometrica **22**, 265–290 (1954)

Aumann, R.J.: Markets with a continuum of traders. Econometrica **32**, 39–50 (1964)

Debreu, G.: Theory of value. New York: Wiley 1959

Glycopantis, D., Muir, A., Yannelis, N.C.: Non-implementation of rational expectations as a perfect Bayesian equilibrium. Mimeo (2003)

Koutsougeras, L., Yannelis, N.C.: Incentive compatibility and information superiority of the core of an economy with differential information. Economic Theory **3**, 195–216 (1993)

Krasa, S., Yannelis, N.C.: The value allocation of an economy with differential information. Econometrica **62**, 881–900 (1994)

Kreps, M.D.: A note on 'fulfilled expectations' equilibria. Journal of Economic Theory **14**, 32–43 (1977)

Myerson, R.: Cooperative games with incomplete information. International Journal of Game Theory **13**, 69–96 (1984)

Prescott, E., Townsend, R.: Pareto optima and competitive equilibria with adverse selection and moral hazard. Econometrica **52**, 21–45 (1984)

Radner, R.: Competitive equilibrium under uncertainty. Econometrica **36**, 31–58 (1968)

Radner, R.: Rational expectation equilibrium: generic existence and information revealed by prices. Econometrica **47**, 655–678 (1979)

Tourky, R., Yannelis, N.C.: Private expectations equilibria. Mimeo (2003)

Wilson, R.: Information, efficiency, and the core of an economy. Econometrica **46**, 807–816 (1978)

Yannelis, N.C.: The core of an economy with differential information. Economic Theory **1**, 183–198 (1991)

D.G. *London, UK*
N.C.Y. *Urbana-Champaign, IL, USA*

February 8, 2004

Table of Contents

Equilibrium concepts in differential information economies ⋆

Dionysius Glycopantis[1] **and Nicholas C. Yannelis**[2]

[1] Department of Economics, City University, Northampton Square, London EC1V 0HB, UK (e-mail: d.glycopantis@city.ac.uk)
[2] Department of Economics, University of Illinois at Urbana-Champaign, IL 61820, USA (e-mail: nyanneli@uiuc.edu)

Summary. We summarize here basic cooperative and noncooperative equilibrium concepts, in the context of differential information economies with a finite number of agents. These, on the one hand, game theoretic, and, on the other hand, Walrasian equilibrium type concepts are explained, and their relation is pointed out, in the context of specific economies with one or two goods and two or three agents. We analyze the incentive compatibility of several cooperative and noncooperative concepts, and also we discuss briefly the possible implementation of these concepts as perfect Bayesian equilibria through the construction of relevant game trees. This possibility is related to whether the allocation is incentive compatible. This depends on whether there is free disposal or not.

Keywords and Phrases: Differential information economy, Walrasian expectations or Radner equilibrium, Rational expectations equilibria, Free disposal, Weak fine core, Private core, Weak fine value, Private value, Coalitional Bayesian incentive compatibility, Game trees, Perfect Bayesian equilibrium, Sequential equilibrium.

JEL Classification Numbers: D5, D82, C71, C72.

1 Introduction

The classical Walrasian equilibrium model as formalized by Arrow - Debreu (1954) and McKenzie (1954) consists of a finite set of agents each of which is characterized

⋆ We are very grateful to A. Muir for his invaluable help and suggestions. We wish to thank A. Hadjiprocopis for his invaluable help with the implementation of Latex in a Unix environment. He also provided us with numerically approximate solutions to Radner equilibrium and weak fine value problems, using a random selection algorithm.

by her preferences and initial endowments. The Walrasian model captures in a deterministic way the trade or contract (redistribution of initial endowments) among the agents and has played a central role in all aspects of economics. For this model significant results have been obtained, i.e. existence and Pareto optimality of the Walrasian equilibrium, equivalence of the Walrasian equilibrium with the core, (see Debreu and Scarf, 1963), and the relation between the core and the Shapley value, (see Emmons and Scafuri, 1985). These results have also been extended in infinite dimensional spaces (see for example Aumann, 1964; and the books of Hildenbrand, 1974; Khan and Yannelis, 1991).

Although infinite dimensional commodity spaces do capture uncertainty, they do not capture trade under asymmetric (or differential) information. On the other hand, it should be noted that most trades in an economy are made by agents who are asymmetrically informed and the need to introduce differential information into the Cournot - Nash model and the Arrow- Debreu - McKenzie model was evident in the seminal works of Harsanyi (1967) and Radner (1968). Their equilibrium concepts are noncooperative and have found extensive applications. In seminal papers, Wilson (1978) and Myerson (1982) introduced private information in the cooperative concepts of the core and the Shapley value respectively.

Briefly, the purpose of this paper is to survey the basic equilibrium concepts in economies with differential information. We employ a set of examples of finite economies which enable us to compare the outcomes that different equilibrium concepts generate. Also, we examine the implementation and the incentive compatibility of different equilibrium concepts.

Our survey differs from the two recent ones by Forges (1998), Forges et al. (2000) and Ichiishi and Yamazaki (2002). These papers follow the Harsanyi type model and focus on the devolopment of cooperative, core concepts. In contrast, we focus on the partition model, examine in detail additional concepts such as the Shapley value and provide an extensive form foundation for the concepts we examine. Furthermore we analyze the incentive compatibility of the different equilibrium concepts and consider their implementation as a perfect Bayesian equilibrium (PBE). These considerations can help us to decide how to choose among the available equilibrium concepts the most appropriate one. We also provide several illuminating examples which enable one to contrast and compare the different equilibrium notions. These examples could be especially useful to those who start work in the area.

A finite economy with differential information consists of a finite set of agents and states of nature. Each agent is characterized by a random utility function, a random consumption set, random initial endowments, a private information set which is a partition of the set of the states of nature, and a prior probability distribution on these states. For such an economy a number of cooperative and non-cooperative equilibrium concepts have been developed.

We believe that the natural and intuitive way to proceed is to analyse concepts in terms of measurability of allocations (Yannelis, 1991). In particular, as it is well known, (e.g. Prescott and Townsend, 1984; Allen, 2003), without measurability, the set of feasible and incentive compatible allocations is not convex and therefore the existence of an incentive compatible core becomes a serious problem. On the other hand certain measurability conditions imply incentive compatibility and they help us to narrow down the set of admissible allocations to a more manageable equilibrium set which is not only incentive compatibility but also exists. It is precisely for this reason that we follow the measurability approach.

We concentrate here mainly on cooperative concepts which allow for different types of measurability of the proposed allocations, i.e. for alternative forms of information sharing among the agents. In particular we consider the private core, (Yannelis, 1991), which is the set of all state-wise feasible and private information measurable allocations that cannot be dominated, in terms of expected utility, by any coalition's feasible and private information measurable net trades, the weak fine core (WFC), defined in Yannelis (1991) and Koutsougeras and Yannelis (1993), and the concepts of private value and the weak fine value (WFV), (Krasa and Yannelis, 1994), which employ the Shapley value.[1]

On the other hand we discuss the noncooperative concepts of the generalized Walrasian equilibrium type ideas of Radner equilibrium, defined in Radner (1968), and rational expectations equilibrium (REE), which is discussed in Radner (1979), Allen (1981), Einy et al. (2000), Kreps (1977) and Laffont (1985) and Grossman (1981), among others. Unlike the Walrasian equilibrium, Radner equilibrium with positive prices or REE may not exist in well behaved economies.

The paper is organized as follows. Section 2 contains the definition of a differentiable information economy. Section 3 defines cooperative equilibrium concepts. Section 4 defines noncooperative equilibrium concepts and makes some comparisons between the various ideas. Section 5 applies the equilibrium ideas in the context of one-good and Section 6 in that of two-good examples. Section 7 visits the incentive compatibility idea and Section 8 discusses implementation or non-implementation properties, in terms of PBE, of various equilibrium notions. Section 9 pays special attention to the relation between REE and weak core concepts and Section 10 concludes the discussion with some remarks. Finally Appendix I discusses some relations between core concepts.

2 Differential information economy (DIE)

In this section we define the notion of a finite-agent economy with differential information for the case where the set of states of nature, Ω and the number of goods, l, per state are finite. I is a set of n players and $\mathbb{R}^l_+$ will denote the set of positive real numbers.

A differential information exchange economy $\mathcal{E}$ is a set

$$\{((\Omega, \mathcal{F}), X_i, \mathcal{F}_i, u_i, e_i, q_i) : i = 1, \dots, n\}$$

where

1. $\mathcal{F}$ is a σ-algebra generated by a partition of Ω;
2. $X_i : \Omega \to 2^{\mathbb{R}^l_+}$ is the set-valued function giving the *random consumption set* of Agent (Player) i, who is denoted by Pi;

[1] See also Allen and Yannelis (2001) for additional references.

3. $\mathcal{F}_i$ is a partition of Ω generating a sub-σ-algebra of $\mathcal{F}$, denoting the *private information*[2] of Pi; $\mathcal{F}_i$ is a partition of Ω generating a sub-σ-algebra of $\mathcal{F}$, denoting the *private information*[3] of Pi;
4. $u_i : \Omega \times \mathbb{R}^l_+ \to \mathbb{R}$ is the *random utility* function of Pi; for each $\omega \in \Omega$, $u_i(\omega, .)$ is continuous, concave and monotone;
5. $e_i : \Omega \to \mathbb{R}^l_+$ is the *random initial endowment* of Pi, assumed to be $\mathcal{F}_i$-measurable, with $e_i(\omega) \in X_i(\omega)$ for all $\omega \in \Omega$;
6. q_i is an $\mathcal{F}$-measurable probability function on Ω giving the *prior* of Pi. It is assumed that on all elements of $\mathcal{F}_i$ the aggregate q_i is strictly positive. If a common prior is assumed on $\mathcal{F}$, it will be denoted by μ.

We will refer to a function with domain Ω, constant on elements of $\mathcal{F}_i$, as $\mathcal{F}_i$*-measurable*, although, strictly speaking, measurability is with respect to the σ-algebra generated by the partition.

In the first period agents make contracts in the ex ante stage. In the interim stage, i.e., after they have received a signal[4] as to what is the event containing the realized state of nature, they consider the incentive compatibility of the contract.

For any $x_i : \Omega \to \mathbb{R}^l_+$, the *ex ante expected utility* of Pi is given by

$$v_i(x_i) = \sum_{\Omega} u_i(\omega, x_i(\omega)) q_i(\omega).$$

Let $\mathcal{G}$ be a partition of (or σ-algebra on) Ω, belonging to Pi. For $\omega \in \Omega$ denote by $E_i^{\mathcal{G}}(\omega)$ the element of $\mathcal{G}$ containing ω; in the particular case where $\mathcal{G} = \mathcal{F}_i$ denote this just by $E_i(\omega)$. Pi's conditional probability for the state of nature being ω', given that it is actually ω, is then

$$q_i\big(\omega' | E_i^{\mathcal{G}}(\omega)\big) = \begin{cases} 0 & : \quad \omega' \notin E_i^{\mathcal{G}}(\omega) \\ \frac{q_i(\omega')}{q_i\left(E_i^{\mathcal{G}}(\omega)\right)} & : \quad \omega' \in E_i^{\mathcal{G}}(\omega). \end{cases}$$

The *interim expected utility* function of Pi, $v_i(x|\mathcal{G})$, is given by

$$v_i(x|\mathcal{G})(\omega) = \sum_{\omega'} u_i(\omega', x_i(\omega')) q_i\big(\omega' | E_i^{\mathcal{G}}(\omega)\big),$$

which defines a $\mathcal{G}$-measurable random variable.

Denote by $L_1(q_i, \mathbb{R}^l)$ the space of all equivalence classes of $\mathcal{F}$-measurable functions $f_i : \Omega \to \mathbb{R}^l$; when a common prior μ is assumed $L_1(q_i, \mathbb{R}^l)$ will be replaced by $L_1(\mu, \mathbb{R}^l)$. L_{X_i} is the set of all $\mathcal{F}_i$-measurable selections from the random consumption set of Agent i, i.e.,

---

[2] Following Aumann (1987) we assume that the players' information partitions are common knowledge.

[3] Sometimes $\mathcal{F}_i$ will denote the σ-algebra generated by the partition, as will be clear from the context.

[4] A signal to Pi is an $\mathcal{F}_i$-measurable function to all of the possible distinct observations specific to the player; that is, it induces the partition $\mathcal{F}_i$, and so gives the finest discrimination of states of nature directly available to Pi.

$$L_{X_i} = \{x_i \in L_1(q_i, \mathbb{R}^l) : x_i : \Omega \to \mathbb{R}^l \text{ is } \mathcal{F}_i\text{-measurable and } x_i(\omega) \in X_i(\omega)\ q_i\text{-a.e.}\}$$

and let $L_X = \prod_{i=1}^n L_{X_i}$.

Also let

$$\bar{L}_{X_i} = \{x_i \in L_1(q_i, \mathbb{R}^l) : x_i(\omega) \in X_i(\omega)\ q_i\text{-a.e.}\}$$

and let $\bar{L}_X = \prod_{i=1}^n \bar{L}_{X_i}$.

An element $x = (x_1, \ldots, x_n) \in \bar{L}_X$ will be called an *allocation*. For any subset of players S, an element $(y_i)_{i \in S} \in \prod_{i \in S} \bar{L}_{X_i}$ will also be called an allocation, although strictly speaking it is an allocation to S.

In case there is only one good, we shall use the notation $L^1_{X_i}$, L^1_X etc. When a common prior is also assumed $L_1(q_i, \mathbb{R}^l)$ will be replaced by $L_1(\mu, \mathbb{R}^l)$.

Finally, suppose we have a coalition S, with members denoted by i. Their pooled information $\bigvee_{i \in S} \mathcal{F}_i$ will be denoted by $\mathcal{F}_S$ [5]. We assume that $\mathcal{F}_I = \mathcal{F}$.

3 Cooperative equilibrium concepts: Core and Shapley value

We discuss here certain fundamental concepts.[6] First we define the notion of the private core (Yannelis, 1991).

Definition 3.1. An allocation $x \in L_X$ is said to be a *private core allocation* if

(i) $\sum_{i=1}^n x_i = \sum_{i=1}^n e_i$ and
(ii) there do not exist coalition S and allocation $(y_i)_{i \in S} \in \prod_{i \in S} L_{X_i}$ such that $\sum_{i \in S} y_i = \sum_{i \in S} e_i$ and $v_i(y_i) > v_i(x_i)$ for all $i \in S$.

The private core is an ex ante concept and under mild conditions it is not empty, as shown in Yannelis (1991) and Glycopantis et al. (2001). If the feasibility condition (i) is replaced by (i)′ $\sum_{i=1}^n x_i \leq \sum_{i=1}^n e_i$ then free disposal is allowed.

Next we define the *weak fine core* (WFC) (Yannelis, 1991; Koutsougeras and Yannelis, 1993). This is a refinement of the fine core concept of Wilson (1978) or Srivastava (1984). The fine core notion of Wilson as well as that in Koutsougeras and Yannelis may be empty in well behaved economies. This is why we are working with a different concept.

Definition 3.2. An allocation $x = (x_1, \ldots, x_n) \in \bar{L}_X$ is said to be a *WFC allocation* if

(i) each $x_i(\omega)$ is F_I-measurable;
(ii) $\sum_{i=1}^n x_i(\omega) = \sum_{i=1}^n e_i(\omega)$, for all $\omega \in \Omega$;
(iii) there do not exist coalition S and allocation $(y_i)_{i \in S} \in \prod_{i \in S} \bar{L}_{X_i}$ such that $y_i(\cdot) - e_i(\cdot)$ is $\mathcal{F}_S$-measurable for all $i \in S$, $\sum_{i \in S} y_i = \sum_{i \in S} e_i$ and $v_i(y_i) > v_i(x_i)$ for all $i \in S$.

[5] The "join" $\bigvee_{i \in S} \mathcal{F}_i$ denotes the smallest σ-algebra containing all $\mathcal{F}_i$, for $i \in S$.
[6] The interim weak fine core (IWFC) is discussed in a later section.

As comparisons are made on the basis of expected utility, the weak fine core is also an ex ante concept. It captures the idea of an allocation which is ex ante "full information" Pareto optimal. As with the private core the feasibility condition can be relaxed to (ii)′ $\sum_{i=1}^{n} x_i(\omega) \leq \sum_{i=1}^{n} e_i(\omega)$, for all $\omega \in \Omega$.

Finally we define the concept of *weak fine value* (WFV) (see Krasa and Yannelis, 1994, 1996). We must first define a transferable utility (TU) game in which each agent's utility is weighted by a non-negative factor λ_i, $(i = 1, ..., n)$, which allows for interpersonal comparisons. In a TU-game an outcome can be realized through transfers of payoffs among the agents. On the other hand a (weak) fine value allocation is more specific. It is realizable through a redistribution of payoffs among the agents and, following this, no side payments are necessary.[7] The WFV set is also non-empty.

A game with side payments is defined as follows.

Definition 3.3. A game with side payments $\Gamma = (I, V)$ consists of a finite set of agents $I = \{1, ..., n\}$ and a superadditive[8] , real valued function V defined on 2^I such that $V(\emptyset) = 0$. Each $S \subseteq I$ is called a coalition and $V(S)$ is the 'worth' of the coalition S.

The Shapley value of the game Γ (Shapley, 1953) is a rule that assigns to each Agent i a payoff, $Sh_i(V)$, given by the formula[9]

$$Sh_i(V) = \sum_{\substack{S \subseteq I \\ S \supseteq \{i\}}} \frac{(\mid S \mid -1)!(\mid I \mid - \mid S \mid)!}{\mid I \mid!} [V(S) - V(S \backslash \{i\})]. \qquad (1)$$

The Shapley value has the property that $\sum_{i \in I} Sh_i(V) = V(I)$, i.e. the implied allocation of payoffs is Pareto efficient.

We now define for each DIE, $\mathcal{E}$, with common prior μ, which is assumed for simplicity, and for each set of weights, $\lambda = \{\lambda_i \geq 0 : i = 1, \ldots, n\}$, the associated game with side payments (I, V_λ). We also refer to this as a *transferable utility* (TU) game.

Definition 3.4. Given $\{\mathcal{E}, \lambda\}$ an associated game $\Gamma_\lambda = (I, V_\lambda)$ is defined as follows: For every coalition $S \subset I$ let

$$V_\lambda(S) = \max_x \sum_{i \in S} \lambda_i \sum_{\omega \in \Omega} u_i(\omega, x_i(\omega))\mu(\omega) \qquad (2)$$

subject to

(i) $\sum_{i \in S} x_i(\omega) = \sum_{i \in S} e_i(\omega)$, for all $\omega \in \Omega$, and
(ii) $x_i - e_i$ is $\bigvee_{i \in S} \mathcal{F}_i-$measurable.

We are now ready to define the WFV allocation.

---

[7] See Emmons and Scafuri (1985, p. 60) and the examples in Section 6 below for further discussion.

[8] This means that given disjoint $S, T \subset I$ then $V(S) + V(T) \leq V(S \cup T)$.

[9] The Shapley value measure is the sum of the expected marginal contributions an agent can make to all the coalitions of which he/she can be a member (see Shapley, 1953).

Definition 3.5. An allocation $x = (x_1, \ldots, x_n) \in \bar{L}_X$ is said to be a *WFV allocation* of the differential information economy, $\mathcal{E}$, if the following conditions hold

(i) Each net trade $x_i - e_i$ is $\bigvee_{i=1}^{n} \mathcal{F}_i$-measurable,
(ii) $\sum_{i=1}^{n} x_i = \sum_{i=1}^{n} e_i$ and
(iii) There exist $\lambda_i \geq 0$, for every $i = 1, ..., n$, which are not all equal to zero, with $\sum_{\omega \in \Omega} \lambda_i u_i(\omega, x_i(\omega))\mu(\omega) = Sh_i(V_\lambda)$ for all i, where $Sh_i(V_\lambda)$ is the Shapley value of Agent i derived from the game (I, V_λ), defined in (2) above.[10]

Condition (i) requires the pooled information measurability of net trades. Condition (ii) is the market clearing condition and (iii) says that the expected utility of each agent multiplied by his/her weight, λ_i, must be equal to his/her Shapley value derived from the TU game (I, V_λ). Obviously for the actual utility that the agent will obtain the weight must not be taken into account. Therefore an agent could obtain the utility of a positive allocation even if λ_i were zero.

If condition (ii) in Definitions 3.4 and (i) in 3.5 are replaced by $x_i - e_i$ is $\mathcal{F}_i$-measurable, for all i, then we obtain the definition of the *private value allocation.*

An immediate consequence of Definition 3.4 is that $Sh_i(V_\lambda) \geq \lambda_i \sum_{\omega \in \Omega} u_i(\omega, e_i(\omega))\mu(\omega)$ for every i, i.e. the value allocation is individually rational. This follows immediately from the fact that the game (V_λ, I) is superadditive for all weights λ. Similarly, efficiency of the Shapley value implies that the weak-fine (private) value allocation is weak-fine (private) Pareto efficient.

Note 3.1. The core of an economy with differential information was first defined by Wilson (1978) and the Shapley value with differential information by Myerson (1982). The above analysis is based on the measurability approach introduced by Yannelis (1991). This approach enables one to prove readily the existence of alternative core and value concepts. Furthermore, as we will see in subsequent sections, certain measurability restrictions, as for example the private information measurability of an allocation, ensure incentive compatibility. General existence results for the core and value can be found in Yannelis (1991), Allen (1991a, 1991b), Krasa - Yannelis (1994), Lefebvre (2001) and Glycopantis et al. (2001). The reader is referred to the Appendix for a more complete list of core concepts.

4 Noncooperative equilibrium concepts: Walrasian expectations (or Radner) equilibrium and REE

In order to define a competitive equilibrium in the sense of Radner we need the following. A *price system* is an $\mathcal{F}$-measurable, non-zero function $p : \Omega \to \mathbb{R}^l_+$ and the *budget set* of Agent i is given by

$$B_i(p) = \Big\{ x_i : \; x_i : \Omega \to \mathbb{R}^l \text{ is } \mathcal{F}_i\text{-measurable } x_i(\omega) \in X_i(\omega)$$

[10] By replacing the join measurability with private information measurability we can define the private value allocation.

$$\text{and} \sum_{\omega\in\Omega} p(\omega)x_i(\omega) \leq \sum_{\omega\in\Omega} p(\omega)e_i(\omega)\Big\}.$$

Notice that the budget constraint is across states of nature.

Definition 4.1. A pair (p, x), where p is a price system and $x = (x_1, \ldots, x_n) \in L_X$ is an allocation, is a *Walrasian expectations or Radner equilibrium* if

(i) for all i the consumption function maximizes v_i on $B_i(p)$
(ii) $\sum_{i=1}^n x_i \leq \sum_{i=1}^n e_i$ (free disposal), and
(iii) $\sum_{\omega\in\Omega} p(\omega)\sum_{i=1}^n x_i(\omega) = \sum_{\omega\in\Omega} p(\omega)\sum_{i=1}^n e_i(\omega)$.

This is an ex ante concept. We allow for free disposal, because otherwise a Radner equilibrium with positive prices might not exist. This is demonstrated below through Example 5.2 in which a price becomes negative. In general, for purposes of comparison we consider also the case with $\sum_{i=1}^n x_i = \sum_{i=1}^n e_i$.

Proposition 4.1. A (free disposal) Radner equilibrium is in the (free disposal) private core.

The proof parallels the usual one of the complete information case.

We note that a Radner equilibrium with free disposal may not be in the non-free disposal private core. The point can be made using Example 5.2 below, in which the Radner equilibrium with free disposal and private core without free disposal consist of completely different allocations. The question arises why the proposition immediately above fails. The argument cannot be pushed through because under different free disposal assumptions the feasibility condition is different.

Next we turn our attention to the notion of REE. We shall need the following. Let $\sigma(p)$ be the smallest sub-σ-algebra of $\mathcal{F}$ for which a price system $p : \Omega \to \mathbb{R}^l_+$ is measurable and let $\mathcal{G}_i = \sigma(p) \vee \mathcal{F}_i$ denote the smallest σ-algebra containing both $\sigma(p)$ and $\mathcal{F}_i$. We shall also condition the expected utility of the agents on $\mathcal{G}$ which produces a random variable.

Definition 4.2. A pair (p, x), where p is a price system and $x = (x_1, \ldots, x_n) \in \bar{L}_X$ is an allocation, is a *REE* if

(i) for all i the consumption function $x_i(\omega)$ is $\mathcal{G}_i$-measurable;
(ii) for all i and for all ω the consumption function maximizes $v_i(x_i|\mathcal{G}_i)(\omega)$ subject to the budget constraint at state ω,

$$p(\omega)x_i(\omega) \leq p(\omega)e_i(\omega);$$

(iii) $\sum_{i=1}^n x_i(\omega) = \sum_{i=1}^n e_i(\omega)$ for all $\omega \in \Omega$.

REE is an interim concept because we condition on information from prices as well. An REE is said to be *fully revealing* if $\mathcal{G}_i = \mathcal{F} = \bigvee_{i\in I} \mathcal{F}_i$ for all $i \in I$. Although in the definition we do not allow for free disposal, we comment briefly on such an assumption in the context of Example 5.2.

Note 4.1. The concept of Radner equilibrium is due to Radner (1968) and it extends the Arrow-Debreu contingent claims model, (see Debreu, 1959, Ch. 7), to allow for differential information. The existence of a free disposal Radner equilibrium can be found in Radner (1968). The definition of REE is taken from Radner (1979) and Allen (1981). The REE does not exist always, may not be fully Pareto optimal, or incentive compatible and may not be implementable as a PBE (Glycopantis et al., 2003b). The Radner equilibrium without free disposal is always incentive compatible, as it is contained in the private core. Moreover, under standard assumptions, it exists, as shown by Radner (1968). An example illustrating both concepts can be found below.

5 Illustrations of equilibrium concepts and comparisons to each other: One-good examples

We now offer some comments on and make comparisons between the various equilibrium notions. In many instances we will use the same example to compute different equilibrium concepts. Hence the outcomes that different equilibrium concepts generate will become clear.

As we saw in Proposition 4.1 the Radner equilibrium allocations are a subset of the corresponding private core allocations. Of course it is possible that a Radner equilibrium allocation with positive prices might not exist. In the two-agent economy of Example 5.2 below, assuming non-free disposal the unique private core is the initial endowments allocation while no Radner equilibrium exists. On the other hand, assuming free disposal the REE coincides with the initial endowments allocation which does not belong to the private core. It follows that *the REE allocations need not be in the private core*. Therefore *a REE need not be a Radner equilibrium* either. In Example 5.1 below, without free disposal no Radner equilibrium with positive prices exists but REE does. It is unique and it implies no-trade.

As for *the comparison between private core and WFC allocations* the two sets could intersect but there is no definite relation. Indeed the measurability requirement of the private core allocations separates the two concepts. Finally notice that no allocation which does not distribute the total resource could be in the WFC.

For $n = 2$ one can easily verify that the WFV belongs to the weak fine core. However it is known (see for example Scafuri and Yannelis, 1984) that for $n \geq 3$ *a value allocation may not be a core allocation, and therefore may not be a Radner equilibrium*. Also a value allocation might not belong to any Walrasian type set.

In a later section we shall discuss whether core and Walrasian type allocations have certain desirable properties, from the point of view of incentive compatibility. We shall then turn our attention to the implementation of such allocations.

In this and the following sections we indicate, by putting dates, whether we have already discussed in Glycopantis et al. (2001, 2003a, 2003b), at least partly, the various examples. Where both types are calculated we find it more convenient to start with the non-cooperative concepts.

Example 5.1. (2001, 2003a) Consider the following three agents economy, $I = \{1, 2, 3\}$ with one commodity, i.e. $X_i = \mathbb{R}_+$ for each i, and three states of nature $\Omega = \{a, b, c\}$.

The endowments and information partitions of the agents are given by

$$
\begin{aligned}
e_1 &= (5,5,0), & \mathcal{F}_1 &= \{\{a,b\},\{c\}\};\\
e_2 &= (5,0,5), & \mathcal{F}_2 &= \{\{a,c\},\{b\}\};\\
e_3 &= (0,0,0), & \mathcal{F}_3 &= \{\{a\},\{b\},\{c\}\}.
\end{aligned}
$$

$u_i(\omega, x_i(\omega)) = x_i^{\frac{1}{2}}$ and every player has the same prior distribution $\mu(\{\omega\}) = \frac{1}{3}$, for $\omega \in \Omega$.

It was shown in Appendix II of Glycopantis et al. (2001) that, without free disposal, the redistribution

$$
\begin{pmatrix} 4 & 4 & 1 \\ 4 & 1 & 4 \\ 2 & 0 & 0 \end{pmatrix}
$$

is a private core allocation, where the ith line refers to Player i and the columns from left to right to states a, b and c.

If the private information set of Agent 3 is the trivial partition, i.e., $\mathcal{F}_3' = \{a,b,c\}$, then no trade takes place and clearly in this case he gets zero utility. Thus the private core is sensitive to information asymmetries. On the other hand in a Radner equilibrium or a REE, Agent 3 will always receive zero quantities as he has no initial endowments, irrespective of whether his private information partition is the full one or the trivial one.

Example 5.2. (2001, 2003a) We now consider Example 5.1 without Agent 3.

For the various types of allocations below, we distinguish between the cases without and with free disposal. We denote the prices by $p(a) = p_1, p(b) = p_2, p(c) = p_3$. Throughout $\varepsilon, \delta \geq 0$.

A. REE

Now, a price function, $p(\omega)$, known to both agents, is defined on Ω. Apart from his own private $E_i \subseteq \mathcal{F}_i$, each agent also receives a price signal which is a value in the range of the price function. Combining the two types of signals he deduces the event from Ω that has been realized, $E_{p,E_i} = \{\omega : p(\omega) = p \text{ and } \omega \in E_i\}$. He then chooses a constant consumption on E_{p,E_i} which maximizes his interim expected utility subject to the budget set at state ω.

We now distinguish between:

Case 1. All prices positive and $p_1 \neq p_2 \neq p_3$.
Then, as soon as the price signal is announced every agent knows the exact state of nature and simply demands his initial endowment in that state.

Case 2. All prices positive and $p_1 = p_2 \neq p_3$.
Then Agent 2 will always realize which is the state of nature and will demand his initial endowment. On the other hand Agent 1 will not be able to distinguish between states a and b. However given the fact that his utility function is the same across states, he will also demand his initial endowment in all states of nature.

Case 3. All prices positive and $p_1 = p_3 \neq p_2$.
This is identical to Case 2 with the roles of the two agents interchanged.

Case 4. The positive prices are constant on Ω and hence non-revealing. Each agent relies exclusively on his private information and will demand in each state his initial endowment.
In all cases the rational expectations price function can be any such that its range of values is a positive vector and it will confirm the initial endowments as equilibrium allocation. Furthermore it makes no difference to the above reasoning whether free disposal is allowed or not.

We can also argue in general that with one good per state and monotonic utility functions, the measurability of the allocations implies that REE, fully revealing or not, simply confirms the initial endowments.

B. Radner equilibrium

The measurability of allocations implies that we require consumptions $x_1(a) = x_2(b) = x$ and $x_1(c)$ for Agent 1, and $x_2(a) = x_2(c) = y$ and $x_2(b)$ for Agent 2. We can also write $x = 5 - \varepsilon, x_1(c) = \delta, y = 5 - \delta$ and $x_2(b) = \varepsilon$.

We now consider,

Case 1. Without free disposal
There is no Radner equilibrium with prices in $\mathbb{R}^3_+$.

Case 2. With free disposal.
The prices are $p_1 = 0, p_2 = p_3 > 0$ and the allocation is

$$\begin{pmatrix} 4 & 4 & 1 \\ 4 & 1 & 4 \end{pmatrix}.$$

It corresponds to $\varepsilon, \delta = 1$ which means that in state a each of the agents throws away one unit of the good.

C. WFC

The agents pool their information and therefore any feasible consumption vector to either agent will be measurable. Hence we do not need to distinguish between free disposal and non-free disposal. All WFC allocations will exhaust the resource in each state of nature.

There are uncountably many such allocations, as for example

$$\begin{pmatrix} 5 & 2.5 & 2.5 \\ 5 & 2.5 & 2.5 \end{pmatrix}.$$

This allocation is $\bigvee_{i=1}^{2} \mathcal{F}_i$-measurable and cannot be dominated by any coalition of agents using their pooled information.

Referring back to Example 5.1 we can note that a private core allocation is not necessarily a WFC allocation. For example the division (4, 4, 1), (4, 1, 4) and (2, 0, 0), to Agents 1, 2 and 3 respectively, is a private core but not a weak fine core allocation. The first two agents can get together, pool their information and do

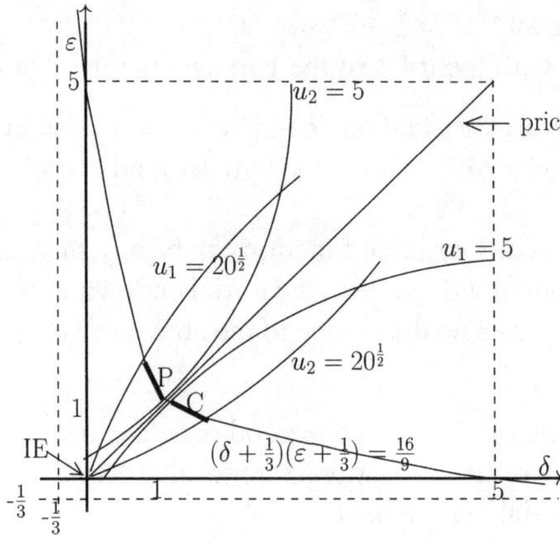

PC: private core; IE: Initial Endowments

Figure 1

better. They can realize the WFC allocation, (5, 2.5, 2.5), (5, 2.5, 2.5) and (0, 0, 0) which does not belong to the private core because of lack of measurability.

D. Private core

Case 1. Without free disposal.
No individual can increase his allocation and retain measurability. Therefore, in this case the only allocation in the core is the initial endowments.

Case 2. With free disposal.
Free disposal can take the form:

$$\begin{pmatrix} 5-\varepsilon & 5-\varepsilon & \delta \\ 5-\delta & \varepsilon & 5-\delta \end{pmatrix}$$

where $\varepsilon, \delta > 0$.

The private core is the section of the curve $(\delta + \frac{1}{3})(\varepsilon + \frac{1}{3}) = \frac{16}{9}$ between the indifference curves corresponding to $\mathcal{U}_1 = 20^{\frac{1}{2}}$ and $\mathcal{U}_2 = 20^{\frac{1}{2}}$. Notice that the allocation

$$\begin{pmatrix} 4 & 4 & 1 \\ 4 & 1 & 4 \end{pmatrix}$$

corresponds to $\delta, \varepsilon = 1$ and is in the private core. The private core and the Radner equilibrium are shown in Figure 1.

E. WFV

Here we shall show that $x_1 = x_2 = (5, 2.5, 2.5)$ is a weak fine value allocation. First we note that the "join" $\mathcal{F}_1 \vee \mathcal{F}_2 = \{\{a\}\{b\}\{c\}\}$. So every allocation is $\mathcal{F}_1 \vee \mathcal{F}_2$-measurable and condition (i) of Definition 3.5 is satisfied. Condition (ii) is also immediately satisfied.

First V_λ is calculated to be

$$V_\lambda(\{1\}) = \frac{2\times 5^{\frac{1}{2}}}{3}\lambda_1, V_\lambda(\{2\}) = \frac{2\times 5^{\frac{1}{2}}}{3}\lambda_2 \quad \text{and}$$
$$V_\lambda(\{1,2\}) = \frac{10^{\frac{1}{2}}+2\times 5^{\frac{1}{2}}}{3}(\lambda_1^2+\lambda_2^2)^{\frac{1}{2}}$$

It is easy to see that

$$Sh_1(V_\lambda) = \frac{1}{2}\left\{\frac{2\times 5^{\frac{1}{2}}}{3}\lambda_1 + \frac{10^{\frac{1}{2}}+2\times 5^{\frac{1}{2}}}{3}(\lambda_1^2+\lambda_2^2)^{\frac{1}{2}} - \frac{2\times 5^{\frac{1}{2}}}{3}\lambda_2\right\}. \tag{3}$$

Definition 3.5 gives

$$2(2.5)^{\frac{1}{2}}\lambda_1 = \frac{10^{\frac{1}{2}}+2\times 5^{\frac{1}{2}}}{2}(\lambda_1^2+\lambda_1^2)^{\frac{1}{2}} - 5^{\frac{1}{2}}\lambda_2. \tag{4}$$

Similarly the condition on player 2's allocation gives

$$2(2.5)^{\frac{1}{2}}\lambda_2 = \frac{10^{\frac{1}{2}}+2\times 5^{\frac{1}{2}}}{2}(\lambda_1^2+\lambda_1^2)^{\frac{1}{2}} - 5^{\frac{1}{2}}\lambda_2. \tag{5}$$

Subtracting we get $2\times 2^{\frac{1}{2}}(\lambda_1-\lambda_2) = 5^{\frac{1}{2}}(\lambda_1-\lambda_2)$.

It follows that $\lambda_1 = \lambda_2$. Substituting this common value λ not equal to 0 back into one of the conditions, λ cancels leaving $2(2.5)^{\frac{1}{2}} = \frac{10^{\frac{1}{2}}+2\times 5^{\frac{1}{2}}}{2}\times 2^{\frac{1}{2}} - 5^{\frac{1}{2}}$ which is an identity. It follows that Definition 3.5 is satisfied.

Next we investigate whether there are any other WFV. The conditions are $\lambda_1\left[x^{\frac{1}{2}}+y^{\frac{1}{2}}+z^{\frac{1}{2}}\right] = 5^{\frac{1}{2}}(\lambda_1-\lambda_2)+k(\lambda_1^2+\lambda_2^2)^{\frac{1}{2}}$ and $\lambda_2\left[(10-x)^{\frac{1}{2}}+(5-y)^{\frac{1}{2}}+(5-z)^{\frac{1}{2}}\right] = 5^{\frac{1}{2}}(\lambda_2-\lambda_1)+k(\lambda_1^2+\lambda_2^2)^{\frac{1}{2}}$ where $k = \frac{10^{\frac{1}{2}}+2\times 5^{\frac{1}{2}}}{2}$.

There is an obvious symmetry here: if $\lambda_1, \lambda_2, x, y, z$ is a solution then so is $\lambda_2, \lambda_1, 10-x, 5-y, 5-z$, so that we may assume, without loss of generality, that λ_2 is different from zero, since both λ's cannot be zero, and write $\theta = \frac{\lambda_1}{\lambda_2}$. Subtracting the two equations we obtain $\theta S_1 - S_2 = 2\times 5^{\frac{1}{2}}(\theta-1)$, where $S_1 = (x)^{\frac{1}{2}}+(y)^{\frac{1}{2}}+(z)^{\frac{1}{2}}$, $S_2 = (10-x)^{\frac{1}{2}}+(5-y)^{\frac{1}{2}}+(5-z)^{\frac{1}{2}}$, which implies $\theta = \frac{S_2-2\times 5^{\frac{1}{2}}}{S_1-2\times 5^{\frac{1}{2}}}$.

We also have $\theta S_1 = 5^{\frac{1}{2}}(\theta-1)+k(\theta^2+1)^{\frac{1}{2}}$ which implies $[\theta(S_1-5^{\frac{1}{2}})+5^{\frac{1}{2}}]^2 = k(\theta^2+1)^{\frac{1}{2}}$. This in turn implies $\{(S_1-5^{\frac{1}{2}})^2-k^2\}\theta^2+2\times 5^{\frac{1}{2}}(S_1-5^{\frac{1}{2}})\theta+5-k^2 = 0$. This has real roots iff $5(S_1-5^{\frac{1}{2}})^2 \geq (5-k^2)\{(S_1-5^{\frac{1}{2}})^2-k^2\}$, or, equivalently, $(S_1-5^{\frac{1}{2}})^2 \geq k^2-5$, or $S_1 \geq 5^{\frac{1}{2}}+(k^2-5)^{\frac{1}{2}}$, which implies the root $S_1 = 5.32978$. By symmetry we also need $S_2 \geq 5.32978$. The symmetric case $\theta = 1$ gives $S_1 = S_2 = 2^{\frac{1}{2}}k$ which has an approximate value of 5.39835. It corresponds to $x_1 = x_2 = (5, 2.5, 2.5)$.

Clearly there is not much room to move away from the symmetric case. On the other hand if S_1 goes up then S_2 goes down. This follows from the fact that the sum of the payoffs to the players is equal to $V_\lambda(\{1,2\})$. This suggests the problem

Maximize S_1 subject to $S_2 = \varrho$.

The First Order Conditions are: $(10-x)^{\frac{1}{2}} = \frac{1}{2}\eta x^{\frac{1}{2}}$, $(5-y)^{\frac{1}{2}} = \frac{1}{2}\eta y^{\frac{1}{2}}$ and $(5-z)^{\frac{1}{2}} = \frac{1}{2}\eta z^{\frac{1}{2}}$.

From these we obtain $\frac{y^{\frac{1}{2}}}{x^{\frac{1}{2}}} = \frac{(5-y)^{\frac{1}{2}}}{(10-x)^{\frac{1}{2}}}$ and $\frac{z^{\frac{1}{2}}}{x^{\frac{1}{2}}} = \frac{(5-z)^{\frac{1}{2}}}{(10-x)^{\frac{1}{2}}}$, which imply $x=2y=2z$.

Re-substituting in $S_2 = \varrho$ we derive $\varrho = (10-2z)^{\frac{1}{2}} + (5-z)^{\frac{1}{2}} + (5-z)^{\frac{1}{2}} = (2+2^{\frac{1}{2}})(5-z)^{\frac{1}{2}}$ which for $\varrho = 5.32978$ implies, approximately, $y = z = 5 - (\frac{\varrho}{2+2^{\frac{1}{2}}})^2 = 2.56310$, $x = 5.12621$, $S_1 = 5.46605$, and $\theta = 0.86290$. It follows that the WFV allocations correspond to $\theta \in [0.86290, 1.158882837]$, where the two numbers are the inverse of each other.

Example 5.3. The problem is a two-state, $\Omega = \{a, b\}$, three-player game with utilities and initial endowments given by:

$$u_1(x_{1j}) = x_{1j}^{\frac{1}{2}}; \quad e_1 = (4,0), \quad F_1 = \{\{a\},\{b\}\}$$
$$u_2(x_{2j}) = x_{2j}^{\frac{1}{2}}; \quad e_2 = (0,4), \quad F_2 = \{\{a\},\{b\}\}$$
$$u_3(x_{3j}) = x_{3j}^{\frac{1}{2}}; \quad e_3 = (0,0), \quad F_3 = \{a,b\},$$

where x_{ij} denotes consumption of Player i in state j, (a is identified with 1 and b with 2). Every player has the same prior distribution $\mu(\omega) = \frac{1}{2}$ for $\omega \in \Omega$.

The associated TU game has value function

$$V_\lambda(\{1\}) = \lambda_1, V_\lambda(\{2\}) = \lambda_2, V_\lambda(\{3\}) = 0,$$
$$V_\lambda(\{1,2\}) = 2(\lambda_1^2+\lambda_2^2)^{\frac{1}{2}}, V_\lambda(\{1,3\}) = (\lambda_1^2+\lambda_3^2)^{\frac{1}{2}}, V_\lambda(\{2,3\}) = (\lambda_2^2+\lambda_3^2)^{\frac{1}{2}},$$
$$V_\lambda(\{1,2,3\}) = 2(\lambda_1^2+\lambda_2^2+\lambda_3^2)^{\frac{1}{2}}.$$

The equations for a value allocation are then:

$$\frac{2}{3}\lambda_1 + \frac{1}{3}\Big(2(\lambda_1^2+\lambda_2^2)^{\frac{1}{2}} - \lambda_2\Big) + \frac{1}{3}(\lambda_1^2+\lambda_3^2)^{\frac{1}{2}}$$
$$+\frac{2}{3}\left(2(\lambda_1^2+\lambda_2^2+\lambda_3^2)^{\frac{1}{2}} - (\lambda_2^2+\lambda_3^2)^{\frac{1}{2}}\right) = \lambda_1\left(x_{11}^{\frac{1}{2}} + x_{12}^{\frac{1}{2}}\right),$$
$$\frac{2}{3}\lambda_2 + \frac{1}{3}\Big(2(\lambda_1^2+\lambda_2^2)^{\frac{1}{2}} - \lambda_1\Big) + \frac{1}{3}(\lambda_2^2+\lambda_3^2)^{\frac{1}{2}}$$
$$+\frac{2}{3}\left(2(\lambda_1^2+\lambda_2^2+\lambda_3^2)^{\frac{1}{2}} - (\lambda_1^2+\lambda_3^2)^{\frac{1}{2}}\right) = \lambda_2\left(x_{21}^{\frac{1}{2}} + x_{22}^{\frac{1}{2}}\right),$$
$$\frac{1}{3}\Big((\lambda_1^2+\lambda_3^2)^{\frac{1}{2}} - \lambda_1\Big) + \frac{1}{3}\Big((\lambda_2^2+\lambda_3^2)^{\frac{1}{2}} - \lambda_2\Big)$$
$$+\frac{4}{3}\left((\lambda_1^2+\lambda_2^2+\lambda_3^2)^{\frac{1}{2}} - (\lambda_1^2+\lambda_2^2)^{\frac{1}{2}}\right) = \lambda_3\left(x_{31}^{\frac{1}{2}} + x_{32}^{\frac{1}{2}}\right),$$

subject to $x_{11} + x_{21} + x_{31} = 4.x_{12} + x_{22} + x_{32} = 4$.

The left-hand side are just numbers which we can calculate. General solution of these equations seems difficult, but we would hope to get a symmetric solution, in the following sense: the economy is symmetric under the interchange of Agent

1 with Agent 2, together with interchange of the good in state 1 and the good in state 2; so we might expect a solution in which

$$x_{11} = x_{22},\ x_{12} = x_{21},\ x_{31} = x_{32},\ \lambda_1 = \lambda_2.$$

We will write, for simplicity

$$x_{11}^{\frac{1}{2}} = x_{22}^{\frac{1}{2}} = x,\ x_{12}^{\frac{1}{2}} = x_{21}^{\frac{1}{2}} = y,$$
$$\text{and hence } x_{31} = x_{32} = (4 - (x^2 + y^2))^{\frac{1}{2}},\ \ \lambda_1 = \lambda_2 = \lambda.$$

We will treat two cases. Firstly, if $\lambda_3 = 0$, the last equation is identically satisfied and the first two equations (which are the same) give $2 \times 2^{\frac{1}{2}}\lambda = \lambda(x + y)$. So λ is arbitrary and $x + y = 2 \times 2^{\frac{1}{2}}$. If we suppose $x = 2^{\frac{1}{2}} + \delta,\ y = (-\delta)^{\frac{1}{2}}$, then $x_{11} + x_{21} = 4 + \delta^2$, so we have $\delta = 0$ and hence

$$x_{11} = x_{12} = x_{21} = x_{22} = 2,\ x_{31} = x_{32} = 0,$$

with $\lambda_1 = \lambda_2 > 0$ arbitrary and $\lambda_3 = 0$.

Now consider the possibility that $\lambda_3 > 0$ and we may normalise it to be equal to 1. The first two equations are the same and they state:

$$\frac{1}{3}\left(2(2)^{\frac{1}{2}} + 1\right)\lambda - \frac{1}{3}(\lambda^2 + 1)^{\frac{1}{2}} + \frac{4}{3}(2\lambda^2 + 1)^{\frac{1}{2}} = \lambda(x + y). \tag{6}$$

The third equation becomes

$$-\frac{2}{3}\left(2(2)^{\frac{1}{2}} + 1\right)\lambda + \frac{2}{3}(\lambda^2 + 1)^{\frac{1}{2}} + \frac{4}{3}(2\lambda^2 + 1)^{\frac{1}{2}} = 2\left[4 - (x^2 + y^2)\right]^{\frac{1}{2}}. \tag{7}$$

It is a matter of tedious calculations on equations (16) and (17) to show that there are no value allocations with $\lambda_3 \neq 0$ which are symmetric.

We now consider approximate equilibria, using the random algorithm. First we look into the case where in the equations for a value allocation we insert $\lambda_i = 1,\ \ \forall i$. The system does not perform very well. Approximate values can be found but the total error, the square root of the sum of squares of RHS-LHS of the equations, is 0.21098557 which is rather large. On the other hand variations in the total resource improve the approximation.

If we allow in the system above for the λ_i's also to be chosen then a rather satisfactory approximate solution emerges:

$x_{11} = 1.9999,\ x_{12} = 2.0001,\ x_{21} = 2{,}0000,\ x_{22} = 1.9998,\ x_{31} = x_{32} = 0$ (approximately), $\lambda_1 = 1,\ \lambda_2 = 1.00009,\ \lambda_3 = 0.0129$, with total error 0.000000007.

In the example we have examined Agent 3 has zero endowments and bad information. As a result, when all the λ_i's can be chosen the solution of the equations of the value allocation are approximately the same as when no weight is attached to Agent 3.

6 Two-good examples

We note that with one good per state and monotone utility functions there is a direct relation between allocations and utilities, i.e. $x \geq y$ iff $u(x) \geq u(y)$. This allows one to prove results which do not hold in general. This is the reason why we present also examples with two goods. We also note that in the one good case the unique REE allocation exists always and it coincides with no trade. Thus it exists, it is incentive compatible and Pareto optimal. However, as it is shown below, this is not the case when there are two goods.

Example 6.1. (2003b) We consider a two-agent economy, $I = \{1, 2\}$ with two commodities, i.e. $X_i = \mathbb{R}^2_+$ for each i, and three states of nature $\Omega = \{a, b, c\}$.

The endowments, per state a, b, and c respectively, and information partitions of the agents are given by

$$e_1 = ((7,1),(7,1),(4,1)), \qquad \mathcal{F}_1 = \{\{a,b\},\{c\}\};$$
$$e_2 = ((1,10),(1,7),(1,7)), \qquad \mathcal{F}_2 = \{\{a\},\{b,c\}\}.$$

We shall denote $A_1 = \{a, b\}$, $c_1 = \{c\}$, $a_2 = \{a\}$, $A_2 = \{b, c\}$. $u_i(\omega, x_{i1}(\omega), x_{i2}(\omega)) = x_{i1}^{\frac{1}{2}} x_{i2}^{\frac{1}{2}}$, and for all players $\mu(\{\omega\}) = \frac{1}{3}$, for $\omega \in \Omega$. We have that $u_1(7,1) = 2.65$, $u_1(4,1) = 2$, $u_2(1,10) = 3.16$, $u_2(1,7) = 2.65$ and the expected utilities of the initial allocations, multiplied by 3, are given by $\mathcal{U}_1 = 7.3$ and $\mathcal{U}_2 = 8.46$.

A. REE

Case 1.
First, we are looking for a fully revealing REE. Prices are normalized so that $p_1 = 1$ in each state. In effect we are analyzing an Edgeworth box economy per state.

State a. We find that

$$(p_1, p_2) = \left(1, \frac{8}{11}\right); \; x^*_{11} = \frac{85}{22}, \; x^*_{12} = \frac{85}{16},$$
$$x^*_{21} = \frac{91}{22}, \; x^*_{22} = \frac{91}{16}; \; u^*_1 = 4.53, \; u^*_2 = 4.85.$$

State b. We find that

$$(p_1, p_2) = (1, 1); \; x^*_{11} = 4, \; x^*_{12} = 4, \; x^*_{21} = 4, \; x^*_{22} = 4; \; u^*_1 = 4, \; u^*_2 = 4.$$

State c. We find that

$$(p_1, p_2) = \left(1, \frac{5}{8}\right); \; x^*_{11} = \frac{37}{16}, \; x^*_{12} = \frac{37}{10}, \; x^*_{21} = \frac{43}{16},$$
$$x^*_{22} = \frac{43}{10}; \; u^*_1 = 2.93, \; u^*_2 = 3.40.$$

The normalized expected utilities of the equilibrium allocations are $\mathcal{U}_1 = 11.46$, $\mathcal{U}_2 = 12.25$. This completes the analysis of the fully revealing REE.

We now look into whether there is a partially revealing or a non-revealing REE as well.

Case 2. Referring to the three states, we consider price vectors $p^1 = p^2 \neq p^3$ or $p^1 \neq p^2 = p^3$ or $p^1 = p^3 \neq p^2$.

We find that in all these cases no REE exists.

Case 3. We consider the price vectors to be equal, i.e. $p^1 = p^2 = p^3$, which means that the Agents get no information from the prices.

We find that no such equilibrium exists.

The above analysis shows that there is only a fully revealing REE. The equilibrium quantities are different in each state and therefore the REE allocations do not belong to either the private core or Radner equilibria.

Next we characterize the Radner equilibria. Apart from the analysis in the context of Example 6.1, (Radner equilibria 1), we also consider a modified model, in Example 6.2, in which every agent can distinguish between all states of nature, (Radner equilibria 2). The calculations in the latter case can be contrasted to the ones for the fully revealing equilibria.

Existence arguments in the case of correspondences can be advanced. However the actual calculation of such equilibria is not always straightforward.

B. Radner equilibria 1

The price vectors are $p(a) = p^1 = (p_1^1, p_2^1)$, $p(b) = p^2 = (p_1^2, p_2^2)$ and $p(c) = p^3 = (p_1^3, p_2^3)$. On the other hand we require measurability of allocations with respect to the private information of the agents.

The problems of the agents are:

Agent 1.
Maximize $\mathcal{U}_1 = 2(AB)^{\frac{1}{2}} + (x_{11}^3 x_{12}^3)^{\frac{1}{2}}$
Subject to

$$A(p_1^1 + p_1^2) + B(p_2^1 + p_2^2) + p_1^3 x_{11}^3 + p_2^3 x_{12}^3 = 7(p_1^1 + p_1^2) + (p_2^1 + p_2^2) + 4p_1^3 + p_2^3$$

and

Agent 2.
Maximize $\mathcal{U}_2 = (x_{21}^1 x_{22}^1)^{\frac{1}{2}} + 2(CD)^{\frac{1}{2}}$
Subject to

$$p_1^1 x_{21}^1 + p_2^1 x_{22}^1 + C(p_1^2 + p_1^3) + D(p_2^2 + p_2^3) = p_1^1 + 10p_2^1 + (p_1^2 + p_1^3) + 7(p_2^2 + p_2^3).$$

Applying a Gorman (1959) type argument we see that the demands of the agents will be of the form: $A = \frac{M_1}{2(p_1^1 + p_1^2)}$, $B = \frac{M_1}{2(p_2^1 + p_2^2)}$, $x_{11}^3 = \frac{M_2}{2p_1^3}$, $x_{12}^3 = \frac{M_2}{2p_2^3}$, $x_{21}^1 = \frac{m_1}{2p_1^1}$, $x_{22}^1 = \frac{m_1}{2p_2^1}$, $C = \frac{m_2}{2(p_1^2 + p_1^3)}$ and $D = \frac{m_2}{2(p_2^2 + p_2^3)}$.

It follows that a Radner equilibrium with non-negative prices exists if the following system of equations has a non-negative solution.

$$\frac{2}{((p_1^1 + p_1^2)(p_2^1 + p_2^2))^{\frac{1}{2}}} = \frac{1}{(p_1^3 p_2^3)^{\frac{1}{2}}},$$

$$M_1 + M_2 = 7(p_1^1 + p_1^2) + (p_2^1 + p_2^2) + 4p_1^3 + p_2^3$$
$$\frac{1}{(p_1^1 p_2^1)^{\frac{1}{2}}} = \frac{2}{((p_1^2 + p_1^3)(p_2^2 + p_2^3))^{\frac{1}{2}}}$$
$$m_1 + m_2 = p_1^1 + 10p_2^1 + (p_1^2 + p_1^3) + 7(p_2^2 + p_2^3)$$
$$\frac{M_1}{2(p_1^1 + p_1^2)} + \frac{m_1}{2p_1^1} = 8, \quad \frac{M_1}{2(p_2^1 + p_2^2)} + \frac{m_1}{2p_2^1} = 11$$
$$\frac{M_2}{2p_1^3} + \frac{m_2}{2(p_1^2 + p_1^3)} = 5, \quad \frac{M_2}{2p_2^3} + \frac{m_2}{2(p_2^2 + p_2^3)} = 8$$
$$\frac{M_1}{2(p_1^1 + p_1^2)} + \frac{m_2}{2(p_1^2 + p_1^3)} = 8, \quad \frac{M_1}{2(p_2^1 + p_2^2)} + \frac{m_2}{2(p_2^2 + p_2^3)} = 8.$$

The above system of equations is homogeneous of degree zero in the p_j^i's, the M_i's and the m_i's. Therefore some price, for example, p_1^1 could be fixed which reduces by one the number of unknowns. However the market equilibrium equations have one degree of redundancy as a consequence of Walras' law,

$$p_1^1(A + x_{21}^1 - 8) + p_2^1(B + x_{22}^1 - 11) + p_1^2(A + C - 8) + p_2^2(B + D - 8)$$
$$+ p_1^3(x_{11}^3 + C - 5) + p_2^3(x_{12}^3 + D - 8) = 0.$$

One can prove the existence of a Radner equilibrium by modifying the usual argument in general equilibrium theory, to take into account the fact that for Cobb-Douglas utility functions the demands are not defined on the whole boundary of the simplex. It is a rather tedious argument and we do not include it.

Approximate values for the equilibrium were obtained from the application of the random selection algorithm. A succession of random variables was appraised using a criterion consisting of the square root of the sum of squares of errors, the best selection so far being retained at each step. We did not normalize prices and all equations were used.

We obtained $p_1^1 = 1.1566$, $p_2^1 = 0.5876$, $p_1^2 = 0.3979$, $p_2^2 = 1.08597$, $p_1^3 = 1.3272$, $p_2^3 = 0.49009$, $M_1 = 14.1971$, $M_2 = 4.1574$, $m_1 = 7.9433$, and $m_2 = 11.8474$, which satisfy the equations to three decimal places. We have also checked the accuracy to more decimal places. If an error implies infeasibility in the sense that demand is larger than the resource then the implication is that a small quantity is not forthcoming. In the calculations we did not normalize prices, in order to allow for the maximum flexibility in the algorithm.

The same approximate solution can be obtained using Newton's method, starting the iteration from a suitable initial set of values. In order to avoid the problems arising from the need to invert a singular matrix, we normalized $p_1^2 = 1$ and, invoking Walras' law, we left out the 4th market equilibrium equation.

However there are dangers which may be illustrated by leaving out the 6th market equation. For the same initial values we approach a different point, where p_2^2 is essentially zero but the sixth equation is not satisfied. This is possible because in the Walras equation the contribution from the 6th equation has coefficient p_2^2 and thus can take any value. This means that a particular limit point cannot be a Radner equilibrium.

We also note that, of course, approximate solutions are not necessarily near the true solution. Even with continuity of functions the changes in the values corresponding to small changes in the variables might be very large.

We now have a digression the purpose of which is to explain that the full information, deterministic Radner equilibrium is not the same as the fully revealing REE.

C. Radner equilibria 2

Example 6.2. We shall now calculate the Radner equilibrium for the case with $\mathcal{F}_1 = \mathcal{F}_2 = \{\{a\}\{b\}\{c\}\}$. All other data are as in Example 6.1.

The problems of the two agents are:

Agent 1.
Maximize $\mathcal{U}_1 = (x^1_{11}x^1_{12})^{\frac{1}{2}} + (x^2_{11}x^2_{12})^{\frac{1}{2}} + (x^3_{11}x^3_{12})^{\frac{1}{2}}$
Subject to

$$p^1_1x^1_{11}+p^1_2x^1_{12}+p^2_1x^2_{11}+p^2_2x^2_{12}+p^3_1x^3_{11}+p^3_2x^3_{12} = 7(p^1_1+p^2_1)+(p^1_2+p^2_2)+4p^3_1+p^3_2$$

and

Agent 2.
Maximize $\mathcal{U}_2 = (x^1_{21}x^1_{22})^{\frac{1}{2}} + (x^2_{21}x^2_{22})^{\frac{1}{2}} + (x^3_{21}x^3_{22})^{\frac{1}{2}}$
Subject to

$$p^1_1x^1_{21}+p^1_2x^1_{22}+p^2_1x^2_{21}+p^2_2x^2_{22}+p^3_1x^3_{21}+p^3_2x^3_{22} = p^1_1+10p^1_2+(p^2_1+p^3_1)+7(p^2_2+p^3_2).$$

Applying a Gorman type argument we obtain $x^i_{1j} = \frac{M_i}{2p^i_j}$ and $x^i_{2j} = \frac{m_i}{2p^i_j}$. These demands imply $\mathcal{U}_1 = \frac{1}{2(p^1_1p^1_2)^{\frac{1}{2}}}M_1 + \frac{1}{2(p^2_1p^2_2)^{\frac{1}{2}}}M_2 + \frac{1}{2(p^3_1p^3_2)^{\frac{1}{2}}}M_3$ and $\mathcal{U}_2 = \frac{1}{2(p^1_1p^1_2)^{\frac{1}{2}}}m_1 + \frac{1}{2(p^2_1p^2_2)^{\frac{1}{2}}}m_2 + \frac{1}{2(p^3_1p^3_2)^{\frac{1}{2}}}m_3$.

The above $\mathcal{U}_1$ and $\mathcal{U}_2$ have to be maximized, each subject to the Agent's constraint cast in terms of M_i's for Agent 1 and m_i's for Agent 2, which is done below.

Notice that no price could be zero because both agents would seek infinite utility. Conditions for Radner equilibrium, such that each agent buys every good, are:

$$p^1_1p^1_2 = p^2_1p^2_2, \quad M_1 + M_2 + M_3 = 7(p^1_1 + p^2_1) + (p^1_2 + p^2_2) + 4p^3_1 + p^3_2$$

$$p^1_1p^1_2 = p^3_1p^3_2, \quad m_1 + m_2 + m_3 = p^1_1 + 10p^1_2 + (p^2_1 + p^3_1) + 7(p^2_2 + p^3_2)$$

$$\frac{M_1}{2p^1_1} + \frac{m_1}{2p^1_1} = 8, \quad \frac{M_1}{2p^1_2} + \frac{m_1}{2p^1_2} = 11$$

$$\frac{M_2}{2p^2_1} + \frac{m_2}{2p^2_1} = 8, \quad \frac{M_2}{2p^2_2} + \frac{m_2}{2p^2_2} = 8$$

$$\frac{M_3}{2p^3_1} + \frac{m_3}{2p^3_1} = 5, \quad \frac{M_3}{2p^3_2} + \frac{m_3}{2p^3_2} = 8.$$

The solution is obtained as follows: We normalize prices by setting $p_1^1 = 1$. From the 5th and 6th equation we obtain $p_2^1 = \frac{8}{11}$ and the 7th and 8th equation imply $p_2^1 = p_2^2$. The 9th and 10th equation imply $p_2^3 = \frac{5}{8}p_1^3$. Putting the last relations into the 1st and 2nd we get the remaining prices. Putting all the information together we have $p_1^1 = 1$, $p_2^1 = \frac{8}{11}$, $p_1^2 = p_2^2 = (\frac{8}{11})^{\frac{1}{2}}$, $p_1^3 = (\frac{64}{55})^{\frac{1}{2}}$, and $p_2^3 = \frac{5}{8} \times (\frac{64}{55})^{\frac{1}{2}}$.

Employing the above values for p_j^i we obtain for M_i and m_i the following relations:

$$M_1 + M_2 + M_3 = 7 \times \frac{8}{11} + 8 \times \left(\frac{8}{11}\right)^{\frac{1}{2}} + 4\frac{5}{8} \times \left(\frac{64}{55}\right)^{\frac{1}{2}},$$

$$m_1 + m_2 + m_3 = 8\frac{3}{8} + 8 \times \left(\frac{8}{11}\right)^{\frac{1}{2}} + 5\frac{3}{8} \times \left(\frac{64}{55}\right)^{\frac{1}{2}},$$

$$M_1 + m_1 = 16, \quad M_2 + m_2 = 16 \times \left(\frac{8}{11}\right)^{\frac{1}{2}} \quad \text{and } M_3 + m_3 = 10 \times \left(\frac{64}{55}\right)^{\frac{1}{2}},$$

which imply a possible solution $M_1 = 7 \times \frac{8}{11}, m_1 = 8 \times \frac{3}{11}, M_2 = m_2 = 8 \times (\frac{8}{11})^{\frac{1}{2}}$, $M_3 = 4\frac{5}{8} \times (\frac{64}{55})^{\frac{1}{2}}$ and $m_3 = 5\frac{3}{8} \times (\frac{64}{55})^{\frac{1}{2}}$. An obvious solution is $M_1 = 7\frac{8}{11}$, $m_1 = 8\frac{3}{11}$, $M_2 = m_2 = 8 \times (\frac{8}{11})^{\frac{1}{2}}$, $M_3 = 4\frac{5}{8} \times (\frac{64}{11})^{\frac{1}{2}}$ and $m_3 = 5\frac{3}{8} \times (\frac{64}{11})^{\frac{1}{2}}$.

However this solution is not unique. For example, we can add to the value for M_1 a small $\epsilon > 0$ and subtract it from m_1, and then adjust in the opposite direction M_2 and m_2. We obtain then a new solution to the system with the same maximum value for the utilities.

It follows that the normalized prices for an interior solution are unique, and so are the maximum utilities, but the M_i's and the m_i's can assume a number of values. The explanation of the last observation is as follows. The product of the two goods to the power $\frac{1}{2}$ becomes one good and given the equilibrium prices the structure of the problem is such that the agents are as well off with $\epsilon > 0$ as with $\epsilon = 0$.

One can ask why is it that the same argument would not apply to the previous formulation of Radner equilibria 1. There we seemed to be getting locally unique values of M_i's and m_i's. The reason was that we did not have the property that rearranging incomes between the agents in Period 1 can be fully compensated by doing so also in, for example, Period 2. In the present case the periods are among themselves separated. This was not the case in the previous formulation.

In that case, if we increase the composite commodity $(AB)^{\frac{1}{2}}$, where the M_i's have been calculated and decrease$(x_{21}^1 x_{22}^1)^{\frac{1}{2}}$, by adjusting M_1's and m_1's, then we have to decrease the commodity $(x_{11}^3 x_{12}^3)^{\frac{1}{2}}$, and increase $(CD)^{\frac{1}{2}}$, which requires a reduction in $(AB)^{\frac{1}{2}}$. Everything was finally balanced there.

There are also approximate equilibria from the random algorithm, which approach the true equilibrium above. Its application gives:

$$p_1^1 = 1, \; p_2^1 = 0.7272, \; p_1^2 = 0.8528, \; p_2^2 = 0.8528, \; p_1^3 = 1.0787, \; p_2^3 = 0.6742$$

and, approximately, $M_1 + m_1$ is 16.000051, $M_2 + m_2$ is 13.6448, and $M_3 + m_3$ is 10.7871. The algorithm also captures the fact that the values of the M_i's and m_i's are not fully determined.

On the basis of the above analysis, we see that full information Radner equilibrium is not the same as fully revealing REE because in the latter case a monotonic, nonlinear transformation can be applied, such as replacing $(x_{11}^i x_{12}^i)^{\frac{1}{2}}$ by $(x_{11}^i x_{12}^i)$, without affecting the results as the calculations are per period. This is not the case in Radner equilibrium where the calculations are on the sum over all the periods.

We return now to the characterization of equilibrium concepts in Example 6.1.

D. WFC

With respect to the cooperative equilibrium concepts, first we show that in this example the fully revealing REE is in the WFC. These allocations are obtained by solving the following problem, where we use superscripts to characterize the states. Superscripts 1, 2 and 3 correspond to states a, b and c respectively. The WFC is characterized as follows:

Problem

Maximize $\mathcal{U}_1 = (x_{11}^1 x_{12}^1)^{\frac{1}{2}} + (x_{11}^2 x_{12}^2)^{\frac{1}{2}} + (x_{11}^3 x_{12}^3)^{\frac{1}{2}}$
Subject to

$$((8-x_{11}^1)(11-x_{12}^1))^{\frac{1}{2}}+((8-x_{11}^2)(8-x_{12}^2))^{\frac{1}{2}}+((5-x_{11}^3)(8-x_{12}^3))^{\frac{1}{2}}=\mathcal{U}_2 \text{ (fixed)}$$
$$\mathcal{U}_1 \geq 7.3, \quad \mathcal{U}_2 \geq 8.46.$$

The conditions on the utility functions imply that there is a unique interior maximum per $\mathcal{U}_2$. Setting up the Lagrangean function we obtain the first order conditions:

$$\frac{{x_{12}^1}^{\frac{1}{2}}}{{x_{11}^1}^{\frac{1}{2}}} = \ell \frac{(11-x_{12}^1)^{\frac{1}{2}}}{(8-x_{11}^1)^{\frac{1}{2}}}$$
$$\frac{{x_{12}^2}^{\frac{1}{2}}}{{x_{11}^2}^{\frac{1}{2}}} = \ell \frac{(8-x_{12}^2)^{\frac{1}{2}}}{(8-x_{11}^2)^{\frac{1}{2}}}$$
$$\frac{{x_{12}^3}^{\frac{1}{2}}}{{x_{11}^3}^{\frac{1}{2}}} = \ell \frac{(8-x_{12}^3)^{\frac{1}{2}}}{(5-x_{11}^3)^{\frac{1}{2}}}$$
$$\left(\frac{{x_{12}^1}^{\frac{1}{2}}}{{x_{11}^1}^{\frac{1}{2}}}\right)^{-1} = \ell \left(\frac{(11-x_{12}^1)^{\frac{1}{2}}}{(8-x_{11}^1)^{\frac{1}{2}}}\right)^{-1}$$
$$\left(\frac{{x_{12}^2}^{\frac{1}{2}}}{{x_{11}^2}^{\frac{1}{2}}}\right)^{-1} = \ell \left(\frac{(8-x_{12}^2)^{\frac{1}{2}}}{(8-x_{11}^2)^{\frac{1}{2}}}\right)^{-1}$$
$$\left(\frac{{x_{12}^3}^{\frac{1}{2}}}{{x_{11}^3}^{\frac{1}{2}}}\right)^{-1} = \ell \left(\frac{(8-x_{12}^3)^{\frac{1}{2}}}{(5-x_{11}^3)^{\frac{1}{2}}}\right)^{-1}$$
$$\left(\left(8-x_{11}^1\right)\left(11-x_{12}^1\right)\right)^{\frac{1}{2}} + ((8-x_{11}^2)(8-x_{12}^2))^{\frac{1}{2}} + \left(\left(5-x_{11}^3\right)\left(8-x_{12}^3\right)\right)^{\frac{1}{2}}$$
$$= \mathcal{U}_2 \text{ (fixed).}$$

It is easy to see that these conditions are satisfied by the REE allocations with the Lagrange multiplier $\ell = 1$.

E. Private core

Next we look at the way we can obtain the private core allocations and then we shall have to find the WFV allocations. We allow for free disposal and see what happens. For the private core allocations we impose private information measurability and solve the following:

Problem

Maximize $\mathcal{U}_1 = 2A^{\frac{1}{2}}B^{\frac{1}{2}} + (x_{11}^3 x_{12}^3)^{\frac{1}{2}}$
Subject to

$$\begin{aligned}&(x_{21}^1 x_{22}^1)^{\frac{1}{2}} + 2C^{\frac{1}{2}}D^{\frac{1}{2}} \geq \mathcal{U}_2 \text{ (fixed)}\\ &A + x_{21}^1 \leq 8,\ \ B + x_{22}^1 \leq 8,\ \ A + C \leq 8,\ \ B + D \leq 8,\\ &A + C \leq 8,\ \ B + D \leq 8,\ \ x_{11}^3 + C \leq 5,\ \ x_{12}^3 + D \leq 8,\\ &\mathcal{U}_1 \geq 7.3, \mathcal{U}_2 \geq 8.46.\end{aligned}$$

We operate with equality constraints eliminating $x_{21}^1,\ x_{22}^1,\ x_{11}^3,\ x_{12}^3,\ A$ and D and forming the Lagrangean $L = 2(8-C)^{\frac{1}{2}}(B)^{\frac{1}{2}} + \lambda\{(C)^{\frac{1}{2}}(11-B)^{\frac{1}{2}} + 2(C)^{\frac{1}{2}}(8-B)^{\frac{1}{2}} - \mathcal{U}_2\}$.

First order conditions are

$$\left(\frac{8-C}{B}\right)^{\frac{1}{2}} + \frac{1}{2}\left(\frac{5-C}{B}\right)^{\frac{1}{2}} = \ell\left\{\frac{1}{2}\left(\frac{C}{11-B}\right)^{\frac{1}{2}} + \left(\frac{C}{8-B}\right)^{\frac{1}{2}}\right\} \tag{8}$$

and

$$\left(\frac{B}{8-C}\right)^{\frac{1}{2}} + \frac{1}{2}\left(\frac{B}{5-C}\right)^{\frac{1}{2}} = \ell\left\{\frac{1}{2}\left(\frac{11-B}{C}\right)^{\frac{1}{2}} + \left(\frac{8-B}{C}\right)^{\frac{1}{2}}\right\} \tag{9}$$

which we can rewrite as

$$\frac{1}{C^{\frac{1}{2}}}\left\{(8-C)^{\frac{1}{2}} + \frac{1}{2}(5-C)^{\frac{1}{2}}\right\} = \ell B^{\frac{1}{2}}\left\{\frac{1}{2}\left(\frac{1}{11-B}\right)^{\frac{1}{2}} + \left(\frac{1}{8-B}\right)^{\frac{1}{2}}\right\} \tag{10}$$

and

$$C^{\frac{1}{2}}\left\{\left(\frac{1}{8-C}\right)^{\frac{1}{2}} + \frac{1}{2}\left(\frac{1}{5-C}\right)^{\frac{1}{2}}\right\} = \ell\frac{1}{B^{\frac{1}{2}}}\left\{\frac{1}{2}(11-B)^{\frac{1}{2}} + (8-B)^{\frac{1}{2}}\right\}. \tag{11}$$

Dividing gives

$$\frac{1}{C}\left\{\frac{(8-C)^{\frac{1}{2}} + \frac{1}{2}(5-C)^{\frac{1}{2}}}{\frac{1}{2}\frac{1}{(8-C)^{\frac{1}{2}}} + \frac{1}{(5-C)^{\frac{1}{2}}}}\right\} = B\left\{\frac{\frac{1}{2}\frac{1}{(11-B)^{\frac{1}{2}}} + \frac{1}{(8-B)^{\frac{1}{2}}}}{\frac{1}{2}(11-B)^{\frac{1}{2}} + (8-B)^{\frac{1}{2}}}\right\}. \tag{12}$$

It is a matter of routine substitutions to show that the allocation $x_1 = ((5.5, 5.5), (5.5, 5.5), (2.5, 5.5))$, $x_2 = ((2.5, 5.5), (2.5, 2.5), (2.5, 2.5))$ is in the private core, with normalized expected utilities $\mathcal{U}_1 = 14.70$ and $\mathcal{U}_2 = 8.70$.

Next we show that this allocation cannot be obtained as a Radner equilibrium with positive prices. We are looking for equality in all the conditions stated in the section Radner equilibria 1. A corner solution would require some zero quantities.

Substituting into the conditions for the demand functions we obtain $M_1 = 11(p_1^1 + p_1^2)$, $m_1 = 5p_1^1$, $M_1 = 11(p_2^1 + p_2^2)$, $m_1 = 11p_2^1$, $M_2 = 5p_1^3$, $m_2 = 5(p_1^1 + p_1^2)$, $M_2 = 11p_3^2$ and $m_2 = (p_2^2 + p_2^3)$. We normalize and set $p_1^1 = 1$. Then we obtain $m_1 = 5$, $p_2^1 = \frac{5}{11}$, $p_2^3 = 1$, $p_1^3 = \frac{11}{5}$, and we require further that $4p_1^3p_2^3 = (p_1^1 + p_1^2)(p_2^1 + p_2^2)$ and $4p_1^1p_2^1 = (p_1^2 + p_1^3)(p_2^2 + p_2^3)$. These equations cannot be satisfied by nonnegative prices because they imply $-3.890 = \frac{6}{11}p_1^2 + \frac{6}{5}p_2^2$.

Obviously there are measurable allocations which are not in the private core, such as

$$x_1 = ((5,5),(5,5),(2,5)), \quad \text{and} \quad x_2 = ((3,6),(3,3),(3,3)),$$
$$x_1 = ((4,4),(4,4,(1,4)), \qquad x_2 = ((4,7),(4,4),(4,4))$$

as can be seen through routine calculations.

On the other hand we can show directly that a Radner equilibrium is in the private core. Taking into account the constraints for demand to be equal to supply, the first order conditions for the agents' maximization of utilities can be cast as follows.

For Agent 1:

$$\frac{B^{\frac{1}{2}}}{(8-C)^{\frac{1}{2}}} - \ell'(p_1^1 + p_1^2) = 0, \qquad \frac{(8-C)^{\frac{1}{2}}}{B^{\frac{1}{2}}} - \ell'(p_2^1 + p_2^2) = 0, \tag{13}$$

$$\frac{1}{2}\frac{B^{\frac{1}{2}}}{(5-C)^{\frac{1}{2}}} - \ell' p_1^3 = 0, \quad \text{and} \quad \frac{1}{2}\frac{(5-C)^{\frac{1}{2}}}{B^{\frac{1}{2}}} - \ell' p_2^3 = 0, \tag{14}$$

and for Agent 2:

$$\frac{1}{2}\frac{(11-B)^{\frac{1}{2}}}{C^{\frac{1}{2}}} - \psi p_1^1 = 0, \qquad \frac{1}{2}\frac{C^{\frac{1}{2}}}{(11-B)^{\frac{1}{2}}} - \psi p_2^1 = 0, \tag{15}$$

$$\frac{(8-B)^{\frac{1}{2}}}{C^{\frac{1}{2}}} - \psi(p_1^2 + p_1^3) = 0, \quad \text{and} \quad \frac{C^{\frac{1}{2}}}{(8-B)^{\frac{1}{2}}} - \psi(p_2^2 + p_2^3) = 0. \tag{16}$$

Substituting (14), (15), (16) and (17) into (9) and (10) we obtain in both instances the relation $\ell' = \ell\psi$ which shows that the Radner equilibrium is in the private core.

F. WFV

Routine calculations imply $V_\lambda(\{1\}) = \frac{1}{3}\lambda_1 A$, $V_\lambda(\{2\}) = \frac{1}{3}\lambda_2 B$, where $A = (2(7)^{\frac{1}{2}} + 2)$ and $B = (2(7)^{\frac{1}{2}} + 10^{\frac{1}{2}})$.

Next we have, $V_\lambda(\{1,2\}) = \frac{1}{3}\max_x \{\lambda_1(x_{11}^1 x_{12}^1)^{\frac{1}{2}} + \lambda_2(8 - x_{11}^1)^{\frac{1}{2}}(11 - x_{12}^1)^{\frac{1}{2}} + \lambda_1(x_{11}^2 x_{12}^2)^{\frac{1}{2}} + \lambda_2(8 - x_{11}^2)^{\frac{1}{2}}(8 - x_{12}^2)^{\frac{1}{2}} + \lambda_1(x_{11}^3 x_{12}^3)^{\frac{1}{2}} + \lambda_2(5 - x_{11}^3)^{\frac{1}{2}}(8 - x_{12}^3)^{\frac{1}{2}}\}$.

We define the per period terms of the sum by $\mathcal{U}^1, \mathcal{U}^2$ and $\mathcal{U}^3$. We assume that both λ's are positive. Otherwise all the weight is put on one agent. We can do separate maximization and defining $\Lambda_1 = \lambda_1^2$, $\Lambda_2 = \lambda_2^2$ we obtain the conditions

(i) $\Lambda_1 x_{12}^1(8 - x_{11}^1) = \Lambda_2 x_{11}^1(11 - x_{12}^1)$ and $\Lambda_1 x_{11}^1(11 - x_{12}^1) = \Lambda_2 x_{12}^1(8 - x_{11}^1)$
(ii) $\Lambda_1 x_{12}^2(8 - x_{11}^2) = \Lambda_2 x_{11}^2(8 - x_{12}^2)$ and $\Lambda_1 x_{11}^2(8 - x_{12}^2) = \Lambda_2 x_{12}^2(8 - x_{11}^2)$
(iii) $\Lambda_1 x_{12}^3(5 - x_{11}^3) = \Lambda_2 x_{11}^3(8 - x_{12}^3)$ and $\Lambda_1 x_{11}^3(11 - x_{12}^3) = \Lambda_2 x_{12}^3(5 - x_{11}^3)$

From (i), (ii) and (iii) we obtain, respectively, $x_{12}^1 = \frac{11}{8}x_{11}^1$, $x_{12}^2 = x_{11}^2$ and $x_{12}^3 = \frac{8}{5}x_{11}^3$. Which means that the maximum will be sought on these flats.

From the above we obtain $\mathcal{U}^1 = (\frac{11}{8})^{\frac{1}{2}}(\lambda_1 x_{11}^1 + \lambda_2(8 - x_{11}^1))$, $\mathcal{U}^2 = \lambda_1 x_{11}^2 + \lambda_2(8 - x_{11}^2))$ and $\mathcal{U}^3 = (\frac{8}{5})^{\frac{1}{2}}(\lambda_1 x_{11}^3 + \lambda_2(5 - x_{11}^3))$. It follows, that $V_\lambda(\{1,2\}) = \frac{1}{3}\max_{x_1}[(\frac{11}{8})^{\frac{1}{2}}(\lambda_1 x_{11}^1 + \lambda_2(8 - x_{11}^1)) + (\lambda_1 x_{11}^2 + \lambda_2(8 - x_{11}^2)) + (\frac{8}{5})^{\frac{1}{2}}(\lambda_1 x_{11}^3 + \lambda_2(5 - x_{11}^3))]$. I.e. $V_\lambda(\{1,2\}) = \frac{1}{3}\max_{x_1}[8(\frac{11}{8})^{\frac{1}{2}}\{\max(\lambda_1, \lambda_2)\} + 8\{\max(\lambda_1, \lambda_2)\} + 5(\frac{8}{5})^{\frac{1}{2}}\{\max(\lambda_1, \lambda_2)\}]$, which we can write as $V_\lambda(\{1,2\}) = C\max(\lambda_1, \lambda_2)$, where $C = (88)^{\frac{1}{2}} + 8 + (40)^{\frac{1}{2}}$. The significance of the flats is clear. For maximization the choice from the extreme values of the variable x_1 depends on the values of λ_1 and λ_2. In particular for $\lambda_1 > \lambda_2$ all endowments are allocated to the utility function of Agent 1, for $\lambda_1 < \lambda_2$ the one of Agent 2, and for $\lambda_1 = \lambda_2$ the allocation can be arbitrary. This can be seen by obtaining $V_\lambda(\{1,2\})$ through the per period maximization of the utility of Agent 1 subject to the utility of Agent 2 being fixed.

For WFV allocations we require solutions to

$$\lambda_1 \sum_\omega (x_{11}(\omega)x_{12}(\omega))^{\frac{1}{2}} = \frac{1}{2}\{C\max\{\lambda_1, \lambda_2\} + A\lambda_1 - B\lambda_2\} \tag{17}$$

$$\lambda_2 \sum_\omega (x_{21}(\omega)x_{22}(\omega)) = \frac{1}{2}\{C\max\{\lambda_1, \lambda_2\} - A\lambda_1 + B\lambda_2\}$$

subject to

$$x_1 + x_2 \le e_1 + e_2,$$

relaxing the feasibility condition. The right-hand sides of the equations above are the $Sh_i(V_\lambda)$'s.

The set of WFV allocations is not empty. It can be checked that for $\lambda_1 = \lambda_2$ the allocation in which P1 gets $((4, \frac{11}{2}), (5,5), (\frac{5}{4}, 2))$ and P2 gets $((4, \frac{11}{2}), (3,3), (\frac{15}{4}, 6))$ is a WFV allocation. We see this by inserting these allocations and $\lambda_1 = \lambda_2$ into the relations above to obtain

$$(22)^{\frac{1}{2}} + 5 + (2.5)^{\frac{1}{2}} = \frac{1}{2}\left((88)^{\frac{1}{2}} + 8 + (40)^{\frac{1}{2}} + 2(7)^{\frac{1}{2}} + 2 - (10)^{\frac{1}{2}} - 2(7)^{\frac{1}{2}}\right)$$

$$2((22)^{\frac{1}{2}} + 3 + (7.5)^{\frac{1}{2}}) = \frac{1}{2}\left((88)^{\frac{1}{2}} + 8 + (40)^{\frac{1}{2}} - 2(7)^{\frac{1}{2}} - 2 + (10)^{\frac{1}{2}} + 2(7)^{\frac{1}{2}}\right)$$

which can be checked that they are satisfied.

On the other hand, it is a matter of tedious calculations to show that the fully revealing REE is not a WFV allocation although it belongs to the weak fine core.

Performing the calculations we obtain the relations

$$11.46\lambda_1 = \frac{1}{2}\left[23.71\{\max(\lambda_1, \lambda_2)\} + 7.291\lambda_1 - 8.45378\lambda_2\right] \text{ and}$$
$$12.25\lambda_2 = \frac{1}{2}\left[23.71\{\max(\lambda_1, \lambda_2)\} - 7.291\lambda_1 + 8.45378\lambda_2\right].$$

We distinguish between two cases and we examine whether the REE is in a weak fine value allocation.

Case 1. $\lambda_1 \geq \lambda_2$
We require

$$11.46\lambda_1 = \frac{1}{2}\left[23.71\lambda_1 + 7.291\lambda_1 - 8.45378\lambda_2\right] \text{ and}$$
$$12.25\lambda_2 = \frac{1}{2}\left[23.71\lambda_1 - 7.291\lambda_1 + 8.45378\lambda_2\right]$$

which imply $4.04\lambda_1 = 4.23\lambda_2$ and $8.21\lambda_1 = 8.22\lambda_1$ both of which cannot be satisfied.

Case 2. $\lambda_2 \geq \lambda_1$
We require now

$$22.92\lambda_1 = 23.71\lambda_2 + 7.291\lambda_1 - 8.45378\lambda_2 \text{ and}$$
$$24.50\lambda_2 = 23.71\lambda_2 - 7.291\lambda_1 + 8.45378\lambda_2$$

which imply $15.63\lambda_1 = 15.26\lambda_2$ and $7.66\lambda_2 = 7.29\lambda_1$ which again cannot be satisfied.

The question arises why is the set of WFV allocations smaller than the WFC, although this of course is only true in the case of two agents. An intuitive explanation is that for the WFV allocations the conditions are more stringent because of the homogeneity of equations in λ_1, and λ_2. We need to get from both equations in (18) the same $\frac{\lambda_1}{\lambda_2}$, and if we are given $x(\omega)$ this is highly unlikely to happen.

Now we show that a WFV equilibrium exists only for $\lambda_1 = \lambda_2$.

Adding side by side the equations (18), we get on the RHS $C\max\{\lambda_1, \lambda_2\}$ which is equal to $V_\lambda(\{1,2\})$. Therefore the sum on the LHS must be also equal to $V_\lambda(\{1,2\})$ and therefore a maximum, and we have seen how this depends on the weights λ_1 and λ_2.

Putting all the information together leads to the following possibilities. $\lambda_1 > \lambda_2$ requires

$$\lambda_1 C = \frac{1}{2}\left\{\lambda_1 C + A\lambda_1 - B\lambda_2\right\}$$
$$0 = \frac{1}{2}\left\{\lambda_1 C - A\lambda_1 + B\lambda_2\right\}.$$

Either of these leads to

$$B\lambda_2 = (A - C)\lambda_1 < 0$$

which is impossible. Similarly $\lambda_1 < \lambda_2$ is impossible.

Finally with $\lambda_1 = \lambda_2$ the equations for a weak-fine-value become, on writing $y_\alpha = x_{12}(\omega_\alpha)$ and recalling that $x_\alpha = \frac{8}{11} y_\alpha$,

$$\left(\frac{8}{11}\right)^{\frac{1}{2}} (2y_1 - 11) + (2y_2 - 8) + \left(\frac{5}{8}\right)^{\frac{1}{2}} (2y_3 - 8) = 2 - 10^{\frac{1}{2}}$$

which is satisfied by the previous specified allocation.

Example 6.3. The problem is a two-state, $\Omega = \{a, b\}$, three-player, two-good game with utilities and initial endowments given by:

$$\begin{aligned} u_1(x_{11}^j, x_{12}^j) &= \min(x_{11}^j, x_{12}^j); \quad e_1 = ((1,0),(1,0)), \quad F_1 = \{\{a\},\{b\}\} \\ u_2(x_{21}^j, x_{22}^j) &= \min(x_{21}^j, x_{22}^j); \quad e_2 = ((0,1),(0,1)), \quad F_2 = \{\{a\},\{b\}\} \\ u_3(x_{31}^j, x_{32}^j) &= \frac{x_{31}^j + x_{32}^j}{2}; \quad e_3 = ((0,0),(0,0)), \quad F_3 = \{\{a,b\}\}, \end{aligned}$$

where x_{ik}^j denotes consumption of Player i of Good k, in state j. Every player has the same prior distribution $\mu(\omega) = \frac{1}{2}$ for $\omega \in \Omega$.

The weights of the agents are λ_i for $i = 1, 2, 3$. First we calculate the characteristic function V_λ.

For $S = \{1\}$, $\{2\}$ or $\{3\}$ we have $e_i = x_i$ and so $u_i = 0$. Therefore $V_\lambda(\{i\}) = 0$. Next consider $S = (\{1,2\})$. The sum of the weighted utilities

$$\sum_{j \in \Omega} \frac{1}{2}[\lambda_1 \min(x_{11}^j, x_{12}^j) + \lambda_2 \min(x_{21}^j, x_{22}^j)]$$

must be maximized subject to $x_{11}^j + x_{21}^j = 1$ and $x_{12}^j + x_{22}^j = 1$ for $j \in \Omega$. It is straightforward that for a maximum we must have $x_{11} = x_{12}$ and $x_{21} = x_{22}$ and then that $V_\lambda(\{1,2\}) = \max(\lambda_1, \lambda_2)$. It is also straigtforward that $V_\lambda(\{1,3\}) = V_\lambda(\{2,3\}) = \frac{\lambda_3}{2}$.

We now turn our attention to $S = \{1,2,3\}$. The expression

$$\sum_{j \in \Omega} \frac{1}{2} \left[\lambda_1 \min(x_{11}^j, x_{12}^j) + \lambda_2 \min(x_{21}^j, x_{22}^j) + \lambda_3 \frac{x_{31}^j + x_{32}^j}{2} \right]$$

must be maximized subject to $x_{11}^j + x_{21}^j + x_{31}^j = 1 \quad x_{12}^j + x_{22}^j + x_{32}^j = 1$, for $j \in \Omega$.

Again from the first two terms we get $\max(\lambda_1, \lambda_2)$ and for the whole constraint sum $V_\lambda(\{1,2,3\}) = \max(\lambda_1, \lambda_2, \lambda_3)$.

Consider now the special case $\lambda_i = 1$ for $i = 1, 2, 3$. Replacing in the above λ_i by 1 we obtain

$$\begin{aligned} &V_\lambda(\{i\}) = 0, \quad \text{for } i = 1,2,3 \\ &V_\lambda(\{1,2\}) = 1, \quad V_\lambda(\{1,3\}) = V_\lambda(\{2,3\}) = \frac{1}{2} \quad \text{for } i = 2,3 \\ &V_\lambda(\{1,2,3\}) = 1. \end{aligned}$$

For this particular case, $\lambda_i = 1$, the Shapley values are given by

$$Sh_1(V) = 0 + \frac{1}{6}(1-0) + \frac{1}{6}\left(\frac{1}{2} - 0\right) + \frac{2}{6}\left(1 - \frac{1}{2}\right) = \frac{5}{12}$$
$$Sh_2(V) = \frac{5}{12}, \text{ and } Sh_3(V) = \frac{2}{12}.$$

Hence the value allocation is, per state,

$$(x_{11}, x_{12}) = (x_{21}, x_{22}) = \left(\frac{5}{12}, \frac{5}{12}\right) \text{ and } (x_{31}, x_{32}) = \left(\frac{2}{12}, \frac{2}{12}\right).$$

On the other hand any Walrasian type allocation will award zero quantities to Player 3, as he has no initial endowments. Therefore the point that this example is making is that with the number of agents $n \geq 3$, it is possible that there is a value allocation which does not belong to a Walrasian type set, (i.e. it is not a REE or Radner equilibrium).

However it can also be used to make one more point that is equally important. It can be seen that for $\lambda_1 = \lambda_2 = 1$ and $\lambda_3 = 0$ or for the case where there is no third agent and with $\lambda_1 = \lambda_2 = 1$ we have $Sh_1(V) = Sh_2(V) = \frac{1}{2}$. This says that it is possible that a group which includes all the agents can do better for its members than each one of them in isolation, but this is not the end of the story. A sub-group might do even better.

With respect to offering an interpretation of the distinction between side payments and a WFV allocation, we look at the following situation. Two agents have some initial endowments, the same weights, and their utilities are really revenues from selling these quantities in a non-competitive market. We can hand over all quantities to one agent, ask him to sell them on the market, keep his Shapley share and hand the other agent his own. With respect to the weak fine value it corresponds to the case when only a redistribution of the endowments is allowed, in which case we might only be able to do it when specific weights are given to the individuals.

Non-existence of REE:
Finally we discuss a specific version of the well known Kreps (1977) example of a non-existent REE. On the other hand, in the same example, the private core exists which suggests that the latter concept has an advantage over that of REE.[11]

Example 6.4 (2003b). There are two agents $I = \{1, 2\}$, two commodities, i.e. $X_i = \mathbb{R}^2_+$ for each Agent, i, and two states of nature $\Omega = \{\omega_1, \omega_2\}$, considered by the agents as equally probable. In x_{ij} the first index will refer to the agent and the second to the good.

We assume that the endowments, per state ω_1 and ω_2 respectively, and information partitions of the agents are given by

$$e_1 = ((1.5, 1.5), (1.5, 1.5), \qquad \mathcal{F}_1 = \{\{\omega_1\}, \{\omega_2\}\};$$
$$e_2 = ((1.5, 1.5), (1.5, 1.5), \qquad \mathcal{F}_2 = \{\{\omega_1, \omega_2\}\}.$$

[11] Example 6.4 is also discussed in Glycopantis et al. (2003b).

The utility functions of Agents 1 and 2 respectively, are for ω_1 given by $u_1 = log\ x_{11} + x_{12}$ and $u_2 = 2\ log\ x_{21} + x_{22}$ and for state ω_2 by $u_1 = 2\ log\ x_{11} + x_{12}$ and $u_2 = \ log\ x_{21} + x_{22}$.

We consider first the possibility of REE.

Case 1. Fully revealing REE.
Suppose that there exist, after normalization, prices $(p_1(1), p_2(1)) \neq (p_1(2), p_2(2))$, where $p_i(j)$ denotes the price of good i in state j. The problems that the two agents solve are as follows.

State ω_1.

Agent 1:
Maximize $u_1 = \ log\ x_{11} + x_{12}$
Subject to

$$p_1(1)x_{11} + p_2(1)x_{12} = 1.5(p_1(1) + p_2(1))$$

and

Agent 2:
Maximize $u_2 = 2\ log\ x_{21} + x_{22}$
Subject to

$$p_1(1)x_{21} + p_2(1)x_{22} = 1.5(p_1(1) + p_2(1)).$$

The agents solve analogous problems in state ω_2. However it is not possible to find $(p_1(1), p_2(1)) \neq (p_1(2), p_2(2))$. In the two problems the demands of the agents are interchanged so that the total demand stays the same while the total supply is fixed. It is also straightforward to check that there is no multiplicity of equilibria per state.

Case 2. Non-revealing REE.
Now we consider the possibility of $p_1(1) = p_1(2) = p_1$ and $p_2(1) = p_2(2) = p_2$. The two agents would act as follows.

Agent 1:
He can tell the states of nature and obtains the demand functions for $\omega_1, x_{11} = \frac{p_2}{p_1}$ and $x_{12} = \frac{1.5p_1}{p_2} + 0.5$ and for $\omega_2, x_{11} = \frac{2p_2}{p_1}$ and $x_{12} = \frac{1.5p_1}{p_2} - 0.5$ for $3p_1 \geq p_2$.

It is clear that the demands differ per state of nature.

Agent 2:
He sets $x_{21}(\omega_1) = x_{21}(\omega_2) = x_{21}$ and $x_{22}(\omega_1) = x_{22}(\omega_2) = x_{22}$ and solves the problem:

Maximize $u_2 = \frac{1}{2}(2\ log\ x_{21} + x_{22}) + \frac{1}{2}(\ log\ x_{21} + x_{22}) = 1.5\ log\ x_{21} + x_{22}$
Subject to

$$p_1 x_{21} + p_2 x_{22} = 1.5(p_1 + p_2).$$

So the highest indifference curve touches the budget constraint only once. On the other hand the demands of Agent 1 differ per ω. It follows that the markets cannot be cleared with common prices in both states of nature.

The above analysis shows that there is no REE in this model.

Next we consider, in the same example, the existence of private core allocations. These are obtained as solutions of the problem:

Maximize $E_2 = 1.5\ log\ x_{21} + x_{22}$
Subject to

$$\frac{1}{2}(\ log\ x_{11}(\omega_1) + x_{12}(\omega_1)) + \frac{1}{2}(\ log\ x_{11}(\omega_2) + x_{12}(\omega_2)) \geq E_1 \text{ (fixed)},$$

$$x_{1j}(\omega_1), x_{1j}(\omega_2) \geq 0, E_1,\ E_2 \geq 1.5\ log\ 1.5 + 1.5$$

$$x_{21} + x_{11}(\omega_1) \leq 3, \quad x_{21} + x_{11}(\omega_2) \leq 3,$$

$$x_{22} + x_{12}(\omega_1) \leq 3, \quad x_{22} + x_{12}(\omega_2) \leq 3.$$

The structure of the problem, i.e. the continuity of the objective function and the compactness of the feasible set, implies that it has always a solution. In particular, if we set the quantity constraints equal to 3 and $1.5\ log\ 1.5 + 1.5 = E_1$ then the initial allocation is in the private core.

The discussion of Example 6.4 indicates that the REE may not be an appropriate concept to explain trades in DIE. The agents here receive no instructions as to what they should be doing.

7 Incentive compatibility

There are alternative formulations of the notion of incentive compatibility. The basic idea is that an allocation is incentive compatible if no coalition can misreport the realized state of nature and have a distinct possibility of making its members better off.

Suppose we have a coalition S, with members denoted by i, and the complementary set $I \setminus S$ with members j. Let the realized state of nature be ω^*. Each member $i \in S$ sees $E_i(\omega^*)$. Obviously not all $E_i(\omega^*)$ need be the same, however all Agents i know that the actual state of nature could be ω^*.

Consider a state ω' such that for all $j \in I \setminus S$ we have $\omega' \in E_j(\omega^*)$ and for at least one $i \in S$ we have $\omega' \notin E_i(\omega^*)$. Now the coalition S decides that each member i will announce that she has seen her own set $E_i(\omega')$ which, of course, contains a lie. On the other hand we have that $\omega' \in \bigcap_{j \notin S} E_j(\omega^*)$.

The idea is that if all members of $I \setminus S$ believe the statements of the members of S then each $i \in S$ expects to gain. For *coalitional Bayesian incentive compatibility* (CBIC) of an allocation we require that this is not possible. This is the incentive compatibility condition we used in Glycopantis et al. (2001).[12]

We showed in Example 5.1 that in the three-agent economy without free disposal the private core allocation $x_1 = (4, 4, 1)$, $x_2 = (4, 1, 4)$ and $x_3 = (2, 0, 0)$ is incentive compatible. This follows from the fact that Agent 3 who would potentially cheat in state a has no incentive to do so. It has been shown in Koutsougeras and Yannelis (1993) that if the utility functions are monotone and continuous then private core allocations are *always* CBIC.

[12] See Krasa and Yannelis (1994) and Hahn and Yannelis (1997) for related concepts.

On the other hand the WFC allocations are not always incentive compatible, as the proposed redistribution $x_1 = x_2 = (5, 2.5, 2.5)$ in Example 5.2 shows. Indeed, if Agent 1 observes $\{a, b\}$, he has an incentive to report c and Agent 2 has an incentive to report b when he observes $\{a, c\}$.

CBIC coincides in the case of a two-agent economy with the concept of *Individually Bayesian Incentive Compatibility* (IBIC), which refers to the case when S is a singleton.

We consider here explicitly the concept of *Transfer Coalitionally Bayesian Incentive Compatible* (TCBIC) allocations. This allows for transfers between the members of a coalition, and is therefore a strengthening of the concept of Coalitionally Bayesian Incentive Compatibility (CBIC).

Definition 7.1. An allocation $x = (x_1, \ldots, x_n) \in \bar{L}_X$, with or without free disposal, is said to be TCBIC if it is not true that there exists a coalition S, states ω^* and ω', with ω^* different from ω' and $\omega' \in \bigcap_{i \notin S} E_i(\omega^*)$ and a random, net-trade vector, $z = (z_i)_{i \in S}$ among the members of S,

$$(z_i)_{i \in S}, \sum_S z_i = 0$$

such that for all $i \in S$ there exists $\bar{E}_i(\omega^*) \subseteq Z_i(\omega^*) = E_i(\omega^*) \cap (\bigcap_{j \notin S} E_j(\omega^*))$, for which

$$\sum_{\omega \in \bar{E}_i(\omega^*)} u_i(\omega, e_i(\omega) + x_i(\omega') - e_i(\omega') + z_i) q_i\left(\omega | \bar{E}_i(\omega^*)\right) \tag{18}$$
$$> \sum_{\omega \in \bar{E}_i(\omega^*)} u_i(\omega, x_i(\omega)) q_i\left(\omega | \bar{E}_i(\omega^*)\right).$$

Notice that $e_i(\omega) + x_i(\omega') - e_i(\omega') + z_i(\omega) \in X_i(\omega)$ is not necessarily measurable. The definition implies that no coalition can hope that by misreporting a state, every member will become better off if they are believed by the members of the complementary set.

Returning to Definition 7.1, one can define CBIC to correspond to $z_i = 0$ and then IBIC to the case when S is a singleton. Thus we have (not IBCI) $\Rightarrow$ (not CBIC) $\Rightarrow$ (not TCBIC). It follows that TCBIC $\Rightarrow$ CBIC $\Rightarrow$ IBIC.

We now provide a *characterization* of TCBIC:

Proposition 7.1. Let $\mathcal{E}$ be a one-good DIE, and suppose each agent's utility function, $u_i = u_i(\omega, x_i(\omega))$ is monotone in the elements of the vector of goods x_i, that $u_i(., x_i)$ is $\mathcal{F}_i$-measurable in the first argument, and that an element $x = (x_1, \ldots, x_n) \in \bar{L}_X^1$ is a feasible allocation in the sense that $\sum_{i=1}^n x_i(\omega) = \sum_{i=1}^n e_i(\omega)\ \forall \omega$. Consider the following conditions:

(i) $x \in L_X^1 = \prod_{i=1}^n L_{X_i}^1$. and
(ii) x is TCBIC.

Then (i) is equivalent to (ii).

Proof. See Glycopantis et al. (2003a).

Next we state conditions under which the private core allocation is CBIC.

Proposition 7.2. let $\mathcal{E}$ be an arbitrary differential information economy with monotone and continuous utility functions. The private core and the private value are CBIC.

Proof. See Koutsougeras and Yannelis (1993), Krasa and Yannelis (1994), and Hahn and Yannelis (2001).

Corollary 7.1. A no-free disposal Radner equilibrium is CBIC.[13]

Proof. It can be easily shown that any no-free disposal Radner equilibrium belongs to the private core. Therefore by Proposition 7.2 it follows that the Radner equilibrium is CBIC.

Proposition 7.1 characterizes TCBIC and CBIC in terms of private individual measurability of allocations. It will enable us to conclude whether or not, in case of non-free disposal, any of the solution concepts will be TCBIC, whenever feasible allocations are $\mathcal{F}_i$-measurable.

It follows also that the redistribution

$$\begin{pmatrix} 5 & 2.5 & 2.5 \\ 5 & 2.5 & 2.5 \end{pmatrix}$$

is not CBIC because it is not $\mathcal{F}_i$-measurable.

On the other hand the proposition implies, again in Example 5.1, that the allocation

$$\begin{pmatrix} 5 & 5 & 0 \\ 5 & 0 & 5 \end{pmatrix}$$

is incentive compatible. As we have seen this is a non-free disposal REE, and a private core allocation.

We note that the above propositions are not true if we assume free disposal. In that case $\mathcal{F}_i$-measurability *does not* imply incentive compatibility. In the case with *free disposal, private core and Radner equilibrium need not be incentive compatible*. In order to see this we notice that in Example 5.2 the (free disposal) Radner equilibrium is $x_1 = (4, 4, 1)$ and $x_2 = (4, 1, 4)$. The above allocation is clearly $\mathcal{F}_i$-measurable and it can be checked directly that it belongs to the (free disposal) private core. However it is not TBIC since if state a occurs Agent 1 has an incentive to report state c and become better off.

Next we consider Example 6.1. We define $A_1 = \{a, b\}$ and $A_2 = \{b, c\}$. We assume that P1 acts first and that when P2 is to act he has heard the declaration of P1.

As shown in Section 6 the fully revealing REE allocations and corresponding utilities are:

In state a, $x_{11}^* = \dfrac{85}{22}$, $x_{12}^* = \dfrac{85}{16}$, $x_{21}^* = \dfrac{91}{22}$, $x_{22}^* = \dfrac{91}{16}$; $u_1^* = 4.53$, $u_2^* = 4.85$.

[13] A direct proof of the CBIC of the non-free disposal Radner equilibrium with infinitely many commodities has been given in Herves et al. (2003).

In state b, $x_{11}^* = 4,\ x_{12}^* = 4,\ x_{21}^* = 4,\ x_{22}^* = 4;\ u_1^* = 4,\ u_2^* = 4.$

In state c, $x_{11}^* = \frac{37}{16},\ x_{12}^* = \frac{37}{10},\ x_{21}^* = \frac{43}{16},\ x_{22}^* = \frac{43}{10};\ u_1^* = 2.93,\ u_2^* = 3.40.$

The normalized expected utilities of the REE are $\mathcal{U}_1 = 11.46,\ \mathcal{U}_2 = 12.25$.

We can see that the REE redistribution, which belongs also to the WFC, is not CBIC as follows.[14] Suppose that P1 sees $\{a, b\}$ and P2 sees$\{a\}$ but misreports $\{b, c\}$. If P1 believes the lie then state b is believed. So P1 agrees to get the allocation $(4, 4)$. P2 receives the allocation $e_2(a) + x_2(b) - e_2(b) = (1, 10) + (4, 4) - (1, 7) = (4, 7)$ with $u_2(4, 7) = 5.29 > u_2(\frac{91}{22}, \frac{91}{16}) = 4.85$. Hence P2 has a possibility of gaining by misreporting and therefore the REE is not CBIC. On the other hand if P2 sees $\{b, c\}$ and P1 sees $\{c\}$, the latter cannot misreport $\{a, b\}$ and hope to gain if P2 believes it is b.

In employing game trees in the analysis we adopt the definition of IBIC. The game-theoretic equilibrium concept employed will be that of PBE. A play of the game will be a directed path from the initial to a terminal node.

In terms of the game trees, a core allocation will be IBIC if there is a profile of optimal behavioral strategies along which no player misreports the state of nature he has observed. This allows for the possibility that players have an incentive to lie from information sets which are not visited by an optimal play.

In view of the analysis using game trees we comment further on the general idea of CBIC. First we look at it again, in a similar manner to the one in the beginning of Section 4.

Suppose the true state of nature is $\bar{\omega}$. Any coalition can only see together that the state lies in $\bigcap_{i \in S} E_i(\bar{\omega})$. If they decide to lie they must first guess at what is the true state and they will do so at some $\omega^* \in \bigcap_{i \in S} E_i(\bar{\omega})$. Having decided on ω^* as a possible true state, they pick some $\omega' \in \bigcap_{j \notin S} E_j(\omega^*)$ and assuming the system is not CBIC they hope, by each of them announcing $E_i(\omega')$ to secure better payoffs.

This is all contingent on their being believed by $I \setminus S$, which depends on having been correct in guessing that $\omega^* = \bar{\omega}$. If $\omega^* \neq \bar{\omega}$, i.e they guess wrongly, then since $\bigcap_{j \notin S} E_j(\omega^*) \neq \bigcap_{j \notin S} E_j(\bar{\omega})$ the lie may be detected, since possibly $\omega' \notin \bigcap_{j \notin S} E_j(\bar{\omega})$.

Therefore the definition of CBIC can only be about situations where a lie might be beneficial. On the other hand the extensive form forces us to consider the alternative of what happens if the lie is detected. It requires statements concerning earlier decisions by other players to lie or tell the truth and what payoffs will occur whenever a lie is detected, through observations or incompatibility of declarations. Only in this fuller description can players make a decision whether to risk a lie. Such considerations probably open the way to an incentive compatibility definition based on expected gains from lying.

The issue is whether cooperative and noncooperative static solutions can be supported through an appropriate noncooperative solution concept. The analysis below shows that CBIC allocations can be supported by a PBE while absence of

[14] Palfrey and Srivastava (1986) have also shown that the REE may not be incentive compatible.

incentive compatibility implies lack of such support. It is also shown how implementation of allocations becomes possible by introducing an exogenous third party or an endogenous intermediary.

We recall that a PBE consists of a set of players' optimal behavioral strategies, and consistent with these, a set of beliefs which attach a probability distribution to the nodes of each information set (Tirole, 1988). It is a variant of the idea of a sequential equilibrium (Kreps and Wilson, 1982).

Note 7.1. Different notions of incentive compatibility for differential information economies were first introduced by in Krasa and Yannelis (1994). It should be noted that the framework for differential information economies is different than the one in the Harsanyi type models and the notions of incentive compatibility which they use. These models assume that the initial endowments are independent of the state of nature and therefore uncertainty comes only from the utility functions.

Notice that if the initial endowments are assumed to be constant, then most of the examples in this paper cannot be analysed by a Harsanyi type model. A comparison between the DIE model and the Harsanyi type models can be found in Hahn and Yannelis (1997). In particular this paper contains a comparison of some of the Holmström and Myerson (1983) incentive compatibility notions and the ones in the DIE literature.

Finally it is important to notice that in a multilateral contracts model, it appears more appropriate to ensure CBIC rather than IBIC. Obviously CBIC implies IBIC but the reverse is not true, as an example in the preface of this volume demonstrates. Therefore lack of CBIC may make a contract unstable or not viable.

8 Non-implementation of Radner equilibria, of WFC and WFV allocations

We examine here the implementation, as a PBE of different equilibrium concepts. This section is closely related to the previous one. The fundamental issue is to connect, in the context of the partition model, the idea of implementation, in the form of a PBE of an extensive form game, to the CBIC property. Namely, we wish to check whether an allocation can be realized as a PBE in an incomplete information, dynamic game, in the form of a tree, and how this is connected to the CBIC property.

The static concept of the CBIC implies that no agent has an incentive to lie with respect to the state(s) he has observed and the PBE satisfies basic rationality criteria in a game tree in which the agents are asymmetrically informed.

We examine whether cooperative or Walrasian, noncooperative, static equilibrium allocations, can be supported as the outcome of a dynamic, noncooperative solution concept. We also examine the role that a third party can play in supporting an equilibrium.

A general conclusion is that static equilibrium allocations with the CBIC property can be supported, under reasonable rules, as PBE outcomes. This discussion helps us to reach a conclusion as to which equilibrium concept can be considered as appropriate. We find that private core allocations have distinct advantages.[15]

8.1 Non-implementation of Radner equilibria, of WFC and WFV allocations

We consider Example 5.2. We show here that lack of IBIC implies that two agents do not sign a proposed contract because they have an incentive to cheat. Therefore PBE leads to no-trade.

We shall investigate the possible implementation of the allocation

$$\begin{pmatrix} 4 & 4 & 1 \\ 4 & 1 & 4 \end{pmatrix}$$

of Example 5.2, contained in a proposed contract between P1 and P2. As we have seen, with free disposal this is a Radner equilibrium allocation.

This allocation is not IBIC because, as we explained in Section 8, if Agent 1 observes $A_1 = \{a, b\}$, he has an incentive to report c and Agent 2 has an incentive to report b when he observes $A_2 = \{a, c\}$.

We construct a game tree and employ reasonable rules for calculating payoffs. In fact we look at the contract

$$\begin{pmatrix} 5 & 4 & 1 \\ 5 & 1 & 4 \end{pmatrix}.$$

The proposed allocation can be obtained by invoking free disposal in state a. Of course to impose free disposal causes certain problems, because the question arises as to how it will be verified that the agents have actually thrown away 1 unit. However we assume that this is possible. In the analysis below we assume that the players move sequentially.

The *rules* for calculating the payoffs in terms of quantities, i.e. the terms of the contract, are:

(i) If the declarations by the two players are incompatible, that is (c_1, b_2) then no-trade takes place and the players retain their initial endowments.
That is the case when either state c, or state b occurs and Agent 1 reports state c and Agent 2 state b. In state a both agents can lie and the lie cannot be detected by either of them. They are in the events A_1 and A_2 respectively, they get 5 units of the initial endowments and again they are not willing to cooperate. Therefore whenever the declarations are incompatible, no trade takes place and the players retain their initial endowments.

(ii) If the declarations are (A_1, A_2) then even if one of the players is lying, this cannot be detected by his opponent who believes that state a has occurred and both players have received endowment 5. Hence no-trade takes place.

[15] For a thorough analysis in this section the reader is referred also to Glycopantis et al. (2001, 2003a, 2003b).

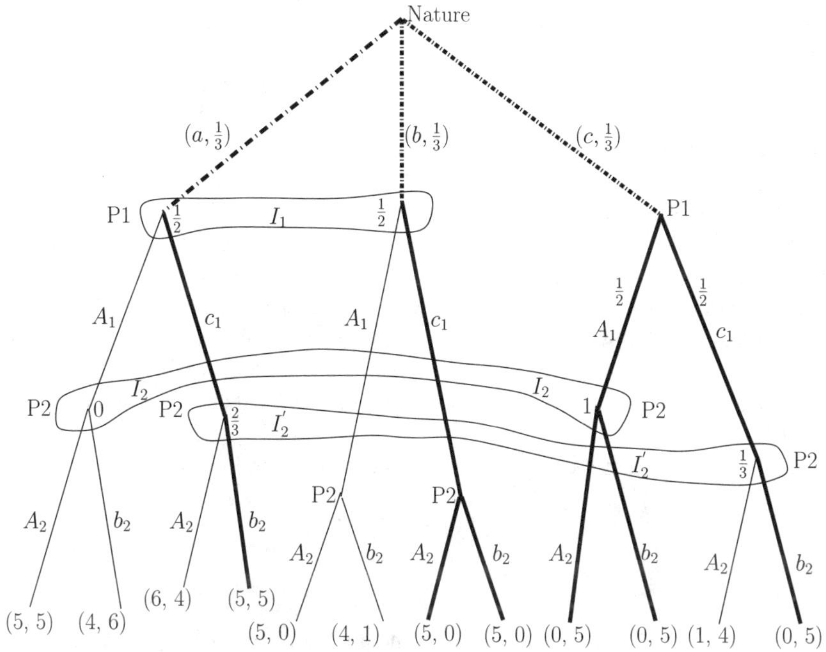

Figure 2

(iii) If the declarations are (A_1, b_2) then a lie can be beneficial and undetected. P1 is trapped and must hand over one unit of his endowment to P2. Obviously if his initial endowment is zero then he has nothing to give.

(iv) If the declarations are (c_1, A_2) then again a lie can be beneficial and undetected. P2 is now trapped and must hand over one unit of his endowment to P1. Obviously if his initial endowment is zero then he has nothing to give.

For the calculations of payoffs the revelation of the actual state of nature is not required. We could specify that a player does not lie if he cannot get a higher payoff by doing so. We assume that each player, given his beliefs, chooses optimally from his information sets.

In Figure 2 we indicate, through heavy lines, plays of the game, obtained through backward induction, which are the outcome of the choices by nature and the optimal behavioral strategies by the players. The interrupted lines signify that nature simply chooses among three alternatives, with equal probabilities. The fractions next to the nodes of the information sets are obtained, whenever possible through Bayesian updating. That is they are consistent with the choice of a state of nature and the optimal behavioral strategies of the players.

For all choices by nature, at least one of the players tells a lie on the optimal play. The players, by lying, avoid the possibility of having to make a payment and the PBE confirms the initial endowments. The decisions to lie imply that the players will not sign the contract $(5, 4, 1)$ and $(5, 1, 4)$. A similar conclusion would have been reached if we investigated directly the allocation (4, 4, 1) and (4, 1, 4).

Finally suppose we were to modify (iii) and (iv) of the *rules* i.e.: (iii) If the declarations are (A_1, b_2) then a lie can be beneficial and undetected, and P1 is trapped and must hand over half of his endowment to P2. Obviously if his endowment is zero then he has nothing to give.
(iv) If the declarations are (c_1, A_2) then again a lie can be beneficial and undetected. P2 is now trapped and must hand over half of his endowment to P1. Obviously if his endowment is zero then he has nothing to give.

The new rules would imply the following changes in the payoffs in Figure 2, from left to right. The second vector would now be (2.5, 7.5), the third vector (7.5, 2.5), the sixth vector (2.5, 2.5) and the eleventh vector (2.5, 2.5). The analysis in Glycopantis et al. (2001) shows that the weak fine core allocation in which both agents receive (5, 2.5, 2.5) cannot be implemented as a PBE. Again this allocation is not IBIC. The same allocation belongs, for equal weights to the agents, also to the WFV.

Finally we note that the PBE implements the initial endowments allocation

$$\begin{pmatrix} 5 & 5 & 0 \\ 5 & 0 & 5 \end{pmatrix}$$

which in the case of non-free disposal, coincides with the REE. However as it is shown in Glycopantis et al. (2003b) a REE is not in general implementable.

8.2 Implementation of Radner equilibria and of WFC allocations through the courts

We shall show briefly that the allocation

$$\begin{pmatrix} 4 & 4 & 1 \\ 4 & 1 & 4 \end{pmatrix}$$

of Example 5.2 can be implemented as a PBE through an exogenous third party. This can be interpreted as a court which imposes penalties when an agent lies.

Nature chooses states a, b and c with equal probabilities. P1 acts first and cannot distinguish between a and b. When P2 is to act we assume that not only he cannot distinguish between a and c but also he does not know what P1 has chosen before him.

The *rules* are:

(i) If a player lies about his observation, then he is penalized by 1 unit of the good. If both players lie then they are both penalized. For example if the declarations are (c_1, b_2) and state a occurs both are penalized. If they choose (c_1, A_2) and state a occurs then the first player is penalized. If a player lies and the other agent has a positive endowment then the court keeps the quantity subtracted for itself. However, if the other agent has no endowment, then the court transfers to him the one unit subtracted from the one who lied.

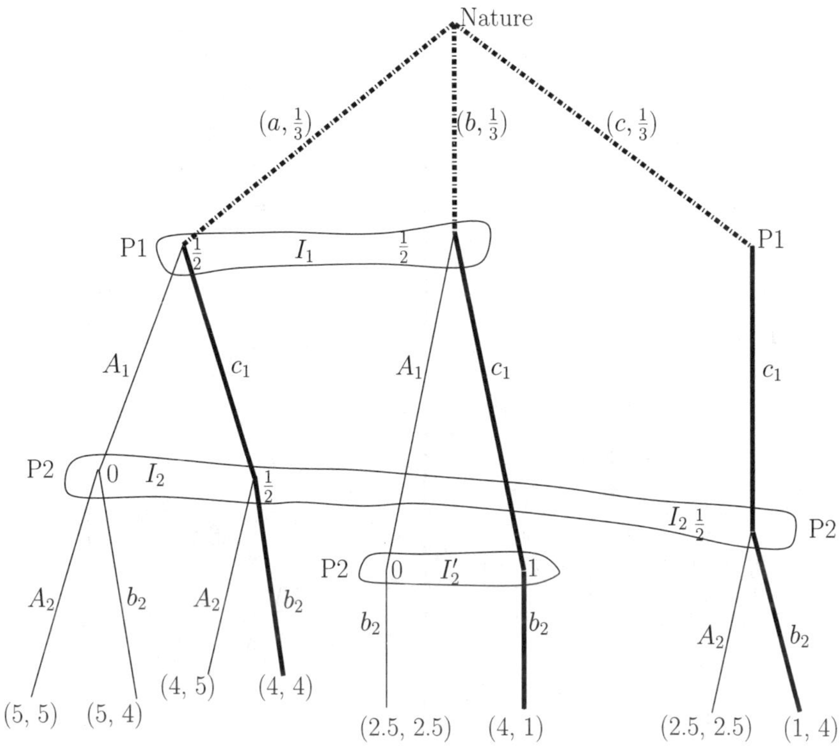

Figure 3

(ii) If the declarations of the two agents are consistent, that is (A_1, A_2) and state a occurs, (A_1, b_2) and state b occurs, (c_1, A_2) and state c occurs, then they divide equally the total endowments in the economy.

We obtain through backward induction the equilibrium strategies by assuming that each player chooses optimally, given his stated beliefs.

Figure 3 indicates, through heavy lines, optimal plays of the game. The fractions next to the nodes of the information sets are obtained through Bayesian updating.

Finally, suppose that the penalties are changed as follows. The court is extremely severe when an agent lies while the other agent has no endowment. It takes all the endowment from the one who is lying and transfers it to the other player.

Now P2 will play A_2 from I_2 and P1 will play A_1 from I_1. Therefore invoking an exogenous agent implies that the PBE will now implement the WFC allocation

$$\begin{pmatrix} 5 & 2.5 & 2.5 \\ 5 & 2.5 & 2.5 \end{pmatrix}.$$

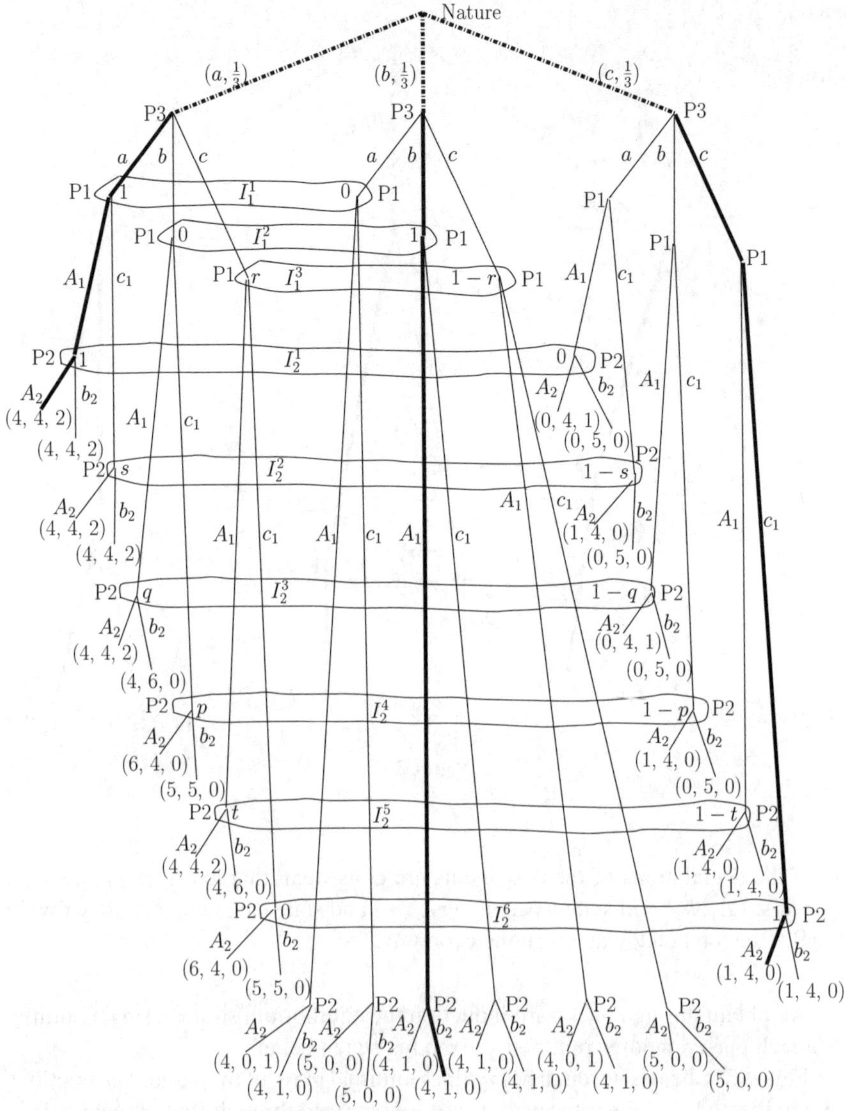

Figure 4

8.3 Implementation of private core allocations

Here we draw upon the discussion in Glycopantis et al. (2001, 2003a). In the case we consider now there is no court and therefore the agents in order to decide must listen to the choices of the other agents before them. P3 is one of the agents and we investigate his role in the implementation of private core allocations. Again we define $A_1 = \{a, b\}$ and $A_2 = \{a, c\}$.

Private core without free disposal seems to be the most satisfactory concept. The third agent, who has superior information, acting as an intermediary, implements the contract and gets rewarded in state a.

We shall consider the private core allocation

$$\begin{pmatrix} 4 & 4 & 1 \\ 4 & 1 & 4 \\ 2 & 0 & 0 \end{pmatrix}$$

of Example 5.1.

We know from Proposition 7.1 that such core allocations are CBIC and we shall show now how they can be supported as PBE of a noncooperative game.

P1 cannot distinguish between states a and b and P2 between a and c. P3 sees on the screen the correct state and moves first. He can either announce exactly what he saw or he can lie. Obviously he can lie in two ways. When P1 comes to decide he has his information from the screen and also he knows what P3 has played. When it is the turn of P2 to decide he has his information from the screen and he also knows what P3 and P1 played before him. Both P1 and P2 can either tell the truth about the information they received from the screen or they can lie.

The *rules* of calculating payoffs, i.e. the terms of the contract, are as follows: If P3 tells the truth we implement the redistribution in the matrix above which is proposed for this particular choice of nature.
If P3 lies then we look into the strategies of P1 and P2 and decide as follows:

(i) If the declaration of P1 and P2 are incompatible we go to the initial endowments and each player keeps his.

(ii) If the declarations are compatible we expect the players to honour their commitments for the state in the overlap, using the endowments of the true state, provided these are positive. If a player's endowment is zero then no transfer from that agent takes place as he has nothing to give.

In Figure 4 we indicate through heavy lines the equilibrium paths. The directed paths (a, a, A_1, A_2) with payoffs (4, 4, 2), (b, b, A_1, b_2) with payoffs (4, 1, 0) and (c, c, c_1, A_2) with payoffs (1, 4, 0) occur, each, with probability $\frac{1}{3}$. It is clear that nobody lies on the optimal paths and that the proposed reallocation is incentive compatible and hence it will be realized.

Further we can show that the PBE in Figure 4 can also be obtained as a sequential equilibrium in the sense of Kreps - Wilson (1982). Now, it is also required that the optimal behavioral strategies, and the beliefs consistent with these, are the limit of a sequence consisting of completely mixed behavioral strategies, and the implied beliefs. Throughout the sequence it is only required that beliefs are consistent with the strategies. The latter are not expected to be optimal.

8.4 Non-implementation of REE

We show here, in the context of an economy with two agents, three states of nature and two goods per state, that a fully revealing REE is not implementable. In fact we consider Example 6.1. We recall that $A_1 = \{a, b\}$, $A_2 = \{b, c\}$, and assume that P1 acts first and that when P2 is to act he has heard the declaration of P1. We have seen in Section 7 that the REE is not CBIC.

Next we show using the sequential decisions approach that the REE is not implementable. We specify the *rules* for calculating payoffs, i.e. the terms of the contract:

(i) If the declarations of the two players are incompatible, that is (c_1, a_2), then this implies that no trade takes place.

(ii) If the declarations of the two players are (A_1, A_2) then this implies that state b is believed. The player who believes it gets his REE allocation (4, 4) and the other player gets the rest. So aA_1A_2 means that P2 has lied but P1 believes it is state b and and gets (4, 4). P2 gets the rest under state a that is (4, 7); bA_1A_2 means that both believe that it is the (actual) state b and each gets (4, 4); cA_1A_2 means that P2 believes it is state b and gets (4, 4) and P1 gains nothing from his lie as he gets (1, 4).

(iii) aA_1a_2, bA_1A_2, cc_1A_2 imply that everybody tells the truth and the contract implements the REE allocation under state a, b, and c respectively. (bA_1A_2 in (ii) and (iii) give of course an identical result).

(iv) ac_1A_2 implies that both lie but their declarations are not incompatible. Each gets his REE under c and there is free disposal.

(v) cA_1a_2 means that both lie and stay with their initial endowments as they cannot get the REE allocations under state a which is the intersection of A_1 and a_2.

(vi) bA_1a_2 implies that P2 misreports and P1 believes and gets his REE under a; P2 gets the rest under b.

(vii) bc_1A_2 means that P1 lies and P2 believes that it is state c. P2 gets his REE allocation under c and P1 gets the rest under b, that is the allocation (5.31, 3.7).

On the game tree of consecutive decisions, the payoffs are translated in terms of utility. The complete optimal paths are shown in Figure 5, through heavy lines. We assume that each player chooses optimally from his information set. Probabilities next to the nodes of the information sets denote the players' beliefs. Strategies and beliefs satisfy the condition of a PBE. Our analysis shows that it is unique[16]. The corresponding normalized expected payoffs of the players are $\mathcal{U}_1 = 10.93$ and $\mathcal{U}_2 = 12.69$.

The equilibrium paths imply that REE is not implementable which matches up with the fact that it is not CBIC. However comparing the normalized expected utilities of the Bayesian equilibrium with those corresponding to the initial allocation we conclude that the proposed contract will be signed. On the other hand P2, because it is not advantageous to him, stops P1 from realizing his normalized REE utility. He ends up with $\mathcal{U}_2 = 12.69$ rather than $\mathcal{U}_2 = 12.25$.

Further, it is shown in Glycopantis et al. (2003b) that if we modify the model into one with simultaneous decisions of the agents again the REE is not implementable.

[16] Notice that as explained in the more detailed analysis, reversing the order of the play between the agents results in more than one PBE.

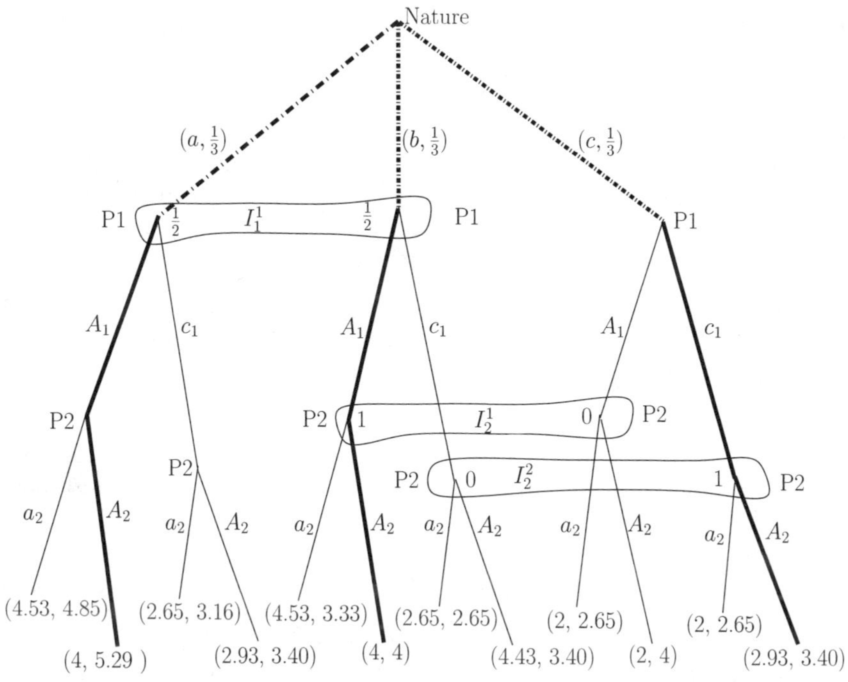

Figure 5

9 REE and weak core concepts

In view of the significance of the REE as an equlibrium concept we look in this section closer at the relation between REE and weak core concepts, which allow for sharing of information among the agents.[17] It is this sharing of information which makes the conditions different and therefore the comparison interesting, as REE is a Walrasian notion. The relation between REE and the private core, in which every agent keeps their own information, has been examined above.

We show here that for state independent utilities, no coalition of agents can block a fully revealing REE. Therefore in this case the REE is always a subset of IWFC and therefore it is interim "fully" Pareto optimal. However for state dependent utility functions the REE is not necessarily in the IWFC as we show below.

We also show that in general a REE does not belong to the WFC. If it so happens that REE does belong to this set then a slight modification of the utility functions implies that the two sets do not overlap anymore.

9.1 REE and IWFC

First we define the cooperative concept of the IWFC concept which is conditional on some information already obtained and shared by coalitions of agents.

[17] This section uses results and statements from Glycopantis et al. (2003b).

Definition 9.1.1. An allocation $x = (x_1, \ldots, x_n) \in \bar{L}_X$ is said to be a IWFC allocation if

(i) each $x_i(\cdot)$ is $\mathcal{F}_I$-measurable;[18]
(ii) $\sum_{i=1}^n x_i(\omega) = \sum_{i=1}^n e_i(\omega)$ for all $\omega \in \Omega$;
(iii) there do not exist state of nature $\omega^* \in \Omega$, coalition S and allocation $(y_i)_{i\in S} \in \prod_{i\in S} \bar{L}_{X_i}$ such that $y_i(\cdot) - e_i(\cdot)$ is $\mathcal{F}_S$-measurable for all $i \in S$, $\sum_{i\in S} y_i(\omega) = \sum_{i\in S} e_i(\omega)$, for all ω and $v_i(y_i|\mathcal{F}_S)(\omega^*) > v_i(x_i|\mathcal{F}_{x_i})(\omega^*)$ for all $i \in S$, where $\mathcal{F}_{x_i}$ denotes the information connected with x_i.

The definition, (see Yannelis, 1991), implies that no coalitions of agents can pool their own information and make each of its members better off.

Proposition 9.1.1. For state independent utility functions, a fully revealing REE allocation belongs to the IWFC.

Proof. Let (x, p) be a fully revealing REE, so that the state of nature that has occurred is known to everybody and x be feasible and measurable with respect to $\mathcal{F}_I$. Suppose now that x is not an element of IWFC. Then there exists $\omega^* \in \Omega$, a coalition S and feasible $(y_i)_{\in S} \in \prod_{i\in S} \bar{L}_{X_i}$ which is $\mathcal{F}_S$-measurable $\forall i \in S$, such that $\sum_{i\in S} y_i(\omega) = \sum_{i\in S} e_i(\omega)$ $\forall \omega \in \Omega$ and

$$v_i(y_i|\mathcal{F}_S)(\omega^*) > v_i(x_i|\mathcal{G}_i)(\omega^*). \tag{19}$$

On the right-hand side of (6) we have that $\mathcal{G}_i = \mathcal{F}$ which in this case is generated by singletons.

We consider the two terms in relation to the Definition 9.1.1. The right-hand side is $v_i(x_i|\mathcal{G}_i)(\omega^*) = u_i(x_i(\omega^*))$, i.e. one single term with probability one. This follows from the fact that x is fully revealing and therefore $E_i^{\mathcal{G}_i}(\omega^*) = \{\omega^*\}$.

On the other hand the left-hand side is

$$v_i(\omega^*, y_i(\omega^*)) = \sum_{\omega'} u_i(y_i(\omega'))q_i(\omega'|E_i^{\mathcal{F}_S}(\omega^*)), \tag{20}$$

where in (7)

$$q_i(\omega'|E_i^{\mathcal{F}_S}(\omega^*)) = \begin{cases} 0 & : \quad \omega' \notin E_i^{\mathcal{F}_S}(\omega^*) \\ \dfrac{q_i(\omega')}{q_i(E_i^{\mathcal{F}_S}(\omega^*))} & : \quad \omega' \in E_i^{\mathcal{F}_S}(\omega^*). \end{cases}$$

and $E_i^{\mathcal{F}_S}(\omega^*)$ is a subset of $\mathcal{F}_S$ on which y_i is constant.

This allows us to take the utility term out of the sum[19] and deduce that $u_i(y_i(\omega^*)) > u_i(x_i(\omega^*))$. This implies that when x_i was chosen y_i was too expensive and therefore $p(\omega^*)y_i(\omega^*) > p(\omega^*)x_i(\omega^*) = p(\omega^*)e_i(\omega^*)$ $\forall\, i \in S$. Then summing up with respect to $i \in S$ we obtain

$$p(\omega^*)\sum_{i\in S} y_i(\omega^*) = \sum_{i\in S} p(\omega^*)y(\omega^*) > \sum_{i\in S} p_i(\omega^*)e_i(\omega^*) = p(\omega^*)\sum_{i\in S} e_i(\omega^*). \tag{21}$$

[18] Recall that for $S \subseteq I$, $\mathcal{F}_S$ denotes the "join" of coalition S, i.e. $\bigvee_{i\in S} \mathcal{F}_i$.

[19] Notice that if $u_i(\omega', x_i(\omega'))$ depended separately on ω' then, in general, it would not have been possible to take $u_i(\omega', y_i(\omega'))$ out of the sum. On the other hand measurability of u_i with respect to its first argument would rescue the proof.

Relation (21) is a contradiction to $\sum_{i \in S} y_i(\omega) = \sum_{i \in S} e_i(\omega)$ because in order to obtain the inequality $p(\omega^*) \sum_{i \in S} y_i(\omega^*) > p(\omega^*) \sum_{i \in S} e_i(\omega^*)$ at least one element of the vector $\sum_{i \in S} y_i(\omega)$ must be larger than the corresponding element of $\sum_{i \in S} e_i(\omega)$.

Remark 9.1.1. With state independent utilities, Proposition 9.1.1 can be proven even if x is a partially revealing or non-revealing REE. It does not matter whether the information of the coalition is finer or not than the one of the REE. Also with state dependent utilities the proposition can be proven for general REE and an appropriately defined WFC concept if coalitions are only allowed to form which have the same information as REE. Then there is no need to take the utility expressions out of the relation $v_i(y_i|\mathcal{F}_S)(\omega^*) > v_i(x_i|\mathcal{G}_i)(\omega^*)$. An interpretation of what the proposition implies is that, under certain conditions, allowing all possible coalitions to share their information will not block the REE allocations.

Kwasnica (1998) has discussed a related result for a different core concept which is not interim fully Pareto optimal.

The conditions under which Proposition 9.1.1 holds are limited. We now construct examples to show that it does not necessarily hold when we have state dependent utilities.

In the examples below the introduction of Agent 3 is done so that the REE satisfy (i) in the definition of the IWFC. Alternatively, without introducing a third agent we can argue that given a REE there exists an IWFC allocation which improves the conditional utility of an agent given some particular state.

Example 9.1.1. There are only two, equally probable, from the point of view of the agents, states of nature, (one can add more states to make the model richer but this is not important), and two goods. Players 1 and 2 cannot distinguish between states a and b. On the other hand their utility functions differ per state. Player 3 can distinguish between all states of nature, has no initial endowments and has some utility function. His role is to ensure that the vector x described below satisfies condition (i) of IWFC. We turn our attention to the other players.

We are assuming the following. In state a: $u_1 = \min\{\epsilon x_{11}, x_{12}\}$, where $\epsilon > 1$, and $e_1=(2,0)$; $u_2=\min\{x_{21}, x_{22}\}$, and $e_2=(0,2)$. In state b: $u_1=\min\{x_{11}, x_{12}\}$, and $e_1=(2,0)$ $u_2=(x_{21}x_{22})^c$, where $c > 0$ will be determined later, and $e_2 = (0,2)$.

We construct two Edgeworth boxes and find the fully revealing REE, and hence our vector x, to be as follows. In state a: $p_1 = 0, p_2 = 1$; Agent 1 gets zero quantities and Agent 2 gets everything; $u_1 = 0$ and $u_2 = 2$. In state b: $p_1 = 1, p_2 = 1$; every agent gets 1 unit of each good; $u_1 = 1$ and $u_2 = 1$. In both states, Player 3 receives no quantities.

We will now show that this REE is not in the IWFC. Since, when the two players share their information, they still cannot distinguish between the two states we still require measurability of the feasible allocation to satisfy condition (iii) of IWFC.

The proposed allocation is that Agent 1 gets $y_1(a) = y_1(b) = (0.75, 0.75)$ and Agent 2 gets $y_2(a) = y_2(b) = (1.25, 1.25)$. The utility levels are as follows. In state a: $u_1 = 0.75$, and $u_2 = 1.25$ and in state b: $u_1 = 0.75$, $u_2 = (1.25 \times 1.25)^c$.

We choose state a for the condition (iii) of IWFC. For agent 1 we have that $v_1(y_1)(a)$ is larger than his REE utility which is zero. Also, for sufficiently large c, we have for agent 2 that $v_2(y_2)(a) = (\frac{1}{2})1.25 + (\frac{1}{2})(1.25 \times 1.25)^c > u_2 = 2$ (REE utility under a).

As for the alternative approach, without introducing a third agent we can argue that, given a REE, there exists an IWFC allocation which does better for some agent. First we use the above y_i allocation to show that it does better under a. Then we can argue that there exists an IWFC allocation which for some agent does even better in terms of utility conditioned on state a.

Example 9.1.2 (2003b). There are two, equally probable, from the point of view of the agents, states $\Omega = \{a, b\}$ and three players $I = \{1, 2, 3\}$. Player 3 can detect all states, but he has no initial endowments; his only role is to ensure that the x_i calculated below satisfy condition (i) of IWFC. Players 1 and 2 cannot distinguish between the states.

We are assuming that in state a: $u_1 = x_{11}^2 x_{12}$, $u_2 = x_{21}^2 x_{22}^2$, $e_1 = (\frac{9}{13}, \frac{9}{13})$, $e_2 = (\frac{4}{13}, \frac{4}{13})$, and in state b: $u_1 = x_{11}^{0.5} x_{12}$, $u_2 = x_{21} x_{22}$, $e_1 = (\frac{9}{13}, \frac{9}{13})$, $e_2 = (\frac{4}{13}, \frac{4}{13})$.

The REE is given by $p(a)=(8,5)$, $x_1(a)=(0.75,0.6)$, $x_2(a)=(0.25,0.4)$, and $p(b)=(5,8)$, $x_1(b)=(0.6,0.75)$, $x_2(b)=(0.4,0.25)$.

In the IWFC definition choose $\omega^*=a$, $S=\{1,2\}$, $y_1(a)=y_1(b)=(0.6,0.8)$, and $y_2(a) = y_2(b) = (0.4, 0.2)$.

Then $v_1(y_1)(a)=0.454$, $u_1(a, x_1(a))=0.337$, $v_2(y_2)(a)=0.043$, $u_2(a,x_2(a)) = 0.01$.

9.2 REE and WFC

Next we consider the relation between REE and the WFC in the context of a more general model than Example 6.1 which was considered above. We find that an REE allocation is not necessarily in the WFC.

Example 9.2.1 (2003b). For simplicity, we treat originally a case with two players, two goods and two states. We also assume, in the beginning, that the players are, in all states, endowed with strictly positive endowments of both goods and that for both players all states are equally probable. We assume that all states, $j \in \Omega$, are distinguishable by the two players when they pool their information.

The normalized expected utility functions of the two players are $\mathcal{U}_1 = \sum_j (x_{11}^j)^\alpha (x_{12}^j)^\beta$ and $\mathcal{U}_2 = \sum_j (x_{21}^j)^\alpha (x_{22}^j)^\beta$ where $\alpha, \beta > 0$. Namely we assume that they have identical, state independent utility functions. These assumptions can be relaxed. In summary, the result of the analysis is that in general the REE does not belong to the WFC.

The WFC allocations are characterized through the following problem:

Maximize $\sum_j (x_{11}^j)^\alpha (x_{12}^j)^\beta$
Subject to

$$\sum_j (S_1^j - x_{11}^j)^\alpha (S_2^j - x_{12}^j)^\beta = \bar{\mathcal{U}}_2 \text{ (fixed)},$$
$$0 \le x_{11}^j \le S_1^j, \quad 0 \le x_{12}^j \le S_2^j \quad \forall j,$$

where S_i^j denotes the total quantity of Good i in state j. Note that $0 < \mathcal{U}_2 < \sum_j (S_1^j)^\alpha (S_2^j)^\beta$.

Because of the feasibility constraints on quantities, the Lagrange theory cannot be applied in general in order to obtain the solution. However we can comment on the relation between REE and WFC allocations by arguing through another route.

We apply a Gorman type separation argument (see Gorman, 1959). We consider the contract curve per state. First we consider the following problem.

Maximize $(x_{11}^j)^\alpha (x_{12}^j)^\beta$
Subject to

$$(S_1^j - x_{11}^j)^\alpha (S_2 - x_{12}^j)^\beta = u_2{}^j \text{ (fixed)},$$
$$0 \le x_{11}^j \le S_1^j, \quad 0 \le x_{12}^j \le S_2^j.$$

The solution implies $S_2^j x_{11}^j = S_1^j x_{12}^j$, which is the diagonal of the Edgeworth box. All WFC allocations are on contract curve in each state, for otherwise we can move to a Pareto superior point on the contract curve. It is also true that a REE, fully revealing or not, will be on the diagonal with every agent receiving positive quantities from both goods. This follows from the fact that otherwise, in at least one state, the markets will not clear.

The actual solution is

$$x_{11}^j = \left(\frac{S_1^j}{S_2^j}\right)^{\frac{\beta}{\alpha+\beta}} \left[(S_1^j)^{\frac{\alpha}{\alpha+\beta}} (S_2^j)^{\frac{\beta}{\alpha+\beta}} - (u_2^j)^{\frac{1}{\alpha+\beta}}\right],$$
$$x_{12}^j = \left(\frac{S_2^j}{S_1^j}\right)^{\frac{\alpha}{\alpha+\beta}} \left[(S_1^j)^{\frac{\alpha}{\alpha+\beta}} (S_2^j)^{\frac{\beta}{\alpha+\beta}} - (u_2^j)^{\frac{1}{\alpha+\beta}}\right].$$

We write $(S_1^j)^{\frac{\alpha}{\alpha+\beta}} (S_2^j)^{\frac{\beta}{\alpha+\beta}} = T^j$ and $(u_2^j)^{\frac{1}{\alpha+\beta}} = W^j$, and substitute into the objective function to get $\sum_j [T^j - W^j]^{(\alpha+\beta)}$ which is to be maximized subject to the constraints $\sum_j u_2^j = \bar{\mathcal{U}}_2$ and $u_2^j \ge 0$ which are equivalent to $\sum_j (W^j)^{(\alpha+\beta)} = \bar{\mathcal{U}}_2$ and $W^j \ge 0$. Considering the solution for the $x's$ we also have $0 \le W^j \le T^j$. So in summary we are solving:

Maximize $\sum_j [T^j - W^j]^\gamma$
Subject to

$$\sum_j (W^j)^\gamma = \bar{\mathcal{U}}_2 \text{ (fixed), and}$$
$$0 \le W^j \le T^j.$$

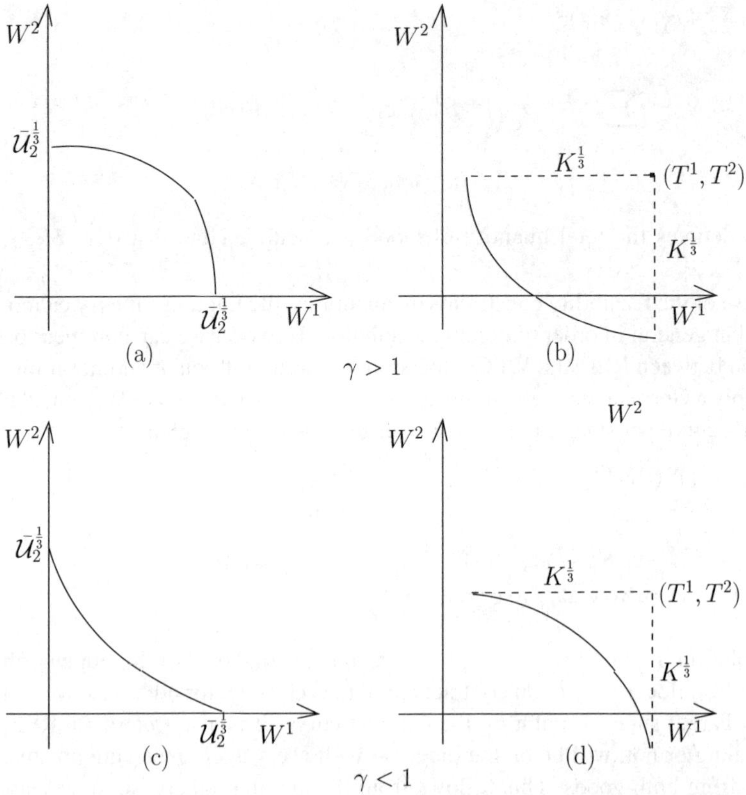

Figure 6

where $\gamma = \alpha + \beta$.

We now look at the form of the functions. Consider $\sum_j (W^j)^\gamma = 1$ for any $\gamma > 0$.
For $\gamma = 1$ this is a hyperplane. In the positive orthant, $\gamma > 1$ causes the surface to bulge away from the hyperplane so as to enclose a convex set including the origin ($\gamma = 2$ is the exemplary case, which is a hypersphere). Conversely for $\gamma < 1$ it produces a surface which bulges in towards the origin. $\sum_j (W^j)^\gamma = \bar{\mathcal{U}}_2$ is similar in shape but scaled by a factor $\bar{\mathcal{U}}_2^{\frac{1}{\gamma}}$.

Finally the shape of $\sum_j [T^j - W^j]^\gamma = K$ (fixed) can be derived from the above. The origin has been shifted to the point with coordinates (T^j) after the surface has been reflected along each coordinate axis.

Now we look at the solution of the overall Gorman problem. We distinguish between:

(i) $\gamma > 1$; the constraint is concave, in the nonnegative area, with perpendicular intersections with the axes. The indifference curves of the objective function are convex, with nonnegative coordinates, (see Fig. 6), and increase in value as we move in the direction of the origin. It follows that the maximum will be at one or both of the corner points. This means that the REE is not in the WFC.

(ii) $\gamma < 1$; in this case the constraint is convex and the indifference curves are concave, (see Fig. 6), and increase in value as we move in the direction of the origin. The solution is away from the corner points at a point of tangency. Even under symmetric conditions there is no reason why the REE should be in the WFC.

(iii) $\gamma = 1$; inspection of the objective function and the constraint shows that the WFC coincides with the linear constraint. It follows that the REE allocation is in the WFC and this is the case in Example 6.1. However, attaching a weight to the utility of Player 1 in one state implies a corner solution and therefore the REE is not in the WFC.

10 Bayesian learning with cooperative solution concepts

As we indicated in the previous sections, the private core and the private value outcomes are sensitive to changes in the private information of the agents. In this section we sketch out how information available to the agents can change through time.

The idea of learning introduces changes in the information structure of the agents. We consider a DIE that extends over many periods. The agents have initially private information which reflects their own personal characteristics, i.e. the random initial endowments and preferences. However, in each period they draw new information from the realized core or value allocation. Hence we consider an economy $\mathcal{E}$ in a dynamic framework.

One way of explaining how the agents refine their private information over time is as follows. Suppose, for example, that the same utility functions and endowments are repeated at each point in time. The chances are that over a long period all states of nature will occur. Suppose now that Agent i knows exactly what this state is, say a, but Agent j observes an element of his information partition with more than one state. Agent j cannot distinguish between the various states in his information set. However he can start slowly associating state a with signals which he originally considered as unimportant or irrelevant and which now he sees coincide with the announcement, through his private core or value allocation, of this state by Agent i. At no stage is it assumed that the agents get together to share their information.

Let $T = \{1, 2, ...\}$ denote the set of time periods and $\sigma(e_i^t, u_i^t)$ the σ-algebra that the random initial endowments and utility function of Agent i generated at time t. At any given point in time $t \in T$, the private information of Agent i is defined as:

$$\mathcal{F}_i^t = \sigma(e_i^t, u_i^t, (x^{t-1}, x^{t-2}, ...)) \tag{22}$$

where x^{t-1}, x^{t-2}, ... are past periods private core or value allocations.

Relation (22) says that at any given point in time t, the private information which becomes available to Agent i is $\sigma(e_i^t, u_i^t)$ together with the information that the private core (value) allocations generated in all previous periods. In this scenario, the private information of Agent i in period $t+1$ will be $\mathcal{F}_i^t$ together with the information the private core (value) allocation generated at period t, i.e. $\sigma(x^t)$. More explicitly, the assumption is that the private information of Agent i at time $t+1$ will be $\mathcal{F}_i^{t+1} = \mathcal{F}_i^t \vee \sigma(x^t)$, which denotes the "join", that is the smallest σ-algebra containing $\mathcal{F}_i^t$ and $\sigma(x^t)$.

Therefore for each Agent i we have that

$$\mathcal{F}_i^t \subseteq \mathcal{F}_i^{t+1} \subseteq F_i^{t+2} \subseteq \dots . \tag{23}$$

Relation (23) represents a learning process for Agent i and it generates a sequence of differential information economies $\{\mathcal{E}^t : t \in T\}$ where now the corresponding private information sets are given by $\{\mathcal{F}_i^t : t \in T\}$.

The agents are myopic, in the sense that they do not form expectations over the entire horizon but only for the current period, i.e. each agent's interim expected utility is based on his/her current period private information. Obviously, since the private information set of each agent becomes finer over time, the interim expected utility of each agent is changing as well. The information gathered at a given time t, will affect the private core (or value) outcome in periods $t+1,\ t+2,\ \dots$. The example below attempts to explain the idea of learning.

Example 10.1. Consider the following DIE with two agents $I = \{1, 2\}$ three states of nature $\Omega = \{a, b, c\}$ and goods, in each state, the quantities of which are denoted by x_{i1}, x_{i2}, where i refers to the agent. The utility functions are given by $u_i(\omega, x_{i1}, x_{i2}) = x_{i1}^{\frac{1}{2}} x_{i2}^{\frac{1}{2}}$, and states are equally probable, i.e. $\mu(\{\omega\}) = \frac{1}{3}$, for $\omega \in \Omega$. Finally the measurable endowments and the private information of the agents is given by

$$\begin{aligned} e_1^t &= ((10,0),(10,0),(0,0)), \quad && \mathcal{F}_1 = \{\{a,b\},\{c\}\}; \\ e_2^t &= ((10,0),(0,0),(10,0)), \quad && \mathcal{F}_2 = \{\{a,c\},\{b\}\}. \end{aligned} \tag{24}$$

The structure of the private information of the agents implies that the private core allocation, (x_1^t, x_2^t), in $t = 1$ consists of the initial endowments.

Notice also that the information generated in Period 2 is the full information $\sigma(x_1^t, x_2^t) = \{\{a\},\{b\},\{c\},\{a,b\},\{a,c\},\{a,b,c\},\emptyset\}$. It follows that the private information of each agent in periods $t \geq 2$ will be

$$\begin{aligned} \mathcal{F}_1^{t+1} &= F_1^t \vee \sigma(x_1^t, x_2^t) = \{\{a\},\{b\},\{c\}\}; \\ \mathcal{F}_2^{t+1} &= F_2^t \vee \sigma(x_1^t, x_2^t) = \{\{a\},\{b\},\{c\}\}. \end{aligned} \tag{25}$$

Now in $t = 2$ the agents will make contracts on the basis of the private information sets in (25). It is straightforward to show that a private core allocation in period $t \geq 2$ will be

$$\begin{aligned} x_1^{t+1} &= ((5,5),(10,0),(0,0)); \\ x_2^{t+1} &= ((5,5),(0,0),(0,10)). \end{aligned} \tag{26}$$

Notice that the allocation in (26) makes both agents better off than the one given in (24). In other words, by refining their private information using the private core allocation they have observed, the agents realized a Pareto improvement.

Of course, in a generalized model with more than two agents and a continuum of states, unlike the above example, there is no need that the full information private core or value will be reached in two periods. The main objective of learning is to examine the possible convergence of the private core or value in an infinitely repeated DIE. In particular, let us denote the one shot limit full information economy by $\bar{\mathcal{E}} = \{(X_i, u_i, \bar{\mathcal{F}}_i, e_i, q_i : i = 1, 2, ..., n)\}$ where $\bar{\mathcal{F}}_i$ is the pooled information of Agent i over the entire horizon, i.e. $\bar{\mathcal{F}}_i = \bigvee_{i=1}^{\infty} \mathcal{F}_i^t$.

The questions that learning addresses itself to are the following:

(i) If $\left\{\mathcal{E}^t : t \in T\right\}$ is a sequence of DIE and x^t is a corresponding private core or value allocation, can we extract a subsequence which converges to a limit full information private core or value allocation for $\bar{\mathcal{E}}$?

(ii) Is the answer to (i) above affirmative, if we allow for bounded rationality in the sense that x^t is now required to be an approximate, ϵ-private core or ϵ-value allocation for $\mathcal{E}^t$, but nonetheless it converges to an exact private core or value allocation for $\bar{\mathcal{E}}$?

(iii) Given a limit full information private core or value allocation say $\bar{x}$ for $\bar{\mathcal{E}}$, can we construct a sequence of ϵ-private core or ϵ-value allocation x^t in $\mathcal{E}^t$ which converges to $\bar{x}$? In other words, can we construct a sequence of bounded rational plays, such that the corresponding ϵ-private core or ϵ-value allocations converge to the limit full information private core or value allocation.

The above questions have been affirmatively answered in Koutsougeras and Yannelis (1999).

It should be noted that in the above framework it may be the case that in the limit incomplete information may still prevail. In other words, it could be the case that

$$\bar{\mathcal{F}}_i = \bigvee_{i=1}^{\infty} \mathcal{F}_i^t \subset \bigvee_{i=1}^{n} \mathcal{F}_i^t.$$

Hence in the limit a private core or value allocation may not be a fully revealing allocation of the same kind. However, if learning in each period reaches the complete information in the limit, i.e. $\bar{\mathcal{F}}_i \supset \bigvee_{i=1}^{n} \mathcal{F}_i^t$ the private core or value allocation is indeed fully revealing.

Learning applied to cooperative solution concepts was first discussed in Koutsougeras and Yannelis (1999). A generalization of their results to non-myopic learning which allows agents to discount the future can be found in Serfes (2001).

11 Concluding remarks

We have reviewed here relations between some of the main cooperative and noncooperative equilibrium concepts in the area of finite economies with asymmetric information. It is precisely the asymmetry in the information of the agents which leads to a variety of cooperative and noncooperative equilibrium concepts. It is then appropriate that their properties be compared. As explained in Glycopantis and Yannelis in this volume, the example of Wilson (1978) shows that even the list of noncooperative concepts employed is not exhaustive.

Notice that we have not examined large economies or economies with infinite dimensional commodity spaces. There is a growing literature on such economies but we decided to focus mainly on finite economies. This was for the sake of simplicity, and also for focusing on conceptual issues rather than proving powerful theorems.

In modeling a DIE, we followed the partition approach. Alternative concepts are defined depending mainly on whether the calculations are in the ex ante or the interim state, the degree of information sharing among the agents, the free disposability or not of goods.

A number of examples calculate in detail equilibria, which makes their comparison transparent. Relations are obtained and the significance of superior information is brought out.

Given the variety of equilibrium concepts, the question arises which ones have satisfactory properties. Two such properties are the static Bayesian incentive compatibility and the dynamic PBE implementability of an equilibrium. We have also exhibited here some of the results obtained earlier which examined the connection between these ideas.

The discussion considered both cooperative and Walrasian type equilibrium concepts. The presentation here points out the positive association between Bayesian incentive compatibility of a concept and its implementability as a PBE. This investigation is wider than the Nash (1953) programme which concentrates in providing support to cooperative, static concepts through noncooperative, extensive form constructions.

A main conclusion is that equilibrium notions which may not be incentive compatible, cannot easily be supported as a PBE, e.g. REE and Radner equilibrium. On the contrary notions which are incentive compatible can be supported as a PBE, e.g. private core and private value.

We consider the area of incomplete and differential information and its modelling important for the development of economic theory. We believe that the introduction of game trees, which give a dynamic dimension to the analysis by making the individual decisions transparent, helps in the development of ideas. The partition model is, in our view, a natural way to analyze DIE and the use of game trees provides a noncooperative foundation of the equilibrium concepts.

Appendix I: On core concepts

We construct here a table containing a number of core concepts, taken as a starting point Yannelis (1991) and Koutsougeras and Yannelis (1993). We assume non-free disposal and that the utility function with which comparisons will be made is the ex ante one. First we cast the definition of a private core allocation in a form which will facilitate the comparison with other concepts.

Definition I.1. An allocation $x(\omega) = (x_1(\omega), x_2(\omega), ..., x_n(\omega))$ with $x_i(\omega) \in X_i(\omega)$ for all $\omega \in \Omega$ and $i = 1, ..., n$, is a *private core allocation* if

(i) x_i is $\mathcal{F}_i$-measurable, for all i,

(ii) $\sum_{i=1}^{n} x_i(\omega) = \sum_{i=1}^{n} e_i(\omega)$ for all ω, and
(iii) there do not exist coalition S and allocation to S given by $y(\omega) = (y_1(\omega), y_2(\omega), ..., y_n(\omega))$ with $y_i(\omega) \in X_i(\omega)$ for al $\omega \in \Omega$ and $i \in S$ such that
  (a) $y_i - e_i$ is $\mathcal{F}_i$-measurable for all i,
  (b) $\sum_{i \in S} y_i(\omega) = \sum_{i \in S} e_i(\omega)$ for all ω, and
  (c) $v_i(y_i) = \sum_{\omega \in \Omega} u_i(y_i(\omega))\mu(\omega) > v_i(x_i) = \sum_{\omega \in \Omega} u_i(y_i(\omega))u_i(x_i(\omega))\mu(\omega)$ for $i \in S$.

We can now proceed to the following classification:

A1: If in (iii) (a) is replaced by $\bigwedge_{i \in S} \mathcal{F}_i$-measurable[20], it is a *coarse core allocation*
A2: If also (i) is replaced by $\bigwedge_{i \in I} \mathcal{F}_i$-measurable, it is a *strong coarse core allocation*
B1: If in (iii) (a) is replaced by $\bigvee_{i \in S} \mathcal{F}_i$-measurable, it is a *fine core allocation*
B2: If also (i) is replaced by $\bigvee_{i \in I} \mathcal{F}_i$-measurable, it is a *a WFC allocation.*

Therefore when we use the terms *coarse* or *fine* we are referring to the measurability of y_i in (iii) (a). The terms *strong* or *weak* refer to the measurability of x_i in (i).

Next we note that if $\mathcal{F} \subseteq \mathcal{G}$ then x is $\mathcal{F}$-measurable $\Longrightarrow$ x is $\mathcal{G}$-measurable. Thus if we make the σ-algebra in (i) finer, we make it easier to find a core element. Conversely, in (iii), where we ask that a certain function should not exist, making the σ-algebra coarser makes it easier to find a core element.

We note the relation between the sets, Fine Core (possibly $\emptyset$) $\subseteq$ Private Core $\subseteq$ Coarse Core. The latter consists of individually rational Pareto optimal allocations. We also have that the strong coarse core is possibly empty, while the WFC exists.

We have that $\bigwedge_{i \in S} \mathcal{F}_i \subseteq \mathcal{F}_i \subseteq \bigvee_{i \in S} \mathcal{F}_i$. Therefore, theoretically, we could have nine core concepts, shown in the table below.

$(i) \backslash (iii)$	$\bigwedge \mathcal{F}_i$	$\mathcal{F}_i$	$\bigvee \mathcal{F}_i$
$\bigwedge \mathcal{F}_i$	Strong Coarse	α	β
$\mathcal{F}_i$	Coarse	Private	Fine
$\bigvee \mathcal{F}_i$	γ	δ	Weak Fine

The set inclusion sign $\supseteq$ applies in each row of the table from left to right, and in each column as we go down.

Note also that since WFC exists so do γ and δ. In the context of measurability the private core concept is important. It has good properties: CBIC and it exists. It is the smallest set which exists and is incentive compatible.

Obviously there are classifications as well, such as producing a table for free disposal and one with interim utility functions. Some comparisons between entries across tables can be made.

---

[20] The "meet" is the largest σ-algebra which is contained in each $\mathcal{F}_i$. It is in a sense the intersection of these algebras.

It is of interest to make a comparison between Definition I.1 of the private core above, (Koutsougeras and Yannelis, 1993), and the definition below, (Yannelis, 1991), which is cast in a positive formulation.

Definition I.2. An allocation $x \in L_X$ is said to be an *interim private core allocation* (IPC) if

(i) $\sum_{i=1}^{n} x_i = \sum_{i=1}^{n} e_i$ and
(ii) for all S and all $(y_i)_{i \in S} \in \prod_{i \in S} L_{X_i}$ such that $\sum_{i \in S} y_i = \sum_{i \in S} e_i$, $\exists i \in S$ such that $v_i(\omega, x_i) \geq v_i(\omega, y_i)$ for some ω with $\mu(\omega) > 0$.

Despite the fact that in Definition I.2 interim expected utility functions were used, one can show that IPC contains the ex ante private core in Definition I.1, i.e. PC$\subseteq$ IPC but not the other way round.

References

Allen, B.: Generic existence of completely revealing equilibria with uncertainty, when prices convey information. Econometrica **49**, 1173–1199 (1981)

Allen, B.: Market games with asymmetric information: the value. University of Pennsylvania, Working paper, ♯ 91-08 (1991a)

Allen, B.: Market games with asymmetric information and non-transferable utility: Representation results and the core. University of Pennsylvania, Working paper, ♯ 91-09 (1991b)

Allen, B.: Incentives and market games with asymmetric information. Economic Theory **21**, 527–544 (2003)

Allen, B., Yannelis, N.C.: Differential information economies. Economic Theory **18**, 263–273 (2001)

Arrow, K.J., Debreu, G.: Existence of an equilibrium in a competitive economy. Econometrica **22**, 265–290 (1954)

Aumann, R.J.: Markets with a continuum of traders. Econometrica **32**, 39–50 (1964)

Aumann, R.J.: Correlated equilibria as an expression of Bayesian rationality. Econometrica **55**, 1–18 (1987)

Debreu, G.: Theory of value. New York: Wiley 1959

Debreu, G., Scarf H.: A limit theorem on the core of an economy. International Economic Review **4**, 235–246 (1963)

Emmons, D., Scafuri, A.J.: Value allocations: an exposition. In: Aliprantis, C.D., Burkinshaw, O., Rothman, N.J. (eds): Advances in equilibrium theory. Lecture notes in economics and mathematical systems, pp. 55–78. Berlin Heidelberg New York: Springer 1985

Einy, E., Moreno, D., Shitovitz, B.: Rational expectations equilibria and the ex post core of an economy with asymmetric information. Journal of Mathematical Economics **34**, 527–535 (2000)

Forges, F.: Le coeur d'une économie d'échange en information asymétrique. Mimeo (1998)

Forges, F., Minelli, E., Vohra, R.: Incentives and the core of an exchange economy: a survey. Mimeo (2000)

Glycopantis, D., Muir, A., Yannelis, N.C.: An extensive form interpretation of the private core. Economic Theory **18**, 293–319 (2001)

Glycopantis, D., Muir, A., Yannelis, N.C.: On extensive form implementation of contracts in differential information economies. Economic Theory **21**, 495–526 (2003a)

Glycopantis, D., Muir, A., Yannelis, N.C.: Non-implementation of rational expectations as a perfect Bayesian equilibrium. Mimeo (2003b)

Glycopantis, A., Yannelis, N.C.: Information, efficiency and the core of the economy: Comments on Wilson's paper. This volume.

Gorman, W.M.: Separable utility and aggregation. Econometrica **27**, 469–481 (1959)

Grossman, S.J.: An introduction to the theory of rational expectations under asymmetric information economies. Review of Economic Studies **48**, 541–560 (1981)

Hahn, G., Yannelis, N.C.: Efficiency and incentive compatibility in differential information economies. Economic Theory **10**, 383–411 (1997)
Hahn, G., Yannelis, N.C.: Coalitional Bayesian Nash implementation in differential information economies. Economic Theory **18**, 485–509 (2001)
Harsanyi, J.C.: Games with incomplete information played by 'Bayesian' players, Part I. Management Science **14**, 159–182 (1967)
Herves, C., Moreno, E., Yannelis, N.C.: Equivalence and incentive compatibility of the Walrasian expectations equilibrium in infinite dimensional commodity spaces. Mimeo (2003). Economic Theory (forthcoming)
Hildenbrand, W.: Core and equilibria in large economies. Princeton, NJ: Princeton University Press 1974
Holmström, B., Myerson, R.: Efficient and durable decision rules with incomplete information. Econometrica **51**, 1799–1819 (1983)
Ichiishi, T., Yamazaki, A.: Preliminary results for cooperative extensions of the Bayesian game. Discussion paper , Graduate School of Economics, Hitotsubashi University, 1-89 (2001-2)
Khan, M.A., Yannelis, N.C. (eds): Equilibrium theory in infinite dimensional spaces, pp. 233–248. Berlin Heidelberg New York: Springer 1991
Koutsougeras, L., Yannelis, N.C.: Incentive compatibility and information superiority of the core of an economy with differential information. Economic Theory **3**, 195–216 (1993)
Koutsougeras, L., Yannelis, N.C.: Bounded rational learning in differential information economies: Core and value. Journal of Mathematical Economics **31**, 373–391 (1999)
Krasa, S., Yannelis, N.C.: The value allocation of an economy with differential information. Econometrica **62**, 881–900 (1994)
Krasa, S., Yannelis, N.C.: Existence and properties of a value allocation for an economy with differential information. Journal of Mathematical Economics **25**, 165–179 (1996)
Kreps, M.D.: A note on 'fulfilled expectations' equilibria. Journal of Economic Theory **14**, 32–43 (1977)
Kreps, M.D., Wilson, R.: Sequential equilibrium. Econometrica **50**, 889–904 (1982)
Kwasnica, A.M.: Welfare properties of rational expectations equilibria: The core. Mimeo. Caltech (1998)
Laffont, J.- J.: On the welfare analysis of rational expectation with asymmetric information. Econometrica **53**, 1–29 (1985)
Lefebvre, I.: An alternative proof of the nonemptiness of the private core. Journal of Economic Theory **14**, 275–291 (2001)
McKenzie, L.W.: On equilibrium in Graham's model of world trade and other competitive systems. Econometrica **22**, 147–161 (1954)
Myerson, R.: Cooperative games with incomplete information. International Journal of Game Theory **13**, 69–96 (1982)
Nash, J.F.: Two-person cooperative games. Econometrica **21**, 128–140 (1953)
Palfrey, T.R., Srivastava, S.: Private information in large economies. Journal of Economic Theory **39**, 34–58 (1986)
Prescott, E., Townsend, R.: Pareto optima and competitive equilibria with adverse selection and moral hazard. Econometrica **52**, 21–45 (1984)
Radner, R.: Competitive equilibrium under uncertainty. Econometrica **36**, 31–58 (1968)
Radner, R.: Rational expectation equilibrium: generic existence and information revealed by prices. Econometrica **47**, 655–678 (1979)
Scafuri, A.J., Yannelis, N.C.: Non-symmetric cardinal value allocations. Econometrica **52**, 1365–1368 (1984)
Serfes, K.: Non-myopic learning in differential information economies: the core. Economic Theory **18**, 333–348 (2001)
Shapley, L.S.: A value for n-person games. In: Kuhn, H.W., Tucker, A.W. (eds) Contributions to the theory of games, II. Princeton: Princeton University Press 1953. Annals of Mathematical Studies **28**, 307–317 (1953)
Srivastava, S.: A limit theorem in the core with differential information. Economic Letters **14**, 111–116 (1984)
Tirole, J.: The theory of industrial organization. Cambridge, MA, London, UK: The MIT Press 1988
Wilson, R.: Information, efficiency, and the core of an economy. Econometrica **46**, 807–816 (1978)
Yannelis, N.C.: The core of an economy with differential information. Economic Theory **1**, 183–198 (1991)

PART 1

CORE NOTIONS, EXISTENCE RESULTS

Information, efficiency, and the core of an economy*

Robert Wilson**

Stanford University

Received: September, 1976; revision version: May, 1977

Summary. The meaning of exchange efficiency is examined in the context of an economy in which agents differ in their endowments of information. Definitions of efficiency, and of the core, are proposed which emphasize the role of communication. Opportunities for insurance are preserved by restricting communication, or in a market system by restricting insider trading, prior to the pooling of information for the purposes of production.

My subject is an economy in which different agents have different information. I propose a definition of exchange efficiency and I characterize the efficient allocations. I then examine an analogous definition of the core and I demonstrate that the core is not empty if the usual regularity conditions are satisfied. An example, however, illustrates that a market process may fail to yield an efficient allocation. In fact, in this example the market allocation is not even individually rational for the agents. Also, in this example the core is empty if there are opportunities for communication which disrupt arrangements for mutual insurance.

1. Formulation

S denotes the set of possible states. For simplicity 1 suppose that the cardinality of S is finite. Some one state s^* in S is the prevailing state. An event is a subset of S.

N denotes the finite set of agents. The information of the ith agent is described by the field F_i of events which he can discern. An event E is in the field F_i iff

* This study was supported in part by the National Science Foundation, Grant SOC76-11446-A01, at the Institute for Mathematical Studies in the Social Sciences, and in part by the Energy Research and Development Administration, Contract E(04-3)-326 PA # 18, at the Systems Optimization Laboratory, Stanford University.

** For discussions and suggestions I am indebted to Robert Aumann, Jonathan Cave, Bengt Holmström, Takao Kobayashi, David Kreps, and John Riley. Kobayashi [3] presents extensions of this work.

he knows whether the prevailing state is in the event E or in the complementary event $S \backslash E$. For example, if agent i observes the value of a random variable y_i, then F_i is the smallest field containing the events in the inverse image of y_i. The minimal nonempty events in the field F_i form a partition of the states denoted by PF_i. Precisely one member of the partition is known by the agent to contain the prevailing state. $PF_i(s)$ denotes the unique member of the partition containing the state s.

The field of events discernable by every agent is the "coarse" field $\bigwedge_N F = \cap_{i \in N} F_i$. By pooling their information they could discern the events in the "fine" field $\bigvee_N F$ for which $P \bigvee_N F(s) = \cap_{i \in N} PF_i(s)$. More generally, the result of a communication system (c.s.) is a collection $(H_i)_{i \in N}$ of fields such that $H_i \supseteq F_i$ for each agent and $\bigvee_N H = \bigvee_N F$. Communication enlarges the field of events an agent can discern but it does not produce new information. The null c.s. is $(F_i)_{i \in N}$ and the full c.s. is $(\bigvee_N F)_{i \in N}$.

A commodity bundle is a member of the Euclidean space with coordinates indexed by the commodities. An agent i has for each state s a set $X_i(s)$ of commodity bundles which are feasible for consumption. One member of $X_i(s)$ is agent i's endowment $e_i(s)$ which he obtains if s is the prevailing state and he engages in no trade. A consequence of trade is an allocation $x = (x_i(s))$ which provides agent i with the consumption $x_i(s) \in X_i(s)$ if state s prevails, provided that $\sum_{i \in N} x_i(s) = \sum_{i \in N} e_i(s)$. One may also require that the consumption plan x_i is measurable with respect to a field F_i'', namely $x_i(s) = x_i(\bar{s})$ if $s \in PF_i''(\bar{s})$. In this case I assume that X_i and e_i are F_i'-measurable for some field $F_i' \supseteq F_i$; and that $F_i'' \supseteq \bigvee_N F'$.

When agent i knows that the prevailing state is in the event $A \in PF_i$ of his partition, or a finer event $A \in PH_i$ discernable from communication, he has a relation $\succ_{iA}$ of preference between feasible consumption plans. For any coarser event $E \in H_i$ the relation $x_i \succ_{iA} \bar{x}_i$ means that $x_i \succ_{iA} \bar{x}_i$ for every event $A \in PH_i$ in the partition for which $A \subseteq E$. It will suffice here to assume that this preference relation is represented by a probability assessment (S, F_i'', μ_i) and by an F_i'-measurable utility function u_i which assigns to each feasible consumption $x_i(s) \in X_i(s)$ in state s a real value $u_i(s, x_i(s)) \equiv u_i[x_i](s)$. If H is a subfield of F_i'' then the conditional expectation of an F_i''-measurable random variable u defined on S is an H-measurable random variable $v \equiv \mathcal{E}_i\{u|H\}$ for which $\int_E u(s)d\mu_i(s) = \int_E v(s)d\mu_i(s)$ for each event $E \in H$. In particular, $x_i \succ_{iE} \bar{x}_i$ for an event $E \in H_i$ iff $\mathcal{E}_i\{u_i[x_i]|H_i\}(s) > \mathcal{E}_i\{u_i[\bar{x}_i]|H_i\}(s)$ for each state $s \in E$. Note that the conditional expectation has a common value $\mathcal{E}_i\{u_i[x_i]|H_i\}(A)$ for $s \in A \in PH_i$ if $\mu_i(A) > 0$. For simplicity I assume that the measure μ_i assesses a positive probability for each nonempty event in $\bigvee_N F$.

For the propositions in Sections 2–4 I impose the usual regularity assumptions which ensure that the sets of feasible allocations and attainable utilities are compact and convex. Namely, for each agent i and each state s the set $X_i(s)$ of feasible consumptions is closed, convex, and bounded below; and the utility function $u_i(s, \cdot)$ defined on this set is continuous and concave.

Table 1

Agent (i)	PF_i	State (s):	Endowments ($e_i(s)$) a	b	Allocation ($x_i(s)$) a	b
1	$\{a\}, \{b\}$		2	0	1	1
2	$\{a, b\}$		0	2	1	1

Table 2

Agent (i)	PF_i	State (s):	Endowments ($e_i(s)$) a	b	c	Allocation ($x, (s)$) a	b	c
1	$\{a\}, \{b, c\}$		5	1	3	5	2	2
2	$\{b\}, \{a, c\}$		3	5	1	2	5	2
3	$\{c\}, \{a, b\}$		1	3	5	2	2	5

2. Efficiency

It is useful to recognize that no single definition of efficiency will suffice for all purposes. The fact that different agents have different information must necessarily eliminate some of the opportunities for mutual insurance. Moreover, the possibility of communication raises the prospect that additional opportunities will be eliminated. My aim here is to identify that definition of efficiency which retains the greatest opportunities for insurance subject to the limitation inherent in the agents' information. In addition, I seek a definition of efficiency which is consistent with a viable definition of the core.

Two simple examples illustrate the primary considerations. There is a single desired commodity; each agent has a utility function which is independent of the state and strictly concave, reflecting aversion to risk; and each agent assigns equal probabilities to the states.

Example 1. There are two agents and two states. The agents' endowments and information are displayed in Table 1. Also shown is an allocation x which would be a favorable arrangement for mutual insurance in the absence of a difference in information. As it is, however, agent 1 has superior information. If the prevailing state is $s^* = a$ he would surely reject the proposed allocation x. That is, the allocation is not individually rational for agent 1, nor is any other allocation which partially insures agent 2 against his perceived risk. Indeed, realizing this, agent 2 has no incentive to offer or accept a contract since it could be advantageous to agent 1 only in state b when it is to his own disadvantage. I conclude that a useful definition of efficiency must include the endowment as an efficient outcome. This example illustrates the phenomenon of adverse selection which often vitiates opportunities for insurance.

Example 2. There are three agents and three states. The agents' endowments and information are displayed in Table 2. As in the previous example there is an allocation which provides an equal amount (3 units) to each agent in each state but which in each state is not individually rational for the agent with superior information. Also shown in Table 2 is an allocation x which escapes this feature and which

Table 3

		Allocation of Claims ($s^* = a$)			Prevailing	Market Allocation (x115)		
Agent (i)	State (s):	a	b	c	state (s^*):	a	b	c
1		666/115	0	0		666	144	225
2		225/115	0	9		225	666	144
3		144/115	9	0		144	225	666

provides complete insurance for the poorly informed agents. If the prevailing state is $s^* = a$, then agent 2 or 3 perceives equal chances that his endowment is 1 or 3 units and therefore he prefers an insured consumption of 2 units.

Another allocation of interest is the one derived from a market for state-contingent claims. Assume that each agent has the utility function $u(s, x) = \log x$ and that the prevailing state is $s^* = a$. Then the equilibrium prices are $(p_a, p_b, p_c) = (1, 16/115, 25/115)$, where p_s is the price of one unit payable in state s, and the resulting allocation of claims is shown on the left in Table 3. The construction of this equilibrium depends on the assumption that an agent cannot sell more claims than his endowment and that no agent infers the prevailing state from the prevailing prices. Proceeding symmetrically for each of the other two states that might prevail yields the actual market allocation shown on the right in Table 3. A short computation reveals that this market allocation violates individual rationality. For instance, in the event $\{a, b\}$ agent 3 is worse off with the market allocation than with his endowment. A rational-expectations model would eliminate this difficulty, of course, since each of the two poorly informed agents could infer the state from the prices. In this case there would be no trade at the prices $p = (1, 0, 0)$ when $s^* = a$, and the market allocation would be the endowment. Implicit communication via the market process preserves individual rationality but still it eliminates the kind of favorable insurance arrangement provided by the allocation x in Table 2. Notice that the market allocation in Table 3 could be improved by equalizing the consumptions of the two poorly informed agents in each state.

The allocation in Table 2 is not immune to criticism. The insurance plan for the two poorly informed agents appears to require the cooperation of the perfectly informed agent regarding states which he knows do not prevail. The possibility of communication raises the prospect that if $s^* = a$ then the coalition of agents 1 and 2 could do better by retaining their endowments, perhaps with agent 2 rewarding agent 1 for saving him the cost of insurance. These matters will be examined further when we study the core in Section 3. I defer the question of what institutionalized process, market or nonmarket, could achieve the allocation in Table 2.[1] For now it suffices to observe that the greatest opportunities for insurance are obtained by restricting communication to the null c.s. I conclude, therefore, that a viable definition of

[1] The allocation in Table 2 can be achieved by a market in state-contingent claims with the fixed prices $p = (1, 1, 1)$ if each agent is prohibited from trading in the claims for which he has perfect, or "inside," information; i.e., agents 1, 2, and 3 are prohibited from trading claims payable in states a, b, and c, respectively.

efficiency without communication should allow the allocation in Table 2 to be efficient.

With these examples in mind I turn to a definition of efficiency. I propose that an allocation is efficient iff in each event which every agent can discern there is no other allocation which each agent prefers given his own information. That is, an allocation is efficient iff there is not an event $E \in \bigwedge_N F$ and another allocation $\bar{x}$ such that $\bar{x}_i \succ_{iE} x_i$, namely $\mathcal{E}_i\{u_i[\bar{x}_i]|F_i\} > \mathcal{E}_i\{u_i[x_i]|F_i\}$ on E, for every agent $i \in N$. Note that the null c.s. is imposed. The origin of the requirement that the contingency must be recognized by every agent is evident in Example 2. There we saw that a reallocation of the endowment may extend over states known by some agents not to prevail.

This notion of efficiency is also called "coarse" efficiency to distinguish it from the weaker concept of "fine" efficiency which admits the full c.s. and allows $E \in \bigvee_N F$. Thus fine efficiency excludes another allocation $\bar{x}$ for which $\mathcal{E}_i\{u_i[\bar{x}_i]|\bigvee_N F\} > \mathcal{E}_i\{u_i[x_i]|\bigvee_N F\}$ on $E \in \bigvee_N F$ for every agent $i \in N$. An allocation which is fine *in*efficient on a coarse event $E \in \bigwedge_N F$ is also coarse *in*efficient. In this sense coarse efficiency is a strong requirement. The corresponding notion of strict efficiency is slightly stronger: an allocation x is strictly efficient iff there is not an event $E \in \bigwedge_N F$ and another allocation $\bar{x}$ such that $\mathcal{E}_i\{u_i[\bar{x}_i]|F_i \geqq \mathcal{E}_i\{u_i[x_i]|F_i\}$ on E for every agent $i \in N$, with strict preference for at least one agent i on at least one event $A \in PF_i$, $A \subseteq E$. I omit the obvious generalization of the definitions of efficiency to include arbitrary communication systems other than the null and full c.s.

For Example 1 the endowment is both strictly efficient and fine efficient. For Example 2 the allocation in Table 2 is both strictly efficient and fine efficient. The role of the distinction between coarse and fine events for this allocation will not be apparent until we study the coarse and fine cores in Section 3. This distinction is evident in the market allocation in Table 3, however. The market allocation is fine efficient but not coarse efficient; and in fact this is true also of the endowment, which is the market allocation resulting from rational expectations.[2] We see here that fine efficiency is compatible with the "informational efficiency" of market processes (e.g., S. Grossman [1] or S. Grossman and J. Stiglitz [2]). In contrast, coarse or strict efficiency emphasizes the advantages of insurance, and therefore the disadvantages of direct or implicit communication.

The existence of efficient allocations is easily verified. Consider nonnegative weights $\lambda_i(s)$ for each agent $i \in N$ and each state $s \in S$, and an allocation that maximizes $\sum_{i \in N} \mathcal{E}_i\{\lambda_i u_i[x_i]\}$ among the set of feasible allocations. If agent i's weighting function λ_i is F_i-measurable, and not all the weights are zero on any coarse event in $P\bigwedge_N F$, then the allocation is efficient; or if additionally all the weights are positive, strictly efficient. Similarly, a fine-efficient allocation is obtained from $\bigvee_N F$-measurable weights, not all zero on any event in $P\bigvee_N F$. For the coarse-efficient allocation shown in Table 2 such a set of weights has $\lambda_i(s) = 15$ if s is the state in which agent i has superior information, and $\lambda_i(s) = 6$ otherwise. The extreme form of "ex ante" efficiency which emphasizes insurance

[2] This is proved in detail in Section 3. See Table 5.

Table 4

Agent (i)	State (s):	Reallocation of x a	b	c
1		5	$2+\beta$	$2-\gamma$
2		$2-\alpha$	5	$2+\gamma$
3		$2+\alpha$	$2-\beta$	5

to the exclusion of informational considerations is reflected in the allocation shown in Table 1 for Example 1 and the similar one for Example 2: these allocations correspond to weights which are not only F_i-measurable but in fact constant over the whole set of states.

In most models of market processes the imputed weights are the reciprocals of the agents' marginal utilities of income. Strict efficiency requires, therefore, that each agent's marginal utility of income is measurable with respect to his information. This is just another way of stating the requirement for optimal insurance. In this case the insurance is against what other agents know about the prevailing state. A market process vitiates the opportunities for this insurance. In Example 2 the perfectly informed agent is an "insider" in the market for state-contingent claims purchased for insurance purposes by the other two agents, and this distorts the prices and their incomes to the evident advantage of the insider.[3]

Provided the utility functions are differentiable one can state the necessary condition for an interior allocation x to be strictly efficient in terms of the marginal rates of substitution (MRS). Considering only a single commodity and assuming $PF'_i(s) = \{s\}$ for simplicity, agent i's MRS between incomes in states s and $\bar{s} \in PF_i(s)$ is $MRS_i(s,\bar{s}) = v_i(\bar{s})/v_i(s)$, where $v_i(s) = u'_i[x_i](s)\mu_i(\{s\})$. Consider a small reallocation such as the cyclic one shown in Table 4 for the allocation of Example 2. If x is to be strictly efficient it must be for $(\alpha,\beta,\gamma) > 0$ that if $\beta/\gamma \geqq MRS_1(b,c)$ and $\gamma/\alpha \geqq MRS_2(c,a)$ so that a marginal reallocation is not unfavorable for agents 1 and 2, then $\alpha/\beta \leqq MRS_3(a,b)$ so that it is not favorable for agent 3. Allowing negative variations as well yields the necessary condition for strict efficiency that $MRS_1(b,c) \cdot MRS_2(c,a) \cdot MRS_3(a,b) = 1$. In general, consider a finite cycle of states $s_1, ..., s_K, s_{K+1} = s_1$ such that $s_{k+1} \in PF_{i(k)}(s_k)$ for some agent $i(k)$. Then an interior allocation is strictly efficient only if $\Pi_k MRS_{i(k)}(s_k, s_{k+1}) = 1$. This condition is a generalization of the equality of agents' MRS's which is the familiar condition for "ex ante" efficiency in an economy without differences in information.

A substantial part of economic theory is the consequence of the observation that bilateral trade suffices to obtain the equality of the agents' MRS's. Here, it is clear that multilateral trade is necessary, though the institutionalized form that this trading might take is ambiguous. As we saw earlier the missing ingredient of ordinary market processes is some form of "income insurance" which enables each agent i to achieve a marginal utility of income which is F_i-measurable, namely the same for each state $s \in PF_i(s^*)$. There are now a number of well-known examples

[3] See Footnote 1.

where the absence of this ingredient has adverse effects; e.g., the study of signalling in labor markets by M. Spence [1] and the study of screening in insurance markets by M. Rothschild and J. Stiglitz [4]. The substance of the matter is whether institutional arrangements to remedy these effects are possible in principle. In the next section I examine the question by studying the core of an economy with differences in information, and I demonstrate the affirmative answer that the core is not empty.

3. The core

In choosing a definition of the core my motive is to identify those allocations having the property that if one is proposed then no subset of the agents has the opportunity and incentive to opt for an alternative allocation. That is, no coalition can block the proposed allocation. When different agents in a coalition have different information their opportunities to take blocking actions jointly are necessarily contingent upon events which they all can discern.

I suggest the following definition of contingent blocking. An allocation is blocked if some coalition can enforce an alternative allocation which they prefer in an event which they all can discern. Specifically, a (nonempty) coalition $M \subseteq N$ can enforce an allocation $\bar{x}$ in an event $E \in \bigwedge_M F$, which its members all can discern, iff $\sum_{i\in M} \bar{x}_i(s) = \sum_{i\in M} e_i(s)$ for each state $s \in E$; and if $\mathcal{E}_i\{u_i[\bar{x}_i]|F_i\} > \mathcal{E}_i\{u_i[x_i]|F_i\}$ on E for each member $i \in M$ then the proposed allocation x is blocked. The core is then the set of unblocked allocations. Note that this definition confines a blocking coalition to its null c.s.

This can also be called the coarse core to distinguish it from the fine core for which a blocking coalition can also use its full c.s. If each coalition M has a specified set $C(M)$ of feasible communication systems then a general definition can be phrased as follows: an allocation x is blocked iff there is a coalition $M \subseteq N$ having a feasible c.s. $(H_i)_{i\in M} \in C(M)$, and event $E \in \bigwedge_M H$ which its members can all discern using the c.s., and an alternative allocation $\bar{x}$ which it can enforce in the event E and which every member $i \in M$ prefers given the information H_i in the event E, namely $\mathcal{E}_i\{u_i[\bar{x}_i]|H_i\} > \mathcal{E}_i\{u_i[x_i]|H_i\}$ on E for each member $i \in M$.

The definition of blocking invokes three considerations, of which the first is peculiar to an economy with differences in information among the agents. A coalition can block only in an event which every member can discern using some one of its feasible communication systems, since otherwise joint action is not possible. Moreover it can object only with an alternative allocation which it can enforce given that the specified event is known to obtain. And lastly, each member must prefer the alternative allocation based on his information derived from the c.s. in whatever finer event he knows or learns to obtain.

The requirement that a proposed allocation be unblocked is postulated as a minimal desideratum for its stability as a candidate in a negotiating process. One can envision that the agents negotiate the terms of an enforceable contract. Each agent has his private information but in an institutionalized setting he may be unable or unwilling to reveal it. The proposal of an unblocked allocation offers no coalition an opportunity and incentive to object in any contingency. Any other allocation is

Table 5

		Alternative allocation		
Agent (i)	State (s):	a	b	c
1		$5+2\varepsilon$	$2-\varepsilon$	$2-\varepsilon$
2		$2-\varepsilon$	$5+2\varepsilon$	$2-\varepsilon$
3		$2-\varepsilon$	$2-\varepsilon$	$5+2\varepsilon$

unlikely to be sustained against counterproposals and ultimately adopted in an event in which a coalition can block; for, each member recognizes the possibility of the event (and its certainty if it is discernable from their null c.s.) and together they have an incentive to opt in favor of their alternative allocation which they can enforce.

For Example 1 the coarse and fine cores consist only of the endowment. For Example 2 the allocation in Table 2 is in the coarse core. The market allocation in Table 3, however, is blocked by agent 3 in the event (a, b).

The fine core for Example 2 is actually empty, as I shall now demonstrate. In each state the perfectly informed agent and each two-agent coalition must get at least their endowments since the full c.s. allows them to identify the prevailing state. Thus the endowment is the only candidate for an unblocked allocation in the fine core. But the endowment is blocked by the whole coalition N using its null c.s. in the whole event S by proposing the alternative allocation displayed in Table 5, provided that $\varepsilon > 0$ is sufficiently small. Thus the fine core is empty for Example 2. The apparent source of this difficulty is the conflict between the smaller coalitions' use of communication to gain advantages, with the whole coalition's opportunity to provide insurance. It is true in general that the more communication is allowed the smaller is the resulting core, but we see in this example that the tension between the null c.s. and the full c.s. is sufficient to eliminate the core. An analogous conclusion is obtained by M. Rothschild and J. Stiglitz [3] in their study of insurance markets with differential information, where full communication occurs implicitly when an insurer infers a buyer's risk class from the type of contract he purchases.

It is easy to verify that the coarse core is never empty. The proof is obtained by constructing another cooperative game for which the players are the pairs (i, A) in which $i \in N$ and $A \in PF_i$. A player (i, A) prefers one allocation x to another $\bar{x}$ iff agent i prefers x to $\bar{x}$ given F_i in the event A. The admissible coalitions are those of the form $(M, E) \equiv \{(i, A) | i \in M, A \in PF_i, A \subseteq E\}$ for $M \subseteq N$ and $E \in \bigwedge_M F$. Such a coalition can enforce the allocation x iff M can enforce it in the event E. It is straightforward to verify that this newly constructed cooperative game is a balanced game as defined by Scarf [5]. Consequently, there exists an unblocked allocation in the ordinary core of this game. Such an unblocked allocation is also unblocked in the economy with differential information. Thus the coarse core is not empty.[4]

[4] An alternative proof consists of showing that in the coarse core is an allocation resulting from a constrained market process. For fixed prices $(p(s))$ which equate demands and supplies in each state, each player (i, A) chooses a feasible consumption plan to maximize $\mathcal{E}_i\{u_i[x_i]|F_i\}(A)$ subject to the budget constraint $\sum_{s\in A} p(s)[x_i(s) - e_i(s)] \leqq 0$.

The substance of this argument is merely the observation that each agent can wear several hats in the negotiating process; or possibly he can delegate responsibility to subordinates, one for each event in his informational partition, to whom he confers responsibility in that event. This approach will not work, of course, whenever any coalition has access to a non-null c.s. In Example 1, for instance, the economy is usefully regarded as a game among the three players $(1, \{a\}), (1, \{b\}), (2, \{a, b\})$. It is then clear that the endowment is the only allocation in the coarse core, since the first two players will invariably insist on getting their endowments. A similar viewpoint in Example 2 motivates the allocation in Table 2, though it is not the only unblocked allocation; and especially, compared to the market allocation in Table 3, it motivates the requirement that the outcome of the game be efficient in the coarse sense.

4. An extension

A natural objection to the definition of the coarse core is the conjecture that either strategic considerations or the usefulness of information in production might favor the "informational efficiency" of market processes. I conclude briefly, therefore, with a more elaborate construction which addresses the matter to the extent that a version of the coarse core remains nonempty.

Assume that each agent i has a set of feasible decisions which is a compact and convex subset of a finite-dimensional Euclidean space (or more generally, a complete separable metric space). For a coalition M a strategy $d^M = (d_i^M(s))$ specifies for each member $i \in M$ a decision rule d_i^M as a function of the state. Given a strategy d^M the coalition M obtains from production the commodity bundle $y^M(s, d^M(s)) = y^M[d^M](s)$ in state s, and each member i has the endowment $e_i(s; d^M(s), d^{-M}(s)) = e_i[d^M, d^{-M}](s)$ depending on the strategy of the complementary coalition $-M = N \backslash M$. The coalition can enforce the allocation x in an event $E \in \bigwedge_M F$ iff $\sum_{i \in M} x_i[d^M, d^{-M}](s) \leqq \sum_{i \in M} e_i[d^M, d^{-M}](s) + y^M[d^M](s)$ for each state $s \in E$ and each strategy d^{-M} of the complementary coalition. Note that the allocation depends upon the strategies of both coalitions.

An outcome is a pair (d^N, x) consisting of a strategy for the whole coalition, such that d_i^N is $\bigvee_N F$-measurable for each agent i, and an allocation that it can enforce in the whole event S. Such an outcome is blocked by a coalition M in an event $E \in \bigwedge_M F$ proposing one of its feasible strategies d^M and an allocation $\bar{x}$ which it can enforce in the event E iff for each member $i \in M$ the decision rule d_i^M is $\bigvee_M F$-measurable and

$$\mathcal{E}_i\{u_i[\bar{x}_i[d^M, d^{-M}]] | F_i\} > \mathcal{E}_i\{u_i[x_i[d^N]] | F_i\}$$

on E for every strategy d^{-M} of the complementary coalition which is $\bigvee_{N \backslash \{i\}} F$-measurable in each component. The core then consists of the unblocked outcomes. (The weak measurability requirement on the complementary coalition's strategy is perhaps unsatisfactory. It envisions that each member of a blocking coalition trusts only that his own information does not leak out to the complementary coalition, since his colleagues may not have motives to withhold.)

Table 6

Decision	State:	Left	Right
Up		2,2	0,2
Down		0,0	4,0

Assume that the endowments $e_i[\cdot]$ and the production functions $y^M[\cdot]$ are each continuous and concave, and F_i' and $\bigvee_M F'$-measurable, respectively. Also, if B is a balanced collection of coalitions, namely there exist weights $a_M \geqq 0$ such that $\sum_{i \in M \in B} \alpha_M = 1$ for each agent $i \in N$, and $d_i^N = \sum_{i \in M \in B} \alpha_M d_i^M$, then $y^N[d^N] \geqq \sum_{M \in B} \alpha_M y^M[d^M]$. This condition is a consequence of the concavity and homogeneity of y^N if $y^M[d^M] = y^N[d^M, 0]$.

The proof that there exists an unblocked outcome follows the previous argument, supplemented by H. Scarf's [6] construction for cooperative games in normal form.

The following example illustrates some of the features of this formulation.

Example 3. Two agents named Row and Column are to play one of two noncooperative games called Left and Right, each equally likely. Only Column knows which game is to be played. Column has no decision to make but Row must choose between two decisions Up and Down. The payoffs (in the single commodity) to Row and Column are shown as ordered pairs in Table 6. The Nash equilibria of this game lead Row to choose Down. This is true also if Column is allowed first to send a message to Row, since Column's incentive is to induce Row to choose Up in either game. In the coarse core of the corresponding cooperative game is the strategy which chooses Up or Down as the state is Left or Right, and which gives to Row an insured payoff of 3 units in either case; indeed Row can let Column make the decision to obtain an insured payoff of 1 unit in either case.

This example illustrates the legal maxim that favors vesting the better-informed agent with the power and consequences of decisions in situations afflicted with moral hazard. The other side of the coin, of course, is the need to insure the poorly informed agent.

References

1. Grossman, S.: On the efficiency of competitive stock markets where traders have diverse information. Journal of Finance **31**, 573–585 (1976)
2. Grossman, S., Stiglitz, I.: Information and competitive price systems. American Economic Review **66**, 246–253 (1976)
3. Kobayashi, T.: Efficiency, equilibrium, and the theory of syndicates with differential information. Ph.D. dissertation, Stanford University, 1977
4. Rothschild, M., Stiglitz, J.: Equilibrium in competitive insurance markets: an essay an the economics of imperfect information. Quarterly Journal of Economics **90**, (629–649 (1976)
5. Scarf, H.: The core of an N-person game. Econometrica **35**, 50–69 (1967)
6. Scarf, H.: On the existence of a cooperative solution for a general class of N-person games. Journal of Economic Theory **3**, 169–181 (1971)
7. Spence, M.: Market signalling. Cambridge: Harvard University Press 1974

Information, efficiency and the core of an economy: Comments on Wilson's paper*

Dionysius Glycopantis[1] **and Nicholas C. Yannelis**[2]

[1] Department of Economics, City University, Northampton Square, London EC1V 0HB, UK (e-mail: d.glycopantis@city.ac.uk)
[2] Department of Economics, University of Illinois at Urbana-Champaign, IL 61820, USA (e-mail: nyanneli@uiuc.edu)

1 Introduction

In his seminal paper, Wilson (1978) discusses the issues of exchange efficiency and the core in the context of differential information economies. His work has attracted a lot of attention and had widespread influence in the development of economic theory.

In this note we offer some explanations and make comments on Wilson's paper which we believe will help to clarify further his ideas. In particular we discuss two of his examples which he uses to make a number of points centered around feasibility, efficiency and his coarse core. We assume that the reader is acquainted with this paper. Examples and tables are numbered as they appear there. The main point is that Wilson's analysis and use of the term rational expectations equilibria (REE) are not the same as in the Radner-Allen approach.

2 Explanations of Examples 2 and 3

In Example 2 we are explaining Wilson's calculations of prices and quantities in a market of state-contingent claims. We also point out that when Wilson uses the term rational expectations equilibria (REE) his definition is not the same as REE given by Radner (1979) and Allen (1981), and used in Glycopantis-Yannelis in this volume. It is precisely the differential information of the agents which accounts for the variety of possible equilibrium ideas.

In Example 3 we give a detailed analysis, confirming Wilson's results, through the interpretation of the game in a tree form. The use of extensive form games lends itself naturally in situations where one of the players sends or does not send a signal to the other one.

---

* We wish to thank A. Muir and R. Wilson for their very helpful comments and suggestions.

Example 2. We explain here how Wilson obtains the prices $(p_a, p_b, p_c) = (1, \frac{16}{115}, \frac{25}{115})$ under state $s^* = a$. Notice that this is not rational expectations equilibria and we will return to this point below. Wilson says that state a has been chosen by nature. P1 sees a, P2 observes $\{a, c\}$ and P3 the event $\{a, b\}$. However each player will also try to sell his endowments in the remaining states because he knows that they are valuable to somebody else.

The calculations below show how to obtain the prices $p_a = 1,\ p_b = \frac{16}{115}\ p_c = \frac{25}{115}$. The agents receive price signals from the auctioneer and maximixe their interim expected utility subject to their budget constraints. We have in effect the following problems.

P1:
Problem

Maximize $u_1 = log x_{1a}$
Subject to

$$p_a x_{1a} = 5p_a + 1p_b + 3p_c$$

P2:
Problem

Maximize $u_2 = log x_{2a} + log x_{2c}$
Subject to

$$p_a x_{2a} + p_c x_{2c} = 3p_a + 5p_b + 1p_c$$

P3:
Problem

Maximize $u_3 = log x_{3a} + log x_{3b}$
Subject to

$$p_a x_{3a} + p_b x_{3b} = p_a + 3p_b + 5p_c$$

The agents send back to the Walrasian auctioneer their quantities demanded as signals. The equilibrium conditions are:
For the quantities, in State a: $\frac{5p_a+1p_b+3p_c}{p_a} + \frac{3p_a+5p_b+1p_c}{2p_a} + \frac{p_a+3p_b+5p_c}{2p_a} = 9$, in State b: $\frac{p_a+3p_b+5p_c}{2p_b} = 9$ and in State c: $\frac{3p_a+5p_b+1p_c}{2p_c} = 9$.

These relations are satisfied by $p_a = 1,\ p_b = \frac{16}{115},\ p_c = \frac{25}{115}$ and the implied allocation of claims is $x_{1a} = \frac{666}{115},\ x_{2a} = \frac{225}{115},\ x_{3a} = \frac{144}{115},\ x_{1b} = x_{2b} = 0,\ x_{3b} = 9,\ x_{1c} = x_{3c} = 0,\ \ x_{2c} = 9$.

We have obtained in these calculations the prices above and the Allocations of Claims in Table III. The first column in the Market Allocation corresponds to the first column of Allocation of claims. It says that when it is revealed that state a has been realized, then the first column of Allocation of Claims is what is relevant. Given the endowments in Table II we can do the calculations per prevailing state and arrive eventually at columns b and c of Market Allocation in Table III.

The above analysis is not in the area of rational expectations in the usual sense. First the endowments in Wilson's formulation are not private information measurable and also, which is probably more significant, we now have a function

$p : \Omega \to \mathbb{R}^l$ per prevailing state. In REE there is only one price function defined on Ω. Hence, although REE is itself an interim concept, here we have an alternative interim concept.

Suppose we were to define a REE in the context of Wilson. It could go as follows. We are looking for $p : \Omega \to \mathbb{R}^l$ such that each agent, given the element of his information set which he observes and the signal that he receives from prices, maximizes his interim expected utility subject to his relevant budget constraint and such that when the state is revealed the markets clear. Now we would not insist on any kind of measurability of the allocations.

According to the above definition $(p_a, p_b, p_c) = (1, 1, 1)$ is a non-revealing REE set of prices and the Allocation in Table II is the corresponding REE quantities. The agents rely only on their information sets and maximize interim expected utility subject to their budget constraints without insisting on measurability of their choices. This is consistent with the fact that the endowments are not measurable.

With respect to Wilson's Footnote 3, what is meant there is that given prices $p = (1, 1, 1)$, the agents are only maximizing interim expected utility, when the state of nature is uncertain, and otherwise they keep their endowment. The outcome is again the Allocation in Table II. Perhaps defining a new REE notion is more satisfactory than prohibiting traders, under some circumstances from trading.

Below Table III, Wilson refers to REE but in a different sense to the one described above. This follows from the nature of the prices he proposes. The vector $p = (1, 0, 0)$ clears the market only under the condition $s^* = a$. Indeed it can be replaced by any non-negative price vector $p = (1, k_1, k_2)$ with $k_i \neq 1$ and analogous prices identifying the other states. Everybody demands his own endowment. Also, if all prices are positive and different, and again ignoring the lack of measurability of the initial allocation, we have a fully revealing REE, and these initial endowments are confirmed as an equilibrium.

Example 3. This example calculates a Nash equilibrium in the context of a normal form game in which players have differential information. Originally the player of the columns is not allowed to send a signal and in the second instance he can signal to the player of the rows. The payoffs in Table VI are in a single commodity. Agents 1 and 2 have strictly concave and increasing utility functions u_1 and u_2 on this good. As there is no confusion, the payoffs in the trees below are given in terms of the commodity.

In order to make the analysis clearer, we cast it in a tree form attaching to "Nature", as the third agent, the possibility to choose in the beginning between states left (L) and right (R) with equal probabilities. We call P2 the Column player and P1 the Row player.

The information sets are given by $\mathcal{F}_1 = \{\{L, R\}\}$ and $\mathcal{F}_2 = \{\{L\}, \{R\}\}$. P1 cannot distinguish between L and R that nature chose, but P2 can do so.

Case 1. P2 makes no announcement.

Payoffs are determined from nature and the decision of P1 who, given the probabilities of choices, maximizes his expected utility. The Nash equilibrium, indicated in Figure 1, is for P1 to play d. P2 is completely passive and does nothing.

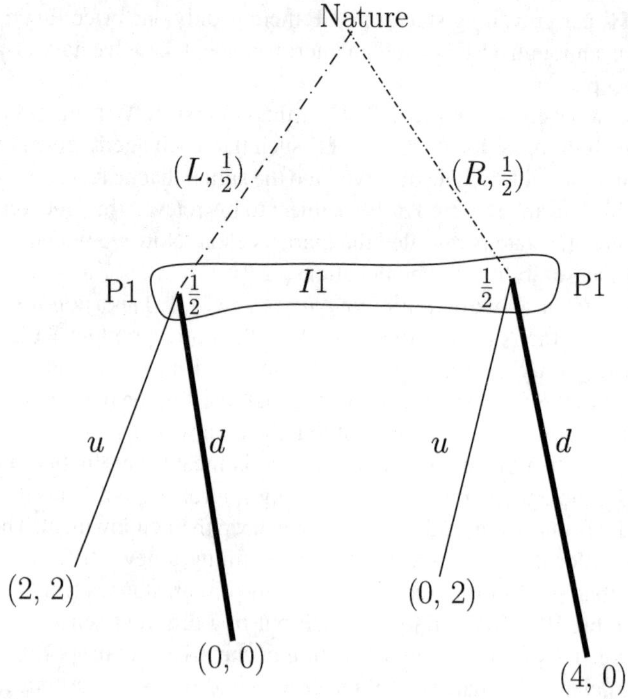

Figure 1

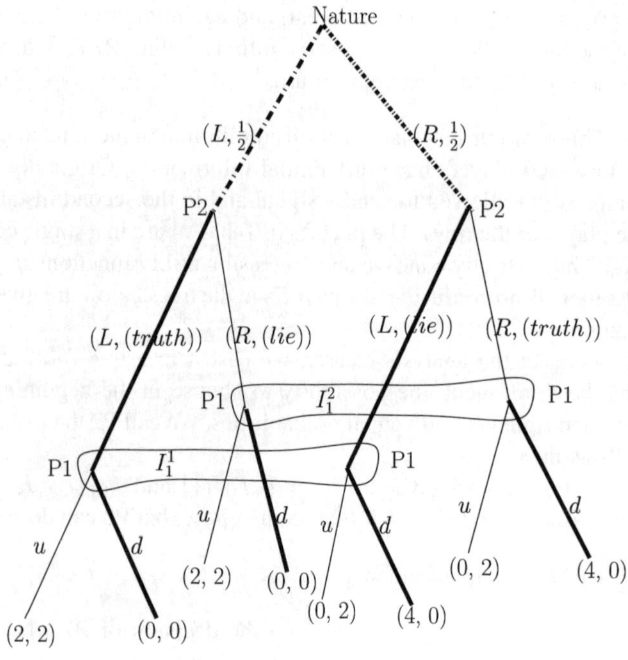

Figure 2

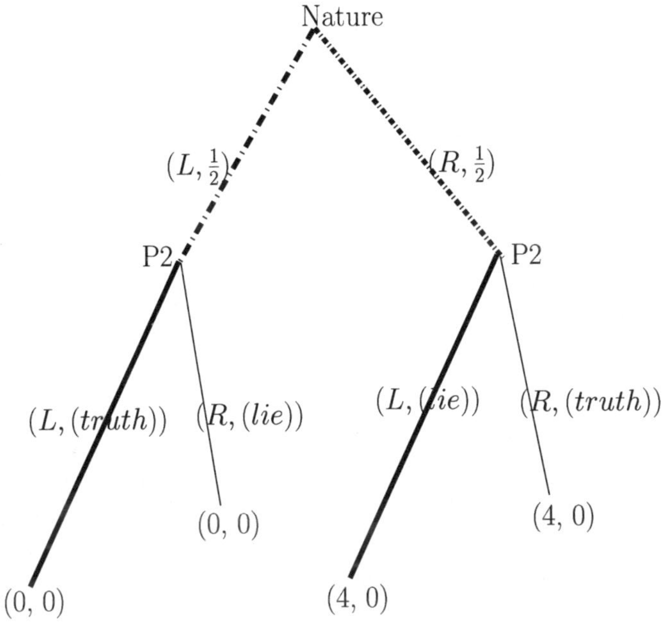

Figure 3

Case 2. P2 is allowed to send a signal to P1.
We now construct Figure 2. Payoffs are determined from nature and the decision of P1. He cannot distinguish between L and R that nature chose but can hear the announcement of P2 who can either tell the truth or choose to lie. We consider various possibilities of pure strategies, where the first choice of each player refers to his first information set, from left to right:

P2, $(L, R) \Rightarrow (u, d)$ for P1, not Nash;
P2, $(L, L) \Rightarrow (d, d)$ for P1, Nash;
P2, $(L, L) \Rightarrow (d, u)$ for P1, not Nash;
P2, $(R, L) \Rightarrow (u, d)$ for P1, not Nash;
P2, $(R, R) \Rightarrow (d, d)$ for P1, Nash;
P2, $(R, R) \Rightarrow (u, d)$ for P1, not Nash.

One of the possible Nash equilibria is indicated on Figure 2 with heavy lines. It can be obtained by folding up the tree to the one in Figure 3. The above confirms the statement of Wilson that in this case also P1 will play d.

There are other Nash equilibria as well. In these P1 will always play d but P2 can use mixed strategies as well. Furthermore all these Nash equilibria, with appropriate probabilities (beliefs) attached to the nodes of the information sets, are also perfect Bayesian equilibria (PBE).

In the same example Wilson says: "In the coarse core of the corresponding cooperative game (an equilibrium) is the strategy which chooses Up or Down as

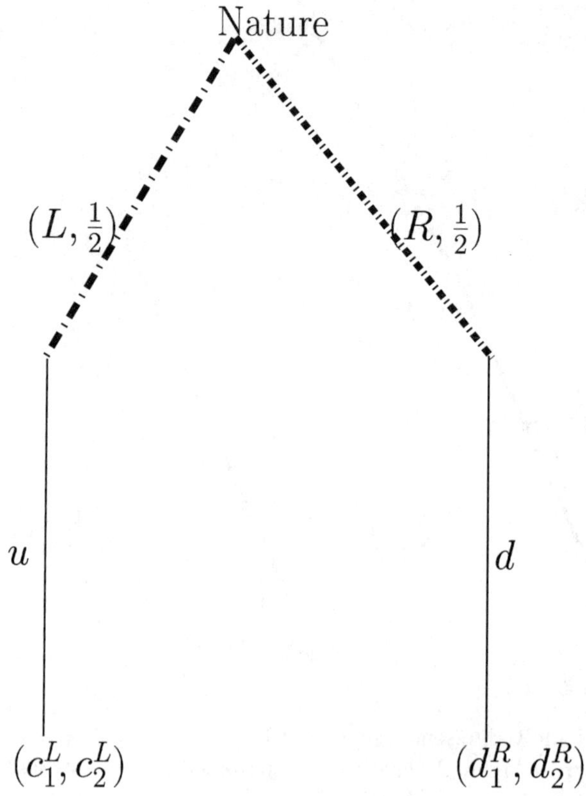

Figure 4

the state is Left or Right, and which gives to Row an insured payoff of 3 units in either case;...". An interpretation is as follows.[1] Cooperation allows us to discard the payoff vectors $(0,0)$ and $(0,2)$, and $e_1^L = 2$, $e_1^R = 4$, $e_2^L = 2$ and $e_2^R = 0$ can be thought of as the players' endowments in the single commodity. The agents act on the basis of the "meet" of the information algebras, $\mathcal{F}_1 = \{\{L, R\}\}$. The corresponding tree is shown in Figure 4. The players know the probabilities and have to reach a decision concerning both nodes. The commodity payoff constraints are $0 \leq c_1^L + c_2^L \leq 4$ and $0 \leq d_1^R + d_2^R \leq 4$.

The expected utilities of the players corresponding to their random endowments are

$$E_1 = \frac{1}{2}u_1(2) + \frac{1}{2}u_1(4), \quad \text{and } E_2 = \frac{1}{2}u_2(2) + \frac{1}{2}u_2(0)$$

The coarse core allocations are obtained as solutions to the problem:

[1] We recall that Wilson defines the coarse core to consist of allocations which cannot be blocked by any coalition of agents who act on the basis of the intersection of their algebras of information sets. Wilson imposes no measurability conditions on allocations.

Problem

Maximize $E_1 = \frac{1}{2}u_1(u) + \frac{1}{2}u_1(d)$
Subject to

$$\frac{1}{2}u_2(u) + \frac{1}{2}u_2(d) \geq E_2 \quad \text{(fixed)}$$

$$E_1 \geq \frac{1}{2}u_1(2) + \frac{1}{2}u_1(4) \quad \text{and } E_2 \geq \frac{1}{2}u_2(2) + \frac{1}{2}u_2(0).$$

Then there is a coarse core allocation in which P1 gets $(3, 3)$ and P2 the allocation $(1, 1)$. We can see this as follows. We maximize the expected utility of one agent subject to a given expected value for the other. A particular solution is the one above.

We note that Example 1, which we did not discuss, illustrates the problem of adverse selection, and requires no special interpretation.

References

Allen, B.: Generic existence of completely revealing equilibria with uncertainty, when prices convey information. Econometrica **49**, 1173–1199 (1981)

Glycopantis, D., Yannelis, N.C.: Equilibrium concepts in differential information economies. This volume.

Radner, R.: Rational expectation equilibrium: generic existence and information revealed by prices. Econometrica **47**, 655–678 (1979)

Wilson, R.: Information, Efficiency, and the core of an economy. Econometrica **46**, 807–816 (1978)

The core of an economy with differential information*

Nicholas C. Yannelis

Department of Economics, University of Illinois at Urbana-Champaign, Champaign, IL 61820, USA

Received: February 12, 1990; revised version

Summary. We introduce a new core concept for an exchange economy with differential information which is contained in the coarse core concept of Wilson (1978). We prove the existence of (i) a core allocation for an exchange economy with differential information and; (ii) an α-core strategy for a game in normal form with differential information.

1. Introduction

An exchange economy with differential information consists of a finite set of agents each of whom is characterized by a random utility function, a random initial endowment, a private information set and a prior.

The purpose of this paper is to study the following questions: How does one define the notion of the core in an exchange economy with differential information? What is the appropriate core concept? Under what conditions an agent's characteristics is the core nonempty?

With finitely many states of nature, the existence of a coarse core allocation for an economy with differential information follows easily from the well known result of Scarf (1967), as first shown in a seminal paper by Wilson (1978). However, with a continuum of states even if there is symmetric information (i.e., the information set of each agent is the same) the domain of the expected utility becomes infinite dimensional (even if there is only one good in the economy), and consequently

* On different occasions I have benefited from discussions, comments and suggestions by C.D. Aliprantis, Kim Border, Don Brown, Baskar Chacravorti, Mark Feldman, Leo Hurwicz, Charlie Kahn, John Ledyard, Andreu Mas-Colell, Flavio Menezes, Tom Palfrey, Ed Prescott, Aldo Rustichini, David Schmeidler and Sanjay Srivastava. Mark. Feldman and Aldo Rustichini both independently brought to my attention the related work of Wilson (1978). My thanks are extended to all the above individuals as well as to a careful referee. Of course, I am responsible for any remaining shortcomings.

Scarf's theorem is not directly applicable. It turns out that in the presence of a continuum of states, functional analytic methods as well as several measure theoretic results seem to be required.

The paper is organized as follows: Sect. 2 contains notation and definitions. The model and the main results are presented in Sect. 3. Sections 4 and 5 contain the proofs of our main theorems. Finally Sect. 6 contains some concluding remarks.

2. Notation and definitions

2.1. Notation

$\mathbf{R}^l$ denotes the l-fold Cartesian product of the set of real numbers $\mathbf{R}$.
$\mathbf{R}^l+$ denotes the positive cone of $\mathbf{R}^l$.
$\mathbf{R}^l_{++}$ denotes the strictly positive elements of $\mathbf{R}^l$.
2^A denotes the set of all nonempty subsets of the set A.
$\varnothing$ denotes the empty set.
$/$ denotes the set theoretic subtraction.
If X is a linear topological space, its dual is the space X^* of all continuous linear functionals on X, and if $p \in X*$, and $x \in X$ the value of p at x is denoted by $p \cdot x$.

2.2. Definitions

If X and Y are sets, the *graph* of the set-valued function (or correspondence), $\phi : X \to 2^Y$ is denoted by $G_\phi = \{(x, y) \in X \times Y : y \in \phi(x)\}$. Let $(T, \mathbf{T}, \mu)$ be a complete, finite measure space, and X be a separable Banach space. The set-valued function $\phi : T \to 2^X$ is said to have a *measurable graph* if $G_\phi \in \mathbf{T} \otimes \beta(X)$, where $\beta(X)$ denotes the Borel σ-algebra on X and $\otimes$ denotes the product σ-algebra. The set-valued function $\phi : T \to 2^X$ is said to be *lower measurable* or just *measurable* if for every open subset V of X, the set $\{t \in T : \phi(t) \cap V \neq \varnothing\}$ is an element of $\mathbf{T}$. A well-known result of Debreu [(1966), p. 359] says that if $\phi : T \to 2^X$ has a measurable graph, then ϕ is lower measurable. Furthermore, if $\phi(\cdot)$ is closed valued and lower measurable then $\phi : T \to 2^X$ has a measurable graph. A theorem of Aumann (1967) which will be of fundamental importance in this paper tells us, that if $(T, \mathbf{T}, \mu)$ is a complete, finite measure space, X is a separable metric space and $\phi : T \to 2^X$ is a nonempty valued correspondence having a measurable graph, then $\phi(\cdot)$ admits a *measurable selection*, i.e., there exists a measurable function $f : T \to X$ such that $f(t) \in \phi(t) \mu$-*a.e.*

Let $(T, \mathbf{T}, \mu)$ be a finite measure space and X be a Banach space. Following Diestel-Uhl (1977) the function $f : T \to X$ is called *simple* if there exist $x_1, x_2, \ldots, x_n$ in X and $\alpha_1, \alpha_2, \ldots, \alpha_n$ in $\mathbf{T}$ such that $f = \sum_{i=1}^n x_i \chi_{\alpha_i}$, where $\chi_{\alpha_i}(t) = 1$ if $t \in \alpha_i$ and $\chi_{\alpha_i}(t) = 0$ if $t \notin \alpha_i$. A function $f : T \to X$ is said to be μ-*measurable* if there exists a sequence of simple functions $f_n : T \to X$ such that $\lim_{n\to\infty} \|f_n(t) - f(t)\| = 0$ for almost all $t \in T$. A μ-measurable function $f : T \to X$ is said to be *Bochner integrable* if there exists a sequence of simple functions $\{f_n : n = 1, 2, ...\}$ such that

$$\lim_{n\to\infty} \int_T \|f_n(t) - f(t)\| d\mu(t) = 0$$

In this case we define for each $E \in \mathbf{T}$ the integral to be $\int_E f(t)d\mu(t) = \lim_{n\to\infty} \int_E f_n(t)d\mu(t)$.

It can be shown [see Diestel-Uhl (1977), Theorem 2, p. 45] that if $f : T \to X$ is a μ-measurable function then, f is Bochner integrable if and only if $\int_T \|f(t)\|d\mu(t) < \infty$.

It is important to note that the *Dominated Convergence Theorem* holds for Bochner integrable functions. In particular, if $f_n : T \to X$, $(n = 1, 2, \ldots)$ is a sequence of Bochner integrable functions such that $\lim_{n\to\infty} f_n(t) = f(t)\mu$-*a.e.*, and $\|f_n(t)\| \leqq g(t)\mu$-*a.e.*, (where $g : T \to \mathbf{R}$ is an integrable function), then f is Bochner integrable and $\lim_{n\to\infty} \int_T \|f_n(t) - f(t)\|d\mu(t) = 0$.

For $1 \leqq p < \infty$, we denote by $L_p(\mu, X)$ the space of equivalence classes of X-valued Bochner integrable functions $x : T \to X$ normed by

$$\|x\|_p = \left(\int_T \|x(t)\|^p d\mu(t)\right)^{1/p}.$$

It is a standard result that normed by the functional $\|\cdot\|_p$ above, $L_p(\mu, X)$ becomes a Banach space [see Diestel-Uhl (1977), p. 50]. Recall that a correspondence $\phi : T \to 2^X$ is said to be *integrably bounded* if there exists a map $h \in L_1(\mu, R)$ such that $\sup\{\|x\| : x \in \phi(t)\} \leqq h(t)\mu$-*a.e.*

A Banach space X has the *Radon-Nikodym Property with respect to the measure space* $(T, \mathbf{T}, \mu)$ if for each μ-continuous measure $G : \mathbf{T} \to X$ of bounded variation there exists $g \in L_1(\mu, X)$ such that $G(E) = \int_E g(t)d\mu(t)$ for all $E \in \mathbf{T}$. A Banach space X has the *Radon-Nikodym Property* (RNP) if X has the RNP with respect to every finite measure space. Recall now [see Diestel-Uhl (1977, Theorem 1, p. 98)] that if $(T, \mathbf{T}, \mu)$ is a finite measure space $1 \leqq p < \infty$, and X is a Banach space, then X^* has the RNP if and only if $(L_p(\mu, X))^* = L_q(\mu, X^*)$ where $\frac{1}{p} + \frac{1}{q} = 1$.

We will close this section by collecting some basic results on Banach lattices [for an excellent treatment see Aliprantis-Burkinshaw (1985)]. Recall that a *Banach lattice* is a Banach space L equipped with an order relation $\geqq$ (i.e., $\geqq$ is a reflexive, antisymmetric and transitive relation) satisfying:

(i) $x \geqq y$ implies $x + z \geqq y + z$ for every z in L,
(ii) $x \geqq y$ implies $\lambda x \geqq \lambda y$ for all $\lambda \geqq 0$,
(iii) for all x, y in L there exists a supremum (least upper bound) $x \vee y$ and an infimum (greatest lower bound) $x \wedge y$,
(iv) $|x| \geqq |y|$ implies $\|x\| \geqq \|y\|$ for all x, y in L.

As usual $x^+ = x \vee 0$, $x^- = (-x) \vee 0$ and $|x| = x \vee (-x) = x^+ + x^-$; we call x^+, x^- the *positive* and *negative parts* of x, respectively and $|x|$ the *absolute value* of x. The symbol $\|\cdot\|$ denotes the norm on L. If x, y are elements of the Banach lattice L, then we define the *order interval* $[x, y]$ as follows:

$$[x, y] = \{z \in L : x \leqq z \leqq y\}.$$

Note that $[x, y]$ is norm closed and convex (hence weakly closed). A Banach lattice L is said to have an *order continuous norm* if, $x_\alpha \downarrow 0$ in L implies $\|x_\alpha\| \downarrow 0$. A very useful result which will play an important role in the sequel is that if L is a Banach lattice then the fact that L has an order continuous norm is equivalent to weak compactness of the order interval $[x, z] = \{y \in L : x \leqq y \leqq z\}$ for every x, z in L [see for instance Aliprantis-Brown-Burkinshaw (1989), Theorem 2.3.8, p. 104 or Lindenstrauss-Tzafriri (1979, p. 28)].

We finally note that Cartwright (1974) has shown that if X is a Banach lattice with order continuous norm (or equivalently X has weakly compact order intervals) then $L_1(\mu, X)$, has weakly compact order intervals, as well. *Cartwright's theorem* will play a crucial role in the proof of our main results.

3. Model and results

3.1. The core of an exchange economy with differential information

Let Y be a separable Banach lattice with an order continuous norm, whose dual Y^* has the RNP.[1] Let $(\Omega, \mathbf{F}, \mu)$ be a complete finite measure space.

An *exchange economy with differential information* $\Gamma = \{(X_i, u_i, e_i, F_i, q_i) : i = 1, 2, \ldots, n\}$ is a set of quintuples $(X_i, u_i, e_i, F_i, q_i)$ where,

(1) $X_i : \Omega \to 2^{Y_+}$ is the *random consumption set* of agent i,
(2) $u_i : \Omega \times X_i \to \mathbf{R}$ is the *random utility function* of agent i,
(3) F_i is a (measurable) partition[2] of $(\Omega, \mathbf{F})$ denoting the *private information of* agent i,
(4) $e_i : \Omega \to Y_+$ is the *random initial endowment of* agent i, $e_i(\cdot)$ is F_i-measurable, Bochner integrable and $e_i(\omega) \in X_i(\omega)$ for all i, μ-*a.e.*,
(5) $q_i : \Omega \to \mathbf{R}_{++}$ is the *prior* of agent i, (i.e., q_i is a Radon-Nikodym derivative having the property that $\int_{t\in\Omega} q_i(t)d\mu(t) = 1$).

Denote by L_{X_i} the *set of all Bochner integrable and F_i-measurable selections from the consumption set* X_i of agent i, i.e.,

$$L_{X_i} = \{x_i \in L_1(\mu, Y_+) : x_i : \Omega \to Y_+ \text{ is } F_i\text{-measurable and } x_i(\omega) \in X_i(\omega) \mu\text{-}a.e\}.$$

For each i, $(i = 1, 2, \ldots, n)$, denote by $E_i(\omega)$ the event in F_i containing the realized state of nature $\omega \in \Omega$ and suppose that $\int_{t\in E_i(\omega)} q_i(t)d\mu(t) > 0$ for all i. Given $E_i(\omega)$ in F_i define the *conditional expected utility* of agent i, $V_i : \Omega \times L_{X_i} \to \mathbf{R}$ by

$$V_i(\omega, x_i) = \int_{t\in E_i(\omega)} u_i(t, x_i(t)) q_i(t|E_i(\omega)) d\mu(t) \tag{3.1}$$

[1] A basic example of a space which satisfies all these conditions is the Euclidean space $\mathbf{R}^l$. Remark 6.1 in Section 6 presents some more examples.

[2] In the sequel by an abuse of notation, we will still denote by an F_i the σ-algebra that the partition F_i generates.

where

$$q_i(t|E_i(\omega)) = \begin{cases} 0 & \text{if } t \notin E_i(\omega) \\ \dfrac{q_i(t)}{\int_{t\in E_i(\omega)} q_i(t)d\mu(t)} & \text{if } t \in E_i(\omega). \end{cases} \tag{3.2}$$

We are now ready to define the central notions of the paper.

Definition 3.1.1. *We say that* $x = (x_1, x_2, \dots, x_n) \in \prod_{i=1}^n L_{X_i}$ *is a* core *allocation for* Γ, *if*

(i) $\sum_{i=1}^n x_i = \sum_{i=1}^n e_i$, *and*
(ii) *it is not true that there exist* $S \subset \{1, 2, \dots, n\}$ *and* $(y_i)_{i\in S} \in \prod_{i\in S} L_{X_i}$ *such that* $\sum_{i\in S} y_i = \sum_{i\in S} e_i$ *and* $V_i(\omega, y_i) > V_i(\omega, x_i)$ *for all* $i \in S$ *for* μ*-almost all* $\omega \in \Omega$ *(where* V_i *is given by 3.1).*

A couple of comments are in order: Note that $x \in \prod_{i=1}^n L_{X_i}$ implies that each $x_i(\cdot)$ is F_i-measurable and therefore the vector $x(\omega) = (x_i(\omega), x_2(\omega), \dots, x_n(\omega)) \in \prod_{i=1}^n X_i(\omega)$ is $\bigvee_{i=1}^n F_i$-measurable (where $\bigvee_{i=1}^n F_i$ denotes the join, the smallest partition containing $F_1, F_2, \dots F_n$). Condition (i) above implies that the markets are cleared in each state of nature, i.e., $\sum_{i=1}^n x_i(\omega) = \sum_{i=1}^n e_i(\omega)\mu$*-a.e.* Condition (ii) shows that no coalition of agents (while each agent in the coalition uses his/her own private information) can redistribute their initial endowments among themselves for any state of nature and make the conditional expected utility of each agent in the coalition better off. Note that Condition (ii) of Definition 3.1.1) implies the following condition:

(ii)′ It is not true that there exist $S \subset \{1, 2, \dots, n\}$ and $y : \Omega \to \prod_{i\in S} X_i$, $y_i(\cdot)$ if $\bigwedge_{i\in S} F_i$-measurable (where $\bigwedge_{i\in S} F_i$ denotes the meet, i.e., the maximal partition contained in all of them) such that $\sum_{i\in S} y_i(\omega) = \sum_{i\in S} e_i(\omega)\mu$*-a.e.* and $V_i(\omega, y_i) > V_i(\omega, x_i)$ for all $i \in S$ for μ-almost all $\omega \in \Omega$.

The above blocking notion is the one adopted by Wilson (1978) to define his *coarse core* concept.[3] Note that since each $y_i(\cdot)$ is $\bigwedge_{i\in S} F_i$-measurable, the information is verifiable by each member of the coalition. For instance, if we imagine that agents negotiate the terms of a contract, then Wilson's definition tells us that a coarse core allocation has the property that no coalition of agents can exchange their own information (in fact, information is verifiable by each member of the coalition) and make each agent in the coalition better off. In other words, contracts are realizable because information is verifiable. However, according to our Condition (ii) of Definition 3.1.1, information is not necessarily verifiable by all the members of the coalition (it is only privately verifiable). The latter makes the core smaller, i.e., any core allocation satisfying the Definition 3.1.1 is a coarse core allocation as well. (Recall that if $y_i(\cdot)$ is $\bigwedge_{i\in S} F_i$-measurable, it is also F_i-measurable; of coarse the reverse is not true). Hence, the theorems that we will prove on the existence of core allocations will imply the existence of coarse core allocations as well.

[3] See also Kobayashi (1980) who has also used the coarse core.

Note that if we were to narrow the set of core allocations by replacing the F_i-measurability of $y_i(\cdot)$ in (ii) of Definition 3.1.1 with the $\bigvee_{i \in S} F_i$-measurability of $y : \Omega \to \prod_{i \in S} X_i$, then it is easy to construct examples which satisfy all the assumptions of Theorem 3.1 below, but the core is empty [see Wilson (1978) or Berliant (1990) for examples to that effect]. We are not aware of any natural set of assumptions on utility functions and initial endowments which will guarantee the existence of such a core. Finally, it is worth pointing out that a core notion which allows for complete exchange of information among agents in each coalition may not be an appropriate concept since in most applications, agents do not have an incentive to reveal their own private information (think of situations of moral hazard or adverse selection).

Definition 3.1.2. *We say that $x \in \prod_{i=1}^n L_{X_i}$ is (interim)* Pareto optimal *if:*[4]

(i) $\sum_{i=1}^n x_i = \sum_{i=1}^n e_i$, *and*
(ii) it is not true that there exists $y \in \prod_{i=1}^n L_{X_i}$ such that $\sum_{i=1}^n y_i = \sum_{i=1}^n e_i$ and $V_i(\omega, y_i) > V_i(\omega, x_i)$ for all i for μ-almost all $\omega \in \Omega$ (where V_i is given by (3.1)).

Definition 3.1.3. *We say that $x \in \prod_{i=1}^n L_{X_i}$ is individually rational if:*

(i) $\sum_{i=1}^n x_i = \sum_{i=1}^n e_i$ *and*
(ii) $V_i(\omega, x_i) \geqq V_i(\omega, e_i)$ for all i and for some $\omega \in \Omega$ (where V_i is given by (3.1)).

Finally, if the private information set of each agent, is the same (i.e., there is symmetric information so $F_i \equiv F$ for all i) we call any $x \in \prod_{i=1}^n L_{X_i}$ satisfying (i) and (ii) of Definition 3.1.1 a *symmetric core allocation* for Γ.

We are now ready to state our first main result:

Theorem 3.1. *Let $\Gamma = \{(X_i, u_i, e_i, F_i, q_i) : i = 1, 2, \dots, n\}$ be an exchange economy with differential information satisfying the following assumptions, for each i ($i = 1, 2, \dots, n$),*

(a:3.1) $X_i : \Omega \to 2^{Y_+}$ is an integrably bounded, convex, closed, nonempty valued and F_i-measurable correspondence,
(a.3.2) for each $\omega \in \Omega$, $u_i(\omega, \cdot)$ is weakly continuous and integrably bounded, and
(a.3.3) for each $\omega \in \Omega$, $u_i(\omega, \cdot)$ is concave.

Then a core allocation exists in Γ.

The following corollaries follow directly from Theorem 3.1.

Corollary 3.1. *Let $\Gamma = \{(X_i, u_i, e_i, F_i, q_i) : i = 1, 2, \dots, n\}$ be an exchange economy with differential information satisfying all the assumptions of Theorem 3.1. Then an individually rational and Pareto optimal allocation exists in Γ.*

Corollary 3.2. *Let $\Gamma = \{(X_i, u_i, e_i, F_i, q_i) : i = 1, 2, \dots, n\}$ be an exchange economy with symmetric information (i.e., $F_i \equiv F$ for all i), satisfying all the assumptions of Theorem 3.1. Then a symmetric core allocation exists in Γ.*

[4] A similar notion is defined by Palfrey and Srivastava (1987).

3.2. The α-core of a game in normal form with differential information

A *game in normal form with differential information* $B = \{(X_i, u_i, F_i, q_i) : i = 1, 2, \dots, n\}$ is a set of quadruples (X_i, u_i, F_i, q_i) where

(1) $X_i : \Omega \to 2^Y$ is the *strategy set-valued function* of player i,
(2) $u_i : \Omega \times \prod_{i=1}^n X_i \to \mathbf{R}$ is the *random payoff function* of player i,
(3) F_i is a (measurable) partition of $(\Omega, \mathbf{F})$ denoting the *private information* of player i, and
(4) $q_i : \Omega \to \mathbf{R}_{++}$ is the *prior* of player i (i.e., q_i is a Radon-Nikodym derivative having the property that $\int_{t\in\Omega} q_i(t)d\mu(t) = 1$).

For each i ($i = 1, 2, \dots, n$) denote by $E_i(\omega)$ the event in F_i containing the true state of nature $\omega \in \Omega$ and suppose that $\int_{t\in E_i(\omega)} q_i(t)d\mu(t) > 0$. Given $E_i(\omega)$ in F_i define the *conditional expected payoff* of player i, $V_i : \Omega \times \prod_{i=1}^n L_{X_i} \to \mathbf{R}$ by

$$V_i(\omega, x) = \int_{t\in E_i(\omega)} u_i(t, x(t))q_i(t|E_i(\omega))d\mu(t), \tag{3.3}$$

where $q_i(t|E_i(\omega))$ is defined as in (3.2).

Before we defne the notion of an *α-core strategy for the game* $\mathbf{B}$ we need to introduce some notation. Denote by I the set of players $\{1, 2, \dots, n\}$. If $S \subset I$ then $(y^S, x^{I/S})$ denotes the vector z in $\prod_{i=1}^n L_{X_i}$ where $z_i = y_i$ if $i \in S$ and $z_i = x_i$ if $i \notin S$.

Definition 3.2.1. We say that $x \in \prod_{i=1}^n L_{X_i} = L_X$ is an *α-core strategy* for B if:

(i) It is not true that there exist $S \subset I$ and $(y_i)_{i\in S} \in \prod_{i\in S}^n L_{X_i}$ such that for any $z^{I/S} \in \prod_{i\notin S} L_{X_i}$, $V_i(\omega, (y^S, z^{I/S})) > V_i(\omega, x)$ for all $i \in S$ for μ-almost all $\omega \in \Omega$ (where V_i is given by (3.3)).

Note that as before $x \in \prod_{i=1}^n L_{X_i}$ implies that $x_i(\cdot)$ is F_i-measurable and consequently the vector $x(\omega) = (x_1(\omega), \dots, x_n(\omega))$ is $\bigvee_{i\in I} F_i$-measurable. Condition (i) in Definition 3.2.1 indicates that no coalition of players is able to change its strategy (while each player in the coalition uses his/her own private information) and make the expected utility of each member in the coalition better off, no matter what the complementary coalition chooses to do (each member in the complementary coalition is also allowed to take advantage of his/her own private information). Following the previous definition of a coarse core allocation for an economy wich differential information, we can define an α-coarse strategy for the game B, and show that the set of α-coarse core strategies contains the set of α-core strategies for the game B.

Since there is no exchange of information among players in each coalition one may suggest that it is possible to analyze games in normal form wich differential information (or economies with different information) in a noncooperative setting adopting the notion of a Bayesian Nash equilibrium or correlated equilibrium. However, the latter concepts do not yield Pareto optimal outcomes, contrary to the core or α-core. It seems to us that selecting outcomes out of the Pareto frontier is an attractive property for an allocation mechanism to have. The latter makes the core concept appealing in an economy with differential information.

We can now state our second main result.

Theorem 3.2. *Let $B = \{(X_i, u_i, F_i, q_i) : i = 1, 2, \dots, n\}$ be a game in normal form with differential information satisfying the following assumptions for each player i ($i = 1, 2, \dots, n$),*

(a.3.2.1) $X_i : \Omega \to 2^{Y_+}$ is an integrably bounded, nonempty convex weakly compact valued and F_i-measurable correspondence,[5]

(a.3.2.2) for each $\omega \in \Omega$, $u_i(\omega, \cdot)$ is weakly continuous and integrably bounded, and

(a.3.2.3) for each $\omega \in \Omega$, $u_i(\omega, \cdot)$ is concave.

Then an α-core strategy exists in B.

4. Proof of Theorem 3.1

We first state the well-known core existence result of Scarf (1967) [see also Border (1984) or Yannelis (1990) for recent generalizations] which is going to play a crucial role in the proof of Theorem 3.1. We will first need some notation.

Let $\mathbf{E} = \{(X_i, u_i, e_i) : i = 1, 2, \dots, n\}$ be an exchange economy, where

(1) $X_i \subset \mathbf{R}^l$ is the *consumption* set of agent i,
(2) $u_i : X_i \to \mathbf{R}$ is the *utility function* of agent i, and
(3) $e_i \in X_i$ is the *initial endowment* of agent i.

Define the set-valued function $P_i : X_i \to 2^{X_i}$ by $P_i(x_i) = \{y \in X_i : u_i(y_i) > u_i(x_i)\}$. *Scarf's result* asserts that if X_i is a nonempty, closed, convex and bounded from below subset of $\mathbf{R}^l$, u_i is quasi concave and continuous (i.e., if P_i is convex valued and has an open graph in $X_i \times X_i$), then core allocations exist in $\mathbf{E}$, i.e., there exists $x \in \prod_{i=1}^n X_i$ such that:

(i) $\sum_{i=1}^n x_i = \sum_{i=1}^n e_i$, and
(ii) it is not true that there exist $S \subset \{1, 2, \dots, n\}$ and $(y_i)_{i \in S} \in \prod_{i \in S} X_i$ such that $\sum_{i \in S} y_i = \sum_{i \in S} e_i$ and $y_i \in P_i(x_i)$ for all $i \in S$.

We begin the proof of Theorem 3.1 by constructing a new economy $G = \{(L_{X_i}, P_i, e_i) : i = 1, 2, \dots, n\}$, where

(i) L_{X_i} is the *consumption set* of agent i,
(ii) $P_i : L_{X_i} \to 2^{L_{X_i}}$ is the *preference correspondence* of agent i defined by $P_i(x_i) = \{y_i \in L_{X_i} : V_i(\omega, y_i) > V_i(\omega, x_i)$ for μ-almost all $\omega \in \Omega\}$ and
(iii) $e_i \in L_{X_i}$ for all i, is the *initial endowment* of agent i.

Note the existente of a core allocation for G implies the existence of a core allocation for the original economy $\Gamma = \{(X_i, u_i, e_i, F_i, q_i) : i = 1, 2, \dots, n\}$. Hence, all

[5] The assumption that $X_i(\cdot)$ takes values in the positive cone of Y, is not needed for the proof of this theorem.

we need to show is that a core allocation exists in the economy G. To this end we first show that for each i, L_{X_i} is closed, bounded, convex, nonempty and that $P_i : L_{X_i} \to 2^{L_{X_i}}$ is convex valued having a weakly open graph (i.e., the set $G_{P_i} = \{(x,y) \in L_{X_i} \times L_{X_i} : y \in P_i(x)\}$ is weakly open in $L_{X_i} \times L_{X_i}$).

Note the fact that L_{X_i} is convex, closed and bounded follows directly from assumption (a.3.1). To prove that L_{X_i} is nonempty, recall that $X_i : \Omega \to 2^{Y_+}$ is F_i-measurable, nonempty, closed valued and therefore $G_{X_i} \in F_i \otimes \beta(Y_+)$. By the Aumann (1967) measurable selection theorem, we can obtain an F_i-measurable function $f_i : \Omega \to Y_+$ such that $f_i(\omega) \in X_i(\omega) \mu\text{-}a.e.$ Since X_i is integrably bounded, we can conclude that $f_i \in L_1(\mu, Y_+)$. Hence, $f_i \in L_{X_i}$ and this proves that L_{X_i} is nonempty.

In order to show that for each i, P_i has a weakly open graph, we will first need the following claim[6]:

Claim 4.1. For each i ($i = 1, 2, \dots$) and for each $\omega \in \Omega$, $V_i(\omega, \cdot)$ is weakly continuous.

Proof. Fix i ($i = 1, 2, \dots, n$) and $\omega \in \Omega$ and let $E_i(\omega)$ be an event in F_i. Consider the sequence $\{x_i^m : m = 1, 2 \dots\}$ in $L_{X_i} \subset L_1(\mu, Y)$, which converges weakly to $x_i \in L_{X_i}$, i.e., $p \cdot x_i^m$ converges to $p \cdot x_i$ for any $p \in L_\infty(\mu, Y^*) = (L_1(\mu, Y))^*$ (recall that Y^* has the RNP). Note that x_i^m converges weakly to x_i is equivalent to the fact that $p \cdot x_i^m \chi_A = p\chi_A \cdot x_i^m$ converges to $p \cdot x_i \chi_A = p\chi_A \cdot x_i$ for any $p \in L_\infty(\mu, Y^*)$, $A \in \mathbf{F}$ and each condition above implies that $y^* \cdot x_i^m \chi_A = y^* \chi_A \cdot x_i^m$ converges to $y^* \cdot x_i \chi_A = y^* \chi_A \cdot x_i$ for any $y^* \in Y^*$, $A \in \mathbf{F}$. If we show that $x_i^m \chi_{E_i(\omega)}$ converges pointwise in the weak topology of X_i to $x_i \chi_{E_i(\omega)}$, then since for each $\omega \in \Omega$, $u_i(\omega, \cdot)$ is weakly continuous and integrably bounded the weak continuity of $V_i(\omega, \cdot)$ will follow from the Lebesgue dominated convergence theorem. Now if $F_i = \{E_i^1, E_i^2, \dots\}$ is the partition of agent i, then the fact that x_i^m and x_i are elements of L_{x_i} implies that $x_i^m = \sum_{k=1}^\infty x_{i,k}^m \chi_{E_i^k}$, $x_i = \sum_{k=1}^\infty x_{i,k} \chi_{E_i^k}$, for $x_{i,k}^m, x_{i,k}$ in X_i and consequently we can conclude that $x_i^m \chi_{E_i(\omega)} = \sum_{k=1}^\infty x_{i,k}^m \chi_{E_i^k \cap E_i(\omega)}$ converges weakly to $x_i \chi_{E_i(\omega)} = \sum_{k=1}^\infty x_{i,k} \chi_{E_i^k \cap E_i(\omega)}$. This completes the proof of the claim.

In view of Claim 4.1 we can now conclude that for each i, P_i has a weakly open graph. Moreover, since for each $\omega \in \boldsymbol{\Omega}$, $u_i(\omega, \cdot)$ is concave so is $V_i(\omega, \cdot)$ and therefore, P_i is convex valued. We will now construct a suitable family of truncated subeconomies in a finite dimensional commodity space, each of which satisfies the assumptions of Scarf's theorem. Applying Scarf's theorem, we will obtain a net of core allocations for each subeconomy. By taking limits we will show that the existence of a core allocation for each subeconomy implies the existence of a core allocation for the original economy G.

Let $\mathbf{A}$ be the set of all finite dimensional subspaces of $L_1(\mu, Y_+)$ containing the initial endowments. For each $\alpha \in A$ define the consumption set of agent i, $L_{X_i}^\alpha$ by $L_{X_i}^\alpha = L_{X_i} \cap \alpha$ and the preference correspondence of agent i, $P_i^\alpha : L_{X_i}^\alpha \to 2^{L_{X_i}^\alpha}$ by $P_i^\alpha(x_i) = P_i(x_i) \cap L_{X_i}^\alpha$. We now have an economy $G^\alpha = \{(L_{X_i}^\alpha, P_i, e_i) : i =$

[6] A similar result is proved in Yannelis & Rustichini (1991).

$1, 2, \dots, n\}$ in a finite dimensional commodity space, where,

$$L^{\alpha}_{X_i} \text{ is the consumption set of agent } i, \tag{4.1}$$

$$P^{\alpha}_i : L^{\alpha}_{X_i} \to 2^{L_{X^{\alpha}_i}} \text{ is the preference correspondence of agent } i, \text{ and} \tag{4.2}$$

$$e_i \in L^{\alpha}_{X_i} \text{ is the initial endowment of agent } i. \tag{4.3}$$

It can be easily checked that each economy G^{α} satisfes all the assumptions of Scarf's theorem and therefore there exists $x^{\alpha} = (x^{\alpha}_1, \dots, x^{\alpha}_n) \in \prod_{i=1}^{n} L^{\alpha}_{X_i} = L^{\alpha}_X$ such that:

$$\sum_{i=1}^{n} x^{\alpha}_i = \sum_{i=1}^{n} e_1, \text{ and} \tag{4.4}$$

it is not true that there exist $S \subset \{1, 2, \dots, n\}$ and $(y_i)_{i \in S} \in \prod_{i \in S} L^{\alpha}_{X_i}$ such that

$$\sum_{i \in S} y_i = \sum_{i \in S} e_i \text{ and } y_i \in P^{\alpha}_i(x^{\alpha}_i) \text{ for all } i \in S. \tag{4.5}$$

From (4.4) it follows that for each $\alpha \in \mathbf{A}$

$$0 \leqq \sum_{i=1}^{n} x^{\alpha}_i = \sum_{i=)1}^{n} e_i = e.$$

Hence for each $\alpha \in \mathbf{A}$ the vectors x^{α}_i lie in the order interval $[0, e]$. Since by assumption order intervals in Y are weakly compact, by Cartwright's theorem the order interval $[0, e]$ in $\sum_{i=1}^{n} L_{X_i}$ is weakly compact. Direct the set $\mathbf{A}$ by inclusion so that $\{(x^{\alpha}_i, x^{\alpha}_2, \dots, x^{\alpha}_n) : \alpha \in \mathbf{A}\}$ forms a net in $\prod_{i=1}^{n} L_{X_i}$. Since all the vectors x^{α}_i lie in the order interval $[0, e]$ which is weakly compact, the net $\{(x^{\alpha}_i, \dots, x^{\alpha}_n) : \alpha \in \mathbf{A}\}$ has a subnet which converges weakly to some vector $x_1, x_2, \dots, x_n$ in $[0, e]$. We will show that the vector $x_1, \dots, x_n$ is a core allocation for the economy G. Denote the convergent subnet by $\{(x^{\alpha(m)}_i, \dots, x^{\alpha(m)}_n) : m \in M\}$ where M is a set directed by "$\geqq$". Since for all $m \in M$, $\sum_{i=1}^{n} x^{\alpha(m)}_i = \sum_{i=1}^{n} e_i$ and $x^{\alpha(m)}_i$ converges weakly to $x_i \in L_{X_i}$, we conclude that $\sum_{i=1}^{n} x_i = \sum_{i=1}^{n} e_i$. We will now complete the proof by showing that:

It is not true that there exist $S \subset \{1, 2, \dots, n\}$ and $(y_i)_{i \in S} \in \prod_{i \in S} L_{X_i}$ such that

$$\sum_{i \in S} y_i = \sum_{i \in S} e_i \text{ and } y_i \in P_i(x_i) \text{ for all } i \in S. \tag{4.6}$$

Suppose that (4.6) is false, then there exist $S \subset \{1, 2, \dots, n\}$ and $(y_i)_{i \in S} \in \prod_{i \in S} L_{X_i}$ such that $\sum_{i \in S} y_i = \sum_{i \in S} e_i$ and $y_i \in P_i(x)$ for all $i \in S$. Since $x^{\alpha(m)}_i$ converges weakly to x_i and P_i has a weakly open graph, there exists $m_0 \in M$ such that $y_i \in P_i(x^{\alpha(m)}_i)$ for all $m \geqq m_0$ and for all $i \in S$. Choose $m_1 \geqq m_0$ so that, if $m \geq m_1$, $y_i \in L^{\alpha(m)}_{X_i}$ for all $i \in S$. Then $y_i \in P^{\alpha(m)}_i(x^{\alpha(m)}_i)$, for all $m \geqq m_1$ and for all $i \in S$. But this contradicts (4.5). Hence (4.6) holds and this completes the proof of the theorem.

5. Proof of Theorem 3.2

We begin by stating the α-core existence result of Scarf (1971) which is going to be used in the proof of Theorem 3.2.

Let $N = \{(X_i, u_i) : i = 1, 2, \ldots, n\}$ be a *game in normal form* where,

(1) X_i is a compact, convex and nonempty subset of $\mathbf{R}^l$, denoting the *strategy set* of player i, and
(2) $u_i : \prod_{i=1}^n X_i \to \mathbf{R}$ is a quasi-concave function on $\prod_{i=1}^n X_i$, denoting the *payoff* of player i.

The strategy vector $x \in \prod_{i=1}^n X_i$ is said to be an *α-core strategy* for N if:

It is not true that there exist $S \subset \{1, 2, \ldots, n\}$ and $(y_i)_{i \in S} \in \prod_{i \in S} X_i$ such that for any $z^{I/S} \in \prod_{i \notin S} X_i$, $u_i(y^S, z^{I/S}) > u_i(x)$ for all $i \in S$.

As in the proof of Theorem 3.1 we will construct a new game $\bar{\mathbf{B}} = \{(L_{X_i}, V_i) : i = 1, 2, \ldots, n\}$, where

(a) L_{X_i} is the *strategy set* of player i, and
(b) $V_i : \Omega \times \prod_{i=1}^n L_{X_i} \to \mathbf{R}$ is the *payoff function* of player i (defined as in (3.3)).

It is easy to see that the existence of an α-core strategy for $\bar{\mathbf{B}}$ implies the existence of an α-core strategy for the original game $\mathbf{B} = \{(X_i, u_i, F_i, q_i) : i = 1, 2, \ldots, n\}$. Our goal is to construct a suitable family of truncated subgames in a finite dimensional strategy spare, each of which satisfies all the conditions of the Scarf (1971) theorem. Therefore we will obtain a net of α-core strategies for each subgame. As in the proof of Theorem 3.1, operating a limiting argument we can show that the existence of an α-core strategy for each subgame implies the existence of an α-core strategy for the original game $\mathbf{B}$. Before we start the outlined construction of the family of truncated subgames, we need to make some observations.

Note that for each $\omega \in \Omega$, $V_i(\omega, \cdot)$ is weakly continuous (recall Claim 4.1) and by virtue of assumption (a.3.2.3) concave on $\prod_{i=1}^n L_{X_i}$. Moreover, note that each L_{X_i} is convex and nonempty. However, since Scarf's theorem requires the compactness of each strategy set we will need to prove the following claim which is known as Diestel's theorem.

Claim 5.1. The set L_{X_i} is weakly compact in $L_1(\mu, Y)$.

Proof. The proof is based on the celebrated theorem of James (1964) and it is patterned after that Khan (1982). Note that the dual of $L_1(\mu, Y)$ is $L_\infty(\mu, Y^*_{w^*})$ (where w^* denotes the w^*-topology), i.e., $(L_1(\mu, Y))^* = L_\infty(\mu, Y^*_{w^*})$ [see, for instance, Tulcea-Tulcea (1969)]. Let x be an arbitrary element of $L_\infty(\mu, Y^*_{w^*})$. If we show that x attains its supremum on L_{X_i} the result will follow from James' theorem [James (1964)]. Let,

$$\operatorname*{Sup}_{\psi_i \in L_{X_i}} \psi \cdot x = \operatorname*{Sup}_{\psi_i \in L_{X_i}} \int_{\omega \in \Omega} (\psi_i(\omega) \cdot x(\omega)) d\mu(\omega).$$

Note that by Theorem 2.2 in Hiai-Umegaki (1977),

$$\operatorname*{Sup}_{\psi_i \in L_{X_i}} \int_{\omega \in \Omega} (\psi_i(\omega) \cdot x(\omega)) d\mu(\omega) = \int_{\omega \in \Omega} \operatorname*{Sup}_{\phi_i \in X_i(\omega)} (\phi_i \cdot x(\omega)) d\mu(\omega).$$

For each i, define the set-valued function $g_i : \Omega \to 2^Y$ by $g_i(\omega) = \{y \in X_i(\omega) : y \cdot x = \text{Sup}_{\phi_i \in X_i(\omega)} \phi_i \cdot x\}$. It follows from the weak compactness of X_i that for all $\omega \in \Omega$, $g_i(\omega)$ is nonempty. For each i, define $f_i : \Omega \times Y \to [-\infty, \infty]$ by $f_i(\omega, y) = y \cdot x - \text{Sup}_{\phi_i \in X_i(\omega)} \phi \cdot x$. It is easy to see that for each fixed $\omega \in \Omega$, $f_i(\omega, \cdot)$ is continuous and for each fixed $y \in Y$, $f_i(\cdot, y)$ is F_i-measurable and hence $f_i(\cdot, \cdot)$ is jointly F_i-measurable, i.e., for every closed subset V of $[-\infty, \infty]$, $f_1^{-1}(V) = \{(\omega, z) \in \Omega \times Y : z \in X_i(\omega)\}$ belongs to $F_i \otimes \mathbf{B}(Y)$. Since X_i is F_i-measurable the set $G_{X_i} = \{(\omega, x) : x \in X_i(\omega)\}$ is an element of $F_i \otimes \mathbf{B}(Y)$. Moreover, note that $G_{g_i} = f_i^{-1}(0) \cap G_{X_i}$ and since $f_i^{-1}(0)$ and G_{X_i} belong to $F_i \otimes \mathbf{B}(Y)$ so does G_{g_i}. It follows from the Aumann measurable selection theorem that there exists an F_i-measurable function $z_i : \Omega \to Y$ such that $z_i(\omega) \in g_i(\omega)$ μ-*a.e.* Thus, $z_i \in L_{X_i}$ and $\text{Sup}_{\phi_i \in L_{Xi}} \phi_i \cdot x = \int_{\omega \in \Omega} (z_i(\omega) \cdot x(\omega)) d\mu(\omega) = z_i \cdot x$. Since $x \in L_\infty(\mu, Y^*_{w^*})$ was arbitrarily chosen, we conclude that every element of $(L_1(\mu, Y))^*$ attains its supremum on L_{X_i}, and this completes the proof of the fact that L_{X_i} is weakly compact.

We are now ready to construct a suitable family of truncated subgames. To this end let Λ be a family of all finite subsets of L_{X_i}. For each $\lambda \in \Lambda$ let $L^\lambda_{X_i}$ denote the closed convex hull of λ. Then each $L^\lambda_{X_i}$ is a compact, convex, nonempty subset of a finite dimensional Euclidean space and $\bigcup_{\lambda \in \Lambda} L^\lambda_{X_i} = L_{X_i}$ Moreover, the set $\{L^\lambda_{X_i} : \lambda \in \Lambda\}$ is directed upwards by inclusion. For each $\lambda \in \Lambda$ we have a game $\bar{\mathbf{B}}^\lambda = \{(L^\lambda_{X_i}, V^\lambda_i) : i = 1, 2, \ldots, n\}$ where,

$$L^\lambda_{X_i} \text{ is the } \textit{strategy set} \text{ of player } i, \text{ and} \tag{5.1}$$

$$V^\lambda_i : \Omega \times \prod_{i=1}^n L^\lambda_{X_i} \to \mathbf{R} \text{ is the } \textit{payoff function} \text{ of player } i. \tag{5.2}$$

Each $\bar{\mathbf{B}}^\lambda$ satisfies the assumptions of Scarf's α-core existence theorem and therefore there exists $x^\lambda \in \prod_{i=1}^n L^\lambda_{X_i}$ satisfying the following property:

It is not true that there exist $S \subset \{1, 2, \ldots, n\}$ and $(y_i)_{i \in S} \in \prod_{i=1}^n L^\lambda_{X_i}$ such that for each $z^{I/S} \in \prod_{i \notin S} L^\lambda_{X_i}$, $V^\lambda_i(\omega, (y^S, z^{I/S})) > V^\lambda_i(\omega, x^\lambda)$ for all $i \in S$ for μ-almost all $\omega \in \Omega$.

Since the set Λ is directed by inclusion we have constructed a net $\{(x^\lambda_1, x^\lambda_2, \ldots, x^\lambda_n) : \lambda \in \Lambda\}$ of α-core strategies in $\prod_{i=1}^n L_{X_i}$. Since by Claim 5.1 each L_{X_i} is weakly compact so is $\prod_{i=1}^n L_{X_i}$. Hence the net $\{(x^\lambda_1, x^\lambda_2, \ldots, x^\lambda_n) : \lambda \in \Lambda\}$ has a subset which converges weakly to $(x_1, x_2, \ldots, x_n)$ in $\prod_{i=1}^n L_{X_i}$. We must show that $x_1, x_2, \ldots, x_n$ is an α-core strategy for $\mathbf{B}$. Adopting a similar argument with that used in the proof of Theorem 3.1, one can now complete the proof of Theorem 3.2.

6. Concluding remarks

Remark 6.1. In Theorems 3.1 and 3.2, Y is assumed to be a separable Banach lattice with order continuous norm whose dual Y^* has the RNP. Basic examples of spaces which satisfy the above properties are:

(i) the Euclidean space $\mathbf{R}^l$,
(ii) the space l^p $(1 < p < \infty)$ of real sequences $\{a_n : n = 1, 2, \dots\}$ for which the norm $\|a_n\|_p = \left(\sum_{n=1}^{\infty} |a_n|^p\right)^{1/p}$ is finite,
(iii) the space $L^p(\Omega, \mathbf{F}, \mu)$ $(1 < p < \infty)$ of measurable furlctions f on the measure space $(\Omega, \mathbf{F}, \mu)$ for which the norm $\|f\|_p = \left(\int_{\omega \in \Omega} |f(\omega)|^p d\mu(\omega)\right)^{1/p}$ is finite.

It is important to give examples of spaces that Theorems 3.1 and 3.2 do not cover:

(iv) $L_1[0, 1]$ or $L_1(\mu)$, if μ is not purely atomic, c_0, l_∞, $L_\infty[0, 1]$ and
(v) the space $C(X)$ of continuous real-valued functions on the infinite compact Hausdorff space X (with the supremum norm).

Recall that the spaces in (iv) and (v) do not have the RNP moreover, order intervals are not weakly compact in $L_\infty[0, 1]$ and $C(X)$.

Remark 6.2. The separability assumption on Y was used in order to make the Aumann measurable selection theorem applicable. The latter result was used in several steps in the proofs of Theorems 3.1 and 3.2. The relaxation of the separability of Y is possible. In this case however, the consumption set L_{X_i} will be the set of all *Gel'fand integrable selections from the set-valued function* $X_i : \Omega \to 2^{Y^*}$, and one will need to appeal to results on the existence of weak* measurable selections.

Remark 6.3. Theorem 3.1 and its corollaries can be easily extended to coalition production economies provided that the production technology is assumed to be balanced. The proof remains essentially unchanged.

Remark 6.4. Kahn and Mookerjee (1989), have introduced a core-like concept in order to analyse games in normal form with differential information. Their concept in a two-person game, coincides with the coalitional Nash equilibrium. No existence results are given in their paper. However, it is known [see, for instance, Scarf (1971)] that even if preferences are strictly convex and continuous the set of coalitional Nash equilibrium strategies may be empty.

Remark 6.5. We conjecture that the core of a large finite private information economy will converge to the standard Debreu-Scarf (1963) core notion, with the approximation getting finer the larger the private information economy (this will follow from the law of large numbers provided there is some kind of independence among agents). Hence, we can conclude that core allocations in large private information economy will become Walrasian. We also conjecture that without the independence assumption among agents, core allocations in a large private information economy will characterize some kind of rational expectations equilibrium.[7]

---

[7] Kobayashi (1980, p. 1647) has made a related conjecture for the syndicate problem. Moreover, Srivastava (1984) has shown that a Wilson-type core allocation in a differential information economy becomes a full information core allocation as the number of agents in the economy tends to infinity. Finally, Allen (1983) has treated information as a differentiated commodity and she has shown the equivalence of the core and the competitive equilibrium for an economy with a continuum of agents.

References

Allen, B.: General equilibrium with information sales. CARES working paper no. 83–25 (1983). A shortened version appeared in Theory and Decision 21, 1–33 (1986)

Aliprantis, C.D., Brown, D., Burkinshaw, O.: Existence and optimality of competitive equilibria. Berlin Heidelberg New York: Springer 1989

Aliprantis, C.D., Burkinshaw, O.: Positive operators. New York: Academic Press 1985

Aumann, R.J.: Measurable utility and the measurable choice theorem. In: La decision. C.N.R.S., Aix-en-Provence (1967)

Berliant, M: On income taxation and the core. Working paper, University of Rochester (1990)

Border, K.: A core existence theorem for games without ordered preferences. Econometrica **52**, 1537–1542 (1984)

Cartwright, D.: The order completeness of some spaces of vector-valued functions. Bull Aust Math Soc **11**, 57–61 (1974)

Debreu, G.: Integration of correspondences. Proc Fifth Berkeley Symp Stat Probability **11**, 351–372 (1967)

Debreu, G., Scarf, H.: A limit theorem an the core of an economy. Econ Rev **4**, 235–246 (1963)

Diestel, J., Uhl, J.J.: Vector measures. Mathematical surveys, no. 15. American Mathematical Society, Providence Rhode Island (1977)

Hiai, F., Umegaki, H.: Integrals, conditional expectation and martingales of multivalued functions. J Multivar Anal **7**, 149–182 (1977)

James, R.: Weakly compact sets. Transact Am Math Soc **113**, 129–140 (1964)

Kahn, C., Mookerjee, D.: Decentralized exchange and efficiency under moral hazard. Working paper, University of Illinois and Stanford University (1989)

Khan, M.A.: Ort the integration of sec-valued mappings in a nonrellexive Banach space. Working Paper, The Johns Hopkins University (1982)

Kobayashi, T.: Equlibrium contracts for syndicates with differential information. Econometrica **48**, 1635–1665 (1980)

Lindenstrauss, J., Tzafriri, L.: Classical banach spaces II. Berlin Heidelberg New York: Springer 1979

Palfrey, T.R., Srivastava, S.: On Bayesian implementable allocations. Rev Econ Stud **178**, 193–208 (1986)

Scarf, H.: The core of an n-person game. Econometrica **35**, 50–69 (1967)

Scarf, H.: On the existence of a cooperative solution for a general class of n-person games. J Econ Theory **3**, 169–171 (1971)

Srivastava, S.: A limit theorem an the core wich differential information. Econ Lett **134**, 111–116 (1984)

Tulcea, I.A., Tulcea, C.I.: Topics in the theory of liftings. New York: Springer 1968

Wilson, R.: Information, efficiency, and the core of an economy. Econometrica **46**, 807–816 (1978)

Yannelis, N.C.: The core of an economy without ordered preferences. In: Khan, M.A., Yannelis, N.C. (eds.) Equilibrium theory wich infinitely many commodities. Berlin Heidelberg New York: Springer 1991

Yannelis, N.C., Rustichini, A.: Equilibrium points of non-cooperative random and Bayesian games. In: Alipranties, C.D., Border, K.C., Luxemburg, W.A.J. (eds.) Positive operators, Riesz spaces and economics. Berlin Heidelberg New York: Springer 1991

An alternative proof of the nonemptiness of the private core*

Isabelle Lefebvre

CERMSEM, Maison des Sciences Economiques, Université de Paris 1,
106-112 Boulevard de l'Hôpital, 75647 Paris Cedex 13, FRANCE (e-mail: ilefebvr@univ-paris1.fr)

Received: October 4, 1999; revised version: June 7, 2000

Summary. We focus on the private core (Yannelis [19]) of an economy with a finite number of agents with differential information, a continuum of states and an infinite number of commodities. We state a nonemptiness result for the private core and provide a proof based on a fixed-point argument.

Keywords and Phrases: Economy with differential information, Coarse core, Private core, Weak fine core, Nonemptiness, Fixed-point theorem.

JEL Classification Numbers: D51, D82, C71.

1 Introduction

This paper deals with the nonemptiness of the private core notion due to Yannelis [19], a notion which has desirable properties (see for example Koutsougeras-Yannelis [13]). This question has already been treated in different papers and various frameworks commodities-states of nature, for example, by Yannelis [19], Koutsougeras-Yannelis [13], and Allen [3].

Actually, we here consider an economy with differential information in the framework of a finite number of agents, an infinite number of commodities and a continuum of states of nature. Formally, the information of an agent is modelled as a sub-σ-field of the set Ω of states of nature. If the information is incomplete, it restricts the transactions an agent can make.

The physical commodity space is given by a Banach lattice Z, while the space of contingent commodities is given by the set of measurable Bochner integrable

* I wish to thank Prof. J.-M. Bonnisseau for his supervision, and Nicholas Yannelis and the anonymous referee for helpful comments.

functions from Ω to Z. Hence, the consumption sets of the agents will be subsets of the space[1] $L^1(Z)$.

The preferences of each agent i are described by her/his preference correspondence P_i which associates to every allocation $x = (x_i, x_{-i})$ of the economy the set of allocations which are strictly preferred by her/him to x_i, given the allocations x_{-i} of the other agents. Thus, agent's preferences do not need to be neither transitive nor complete, and may be interdependent. Clearly, those cannot in general be representable by utility functions. But, the general feature of the preferences allows us to take into account particular preferences, for example, preferences defined from an "expected utility" as in Allen [4] or in Koutsougeras-Yannelis [13], or defined from a "conditional expected utility" as in Yannelis [19].

This paper's contribution to the economics of information is technical. We provide an alternative proof of the nonemptiness of the private core due to Yannelis [19]. Actually, the method of the proof relies on the deterministic core result of Florenzano [11] and the Banach lattice technique of Yannelis [19]. More precisely, the first main core existence result for an economy with differential information using Banach space methods was proved by Yannelis [19]. The idea there was to go to finite dimension use Scarf's result [17] and operate a limit argument.

There is an another method of proof (Allen [3], Schwalbe [18], Page [15], Balder-Yannelis [5]...) which consists in showing that the game derived from an economy with differential information is balanced and appealing to Scarf's result [17].

In contrast, as in Lefebvre [14], the proof of the nonemptiness of the private core rests here on a fixed-point argument following the approach of Florenzano [11] in her proof of a finite dimensional core nonemptiness result. Precisely here, we first construct a suitable family of abstract applications defined on suitable finite-dimensional sets derived from the economy. Then, by a fixed-point theorem applied to a particular correspondence defined from the primitives of the economy, we deduce the existence of maximal element for those applications. We thus obtain a sequence of maximal element for each abstract application. Then by a compactness argument, we know that the sequence of maximal element for each abstract application converges to an attainable allocation. Finally, we show, by contraposition, that this limit belongs to the core of the economy.

Moreover, we show that the nonemptiness results of the (interim or ex-ante) private core given in Yannelis [19], Allen [4] and Koutsougeras-Yannelis [13] are corollaries of our main theorem. Indeed, we here assume weakest assumptions on the preferences, the initial endowments and the consumption sets. Hence, this paper provides a single proof of those three nonemptiness results. It will be furthermore noticed that the main result implies directly the nonemptiness of the *coarse core* of Yannelis [19] and Koutsougeras-Yannelis [13]. Let note also that the techniques of the proof remain valid to obtain the nonemptiness of the *weak fine core* of Koutsougeras-Yannelis [13], and allow to cover the case of general information-rules of Allen [3].

[1] Confer Appendix for definition.

Let moreover note that if the consumption sets of the agents are subsets of $L^\infty(\mathbb{R}^\ell)$, there exists a short proof of the nonemptiness of the core due to Lefebvre [14]. This proof uses a quasi fixed-point theorem in infinite-dimensional spaces (Thm.2.1 of [14]) applied to a suitable correspondence directly defined from the primitives of the economy. But here, we can not use this theorem and so this short proof because of the non-metrizability of the order intervals in $L^1(Z)$.

The remainder of this paper is organized as follows. Section 2 describes the model. An agent is specified by initial endowments of both physical commodities and information. It then recalls the definition of the private core of an exchange economy with differential information. Then, we state the nonemptiness result. Section 3 presents the proof of the nonemptiness result divived into several steps. Then, we provide an extension for production economies. Section 4 gives concluding remarks. We end this paper by an appendix which contains mathematical definitions and the proof of intermediary claims.

2 The model and the main result

We consider an exchange economy with a finite number n of consumers. We note $N = \{1, \ldots, n\}$ the set of agents and $\mathcal{N} = 2^N \setminus \emptyset$ the family of all nonempty subsets of N, called coalitions in the following.

The uncertainty and initial information are modelled in the usual way. The set of states of nature is denoted by Ω, with typical element ω. Let $\mathbf{F}$ be a σ-field of measurable subsets of Ω, interpreted as events. Let μ be a σ-additive probability measure defined on $(\Omega, \mathbf{F})$. The initial information of the agent $i \in N$ is described by a sub-σ-field $\mathbf{F}_i$ of $\mathbf{F}$.

The physical commodity space is given by a Banach lattice[2] Z (in each state $\omega \in \Omega$). The space of contingent commodities is given by the space $L^1(\Omega, \mathbf{F}, \mu; Z)$ of equivalence classes of Bochner μ-integrable[3] $\mathbf{F}$-measurable[4] functions. Let us denote the consumption set of the agent i by $X_i \subset L^1(\Omega, \mathbf{F}, \mu; Z)$ and her/his random (state dependent) initial endowment by $e_i(\cdot) \in L^1(\Omega, \mathbf{F}, \mu; Z)$.

For all $x = (x_i)_{i \in N} \in \prod_{i \in N} X_i$, we let $P_i(x) \subset X_i$ be the set of allocations which are strictly preferred[5] to x_i by the i-th consumer, given the allocations $(x_j)_{j \neq i}$ of the other consumers. Thus, agent's preferences do not need to be neither transitive nor complete, and may be interdependent.

The exchange economy with differential information that we consider is thus described by the collection $\mathcal{E} = (X_i, e_i, P_i, \mathbf{F}_i)_{i=1}^n$.

---

[2] Definition and basic results are given in appendix.

[3] Definition and basic results are given in appendix.

[4] Usually, we write μ-measurable instead of $\mathbf{F}$-measurable, with μ a measure on $(\Omega, \mathbf{F})$ (see [9], [10]). But, writing the measurability with respect to the σ-field is here more suitable since in the following we will encounter notions of measurability with respect to particular sub-σ-fields of $\mathbf{F}$.

[5] This above framework, describing the tastes of the consumers, encompasses the case where the i-th consumer has a preference relation $\preceq_i$ which is a binary relation on X_i. In this latter case, we associate to $\preceq_i$ for all $x \in \prod_{k \in N} X_k$, the preferred set $P_i(x) = \{x'_i \in X_i \mid x_i \prec_i x'_i\}$ where the strict preference relation $\prec_i$ is defined by $x_i \prec_i x'_i$ if $[x_i \preceq_i x'_i$ and not $x'_i \preceq_i x_i]$.

2.1 Definition of the core

Before defining the private core of an exchange economy with differential information, it is necessary to define firstly the set of allocations of the economy which are attainable, and secondly, the notion of improvness. Obviously, attainable allocations must be at least physically attainable-Condition (1) below, but must also depend on the information-Condition (2), here described by the agent's individual informations $(\mathbf{F}_i)_{i\in N}$.

Definition 2.1 *The allocation* $x^*=(x_i^*)_{i\in N}\in\prod_{i\in N}X_i$ *is attainable for the economy* $\mathcal{E}$ *if*

(1) $\sum_{i\in N} x_i^*(\omega)=\sum_{i\in N} e_i(\omega)$ *μ-a.e.,*
(2) for every $i\in N$, $x_i^*-e_i$ *is* $\mathbf{F}_i$*-measurable.*

We denote by $\mathcal{A}$ *the set of the attainable allocations of the economy* $\mathcal{E}$.

Definition 2.2 *The allocation* $x^*=(x_i^*)_{i\in N}\in\prod_{i\in N}X_i$ *is improved upon by the coalition* $S\in\mathcal{N}$ *if there exists* $(x_i)_{i\in S}\in\prod_{i\in S}X_i$ *such that* $x_i\in P_i(x^*)$ *for every* $i\in S$ *and*

(1S) $\sum_{i\in S} x_i(\omega)=\sum_{i\in S} e_i(\omega)$ *μ-a.e.,*
(2S) for every $i\in S$, x_i-e_i *is* $\mathbf{F}_i$*-measurable.*

The private core of the economy $\mathcal{E}$*, noted* $\mathcal{C}(\mathcal{E})$*, is the set of attainable allocations that no coalition can improve upon.*

In other words, an attainable allocation belongs to the private core $\mathcal{C}(\mathcal{E})$ if it is not possible for agents to join a coalition, reallocate their endowments among themselves -Condition (1S) (while each member of the coalition uses his/her own private information -Condition (2S)), and obtain a strictly preferred allocation for each member of the coalition.

2.2 The nonemptiness result

For every coalition $S\in\mathcal{N}$, we define the set $\mathcal{A}(S)$ of attainable allocations of the coalition S as follows

$$\mathcal{A}(S)=\left\{(x_i)_{i\in S}\in\prod_{i\in S}X_i\;\middle|\;\begin{array}{l}(1S)\ \sum_{i\in S}x_i(\omega)=\sum_{i\in S}e_i(\omega)\ \mu-a.e.\\ (2S)\ \forall i\in S, x_i-e_i \text{ is } \mathbf{F}_i-\text{measurable}\end{array}\right\}.$$

Note that the set $\mathcal{A}$ of the attainable allocations of the economy $\mathcal{E}$ is in fact the set $\mathcal{A}(N)$. Let endow the space $L^1(\Omega,\mathbf{F},\mu;Z)$ by the weak topology $\sigma(L^1,L^\infty)$. Now, in addition to the fact that Z is a Banach lattice, we posit the following assumptions which describe the general framework of this paper, and then state the main result.

Assumption C1 [Consumption side] *For all* $i\in N$,

- $X_i\subset L^1(\Omega,\mathbf{F},\mu;Z)$ *is convex;*

- *[irreflexivity] for all* $x=(x_i)_{i\in N}\in\mathcal{A}$, $x_i \notin coP_i(x)$[6]*;*
- *[open lower sections] for all* $z \in X_i$*, the set* $P_i^{-1}(z) = \{x \in \prod_{k\in N} X_k \mid z \in P_i(x)\}$ *is weakly open in* $\prod_{k\in N} X_k$.

Assumption C2 [Compactness] $\mathcal{A}$ *is weakly compact.*

Assumption S [Survival assumption] *For all* $i \in N$, $\mathcal{A}(\{i\}) \neq \emptyset$.

Let remark that Assumption S implies the nonemptiness of the sets $\mathcal{A}(S)$ for all $S\in\mathcal{N}$ and so of the set $\mathcal{A}=\mathcal{A}(N)$.

Moreover, let note that if, for all $i\in N$ P_i is representable by an utility function $U_i : X_i \to \mathrm{I\!R}$, i.e. $P_i(x) := \{x_i' \in X_i \mid U_i(x_i') > U_i(x_i)\}$, for all $x=(x_i)_{i\in N} \in \prod_{i\in N}X_i$, then assuming that P_i has open lower sections is equivalent to assume that U_i is weakly upper semicontinuous.

Let also note that Yannelis [19] and Koutsougeras-Yannelis [13] consider a stronger assumption than C1 since this implies that the preference correspondences P_i have a weakly open graph.

Theorem 2.1 *Under Assumptions* C1, C2 *and* S, *the private core* $\mathcal{C}(\mathcal{E})$ *of the exchange economy with differential information* $\mathcal{E} = (X_i, e_i, P_i, \mathbf{F}_i)_{i=1}^n$ *is nonempty.*

3 Proof of Theorem 2.1: $\mathcal{C}(\mathcal{E}) \neq \emptyset$

We will here construct a suitable family of abstract applications on finite-dimensional commodity spaces, each of which has a maximal element. We will then obtain a net of maximal element for each application. By taking limits we will show that the existence of maximal element for each application implies the existence of a core allocation for the economy $\mathcal{E} = (X_i, e_i, P_i, \mathbf{F}_i)_{i=1}^n$. To simplify the notations, we note L^1 for $L^1(\Omega, \mathbf{F}, \mu; Z)$ in the following. We now give the proof of Theorem 2.1.
For every $x=(x_i)_{i\in N}\in\prod_{i\in N}X_i$, we define the preferred set $P_S(x)$ as follows

$$P_S(x) = \prod_{i\in S} P_i(x) = \left\{(z_i)_{i\in S} \in \prod_{i\in S} X_i \mid z_i \in P_i(x), \forall i \in S\right\}.$$

Let remark that one has

$$\mathcal{C}(\mathcal{E}) = \{x \in \mathcal{A} \mid \forall S\in\mathcal{N}, \mathcal{A}(S) \cap P_S(x) = \emptyset\}.$$

For every $x \in \mathcal{A}$, we note $\mathcal{I}(x)$ the set of coalitions which can improve upon the attainable allocation x, formally

$$\mathcal{I}(x) = \{S\in\mathcal{N} \mid \mathcal{A}(S) \cap P_S(x) \neq \emptyset\}.$$

Let then remark that $x\in\mathcal{A}$ belongs to $\mathcal{C}(\mathcal{E})$ if and only if $\mathcal{I}(x) = \emptyset$.

[6] We denote co$E$ the convex hull of the set E.

3.1 Abstract applications

For each $S \in \mathcal{N}$, let consider an element $a_S = (a_{S,i})_{i\in S}$ of $\mathcal{A}(S) \subset (L^1)^{\mathrm{card} S}$. To simplify the notations, we will note a_i for $a_{\{i\},i} \in \mathcal{A}(\{i\})$ for all $i \in N$, in the following. Now let $\mathcal{F}$ be the set of finite-dimensional subspaces of L^1 containing the initial endowments $\{e_i, i \in N\}$ and the set $\{a_{S,i} \mid S \in \mathcal{N}, i \in S\}$.

Let fix $F \in \mathcal{F}$. Let endow the finite-dimensional space F with the topology induced by the topology of L^1. For each $i \in N$, let define the set $X_i^F := X_i \cap F$, and the correspondence $P_i^F : \prod_{i\in N} X_i^F \to 2^{X_i^F}$ by $P_i^F(x) := P_i(x) \cap F$, and $P_S^F(x) := \prod_{i\in S} P_i^F(x) = [\prod_{i\in S} P_i(x)] \cap F^{\mathrm{card} S}$. Remark that one has for all $S \in \mathcal{N}, \mathcal{A}(S) \cap F^{\mathrm{card} S} \subset \prod_{i\in S} X_i^F$. For every $x \in \mathcal{A} \cap F^n$, let define the set $\mathcal{I}^F(x)$ by

$$\mathcal{I}^F(x) = \{S \in \mathcal{N} \mid [\mathcal{A}(S) \cap F^{\mathrm{cardS}}] \cap P_S^F(x) \neq \emptyset\}.$$

Lemma 3.1 *For each $F \in \mathcal{F}$, there exists $x^F \in \mathcal{A} \cap F^n$ such that $\mathcal{I}^F(x^F) = \emptyset$.*
The proof of Lemma 3.1, given in Section 3.2, is divided into three steps and rests on a fixed-point argument.

3.2 Proof of Lemma 3.1

3.2.1 Preliminaries

Let fix $F \in \mathcal{F}$. We let

$$\Delta = \left\{ \lambda = (\lambda_S)_{S\in\mathcal{N}} \in \mathbb{R}_+^{\mathcal{N}} \mid \sum_{\{S\in\mathcal{N} \mid i\in S\}} \lambda_S = 1,\ \forall\, i \in N \right\},$$

$$\Sigma = \left\{ \mu = (\mu_S)_{S\in\mathcal{N}} \in \mathbb{R}_+^{\mathcal{N}} \mid \sum_{S\in\mathcal{N}} \mu_S = 1 \right\}.$$

We now define the correspondence ϕ from $[\mathcal{A} \cap F^n] \times \prod_{S\in\mathcal{N}} [\overline{\mathrm{co}}\mathcal{A}(S) \cap F^{\mathrm{cardS}}] \times \Delta \times \Sigma$ to itself[7] (to which we will apply a fixed-point theorem) as follows

$$\phi(x, z, \lambda, \mu) = \phi^1(x, z, \lambda, \mu) \times \prod_{S\in\mathcal{N}} \phi_S^2(x, z, \lambda, \mu) \times \phi^3(x, z, \lambda, \mu) \times \phi^4(x, z, \lambda, \mu),$$

with
$\phi^1(x, z, \lambda, \mu) = \{(\varphi_i^1(x, z, \lambda, \mu))_{i\in N}\}$
where $\varphi_i^1(x, z, \lambda, \mu) = \sum_{\{S\in\mathcal{N} \mid i\in S\}} \lambda_S z_{S,i}$, for $i \in N$;

$\phi_S^2(x, z, \lambda, \mu) = [\overline{\mathrm{co}}\mathcal{A}(S) \cap F^{\mathrm{cardS}}] \cap \mathrm{co} P_S^F(x)$, for each $S \in \mathcal{N}$;

$\phi^3(x, z, \lambda, \mu) = \{\bar{\lambda} \in \Delta \mid \mu \cdot \bar{\lambda} = \max_{\delta\in\Delta} \mu \cdot \delta\}$;

$\phi^4(x, z, \lambda, \mu) = \{\bar{\mu} \in \Sigma \mid \bar{\mu}_S = 0, \forall S \notin \mathcal{I}^F(x)\}$.

[7] We denote $\overline{\mathrm{co}}E$ the convex strongly closed hull of the set $E \subset (L^1)^r$ with r integer.

3.2.2 The fixed-point argument

Lemma 3.2 *ϕ satisfies the assumptions of Theorem 4.4 of [12].*

Proof of Lemma 3.2. • It is easily seen that Δ and Σ are convex and compact subsets of the finite-dimensional Euclidean space $\mathbb{R}^{\mathcal{N}}$. The set Σ is obviously nonempty. And the set Δ contains the vector $\lambda = (\lambda_S)_{S\in\mathcal{N}}$ defined by $\lambda_{\{i\}} = 1$ for all $i \in N$ and $\lambda_S = 0$ if card$S > 1$.

• The set $\mathcal{A}\cap F^n$ is a nonempty, convex and compact subset of a finite-dimensional space, since $\mathcal{A}$ contains $(a_i)_{i\in N}$ where $a_i \in \mathcal{A}(\{i\})\cap F$ for all $i \in N$, and is a convex and weakly compact subset of $(L^1)^n$ (see Claim 5.1 Appendix).

• The set $\prod_{S\in\mathcal{N}}[\overline{\text{co}}\mathcal{A}(S) \cap F^{\text{card}S}]$ is a nonempty, convex and compact subset of a finite-dimensional space. Indeed, for all $S \in \mathcal{N}$, the set $\overline{\text{co}}\mathcal{A}(S) \cap F^{\text{card}S}$ is a nonempty, convex and compact subset of a finite-dimensional space, since the set $\overline{\text{co}}\mathcal{A}(S)$ contains $a_S = (a_{S,i})_{i\in S} \in F^{\text{card}S}$, and is a convex and weakly compact subset of $(L^1)^{\text{card}S}$ (see Claim 5.2 Appendix).

• The mapping ϕ^1 is clearly a continuous mapping from $[\mathcal{A} \cap F^n] \times \prod_{S\in\mathcal{N}} [\overline{\text{co}}\mathcal{A}(S) \cap F^{\text{card}S}] \times \Delta \times \Sigma$ to F^n (hence as a correspondence it is lower semicontinuous with convex values). It remains to check that ϕ^1 $([\mathcal{A} \cap F^n] \times \prod_{S\in\mathcal{N}}[\overline{\text{co}}\mathcal{A}(S) \cap F^{\text{card}S}] \times \Delta \times \Sigma) \subset \mathcal{A}\cap F^n$. But, since ϕ^1 is continuous and $\mathcal{A}\cap F^n$ is compact, it suffices to show that $\phi^1([\mathcal{A}\cap F^n] \times \prod_{S\in\mathcal{N}}[\text{co}\mathcal{A}(S) \cap F^{\text{card}S}] \times \Delta \times \Sigma) \subset \mathcal{A} \cap F^n$. Now, since ϕ^1 is linear in the second variable (i.e. z) and $\mathcal{A}$ is convex, it is sufficient to show $\phi^1([\mathcal{A} \cap F^n] \times \prod_{S\in\mathcal{N}}[\mathcal{A}(S) \cap F^{\text{card}S}] \times \Delta \times \Sigma) \subset \mathcal{A} \cap F^n$. Let then consider $(x, z = (z_S)_{S\in\mathcal{N}}, \lambda, \mu) \in [\mathcal{A} \cap F^n] \times \prod_{S\in\mathcal{N}}[\mathcal{A}(S) \cap F^{\text{card}S}] \times \Delta \times \Sigma$, and let show that $x^* = (x_i^*)_{i\in N} := \phi^1(x, z, \lambda, \mu)$ belongs to $\mathcal{A}\cap F^n$. For all $i \in N$, $x_i^* = \varphi_i^1(x, z, \lambda, \mu) = \sum_{\{S\in\mathcal{N}|i\in S\}} \lambda_S z_{S,i}$ belongs to the convex set X_i^F since $z_{S,i} \in X_i^F$, for all $S \in \mathcal{N}$ and $\lambda \in \Delta$. Hence, x^* belongs to $\prod_{i\in N} X_i^F = (\prod_{i\in N} X_i)\cap F^n$. It then remains to check that $x^* \in \prod_{i\in N} X_i$ satisfies the conditions of the definition of $\mathcal{A}$. Let then fix $S \in \mathcal{N}$. Since $z_S = (z_{S,i})_{i\in S}$ belongs to $\mathcal{A}(S)$ one has: (1S) $\sum_{i\in S} z_{S,i}(\omega) = \sum_{i\in S} e_i(\omega)$ $\mu - a.e.$; and (2S) for all $i \in S$, $z_{S,i} - e_i$ is $\mathbf{F}_i$-measurable. Hence, it holds:
(1) From (1S), we obtain for μ-almost $\omega \in \Omega$,

$$\sum_{i\in N} x_i^*(\omega) = \sum_{i\in N} \sum_{\{S\in\mathcal{N}|i\in S\}} \lambda_S z_{S,i}(\omega) = \sum_{S\in\mathcal{N}} \lambda_S \sum_{i\in S} z_{S,i}(\omega)$$

$$= \sum_{S\in\mathcal{N}} \lambda_S \sum_{i\in S} e_i(\omega) = \sum_{i\in N} \left(\sum_{\{S\in\mathcal{N}|i\in S\}} \lambda_S \right) e_i(\omega) = \sum_{i\in N} e_i(\omega),$$

since λ belongs to Δ.
(2) For all $S \in \mathcal{N}$, for all $i \in S$, $(z_{S,i} - e_i)$ is $\mathbf{F}_i$-measurable. We then deduce that for all $i \in N$, $x_i^* - e_i$ is $\mathbf{F}_i$-measurable since we have $x_i^* - e_i = (\sum_{\{S\in\mathcal{N}|i\in S\}} \lambda_S z_{S,i}) - e_i = \sum_{\{S\in\mathcal{N}|i\in S\}} \lambda_S (z_{S,i} - e_i)$.

• Let fix $S \in \mathcal{N}$. The correspondence ϕ_S^2 is clearly convex valued. To show that ϕ_S^2 is lower semicontinuous, it is sufficient to show that it has open lower sections (confer [20]). Indeed, for every $\bar{z}_S \in \overline{\mathrm{co}}\mathcal{A}(S) \cap F^{\mathrm{card}S}$,

$$(\phi_S^2)^{-1}(\bar{z}_S)=\Big\{(x,z,\lambda,\mu) \in [\mathcal{A} \cap F^n] \times \prod_{S \in \mathcal{N}} [\overline{\mathrm{co}}\mathcal{A}(S) \cap F^{\mathrm{card}S}] \times \Delta \times \Sigma \mid \bar{z}_S \in \mathrm{co}P_S^F(x)\Big\}$$

$$=\Big\{(x,z,\lambda,\mu) \in [\mathcal{A} \cap F^n] \times \prod_{S \in \mathcal{N}} [\overline{\mathrm{co}}\mathcal{A}(S) \cap F^{\mathrm{card}S}] \times \Delta \times \Sigma \mid \bar{z}_{S,i} \in \mathrm{co}P_i^F(x), \forall i \in S\Big\}$$

$$=[\mathcal{A} \cap F^n] \cap [\cap_{i \in S}(\mathrm{co}P_i^F)^{-1}(\bar{z}_{S,i})] \times \prod_{S \in \mathcal{N}} [\overline{\mathrm{co}}\mathcal{A}(S) \cap F^{\mathrm{card}S}] \times \Delta \times \Sigma$$

is an open subset of $[\mathcal{A} \cap F^n] \times \prod_{S \in \mathcal{N}}[\overline{\mathrm{co}}\mathcal{A}(S) \cap F^{\mathrm{card}S}] \times \Delta \times \Sigma$, since from Assumption C1, for all $i \in N$, P_i and so P_i^F and $\mathrm{co}P_i^F$ have open lower sections (confer [20]).

• The correspondence ϕ^3 is clearly convex and compact valued. And it is upper semicontinuous from the Maximum Theorem (Berge 1959 [6]) and the fact that Σ is a nonempty convex compact subset of $\mathbb{R}^{\mathcal{N}}$.

• The correspondence ϕ^4 is clearly convex valued. To show that it is lower semicontinuous, it is sufficient to show that it has open lower sections. Indeed, for every $\bar{\mu} \in \Sigma$, one has

$$(\phi^4)^{-1}(\bar{\mu}) = \Big\{(x,z,\lambda,\mu) \in [\mathcal{A} \cap F^n] \times \prod_{S \in \mathcal{N}} [\overline{\mathrm{co}}\mathcal{A}(S) \cap F^{\mathrm{card}S}] \times \Delta \times \Sigma \mid \bar{\mu} \in \phi^4(x,z,\lambda,\mu)\Big\}$$

$$= \{x \in \mathcal{A} \cap F^n \mid \bar{\mu}_S = 0, \forall S \notin \mathcal{I}^F(x)\} \times \prod_{S \in \mathcal{N}} [\overline{\mathrm{co}}\mathcal{A}(S) \cap F^{\mathrm{card}S}] \times \Delta \times \Sigma.$$

To show it is open in $[\mathcal{A} \cap F^n] \times \prod_{S \in \mathcal{N}}[\overline{\mathrm{co}}\mathcal{A}(S) \cap F^{\mathrm{card}S}] \times \Delta \times \Sigma$, it is equivalent to show that its first component $\Omega = \{x \in \mathcal{A} \cap F^n \mid \bar{\mu}_S = 0, \forall S \notin \mathcal{I}^F(x)\}$ is open in $\mathcal{A} \cap F^n$. Indeed, let $\bar{x} \in \Omega$, we notice that

$$\bar{x} \in \nu_{\bar{x}} = \{x \in \mathcal{A} \cap F^n \mid \mathcal{I}^F(\bar{x}) \subset \mathcal{I}^F(x)\} \subset \Omega,$$

and the proof will be complete if we show that $\nu_{\bar{x}}$ is a neighborhood of $\bar{x}$. But $S \in \mathcal{I}^F(\bar{x})$ if and only if there exists $\bar{z}_S \in [\mathcal{A}(S) \cap F^{\mathrm{card}S}] \cap P_S^F(\bar{x})$, or equivalently $[\bar{x} \in U_S := [\mathcal{A} \cap F^n] \cap [\cap_{i \in S}(P_i^F)^{-1}(\bar{z}_{S,i})]$, with $\bar{z}_S \in [\mathcal{A}(S) \cap F^{\mathrm{card}S}]]$. This latter set U_S is open in $\mathcal{A} \cap F^n$, since from Assumption C1, P_i and so P_i^F have open lower sections. One easily checks that for every $x \in U_S$, one has $S \in \mathcal{I}^F(x)$.

Hence, for every x in the open set $U = \cap_{S\in\mathcal{I}^F(\bar{x})} U_S$, one has $\mathcal{I}^F(\bar{x}) \subset \mathcal{I}^F(x)$, and so $U \subset \nu_{\bar{x}}$. Finally, $\bar{x} \in U \subset \nu_{\bar{x}} \subset \Omega$. This ends the proof of Lemma 3.2. □

3.2.3 The end of the proof of Lemma 3.1

From Lemma 3.2 and Theorem 4.4 of [12], there exists $(x^F, z^F = (z^F_S)_{S\in\mathcal{N}}, \lambda^F, \mu^F) \in [\mathcal{A} \cap F^n] \times \prod_{S\in\mathcal{N}}[\overline{\text{co}}\mathcal{A}(S) \cap F^{\text{cardS}}] \times \Delta \times \Sigma$ such that

(1) $x^F = \phi^1(x^F, z^F, \lambda^F, \mu^F)$,
or equivalently
(1′) $x^F_i = \sum_{\{S\in\mathcal{N} | i\in S\}} \lambda^F_S z^F_{S,i}$, for each $i \in N$;

(2) $\forall\ S \in \mathcal{N}, z^F_S \in [\overline{\text{co}}\mathcal{A}(S) \cap F^{\text{card}S}] \cap \text{co}P^F_S(x^F)$ or $[\overline{\text{co}}\mathcal{A}(S) \cap F^{\text{cardS}}] \cap \text{co}P^F_S(x^F) = \emptyset$,
which implies
(2′) $\forall S \in \mathcal{I}^F(x^F),\ z^F_S \in [\overline{\text{co}}\mathcal{A}(S) \cap F^{\text{cardS}}] \cap \text{co}P^F_S(x^F)$;

(3) $\lambda^F \in \phi^3(x^F, z^F, \lambda^F, \mu^F)$ (since $\phi^3(x^F, z^F, \lambda^F, \mu^F) \neq \emptyset$);

(4) $\mu^F \in \phi^4(x^F, z^F, \lambda^F, \mu^F)$ or $\phi^4(x^F, z^F, \lambda^F, \mu^F) = \emptyset$,
or equivalently
(4′) $[\mu^F_S = 0, \forall S \notin \mathcal{I}^F(x^F)]$ or $[\mathcal{I}^F(x^F) = \emptyset]$.

We will now show that $x^F \in \mathcal{A} \cap F^n$ satisfies $\mathcal{I}^F(x^F) = \emptyset$. Let us suppose, on the contrary that $\mathcal{I}^F(x^F) \neq \emptyset$. From (1′), we have

$$\text{for every } i\in N,\ x^F_i = \sum_{\{S\in\mathcal{N} | i\in S\}} \lambda^F_S z^F_{S,i}.$$

Claim 3.1 *There exists $i \in N$ such that $\lambda^F_S = 0$ for every $S \notin \mathcal{I}^F(x^F), i \in S$.*

Consequently, from Claim 3.1, it holds for some $i \in N$,

$$x^F_i = \sum_{\{S\in\mathcal{N} | i\in S\}} \lambda^F_S z^F_{S,i} = \sum_{\{S\in\mathcal{I}^F(x^F) | i\in S\}} \lambda^F_S z^F_{S,i}.$$

From (2′), one has

$$\text{co}\{z^F_{S,i} \mid S \in \mathcal{I}^F(x^F), i \in S\} \subset \text{co}P^F_i(x^F) \subset \text{co}P_i(x^F).$$

Thus, since λ^F belongs to Δ, from the two above assertions we deduce that $x^F_i \in \text{co}P_i(x^F)$ which contradicts the irreflexivity on P_i (Assumption C1). This ends the proof of Lemma 3.1. □

Proof of Claim 3.1. Since $\mathcal{I}^F(x^F) \neq \emptyset$, from Assertion (4′) one has $\mu^F_S = 0$, for every $S \notin \mathcal{I}^F(x^F)$. Assertion (3) can be rewritten equivalently by saying that λ^F is a solution of the following linear programming problem

$$\text{maximize} \sum_{S\in\mathcal{N}} \mu^F_S \lambda_S = \mu^F \cdot \lambda$$

$$(f_i \cdot \lambda =) \sum_{\{S \in \mathcal{N} | i \in S\}} \lambda_S = 1, \ \forall i \in N,$$

$$\lambda_S \geq 0, \ \forall S \in \mathcal{N};$$

where $f_i = (f_{i,S})_{S \in \mathcal{N}} \in \mathrm{I\!R}^{\mathcal{N}}$ is the vector defined by $f_{i,S} = 1$ if $i \in S$ and $f_{i,S} = 0$ if $i \notin S$. From Kuhn and Tucker's Theorem, there exist multipliers $\alpha_i \in \mathrm{I\!R}$, $i \in N$, such that

$$\mu^F - \sum_{i \in N} \alpha_i f_i \leq 0, \ \lambda^F \geq 0, \text{ and } \left(\mu^F - \sum_{i \in N} \alpha_i f_i \right) \cdot \lambda^F = 0,$$

or equivalently such that, for all $S \in \mathcal{N}$

$$\mu_S^F - \sum_{i \in N} \alpha_i f_{i,S} = \mu_S^F - \sum_{i \in S} \alpha_i \leq 0, \lambda_S^F \geq 0, \text{and } \left(\mu_S^F - \sum_{i \in S} \alpha_i \right) \lambda_S^F = 0.$$

We first remark that for every $j \in N, \alpha_j \geq 0$. Indeed, let $j \in N$ and choose $S = \{j\} \in \mathcal{N}$, then since μ^F belongs to Σ, from above we get $0 \leq \mu_{\{j\}}^F \leq \alpha_j$. Now, since μ^F belongs to Σ, there exists $S^F \in \mathcal{N}$ such that $\mu_{S^F}^F > 0$. From the Kuhn and Tucker Conditions we get $\sum_{j \in S^F} \alpha_j \geq \mu_{S^F}^F > 0$, and hence there exists $i \in S^F$ such that $\alpha_i > 0$. We end the proof by showing that, for each $S \notin \mathcal{I}^F(x^F), i \in S$, one has $\lambda_S^F = 0$. In view of the Kuhn and Tucker's Conditions, it is sufficient to show that $\mu_S^F - \sum_{j \in S} \alpha_j < 0$. Since $S \notin \mathcal{I}^F(x^F)$, from (4′) we know that $\mu_S^F = 0$. Recalling that for every $j \in N$ we have $\alpha_j \geq 0$, we obtain $\mu_S^F - \sum_{j \in S} \alpha_j \leq -\alpha_i < 0$. This ends the proof of Claim 3.1. □

3.3 The end of the proof of $\mathcal{C}(\mathcal{E}) \neq \emptyset$

From Lemma 3.1, for each $F \in \mathcal{F}$, there exists $x^F = (x_1^F, \ldots, x_n^F) \in \mathcal{A} \cap F^n \subset \mathcal{A}$ such that $\mathcal{I}^F(x^F) = \{S \in \mathcal{N} \mid [\mathcal{A}(S) \cap F^{\text{card}S}] \cap P_S^F(x^F) \neq \emptyset\}$ is empty.

Now the collection $\mathcal{F}$, ordered by inclusion, is directed. Since $\mathcal{A}$ is weakly compact, by passing to subnets if necessary, we can assume that $(x^F)_{F \in \mathcal{F}}$ weakly converges to some vector $\bar{x} \in \mathcal{A}$. We now show that $\bar{x}$ belongs to $\mathcal{C}(\mathcal{E})$, i.e. $\mathcal{I}(\bar{x}) = \{S \in \mathcal{N} \mid \mathcal{A}(S) \cap P_S(\bar{x}) \neq \emptyset\}$ is empty. By contraposition, let suppose that there exists $S \in \mathcal{N}$ and $(x_i)_{i \in S} \in \mathcal{A}(S) \cap P_S(\bar{x})$. Since for all $i \in N$, x_i^F converges to $\bar{x}_i$, and P_i has (weakly) open lower sections, there exists $F_0 \in \mathcal{F}$ such that $F \supset F_0$ implies $(x_i)_{i \in S} \in P_S(x^F)$. Choose $F_1 \supset F_0$ such that $(x_i)_{i \in S} \in \mathcal{A}(S) \cap F^{\text{card}S}$, for all $F \supset F_1$. Then, it holds that $(x_i)_{i \in S} \in [\mathcal{A}(S) \cap F^{\text{card}S}] \cap P_S^F(x^F)$ for each $F \in \mathcal{F}, F \supset F_1$. This implies that $\mathcal{I}^F(x^F)$ is nonempty, this is a contradiction. This ends the proof of Theorem 2.1. □

3.4 An extension to the production economies

To describe the production sector of the economy, we use a similar approach to the one developed by Boehm [7]. Each coalition $S \in \mathcal{N}$ has a production set $Y_S \subset L^1(\Omega, \mathbf{F}, \mu; Z)$. The set Y_S will reflect all features which are related to the ability of coalition S to make certain net output bundles available to the members of S through a joint action. Apart from purely technological determinants reflecting the technical knowledge of the coalition S, Y_S will also be determinated by organizational and institutional features inherent to the coalition S. Hence, for the general case, the technology of the economy will be given by the collection of nonempty subsets $T = (Y_S)_{S\in\mathcal{N}}$ of the commodity space $L^1(\Omega, \mathbf{F}, \mu; Z)$. The coalitional production economy with differential information that we consider is described by the collection $\mathcal{E}^T = ((X_i, e_i, P_i, \mathbf{F}_i)_{i=1}^n, T)$.

Definition 3.1 *The allocation $x^* = (x_i^*)_{i\in N} \in \prod_{i\in N} X_i$ is attainable for the coalition $S \in \mathcal{N}$ if there exists a production plan $y_S \in Y_S$ such that*

(1S) $\sum_{i\in S} x_i^*(\omega) = \sum_{i\in S} e_i(\omega) + y_S(\omega)$ *μ-a.e.,*
(2S) for every $i \in S$, $x_i^* - e_i$ is $\mathbf{F}_i$-measurable,
(3S) y_S is $(\vee_{i\in S}\mathbf{F}_i)$-measurable[8].

We denote by $\mathcal{A}^T(S)$ the set of the attainable allocations of the coalition S. As previously, the set $\mathcal{A}^T$ of the attainable allocations of the economy $\mathcal{E}^T$ is the set $\mathcal{A}^T(N)$ of the attainable allocations of the grand coalition N.

Since the choice of a production plan is considered as a joint or collective decision, it seems to be natural that the production plans for a coalition have to be compatible with the pooled information of all the agents of the coalition-Condition (3S) above.

Now, as in Definition 2.2, the allocation $x^* = (x_i^*)_{i\in N} \in \prod_{i\in N} X_i$ is *improved upon by the coalition* $S \in \mathcal{N}$ if there exists $(x_i)_{i\in S} \in \mathcal{A}^T(S)$ such that $x_i \in P_i(x^*)$ for all $i \in S$.

Hence, an attainable allocation belongs to the core $\mathcal{C}(\mathcal{E}^T)$ of the coalitional production economy with differential information $\mathcal{E}^T$, if it is not possible for agents to join a coalition, reallocate their endowments among themselves (while each member of the coalition uses his/her own private information-Condition (2S)) or use those for production (while the coalition uses the pool of the information of its member-Condition (3S)) and obtain a strictly preferred allocation for each member of the coalition.

We now give the assumption which describes the technology and establish the nonemptiness result. Since the below assumptions do not imply $\mathcal{A}^T(S) \neq \emptyset$, for all $S \in \mathcal{N}$, we consider in the following the set $\mathcal{S} := \{S \in \mathcal{N} \mid \mathcal{A}^T(S) \neq \emptyset\}$.

Assumption P [Production side]

- *$Y_N \subset L^1(\Omega, \mathbf{F}, \mu; Z)$ is convex;*

[8] For all $S \in \mathcal{N}$, the term $\vee_{i\in S}\mathbf{F}_i$ denotes the "join", i.e., the minimal sub-σ-field (or partition) containing all $\mathbf{F}_i$.

- *[balancedness] the technology* $T^{\mathcal{S}} := (Y_S)_{S \in \mathcal{S}}$ *is balanced, i.e. for every balanced* [9] *family* β *of coalitions of* $\mathcal{S}$ *and associated weights* $(\lambda_S)_{S \in \beta}$*, one has* $\sum_{S \in \beta} \lambda_S Y_S \subset Y_N$.

Theorem 3.1 *If the coalitional production economy* $\mathcal{E}^T = ((X_i, e_i, P_i, \mathbf{F}_i)_{i=1}^n, T)$ *satisfies Assumptions* C1,C2,S *and* P*, then the core* $\mathcal{C}(\mathcal{E}^T)$ *is nonempty.*

Proof of Theorem 3.1. Recall that one has $\mathcal{C}(\mathcal{E}^T) = \{x \in \mathcal{A}^T \mid \forall S \in \mathcal{S}, \mathcal{A}^T(S) \cap P_S(x) = \emptyset\}$. Consequently, the attainable allocation $x \in \mathcal{A}^T$ belongs to $\mathcal{C}(\mathcal{E}^T)$ if and only if $\mathcal{I}(x) = \emptyset$ with $\mathcal{I}(x) = \{S \in \mathcal{S} \mid \mathcal{A}^T(S) \cap P_S(x) \neq \emptyset\}$. Now, it suffices to adapt the proof of Theorem 2.1. In fact, it consists essentially in showing that the correspondence ϕ^1 of Section 3.2.1 satisfies $\phi^1([\mathcal{A}^T \cap F^n] \times \prod_{S \in \mathcal{S}} [\mathcal{A}^T(S) \cap F^{\mathrm{card} S}] \times \Delta \times \Sigma) \subset \mathcal{A}^T \cap F^n$ which comes from Assumption P. □

4 Concluding remarks

4.1 Related literature

We will now show how we can obtain the results of Allen ([4] Prop.6.2), Koutsougeras-Yannelis ([13] Thm.3.1) and Yannelis ([19] Thm.3.1) as corollaries of our Theorem 2.1.

Firstly, it is clear that Definition 2.2 coincides with Definition 6.1 in Allen [4], Definition 3.1.1 in Yannelis [19] and Definition 3.1 in Koutsougeras-Yannelis [13] (who assume that e_i is $\mathbf{F}_i$-measurable). Indeed, we have to consider for all $i \in N$ the consumption set $X_i := \{x_i \in L^1(\Omega, \mathbf{F}, \mu; Z_+) \mid x_i(\omega) \in X_i(\omega)\ \mu - a.e.\}$[10] where Z_+ is the positive cone of Z and Z has an order continuous norm; and the preference correspondence $P_i : \prod_{i \in N} X_i \to X_i$ defined from the (ex-ante or interim) expected utility[11].

Secondly, we assume weakest assumptions on the preferences $(P_i)_{i \in N}$, the initial endowments $(e_i)_{i \in N}$ and the consumption sets $(X_i)_{i \in N}$ than those of the above nonemptiness results. Indeed, let fix $i \in N$.

Firstly, the convexity of X_i comes from the fact that the correspondences $X_i(\cdot) : \Omega \to 2^{Z_+}$ are convex valued.

Moreover, P_i is convex valued. It comes from the concavity of the utility $u_i(\omega, \cdot) : X_i \to \mathbb{R}$ for each $\omega \in \Omega$ in Yannelis [19] and in Allen [4], and of the utility $u_i : Z_+ \to \mathbb{R}$ in Koutsougeras-Yannelis [13]. Hence, by the definition of P_i, it then holds that for all $x = (x_i)_{i \in N} \in \mathcal{A}$, $x_i \notin \mathrm{co} P_i(x) = P_i(x)$.

---

[9] The family $\beta \subset \mathcal{S}$ of coalitions is balanced if for each $S \in \beta$ there exists $\lambda_S \geq 0$, such that $\sum_{\{S \in \beta \mid i \in S\}} \lambda_S = 1$, for all $i \in N$.

[10] In Allen [4], one has $Z = \mathbb{R}^\ell$ and $X_i(\omega) = [0, \sum_{k \in N} e_k(\omega)]$.

[11] More precisely, with the notations of the authors we have for all $x^* = (x_i^*)_{i \in N} \in \prod_{i \in N} X_i$:
- $P_i(x^*) := \{x_i \in X_i \mid V_i(\omega, x_i) > V_i(\omega, x_i^*), \mu - a.e.\}$ in Yannelis [19];and
- $P_i(x^*) := \{x_i \in X_i \mid \int u_i(x_i(\omega); \omega) d\mu(\omega) > \int u_i(x_i^*(\omega); \omega) d\mu(\omega)\}$ in Koutsougeras-Yannelis [13] and in Allen [4].

Now, P_i has open lower sections. Indeed, in Allen [4] it comes from the fact that $u_i : \mathbb{R}^\ell_+ \times \Omega \to \mathbb{R}$ is concave and upper-semicontinuous on $\mathbb{R}^\ell_+$, and μ-integrably bounded on $X_i \times \Omega$ (via Prop.3.2 of [4]). And since in Yannelis [19] for each $\omega \in \Omega$ the utility $u_i(\omega, \cdot)$ is weakly continuous and integrably bounded, and in Koutsougeras-Yannelis [13] the utility u_i is continuous integrably bounded and concave, it holds that P_i has a weakly open graph and so open lower sections. Hence, Assumption C1 is satisfied.

Furthermore, the set X_i is strongly closed. It is obvious in Allen [4]. And it comes from the fact that the correspondences $X_i(\cdot) : \Omega \to 2^{Z_+}$ are closed valued in Yannelis [19] and Koutsougeras-Yannelis [13] (see Lemma 5.1 of Appendix). Now, the convex and strongly closed set X_i is clearly bounded from below by $0 \in L^1(\Omega, \mathbf{F}, \mu; Z)$. Hence, by Lemma 5.2 of Appendix, $\mathcal{A}$ is weakly compact. So Assumption C2 is satisfied.

Endly, Assumption S is satisfied since for each $i \in N$, the initial endowment e_i belongs to X_i and so to $\mathcal{A}(\{i\}) = \{x_i \in X_i \mid x_i(\omega) = e_i(\omega)\ \mu - a.e. \text{ and } x_i - e_i \text{ is } \mathbf{F}_i-\text{measurable}\}$.

4.2 *Other concepts of core*

Let note that, with the techniques of the proof of Theorem 2.1, we may also prove the nonemptiness of the *coarse core* and the *weak fine core* of an exchange economy with differential information as defined in Koutsougeras-Yannelis [13]. Indeed, let consider the initial informations $(\mathbf{F}_i)_{i \in N}$ as partitions of Ω, and choose the consumption sets X_i and the preference correspondences P_i as defined in Section 4.1 for the link with the *private core* of [13].

Then, we obtain the *coarse core* (Definition 3.2 [13]) if for each $S \neq N$, we replace Condition $(2S)$ of the set $\mathcal{A}(S)$ by: $(2S')$ for all $i \in S$, $x_i - e_i$ is $(\cap_{i \in S} \mathbf{F}_i)$-measurable. Now since, the *private core* is a subset of the *coarse core*, the assumptions of the nonemptiness results of the private core and the coarse core are exactly the same, and the nonemptiness result of the *private core* (Theorem 3.1 [13]) is a corollary of Theorem 2.1 (confer Section 4.1), it holds that the nonemptiness result of the *coarse core* (Theorem 3.2 [13]) is a corollary of Theorem 2.1.

Now, we obtain the *weak fine core* (Definition 3.3(i') [13]) by replacing for all $S \in \mathcal{N}$, Condition $(2S)$ in $\mathcal{A}(S)$ by: $(2S')$ for every $i \in S$, $x_i - e_i$ is $(\vee_{i \in S} \mathbf{F}_i)$-measurable. One may then check that the proof of the nonemptiness of $\mathcal{C}(\mathcal{E})$ (Section 3) remains valid with the above sets $\mathcal{A}(S)$. Furthermore, the assumptions of the nonemptiness result of the *weak fine core* (Theorem 3.3 [13]) (which are exactly the same than in Theorem 3.1 [13]) imply Assumptions C1,C2,S of Theorem 2.1 (confer Section 4.1). Consequently, we may thus provide a nonemptiness result of the *weak fine core* with weakest assumptions, and an alternative proof based on a fixed point argument.

Let furthermore remark that the techniques allow also to obtain the nonemptiness of the core defined from general information-rules due to Allen [3]. The proof of Theorem 2.1 remains valid [12].

5 Appendix

5.1 Mathematical definitions

• Let $(\Omega, \mathbf{F}, \mu)$ be a finite measure space and Z be a Banach space. Following Diestel-Uhl (1977) [9], a function $f : \Omega \to Z$ is called *simple* if there exist $x_1, x_2, \ldots, x_k$ in Z and $E_1, E_2, \ldots, E_k$ in $\mathbf{F}$ such that $f = \sum_{i=1}^{k} x_i \chi_{E_i}$, where $\chi_{E_i}(\omega) = 1$ if $\omega \in E_i$ and $\chi_{E_i}(\omega) = 0$ if $\omega \notin E_i$. Then, the integral $\int_\Omega f(\omega) d\mu(\omega)$ is defined by $\int_\Omega f(\omega) d\mu(\omega) = \sum_{i=1}^{k} x_i \mu(E_i)$; and for each $E \in \mathbf{F}$, the integral $\int_E f(\omega) d\mu(\omega)$ is defined by $\int_E f(\omega) d\mu(\omega) = \sum_{i=1}^{k} x_i \mu(E_i \cap E)$. A function $f : \Omega \to Z$ is called *μ-measurable* if there exists a sequence of simple functions $f_n : \Omega \to Z$ such that $\lim_{n \to +\infty} \|f_n(\omega) - f(\omega)\| = 0$ for μ-almost all $\omega \in \Omega$. A μ-measurable function $f : \Omega \to Z$ is called *Bochner integrable* if there exists a sequence of simple functions $f_n : \Omega \to Z$ such that $\lim_{n \to \infty} \int_\Omega \|f_n(\omega) - f(\omega)\| d\mu(\omega) = 0$. In this case, for each $E \in \mathbf{F}$, the integral $\int_E f(\omega) d\mu(\omega)$ is defined by $\int_E f(\omega) d\mu(\omega) = \lim_{n \to \infty} \int_E f_n(\omega) d\mu(\omega)$.

Theorem (Thm.2 p.45 of Diestel-Uhl [9]) *A μ-measurable function $f : \Omega \to Z$ is Bochner integrable if and only if $\int_\Omega \|f(\omega)\| d\mu(\omega) < \infty$.*

For $1 \leq p < \infty$, we note $L^p(\Omega, \mathbf{F}, \mu; Z)$ the space of equivalence classes of μ-Bochner integrable functions $f : \Omega \to Z$ such that $\|f\|_p = (\int_\Omega (\|f(\omega)\|_Z)^p d\mu(\omega))^{1/p}$
$< \infty$. Normed by the functional $\|.\|_p$ defined above, $L^p(\Omega, \mathbf{F}, \mu; Z)$ is a Banach space (see Diestel-Uhl [9]).

• We will close this section by collecting some basic results on Banach lattices (for an excellent treatment see Aliprantis-Burkinshaw [1]). Recall that a *Banach lattice* is a Banach space L equipped with an order relation $\geq$ (i.e., $\geq$ is a reflexive, antisymmetric and translative relation) satisfying the below conditions (1)–(4):

(1) for all $x, y \in L$, $x \geq y$ implies $x + z \geq y + z$ for every $z \in L$;
(2) for all $x, y \in L$, $x \geq y$ implies $\alpha x \geq \alpha y$ for every $\alpha \geq 0$;
(3) for all $x, y \in L$, there exist a supremum (least upper bound) $x \vee y$ and an infimum (greatest below bound) $x \wedge y$;

As usual we define $x^+ = x \vee 0, x^- = (-x) \vee 0, |x| = x \vee (-x) \equiv x^+ + x^-$; we call x^+, x^- respectively the *positive* and *negative parts* of x, and $|x|$ the *absolute value* of x. The symbol $\| \cdot \|$ denotes the norm on L. Then,
(4) $|x| \geq |y|$ implies $\|x\| \geq \|y\|$ for all $x, y \in L$.

[12] It can be found in a previous version of this paper entitled "A general nonemptiness result of the core of a production economy with asymmetric information" in *Cahiers de la MSE*, Université Paris 1, France 1999.

If x, z are elements of the Banach lattice L, we define the *order interval* $[x, z]$ as follows: $[x, z] := \{y \in L \,|\, x \leq y \leq z\}$. Note that $[x, z]$ is norm closed and convex (hence weakly closed).

A Banach lattice L is said to have an *order continuous norm* if, $x_n \downarrow 0$[13] in L implies $\|x_n\| \downarrow 0$. If L is a Banach lattice then, the fact that L has an order continuous norm is equivalent to the weak compactness of the order interval $[x, z]$ for every $x, z \in L$ (see for instance Thm.2.3.8 of Aliprantis-Brown-Burkinshaw [2]).

We endly recall Cartwright's Theorem [8]: *If Z is a Banach lattice with an order continuous norm, then the order intervals of $L^1(\Omega, \mathbf{F}, \mu; Z)$ are weakly compact.*

5.2 Intermediary results

Claim 5.1 *$\mathcal{A}$ contains $(a_i)_{i \in N} \in \prod_{i \in N} \mathcal{A}(\{i\})$ and is a convex and weakly compact subset of $(L^1)^n$.*

Proof of Claim 5.1. First recall that $\mathcal{A} = \mathcal{A}(N)$. The set $\mathcal{A}$ obviously contains the allocation $(a_i)_{i \in N} \in \prod_{i \in N} X_i$, where $a_i \in \mathcal{A}(\{i\})$ for all $i \in N$. Moreover, $\mathcal{A}$ is also convex from the convexity of X_i for all $i \in N$, and because the operation of taking convex combinations preserves equality (1) and measurability (2). Endly, $\mathcal{A}$ is weakly compact by Assumption C2. □

Claim 5.2 *For all $S \in \mathcal{N}, \overline{\text{co}}\mathcal{A}(S)$ is a convex and weakly compact subset of $(L^1)^{\text{card}S}$.*

Proof of Claim 5.2. Fix $S \in \mathcal{N}$. Let recall that one has:

$$\mathcal{A}(S) = \left\{ (x_i)_{i \in S} \in \prod_{i \in S} X_i \;\middle|\; \begin{array}{l} (1S) \sum_{i \in S} x_i(\omega) = \sum_{i \in S} e_i(\omega)\ \mu - a.e. \\ (2S)\, \forall\, i \in S, x_i - e_i \text{ is } \mathbf{F}_i - \text{measurable} \end{array} \right\}.$$

Let first show that $\mathcal{A}(S)$ is a subset of the projection of $\mathcal{A} \subset (L^1)^n$ on $(L^1)^{\text{card}S}$, noted $p_S(\mathcal{A})$. Indeed, let us consider $(x_i)_{i \in S}$ in $\mathcal{A}(S)$. Then, one has (1S) $\sum_{i \in S} x_i(\omega) = \sum_{i \in S} e_i(\omega)$ μ-a.e.; (2S) for all $i \in S$, $x_i - e_i$ is $\mathbf{F}_i$- measurable.

Let now consider the allocation $(x'_i)_{i \in N} \in \prod_{i \in N} X_i$ defined by $x'_i := x_i$ if $i \in S$, and $x'_i \in \mathcal{A}(\{i\})$ if $i \notin S$, i.e. satisfying $x'_i = e_i$ μ-a.e. and $x'_i - e_i$ is $\mathbf{F}_i$-measurable. We then obtain from above: (1) $\sum_{i \in N} x'_i(\omega) = \sum_{i \in S} x_i(\omega) + \sum_{i \notin S} e_i(\omega) = \sum_{i \in N} e_i(\omega)$ for μ-almost $\omega \in \Omega$; and (2) for all $i \in N$, $x'_i - e_i$ is $\mathbf{F}_i$-measurable.

This means that $(x'_i)_{i \in N}$ belongs to $\mathcal{A} \subset \prod_{i \in N} X_i$. Consequently, $(x_i)_{i \in S} = (x'_i)_{i \in S} = p_S((x'_i)_{i \in N})$ belongs to $p_S(\mathcal{A})$. Endly, since $(x_i)_{i \in S} \in \mathcal{A}(S)$ is taken arbitrary, it finally holds $\mathcal{A}(S) \subset p_S(\mathcal{A})$.

Now, since $\mathcal{A}$ is convex and weakly compact, it holds that $p_S(\mathcal{A})$ is convex and weakly compact, so weakly closed and then strongly closed. Hence, one then has $\overline{\text{co}}\mathcal{A}(S) \subset p_S(\mathcal{A})$. Now, since $\overline{\text{co}}\mathcal{A}(S)$ is convex and strongly closed in the

[13] The notation $x_n \downarrow x$ means that x_n is decreasing (formally, whenever $q \geq p$ implies $x_q \leq x_p$; in symbols, $x_n \downarrow$) and inf $x_n = x$ both hold.

locally convex space $(L^1)^{\mathrm{card}S}$, it is weakly closed (see Thm.3.12 of [16]). Finally, $\overline{\mathrm{co}}\mathcal{A}(S)$ is weakly compact as a weakly closed subset of the weakly compact set $p_S(\mathcal{A})$. □

Lemma 5.1 *Let Z be a Banach space and $(\Omega, \mathbf{F}, \mu)$ a measure space with μ a σ-additive measure on $(\Omega, \mathbf{F})$. If $G : \Omega \to Z$ is a closed valued correspondence, then the set $X = \{g \in L^1(\Omega, \mathbf{F}, \mu; Z) \mid g(\omega) \in G(\omega)\ \mu - a.e.\}$ is strongly closed (i.e. for the L^1 norm topology).*

Proof of Lemma 5.1. To simplify the notations, we note L^1 for $L^1(\Omega, \mathbf{F}, \mu; Z)$ in the following. Let consider a sequence of functions (g^n) of X which converges in L^1 to $\bar{g} \in L^1$. Then, by Theorem III.3.6 of [10], (g^n) converges to $\bar{g}$ in μ-measure. Hence, by Corollary III.6.3 of [10], (g^n) admits a subsequence $(g^{\phi(n)})$ which converges to $\bar{g}$ μ-uniformly, and so ponctually μ-almost everywhere. Thus, one has for μ-almost $\omega \in \Omega$, $g^{\phi(n)}(\omega) \to \bar{g}(\omega)(n \to \infty)$. But moreover, one has for all $n \in \mathbb{N}$, $g^{\phi(n)}(\omega) \in G(\omega)$ μ-a.e., where $G(\omega)$ is closed. Hence, it holds for μ-almost $\omega \in \Omega, \bar{g}(\omega) \in G(\omega)$. This finally means that $\bar{g}$ belongs to X, and implies that X is strongly closed. □

Lemma 5.2 *Let Z be a Banach lattice with an order continuous norm*[14]. *In an exchange economy with differential information $\mathcal{E} = (X_i, e_i, P_i, \mathbf{F}_i)_{i=1}^n$ let us suppose that for all $i \in N$, $X_i \subset L^1(\Omega, \mathbf{F}, \mu; Z)$ is convex, bounded from below*[15] *and strongly (i.e. for the L^1 norm topology) closed. Then, the set $\mathcal{A}$ of attainable allocations of $\mathcal{E}$ is a weakly compact subset of $(L^1(\Omega, \mathbf{F}, \mu; Z))^n$.*

Proof of Lemma 5.2. To simplify the notations, we note L^1 for $L^1(\Omega, \mathbf{F}, \mu; Z)$. For all $i \in N$, since X_i is bounded from below, let consider $b_i \in L^1$ such that for all $x_i \in X_i, x_i(\omega) \geq b_i(\omega)\ \mu - a.e.$ Recall that one has

$$\mathcal{A} = \left\{ (x_i)_{i \in N} \in \prod_{i \in N} X_i \;\middle|\; \begin{array}{l} (1)\ \sum_{i \in N} x_i(\omega) = \sum_{i \in N} e_i(\omega)\ \ \mu - a.e. \\ (2)\ \forall\, i \in N,\ x_i - e_i \text{ is } \mathbf{F}_i - \text{measurable} \end{array} \right\}.$$

Firstly, the set $\mathcal{A}$ is clearly convex since X_i is convex for all $i \in N$ and because the operation of taking convex combinations preserves equality (1) and measurability (2). Now, from its definition and since for all $i \in N$, X_i is strongly (i.e., for the L^1 norm topology) closed in the locally convex space L^1, the set $\mathcal{A}$ is strongly closed in the locally convex space $(L^1)^n$. Therefore, the convex set $\mathcal{A}$ is also weakly closed (see Thm.3.12 of [16]). Moreover, it is easily checked that one has $\mathcal{A} \subset \prod_{i \in N} I_i$ where for all $i \in N$, I_i is the order interval of L^1 defined by $I_i := [b_i, \sum_{i \in N} e_i - \sum_{\{j \in N \mid j \neq i\}} b_j]$. Furthermore from Cartwright's Theorem [8],

[14] Basic examples of Banach lattices Z with an order continuous norm are:
- the Euclidean space $\mathbb{R}^\ell$;
- the space l^p $(1 < p < \infty)$ of real sequences $(a_n)_{n \geq 1}$ for which the norm $||a_n||_p = (\sum_{n=1}^{\infty} |a_n|^p)^{1/p}$ is finite;
- the vector space $L^p(\nu)$ $(1 \leq p < \infty)$ of real-valued ν-measurable functions f on X such that $\int_X |f|^p d\nu < \infty$, where (X, σ, ν) is a measure space.
(see Aliprantis-Burkinshaw [1])

[15] This means that there exists $g \in L^1(\Omega, \mathbf{F}, \mu; Z)$ such that for all $f \in X_i, f(\omega) \geq g(\omega)\ \mu - a.e.$

since Z is a Banach lattice with an order continuous norm, it holds that I_i is weakly compact. This finally implies that $\mathcal{A}$ is weakly compact as a weakly closed subset of the weakly compact set $\prod_{i \in N} I_i$. □

References

1. Aliprantis, C.D., Burkinshaw, O.: Positive operators. New York: Academic Press 1985
2. Aliprantis, C.D., Brown, D., Burkinshaw, O.: Existence and optimality of competitive equilibria. Berlin Heidelberg New York: Springer 1989
3. Allen, B.: Market games with asymmetric information and nontransferable utility: representation results and the core. CARESS Working Paper 91–09, University of Pennsylvania (1991)
4. Allen, B.: Market games with asymmetric information: the private information core. CARESS Working Paper 92–04, University of Pennsylvania (1992)
5. Balder, E.J., Yannelis, N.C.: On the continuity of expected utility. Economic Theory **3**, 625–643 (1993)
6. Berge, C.: Espaces topologiques, fonctions multivoques. Paris: Dunod 1959
7. Boehm, V.: The core of an economy with production. The Review of Economic Studies **41**, 429–436 (1974)
8. Cartwright, D.: The order completeness of some spaces of vector-valued functions. Bulletin of the Australian Mathematical Society **11**, 57–61 (1974)
9. Diestel, J., Uhl, J.J.: Vector measures . In: Mathematical surveys, vol.15. Providence, RI: American Mathematical Society 1977
10. Dunford, N., Schwartz, J.: Linear operators. Part I: General theory. New York: Interscience Publishers 1967
11. Florenzano, M.: On the nonemptiness of the core of a coalitional production economy without ordered preferences. Journal of Mathematical Analysis and Applications **141**, 484–490 (1989)
12. Gourdel, P.: Existence of intransitive equilibria in nonconvex economies. Set Valued Analysis **3** (4), 307–337 (1995)
13. Koutsougeras, L., Yannelis, N.C.: Incentive compatibility and information superiority of the core of an economy with differential information. Economic Theory **3**, 195–216 (1994)
14. Lefebvre, I.: A quasi fixed-point theorem for a product of u.s.c or o.l.s correspondences with an economic application. Cahiers de la MSE, Université de Paris 1, France (1999)
15. Page, F.H.: Market games with differential information and infinite dimensional commodity spaces: the core. Economic Theory **9**, 151–159 (1997)
16. Rudin, W.: Functional analysis. New York: McGraw-Hill 1973
17. Scarf, H.: The core of an n-person game. Econometrica **35**, 50–69 (1967)
18. Schwalbe, U.: The core of a production economy with asymmetric information. University of Mannheim, Department of Economics, Mannheim (1996)
19. Yannelis, N.C.: The core of an economy with differential information. Economic Theory **1**, 183–198 (1991)
20. Yannelis, N.C., Prabhakar, N.D.: Existence of maximal elements and equilibria in linear topological spaces. Journal of Mathematical Economics **12**, 233–245 (1983)

On the continuity of expected utility★

Erik J. Balder[1] **and Nicholas C. Yannelis**[2]

[1] Mathematical Institute, University of Utrecht, P.O. Box 80.010, 3508 TA Utrecht, NETHERLANDS
[2] Department of Economics, University of Illinois, Champaign, IL 61820, USA

Received: April 29, 1992; revised version November 5, 1992

Summary. We provide necessary and sufficient conditions for weak (semi)continuity of the expected utility. Such conditions are also given for the weak compactness of the domain of the expected utility. Our results have useful applications in cooperative solution concepts in economies and games with differential information, in noncooperative games with differential information and in principal-agent problems.

1 Introduction

Recent work on cooperative solution concepts in economies and games with differential information (e.g. Yannelis [25], Krasa-Yannelis [16], Allen [2,3], Koutsougeras-Yannelis [17], Page [22]) has necessitated the consideration of conditions that guarantee the (semi)continuity of an agent's expected utility.[1]

Specifically, in this paper $(\Omega, \mathcal{F}, P)$ is a probability space, representing the *states of the world* and their governing distribution, $(V, \|\cdot\|)$ a separable Banach space of *commodities*, and $X : \Omega \to 2^V$ a set-valued function, prescribing for each state w of the world the set $X(w)$ of possible *consumptions*. We define the set $\mathcal{L}^1_X$ of *feasible state contingent consumption plans* to consist of all Bochner integrable a.e. selections of X, that is, the set of all $x \in \mathcal{L}^1_V$ such that

$$x(\omega) \in X(\omega) \quad \text{a.e. in } \Omega.$$

As usual, $\mathcal{L}^1_V$ stands for the (prequotient) set of all Bochner-integrable V-valued functions on $(\Omega, \mathcal{F}, P)$; the $\mathcal{L}^1$-*seminorm* on this space is defined by

$$\|x\|_1 := \int_\Omega \|x(\omega)\| P(d\omega).^2$$

★ Work done while visiting the Department of Economics, University of Illinois at Urbana-Champaign.

[1] This problem also arises naturally in principal-agent problems (see for example Page [19, 21]) and Kahn [14], as well as in noncooperative games with differential information (see for example, Yannelis-Rustichini [27]).

Let $U : \Omega \times V \to [-\infty, +\infty)$ be a given *utility* function. Then the *expected utility* $I_U(x)$ of a consumption plan $x \in \mathcal{L}^1_X$ is given by

$$I_U(x) := \int_\Omega U(\omega, x(\omega)) P(d\omega),$$

assuming that this integral exists. Clearly, if for each $\omega \in \Omega$ the function $U(\omega, \cdot)$ is norm-continuous and if U is integrably bounded, then the $\mathcal{L}^1$-seminorm-continuity of I_U would follow directly from Lebesgue's dominated convergence theorem [11]. However, the corresponding $\mathcal{L}^1$-compactness of $\mathcal{L}^1_X$, on which I_U is defined, is only found under quite heavy conditions, even when X has only *finite* sets as its values:

Example 1.1. *Consider for $(\Omega, \mathcal{F}, P)$ the unit interval cum Lebesgue measure. Let the consumption set $X(\omega)$ be $\{-1, +1\}$ for all ω. Then the sequence (x_k) of* Rademacher *functions $x_k : [0, 1] \to \{-1, +1\}$, defined by*

$$x_k(\omega) := \operatorname{sgn}(\sin(2\pi k\omega)),$$

forms a sequence of consumption plans that does not contain any subsequence which converges in $\mathcal{L}^1$-seminorm; obviously, this implies that the set $\mathcal{L}^1_X$ cannot be compact for the $\mathcal{L}^1$-seminorm. Indeed, if such a subsequence did exist, the corresponding limit consumption plan would have to be a.e. equal to zero (note that $\int_B x_k \to 0$ for every interval $B := [\alpha, \beta]$; start by observing that when $\alpha, \beta \in [0, 1]$ have finite binary expansions this is trivial). But since $\|x_k\|_1 = \int_{[0,1]} |x_k(w)| dw = 1$ for all k, the $\mathcal{L}^1$-norm of the limit consumption plan would have to be equal to 1 at the same time.

Thus, in such situations the attainment of a maximum of the expected utility is not guaranteed. To this end stronger continuity conditions (viz. *weak* continuity in the second variable) must be imposed on U. The corresponding continuity found for I_U in this way is weak continuity. At the same time, imposing weak compactness upon the values of X yields weak compactness of the set $\mathcal{L}^1_X$ (Diestel's theorem [26]). Hence, in this situation attainment of the maximum of I_U is guaranteed.

The purpose of this paper is to investigate the necessary and sufficient conditions for the following properties:

- weak and strong (semi)continuity of I_U on $\mathcal{L}^1_x$,
- weak and strong closedness and weak compactness of $\mathcal{L}^1$.

In view of recent work on cooperative and noncooperative solution concepts in economies and games with differential information, as well as in principal-agent problems, an answer to the above question is of fundamental importance. For this enables us to prove – via the usual forms of analysis – the existence of value and core allocations in economies with differential information, as well as the existence of a correlated equilibrium in games with differential information. The techniques employed in this paper are mostly based on classical developments in the calculus of variations and optimal control theory.

[2] Since all principal results hold modulo sets of measure zero, one could alternatively work with the usual equivalence class structure. One consequence of choosing for the prequotient setup is, of course, that the L_1-norm is traded in for its seminorm analogue.

This paper is organized as follows: First, we state our principal results (Sect. 2), and their economic applications (Sect. 3). Our mathematical tools, their proofs, as well as all other proofs have been collected in Section 4. Some notation to be used below is as follows: V^* stands for the topological dual space of $(V, \|\cdot\|)$. As usual $\|\cdot\|^*$ stands for the dual norm on V^* [i.e., $\|x^*\|^* := \sup\{\langle x, x^*\rangle : x \in V, \|x\| \leq 1\}$, where $\langle x, x^*\rangle := x^*(x)$.

2 Main results

Let us observe that the probability space $(\Omega, \mathcal{F}, P)$ can always be decomposed into an atomless part Ω_1 and a countable union Ω_2 of atoms. Let $U : \Omega \times V \to [-\infty, +\infty)$ be a given *utility function*, which we suppose to be $\mathcal{F} \times \mathcal{B}(V)$-measurable; here $\mathcal{B}(V)$ stands for the Borel σ-algebra on $(V, \|\cdot\|)$. The *expected utility* functional I_U on $\mathcal{L}^1_V$ is given by

$$I_V(x) := \int_\Omega U(\omega, x(\omega))P(d\omega),$$

where we use the following *convention* regarding the integration of any measurable function $\phi : \Omega \to [-\infty, +\infty] : \int \phi := \int \phi^+ - \int \phi^-$, with $+\infty - +\infty := -\infty$. Let $X : \Omega \to 2^V$ be a given set-valued function; we imagine the *consumption* set $X(\omega)$ to comprise all feasible (e.g., budgetary) consumption plans under the state of nature ω. The graph of X is supposed to be $\mathcal{F} \times \mathcal{B}(V)$-measurable. We define the set $\mathcal{L}^1_X$ of all integrable state contingent *consumption plans* by

$$\mathcal{L}^1_X := \{x \in \mathcal{L}^1_V : x(\omega) \in X(\omega)P\text{-a.e. in } \Omega\}.$$

We distinguish between *strong* and *weak* (semi)continuity of the expected utility functional I_U on $\mathcal{L}^1_X$. The first kind of continuity is with respect to the seminorm $\|\cdot\|_1$ (see Sect. 1), and the second kind of continuity is with respect to the weak topology $\sigma(\mathcal{L}^1_V, \mathcal{L}^\infty_{V^*}[V])$, restricted to $\mathcal{L}^1_x$. Here $\mathcal{L}^\infty_{V^*}[V]$ stands for the set of all functions $p : \Omega \to V^*$ that are bounded [i.e., $\sup_{\omega\in\Omega} \|p(\omega)\|^* < +\infty$ and V-scalarly measurable [i.e., $\omega \to \langle x, p(\omega)\langle$ is $\mathcal{F}$-measurable for every $x \in V$]. It is well-known that $\mathcal{L}^\infty_{V^*}[V]$ is the dual of $(\mathcal{L}^1_V, \|\cdot\|_1)[12, VI]$). Recall also that $\sigma(\mathcal{L}^1_V, \mathcal{L}^\infty_{V^*}[V])$ is defined as the weakest topology on $\mathcal{L}^1_V$ for which all functionals

$$x \mapsto \int_\Omega \langle x(\omega), p(\omega)\rangle P(d\omega), \quad p \in \mathcal{L}^\infty_{V^*}[V],$$

are continuous. In other words, this is the weakest topology that one could define for the consumption plans so that at least all the very simple utility functions of the type $U_p(\omega, x) := \langle x, p(\omega)\rangle, p \in \mathcal{L}^\infty_{V^*}[V]$, one would have the corresponding expected utility functionals $I_{U_p}(x)$ depend continuously upon the consumption plan variable x. With the same topologies in mind, we can also distinguish between strong and weak closedness of the set $\mathcal{L}^1_X$ of consumption plans. Similarly, on the commodity space V we make a distinction between the weak topology $\sigma(V, V^*)$ and the strong norm-topology (however, the corresponding σ-algebras on V coincide). Thus, we shall be considering two weak topologies and two strong topologies, respectively on the space $\mathcal{L}^1_V$ (and/or its subsets) and on the space V (and/or its subsets); from the context the reader can always deduce which space is intended.

The following *nontriviality hypothesis* will be adopted in this entire section:

$$\text{there exists at least one } \bar{x} \in \mathcal{L}^1_X \text{ with } -\infty < I_U(\bar{x}).$$

Of course, this hypothesis is extremely mild: it only prevents a completely trivial situation. On some occasions we shall require the only slightly more restrictive *strict nontriviality hypothesis*

$$\text{there exists at least one } \bar{x} \in \mathcal{L}^1_X \text{ with } -\infty < I_U(\bar{x}) < +\infty,$$

but when this reinforcement is needed, it will always be stated explicitly.

Our first result concerns a necessary and sufficient condition for the weak closedness of the set $\mathcal{L}^1_X$ of integrable consumption plans:

Theorem 2.1. *The following statements are equivalent.*

i. $X(\omega)$ is convex and closed a.e. in Ω_1, and weakly closed a.e. in Ω_2.[3]
ii. $\mathcal{L}^1_X$ is weakly closed.

By Mazur's theorem the adjective "closed" for a *convex* subset of V can be interpreted equivalently as weakly closed and as strongly closed; hence "convex and closed" above needs no further specification.

Our second result is similar in nature, but now the strong closedness of the set of integrable consumption plans is addressed:

Theorem 2.2. *The following statements are equivalent.*

i. $X(\omega)$ is strongly closed a.e. in Ω,
ii. $\mathcal{L}^1_X$ is strongly closed.

In this connection it is useful to recall the following related result which has to do with weak compactness of the set of integrable consumption plans. The necessity part comes from [15, Thm. 3.6]; the sufficiency part in the above result – frequently referred to as Diestel's theorem – is better known (see for instance [26]). It has been refined in [8], using *K-convergence*, a Cesaro-type of pointwise convergence (for arithmetic averages).

Theorem 2.3 (Klei). *Suppose that the set $\mathcal{L}^1_X$ of integrable consumption plans is relatively weakly compact. Then*

$$X(\omega) \text{ is relatively weakly compact a.e. in } \Omega.$$

The converse implication holds also, provided that X is integrably bounded.

Recall here that the multifunction X is said to be *integrably bounded* if for some $\psi \in \mathcal{L}^1_{\mathbf{R}}$

$$\sup_{x \in X(\omega)} \|x\| \leq \psi(\omega) \text{ a.e. in } \Omega.$$

[3] Such condensed formulations are used throughout: we mean to say that for P-almost every $\omega \in \Omega_1$ the set $X(\omega) \subset V$ is convex and closed, etc.

Note that this additional condition is essential for the sufficiency part, as is shown by the following counterexample.

Example 2.4. *Consider $\Omega := (0,1)$, equipped with the Borel σ-algebra and the Lebesgue measure P. Define $X(\omega) := [0, 1/\omega]$. Then the the sequence $(x_k) \in \mathcal{L}^1_X$, defined by $x_k(\omega) := 1/\omega$ if $1/k \le \omega < 1$, and $x_k(\omega) := 0$ otherwise, does not have a convergent subsequence, since it is not even uniformly integrable.*

Corollary 2.5. *Suppose that the set$\mathcal{L}^1_x$ of integrable consumption plans is weakly compact. Then*

$$X(\omega) \text{ is convex and weakly compact a.a. in } \Omega_1,$$

$$X(\omega) \text{ is weakly compact a.e. in } \Omega_2.$$

The converse implication holds also, provided that X is integrably bounded.

Proof. Combine Theorems 2.1 and 2.3. □

It is interesting to observe that for the strong topologies the counterpart to the above result fails as far as the sufficiency part is concerned [15, p. 316], even if $V = \mathbf{R}$ (the necessity part has an analogue [15, Prop. 3.12]). Next, we occupy ourselves with necessary conditions for weak upper semicontinuity and weak continuity of the expected utility.

Theorem 2.6. *Suppose that the expected utility I_U is weakly upper semicontinuous and that the set $\mathcal{L}^1_X$ of all integrable consumption plans is weakly closed. Suppose also that for each of the countably many atoms $A \subset \Omega_2$ there exist constants $M_A, K_A > 0$ such that*

$$U(\omega, \cdot) \le K_A + M_A \|\cdot\| \text{ on } X(\omega) \text{ a.e. in } A.$$

Then

i. $U(\omega,\cdot)$ is concave and upper semicontinuous on the convex closed set $X(\omega)$ a.e. in Ω_1,
ii. $U(\omega,\cdot)$ is weakly upper semicontinuous on the weakly closed set $X(\omega)$ a.e. in Ω_2.

Corollary 2.7. *Suppose that the expected utility I_U is weakly continuous and that the set $\mathcal{L}^1_x$ of all integrable consumption plans is weakly closed. Suppose also that for each of the countably many atoms $A \subset \Omega_2$ there exists contains $M_A, K_A > 0$ such that*

$$|U(\omega)| \le K_A + M_A \|\cdot\| \text{ on } X(\omega) \text{ a.e. in } A.$$

Then, under the strict nontriviality hypothesis,

i. $U(\omega,\cdot)$ is affine and continuous on the convex closed set $X(\omega)$ a.e. in Ω_1,
ii. $U(\omega,\cdot)$ is weakly continuous on the weakly closed set $X(\omega)$ a.e. in Ω_2.

The corresponding sufficient conditions for weak upper semicontinuity and weak continuity of the expected utility are as follows:

Theorem 2.8. *Suppose that a.e. in Ω_1*

$$X(\omega) \text{ is convex and closed,}$$

$$U(\omega,\cdot) \text{ is concave and upper semicontinous on } X(\omega),$$

and

$$U(\omega,\cdot) \le \psi(\omega) + M\|\cdot\|$$

for some $M > 0$ and $\psi \in \mathcal{L}^1_{\mathbf{R}}$. Suppose further that a.e. in Ω_2

$$X(\omega) \text{ is weakly closed,}$$

$$U(\omega,\cdot) \text{ is weakly upper semicontinuous on } X(\omega).$$

Then I_U is weakly upper semicontinuous on the weakly closed set $\mathcal{L}^1_X$.

Corollary 2.9. *Suppose that a.e. in Ω_1*

$$X(\omega) \text{ is convex and closed,}$$

$$U(\omega,\cdot) \text{ is affine and continuous on } X(\omega),$$

and

$$|U(\omega,\cdot)| \le \psi(\omega) + M\|\cdot\|$$

for some $M > 0$ and $\psi \in \mathcal{L}^1_{\mathbf{R}}$. Suppose further that a.e. in Ω_2

$$X(\omega) \text{ is weakly closed,}$$

$$U(\omega,\cdot) \text{ is weakly continuous on } X(\omega).$$

Then I_U is weakly continuous on the weakly closed set $\mathcal{L}^1_X$.

For strong continuity of the expected utility we have the following characterization:

Theorem 2.10. *Suppose that there exists a constant $M > 0$ and $\psi \in \mathcal{L}^1_{\mathbf{R}}$ such that*

$$U(\omega,\cdot) \le \psi(\omega) + M\|\cdot\| \text{ on } X(\omega) \text{ a.e. in } \Omega.$$

Then the following statements are equivalent:

i. $U(\omega,\cdot)$ is strongly upper semicontinuous on the strongly closed set $X(\omega)$ a.e. in Ω_2,
ii. $\mathcal{L}^1_X$ is strongly closed and I_U is strongly upper semicontinuous on $\mathcal{L}^1_X$.

Corollary 2.11. *Suppose that there exist a constant $M > 0$ and $\psi \in \mathcal{L}^1_{\mathbf{R}}$ such that*

$$|U(\omega,\cdot)| \le \psi(\omega) + M\|\cdot\| \text{ on } X(\omega) \text{ a.e.in } \Omega_2.$$

Then, under the strict nontriviality hypothesis, the following statements are equivalent:

i. $U(\omega,\cdot)$ is strongly continuous on the strongly closed set $X(\omega)$ a.e. in Ω,
ii. $\mathcal{L}^1_X$ is strongly closed and I_U is strongly continuous on $\mathcal{L}^1_X$.

3 Applications

3.1 Market games with differential information

Consider an *exchange economy with differential information* $\mathcal{E} = \{(X_i, U_i, \mathcal{F}_i, e_i, P) : i \in I\}, I := \{1, ..., n\}$, where

i. $X_i : \Omega \to 2^V$ is a multifunction prescribing agent i's potential *consumption sets* [i.e., $X_i(\omega)$ is i's potential consumption set in state $\omega \in \Omega$],
ii. $U_i : c\Omega \times V \to \mathbf{R}$ is the state dependent *utility function* of agent i,
iii. $\mathcal{F}_i$ is a sub σ-algebra of $(\Omega, \mathcal{F}))$ denoting the *private information* of agent i about the state of nature,
iv. $e_i : \Omega \to V$ is the *initial endowment* of agent i, where e_i is $\mathcal{F}_i$-measurable and $e_i(\omega) \in X_i(\omega) P$-a.e.,
v. P is a probability measure on Ω representing the *common probability beliefs* of the players concerning states of nature.

Suppose that for the economy $\mathcal{E}$ the following assumptions hold for each $i \in I$:

$$X_i(\omega) \text{ is convex, nonempty and weakly compact a.e. in } \Omega, \tag{3.1}$$

$$X_i \text{ is integrably bounded,} \tag{3.2}$$

$$U_i(\omega, \cdot) \text{ is concave and upper semicontinuous on } X_i(\omega) \text{ a.e. in } \Omega, \tag{3.3}$$

$$U_i \text{ is integrably bounded from above.} \tag{3.4}$$

Note that if the commodity space V is assumed to be a Banach lattice with an order continuous norm (which implies that the order intervals are weakly compact [1]), then it is reasonable to assume that the state contingent consumption set $X_i(\omega)$ of each agent i is contained in the order interval $[0, e(\omega)]$; where $e(\omega) : \sum_{i \in I}(\omega)$.

In this case we may replace (3.1)–(3.2) with simple integrable boundedness of e_i for each agent i.

We will now indicate how our results can be used to prove the existence of a Shapley value allocation for an exchange economy with differential information (see for example [16]). For this one associates with the economy $\mathcal{E}$ the following game with side-payments: for each collection $\lambda := \{\lambda_1, ..., \lambda_n\}$ of nonnegative weights $\lambda_i, \sum_{i=1}^n \lambda_i = 1$, define the *side payment game* (I, V_λ) according to the following rule: for each *coalition* $S \in 2^I$, let

$$V_\lambda(S) := \sup_x \sum_{i \in S} \lambda_i \int_\Omega U_i(\omega, x_i(\omega)) P(d\omega),$$

where the supremum is taken over all $x := (x_i)_{i \in S}, x_i \in \mathcal{L}^1_{X_i}$, subject to

$$\sum_{i \in S} x_i(\omega) = \sum_{i \in S} e_i(\omega) \text{ a.e. in } \Omega.$$

Here $\mathcal{L}^1_{X_i}$ stands for the collection of all $x_i \in \mathcal{L}^1_V(\Omega, \mathcal{F}_i, , P)$ such that x_i is $\mathcal{F}_i$-measurable and $x_i(\omega) \in X_i(\omega)$ a.e. in Ω.

First, let us verify that the supremum above is actually attained, by the Weierstrass theorem. By Theorem 2.3 each $\mathcal{L}^1_{X_i}$ is weakly compact, $i \in S$; hence, so is their product. Since $x \mapsto \sum_{i \in S} x_i$ is obviously weakly continuous, we conclude that the above supremum is taken over a weakly compact set. Since each U_i satisfies the conditions in Theorem 2.8, each I_{U_i} is weakly upper semicontinuous, $i \in S$; hence, so is their sum. This proves the attainment of the supremum in the definition of the Shapley value of the game (I, V_λ). The above existence problem arises naturally if one wants either to prove the existence of a Shapley value allocation in an exchange economy with differential information or to show that a TU market game in characteristic function form is well-defined for such an economy (see for instance [25] or [2,3]).

We now examine an application to the core of an exchange economy with differential information. Following Yannelis [25], the *private core* of $\mathcal{E}$ is defined as follows. The vector $x \in \Pi_{i=1}^n \mathcal{L}^1_{X_i}$ is said to be a *private core allocation* for $\mathcal{E}$ if

i. $\sum_{i=1}^n x_i = \sum_{i=1}^n e_i$,
ii. there does not exist $S \subset I$ and $(y_i)_{i \in S} \in \Pi_{i \in S} \mathcal{L}^1_{X_i}$ such that $\sum_{i \in S} y_i = \sum_{i \in S} e_i$ and $I_{U_i}(y_i) > I_{U_i}(x_i)$ for all $i \in S$.

Following Shapley-Shubik [24], we may convert the economy $\mathcal{E}$ to a *market game* (V, I) as follows: Define $V : 2^I \to \mathbf{R}^n$ by

$$V(S) = \left\{ z \in \mathbf{R}^{|S|} : z_i \leq I_{U_i}(x_i), x_i \in \mathcal{L}^1_{X_i}, i \in S, \sum_{i \in S} x_i = \sum_{i \in S} e_i \right\};$$

here $|S|$ stands for the number of elements in S. For $S \in I$, clearly the set $V(S)$ is convex, nonempty and bounded from above. In view of Theorem 2.8 the function I_{U_i} is weakly upper semicontinuous; hence, $V(S)$ must be closed. Hence, the market game (V, I) is balanced, and has therefore a nonempty core (Scarf's theorem [23]). Standard arguments can now be applied to show that nonemptiness of the core of the game (V, I) implies nonemptiness of the core of the economy $\mathcal{E}$. Related arguments have been employed by Allen [3] to show nonemptiness of the private core of an economy with a finite-dimensional commodity space. Using the K-compactness of the $\mathcal{L}^1_{X_i}$ (as introduced in [8, Corollary 4.2]) and the sequential weak upper semicontinuity of expected utilities I_U, Page [22] has shown that the market game (V, I) corresponding to an exchange economy with an infinite dimensional commodity space is well-defined and balanced, and hence has a nonempty core.

3.2 Principal-agent contracting games with adverse selection

Consider a principal-agent contracting game $\mathcal{G} = \{T, X, U_1, U_2, P, Q\}$, where

i. $(T, \mathcal{T})$ is a measurable space of *agent types*,
ii. $X : \Omega \to 2^V$ prescribes the potential payoffs in each state of nature (i.e., $X(\omega)$ is the set of potential contract payoffs in state $\omega \in \Omega$),
iii. $U_1 : T \times \Omega \times V \to \mathbf{R}$ is *the principal's utility function*, type and state dependent,

iv. $U_2 : T \times \Omega \times V \to \mathbf{R}$ is *the agent's utility function*, again type and state dependent,
v. P is a probability measure on Ω, representing the principal's and the agent's common beliefs concerning states of nature,
vi. Q is a probability measure on T, representing the principal's probability beliefs concerning agent types.

Suppose that for the game $\mathcal{G}$ the following assumptions hold:

$$\text{the } \sigma\text{-algebra } \mathcal{F} \text{ is countably generated,} \tag{3.5}$$

$$X(\omega) \text{ is convex, nonempty and weakly compact a.e. in } \Omega, \tag{3.6}$$

$$X \text{ is lower measurable and integrably bounded.} \tag{3.7}$$

As a consequence, $\mathcal{L}^1_X$ forms the set of all (measurable) state contingent contracts. Also, we require:

$$\text{for each } t \in T, U_1(t,\omega,\cdot) \text{ is concave and upper semicontinuous on } X(\omega) \text{ a.e. in } \Omega, \tag{3.8}$$

$$\text{for each } t \in T, U_2(t,\omega,\cdot) \text{ is affine and continuous on } X(\omega) \text{ a.e. in } \Omega \tag{3.9}$$

$$U_1 \text{ is product measurable and integrably bounded from above with respect to } P \times Q. \tag{3.10}$$

$$U_2 \text{ is product measurable and integrably bounded with respect to } P \times Q. \tag{3.11}$$

Note that (3.10)–(3.11) must be understood as follows: there exist $P \times Q$-integrably functions $\gamma_1, \gamma_2 : T \times \Omega \to \mathbf{R}$ with

$$\sup_{x \in V} U_1(t,\omega,x) \le \gamma(t,\omega) \text{ in } T \times \Omega$$

and

$$\sup_{x \in V} |U_2(t,\omega,x)| \le \gamma(t,\omega) \text{ in } T \times \Omega.$$

If the agent is of type $t \in T$ and the principal and agent enter into the contract $x \in \mathcal{L}^1_X$, then

$$I_{U_1}(t,x) := \int_\Omega U_2(t,\omega,x(\omega))P(d\omega).$$

is the principal's expected utility, while the type t agent's expected utility is given by

$$I_{U_2}(t,x) := \int_\Omega U_2(t,\omega,x(\omega))P(d\omega).$$

By Corollary 2.9, $I_{U_2}(t,\cdot)$ is weakly continuous on $\mathcal{L}^1_X$ for each $t \in T$, and by assumption (3.8) above, $I_{U_2}(t,\cdot)$ is also affine on $\mathcal{L}^1_X$ for each $t \in T$. Finally, I_{U_2}

is $\mathcal{T} \times \mathcal{B}_w$-measurable on $T \times \mathcal{L}^1_X$, where $\mathcal{B}_w$ denoes the Borel σ-algebra for the weak topology on $\mathcal{L}^1_V$.

A contract mechanism is a mapping $\xi : T \to \mathcal{L}^1_X$ from agent types into the set of contracts. Let Ξ denote the set of all $(\mathcal{T}, \mathcal{B}_w)$-measurable contract mechanism. The principal's contracting problem, with adverse selection, is now given by

$$\sup_{\xi \in \Xi} J(\xi) := \int_T I_{U_1}(t, \xi(t)) Q(dt) \tag{3.12}$$

subject to

$$I_{U_2}(t, \xi(t)) \geq I_{U_2}(t, \xi(t')) \quad \text{for all } t, t' \text{ in } T, \tag{3.13}$$
$$I_{U_2}(t, \xi(t)) \geq 0 \quad \text{for all } t \text{ in } T. \tag{3.14}$$

Verbally, this contracting problem can be described as follows: The principal chooses a mechanism $\xi \in \Xi$. Given the mechanism ξ chosen by the principal, the agent responds by making a report to the principal concerning his/her type. If a type t agent reports his/her type as t' (i.e., the agent lies about his/her type), then the principal and agent enter into contract $\xi(t') \in \mathcal{L}^1_X$. Constraints (3.13) are *incentive compatibility* constraints; they guarantee that the mechanism chosen by the principal induces truthful reporting by the agent, and constraints (3.14), the *individual rationality* constraints, guarantee that the mechanism chosen by the principal is such that, given truthful reporting by the agent, it is rational for the agent – no matter what his/her type – to enter into a contract with the principal. Let Ξ_0 denote the set of all $\xi \in \Xi$ satisfying (3.13)–(3.14); it is trivial to verify that Ξ_0 is convex.

In order to guarantee that there exists at least one mechanism in Ξ_0, the following nontriviality hypothesis is sufficient:

$$\text{there exists an } \bar{x} \in \mathcal{L}^1 \text{ such that } I_{U_2}(t, \bar{x}) \geq 0 \quad \text{for all } t \in T.$$

Indeed, then the corresponding constant mechanism belongs to Ξ_0. Using the general existence result of [9], to which the above properties precisely apply, one can then conclude the existence of an optimal contract mechanism for the principal (the proof in [9] still depends heavily on the K-convergence results of [8], and thus follows essentially the same line of proofs as [21, 20], but uses the equivalence result in [10, III.2] for compact-valued multifunctions to obtain a slightly better result).

4 Mathematical preliminaries and proofs of the main results

In this section we develop the tools to be used in deriving the main results of this paper. Let $f : \Omega \times V \to [-\infty, +\infty]$ be a given function, which we suppose to be $\mathcal{F} \times \mathcal{B}(v)$-measurable. We define the integral functional $I_f : \mathcal{L}^1_V \to [-\infty, +\infty]$ by

$$I_f(v) := \int_\Omega f(\omega, v(\omega)) P(d\omega),$$

using the opposite of the integration convention introduced in Section 2: for any $\mathcal{F}$-measurable function $\phi : \Omega \to [-\infty, +\infty]$ we still set $\int \phi := \int \phi^+ - \int \phi^-$, but this time with $+\infty - +\infty := +\infty$. Sometimes we shall wish to restrict considerations to a particular integration domain $B \subset \Omega$. We then define $I_f^B : \mathcal{L}_V^1(B) \to [-\infty, +\infty]$ by obvious restriction:

$$I_f^B(v) := \int_B f(\omega, v(\omega))P(d\omega).$$

Throughout this section the following truly minimal *nontriviality hypothesis* will be in force:

$$\text{there exists at least one } \bar{v} \in \mathcal{L}_V^1 \text{ with } I_f(\bar{v}) < +\infty.$$

We start out by giving necessary conditions for weak lower semicontinuity of I_f in the presence of atomlessness. Of course, any necessary condition for strong lower semicontinuity automatically qualifies as a necessary condition for weak lower semicontinuity (but not conversely). The following result, as well as its proof, can be found in [18] (as shown here, the fact that V is finite-dimensional in [18], does not affect the validity of the result in our present context).

Lemma 4.1. *Assume that $(\Omega, \mathcal{F}, P)$ is atomless*[4]*. Suppose that I_f is strongly lower semicontinuous on $\mathcal{L}_V^1$. Then there exist a constant $M > 0$ and a function $\psi \in \mathcal{L}_{\mathbf{R}}^1$ such that*

$$f(\omega, \cdot) \geq \psi(\omega) - M\|\cdot\| \text{ on } V \text{ a.e. in } \Omega. \tag{4.1}$$

Proof. Suppose that (4.1) does not hold. Then for arbitrary $n \in \mathbf{N}$ the function $\psi : \Omega \to [-\infty, +\infty]$, defined by

$$\psi_m(\omega) := \inf_{x \in V} [f(\omega, x) + n\|x\|].$$

and measurable by [10,III.39], satisfies

$$\int_\Omega \psi_n dP = -\infty.$$

Note here that $\psi_n(\omega) \leq f^+(\omega, \bar{v}(\omega) + n\|\bar{v}(\omega)\|$, and by virtue of the nontriviality hypothesis the right side forms a P-integrable function. By the fact that $(\Omega, \mathcal{F}, P)$ is atomless, we can find a measurable partition of Ω, all whose n components have P-measure $P(\Omega)/n$. Now for at least one such component, which we denote by $A_n \in \mathcal{F}$, it must be true that $\int_{A_n} \psi dP = -\infty$, by the above. Hence also $\int_{A_n} (\psi_n + 1)dP = -\infty$, and this implies in turn that $\int_{B_n} (\psi_n + 1)^- dP = +\infty$, where B_n is defined as the set of those $\omega \in A_n$ for which $\psi_n < -1$. By definition of the latter integral, there exists and integrable function $s_n : B_n \to \mathbf{R}, 0 \leq s_n \leq$

[4] I.e. assume that the purely atomic part Ω_2 is a null set.

$(\psi_n + 1)^-$ on B_n (e.g., a step function), such that $i_n := \int_{B_n} s_n dP \geq 1$. Setting $\phi_n := -i_n^{-1} s_n$ now gives

$$P(B_n) \leq P(A_n) = P(\Omega)/n, \quad \int_{B_n} \phi_n dP = -1,$$
$$0 \geq \phi_n \geq -i_n^{-1}(\psi_n + 1)^- \geq -(\psi_n + 1)^- = \psi_n + 1 \text{ on } B_n.$$

The last inequality guarantees that for every $\omega \in B_n$ the set

$$\{x \in V : f(\omega, x) + n\|x\| \leq \phi_n(\omega)\}$$

is nonempty. So by the Von Neumann-Aumann measurable selection theorem [10,III.22] there exists a $\mathcal{F}$-measurable function $v_n : B_n \to V$ such that for a.e. ω in B_n

$$f(\omega, v_n(\omega)) + n\|v_n(\omega)\| \leq \phi_n(\omega).$$

Now either (i) $\int_{B_n} \|v_n\| dP \leq n^{-1}$ or (ii) $\int_{B_n} \|v_n\| dP > n^{-1}$. In case (i) we set $C_n := B_n$, and in case (ii) atomlessness guarantees the existence of a measurable subset C_n of B_n with $\int_{C_n} \|v_n\| dP = n^{-1}$. Outside C_n we set $v_n := \bar{v}$. In this way we end up with

$$\|v_n - \bar{v}\|_1 \leq \int_{C_n} (\|v_n\| + \|\bar{v}\|) dP \leq \frac{1}{n} + \int_{B_n} \|\bar{v}\| dP.$$

In view of $P(C_n) \leq P(B_n) \leq n^{-1}$, this shows that the sequence (v_n) converges in $\|\cdot\|_1$ to $\bar{v}$. But by the above

$$I_f(v_n) \leq \int_{\Omega \setminus C_n} f(\cdot, \bar{v}(\cdot)) dP + \int_{C_n} (\phi_n - n\|v_n\|) dP.$$

By (i)–(ii) above it is easy to see that, either way, the second integral on the right is at most -1. This means $\liminf_n I_f(v_n) \leq I_f(\bar{v}) - 1$, so that a contradiction with the lower semicontinuity hypothesis has been reached. □

Thus, we see that for atomless $(\Omega, \mathcal{F}, P)$ the most obvious condition for the integral functional I_f to be nowhere $-\infty$, is, at the same time, a necessary condition for its strong semicontinuity. In Example 4.4 below we show that atomlessness is essential for this finding.

We shall now discuss some results which specifically address *weak* lower semicontinuity. Let us denote the duality between $\mathcal{L}^1_V$ and its dual $\mathcal{L}^\infty_{V^*}[V]$ (cf. Sect. 2) by

$$\prec v, p \succ := \int_\Omega \langle (v(\omega), p(\omega) \rangle P(d\omega).$$

The following result is well-known; for generalizations, see [13, 18, 7]. It shows that weak lower semicontinuity of I_f forces the integrand not only to be lower semicontinuous in the second variable, but convex as well. Here atomlessness is again an essential ingredient, as borne out by Example 4.4 below.

Proposition 4.2. *Assume that $(\Omega, \mathcal{F}, P)$ is atomless. Suppose that I_f is weakly lower semicontinuous on $\mathcal{L}^1_V$. Then*

i. I_f *is convex on* $\mathcal{L}^1_v$.
ii. $f(\omega, \cdot)$ *is convex and lower semicontinuous on* V *a.e. in* Ω.

Proof. i. Consider the *epigraph* E *of* I_f [5, p. 11], defined by

$$E := \{(v, \alpha) \in \mathcal{L}^1_V \times \mathbf{R} : \alpha \geq I_f(v)\}.$$

Clearly, E is a closed set for the product of the weak topology (on $\mathcal{L}^1_V$) and the ordinary topology (on $\mathbf{R}$), as a consequence of the hypothesis. We must prove that E is a convex set. To do so, we first establish the following convexity criterion: E is convex if and only if for every finite subset $\{p_1, ..., p_N\}$, $N \in \mathbf{N}$, of the dual space $\mathcal{L}^\infty_{V^*}[V]$ one has that

$$C := \{(\prec v, p_1 \succ, ..., \prec v, p_N \succ, \alpha) : (v, \alpha) \in E\} \tag{4.2}$$

is a convex subset of $\mathbf{R}^{N+1}$. Indeed, for arbitrary $0 < \lambda < 1, (v, \alpha), (v', \alpha') \in E$, we have to check that $(w, \gamma) := \lambda(v, \alpha) + (1 - \lambda)(v', \alpha')$ belongs to E, viz., that $I_f(w) \leq \gamma$. If this were not true, then, by closedness of E, there would be a weakly open subset W of $\mathcal{L}^1_V$, containing w, and a $\delta > 0$ such that $(v, \alpha) \notin E$ whenever $v \in W$ and $\gamma - \delta < \alpha < \gamma + \delta$. By definition of the basis of the weak topology, there exists a finite collection $\{p_1, ..., p_N\} \subset \mathcal{L}^\infty_{V^*}[V]$ for some $N \in \mathbf{N}$, such that for every $v \in \mathcal{L}^1_V$

$$| \prec v - w, p_i \succ | \langle 1, i = 1, ..., N, \text{ implies } v \in W.$$

Let C be the convex set of (4.2). Evidently, by convexity, the $N + 1$-vector with coordinates $\prec w, p_i \succ, i = 1, ..., N$, and last coordinate γ, belongs to C. By definition of C, this means that there exists $(\tilde{v}, \tilde{\alpha}) \in E$ such that $\prec \tilde{v}, p_i \succ = \prec w, p_i \succ, i = 1, ..., N$ and $\tilde{\alpha} = \gamma$. But then the above implies $\tilde{v} \in W$ and $(\tilde{v}, \tilde{\alpha}) \notin E$. This contradiction proves the validity of the convexity criterion. Next, it is easy to establish that all sets C of the form (4.2) are indeed convex: Let $0 < \lambda < 1$ and $(v, \alpha), (v', \alpha') \in E$ be arbitrary. Then for $(w, \gamma) := \lambda(v, \alpha) + (1 - \lambda)(v', \alpha')$ to belong to C it is enough to verify the existence of some $\tilde{v} \in \mathcal{L}^1_V$ with $\prec \tilde{v}, p_i \succ = \prec w, p_i \succ, i = 1, ..., N$ and $I_f(\tilde{v}) \leq \gamma$. By Liapunov's theorem (which we may invoke because $(\Omega, \mathcal{F}, P)$ is assumed to be atomless) there exists a measurable subset B of Ω such that

$$\int_B (\langle v, p_1 \rangle, ..., \langle v, p_N \rangle, f(\cdot, v(\cdot)), \langle v', p_1 \rangle, ..., \langle v', p_N \rangle, f(\cdot, v'(\cdot))dP$$
$$= \lambda(\prec v, p_1 \succ, ..., \prec v, p_N \succ, I_f(v), \prec v', p_1 \succ, ..., \prec v', p_N \succ, I_f(v')).$$

(Note that $I_f(v), I_f(v') \in \mathbf{R}$ by Lemma 4.1 and by $I_f(v) \leq \alpha, I_f(v') \leq \alpha'$). Then setting $\tilde{v} := v$ on B and $\tilde{v}' := v'$ on the complement of B gives the desired integrable function. This establishes the convexity of E, which immediately implies convexity of I_f.

ii. By i, I_f is a convex semicontinuous function on $\mathcal{L}^1_V$. Moreover, we find $I_f > -\infty$ (by Lemma 4.1) and $I_f(\bar{v}) < +\infty$ (by nontriviality). By a well-known result from convex analysis this implies

$$I_f^{**}(v) = I_f(v) \quad \text{for all } v \in \mathcal{L}^1_V,$$

where it should be recalled that

$$I_f^{**}(v) := \sup_{p\in\mathcal{L}^\infty_{V^*}[V]} [\prec v,p\succ - I_f^*(p)]$$

with

$$I_f^*(p) := \sup_{v\in\mathcal{L}^1_V} [\prec v,p\succ - I_f(v)]$$

define two successive instances of *Fenchel conjugation*. Now as a consequence of *decomposability* [10, p. 197] of $\mathcal{L}^1_V$ and $\mathcal{L}^\infty_{V^*}[V]$ – a formalization of the fact that these spaces are both rich in measurable functions – and the Von Neumann-Aumann measurable selection theorem one has the following integral functional representation [10, VII.7]:

$$I_f^{**}(v) = I_{f^{**}}(v) := \int_\Omega f^{**}(\omega, v(\omega))P(d\omega).$$

Here

$$I^{**}(\omega,x) = \sup_{x^*\in V^*} [\langle x,x^*\rangle - f^*(\omega,x^*)]$$

with

$$f^*(\omega,x^*) := \sup_{x\in V}[\langle x,x^*\rangle - f(\omega,x)]$$

denote two successive Fenchel-conjugations with respect to the second argument. It should be kept in mind that for every $\omega\in\Omega$

$$f^{**}(\omega,\cdot) \text{ is the convex lower semicontinuous hull of } f(\omega,\cdot).$$

It follows therefore that $I_f(v) = I_{f^{**}}(v)$ for all $v\in\mathcal{L}^1_V$. By decomposability of $\mathcal{L}^1_V$, the nontriviality hypothesis and Lemma 4.1 we may apply [6, Thm. B.2]. This implies that

$$f(\omega,\cdot) = f^{**}(\omega,\cdot) \text{ a.e. in } \Omega.$$

This finishes the proofs. □

We shall now obtain a characterization of *strong* lower semicontinuity of I_f, which will play an essential role in our study of the necessary conditions for *weak* lower semicontinuity on atoms; this result is valid for a general finite measure space.

Proposition 4.3. *Suppose that there exist a constant $M>0$ and $\psi\in\mathcal{L}^1_{\mathbf{R}}$ such that*

$$f(\omega,\cdot) \geq \psi(\omega) - M\|\cdot\| \text{ on } V \text{ a.e. in } \Omega.$$

Then the following statements are equivalent:

i. $f(\omega,\cdot)$ is strongly lower semicontinuous on V a.e. in Ω,
ii. I_f is strongly lower semicontinuous on $\mathcal{L}^1_V$.

Proof. ii $\Rightarrow$ i: From the given inequality for f it follows that

$$I_f(v) \geq \int_\Omega \psi dP - M\|v\|_1 \text{ for all } v \in \mathcal{L}^1_V.$$

Hence, it follows by lower semicontinuity of I_f that for every $v \in \mathcal{L}^1_V$

$$I_f(v) = \sup_{n\in N} \inf_{w\in\mathcal{L}^1_V} [n\|v-w\|_1 + I_f(w)].$$

This follows by [4, p. 391]. In view of the nontriviality hypothesis and the decomposability of $\mathcal{L}^1_V$ (already used in the proof of Proposition 4.2) it follows by [6, Thm. B.1] (or by mimicking the proof of [10, Theorem VII.7]) that

$$I_f(v) = \sup_{n\in N} \int_\Omega \inf_{y\in V} [n\|(\omega) - y\| + f(\omega, y)] P(d\omega).$$

Note that, by our given inequality for f, the monotone convergence theorem can be invoked, giving

$$I_f(v) = \int_\Omega \bar{f}(\omega, v(\omega)) P(d\omega), \tag{4.3}$$

where we define

$$\bar{f}(\omega, x) := \sup_{n\in N} \inf_{y\in V} [n\|x - y\| + f(\omega, y)].$$

By the given inequality for f and easy ad hoc inspection (cf. [4, p. 391]) it follows from this definition that for a.e. ω

$$\bar{f}(\omega, \cdot) \text{ is the strongly lower semicontinuous hull of } f(\omega, \cdot).$$

By the nontriviality hypothesis and (4.3) it follows from [6, Thm. B.2] that

$$f(\omega, \cdot) = \bar{f}(\omega, \cdot) \text{ a.e. in } \Omega,$$

giving i.

i $\Rightarrow$ ii: Let (v_k) be an arbitrary sequence in $\mathcal{L}^1_V$ such that $\|v_k - v_0\|_1 \to 0$. Let $\gamma := \liminf_k I_f(v_k)$. Then for some subsequence (v_{k_i}) we shall actually have $\gamma = \lim_i I_f(v_{k_i})$. By [4, 2.5.3] there exists a further subsequence of (v_{k_i}), say (v_{k_j}), such that for a.e. ω

$$\lim_{j\to\infty} \|v_{k_j}(\omega) - v_0(\omega)\| = 0.$$

Therefore, Fatou's lemma gives

$$\gamma + M\|v_0\|_1 = \lim_j \int_\Omega [f(\omega, v_{k_j}(\omega)) + M\|v_{k_j}(\omega)\|] P(d(\omega) \geq I_f(v_0) + M\|v_0\|_1$$

(the integrand in the middle expression is minorized by the integrable function $\psi(\omega)$). This shows the validity of ii. □

Note the similarity of our proofs of Proposition 4.2 and of the necessity part of the above result. A much more complicated, hybrid version of both results was given in [6], in connection with certain classical notions in the calculus of variations.

Even though Proposition 4.3 captures the semicontinuity aspect of its counterpart Proposition 4.2, there can be no question of emulating the convexity aspect of Proposition 4.2 or the boundedness feature of Lemma 4.1 if atomlessness is no longer satisfied:

Example 4.4. *Let $(\Omega, \mathcal{F}, P)$ be the purely atomic measure space consisting of the singleton $\{\tilde{\omega}\}$ with $P(\{\tilde{\omega}\}) = 1$. Consider as V the separable Banach space formed by all continuous real-valued functions on the unit interval [0,1]; the norm on V is the usual supremum norm. Define $f(\tilde{\omega}, x) := -[x(0)]^2$; this is evidently a non-convex function. However, if $v_l \to v_0$ weakly, then (equivalently) $v_l(\tilde{\omega}) \to v_0(\tilde{\omega})$ weakly in V. Now V^* is known to be identifiable with the set of all bounded signed Borel measures on [0,1]; in particular, V^* contains the point probability concentrated at 0. This immediately implies the convergence of $I_f(v_l) = f(\tilde{\omega}, v_l(\tilde{\omega}))$ to $I_f(v_0) = f(\tilde{\omega}, v_0(\tilde{\omega}))$. Thus, I_f is weakly continuous, but $f(\omega, \cdot)$ is neither convex – let alone affine – nor does it obey the lower bound in Lemma 4.1.*

Necessary conditions for weak lower semicontinuity of I_f take on a particularly easy form on atoms. We shall see how Proposition 4.3 plays an auxiliary role in connection with the following lemma:

Lemma 4.5. *Let A be an atom of $(\Omega, \mathcal{F}, P)$. Then every function $v : \Omega \to V$ which is measurable with respect to $\mathcal{F}$ and $\mathcal{B}(V)$ is constant a.e. on A. More generally, every multifunction which has strongly closed values and for which $\mathrm{gph}\Gamma := \{(\omega, x) \in \Omega \times V : x \in \Gamma(\omega)\}$ is $\mathcal{F} \times \mathcal{B}(V)$-measurable, is equal to a constant set a.e. on A.*

Proof. Let (x_j) be a sequence in V which is strongly dense. For arbitrary $j \in \mathbf{N}$, the function

$$\phi_j : \omega \mapsto \mathrm{dist}(x_j, \Gamma(\omega)) := \inf_{x \in \Gamma(\omega)} \|x - x_j\|$$

is measurable by [10,III.30]. By an elementary property of measurable, real-valued functions on atoms, ϕ_j must be a.e. constant on A for every j. It remains to observe that when two strongly closed subsets C, D of V satisfy $\mathrm{dist}(x_j, C) = \mathrm{dist}(x_j, D)$ for all j, then $C = D$. □

Proposition 4.6. *Let A be an atom of $(\Omega, \mathcal{F}, P)$. Suppose that I_f^A is weakly lower semicontinuous on $\mathcal{L}_V^1(A)$ and that there exist constants $M, K > 0$ such that*

$$f(\omega, \cdot) \geq K - M\|\cdot\| \text{ on } V \text{ a.e. in } A.$$

Then

$$f(\omega, \cdot) \text{ is weakly lower semicontinuous on } V \text{ a.e. in } A.$$

Proof. A *fortiori*, I_f^A is strongly lower semicontinuous on $\mathcal{L}_V^1(A)$, so by Proposition 4.3.

$$f(\omega,\cdot) \text{ is strongly lower semicontinuous on } V \text{ a.e. in } A. \tag{4.4}$$

Therefore, the multifunction $\Gamma : \Omega \to 2^{V \times \mathbf{R}}$, defined by

$$\Gamma(\omega,\cdot) := \{(x,\lambda) \in V \times \mathbf{R} : \lambda \geq f(\omega,x)\},$$

satisfies all conditions of Lemma 4.5. It follows that there exist a null set N and a closed set $C \subset V \times \mathbf{R}$ such that $\Gamma(\omega) = C$ for all $\omega \in -A\backslash N$. It thus follows that there exists a strongly lower semicontinuous function $g : V \to (-\infty, +\infty]$ such that

$$f(\omega,\cdot) = g \text{ for all } \omega \in A\backslash N. \tag{4.5}$$

It remains to show that g is also *weakly* lower semicontinuous. To this end, let (x_l) be a generalized sequence weakly converging to x_0 in V. Define, correspondingly, $v_l \in \mathcal{L}_V^1(A)$ by $v_l(\omega) := x_l$; then (v_l) converges weakly in $\mathcal{L}_V^1(A)$ to v_0, so we get

$$\begin{aligned} P(A)g(x_0) &= \int_A f(\omega, v_0(\omega))P(d\omega) \leq \liminf_l \int_A f(\omega, v_l(\omega))P(d\omega) \\ &= P(A) \liminf_l g(x_l), \end{aligned}$$

thanks to lower semicontinuity of I_f^A. □

The pattern emerging from the aforegoing results is as follows: (a) in the presence of atomlessness, weak lower semicontinuity of the integral functional is associated with lower semicontinuity and convexity of the integrand (in the second variable); (b) on atoms this is associated with weak lower semicontinuity of the integrand (without convexity). This impression is confirmed by the following result.

Proposition 4.7. *Assume that $(\Omega, \mathcal{F}, P)$ is atomless. Suppose that a.e. in Ω*

$$f(\omega,\cdot) \text{ is convex and lower semicontinuous on } V,$$

and

$$f(\omega,\cdot) \geq \psi(\omega) - M\|\cdot\| \text{ on } V$$

for some constant $M > 0$ and $\psi \in \mathcal{L}_{\mathbf{R}}^1$. Then I_f is weakly lower semicontinuous on $\mathcal{L}_V^1$.

Proof. The integral functional I_f is strongly semicontinuous (by Proposition 4.3) and convex (obvious). Therefore, it must also be weakly lower semicontinuous (Mazur's theorem [5, 1.3.5]). □

Remark 4.8. *Combining Lemma 4.1 and Proposition 4.2.ii, we observe that the converse of the implication in Proposition 4.7 is also valid.*

On atoms, on the other hand, the situation is even simpler:

Proposition 4.9. *Let A be an atom of $(\Omega, \mathcal{F}, P)$. Suppose that*

$$f(\omega, \cdot) \text{ is weakly lower semicontinuous on } V \text{ a.e. in } A.$$

Then the integral functional I_f^A is weakly lower semicontinuous on $\mathcal{L}_V^1(A)$.

Proof. Note first that a *fortiori*

$$f(\omega, \cdot) \text{ is strongly lower semicontinuous on } V \text{ a.e. in } A.$$

So we can repeat the part of the proof of Proposition 4.6 leading from (4.4) to (4.5). Using the notation introduced there, we get for every $v \in \mathcal{L}_V^1(A)$

$$I_f^A(v) = \int_A f(\omega, v(\omega))P(d\omega) = P(A)g(x),$$

where x stands for the a.e. constant value taken by v on the atom A (Lemma 4.5), and where g is weakly lower semicontinuous. The proof is now easily finished. □

Proof of Theorem 2.1. ii⇒ i: Define the $\mathcal{F} \times \mathcal{B}(V)$-measurable function $f_X : \Omega \times V \to \{0, +\infty\}$ by setting $f_X(\omega, x) := 0$ if $x \in X(\omega)$ and $f_X(\omega, x) := +\infty$ if not. Clearly, the integral functional $I := I_{f_X}$ is as follows: $I(v) = 0$ if $v \in \mathcal{L}_X^1$ and $I(v) = +\infty$ if not (note in particular that $I(\bar{x}) < +\infty$ by the nontriviality hypothesis). Therefore, weak closedness of $\mathcal{L}_X^1$ is equivalent to I being weakly lower semicontinuous on $\mathcal{L}_V^1$. Because of the obvious identity

$$I(v) = \int_{\Omega_1} f_X(\omega, v(\omega))P(d\omega) + \int_{\Omega_2} f_X(\omega, v(\omega)P(d\omega) =: I_1(v) + I_2(v),$$

we see that this is equivalent to having I_1 weakly lower semicontinuous on $\mathcal{L}_V^1(\Omega_1)$ and I_2 on $\mathcal{L}_V^1(\Omega_2)$ separately. By Proposition 4.2 (note that $f_X \geq 0$) the semicontinuity of I_1 implies

$$f_X(\omega, \cdot) \text{ is convex and lower semicontinuous for a.e. } \omega \in \Omega_1,$$

which in turn is precisely equivalent to the first part of i. Also, semicontinuity of I_2 on $\mathcal{L}_V^1(\Omega_2)$ implies that on every atom A which is part of Ω_2 (note that $f_X \geq 0$)

$$f_X(\omega, \cdot) \text{ is weakly lower semicontinuous on } A,$$

by virtue of Proposition 4.6. Since Ω_2 is the countable union of such atoms, this finishes the proof of i.

i⇒ ii: f_X now cearly satisfies the conditions of Proposition 4.7 on Ω_1 and Proposition 4.9 on Ω_2. Therefore, I is weakly lower semicontinuous; in view of what was said about $I := I_{f_X}$ above, this implies ii. □

Proof of Theorem 2.2. Define f_X and $I := I_{f_X}$ as in the proof of the previous theorem. Then the result follows directly from Proposition 4.3. □

Proof of Theorem 2.6. By Theorem 2.1 we already know the stated facts about the values $X(\omega)$. Define the $\mathcal{F} \times \mathcal{B}(V)$-measurable function $f : \Omega \times V \to [-\infty, +\infty]$

as follows: set $f(\omega,x) := -U(\omega,x)$ if $x \in X(\omega)$ and $f(\omega,x) := +\infty$ if not. Then I_f equals $-I_U$ on $\mathcal{L}^1_X$ (by the integration conventions) and $+\infty$ on $\mathcal{L}^1_V \backslash \mathcal{L}^1_X$ [note how the switch in sign precisely explains the difference in integration conventions and nontriviality hypotheses between Sects. 2 and the present one!]. It follows directly from the hypotheses that I_f is weakly lower semicontinuous on $\mathcal{L}^1_V$, so by splitting I_f over the atomless part Ω_1, and its complement Ω_2, as done in the proof of Theorem 2.1, and successively applying Propositions 4.2 and 4.6, we find that for a.e. $w \in \Omega_1$,

$$f(\omega,\cdot) \text{ is convex and lower semicontinuous on } V, \tag{4.6}$$

and for a.e. $\omega \in \Omega_2$

$$f(\omega,\cdot) \text{ is weakly lower semicontinuous on } V.$$

In view of the already established properties of $X(\omega)$, the former is equivalent to

$$U(\omega,\cdot) \text{ is convex and lower semicontinuous on } X(\omega),$$

and the latter to

$$U(\omega,\cdot) \text{ is weakly lower semicontinuous on } X(\omega). \qquad \square$$

Proof of Theorem 2.8. Define f as in the previous proof. Then our conditions guarantee that Propositions 4.7 and 4.9 may be applied. In view of the already established weak closedness of $\mathcal{L}^1_X$ (Theorem 2.1), the desired weak continuity of $-I_U$, follows from the nature of I_f, established in the previous proof. □

Proof of Theorem 2.10. The proof essentially consists of an application of Proposition 4.3 to the function f used in the previous two proofs. Details are left to the reader. □

Acknowledgements. We wish to express our sincere gratitude to Frank Page and the referee for kindly providing material for the Section 3.2, and for suggesting several improvements in the presentation of our results.

References

1. Aliprantis, C.D., Burkinshaw, O.: Positive operators. New York: Academic Press 1985
2. Allen, B.: Market games with asymmetric information: the value. CARESS Working Paper ##91-08, University of Pennsylvania (1991)
3. Allen, B.: Market games with asymmetric information: the private information core. CARESS Working Paper ##92-04, University of Pennsylvania (1992)
4. Ash, R.B.: Real analysis and probability. New York: Academic Press 1972
5. Aubin, J.P., Ekeland, L: Applied nonlinear analysis. New York: Wiley 1984
6. Balder, E.J.: On seminormality of integral functionals and their integrands. SIAM J. Control Optim. **24**, 95–121 (1986)
7. Balder, E.J.: Necessary and suffcient conditions for L,-strong-weak semicontinuity of integral functionals. Nonlinear Anal. TMA **11**, 1399–1404 (1987)
8. Balder, E.J.: New sequential compactness results for spaces of scalarly integrable functions. J. Math. Anal. Appl. **151**, 1–16 (1990)

9. Balder, E.J.: On the existence of optimal contract mechanisms for the incomplete information principal agent model. Mimeo
10. Castaing, C., Valadier, M.: Convex analysis and measurable multifunctions. In: Lect. Notes Math., vol. 588. Berlin: Springer 1977
11. Diestel, J., Uhl, J.J.: Vector measures. In: Mathematical Surveys No. 15. Providence: American Mathematical Society (1977)
12. Ionescu Tulcea, C., Ionescu Tulcea, A.: Topics in the theory of lifting. In: Ergebnisse der Mathematik und ihre Grenzgebiete, vol. 48. Berlin: Springer 1969
13. Ioffe, A.D.: On lower semicontinuity of integral functionals. SIAM J. Control Optim. **15**, 521–538 (1977)
14. Kahn, C.M.: Existence and characterization of optimal employment contracts an a continuous state space. J. Econ. Theory (forthcoming)
15. Klei, H.A.: A compactness criterion in $L^1(E)$ and Radon-Nikodym theorems for multimeasures. Bull. Sci. Math. **112**, 305–324 (1988)
16. Krasa, S., Yannelis, N.C.: The value allocation of an economy with differential information. Faculty Working Paper BEBR ##91-0149, University of Illinois (1991)
17. Koutsougeras, L., Yannelis, N.C.: Incentive compatibility and information superiority of the core of an economy with differential information. Econ. Theory **3**, 195–216 (1993)
18. Olech, C.: A characterization of L,-weak lower sernicontinuity of integral functionals. Bull. Acad. Pol. Sci. **25**, 135–142 (1977)
19. Page, F.H.: The existence of optimal contraets in the principal-agent model. J. Math. Econ. **16**, 157–167 (1987)
20. Page, F.H.: Incentive compatible strategies for general Stackelberg games with incomplete information. Intern. J. Garne Theory **18**, 409–421 (1989)
21. Page, F.H.: Optimal contract mechanisms for principal-agent problems with moral hazard and averse selection. Econ. Theory **1**, 323–338 (1991)
22. Page, F.H.: Market games with differential information and infinite commodity spaces: the core. Working Paper, Economics, Finance and Legal Studies, University of Alabama (1992)
23. Scarf, H.: The core of an N person game. Econometrica **35**, 50–69 (1967)
24. Shapley, L.S., Shubik, M.: On market games. J. Econ. Theory **1**, 9–25 (1969)
25. Yannelis, N.C.: The core of an economy with differential information. Econ. Theory **1**, 183–198 (1991)
26. Yannelis, N.C.: Integration of Banach-valued correspondences. In: Khan, M.A., Yannelis, N.C. (eds.) Equilibrium theory in infinite dimensional spaces. Berlin: Springer 1991
27. Yannelis, N.C., Rustichini, A.: Ort the existence of correlated equilibria, In: Khan, M.A., Yannelis, N.C. (eds.) Equilibrium theory in infinite dimensional spaces. Berlin: Springer 1991

Market games with differential information and infinite dimensional commodity spaces: the core*

Frank H. Page, Jr.

Department of Finance, University of Alabama, Tuscaloosa, AL 35487, USA

Received: August 21, 1994; revised version: September 5, 1995

Summary. We provide an alternative proof of the existence of core allocations in exchange economies with differential information and infinite dimensional commodity spaces. We also identify a critical feature of information sharing rules that ensures nonemptiness of the core. In essence, the only condition we require on the sharing rules is that profitable "insider trading" be prohibited. In the absence of insider trading, balancedness is guaranteed and core nonemptiness follows.

JEL Classification Numbers: C71, D82.

1 Introduction

There are three main notions of the core for market games induced by exchange economies with differential information: the coarse core and the fine core, introduced by Wilson (1978), and the private core, introduced by Yannelis (1991).[1] Each of these core notions corresponds to assumptions made concerning the extent to which the potential information available to coalition members is shared and used in trading within the coalition. In the coarse core, it is assumed that only the potential information common to all coalition members is used in trading, while in the fine core, it is assumed that all potential information available to coalition members is used in trading. By contrast, in the private core, it is assumed that coalitional trading takes place via bargaining, based only upon each coalition member's potential private information (i.e., without information sharing).

* I thank Dan Acre, Erik Balder, Myrna Wooders, and Nicholas Yannelis for helpful comments. This paper is a greatly revised version of my paper entitled, "A Variational Problem Arising in Market Games with Differential Information," written in August of 1991

[1] The coarse core and fine core are problematic (i.e., existence and ex post incentive compatibility problems arise – see Koutsougeras and Yannelis (1993)). In order to remedy these difficulties, Yannelis (1991) introduced the private core as well as refinements of the coarse and fine cores.

Allen (1991, 1992), recognizing that certain features of information sharing were common to all the sharing arrangements described above, developed the notion of an abstract sharing rule. She showed that when sharing rules are appropriately specified coalitional utility opportunity sets are determined and a game in characteristic form is naturally defined. With certain conditions on the sharing rules, Allen established the nonemptiness of the core of the derived game. Moreover, nonemptiness of the coarse core, the fine core, and the private core follow as special cases of Allen's result.

In this paper, we extend Allen's approach to include infinite dimensional commodity spaces and we identify a critical feature of sharing rules implying the nonemptiness of the core. In essence, the only condition we require on the sharing rules is that profitable "insider trading" be prohibited. The no-insider-trading condition rules out the formation of informationally advantaged small trading groups, so that profitable trades by small groups due to superior information are not possible. In fact, this is precisely the meaning of Allen's boundedness condition (Allen (1991), p. 21). The intuition, then, behind our results as well as Allen's is that small and intermediate size coalitions are not "too powerful" relative to the grand coalition.

Core nonemptiness for exchange economies with differential information and infinite dimensional commodity spaces is particularly interesting for financial models. For example, in asset trading models differential information is an essential ingredient and "commodity spaces" are naturally infinite dimensional, consisting of financial assets with time and state contingent payoffs (e.g., continuous-time asset return processes, see Duffie and Huang (1985)). Yannelis' (1991) introduced an exchange model that permits infinite dimensional commodity spaces. In his model, information sharing arrangements are those consistent with the private and coarse cores. Using a limit argument, Yannelis establishes nonemptiness for these core concepts. Yannelis also remarks that the fine core may be empty. An example by Koutsougeras and Yannelis (1993, Sect. 5.3) bears out this remark. From our analysis we can conclude that the fine core is empty precisely because information sharing arrangements are such that there is an opportunity for insider trading.

Following Yannelis (1991), we assume that the commodity space is a Banach lattice with order continuous norm. This assumption allows us to prove core nonemptiness in a direct manner using K-compactness. As is the case with Allen's results, the nonemptiness of the coarse core, the fine core, and the private core follow as special cases of our results.

We shall proceed as follows. In Section 2, we present basic ingredients and technical details, and we show that, under any set of information sharing rules, a coalition's feasible set of trades is K-compact. In Section 3, we show that the NTU market game in characteristic form is well-defined, and we show that a core allocation exists – provided the information sharing rules prohibit insider trading. In Section 4, we show that by making specific assumptions concerning the nature of information sharing rules, we obtain the three main core notions: the coarse core, the fine core, and the private core.

2 Preliminaries

2.1 Elements of the model

Consider an exchange economy populated by m agents, indexed by $i = 1, 2, ...m$. Let 2^I denote the set of all coalitions (or nonempty subsets) of agents, where $I = \{1, 2, ...m\}$. Suppose the commodity space is given by a Banach lattice Y with positive cone Y_+ and order continuous norm $\| \cdot \|$ (the terminology here is the same as that found in Yannelis (1991)). Suppose also that the uncertainty in the economy is given by the probability space $(\Omega, \mathbf{F}, \mu)$, where $\mathbf{F}$ is a σ-field of payoff relevant events and μ a probability measure representing each agent's ex ante probability beliefs concerning the events in $\mathbf{F}$ (the fact that all agents have the same probability beliefs is not essential for the analysis). Finally, suppose that the set of all possible vectors of state-contingent consumption (or payoffs) is given by $\mathbf{L}_1(\Omega, \mathbf{F}, \mu; Y)$, the space of equivalence classes of Y-valued, $\mathbf{F}$-measurable, Bochner integrable functions, $x(\cdot) : \Omega \to Y$, normed by

$$\|x\|_1 = \int_\Omega \|x(\omega)\| d\mu(\omega).$$

$\mathbf{L}_1(\Omega, \mathbf{F}, \mu; Y)$ is a Banach space under the norm $\| \cdot \|_1$ (see Diestel and Uhl (1977), p. 50).

For each agent $i \in I$, we have the following:

$\mathbf{F}_i$, a sub-σ-field of $\mathbf{F}$ summarizing the ith agent's initial private information about payoff relevant events.

$Y_i(\cdot)$: $\Omega \to 2^{Y_+}$, a random set-valued mapping specifying for each state of nature, the ith agent's consumption set. We will assume that for each $\omega \in \Omega, Y_i(\omega) \subset Y_+$ is nonempty, convex, and sequentially weakly closed.

$\mathbf{e}_i(\cdot)$: $\Omega \to Y_+$, the ith agent's random initial endowment. We will assume that $\mathbf{e}_i(\cdot)$ is $\mathbf{F}_i$-measurable and Bochner integrable and that $\mathbf{e}_i(\omega) \in Y_i(\omega)$ a.e. $[\mu]$.

$u_i(\cdot, \cdot)$: $\Omega \times Y \to R$, the ith agent's random utility function. We will assume that

(i) u_i is $\mathbf{F} \times \mathbf{B}_+$-measurable, where $\mathbf{B}_+$ denotes the Borel σ-field in Y_+ generated by the relative weak topology in Y_+,

(ii) for each $\omega \in \Omega, u_i(\omega, \cdot)$ is concave and sequentially weakly upper semicontinuous on Y_+.

(iii) $u_i(\cdot, \cdot)$ is integrable bounded from above. Thus, for some $\xi(\cdot) \in \mathbf{L}_1(\Omega, \mathbf{F}, \mu; R), u_i(\omega, y) \leq \xi(\omega)$ on $\Omega \times Y$.

We will denote by Γ any exchange economy with differential information satisfying the assumptions above.

2.1.1. Remarks If the positive cone Y_+ is metrizable for the weak topology, then for each $\omega \in \Omega$, $u_i(\omega, \cdot)$ is sequentially weakly upper semicontinuous on Y_+ if and only if $u_i(\omega, \cdot)$ is weakly upper semicontinuous on Y_+ (Munkres (1975), p. 128).

2.2 Information sharing and feasible trades

Now consider an exchange economy Γ with initial private information $(\mathbf{F}_i)_{i\in I}$, and suppose information sharing is specified via a set of rules $\{\mathbf{F}_i(S) : i \in S, S \in 2^I\}$, where $\mathbf{F}_i(S)$ is the σ-field of payoff relevant events (i.e., the information) made available to agent i in coalition S. These information sharing rules can arise in many ways and take many forms. For example, coalition members, $i \in S$, could pool their resources (or be required to pay a fee for membership) and purchase observations of payoff relevant random variables which generate the conditioning σ-fields $\mathbf{F}_i(S)$. Alternatively, coalition members could simply share their initial private information, $(\mathbf{F}_i)_{i\in S}$. For example, in the terminology of Wilson (1978), there could be coarse information sharing: for $i \in S, \mathbf{F}_i(S) = \bigcap_{i\in S}\mathbf{F}_i$ where $\bigcap_{i\in S}\mathbf{F}_i$ is the largest σ-field common to all the σ-fields $\mathbf{F}_i$ – or there could be fine information sharing: for $i \in S, \mathbf{F}_i(S) = \sigma(\bigcup_{i\in S}\mathbf{F}_i)$ where $\sigma(\bigcup_{i\in S}\mathbf{F}_i)$ is the smallest σ-field containing all the σ-fields $\mathbf{F}_i$.

We will make two assumptions concerning information sharing by coalitions in the economy Γ:

(A-1) Honest reporting: As in Wilson (1978) and Kobayashi (1980), we shall assume that "...members of a coalition release their private information [to the coalition] honestly..." [Kobayashi (1980), p. 1639], where the agent's private information is represented by his initial σ-field $\mathbf{F}_i$.

Assumption (A-1) is implicit in the work of Yannelis (1991), Koutsougeras and Yannelis (1993), and Allen (1991, 1992).

(A-2) No insider trading: We shall also assume that the information sharing rules $\{\mathbf{F}_i(S) : i \in S, S \in 2^I\}$, are such that for each coalition $S \in 2^I$,

$$\mathbf{F}_i \subset \mathbf{F}_i(S) \subset \mathbf{F}_i(I) \text{ for } i \in S.$$

(A-2) is similar to Allen's (1991) assumption that information is bounded. Under (A-2), information sharing is such that no agent is made informationally worse off by joining a coalition and membership in the grand coalition makes available to the individual agent more information than does membership in any subcoalition. The condition, $\mathbf{F}_i(S) \subset \mathbf{F}_i(I)$ for $i \in S$ and $S \in 2^I$, rules out the possibility of informationally advantaged subcoalitions, and therefore, eliminates insider trading. As we shall see below, (A-2) guarantees balancedness and therfore nonemptiness of the core.

2.3 K-compactness of feasible trades

2.3.1. Definitions

(i) *K-convergence.* A sequence of functions $\{x_n, (\cdot)\}_n \subset \mathbf{L}_1(\Omega, \mathbf{F}, \mu; Y)$ is said to K-converge a.e. $[\mu]$ to a function $x\hat{}(\cdot)\mathbf{L}_1(\Omega, \mathbf{F}, \mu; Y)$, if for each subsequence $\{x_{nk}(\cdot)\}_k$ of $\{x_n(\cdot)\}_n$ there is a μ-null set $N \in \mathbf{F}$ (i.e., $\mu(N) = 0$) such that for the sequence of function $\{x^k(\cdot)\}_k$, where

$$x^k(\cdot) \equiv (1/k)[x_{n1}(\cdot) + ... + x_{nk}(\cdot)],$$

$\{x^k(\omega)\}_k \subset Y$ converges weakly to $x\hat{}(\omega) \in Y$ for every $\omega \in \Omega \backslash N$. The function $x\hat{}(\cdot)$ is referred to as a K-limit of the sequence $\{x_n(\cdot)\}_n$.

(ii) *K-compactness.* A subset Φ of $\mathbf{L}_1(\Omega, \mathbf{F}, \mu; Y)$ is said to be relatively K-compact $[\mu]$ if every sequence in Φ contains a subsequence which K-converges a.e. $[\mu]$ to some $x\hat{}(\cdot) \in \mathbf{L}_1(\Omega, \mathbf{F}, \mu; Y)$. Φ is said to be K-compact $[\mu]$ if every sequence in Φ contains a subsequence which K-converges a.e. $[\mu]$ to some $x\hat{}(\cdot)$ in Φ.

2.3.2. Remarks The following Theorem due to Diaz (1994) relates relative K-compactness to relative weak compactness:
Assume Y^* has the Radon-Nikodym property and let Φ be a (norm) bounded subset of $\mathbf{L},(\Omega, \mathbf{F}, \mu; Y)$. Then, Φ is weakly relatively compact if and only if Φ is uniformly integrable and every sequence $\{x_n(\cdot)\}_n$ in Φ has a subsequence $\{x_{nk}(\cdot)\}_k$ such that $\{x^k(\cdot)\}_k$ converges weakly a.e. $[\mu]$ (where again $x^k(\cdot) \equiv (1/k)[x_{n1}(\cdot) + ... + x_{nk}(\cdot)]$).

Thus, given that Y^* has the Radon-Nikodym property, if Φ a bounded subset of $\mathbf{L},(\Omega, \mathbf{F}, \mu, Y)$ is uniformly integrable and relatively K-compact $[\mu]$, then Φ is weakly relatively compact.

For any coalition $S \in 2^I$ in exchange economy Γ with initial private information $(\mathbf{F}_i)_{i\in I}$ and information sharing rules $\{\mathbf{F}_i(S) : i \in S, S \in 2^I\}$, the feasible set of trades, $\Phi(S)$, is given by the set of all S-tuples $(x_i(\cdot))_{i\in S}$ such that

(i) $x_i(\cdot)$ is $\mathbf{F}_i(S)$-measurable, $x_i(\omega) \in Y_i(\omega)$ a.e. $[\mu]$, and
(ii)

$$\sum_{i\in S} x_i(\omega) = \sum_{i\in S} e_i(\omega) \text{ a.e. } [\mu] \tag{2.1}$$

2.3.3 Theorem (K-compactness of $\Phi(S)$)
For any coalition S in economy Γ, $\Phi(S)$ is nonempty, convex, and K-compact $[\mu]$.

Proof. $\Phi(S)$ is nonempty since $(e_i(\cdot))_{i\in S}$ is in $\Phi(S)$, and $\Phi(S)$ is clearly convex. For each $\omega \in \Omega$, define

$$B(\omega) = \left\{ x \in Y : 0 \leq x \leq \sum_{i\in I} e_i(\omega) \right\}.$$

Observe that for any S-tuple $(x_i(\cdot))_{i\in S}$ in $\Phi(S), x_i(\omega) \in B(\omega)$ a.e. $[\mu]$ for each $i \in S$. Moreover, observe that:

(i) For each $\omega \in \Omega$ and $x \in B(\omega)$, $|x| \leq |\sum_{i\in I} e_i(\omega)|$, and since Y is a Banach lattice, this implies that $\|x\| \leq \|\sum_{i\in I} e_i(\omega)\|(|x| = x^+ - x^-)$. Thus, since the initial endowment functions, $e_i(\cdot)$, are Bochner integrable, the multifunction $B(\cdot)$ is integrably bounded.
(ii) For each $\omega \in \Omega$, $B(\omega)$ is equal to the order interval $[0, \sum_{i\in I} e_i(\omega)]$. Thus, since the Banach lattice Y has order continuous norm, $B(\omega)$ is weakly compact for each $\omega \in \Omega$ (see Aliprantis, Brown, and Burkinshaw (1990)).

Now let $\{(x_{in}(\cdot))_{i\in S}\}_n$ be a sequence in $\Phi(S)$. Observe that for all $n, x_{in}(\omega) \in B(\omega)$ a.e. $[\mu]$ for $i \in S$. Given observation (ii) above, we have for each $i \in S$

$$\{x_{in}(\omega)\}_n \text{ is relatively weakly compact a.e. } [\mu],$$

and by observation (i), we have for each $i \in S$

$$\{x_{in}(\cdot)\}_n \subset \mathbf{L}_1(\Omega, \mathbf{F}_i(S), \mu; Y),$$

and

$$\sup_n \int_\Omega \|x_{in}(\omega)\| d\mu(\omega) < +\infty.$$

By Theorem B in Balder (1989), there is a subsequence $\{(x_{ink}(\cdot))_{i\in S}\}_k$ of $\{(x_{in}(\cdot))_{i\in S}\}_n$ such that for each $i \in S$, $\{x_{ink}(\cdot)\}_k$ K-converges a.e. $[\mu]$ to a K-limit $x\hat{}_i(\cdot) \in \mathbf{L}_1(\Omega, \mathbf{F}_i(S), \mu; Y)$. Clearly, $\sum_{i\in S} x\hat{}_i(\omega) = \sum_{i\in S} e_i(\omega)$ a.e. $[\mu]$, and since $Y_i(\omega)$ is nonempty, convex, and sequentially weakly closed $x\hat{}_i(\omega) \in Y_i(\omega)$ a.e. $[\mu]$. Thus, the S-tuple $(x\hat{}_i(\cdot))_{i\in S}$ is in $\Phi(S)$. □

2.3.4 Remarks It follows from a generalization of Diestel's Theorem that, for any coalition $S, \Phi(S)$ is also weakly compact (see Balder (1990)).

3 Nontransferable utility (NTU) game

3.1 Characteristic form

A cooperative nontransferable utility (NTU) game in characteristic form consists of a set of agents $I = \{1, 2, ..., m\}$ and a set-valued mapping $V(\cdot)$ defined on 2^I with nonempty, closed values in R^m such that $V(I)$ is bounded from above and $V(\varnothing) = \{0\}$.

3.1.1 Definition For the market economy $\Gamma, V(\cdot)$ is defined as follows: for any coalition $S \in 2^I (w_1, w_2, ..., w_m) \in V(S)$ if and only if there exists an S-tuple of Y-valued functions $(x_i(\cdot))_{i\in S}$ in $\Phi(S)$ (see (2.1)) such that for $i \in S$,

$$w_i \leq \int_\Omega u_i(\omega, x_i(\omega)) d\mu(\omega).$$

3.1.2 Theorem (The induced NTU game is well-defined)
The set-valued mapping $V(\cdot)$, defined above, is such that $V(I)$ is bounded from above, and for any coalition $S \in 2^I, V(S)$ is nonempty, convex, and closed.

Proof. Clearly, $V(I)$ is bounded from above. To prove the rest of the result we need to show that $V(I)$ is nonempty, convex, and closed. Obviously, $V(I) \neq \varnothing$. Moreover, since $u_i(\omega, \cdot)$ is concave for each i and ω, it is clear that $V(I)$ is convex.

Now let $\{(w_{1n}, ..., w_{mn})\}_n \subset V(I)$ and $\{(x_{1n}(\cdot), ..., x_{mn}(\cdot))\}_n \subset (\Phi(I)$ be such that

(i) $\lim_n w_{in} = w_i^*$ for each i, and
(ii) $w_{in} \leq \int_\Omega u_i(\omega, x_{in}(\omega)) d\mu(\omega)$ for each i and n.

Since $\Phi(I)$ is K-compact $[\mu]$, we can assume without loss of generality that $\{(x_{1n}(\cdot), ..., x_{mn}(\cdot))\}_n$ K-converges a.e. $[\mu]$ to some $(\hat{x}_1(\cdot), ..., \hat{x}_m(\cdot)) \in \Phi(I)$. Moreover, since for each $i \in I$, $u_i(\cdot, \cdot)$ is integrably bounded from above and $u_i(\omega, \cdot)$ concave and sequentially weakly upper semicontinuous on Y_+, by Corollary 2.2 of Balder (1990),

$$\limsup_n \int_\Omega u_i(\omega, x_{in}(\omega))d\mu(\omega) \leq \int_\Omega u_i(\omega, \hat{x}_i(\omega))d\mu(\omega) \;\; \text{for each} \;\; i \in I.$$

Given (i) and (ii) above,

$$w_i^* \leq \int_\Omega u_i(\omega, \hat{x}_i(\omega))d\mu(\omega) \;\; \text{for each} \;\; i \in I.$$

Thus, $(w_1^*, ..., w_m^*) \in V(I)$. □

Thus by Theorem 3.1.2, $V(\cdot)$ defines an NTU market game in characteristic form for the exchange economy Γ.

3.2 The NTU core of the market game

3.2.1 Definition An m-tuple $(x_1(\cdot), ..., x_m(\cdot))$ is said to be a NTU core allocation for Γ given initial private information $\{\mathbf{F}_1, ..., \mathbf{F}_m\}$ and information sharing rules $\{\mathbf{F}_i(S) : i \in S, S \in 2^I\}$ if the following conditions hold:

(i) $(x_1(\cdot), ..., x_m(\cdot)) \in \Phi(I)$,
(ii) there does not exist a coalition S and an S-tuple $(y_i(\cdot))_{i \in S}$ contained in $\Phi(S)$ such that for $i \in S$

$$\int_\Omega u_i(\omega, y_i(\omega))d\mu(\omega) > \int_\Omega u_i(\omega, x_i(\omega))d\mu(\omega).$$

3.2.2 Theorem (Nonemptiness of the NTU core)
Suppose there is honest reporting, so that [A-1] holds. Then an NTU core allocation exists for any exchange economy Γ with initial private information $\{\mathbf{F}_1, ..., \mathbf{F}_m\}$ and information sharing rules $\{\mathbf{F}_i(S) : i \in S, S \in 2^I\}$ satisfying the no-insider-trading condition, [A-2].

Before giving a proof, we need to recall the following notions from Scarf (1967). Let $\Lambda \subset 2^I$ be a balanced collection of coalitions with weights $\{\lambda_S : S \in \Lambda\}$. An NTU game $(I, V(\cdot))$ is said to be balanced if $\bigcap_{S \in \Lambda} V(S) \subset V(I)$ for every balanced collection of coalitions Λ.

The importance of the NTU notion of balancedness is made clear by the following classic result:

Scarf's Theorem. A balanced NTU game $(I, V(\cdot))$ has a core.

In order to show that an NTU core allocation exists for Γ, it suffices to show that the induced game in characteristic form, $(I, V(\cdot))$, is balanced.

Proof of Theorem 3.2.2. Let Λ be a balanced collection of coalitions with weights $\{\lambda_S : S \in \Lambda\}$ and let $(w_1, ..., w_m) \in V(S)$ for each coalition $S \in \Lambda$. Thus, for each $S \in \Lambda$, there is an S-tuple $(x_{iS}(\cdot))_{i \in S} \in \Phi(S)$ such that

$$w_i \leq \int_\Omega u_i(\omega, x_{iS}(\omega))d\mu(\omega) \text{ for } i \in S. \tag{*}$$

Now, for each $i \in I$ define $z_i(\cdot) = \sum_{S \in \Lambda, S \supset \{i\}} \lambda_S x_{iS}(\cdot)$. By inequality (*) and the concavity of the $u_i(\omega, \cdot)$, we have for each $i \in I$

$$\begin{aligned} w_i = \sum_{S \in \Lambda, S \supset \{i\}} \lambda_S w_i &\leq \sum_{S \in \Lambda, S \supset \{i\}} \lambda_S \int_\Omega u_i(\omega, x_{iS}(\omega))d\mu(\omega) \\ &\leq \int_\Omega u_i(\omega, z_i(\omega))d\mu(\omega). \end{aligned} \tag{**}$$

By the no-insider-trading condition, $z_i(\cdot)$ is $\mathbf{F}_i(I)$-measurable for all $i \in I$, and given the convexity of $Y_i(\omega)$ for all $i \in I$ and $\omega \in \Omega$, $z_i(\omega) \in Y_i(\omega)$ a.e. $[\mu]$. Finally, for all $\omega \in \Omega$

$$\sum_{i \in I} z_i(\omega) = \sum_{S \in \Lambda} \sum_{i \in S} \lambda_S x_{iS}(\omega) = \sum_{S \in \Lambda} \lambda_S \sum_{i \in S} x_{iS}(\omega),$$

and since $(x_{iS}(\cdot))_{i \in S} \in \Phi(S)$,

$$\begin{aligned} \sum_{S \in \Lambda} \lambda_S \sum_{i \in S} x_{iS}(\omega) &= \sum_{S \in \Lambda} \lambda_S \sum_{i \in S} e_i(\omega) \\ &= \sum_{i \in I} e_i(\omega) \sum_{S \in \Lambda, S \supset \{i\}} \lambda_S \\ &= \sum_{i \in I} e_i(\omega) \text{ a.e. } [\mu]. \end{aligned}$$

Thus, $\sum_{i \in I} z_i(\omega) = \sum_{i \in I} e_i(\omega)$ a.e. $[\mu]$, and we can conclude that

$$(z_i(\cdot))_{i \in I} \in \Phi(I).$$

Thus, by inequality (**), $(w_1, ..., w_m) \in V(I)$, and hence the NTU game $(I, V(\cdot))$ induced by Γ is balanced. By Scarf's Theorem and Theorem 3.1.2, an NTU core allocation exists for Γ. □

4 Discussion

By making specific assumptions concerning the nature of information sharing rules $\{\mathbf{F}_i(S) : i \in S, S \in 2^I\}$, we can obtain the three main core notions: the coarse core, the fine core, and the private core. Consider a market economy Γ with initial private information $\{\mathbf{F}_1, ...\mathbf{F}_m\}$ and information sharing rules $\{\mathbf{F}_i(S) : i \in S, S \in 2^I\}$.

The private core Yannelis (1991): If $\mathbf{F}_i(S) = \mathbf{F}_i$ for all $i \in S$ and $S \in 2^I$, so that information is not shared within any coalition, then the NTU core of Γ is called the private core. Notice that if the information sharing rules, $\{\mathbf{F}_i(S) : i \in S, S \in 2^I\}$, are specified in this way, then the no-insider-trading condition is satisfied, and we can conclude via Theorem 3.2.2 that the private core is nonempty (Allen (1992) refers to this core notion as the private information core).

The coarse core Wilson (1978): If for the grand coalition, $\mathbf{F}_i(I) = \mathbf{F}_i$ for all $i \in I$, while $\mathbf{F}_i(S) = \bigcap_{i \in S} \mathbf{F}_i$, (i.e., the largest σ-field common to all the σ-fields $\mathbf{F}_i$, for $i \in S$) for each possible blocking coalition $S \in 2^I$, then the NTU core of Γ is called the coarse core (see Yannelis (1991)). Again notice that if the information sharing rules, $\{\mathbf{F}_i(S) : i \in S, S \in 2^I\}$, are specified in this way, then the no-insider-trading condition is satisfied, and we can conclude via Theorem 3.2.2 that the coarse core is nonempty.

The fine core Wilson (1978): If for the grand coalition, $\mathbf{F}_i(I) = \mathbf{F}_i$ for all $i \in I$, while for each possible blocking coalition $S \in 2^I$, $\mathbf{F}_i(S) = \sigma(\bigcup_{i \in S} \mathbf{F}_i)$ (i.e., the smallest σ-field containing all the σ-fields $\mathbf{F}_i$, for $i \in S$), then the NTU core of Γ is called the fine core (see Yannelis (1991)). Now notice that if the information sharing rules, $\{\mathbf{F}_i(S) : i \in S, S \in 2^I\}$, are specified in this way, then the no-insider-trading condition fails to hold and the fine core may be empty. This is the case because for any potential blocking coalition S, the Banach space of state contingent consumption functions $\mathbf{L}_1(\Omega, \mathbf{F}_i(S), \mu; Y)$ is larger than the space $\mathbf{L}_1(\Omega, \mathbf{F}_i(I), \mu; Y)$ for all $i \in S$ (see Koutsougeras and Yannelis (1993) for an example). One way around this problem is to assume that $\mathbf{F}_i(I) = \sigma(\bigcup_{i \in I} \mathbf{F}_i)$, so that the no-insider-trading condition is restored. Thus, if the information sharing rules, $\{\mathbf{F}_i(S) : i \in S, S \in 2^I\}$, are modified in this way, we can conclude via Theorem 3.2.2 that the core is nonempty. If we assume that $\mathbf{F}_i(I) = \sigma(\bigcup_{i \in I} \mathbf{F}_i)$, then the resulting core notion corresponds to the *weak fine core* notion in Koutsougeras and Yannelis (1993).

References

Aliprantis, C.D., Brown, D., Burkinshaw, O.: Existence and optimality of competitive equilibria. New York: Springer 1990

Allen, B.: Market games with asymmetric information and nontransferable utility: representation results and the core. Working Paper #91-09, CARESS, University of Pennsylvania 1991

Allen, B.: Market games with asymmetric information: the private information core. Working Paper #92-04, CARESS, University of Pennsylvania 1992

Balder, E. J.: Infinite-dimensional extension of a theorem of Komlos'. Probability Theory and Related Fields **81**, 185–188 (1989)

Balder, E. J.: New sequential compactness results for spaces of scalarly integrable function. Journal of Mathematical Analysis and Applications **151**, 1–16 (1990)

Diaz, S.: Weak compactness in $L^1(\mu, x)$. Working Paper, Departamento de Matematica Aplicada II, Universidad de Sevilla 1994

Diestel, J., Uhl, J.J. Jr.: Vector measures. Mathematical Surveys 15, American Mathematical Society, Providence, Rhode Island 1977

Duffie, D., Huang, C. F.: Implementing Arrow-Debreu equilibria by continuous trading of few long-lived securities. Econometrica **53**, 1337–1356 (1985)

Kobayashi, T.: Equilibrium contracts for syndicates with differential information. Econometrica **48**, 1635–1665 (1980)

Koutsougeras, L., Yannelis, N.C.: Incentive compatibility and information superiority of the core of an economy with differential information. Economic Theory **3**, 195–216 (1993)
Munkres, J.R.: Topology: a first course. Englewood Cliffs, N.J.: Prentice-Hall 1975
Scarf, H.: The core of an N person game. Econometrica **35**, 50–69 (1967)
Wilson, R.: Information, efficiency, and the core of an economy. Econometrica **46**, 807–816 (1978)
Yannelis, N.C.: The core of an economy with differential information. Economic Theory **1**, 183–198 (1991)

A two-stage core with applications to asset market and differential information economies*

Leonidas C. Koutsougeras
School of Economic Studies, University of Manchester, Manchester M139PL, UK

Received: December 20, 1995; revised version: December 6, 1996

Summary. We introduce a new core concept, called the two-stage core, which is appropriate for economies with sequential trade. We prove a general existence theorem and present two applications of the two-stage core: (i) In asset markets economies where we extend our existence proof to the case of consumption sets with no lower bound, in order to capture the case of arbitrary short sales of assets. Further, we show that the two-stage core is non empty in the Hart (1975) example where a rational expectations equilibrium fails to exist. (ii) In differential information economies where we provide sufficient conditions for the incentive compatibility of trades. Namely, that no coalition of agents can misreport the true state and provide improvements to all its members, even by redistributing the benefits from misreporting.

JEL Classification Number: D5.

1 Introduction

The standard Arrow-Debreu model of an exchange economy can be extended to account for uncertainty, by using the idea of contingent plans. In the standard scenario, agents are assumed to plan trades which are contingent upon the occurrence of uncertain events, in a way so that their expected payoff is maximized. The cooperative allocation of risks can be developed along the same line, by considering coalitions of agents who coordinate their contingent plans in order to achieve mutual benefits.

However, the dynamic nature of the uncertainty model allows for the possibility of *transient* formation of coalitions. In such a case, the benefits that a coalition can secure for its members are not well defined, because they

* This paper is an extract from my Ph.D thesis at the University of Illinois at Urbana-Champaign. I am grateful to Professors Wayne J. Shafer and Nicholas C. Yannelis for their patience, encouragement and useful advise. My thanks are also extended to Professors R. Anderson, S. Krasa, J-F. Mertens, H. Polemarchakis and A. Villamil for enlightening discussions.

depend on the behavior of its members in periods where they do not cooperate. In particular, the prospect of sequential formation of coalitions introduces a mutual dependency, between the *ex ante* and the *ex post* trading possibilities among groups of agents. In this way, the *ex ante* (*ex post*) benefits that a coalition can secure for its members becomes conditional on the *ex post* (*ex ante*) activities of its members. These issues trace back to Gale [5] and to Repullo [14], which discuss alternative core notions for economies with a sequence of markets. We leave it to the reader to consult the above sources for a discussion of several problems, associated with the core in a sequential framework. It's worth pointing out that the transient formation of coalitions, introduces non cooperative elements into the analysis. In other words, a game cannot be purely cooperative in such a framework.

In this paper we introduce an alternative core notion, called the *two-stage core*, that takes into account the possibility of temporary cooperation. The key ideas involved in the two-stage core are as follows: Within each coalition agents make future trades only if they are enforceable, i.e., a coalition may have a limited horizon. For example, if no future trades are enforceable then agents trade only in current commodities. Also we take the conservative point of view that a coalition blocks at some point in time only if it can secure improvements for its members in any possible consequence of a deviation. Our core notion can be easily specified in the asset market and the differential information contexts, which provide a natural environment for our analysis.

So far, the incomplete asset markets literature has addressed extensively the basic issues regarding the existence, optimality and regularity of competitive equilibria[1]. In general, rational expectations equilibria are not efficient, so the cooperative approach in these models seems to have been neglected. As a result there has been no counterpart of the core in an incomplete asset markets setting. In this paper we will show that the core in such a framework is quite interesting. As it is shown in Duffie-Shafer [4], in the asset markets framework competitive equilibria exist in a generic sense, i.e., competitive equilibria exist in all but a negligible set of economies. One exceptional case where a competitive equilibrium fails to exist is the Hart (1975) example. We show in the sequel that the core notion which we introduce here is non empty in that example. This suggests that the bargaining approach can be used to study the allocation of risks, in cases where competitive markets fail to do so. In order to capture the case of arbitrary short sales, we show the existence of the core for the case of consumption sets with no lower bounds. We do this by imposing conditions on the structure of preferences as in Page [11] and in Werner [16] for the existence of competitive equilibria. Our result is of independent interest, since it does not rely on the existence of competitive equilibria.

In differential information economies alternative core notions were introduced in Wilson [17] and in Yannelis [18]. However, in the model employed so far in the study of cooperative concepts in differential information econo-

[1] We refer the reader to Magill-Shafer [10] for an excellent survey of this literature.

mies, there is no trading round after uncertainty is resolved. As a result, agents end up consuming the *ex ante* contracted allocations. Optimality in that model is understood to be in a constrained sense and refers to the *ex ante* (or *interim*) trades only. In this paper we proceed one step further and allow for a trading round *ex post*. In our model individuals engage in state contingent trades under differential information, but have the opportunity to retrade (or recontract) after uncertainty is resolved. Thus, in contrast to the existing literature, we provide a core notion that gives rise to *fully optimal* allocations. Viewed from this perspective the core notion we introduce here characterizes the decisions of individuals both before and after some learning of the true state has occurred. We employ the specification of the two-stage core for differential information economies to address some incentive considerations. Specifically, since there is an *ex post* trading round, after misreporting a state individuals have the opportunity to redistribute among themselves the benefits from misreporting[2]. We provide sufficient conditions that guarantee the incentive compatibility of trades under asymmetric information. In this way, we conclude that our core notion characterizes outcomes that are incentive compatible in addition to being fully optimal.

Our analysis is organized as follows: In section two we develop the formal model. In section three we define a core notion appropriate for models with sequential trade and provide an existence result. Next, in section four we define the core of an asset markets economy as a specification of the two-stage core and demonstrate that in the Hart (1975) example the core is non empty. In section five we apply the two-stage core to a differential information economy and derive an incentive compatibility result. Section six features an example where we calculate two-stage core allocations and illustrate our incentive compatibility results. Some concluding remarks follow in section seven. Finally, we have collected most of the proofs along with the mathematical apparatus used in this paper in three appendices, that constitute sections eight, nine and ten.

2 The model

Consider an economy in two periods with uncertainty in the second period. There is a finite set of agents, denoted by $I = \{1, 2, 3, \ldots, n\}$, that engage in sequential trade as follows. Before a state of nature is realized (*ex ante*) individuals arrange for state contingent deliveries of commodities. In the second period, after the realization of a state of nature (*ex post*), the contracted deliveries take place and agents may engage in further trades of commodities[3].

---

[2] Allocations which are immune to such kind of strategic disclosure of information have been termed strongly incentive compatible in Krasa-Yannelis [8].

[3] Notice that in this framework there are no endowments and thus no consumption in the first period. Hence, agents are only concerned with allocating their second period endowments. This is done for simplicity in the exposition and involves no loss of generality.

Formally, let Ω be the set of all possible states of nature. The uncertainty in the model is described by the triple $(\Omega, \mathscr{F}, \mu)$ where $\mathscr{F}$ is the family of all possible events (a σ-field of subsets of Ω) and μ is a probability distribution on the events in $\mathscr{F}$. We will take the commodity space in each state to be $Y = \Re_+^l$ [4]. Let $L_1(\mu, Y)$ be the space of equivalence classes of (Bochner) integrable functions $x : \Omega \to Y$ (see Appendix I).

An *economy* is defined as the pair $\{\mathscr{E}, \mathscr{R}\}$ with $\mathscr{E} = \{(X_i, u_i, e_i, \mu) : i \in I\}$ where:

(i) X_i: $\Omega \to 2^Y$ the consumption set;
(ii) $u_i : \Omega \times Y \to \Re$ the utility function;
(iii) $e_i : \Omega \to Y, e_i(\omega) \in X_i(\omega)$ a.e. and e_i is (Bochner) integrable, the random initial endowment;
(iv) μ the prior[5] of agent i.

and $\mathscr{R} = \{\mathscr{R}_i \subseteq L_1(\mu, Y) : i \in I\}$ is a collection of restrictions that may apply in the first period trades of each individual in this economy[6].

Define for each $i \in I$, the consumption set of agent i in the second period as :

$$L_{X_i} = \{x \in L_1(\mu, Y) : x(\omega) \in X_i(\omega) \ a.e.\}.$$

The consumption set of each individual $i \in I$ in the first period is defined as :

$$L_{X_i}^R = L_{X_i} \cap \mathscr{R}_i.$$

The expected utility of agent i is given by :

$$v_i(x) = \int u_i(\omega, x(\omega)) d\mu(\omega), \quad for\ x \in L_{X_i}^R.$$

A trading plan (or strategy) for agent i is a pair $(x_1, x_2) \in L_{X_i}^R \times L_{X_i}$ i.e., a pair consisting of an *ex ante* and an *ex post* trade. One may think of *ex ante* trades taking the form of state contingent contracts. For the moment we will treat *ex ante* trades in this way. Alternatively first period state contingent trades could be in the form of asset trading. In our application to asset markets we will describe *ex ante* trades in terms of portfolios of assets.

3 The two-stage core

Consider the case where agents arrange, subject to restrictions, state contingent deliveries of commodities, through coalitional bargaining. Once a state has been formed, agents carry out the appropriate trades, and treating

[4] The analysis can be extended to hold for more general consumption spaces.

[5] Although we have considered the same prior for all agents, we may also allow for different priors as in Yannelis [18].

[6] For the moment we will treat $\mathscr{R}$ in an abstract way and save further specification of its nature for later sections.

the resulting allocation as their new endowment they bargain anew the exchange of commodities in that state. The set of core outcomes in such a situation should be a collection of *ex ante* and *ex post* bargaining outcomes that cannot be improved upon by any coalition.

A coalition may reject either a proposed *ex ante* trade, or a proposed allocation of commodities *ex post*. Therefore, the relevant core notion should account for possible deviations of coalitions in either period. Notice that since we will consider all possible coalitions of agents in both periods we allow the possibility that agents may not be committed to adhere *ex post* with a coalition that was formed *ex ante*.

To begin with, *ex post* core allocations are easy to characterize using the usual core notion for an exchange economy, pointwise for each $\omega \in \Omega$.

Definition 3.1 Given $x' \in \prod_{i \in I} L^R_{X_i}$ let $C(\mathscr{E};x')$ denote the following set of allocations: $x \in \prod_{i \in I} L_{X_i}$ such that

(i) $\sum_{i \in I} x_i(\omega) = \sum_{i \in I} x'_i(\omega)$ $\omega - a.e.$

(ii) *For* $\omega - a.e.$ $\nexists\ S \subset I$ and $y \in \prod_{i \in S} X_i(\omega)$ such that $\sum_{i \in S} y_i = \sum_{i \in S} x'_i(\omega)$ and $u_i(\omega, y_i) > u_i(\omega, x_i(\omega))$ for all $i \in S$.

i.e., $C(\mathscr{E};x')$ is the set of statewise core allocations resulting from a given *ex ante* trade x'.

We now turn to characterize the set of *ex ante* trades. Since trade is sequential, *ex ante* trades have to be evaluated on the basis of the final allocation they give rise to. This is the key idea behind the following definition.

Definition 3.2 A two period trading plan $(x_1, x_2) \in \prod_{i \in I} L^R_{X_i} \times \prod_{i \in I} L_{X_i}$ is in the *two-stage core* if the following are true

(i) $\sum_{i \in I} x^1_i(\omega) = \sum_{i \in I} e_i(\omega)$ $\omega - a.e,$ $x^2 \in C(\mathscr{E};x^1)$

(ii) $\nexists\ S \subset I$ and $y \in \prod_{i \in S} L^R_{X_i}$ so that $\sum_{i \in S} y_i(\omega) = \sum_{i \in S} e_i(\omega)$ $\omega - a.e,$ and $v_i(y_i) > v_i(x^2_i)$ for all $i \in S$.

The first requirement states that the *ex ante* contracted allocation (i.e., x^1) must be feasible and that the *ex post* allocation is a core allocation for the given *ex ante* trade. The second requirement is that it must not be possible for a coalition to arrange *ex ante* trades among its members, that would give them a higher expected utility than the given *ex post* allocation (i.e., x^2). In particular, (ii) precludes the possibility of *ex ante* trades feasible for a coalition in which all members of the coalition improve upon the given *ex post* allocation in *every individually rational outcome that would follow in each state.*

Notice that at no point in this definition the utility of x^1 *per se* matters. According to definition 3.2 when a two period agreement (x^1, x^2) is proposed, the crucial matter in accepting the *ex ante* trade (i.e., x^1) is not its payoff but rather the payoff of the final allocation which will follow that (i.e., x^2). The *ex ante* trade x^1 is crucial to the extend that it rationalizes x^2 as an irrefutable allocation in each state. The interpretation of definition 3.2 is the following. When a coalition considers alternative *ex ante* trades the final allocation is not predictable because it depends on the *ex ante* trades of the complementary coalition. The two-stage core takes the conservative point of view that a coalition blocks an *ex ante* trade, only if it can secure improvements to its members, for any possible (that is, individually rational) scenario that might follow as a consequence of a deviation. In this regard the two-stage core describes bargaining outcomes where agents cannot predict the *ex post* trades that will follow after an *ex ante* trade is agreed upon. In view of this uncertainty the best that an individual or a coalition can do, is *to maximize the minimal expected payoff of a first period trade* (the worst case scenario of each member). From a coalitional point of view, the blocking notion (ii) says that it should not be possible for a coalition to increase the minimal expected payoffs of all its members at the same time.

We now turn to the main result of this section. The following theorem demonstrates that core allocations in the sense of definition 3.2 always exist under standard assumptions.

Theorem 3.3. *Let* $\{\mathscr{E}, \mathscr{R}\}$ *be an economy that satisfies the following conditions:*
(a.3.1) X_i *is a convex, closed and* $\mathscr{F}$ *measurable correspondence.*

(a.3.2) $u_i(.,x)$ *is integrally bounded,* $\mathscr{F}$ *measurable, and* $u_i(\omega,.)$ *is continuous and concave on* Y.
(a.3.3) *for each* $i \in I, \mathscr{R}_i$ *is a closed, convex set containing the initial endowments*[7].

Then a two-stage core allocation for this economy exists.

Proof. See Appendix II.

By way of proof of theorem 3.3 we arrive at the following useful conclusion.

Remark 3.4. Notice that the proof of the above theorem suggests a natural procedure through which two-stage core allocations can be reached. First agents reach a restricted core allocation and then treating this allocation as the new endowment, negotiate an unrestricted core allocation. However, it should be noted that the allocations formed by the procedure just described, are not the only two-stage core allocations. In particular if (x^1, x^2) is a two-stage core allocation x^1 need not be a first period core allocation.

---

[7] Any set of linear restrictions gives rise to closed convex sets. The requirement that the endowment of each agent satisfies the restrictions imposed on this agent's *ex ante* trades, i.e. (a.3.3), merely guarantees that there is voluntary trade.

4 Incomplete asset markets economies

In order to specify the two-stage core for the incomplete asset markets framework a few qualifications are necessary. First, we have to specify the *ex ante* choice space of individuals in terms of portfolios of assets, since in this framework *ex ante* agreements take the form of asset trades. Second, we must allow for the possibility that agents can go arbitrarily short in asset trading. In order to do this we must allow for consumption sets which are not bounded below, as some agents may promise to deliver a very large quantity of a commodity in some states and then plan to take some of it back if one of these states occur. In this case agents can certainly promise to deliver an infinite amount of a commodity and extensions of the two-stage core to the case of arbitrary short sales require the use of consumption sets with no lower bound. We will provide here a basic result that shows the existence of core allocations in an economy with unbounded consumption sets.

To this end in this section we will take $Y = \mathfrak{R}^l$ to be the consumption set of each individual in each state. Consider the case where agents arrange *ex ante* trades using a finite number of securities which are available in the first period. A (real) asset is a mapping $r : \Omega \to Y$ and we will require that $r \in L^1(\mu, Y)$. In other words an asset r is a promise of a return $r(\cdot)$, contingent upon the realization of a state of nature. Let $(r_k)_{k=1}^m$ summarize the assets available in the economy—we will refer to this collection as the *asset structure*. An asset trade (or a portfolio of assets) is a vector $\theta_i \in \mathfrak{R}^m$, where θ_i^k specifies the number of the kth asset that agent i holds. If $\theta_i^k > 0$ (respectively $\theta_i^k < 0$) then agent i demands (supplies) the kth asset. Upon realization of a state an individual holding a portfolio $\theta_i \in \mathfrak{R}^m$ commands the net commodity bundle given by $e_i(\cdot) + \sum_{k=1}^{m} \theta_i^k \cdot r_k(\cdot)$. Given an asset structure the allocations that can be contracted in the first period are those that can be attained through trade of the existing assets. In this section we specify $\mathscr{R}$ in the definition of the economy in section 2 as follows. For each $i \in I$ let $\mathscr{R}_i = sp(r_k)_{k=1}^m$. We can specify now an asset markets economy as a pair $\{\mathscr{E}, (r_k)_{k=1}^m\}$.

In the presence of asset markets with an incomplete structure, the first period consumption set of each agent $i \in I$ can be specified as follows:

$$L_{X_i}^A = L_{X_i} \cap \{x \in L^1(\mu, Y) : x - e_i \in sp(r_k)_{k=1}^m\}$$

i.e., the allocations attainable through the exchange of assets.

A remark is in order regarding the incompleteness of the asset structure. Asset markets in this setup can be incomplete in the sense that the asset structure does not allow for full diversification of trades across states. Notice that, in general, full diversification requires a number of assets at least equal to $|\Omega| \times l$. In particular, for any number of real numeraire assets with independent returns across states does not allow for complete spanning of L_{X_i}.

Now, we can define the choice space of each agent in terms of the portfolio choices available to them in the first period as follows:

$$L_{\theta_i} = \left\{ \theta_i \in \Re^m : e_i + \sum_{k=1}^{m} \theta_i^k \cdot r_k(\cdot) \in L_{X_i}^A \right\}$$

Notice that L_{θ_i} has no lower bound.

Using L_{θ_i} in place of $L_{X_i}^R$ in definition 3.2 we derive the core of an asset markets economy as a specification of the two-stage core. In particular, we have the following definition of the core of an asset market economy.

Definition 4.2 A collection of two period trading plans $(\theta, x) \in \prod_{i \in I} L_{\theta_i} \times \prod_{i \in I} L_{X_i}$ is in *the core of an asset market economy* if

(i) $\sum_{i \in I} \theta_i = 0, \; x \in C(\mathscr{E}; \theta)$

(ii) $\nexists \; S \subset I$ and $\phi \in \prod_{i \in S} L_{\theta_i}$ so that $\sum_{i \in S} \phi_i = 0$ and

$$v_i\left(e_i + \sum_{k=1}^{m} \phi_i^k \cdot r_k \right) > v_i(x_i) \text{ for all } i \in S.$$

As in the original definition of the two-stage core (i) requires that the asset trades are feasible and the final commodity allocation is a statewise core allocation for the given asset trades and (ii) requires that no coalition of agents can improve over the final allocation by trading assets among themselves. It can be shown that requirement (ii) is equivalent to the following statement:

(ii)$'$ $\nexists \; S \subset I$ and $\phi \in \prod_{i \in S} L_{\theta_i}$ so that $\sum_{i \in S} \phi_i = \sum_{i \in S} \theta_i$ and

$$v_i\left(e_i + \sum_{k=1}^{m} (\phi_i^k - \theta_i^k) \cdot r_k \right) > v_i(x_i) \text{ for all } i \in S.$$

This last statement (ii)$'$ gives rise to an alternative interpretation of its equivalent statement (ii). Namely, that no coalition of agents can improve over the final allocation x by *redistributing* assets among its members. In other words no group of agents can improve over the final commodity allocation by reallocating the asset gains and liabilities among the members of the group.

The following theorem asserts the existence of core allocations for an economy as the one above. Notice that in view of remark 3.4 we need to show that core allocations in the *ex ante* sense exist. The same argument can be used to show the existence of core allocations in each state.

Define for each $i \in I$ w_i: $L_{\theta_i} \to \Re$, as $w_i(\theta) = v_i(e_i + \sum_{k=1}^{m} \theta_i^k \cdot r_k)$, (i.e. the utility function induced on portfolios from the consumption of their returns). Notice that since $u_i(\omega, \cdot)$ is continuous and concave on Y, $w_i(\cdot)$ is continuous and concave on L_{θ_i}.

Also, define $P_i : L_{\theta_i} \to 2^{L_{\theta_i}}$ as $P_i(\theta) = \{\phi \in L_{\theta_i} : w_i(\phi) > w_i(\theta_i)\}$. Since $w_i(\cdot)$ is concave we have that $P_i(\cdot)$ is convex valued. Let $O^+P_i(\cdot)$ denote the recession cone of $P_i(\cdot)$(see Appendix I). We will adopt the following assumption:

[P] For each $S \subset I$ we have: If $\theta \in \prod_{i\in S} L_{\theta_i}$ with $\sum_{i\in S} \theta_i = 0$

and $\theta_i = 0 \in O^+P_i(0)$ for all i, then it must be $\theta_i = 0$ for all $i \in S$.

This type of assumption has appeared in the literature in many different ways, and is generally referred to as limited arbitrage condition. The one we use here, drawn from Page [11], restricts the *structure* of the preferences existing in the economy rather than individual preferences *per se.* It guarantees that no subset of agents could engage in unbounded and mutually improving trades of assets. Similar conditions have appeared elsewhere such as in Werner [16], in the study of competitive equilibria in a general equilibrium model with unbounded consumption sets. Lately, similar results on the core have appeared in Chichilnisky [3] and in Page-Wooders [12].

We are now ready to state the following theorem.

Theorem 4.3. *Let* $\Lambda = \{(L_{\theta_i}, w_i, 0) : i \in I)\}$ *be an economy where* $w_i(\cdot)$ *is continuous and concave on* L_{θ_i} *for each* $i \in I$ *and preferences in the economy satisfy* [P]. *Then the core of the economy in non empty.*

Proof. See Appendix III.

4.1 Application: The Hart (1975) example

Hart (1975) has provided an example, where a competitive equilibrium fails to exist. We demonstrate in this section that the two-stage core is non empty in that example. Our study of the Hart (1975) example is motivated by the observation that the structure of preferences there satisfies all the assumptions of theorem 4.3. The utility functions in that example are defined in the non-negative orthant only, so that indifference curves of both agents are bounded below. Clearly, if indifference curves are bounded below then the recession cone of the preferred set is the non-negative orthant. Thus, the recession cones of the preferred sets of both agents in that example are the same and each one is contained in a half space. Therefore, [P] is satisfied. According to theorem 4.3 a two-stage core allocation should exist in this example and indeed this is the case. In order to see this consider Hart's (1975) original example which, for convenience, we reproduce below.

There are two states of nature $\Omega = \{a, b\}$ each one occurring with the same probability. In each state there are two commodities available. The asset structure consists of two securities with returns summarized by the following matrix:

	a	b
r_1	$(1, 0)$	$(1, 0)$
r_2	$(0, 1)$	$(0, 1)$

There are two agents $I = \{i, j\}$ with preferences represented by:

$$v_i(x_i, y_i) = 2^{3/2}\{x_i(a)\}^{1/2} + \{y_i(a)\}^{1/2} + 2^{3/2}\{x_i(b)\}^{1/2} + \{y_i(b)\}^{1/2} \text{ for agent } i$$

$$v_j(x_j, y_j) = \{x_j(a)\}^{1/2} + 2^{3/2}\{y_j(a)\}^{1/2} + \{x_j(b)\}^{1/2} + 2^{3/2}\{y_j(b)\}^{1/2} \text{ for agent } j.$$

Finally, the endowments of the agents in each state are summarized by the matrix below:

	a	b
e_i	$(5/2, 50/21)$	$(13/21, 1/2)$
e_j	$(1/2, 13/21)$	$(50/21, 5/2)$

In order to find a two-stage core allocation for this example, by remark 3.4 we need to calculate a first period core allocation and then treating this allocation as the new endowment, calculate a statewise core allocation. We will establish here that a first period core allocation exists for the asset trades. After that a statewise core allocation is only a matter of calculations.

To this effect let (z_i^1, z_i^2) and (z_j^1, z_j^2) denote the portfolio holdings of agents i and j respectively. We solve the following problem:

$$\begin{aligned} \max\ & 2^{3/2}\{5/2 + z_i^1\}^{1/2} + \{50/21 + z_i^2\}^{1/2} + 2^{3/2}\{13/21 + z_i^1\}^{1/2} \\ & + \{1/2 + z_i^2\}^{1/2} + \{1/2 + z_j^1\}^{1/2} + 2/3\{13/21 + z_j^2\}^{1/2} \\ & + \{50/21 + z_j^1\}^{1/2} + 2^{3/2}\{5/2 + z_i^2\}^{1/2} \\ & \text{s.t } z_i^1 + z_j^1 = 0 \text{ and } z_i^2 + z_j^2 = 0 \end{aligned}$$

This problem has the solution $z_i^1 = -z_j^1 = 0.4270$ $z_i^2 = -z_j^2 = -0.4270$. Finally, it can be checked that the allocation:

	a	b
(x_i, y_i)	$(5/2 + 0.4270, 50/21 - 0.4270)$	$(13/21 + 0.4270, 1/2 - 0.4270)$
(x_j, y_j)	$(1/2 - 0.4270, 13/21 + 0.4270)$	$(50/21 - 0.4270, 5/2 + 0.4270)$

is individually rational for both agents. In particular, we have:

$$v_h(x_h, y_h) = 9.3998 > 8.9476 = v(e_h) \text{ for } h = i, j.$$

Thus, a first period core allocation exists. From this point on it is easy to calculate a statewise core allocation using as endowments the allocation just calculated. By remark 3.4 the resulting pair will be a two stage core allocation.

5 Differential information economies

An alternative case for the restrictions on the first period trades is the presence of differential information. In this case agents confine themselves to state contingent trades, which are verifiable by using the information available to them. For example, given two states ω and ω' an agent would not accept a trade in ω which is different than the trade in ω', unless he/she can distinguish between the two states using the information available to him/her.

Formally, for each $i \in I$ let $\mathscr{F}_i$, where $\sigma(u_i, e_i) \subset \mathscr{F}_i$, be a countably generated sub-σ-field of $\mathscr{F}$, denoting the *private* information of agent i. We can specify now the informational constraint of each agent as follows:

$$\text{for each } i \in I, \mathscr{R}_i = \{x : \Omega \to Y |\ x \text{ is } \mathscr{F}_i - \textit{measurable}\}.$$

The first period consumption set of each agent $i \in I$ can be specified as follows:

$$L^P_{X_i} = L_{X_i} \cap \{x : \Omega \to Y |\ x \text{ is } \mathscr{F}_i - \textit{measurable}\}$$

In this case, we have the following extension of the core notion developed in Yannelis [18].

Definition 5.1. A collection of two period trading plans $(x^1, x^2) \in \prod_{i \in I} L^P_{X_i} \times \prod_{i \in I} L_{X_i}$ is in the *private two-stage core* if

(i) $\sum\limits_{i \in I} x^1_i(\omega) = \sum\limits_{i \in I} e^1_i(\omega), \omega - a.e,\ x^2 \in C(\mathscr{E}, x^1)$

(ii) $\nexists\ S \subset I$ and $y \in \prod\limits_{i \in S} L^P_{X_i}$ so that $\sum\limits_{i \in S} y_i(\omega) = \sum\limits_{i \in S} e^1_i(\omega), \omega - a \cdot e,$ and $v_i(y_i) > v_i(x^2_i)$ for all $i \in S$.

The information available to each agent or each coalition may depend on a communication structure available in the economy. By appropriate specifications of the informational restrictions of each individual or groups of individuals one may obtain the natural extensions of the fine and coarse core versions developed in Wilson [17].

Notice that in the private version of the two-stage core agents face 'asymmetric' restrictions in their trades. It is for this reason that in economies where *ex ante* trades are arranged through assets, this version of the two-stage core corresponds to the case of restricted participation in asset markets studied in Balasko-Cass-Siconolfi [2]. Indeed, the presence of differential information provides reasonable grounds for restricted participation in asset markets as we conclude in the following section.

5.1 Incentive compatibility

One might wonder why in the presence of differential information agents would confine themselves to trades compatible with their private information only, instead of pooling their information. The answer to this question is based on incentive considerations. Specifically, if an agent agrees to make trades contingent on states that he/she cannot distinguish, then it might be possible for another agent or group of agents to strategically misreport the state to their advantage. Our objective in this section is to examine the possibility of strategic revelation of the state in differential information models. We first provide a definition of *coalitionally incentive compatible* two period trades.

In the following definition for each $\omega \in \Omega$ we denote by $E_i(\omega)$ the event in the information set of agent i, that contains ω.

Definition 5.1.1. A two period allocation $(x^1, x^2) \in \prod_{i \in I} L^R_{X_i} \times \prod_{i \in I} L_{X_i}$ is *incentive compatible* if for $\omega - a.e.$ it is not true that there exist $S \subset I$ and $y_i \in \prod_{i \in S} X_i(\omega)$ so that

(i) $\omega' \in \bigcap_{i \notin S} E_i(\omega)$

(ii) $\sum_{i \in S} y_i = \sum_{i \in S} [e_i(\omega) + x_i^1(\omega') - e_i(\omega')]$, and $u_i(\omega, y_i) > u_i(\omega, x_i^2(\omega))$ for all $i \in S$

Requirement (i) says that individuals not in the coalition S should not be in a position to distinguish between the true state and the false one announced by the coalition S. In other words the coalition S can misreport the state ω without being detected by agents not in the coalition. Condition (ii) models the idea that misreporting the state does not enable members of the coalition S to achieve an allocation that improves all the members of the coalition, upon the allocation that they would receive by revealing the true state. This definition has the flavor of the core in that it requires that no coalition can cheat in an 'undetectable' way and attain a feasible allocation that improves all its members i.e., there is group compatibility of incentives. The underlying principle to definition 5.1.1 is that the incentive to misreport a state arises from the possible benefits resulting for the members of the 'cheating' coalition. Incentive compatibility requires the truthful revelation of the state to be a dominant strategy i.e., truthful revelation of the state must be a strategy which is at least as advantageous as any false announcement of the state. Thus, definition 5.1.1 precludes the possibility that the members of a coalition can misreport the state, receive the *ex ante* trades that have been agreed upon for the misreported state and make improvements by engaging in transactions 'under the table'.

The incentive compatibility criterion presented above, is referred to as strong coalitional incentive compatibility in Krasa-Yannelis [8]. However, no existence result is provided there for this notion of incentive compatibility. It is quite different than the incentive compatibility criterion appearing in Koutsougeras-Yannelis [7]. In the above papers the model considered does not allow for trades in the second period, after a state is announced. Agents agree *ex ante* in state contingent deliveries and do not meet again *ex post* for spot trades. Thus, in the incentive compatibility criteria developed in the above papers there is no room for retrading after misreporting the state that has occurred. We differ in the following important aspect: We allow for spot trades after agents make the *ex ante* promised deliveries, so we need to evaluate incentives with respect to the allocation *after* spot trades have taken place.

We need to secure first that incentive compatible allocations exist. Next, notice that even though an outcome may guarantee the compatibility of incentives it need not be optimal. The following theorem and its corollary show the existence of allocations that are both fully optimal and incentive compatible.

Theorem 5.1.2. *Let* $(x^1, x^2) \in \prod_{i\in I} L^P_{X_i} \times \prod_{i\in I} L_{X_i}$ *be a collection of two period trading plans, where* $\sum_{i\in I} x_i^1(\omega) = \sum_{i\in I} e_i(\omega)$ $\omega - a \cdot e$ and $x^2 \in \mathscr{C}(\mathscr{E}, x^1)$. *Then,* (x^1, x^2) *is an incentive compatible allocation.*

Proof. Suppose not, then for a set $A \subset \Omega$, where $\mu(A) > 0$, we have that for each $\omega \in A$ there exist $S \subset I, \omega' \in \Omega$ and $y \in \prod_{i\in S} X_i(\omega)$ with $\omega' \in \bigcap_{i\notin S} E_i(\omega)$ $\sum_{i\in S} y_i = \sum_{i\in S} [e_i(\omega) + x_i^1(\omega') - e_i(\omega')]$, and $u_i(\omega, y_i) > u_i(\omega, x_i^2(\omega))$ for all $i \in I$.

Since $\sum_{i\in I} x_i^1(\omega) = \sum_{i\in I} e_i(\omega)$ $\omega - a \cdot e.$ we have $\sum_{i\in I} [x_i^1(\omega) - e_i(\omega)] = 0$ $\omega - a \cdot e.$

Hence, for each $\omega \in A$ $\sum_{i\in S} [x_i^1(\omega) - e_i(\omega)] = -\sum_{i\notin S} [x_i^1(\omega) - e_i(\omega)]$.
By measurability of the net trades and the fact that $\omega' \in \bigcap_{i\notin S} E_i(\omega)$ we have

$$-\sum_{i\notin S} [x_i^1(\omega) - e_i(\omega)] = -\sum_{i\notin S} [x_i^1(\omega') - e_i(\omega')].$$

It follows then that

$$\sum_{i\in S} [x_i^1(\omega) - e_i(\omega)] = \sum_{i\in S} [x_i^1(\omega') - e_i(\omega')].$$

Thus,

$$\begin{aligned} \sum_{i\in S} y_i &= \sum_{i\in S} [e_i(\omega) + x_i^1(\omega') - e_i(\omega')] \\ &= \sum_{i\in S} [e_i(\omega) + x_i^1(\omega) - e_i(\omega)] \\ &= \sum_{i\in S} x_i^1(\omega). \end{aligned}$$

However, this last equality along with $u_i(y_i, \omega) > u_i(\omega, x_i^2(\omega))$ for all $i \in S$ imply that x^2 is blocked for each $\omega \in A$, a contradiction to $x^2 \in \mathscr{C}(\mathscr{E}, x^1)$. □

The following corollary follows directly from definition 3.2.

Corollary 5.1.4. A private two-stage core allocation is an incentive compatible core allocation.

Remark 5.1.5. Theorem 5.1.2 only requires that $x^2 \in \mathscr{C}(\mathscr{E}, x^1)$ i.e. that x^2 is in the core $\omega - a.e.$, given a feasible and $\mathscr{F}_i$-measurable *ex ante* trade x^1. Thus theorem 5.1.2 will still hold for any equilibrium outcome in the second period that is in the core for $\omega - a.e.$ In particular if x^2 is a Walrasian allocation $\omega - a.e.$, it will be incentive compatible in the sense of definition 5.1.1.

Remark 5.1.6. It should be noted that definition 5.1.1 and Theorem 5.1.2 apply also for the case where *ex ante* trades are made through assets. A different incentive compatibility criterion in the context of incomplete asset markets is developed in Younés [20]. In that paper the author studies incentives in a competitive environment and addresses the incentive compatibility properties of competitive equilibria.

This incentive compatibility result can be viewed as a justification for the restriction of the first period trades to those that are compatible with the private information available to each agent. This justification is based on the *incentive structure that this kind of asset trades generate.* The use of private information is a sufficient condition for incentive, compatibility. On the other hand we emphasize at this point that it is not necessary i.e., there may be other information sharing rules for which incentive compatible trades in the sense discussed here exist. However, as we demonstrate below by means of an example, even with very simple information structures, there are incentive problems in any information sharing rule that involves (complete or partial) pooling of information.

Certainly, one could imagine that the *ex ante* contingent trades are made through the exchange of assets and the asset structure may or may not be complete. In that case one only needs to express the first period trades in terms of portfolios of assets, without altering the validity of the analysis. In an asset market economy with differential information, theorem 5.1.2 provides a case for restricted participation in asset markets as in Balasko-Cass-Siconolfi [2]. One may argue that in the presence of differential information agents may choose to restrict themselves to portfolios with privately verifiable returns, because in this way they avoid being cheated by a misreport of the state that occurs, according to theorem 5.1.2. In other words individuals restrict themselves to asset trades that guarantee the compatibility of incentives after a state is realized.

6 Example

In this section we present an example of an economy which is characterized by differential information in addition to an incomplete asset market structure. The purpose here is to demonstrate the incentive compatibility results discussed in section 5.1. The presence of incomplete asset markets in this example merely simplifies calculations. Specifically, we consider an economy which is summarized as follows:

There are three states of nature $\Omega = \{a, b, c\}$ each one occurring with the same probability. In each state there are two commodities available. The asset structure consists of three securities with returns summarized by the following matrix:

	a	b	c
r_1	$(1,0)$	$(0,0)$	$(0,0)$
r_2	$(0,0)$	$(1,0)$	$(0,0)$
r_3	$(0,0)$	$(0,0)$	$(1,0)$

There are three agents $I = \{j, k, l\}$ with preferences represented by

$$u_h(\omega, x_h(\omega), y_h(\omega)) = [x_h(\omega) \cdot y_h(\omega)]^{1/2}, \textit{ for } h = j, k, l, \text{ and } \omega = a, b, c.$$

The endowments of the agents in each state are summarized by the matrix below

	a	b	c
e_j	$(10,4)$	$(10,4)$	$(0,4)$
e_k	$(10,1)$	$(0,4)$	$(10,1)$
e_l	$(0,0)$	$(0,0)$	$(0,0)$

The private information of each agent is $\mathscr{F}_j = \{\{a,b\},\{c\}\}, \mathscr{F}_k = \{\{a,c\},\{b\}\}$, and $\mathscr{F}_1 = \{\{a\},\{b,c\}\}$.

It can be verified that the following is a security-consumption private core allocation

	θ	$x(a)$	$x(b)$	$x(c)$
agent j	$(-2,-2,5)$	$(12,3)$	$(13/2, 26/5)$	$(13/2, 13/4)$
agent k	$(-5,2,-5)$	$(9/2, 9/8)$	$(7/2, 14/5)$	$(7/2, 7/4)$
agent l	$(7,0,0)$	$(7/2, 7/8)$	$(0,0)$	$(0,0)$

First notice that in this allocation no group of agents can, or would like to misreport state (a) whenever it occurs: Suppose that state (a) has occurred. Then the coalition $\{j, l\}$ can instead report that state (c) has occurred. In that case agents j and l receive the returns (15,4) and (0,0) respectively. After some tedious calculations, which we omit, it can be verified that the following relations cannot hold simultaneously.

$$x_j + x_l = 15, y_j + y_l = 4 \text{ (feasibility)}$$

$$(x_j \cdot y_j)^{1/2} > \sqrt{12.3} \text{ and } (x_l . y_l)^{1/2} > \sqrt{7/2.7/8} \text{ (dominance)}$$

A similar argument establishes that the same holds if the coalition $\{k, l\}$ claimed that the state is (b). This demonstrates that the security-commodity trade in the core allocation above is, according to definition 5.1.1, incentive compatible.

Paradoxically, as a result of the asset trade in the first period, due to the agents' j and k desire for risk diversification agent 1 ends up holding a portfolio with positive returns and hence positive consumption in state (a), although this agent started with zero endowments. Agent l acts as an intermediary between agents j and k and as a result of this agent l ends up holding a portfolio with positive returns. This result reveals an important aspect of restricted participation in asset markets, not accounted for in Balasko-Cass-Siconolfi [2], i.e., that the two-stage core rewards agents with privileged access to asset markets that can intermediate among agents with restricted access to asset markets.

In order to make the contrast clear suppose that agents j and k make state contingent trades which are measurable with their pooled information. Then they will both have full information. An asset-commodity core allocation calculated in this case is the following (weight i's utility by 1/3, j's by 2/3 and maximize the sum of the expected utilities)

	θ	$x(a)$	$x(b)$	$x(c)$
agent j	$(0,-8,5)$	$(13,13/4)$	$(7/2,14/5)$	$(13/2,13/4)$
agent k	$(0,8,-5)$	$(7,7/4)$	$(13/2,26/5)$	$(7/2,7/4)$
agent l	$(0,0,0)$	$(0,0)$	$(0,0)$	$(0,0)$

Suppose that (*a*) occurs but $\{k, l\}$ announce that the state is (*b*) instead, in which case this coalition receives the aggregate bundle (18, 1). Then notice that this coalition can achieve the allocation:

$$(x_k, y_k) = (17, 3/4)$$
$$(x_l, y_l) = (1, 1/4)$$

Notice now that $x_k + x_l = 18, y_k + y_l = 1, (x_k.y_k)^{1/2} > 7/2, (x_l.y_l)^{1/2} > 0$.

To recap: Agent k can 'make it worth' to agent l, so that they jointly announce (*b*) rather than (*a*) that has truly occurred.

Finally notice that in this case, once information has been pooled, the privileged access of agent l in asset markets disappears and this agent ends up with zero consumption.

7 Concluding remarks

We believe that the study of cooperative concepts in a sequential setting is interesting because the extensions of these concepts in dynamic models are not straightforward. A dynamic framework allows the possibility of transient cooperation which in turn leads to situations where cooperative and non cooperative elements coexist. In this paper we have introduced a core notion for economies with sequential trade. The notion we discussed here is based on the principle that coalitions block only when they can make trades which are improving in any consequence that may follow. This makes blocking rather difficult. However, one can construct a variety of core concepts by employing different blocking notions, based on how individuals may percieve the future consequences of a deviation. This will be the topic of a different paper. Our approach can be used as a basis to develop other cooperative solution concepts for sequential games.

The possibility of temporary cooperation arises in cases where agents cannot arrange trades for all current and future contingent commodities at once. This is the case in incomplete asset markets and in differential information models. It turns out that the core has some interesting properties when applied to this class of models. However, restrictions on state contingent forward trades arise naturally in a large variety of fields in economics and finance, ranging from simple insurance or wage contracts up to international financial markets. The core notion presented in definition 3.2 above, is general enough to capture any kind of constraints (regional, institutional etc.) that one may wish to incorporate in an economic model. Hence, the underlying idea of the two-stage core developed in this paper, can be applied to a wide variety of economies with uncertainty and limited risk diversification. This makes our approach look promising in further applications.

Another interesting aspect of our analysis is that it may be useful in understanding the nature of the incompleteness of the asset structure. One may argue that in the presence of differential information agents may confine themselves to portfolios with returns that are privately verifiable because in this way the compatibility of the incentives to actually carry out the pre-committed trades is guaranteed. This results in agents trading only a subset of the assets available. Moreover, the example presented here has obvious implications regarding the role of intermediation. Our example reveals how intermediation arises due to restricted participation in asset markets, which directly corresponds to our model. Restricted participation gives rise to arbitrage opportunities in the sense that an agent with zero endowments but privileged access to some asset markets ends up with positive consumption, simply by intermediating in the asset trades of the other agents.

Further work on the theoretical context could involve the study of convergence properties of the core notion presented here. This is interesting because, as we showed here, the asset market version of our core notion exists in cases where rational expectations equilibria do not exist. This is also interesting for the differential information case because we would be able to study which, if any, of the complications arising from the presence of differential information persist in the limit.

8 Appendix I

In this section we have collected some preliminary mathematical facts that will be used in our proofs. Let X, Y be two sets. The graph of a correspondence, $f : X \to 2^Y$ is defined as $G_f = \{(x, y) \in X \times Y : y \in f(x)\}$. Let (T, τ, μ) be a complete, finite measure space and X be a separable Banach space. Let $\mathscr{B}(X)$ denote the Borel σ-algebra of X i.e. the smallest σ-algebra containing all the open subsets of X. The correspondence $\varphi : T \to 2^X$ is said to have a measurable graph if $G_\varphi \in \tau \otimes \mathscr{B}(X)$. We say that $\varphi : T \to 2^X$ is lower measurable if for every open $V \subset X$ we have that $\{t \in T : \varphi(t) \cap V \neq \emptyset\} \in \tau$. It is a known result that if $\varphi : T \to 2^X$ has a measurable graph then it is lower measurable. Further, if $\varphi : T \to 2^X$ is lower measurable and closed valued then it has a measurable graph. A well known result by Aumann that will be useful to us states that if (T, τ, μ) is a complete finite measure space, X is a separable metric space and $\varphi : T \to 2^X$ is non empty valued and has a measurable graph, then there exists a measurable function $f : T \to X$ such that $f(t) \in \varphi(t)$, t-a.e.

A function $f : T \to X$ is called simple if there exist $x_1, x_2, x_3, \ldots, x_n$ in X and $a_1, a_2, a_3, \ldots, a_n$ in T such that $f = \sum_{i=1}^n x_i \cdot \chi_{a_i}$ where $\chi_{a_i} = 1$ for $t \in a_i$ and $\chi_{a_i} = 0$ for $t \notin a_i$. A function $g : T \to X$ is called measurable if there is a sequence of simple functions $g_n : T \to X$ such that $\lim_{n \to \infty} \|g_n(t) - g(t)\| = 0$ t-ae. A measurable function $f : T \to X$ is called Bochner integrable if there exists a sequence of simple functions $g_n : T \to X$ such that $\lim_{n \to \infty} \int \|g_n(t) - f(t)\| d\mu(t) = 0$. In such a case the integral of f is defined to be $\int f(t) = \lim_{n \to \infty} \int g_n(t) d\mu(t)$. We will denote by $L_1(\mu, X)$ the space of equivalence classes of Bochner integrable functions $f: T \to X$ which is normed by $\|f\| = \int \|f(t)\| d\mu(t)$. It is a known result that $L_1(\mu, X)$ normed in this way is a Banach space.

We turn now to introduce some elementary facts about the recession cones of convex sets. A standard reference for this part is Rocafellar [13]. Let $C \subset \Re^m$ be a non empty convex set. We say that C recedes in the direction D if C includes all the half-lines in the direction D, that start at points of C, i.e. C recedes in the direction of y if and only if for every $x \in C$ and $\lambda \geq 0, x + \lambda y \in C$. The set of all vectors $y \in \Re^m$ in the direction of which C recedes is referred to as the directions of recession of C, denoted as $O^+(C)$. Theorem 8.1 in Rockafellar establishes that $O^+(C)$ is a convex cone containing the origin, called the recession cone of C. In fact $O^+(C)$ can be thought of as the largest cone contained in C. Another fact that we use in the sequel is that if $C \subset \Re^m$ is a non empty convex set then $O^+(riC) = O^+(clC)$ (Rockafellar [13] corollary 8.3.1, pp. 63). Moreover, for an arbitrary collection A of closed convex sets with non empty intersection we have $O^+(\bigcap_{i \in A} C_i) = \bigcap_{i \in A} O^+(C_i)$. Finally, for a non empty convex set $C \subset \Re^m$ we have that C is bounded if and only if $O^+(C) = 0$. With these preliminaries out of the way we now proceed to the proofs of our results.

9 Appendix II

For the proof of theorem 3.3 we will make use of the following proposition:

Proposition 9.1. *There is a measurable function $x : \Omega \to Y^I$ such that $x \in C(\mathscr{E}, x)$.*

Before we start with the proof of proposition 9.1 some notation and elementary facts are in order. For each $S \subset I$ let Y^S be the $\|S\|$-fold Cartesian product of Y. Define $F_S : \Omega \times Y^I \to 2^{Y^S}$ as $F_S(\omega, x) = \{y \in Y^S : \sum_{i \in S} y_i = \sum_{i \in S} e_i(\omega)\}$. Moreover, define $P_i : \Omega \times Y^I \to 2^Y$ for each $i \in I$ as $P_i(\omega, x) = \{y \in Y : u_i(\omega, y) > u_i(\omega, x_i(\omega))\}$ (the preference correspondence of each agent) and $P_S = \prod_{i \in S} P_i$ (the preferred set of each coalition). Let now $B_S : \Omega \times Y^I \to 2^{Y^S}$ be defined as $B_S(\omega, x) = P_S(\omega, x) \cap F_S(\omega, x)$.

Lemma 9.2. *The correspondence B_S is lower measurable.*

Proof. Given $y \in Y$ define the functions $f_y^i : \Omega \times Y^I \to \Re$ by

$$f_y^i(\omega, x) = u_i(\omega, y) - u_i(\omega, x_i(\omega))$$

for each $i \in I$. Since $u_i(\omega, \cdot)$ is continuous on Y and $u_i(\cdot, x)$ is measurable on Ω, we have that $u_i(\cdot, \cdot)$ is jointly $\sigma(u_i) \otimes \mathscr{B}(Y)$ measurable. Thus, we conclude that $f_y^i(\cdot)$ is measurable $\sigma(u_i) \otimes \mathscr{B}(Y)$ for each $y \in Y$.

Notice now that $P_i(\omega, x) = \{y \in Y : f_y^i(\omega, x) > 0\}$. Hence for each $y \in Y$ we have that

$$(f_y^i)^{-1}(0,\infty) = \{(\omega,x) \in \Omega \times Y^I : f_y^i(\omega,x) > 0\}$$
$$= \{(\omega,x) \in \Omega \times Y^I : y \in P_i(\omega,x)\} \tag{1}$$

From the measurability of $f_y^i(\cdot)$ we conclude that

$$\{(\omega,x) \in \Omega \times Y^I : y \in P_i(\omega,x)\} \in \sigma(u_i) \otimes \mathcal{B}(Y^I)$$

Therefore, $\{(\omega,x) \in \Omega \times Y^I : y \in P_S(\omega,x)\} \in (\bigvee_{i\in S} \sigma(u_i)) \otimes \mathcal{B}(Y^I)$. Next given $y \in Y_S$ define for each $S \subset I, g_y^S : \Omega \times Y^I \to \Re$ by $g_y^S(\omega,x) = \|\sum_{i\in S} y_i - \sum_{i\in S} e_i(\omega)\|_Y$, where $\|\cdot\|_Y$ is the norm of Y. We have that $g_y^S(\cdot)$ is measurable $(\bigvee_{i\in S} \sigma(e_i))\otimes \mathcal{B}(Y^I)$ for each $y \in Y^S$. Notice now that $F_S(\omega,x) = \{y \in Y^S : g_y^S(\omega,x) = 0\}$. Therefore, for each $y \in Y^S$ we have $\{(\omega,x) \in \Omega \times Y^I : y \in F_S(\omega,x)\} \in (\bigvee_{i\in S} \sigma(e_i)) \otimes \mathcal{B}(Y^I)$. Let now V be an open subset of Y^S and D a countable dense subset of Y^S. We have

$$\begin{aligned}
&\{(\omega,x) \in \Omega \times Y^I : B_S(\omega,x) \textstyle\bigcap V \neq \emptyset\} \\
&\quad = \left\{(\omega,x) \in \Omega \times Y^I : y \in B_S(\omega,x), y \in V\right\} \\
&\quad = \left\{(\omega,x) \in \Omega \times Y^I : d \in B_S(\omega,x), d \in D\right\} \\
&\quad = \left\{(\omega,x) \in \Omega \times Y^I : d \in P_S(\omega,x), d \in D\right\} \\
&\qquad \textstyle\bigcap \left\{(\omega,x) \in \Omega \times Y^I : d \in F_S(\omega,x), d \in D\right\} \\
&\quad = \left(\bigcup_{d\in D} \{(\omega,x) \in \Omega \times Y^I : d \in P_S(\omega,x)\}\right) \\
&\qquad \bigcap \left(\bigcup_{d\in D} \{(\omega,x) \in \Omega \times Y^I : d \in F_S(\omega,x)\}\right)
\end{aligned}$$

Thus, we can conclude that

$$\{(\omega,x) \in \Omega \times Y^I : B_S(\omega,x) \textstyle\bigcap V \neq \emptyset\} \in (\bigvee_{i\in S} \sigma(u_i, e_i)) \otimes \mathcal{B}(Y^I) \; \square$$

We are ready now to proceed with the proof of proposition 9.1.

Proof of Proposition 9.1. For each $S \subset I$ define

$$C_S(\omega) = \{x \in F_I(\omega) : P_S(\omega,x) \textstyle\bigcap F_S(\omega,x) = \emptyset\}$$

and let $C(\omega) = \cap_{S\subset I} C_S(\omega)$.
By the lemma above the correspondence $B_S(\omega,x) = P_S(\omega,x) \cap F_S(\omega,x)$ is lower measurable. Thus, the set

$$\{(\omega,x) \in \Omega \times Y^I : B_S(\omega,x) = \emptyset\} = \{(\omega,x) \in \Omega \times Y^I : B_S(\omega,x) \bigcap Y^S \neq \emptyset\}^c$$
$$= \{(\omega,x) \in \Omega \times Y^I : x \in C_S(\omega)\}$$
$$= G_{C_S}$$

is $(\bigvee_{i \in S} \sigma(u_i, e_i)) \otimes (Y^I)$-measurable. We conclude then that the correspondence $C(\omega) = \cap_{S \subset I} C_S(\omega)$ has a measurable graph. By Scarf's theorem [15], $C(\omega)$ is also non empty valued for each $\omega \in \Omega$. Thus, by the Aumann measurable selection theorem there exists a measurable function $x : \Omega \to Y^I$ such that $x(\omega) \in C(\omega)$ $\omega - a.e.$ □

Proof of Theorem 3.4. Recall that the set consists of allocations that are statewise in the core. Notice that for each $\omega \in \Omega$ all the conditions of Scarf's theorem [15], for the existence of core allocations are satisfied. Hence by appealing to that theorem for each $\omega \in \Omega$, we can establish that for each $x \in \prod_{i \in I} L^R_{X_i}$. Therefore, it suffices to show the existence of $x^1 \in \prod_{i \in I} L^R_{X_i}$ that satisfies (ii) in definition 3.3. From claim 5.1 in Yannelis [18] we know that L_{X_i} is a weakly compact set. By (a.3.3) $L^R_{X_i}$ is a weakly compact set as well. The economy $\Lambda = \{(L^R_{X_i}, v_i, e_i) : i \in I\}$ satisfies all the conditions of theorem 3.1 in Yannelis [18], thus there is an allocation $x^1 \in \prod_{i \in I} L^R_{X_i}$ so that $\sum_{i \in S} x^1_i(\omega) = \sum_{i \in S} e_i(\omega)$ $\omega - a.e.$, and also $\nexists S \subset I$ and $y \in \prod_{i \in S} L^R_{X_i}$ so that $\sum_{i \in S} y_i(\omega) = \sum_{i \in S} e_i(\omega)$ $\omega - a.e.$ and $v_i(y_i) > v_i(x^1_i)$, for all $i \in S$. By proposition 9.1 above there exists a measurable function $x^2 \in C(\mathscr{E}, x^1)$. We claim that (x^1, x^2) is a two-stage core allocation for the original economy.

To start with (x^1, x^2) satisfies condition (i) in definition 3.3. Suppose by way of contradiction that for some $S \subset I$ and $y \in \prod_{i \in S} L^R_{X_i}$ we have $\sum_{i \in S} y_i(\omega) = \sum_{i \in S} e_i(\omega)$ $\omega - a.e.$ and $v_i(y_i) > v_i(x^2_i)$, for all $i \in S$. However, $x^2 \in C(\mathscr{E}, x^1)$ is individually rational so for ω-ae we have $u_i(\omega, x^2_i(\omega)) \geq u_i(\omega, x^1_i(\omega))$. Thus, we have $v_i(x^2_i) \geq v_i(x^1_i)$ for all $i \in I$. But then $v_i(y_i) > v_i(x^2_i) \geq v_i(x^1_i)$ for all $i \in S$ which is a contradiction. □

10 Appendix III

In this section we show how the analysis of the two-stage core can be extended to allow for arbitrary short sales in the first period. Notice that we only need to establish the existence of first period core allocations. Statewise core allocations in the second period exist by Scarf's theorem. Finally by remark 3.4, pairs of allocations that are in the core in each period are two-stage core allocations. Thus, it suffices to prove the existence of first period core allocations when there is no lower bound on L_{θ_i}. The following lemma is the key to the proof of our existence theorem.

Lemma 10.1. [P] implies that for every coalition $S \subset I$ the set of feasible and individually rational allocations is bounded.

Proof. Suppose that assumption [P] is true.
Define $F^S = \{\phi \in \prod_{i \in S} L_{\theta_i} : \sum_{i \in S} \phi_i = 0\}$. Rewrite [P] as $\prod_{i \in S} O^+ P_i(0) \cap F^S = \{0\}$ for each $S \subset I$. Given $\theta \in F^S$ we have :

$$\begin{aligned} \theta + \lambda\phi \in F^S, \forall \lambda \geq 0 &\Leftrightarrow \sum_{i \in S} \theta_i + \lambda \sum_{i \in S} \phi_i = 0, \forall \lambda \geq 0 \\ &\Leftrightarrow \sum_{i \in S} \phi_i = 0 \\ &\Leftrightarrow \phi \in F^S \end{aligned} \tag{2}$$

Thus, $F^S = O^+ F^S$. Moreover we have $O^+(\prod_{i \in S} P_i(0)) \subset \prod_{i \in S} O^+ P_i(0)$. Hence, [P] implies that $O^+(\prod_{i \in S} P_i(0)) \cap O^+ F^S = \{0\}$. By corollary 8.3.3 in Rockafellar [13], we have that $O^+[(\prod_{i \in S} P_i(0)) \cap F^S] = \{0\}$. Finally, by theorem 8.4 in Rockafellar we conclude that $(\prod_{i \in S} P_i(0)) \cap F^S$ is bounded. □

We now turn to the existence proof of core allocations with unbounded short sales.

Proof of Theorem 4.3. Denote by $\mathscr{I}$ the family of all non empty subsets of I. Let $(I, \mathscr{V})$ be the exchange game derived from the economy $\Lambda = \{(L_{\theta_i}, w_i, 0) : i \in I\}$, where $\mathscr{V} : \mathscr{I} \to \in^{\Re^{|\mathscr{I}|}}$ is defined as follows:

$$\begin{aligned} \mathscr{V}'(S) = \{t \in \Re^{|I|} : \quad & \text{For each } i \in S, t_i \leq w_i(\phi_i) \\ & \text{for some } \phi \in \prod_{i \in S} L_{\theta_i} \text{ s.t.} \sum_{i \in S} \phi_i = 0\} \end{aligned}$$

Notice that the value functions of this game, need not be bounded above. Consider now a new game $(\mathscr{I}, \mathscr{V}')$ where:

$$\begin{aligned} \mathscr{V}'(S) = & \{t \in \Re^{|I|} : \text{For } i \in S, t_i \leq w_i(\phi_i) \text{ for some } \phi \in \prod_{i \in S} L_{\theta_i} \\ & \text{s.t.} \sum_{i \in S} \phi_i = 0, w_i(\phi_i) \geq w_i(0)\} \end{aligned}$$

Notice that $\mathscr{V}'(N) \setminus \cup_{S \subset I} \operatorname{int} \mathscr{V}'(S) = \mathscr{V}(N) \setminus \cup_{S \subset I} \operatorname{int} \mathscr{V}(S)$ i.e. $Core(I, \mathscr{V}') = Core(I, \mathscr{V})$. Therefore, it suffices to show that the core of $(I, \mathscr{V}')$ is non empty. Since [P] is satisfied we have by lemma 4.4 that the value functions $\mathscr{V}'(\cdot)$ are bounded above and it can be verified that they satisfy all the other conditions in Scarfs theorem (see Aliprantis et al [1]), i.e. they are non empty, closed and comprehensive. Furthermore, since L_{θ_i} is closed and convex and $w_i(\cdot)$ is continuous and concave it follows that the game is balanced. Therefore, all the conditions of Scarf's theorem are satisfied and we can conclude that the core of $(I, \mathscr{V})$ is non empty, which completes the proof. □

The above proof illuminates the essence of the cone condition [P] (as well as some related cone conditions) for the existence of the core: It helps recover the boundedness of the value functions that is lost when one removes the lower bounds of the consumption sets.

References

1. Aliprantis, C.D., Brown D.J., Burkinshaw, O.: Existence and Optimality of Competitive Equilibria, Berlin Heidelberg New York: Springer 1982
2. Balasko, Y., Cass, D., Siconolfi P.: The Structure of Financial Equilibrium with Exogenous Yields: The Case of Restricted Participation. Journal of Mathematical Economics **19**, 195–216 (1991)
3. Chichilnisky G.: A Unified Perspective on Resource Allocation: Limited Arbitrage is Necessary and Sufficient for the Existence of a Competitive Equilibrium, the Core and Social Choice. C.O.R.E Discussion Paper No. 9527 (1995)
4. Duffie, D., Shafer W.: Equilibrium in Incomplete Markets. I: A Basic Model of Generic Existence. Journal of Mathematical Economics **14**, 285–300 (1985)
5. Gale, D.: The Core of a monetary Economy without Trust. Journal of Economic Theory **19**, 456–491 (1978)
6. Hart, O.: On the Optimality of Equilibrium when the Market Structure is Incomplete. Journal of Economic Theory **11**(3), 418–443 (1975)
7. Koutsougeras, L., Yannelis N.C.: Incentive Compatibility and Information Superiority of the Core of an Economy with Differential Information. Economic Theory **3**, 195–216 (1993)
8. Krasa, S., Yannelis N.C.: The Value Allocation of an Economy with Differential Information. Econometrica **62**, 881–900 (1994)
9. Magill, M., Shafer W.: Characterisation of Generically Complete Real Asset Structures. Journal of Mathematical Economics **19**, 167–194 (1990)
10. Magill, M., Shafer W.: Incomplete Markets. In: Hildenbrand, Sonnenschein (eds.) Handbook of Mathematical Economics, vol. IV. Amsterdam: Elsevier 1991
11. Page, F.H. Jr.: On Equilibrium in Hart's Securities Exchange Model. Journal of Economic Theory **41**, 392–404 (1987)
12. Page, F.H. Jr., Wooders M.: Arbitrage with Price-Dependent Preferences: Equilibrium and Market Stability. Proceedings of the European Workshop on General Equilibrium Theory, Barcelona (1995)
13. Rockafellar, T.: Convex Analysis. Princeton: Princeton University Press 1970
14. Repullo, R.: The Core of an Economy with Transaction Costs. Review of Economic Studies **LV**, 447–458 (1988)
15. Scarf, H. E.: The Core of an N Person Game. Econometrica **35**, 50–69 (1967)
16. Werner, J.: Arbitrage and existence of competitive equilibrium. Econometrica **55**, 1403–1418 (1987)
17. Wilson, R.: Information, Efficiency and the Core of an Economy. Econometrica **46**, 807–816 (1978)
18. Yannelis, N.C.: (a) The Core of an Economy with Differential Information. Economic Theory **1**, 183–195 (1991)
19. Yannelis, N.C.: (b) Set Valued Functions of Two Variables in Economic Theory. In: Equilibrium Theory in Infinite Dimensional Spaces, pp. 36–72. Berlin Heidelberg New York: Springer 1991
20. Younés, Y.: Towards a Theory of Incentive Compatibility for Economies with Incomplete Market Structures. Mimeo (1992)

Balancedness and the core in economies with asymmetric information*

Stefan Maus

Department of Quantitative Economics, University of Maastricht, P.O. Box 616, 6200 MD Maastricht, THE NETHERLANDS (e-mail: s.maus@ke.unimaas.nl)

Received: December 12, 2001; revised version: November 1, 2002

Summary. A condition is given that is equivalent to balancedness of all NTU-games derived from an exchange economy with asymmetric information when endowments are variable. The condition is applicable to the ex-ante model with expected utilities, but also to the more general model of Arrow-Radner type economies without subjective probabilities. Differences in the interpretation of measurability assumptions between these two models are discussed, and another model with information consistent utility functions is developed in which the result would also hold.

Keywords and Phrases: Balancedness, Core, Communication system, Exchange economy with asymmetric information.

JEL Classification Numbers: C71, D82.

1 Introduction

The standard Arrow-Debreu model of an exchange economy can be extended to account for uncertainty by differentiating the commodities over the states of nature, as first proposed already by Debreu [4] in Chapter 7 of his 'Theory of Value'. This approach has been extended to account for asymmetric information in a seminal paper by Radner [11, 12], using economies where agents possess different information. It is assumed, that an agent can carry out only such trades that are compatible with his information structure.

After this development in the fifities and sixties, the seventies and eighties saw a growing literature on Rational Expectations Equilibrium (REE), which is a natural extension of Arrow-Debreu's deterministic model of Walrasian equilibrium

* I thank two anonymous referees whose comments led to an improvement of the paper.

to a differential information framework. However, prices in a fully revealing REE are not able to reflect the informational asymmetries of agents. This left room for criticism and further work.

A new literature emerges from the work of Wilson [15], who considers the core of an economy with differential information. Wilson considered the problem of how agents within a coalition share information. He gave two different scenarios, corresponding to the notion of fine core (pooling information) and coarse core (use only common information) of an economy with asymmetric information structure. He uses an interim core concept, where agents engage in coalitional negotiations after receiving their private information. For both core notions problems with existence and incentive compatibility arise, so that this approach was not pursued further for some time.

In the early nineties, the core was reconsidered. Yannelis [16] considered a new information sharing concept based on measurability constraints, and thus introduced the private core. This core concept exists under quite general assumptions, and it is coalitionally Bayesian incentive compatible, as shown in Koutsougeras-Yannelis [7]. Furthermore, it rewards the better informed agents, and it provides sensible outcomes in some situations where REE fails too exist. Allen [1–3] carries the development further, by introducing more general information sharing rules. Recently, Einy et al. [5] have shown a core equivalence theorem for large economies.

In one of her mentioned articles, Allen [3] considers the ex-ante core, where agents have to form coalitions before the true state of the world is revealed to them. She allows for arbitrary communication systems rather than fine or coarse communication. This raises the question to find conditions on the communication system that assure nonemptiness of the ex-ante core. In this paper such a condition is given. It turns out to be an equivalent to balancedness of all NTU-games derived from the exchange economy with asymmetric information when endowments are viewed as variable. Moreover, the condition is applicable to the model without expected utility, as introduced in Schwalbe [14]. Most articles consider agents' preferences derived from state-dependent preferences by taking expectations with respect to some subjective probability measure over the states of the world. However, such a description of preferences is not necessary. As Debreu [4] explains in Chapter 7 of his 'Theory of Value', and as is the case in Radner [11], preferences under uncertainty can be formulated without referring to probabilities. The case of expected utility functions is then included in this more general approach.

The analysis is organized as follows: In Section 2 the formal model is developed and the core concepts are defined. Section 3 contains the results and some examples to illustrate the relation of the new condition with other conditions. In Section 4 a discussion is included on measurability assumptions and how the implications of these differ for the models with and without expectations. Some concluding remarks follow in Section 5.

2 Preliminaries

In this section the relevant definitions for the notion of an exchange economy with asymmetric information are given. Two different alternatives are considered, one, where the agents have state-dependent utility functions and priors, and another one, where the utility function is defined on the state-commodity space altogether. The state-dependent utility case includes the expected utility model with finitely many states of the world as in Allen [3]. The case of a utility function avoiding the usage of expected utility by defining the utility function on the state-commodity space appears in Schwalbe [14]. There, balancedness for all exchange economies with asymmetric information is shown. But this holds only if one assumes the feasible allocations for the economy to be determined on the maximum amount of information available, that is, the information an agent would have if all coalitions could be formed simultaneously. Here, I will only assume that agents can use the information available to them in the grand coalition. This seems less artificial, and is in agreement with the expected utility model of Allen [3].

2.1 Information

Let Ω be the finite set of states of the world. Let $\mathcal{P}^*$ be the set of partitions of Ω. A $\mathcal{P} \in \mathcal{P}^*$ is called an *information set*. The interpretation is that states contained in an element $S \in \mathcal{P}$ cannot be distinguished under that information set. For each $\omega \in \Omega$ denote by $\mathcal{P}(\omega)$ the element of the partition $\mathcal{P}$ that contains ω.

An information set $\widetilde{\mathcal{P}}$ is called *finer* than $\mathcal{P}$, if every element of $\widetilde{\mathcal{P}}$ is contained in an element of $\mathcal{P}$. $\mathcal{P}$ is then called *coarser* than $\widetilde{\mathcal{P}}$.

Let N be a finite set of agents. Each agent has an initial endowment of information, described by $\mathcal{P}_i \in \mathcal{P}^*$. Then forming coalitions the information of agents may change, e.g. due to communication. Let $\mathcal{P}_i^S$ be the information that agent $i \in S$ has if the coalition S is formed. Throughout I assume that $\mathcal{P}_i^{\{i\}} = \mathcal{P}_i$. A collection $(\mathcal{P}_i^S)_{i \in S, S \subset N}$ is called a *communication system*.

$$\mathcal{P}_i^m := \bigvee_{S \ni i} \mathcal{P}_i^S := \left\{ \bigcap_{S \ni i} \mathcal{P}_i^S(\omega) | \omega \in \Omega \right\}$$

is the maximum amount of information of agent i. Checking that $\mathcal{P}_i^m \in \mathcal{P}^*$ is straightforward.

An information set $\mathcal{P}$ generates a σ-algebra $\sigma(\mathcal{P})$. A communication system $(\mathcal{P}_i^S)_{i \in S, S \subset N}$ is called *nested* if $\sigma(\mathcal{P}_i^S) \subset \sigma(\mathcal{P}_i^T)$ or equivalently $\mathcal{P}_i^S \subset \sigma(\mathcal{P}_i^T)$ for all $i \in S \subset T$. It is called *bounded* if $\sigma(\mathcal{P}_i^S) \subset \sigma(\mathcal{P}_i^N)$ for all $i \in S \subset N$.

Information restricts the possible net trades of an agent. He cannot trade different amounts on events that he cannot distinguish. Formally this is captured by the following. Let $\mathcal{P}$ be the information the agent has. Then his trades of k goods are limited to the following set of functions

$$X_{\mathcal{P}} := \{x | x : \Omega \to \mathbb{R}^k \text{ and } x \text{ is } \sigma(\mathcal{P})\text{-measurable}\}.$$

Hence, $x \in X_{\mathcal{P}}$ if and only if x is constant on elements of $\mathcal{P}$. The characteristic function of the set Ω, denoted by

$$\begin{aligned} \chi_\Omega &: \Omega \to \mathbb{R} \\ &: \omega \longmapsto 1, \end{aligned}$$

is in $X_{\mathcal{P}}$ for every $\mathcal{P}$ in a one good economy for example. When there are $k > 1$ goods

$$\chi_\Omega \underbrace{\begin{pmatrix} 1 \\ \vdots \\ 1 \end{pmatrix}}_{\in \mathbb{R}^k}$$

is in $X_{\mathcal{P}}$ for every $\mathcal{P}$. I will denote $X_{\mathcal{P}_i^S}$ by X_i^S and $X_{\mathcal{P}_i^m}$ by X_i^m. Then, if $S = \{i\}$ I will write X_i. Call $X_{\mathcal{P}}$ the set of *informationally feasible* trades under $\mathcal{P}$, X_i^S the set of informationally feasible trades of agent i in coalition S and $\prod_{i \in S} X_i^S$ the set of informationally feasible trades of the coalition S.

2.2 Exchange economies with asymmetric information

An *exchange economy with asymmetric information* $\mathbb{E}$ is given by

1. a *finite set of agents* $N = \{1, 2, \dots, n\}$,
2. a *finite set* Ω *of states of the economy*,
3. the *initial endowments* $e_i : \Omega \to \mathbb{R}^k$ for every agent $i \in N$,
4. the *communication system* $(\mathcal{P}_i^S)_{i \in N, S \subset N}$,
5. the *utility functions* $u_i : (\mathbb{R}^k)^\Omega \to \mathbb{R}$ for every agent $i \in N$ or

5.' *state-dependent utility functions* $u_i' : \Omega \times \mathbb{R}^k \to \mathbb{R}$ and *strictly positive subjective probabilities* $\mu_i(\omega) > 0$.

When using 5 I will write

$$\mathbb{E} = (N, \Omega, (e_i, u_i, (\mathcal{P}_i^S)_{i \in S \subset N})_{i \in N})$$

and speak of the *model without expectations*, when using $5'$ I will write

$$\mathbb{E} = (N, \Omega, (e_i, u_i', \mu_i, (\mathcal{P}_i^S)_{i \in S \subset N})_{i \in N})$$

and speak of the *expected utility model*.

A vector of net trades $(z_i)_{i \in S}$ satisfying $e_i + z_i \geq 0$ for all $i \in S$ and $\sum_{i \in S} z_i = 0$ is called *physically feasible* for the coalition $S \subset N$.

Moreover, for the results in this paper it is necessary to assume that

6. the utility functions are *quasiconcave* in $(\mathbb{R}^k)^\Omega$

6.' the state-dependent utility functions $u(\omega, \cdot)$ are *concave* in $\mathbb{R}^k$.

By taking expectations

$$u_i(x_i) := E_{\mu_i} u_i'(\omega, x_i(\omega)) := \sum_{\omega \in \Omega} \mu_i(\omega) u_i'(\omega, x_i(\omega)),$$

it becomes clear, that the expected utility model with concave state-dependent utilities is really a subclass of the model without expectations. In Allen [3] it was assumed, that the state-dependent utility functions be concave, which is reflected here in assumption 6'. The proofs in the more general case of assumptions 5 and 6 demand only quasiconcavity of the utility function. As the integral of concave functions is concave, this encompasses the case of state-dependent concave utilities. Nothing similar can be said about the case of state-dependent quasi-concave utilities, since the integral of quasiconcave functions need not be quasiconcave.

Here no assumptions are made with respect to measurability of initial endowments or state-dependent utility functions. It might be regarded as unreasonable, to have initial endowments or utility functions that contain more information than any of the information partitions the agent has in the coalitions of the game. For the decision to form coalitions and trade in these coalitions the information contained in endowments and utility seems essential, and therefore it might be argued that the agent has to know it. I will discuss these matters in Section 4.

2.3 The core

An $NTU\text{-}game\ in\ characteristic\ function\ form$ is a correspondence $V : 2^N \setminus \{\emptyset\} \to \mathbb{R}^N$ satisfying

1. $V(S)$ is nonempty and closed for $S \neq \emptyset$,
2. if $x \in V(S)$ and $y \in \mathbb{R}^N$ is such that $y_i \leq x_i$ for all $i \in S$ then $y \in V(S)$,
3. for every $i \in N$ there is an $m_i \in \mathbb{R}$ with $V(\{i\}) = \{x \in \mathbb{R}^N | x_i \leq m_i\}$, and $V(N) \cap \{x \in \mathbb{R}^N | x_i \geq m_i\ \forall i \in N\}$ is nonempty and compact.

A collection of coalitions $\mathcal{B} \subseteq 2^N \setminus \{\emptyset\}$ is *balanced* if there are positive real numbers λ_S for every $S \in \mathcal{B}$ such that $\sum_{S \in \mathcal{B}: i \in S} \lambda_S = 1$ for every $i \in N$. An NTU-game V is *balanced* if $\bigcap_{S \in \mathcal{B}} V(S) \subseteq V(N)$ for every balanced collection $\mathcal{B}$. Scarf [13] has proved that if V is balanced, then it has a nonempty core.

The $NTU\text{-}game\ associated\ with\ the\ expected\ utility\ model$ is defined by

$$\begin{aligned} V^{eu}_{\mathbb{E}}(S) = \{x \in \mathbb{R}^N; &\text{there exists } (z_i)_{i \in S} \in \textstyle\prod_{i \in S} X_i^S \text{ such that} \\ &e_i + z_i \geq 0,\ \textstyle\sum_{i \in S} z_i = 0 \\ &\text{and } x_i \leq E_{\mu_i} u_i(\omega, (e_i + z_i)(\omega))\} \end{aligned}$$

for each coalition $S \neq \emptyset$.

Allen [3] shows that this defines indeed an NTU-game. Let $\text{int}(X)$ denote the $interior$ of a set $X \subseteq \mathbb{R}^N$ with respect to the usual topology on $\mathbb{R}^N$. The $expected\ utility\ core$ of the exchange economy with asymmetric information is then defined to be the NTU-core of the associated NTU-game:

$$C^{eu}(\mathbb{E}) := C(V^{eu}_{\mathbb{E}}) = V^{eu}_{\mathbb{E}}(N) \setminus \bigcup_{\emptyset \neq S \subset N} \text{int}(V^{eu}_{\mathbb{E}}(S)).$$

The $NTU\text{-}game\ associated\ with\ the\ model\ without\ expectations$ is defined by

$$\begin{aligned} V_{\mathbb{E}}(S) = \{x \in \mathbb{R}^N; &\text{there exists } (z_i)_{i \in S} \in \textstyle\prod_{i \in S} X_i^S \text{ such that} \\ &e_i + z_i \geq 0,\ \textstyle\sum_{i \in S} z_i = 0 \\ &\text{and } x_i \leq u_i(e_i + z_i)\} \end{aligned}$$

for each coalition $S \neq \emptyset$.

The *core of the exchange economy with asymmetric information* is then defined to be the NTU-core of the associated NTU-game:

$$C(\mathbb{E}) := C(V_{\mathbb{E}}) = V_{\mathbb{E}}(N) \setminus \bigcup_{\emptyset \neq S \subset N} \text{int}(V_{\mathbb{E}}(S)).$$

3 Balancedness of the market games with asymmetric information

Schwalbe [14] defines the *maximum information* of an agent i as the information $\mathcal{P}_i^m$ he could gain by joining all coalitions simultaneously. His set of *feasible allocations* E for the whole economy then consists of all physically feasible allocations $(x_i)_{i \in N} \in \prod_{i \in N} X_i^m$. That makes the set E large enough to assure balancedness of the associated market game with asymmetric information for any communication system. As there are no obvious reasons for using this notion of maximum information, I will give here another condition to assure balancedness based on the sets X_i^S, and use P_i^N, rather than P_i^m, for the whole economy. Note that $P_i^m = P_i^N$ if the communication system is bounded.

I give an example to show that the condition contained in Allen [3], to guarantee (total) balancedness of the associated market game in the expected utility model, is not correct, and then present an alternative that is applicable to both models.

The following definitions are taken from Allen [3].

Definition 1. *The communication system $(P_i^S)_{i,S}$ is essentially nested if for all $i \in N$ and all coalitions S and T such that $i \in S \subseteq T \subseteq N$, if $\Omega' \subset \Omega$ is such that $0 < \mu_i(\Omega') < 1$ where $\Omega' \in \sigma(P_i^S)$, and $\Omega' = \Omega_1 \cup \Omega_2 \cup \ldots \cup \Omega_L, L \geq 1$ for some disjoint $\Omega_l \in P_{j(l)}^S$ with $j(l) \in S \setminus \{i\}$, $l = 1, \ldots, L$, then $\Omega' \in \sigma(P_i^T)$ and each $\Omega_l \in \sigma(P_{j(l)}^T)$. The communication system $(P_i^S)_{i,S}$ is essentially bounded if the condition above holds for $T = N$.*

Obviously essentially nested implies essentially bounded. Furthermore, nested implies essentially nested and bounded implies essentially bounded. Now the claim in Allen [3] is, that essential nestedness is equivalent to total balancedness of the associated market games with varying utilities and endowments, and essential boundedness is equivalent to balancedness. I give a counterexample to show that this is not so. The game specified will be essentially nested, but not balanced, hence constituting a counterexample to both claims.

Example 2. Let there be five agents $N = \{1, 2, 3, 4, 5\}$, 4 states $\Omega = \{1, 2, 3, 4\}$ and prior $\mu_i(\omega) = \mu(\omega) = \frac{1}{4}$ for all $i \in N$. The communication system is given by

$$P_1^S = \{\{1, 2\}, \{3, 4\}\} \; 1 \in S, |S| \leq 4$$

$$P_2^S = \{\{1, 4\}, \{2, 3\}\} \; 2 \in S, |S| \leq 4$$

$$P_3^S = \{\{1, 3, 4\}, \{2\}\} \; 3 \in S, |S| \leq 4$$

$$P_4^S = \{\{1, 2, 3\}, \{4\}\} \; 4 \in S, |S| \leq 4$$

$$P_5^S = \quad \{\Omega\} \qquad 5 \in S, |S| \leq 4$$

and $P_i^N = \{\Omega\}$ for all $i \in N$. This communication system is obviously essentially nested, as $\Omega' \subset \Omega$ with $0 < \mu_i(\Omega') < 1, \Omega' \in \sigma(P_i^S)$, and $\Omega' = \Omega_1 \cup \Omega_2 \cup \ldots \cup \Omega_L$ for some disjoint $\Omega_l \in P^S_{j(l)}$ with $j(l) \in S \setminus \{i\}, l = 1, \ldots, L$, does not exist for any $i \in S \subseteq N$.

Now consider the following one good economy

$$e_1 = (1,1,0,0)\ , u_1 : \Omega \times R, u(\omega, x) = \chi_{\{1,2\}}(\omega)x + 3\chi_{\{3,4\}}(\omega)x,$$
$$e_2 = (0,1,1,0)\ , u_2 : \Omega \times R, u(\omega, x) = \chi_{\{2,3\}}(\omega)x + 3\chi_{\{1,4\}}(\omega)x,$$
$$e_3 = (0,0,0,0)\ , u_3 : \Omega \times R, u(\omega, x) = x,$$
$$e_4 = (0,0,0,1)\ , u_4 : \Omega \times R, u(\omega, x) = x,$$
$$e_5 = (1,1,1,1)\ , u_5 : \Omega \times R, u(\omega, x) = x.$$

There will be no exchange of goods in the grand coalition as only $\{\emptyset, \{\Omega\}\}$-measurable trades are allowed there and so only agent 5 could possibly trade as all others have a state with 0 endowment. So the unique candidate for the core is the utility vector $(u_i(e_i))_{i \in N}$ arising from the initial endowment e. The expected utility is then

$$u_1(e_1) = \frac{1}{2}$$
$$u_2(e_2) = \frac{1}{2},$$
$$u_3(e_3) = 0,$$
$$u_4(e_4) = \frac{1}{4},$$
$$u_5(e_5) = 1.$$

Consider the coalition $S = \{1,2,3,4\}$. The net trades

$$z_1 = \left(-1, -1, \frac{1}{2}, \frac{1}{2}\right),$$
$$z_2 = \left(\frac{1}{2}, -1, -1, \frac{1}{2}\right),$$
$$z_3 = \left(0, 1\frac{1}{2}, 0, 0\right),$$
$$z_4 = \left(\frac{1}{2}, \frac{1}{2}, \frac{1}{2}, -1\right)$$

are informationally and physically feasible and lead to the following expected utility

$$u_1(e_1 + z_1) = \frac{3}{4} > u_1(e_1) = \frac{1}{2}$$

$$u_2(e_2 + z_2) = \frac{3}{4} > u_2(e_2) = \frac{1}{2},$$

$$u_3(e_3 + z_3) = \frac{3}{8} > u_3(e_3) = 0,$$

$$u_4(e_4 + z_4) = \frac{3}{8} > u_4(e_4) = \frac{1}{4},$$

and hence there is a deviation to the utiliy vector $(u_i(e_i))_{i \in N}$. So the core is empty and the game cannot be balanced.

Remark 3. The restriction of information in the grand coalition to $\{\Omega\}$ might seem strong. But the essential idea is only that net trades that are possible in a subcoalition are removed in the grand coalition and hence deviations of this subcoalition become possible. Thus, the assumption $P_i^N = \{\Omega\}$ is only made for simplicity.

The idea behind the condition of essential nestedness/boundedness was that the possibility of net trades is not only tied to physical feasibility, but also to informational feasibility. Hence, not all combinations of physically feasible net trades have to be considered as possible deviations. I propose the following definition that captures this idea better than essentially nestedness/boundedness.

Definition 4. *The communication system $(P_i^S)_{i,S}$ is trade nested if for all coalitions $S \subseteq T \subseteq N$ and all $(z_j)_{j \in S} \in \prod_{j \in S} X_j^S$ with $\sum_{j \in S} z_j = 0$, it holds that $(z_j)_{j \in S} \in \prod_{j \in S} X_j^T$. The communication system $(P_i^S)_{i,S}$ is trade bounded if the condition above holds for $T = N$.*

Clearly, the system in Example 2 is neither trade nested nor trade bounded. Now I claim the following for the model without expectations.

Theorem 5. *Fix the number of goods k and the finite sets Ω and N arbitrarily. Consider all exchange economies with asymmetric information and with these parameters fixed. As endowments and utilities vary, all NTU market games with asymmetric information are totally balanced if and only if $(\mathcal{P}_i^S)_{i,S}$ is trade nested.*

Proof. Necessity: Suppose that the communication system $(\mathcal{P}_i^S)_{i,S}$ is not trade nested. Then one can construct an exchange economy with asymmetric information that is not totally balanced, as follows. By assumption there are coalitions $S \subset T$ and $(z_j)_{j \in S} \in \prod_{j \in S} X_j^S$ such that $\sum_{j \in S} z_j = 0$ and $(z_j)_{j \in S} \notin \prod_{j \in S} X_j^T$. Consequently there is an agent $\widehat{j} \in S$ such that $z_{\widehat{j}} \notin X_{\widehat{j}}^T$. Now let K be large enough to ensure that

$$\widetilde{z}_j := z_j - K\chi_\Omega \begin{pmatrix} 1 \\ \vdots \\ 1 \end{pmatrix} < 0$$

for all $j \in S, j \neq \widehat{j}$. Then put $e_j := -\widetilde{z}_j$ for all $j \in S, j \neq \widehat{j}$ and $e_{\widehat{j}} := 0$ for all $j \in N \setminus S$ and for $j = \widehat{j}$. Let the utility functions of the agents $j \neq \widehat{j}$ be the zero-functions and for agent $\widehat{j}$ let $u_{\widehat{j}}(x) := \frac{1}{|\Omega|} \sum_{\omega \in \Omega} \sum_{m=1}^{k} x_m(\omega)$. I show that the subgame (T, V) is not balanced. Consider the balanced collection $\mathcal{B} := \{S, T \setminus S\}$ with constant weight function $\chi_{\mathcal{B}}$. Obviously

$$V(T \setminus S) = \{x \in \mathbb{R}^T | x_i \leq 0 \text{ for all } i \in T \setminus S\}.$$

The best that $\widehat{j}$ can get in coalition S is the sum of all the initial endowments of the others

$$\widetilde{z}_{\widehat{j}} := \sum_{j \in S, j \neq \widehat{j}} e_j = - \sum_{j \in S, j \neq \widehat{j}} \widetilde{z}_j = - \sum_{j \in S, j \neq \widehat{j}} z_j + (|S| - 1)K \begin{pmatrix} 1 \\ \vdots \\ 1 \end{pmatrix} \chi_\Omega$$

$$= z_{\widehat{j}} + (|S| - 1)K \begin{pmatrix} 1 \\ \vdots \\ 1 \end{pmatrix} \chi_\Omega > 0.$$

So $\widetilde{z}_{\widehat{j}}$ is $\mathcal{P}^S_{\widehat{j}}$ measurable and

$$u_{\widehat{j}}(\widetilde{z}_{\widehat{j}}) = \frac{1}{|\Omega|} \sum_{\omega \in \Omega} \sum_{m=1}^{k} \widetilde{z}_{\widehat{j}m}(\omega) > 0.$$

As u_j is the zero function for $j \neq \widehat{j}$ we have

$$V(S) = \{x \in \mathbb{R}^T | x_i \leq 0 \text{ for all } i \in S \setminus \{\widehat{j}\} \text{ and } x_{\widehat{j}} \leq u_{\widehat{j}}(\widetilde{z}_{\widehat{j}})\}.$$

Hence

$$\bigcap_{B \in \mathcal{B}} V(B) = \{x \in \mathbb{R}^T | x_i \leq 0 \text{ for all } i \in T \setminus \{\widehat{j}\} \text{ and } x_{\widehat{j}} \leq u_{\widehat{j}}(\widetilde{z}_{\widehat{j}})\}.$$

But on the other hand, as $(z_j)_{j \in S} \notin \prod_{j \in S} X_j^T$ one has also that $(\widetilde{z}_j)_{j \in S} \notin \prod_{j \in S} X_j^T$. Furthermore, as $\widetilde{z}_{\widehat{j}} = e(T) := \sum_{j \in T} e_j$ and $e(T \setminus S) = 0$, any physically feasible vector of net trades $(z'_j)_{j \in T} \in \prod_{j \in T} X_j^T$ leads to a utility for agent $\widehat{j}$ that is strictly less than $u_{\widehat{j}}(\widetilde{z}_{\widehat{j}})$. So the balancedness condition

$$V(T) \supseteq \bigcap_{B \in \mathcal{B}} V(B)$$

is violated in the subgame (T, V). This shows that the game (N, V) is not totally balanced.

Sufficiency: To show total balancedness of the derived game (N, V), it has to be shown that every subgame (T, V), $T \subseteq N$ is balanced, i.e. $V(T) \supseteq \bigcap_{B \in \mathcal{B}} V(B)$

holds for every balanced collection $\mathcal{B}$ of subsets of T. So let $\mathcal{B}$ be an arbitrary balanced collection of subsets of T with weights $\gamma : \mathcal{B} \to (0, 1]$. Let $x \in \bigcap_{B\in\mathcal{B}} V(B)$. For every $B \in \mathcal{B}$ one has $x \in V(B)$, and hence there are by definition of $V(B)$ net trades z_i^B, $i \in B$, that are informationally and physically feasible in the coalition B such that $x_i \le u_i(e_i + z_i^B)$ for all $i \in B$. Extend $(z^B)_{i\in B}$ to a vector of net trades in the grand coalition T of the subgame by putting $z_i^B := 0$ for all $i \in T \setminus B$. The communication system is trade nested so $(z_i^B)_{i\in T} \in \prod_{i\in B} X_i^T$, and as $0 \in X_i^T$ this implies $(z_i^B)_{i\in T} \in \prod_{i\in T} X_i^T$. Moreover $\sum_{i\in T} z_i^B = \sum_{i\in B} z_i^B = 0$. Consequently $(z_i^B)_{i\in T}$ is an informationally and physically feasible vector of net trades in T. Now let $z_i := \sum_{B\in\mathcal{B}, B\ni i} \gamma(B) z_i^B = \sum_{B\in\mathcal{B}} \gamma(B) z_i^B$ for every $i \in T$. As it follows easily that the sets X_i^T are convex and contain the z_i^B for every $B \in \mathcal{B}$ one obtains that also $z_i \in X_i^T$, i.e. $(z_i)_{i\in T}$ is informationally feasible in the coalition T. It is clear that $e_i + z_i \ge 0$ for all $i \in T$ so

$$\sum_{i\in T} z_i = \sum_{i\in T}\sum_{B\in\mathcal{B}} \gamma(B) z_i^B = \sum_{B\in\mathcal{B}} \gamma(B) \sum_{i\in T} z_i^B$$

$$= \sum_{B\in\mathcal{B}} \gamma(B) \sum_{i\in B} z_i^B = 0$$

shows that $(z_i)_{i\in T}$ is also physically feasible. Quasiconcavity of the utility functions now implies that for all $i \in T$

$$u_i(e_i+z_i) = u_i\left(e_i+\sum_{B\in\mathcal{B}} \gamma(B) z_i^B\right) = u_i\left(\sum_{B\in\mathcal{B},B\ni i} \gamma(B)e_i + \sum_{B\in\mathcal{B},B\ni i} \gamma(B) z_i^B\right)$$

$$= u_i\left(\sum_{B\in\mathcal{B},B\ni i} \gamma(B)(e_i + z_i^B)\right) \ge \min_{B\in\mathcal{B},B\ni i} \{u_i(e_i + z_i^B)\}$$

$$\ge x_i,$$

so $x \in V(T)$, and that shows $V(T) \supseteq \bigcap_{B\in\mathcal{B}} V(B)$ for all subgames (T, V) and balanced collections $\mathcal{B}$ of subsets of T. Hence the derived game is totally balanced. □

Corollary 6. *Fix the number of goods k and the finite sets Ω and N arbitrarily. Consider all exchange economies with asymmetric information and with these parameters fixed. As endowments and utilities vary, all NTU market games with asymmetric information are balanced if and only if $(\mathcal{P}_i^S)_{i,S}$ is trade bounded.*

Proof. For necessity and sufficiency set $T = N$ in the corresponding parts of the proof of Theorem 5. □

Corollary 7. *Fix the number of goods k and the finite sets Ω and N arbitrarily. Consider all exchange economies with asymmetric information and with these parameters fixed. As endowments and utilities vary, all NTU market games with asymmetric information are*

(i) balanced if and only if $(\mathcal{P}_i^S)_{i,S}$ *is trade bounded and*
(ii) totally balanced if and only if $(\mathcal{P}_i^S)_{i,S}$ *is trade nested.*

Proof. The expected utility case is a subclass of the other model recovered by taking $u_i : (\mathbb{R}^k)^\Omega \to \mathbb{R}$ to be the expected utility functions

$$\sum_{\omega \in \Omega} \mu_i(\omega) u_i(\omega, x(\omega))$$

for all $i \in N$. □

As balancedness implies that the core of the NTU-game is not empty and total balancedness implies that the core of all subgames is not empty one obtains another corollary.

Corollary 8. *Fix the number of goods* k *and the finite sets* Ω *and* N *arbitrarily. Consider all exchange economies with asymmetric information and with these parameters fixed. As endowments and utilities vary, a sufficient condition for all exchange economies with asymmetric information to have a non-empty core is trade boundedness of the communication system. A sufficient condition for all subgames to have a non-empty core is trade nestedness of the communication system.*

Note that the game constructed in the proof of Theorem 5 to show that nestedness is also a necessary condition has a non-empty core. In fact zero utility for every agent, corresponding to no trade, is in the core. So the corollary cannot state equivalence of nonemptiness of the core (subcores) and boundedness (nestedness).

Obvously, nestedness (boundedness) of the communication system implies trade nestedness (boundedness).

To see that the conditions of trade nestedness and trade boundedness are really weaker than nestedness and boundedness of the communication system, consider the following example.

Example 9. Let there be three agents $N = \{1, 2, 3\}$ and 4 states $\Omega = \{1, 2, 3, 4\}$. The communication system given by

$$\begin{aligned} \mathcal{P}_1^S &= \{\{1\}, \{2\}, \{3, 4\}\} \; 1 \in S, |S| \leq 2, \; P_1^N = \{\{1, 2\}, \{3, 4\}\}, \\ \mathcal{P}_2^S &= \quad \{\{1, 2, 3\}, \{4\}\} \qquad 2 \in S, \\ \mathcal{P}_3^S &= \quad \{\{1, 2, 4\}\{3\}\} \qquad 3 \in S. \end{aligned}$$

is trade nested but not nested. This is easily seen, as any physically feasible trades that agent 1 can make must necessarily be constant on the set $\{1, 2\}$.

On the other hand trade nestedness (boundedness) is stronger than essentially nestedness (boundedness) in the expected utility model.

Theorem 10. *Any trade nested (bounded) communication rule is essentially nested (bounded).*

Proof. Let the communication system $(\mathcal{P}_i^S)_{i,S}$ be trade nested. Fix arbitrary i, S and T such that $i \in S \subseteq T \subseteq N$. Suppose there is $\Omega' \subset \Omega$ such that $0 < \mu_i(\Omega') < 1$ where $\Omega' \in \sigma(\mathcal{P}_i^S)$, and $\Omega' = \Omega_1 \cup \Omega_2 \cup \ldots \cup \Omega_L$ for some disjoint $\Omega_l \in \mathcal{P}_{j(l)}^S$ with $j(l) \in S \setminus \{i\}$, $l = 1, \ldots, L$. If such sets Ω', Ω_l do not exist for any combination of i, S and T the communication system is by definition trivially essentially nested. If they exist put $z_i = -\chi_{\Omega'}, z_{j(l)} = \chi_{\Omega_l}$ and $z_j = 0$ for all $j \in S \setminus (\{i\} \cup \{j(l); l = 1 \ldots L\})$. Then $(z_j)_{j \in S} \in \prod_{j \in S} X_j^S$ and $\sum_{j \in S} z_j = 0$ hence by trade nestedness $(z_j)_{j \in S} \in \prod_{j \in S} X_j^T$. But this implies that $\Omega' \in \sigma(\mathcal{P}_i^T)$ and each $\Omega_l \in \sigma(\mathcal{P}_{j(l)}^T)$. As i, S and T were arbitrary this shows that the communication system is essentially nested. The boundedness case follows by specialising T to the grand coalition N. □

4 Discussion of measurability assumptions

As far as the initial endowment is concerned, an assumption such that every information partition $\mathcal{P}_i^S, i \in S \subseteq N$ is finer than the information contained in the initial endowment e_i, i.e. the coarsest partition $\mathcal{P}$ that makes e_i $\sigma(\mathcal{P})-$measurable, seems reasonable. After all, an agent must know his initial endowment for planning his net trade, hence he can use that information in every coalition. Observation of the initial endowments by the agents would not be necessary anymore if one would use utility functions defined on the net trades z rather than the commodity bundle $e_i + z$, and assign infinitely low utility $-\infty$ to net trades where $e_i + z \notin (\mathbb{R}_+^k)^\Omega$. In detail, one switches in the general model from $u_i : (\mathbb{R}^k)^\Omega \to \mathbb{R}$ to

$$\widetilde{u}_i : (\mathbb{R}^k)^\Omega \to \mathbb{R}$$

$$: z \longmapsto \left\{ \begin{array}{c} u_i(e_i + z) \text{ if } e_i + z \geq 0 \\ -\infty \text{ otherwise} \end{array} \right\},$$

and in the expected utility model from $u_i : \Omega \times \mathbb{R}^k \to \mathbb{R}$ to

$$\widetilde{u}_i : \Omega \times \mathbb{R}^k \to \mathbb{R}$$

$$: (\omega, z) \longmapsto \left\{ \begin{array}{c} u_i(\omega, e_i(\omega) + z) \text{ if } e_i(\omega) + z \geq 0 \\ -\infty \text{ otherwise} \end{array} \right\}.$$

Note that quasiconcavity of utility functions is not disturbed by this switch. However the "penalty restriction",

$$e_i + z \notin (\mathbb{R}_+^k)^\Omega \Rightarrow \widetilde{u}_i(e_i + z) = -\infty,$$

on the utility functions that can be used is imposed. Now an agent plans his net trade and observes and outcome in $\mathbb{R}$ or $\mathbb{R}^\Omega$, depending on which model is used. Assume an agent observes an outcome of $-\infty$ in the model without expectations. He could then infer that $e_i + z \geq 0$ is violated. But as he can only trade constant on states he cannot discern he will not be able to find out which state(s) lead to infinitely low utility, and therefore he can gain no information that he does not already have.

Unfortunately, in the expected utility model the agent observes a vector of outcomes, and if these differ he can still gain information. Measurability of the outcome vector $(\widetilde{u}_i(\omega, z(\omega)))_{\omega \in \Omega} \in \mathbb{R}^{\Omega}$ plays also a role here. As decisions in this paper are based on the expected utility of that outcome vector one can easily withdraw from a thorough a discussion of these matters by opting for the usage of the model without expectations to circumvent these problems. But in general for state-dependent utility it might be useful to use state-dependent utility functions that depend also on the information partition the agent is using. That makes sense if the agent has to know his utility function to decide. Utility at some state ω and information partition $\mathcal{P}$ must then equal utility in the nondiscernable states $\omega\prime \in \mathcal{P}(\omega)$. A definition of a state-dependent utility function would then be

$$\widetilde{u} : \Omega \times (\mathbb{R}^k)^{\Omega} \times \mathcal{P}^* \to \mathbb{R},$$

such that the outcome vector $(\widetilde{u}(\omega, x, \mathcal{P}))_{\omega \in \Omega}$ is $\mathcal{P}-$measurable for all $x \in (\mathbb{R}^k)^{\Omega}$ that are informationally feasible, i.e $x - e_i$ is $\mathcal{P}-$measurable. In the expected utility model given here this could be reached by letting the utility function $\widetilde{u}$ at $\mathcal{P} \in \mathcal{P}^*$ be a version of the conditional expected utility of the state-dependent utility function $u : \Omega \times \mathbb{R}^k \to \mathbb{R}$. To be precise, let

$$\widetilde{u}(\omega, x, \mathcal{P}) = \frac{1}{\mu(\mathcal{P}(\omega))} \sum_{\omega\prime \in \mathcal{P}(\omega)} \mu(\omega\prime) u(\omega\prime, x(\omega\prime)).$$

Moreover, it might be argued that subjective probabilites should also reflect information. That could easily be incorporated by taking information dependent subjective probabilities $\widetilde{\mu}(\cdot, \mathcal{P})$, that are consistent in the sense that they arise from an underlying probability measure μ independent of the information partition, i.e.

$$\widetilde{\mu} : \Omega \times \mathcal{P}^* \to \mathbb{R}$$

$$: (\omega, \mathcal{P}) \longmapsto \frac{\mu(\mathcal{P}(\omega))}{|\mathcal{P}(\omega)|}.$$

Obviously $\widetilde{u}$ and $\widetilde{\mu}$ at a particular $\mathcal{P}$ reveal no more information than is contained in $\mathcal{P}$ already. Furthermore, this replacement does not change the expected utility function in the expected utility model, as may be seen from

$$\begin{aligned}
\sum_{\omega \in \Omega} \widetilde{\mu}(\omega, \mathcal{P}) \widetilde{u}(\omega, x(\omega), \mathcal{P}) &= \sum_{\omega \in \Omega} \frac{1}{|\mathcal{P}(\omega)|} \sum_{\omega\prime \in \mathcal{P}(\omega)} \mu(\omega\prime) u(\omega\prime, x(\omega\prime)) \\
&= \sum_{\omega \in \Omega} \sum_{\omega\prime \in \mathcal{P}(\omega)} \frac{1}{|\mathcal{P}(\omega\prime)|} \mu(\omega\prime) u(\omega\prime, x(\omega\prime)) \\
&= \sum_{S \in \mathcal{P}} |S| \sum_{\omega \in S} \frac{1}{|S|} \mu(\omega) u(\omega, x(\omega)) \\
&= \sum_{\omega \in \Omega} \mu(\omega) u(\omega, x(\omega)).
\end{aligned}$$

Summarizing, agent i now plans in a particular coalition S a net trade z that is $\mathcal{P}_i^S$−measurable, observes a utility outcome $(\widetilde{u}(\omega, e_i + z, \mathcal{P}_i^S))_{\omega \in \Omega}$ that is also $\mathcal{P}_i^S$−measurable, and uses that outcome to take expectation under an also $\mathcal{P}_i^S$−measurable subjective probability $\mu(\cdot, \mathcal{P}_i^S)$. So the expected utility model of Allen [3] can be rewritten, with an additional restriction on the used utility functions, to reveal no information to the agent at the planning stage.

All results in this paper would remain valid, if one would transform an exchange economy modeled without or with expectations in the way suggested in this section. What can be said in general for an exchange economy arising from this new framework, but not having an underlying representation in the model with or without expectations, is left open here.

Another approach would be to view the information contained in $\mathcal{P}_i^S$ as the information one is allowed to use for trading, and any additional information contained in the initial endowment or utility vectors is information that is e.g. insider information, and hence trading based on this information is forbidden. Page [10] interprets the boundedness restriction on a communication system into that direction, calling it 'no insider trading'. Pooling information may be impossible to certain degrees as well which would lead to coarser information in larger coalitions.

5 Conclusion

In this paper a condition on the communication system was given to assure balancedness of all NTU-games derived from an asymmetric exchange economy, when endowments are variable. The NTU-games can be derived in an ex-ante setting without using state-dependent utility and subjective probabilities. The condition remains valid in the ex-ante expected utility core. It provides an alternative to the condition given in Allen [3]. Possible extensions are to consider trade boundedness in an interim core concept or to study implications for value or equilibrium concepts. Various results are already available for special communication systems leading to the weak fine, fine and private core. Einy, Moreno and Shitovitz [5], for example, show that with a continuum of traders and 'irreducibility' of the economy the set of equilibrium allocations in the sense of Radner [11, 12] coincides with the private core of Yannelis [16]. Moreover, they show that such an equilibrium allocation exists.

The existence of private and fine value allocations in an ex-ante sense was established in Krasa-Yannelis, [8] and [9]. They also argued that there are problems with the coarse value. Again one might consider arbitrary communication systems and investigate whether trade boundedness leads to an existence result for the value.

References

1. Allen, B.: Market games with asymmetric information and non-transferable utility: Representation results and the core. Center for Analytic Research in Economics and the Social Sciences. Working Paper #91-09, University of Pennsylvannia (1991)
2. Allen, B.: Transferable utility market games with asymmetric information: Representation results and the core. Center for Analytic Research in Economics and the Social Sciences. Working Paper #91-11, University of Pennsylvannia (1991)
3. Allen, B.: Market games with asymmetric information: the core with finitely many states of the world. In: Munier, B., Machina, M.J. (eds.) Models and experiments in risk and rationality. Boston Dordrecht London: Kluwer 1994
4. Debreu, G.: Theory of value. Cowles Foundation Monograph, Vol. 17. New York: Wiley 1959
5. Einy, E., Moreno, D., Shitovitz, B.: Competitive and core allocations in large economies with differential information. Economic Theory **18**, 321–332 (2001)
6. Forges, F., Heifetz, A., Minelli, E.: Incentive compatible core and competitive equilibria in differential information economies. Economic Theory **18**, 349–365 (2001)
7. Koutsougeras, L., Yannelis, N.: Incentive compatiblity and information superiority of the core of an economy with differential information. Economic Theory **3**, 195–216 (1993)
8. Krasa, S., Yannelis, N.: The value allocation of an economy with differential information. Econometrica **62**, 881–900 (1994)
9. Krasa, S., Yannelis, N.: Existence and properties of a value allocation for an economy with differential information. Journal of Mathematical Economics **25**, 165–179 (1996)
10. Page, F.: Market games with differential information and infinite dimensional commodity spaces: the core. Economic Theory **9**, 151–159 (1997)
11. Radner, R.: Competitive equilibrium under uncertainty. Econometrica **36**, 31–58 (1968)
12. Radner, R.: Equilibrium under uncertainty. In: Arrow, K.J., Intrilligator, M.D. (eds.) Handbook of Mathematical Economics, Vol. II, pp. 923–1006. Amsterdam New York Oxford: North-Holland 1982
13. Scarf, H.E.: The core of an n-person game. Econometrica **35**, 50–69 (1967)
14. Schwalbe, U.: The core of economies with asymmetric information. In: Lecture Notes in economics and mathematical systems **474**. Berlin Heidelberg New York: Springer 1999
15. Wilson, R.: Information, efficiency and the core of an economy. Econometrica **46**, 807–816 (1978)
16. Yannelis, N.: The core of an economy with differential information. Economic Theory **1**, 183–198 (1991)

PART 2

CORE AND WALRASIAN EXPECTATIONS EQUILIBRIUM EQUIVALENCE

Competitive and core allocations in large economies with differential information★

Ezra Einy[1], **Diego Moreno**[2], **and Benyamin Shitovitz**[3]

[1] Department of Economics, Ben Gurion University of Negev, Beer Sheva, 841105, ISRAEL (e-mail: einy@bgumail.bgu.ac.il);
[2] Departamento de Economía, Universidad Carlos III de Madrid, 28903 Getafe, SPAIN (e-mail: dmoreno@eco.uc3m.es);
[3] Department of Economics, University of Haifa, Haifa, 31905, ISRAEL (e-mail: binya@econ.haifa.ac.il)

Received March 22, 2000; revised version: May 1, 2000

Summary. We study the core and competitive allocations in exchange economies with a continuum of traders and differential information. We show that if the economy is "irreducible", then a competitive equilibrium, in the sense of Radner (1968, 1982), exists. Moreover, the set of competitive equilibrium allocations coincides with the "private core" (Yannelis, 1991). We also show that the "weak fine core" of an economy coincides with the set of competitive allocations of an associated symmetric information economy in which the traders information is the joint information of all the traders in the original economy.

Keywords and Phrases: General equilibrium in large economies with differential information, Private core, Weak fine core.

JEL Classification Numbers: D50, D82.

1 Introduction

The purpose of this paper is to study the properties of core and competitive allocations in large exchange economies with differential information. Radner (1968 and 1982) introduces a model of an exchange economy with differential information in which every trader is characterized by a state dependent utility function, a random

---

★ This work was done while Einy and Shitovitz visited the Department of Economics of the Universidad Carlos III de Madrid. Einy acknowledges the financial support of the Universidad Carlos III de Madrid. Moreno acknowledges the support of the Spanish Ministry of Education, grant PB97-0091. Shitovitz acknowledges the support of the Spanish Ministry of Education, grant SAB98-0059.
Correspondence to: D. Moreno

initial endowment, an information partition, and a prior belief. In this framework, traders arrange contingent contracts for trading commodities before they obtain any information about the realized state of nature. Radner (1968) extends the notion of Arrow-Debreu competitive equilibrium to this model. In the definition of competitive equilibrium (in the sense of Radner), the information of an agent places a restriction on his feasible trades (i.e., his budget set): better information allows for more contingent trades (i.e., enlarges the agent's budget set). Thus, in a Radner competitive equilibrium better informed agents are generally, ceteris paribus, better off (and they are never worse off) than those with worse information; i.e., a competitive equilibrium rewards the information advantage of a trader.

We consider two period Radner-type economies with a finite number of states of nature and a continuum of traders. In these economies there is uncertainty about the state of nature. In the first period traders arrange contracts that may be contingent on the realized state of nature in the second period. Consumption takes place in the second period. We provide a straightforward extension to these economies of the notion of Radner competitive equilibrium, and examine conditions that guarantee its existence. Also we study the set of core allocations in these economies. It is well known that in perfectly competitive economies (that is, when no individual can affect the overall outcome) with complete information the core coincides with the set of competitive allocations (see, e.g., Aumann, 1964). In an economy with differential information, the set of allocations that a coalition can block depends upon the initial information and the communication opportunities of the members of the coalition. Thus, several alternative notions of core can be considered.

Yannelis (1991) introduces the notion of *private core,* and shows that under appropriate assumptions the private core of an economy is non-empty. In the private core the set of feasible allocations for a blocking coalition must involve a net trade of each member of the coalition that is measurable with respect to his information partition. The private core has some interesting properties: Koutsougeras and Yannelis (1993) show that if there is a finite number of traders, the private core is coalitionally incentive compatible – see Section 4 in Koutsougeras and Yannelis (1993). Also the private core rewards the information advantage of a trader – see Section 5 in Koutsougeras and Yannelis (1993), and the discussion below.

Koutsougeras and Yannelis (1993), see also Allen (1991), introduce the notion of *weak fine core*, a version of Wilson's (1978) fine core, and showed that this core is non-empty. In the definition of the weak fine core blocking net trades are measurable with respect to the joint partition of all the members of the coalition, but in addition all net trades are measurable with respect to the joint partition of all the traders.

In our context, the private core is the appropriate notion of core when the traders have no access to any communication system, and therefore cannot exchange information. The weak fine core is the appropriate notion of core when the traders have access to a communication system that allows them to fully share their information, and under the maintained assumption of perfect competition (that is, assuming that individuals are also "small" from the point of view of information). We study the relations of these cores and the set of competitive allocations in the economies described above, with a continuum of traders and differential information.

First we establish conditions under which a Radner competitive equilibrium exists: we show that if an economy is *irreducible*, and if the traders' utility functions are continuous and increasing, then an equilibrium exists. The existence of competitive equilibrium in economies with a continuum of traders and complete information was studied in Aumann (1966) and Hildenbrand (1970). We establish existence of equilibrium using general results from Hildenbrand (1974). The irreducibility condition was introduced by Mckenzie (1959) for exchange economies with a finite number of traders, and it has been extended to economies with a continuum of traders by Hildenbrand (1974). It expresses the idea that the endowment of every coalition, if added to the allocation of the complementary coalition, can be used to improve the welfare of every member of the complementary coalition. We show that an economy is irreducible if, for example, the initial endowment of every trader is strictly positive at each state of nature.

Next we show that if an economy is irreducible, then the private core of the economy coincides with the set of Radner competitive equilibrium allocations. Thus, private core allocations reward the information advantage of a trader. We provide simple examples which show that this result may not hold for reducible economies. For the weak fine core, we show that under mild assumptions it coincides with the set of competitive allocations of an associated symmetric information economy. This associated economy is identical to the original economy, except for the traders' information, which is the joint information of all the traders in the original economy. Specifically, this result holds if the traders' utility functions are continuous and strictly increasing, and if for every trader there is a state of nature at which his initial endowment is non-zero. Moreover, the result holds whether or not the economy is irreducible.

Thus, whereas private core allocations reward the information advantage of a trader, when the possibility of sharing information is introduced the information advantage is worthless; e.g., if two traders A and B have identical characteristics, except that A is better informed than B (i.e., A's information partition is finer than B's) then in a private core allocation trader A may be better off than trader B; in a weak fine core allocation, however, both traders are equally well off (because a weak fine core allocation is a competitive allocation of the associated symmetric information economy). (In sharp contrasts to this result, Einy, Moreno and Shitovitz (1999) have shown that the weak fine bargaining set contains allocations that are not competitive in the associated symmetric information economy, which suggests that the weak fine bargaining set discriminates in favor of better informed agents.)

The paper is organized as follows. In Section 2 we describe the model. In Section 3 we discuss the existence of competitive equilibrium (in the sense of Radner). In Section 4 we prove the equivalence between competitive and private core allocations. Finally, in Section 5 we establish the equivalence of the weak fine core and the set of competitive allocations of the associated symmetric information economy.

2 The model

We consider a Radner-type exchange economy $\mathcal{E}$ with differential information (e.g., Radner, 1968, 1982). The commodity space is $\Re^l_+$. The space of traders is a measure space (T, Σ, μ), where T is a set (the set of *traders*), Σ is a σ-field of subsets of T (the set of *coalitions*), and μ is a non-atomic measure on Σ. The economy extends over two time periods, $\tau = 0, 1$. Consumption takes place at $\tau = 1$. At $\tau = 0$ there is uncertainty over the state of nature; in this period traders arrange contracts that may be contingent on the realized state of nature at $\tau = 1$. There is a finite space of states of nature, denoted by Ω. At $\tau = 1$ traders do not necessarily know which state of nature $\omega \in \Omega$ actually occurred, although they know their own endowments, and may also have some additional information about the state of nature. We do not assume, however, that traders know their utility function (see below).

The information of a trader $t \in T$ is described by a partition Π_t of Ω. We denote by $\mathcal{F}_t$ the field generated by Π_t. If ω is the true state of the economy at $\tau = 1$, trader t observes the member of Π_t which contains ω. Every trader $t \in T$ has a probability measure q_t on Ω which represents his *prior beliefs*. The preferences of a trader $t \in T$ are represented by a *state dependent utility function,* $u_t : \Omega \times \Re^l_+ \to \Re$ such that for every $(t, x) \in T \times \Re^l_+$, the mapping $(t, x) \to u_t(\omega, x)$ is $\Sigma \times \mathcal{B}$ measurable, where ω is a fixed member of Ω, and $\mathcal{B}$ is the σ-field of Borel subsets of $\Re^l_+$. If x is a random bundle (i.e., a function from Ω to $\Re^l_+$) we denote by $h_t(x)$ the *expected utility from* x of trader $t \in T$. That is

$$h_t(x) = \sum_{\omega \in \Omega} q_t(\omega) u_t(\omega, x(\omega)).$$

An *assignment* is a function $\mathbf{x} : T \times \Omega \to \Re^l_+$ such that for every $\omega \in \Omega$ the function $\mathbf{x}(\cdot, \omega)$ is μ-integrable on T. There is a fixed *initial assignment* $\mathbf{e}$; $\mathbf{e}(t, \omega)$ represents the *initial endowment density* of trader $t \in T$ in the state of nature $\omega \in \Omega$. We assume that for almost every $t \in T$ the function $\mathbf{e}(t, \cdot)$ is $\mathcal{F}_t$-measurable.

Throughout the paper we use the following notations. For two vectors $x = (x_1, \dots, x_l)$ and $y = (y_1, \dots, y_l)$ in $\Re^l$ we write $x \geq y$ when $x_k \geq y_k$ for all $1 \leq k \leq l$, $x > y$ when $x \geq y$ and $x \neq y$, and $x \gg y$ when $x_k > y_k$ for all $1 \leq k \leq l$.

3 Competitive equilibrium

In this section we extend Radner's (1982) definition of competitive equilibrium to our model (see Radner, 1982, Sect. 3.4), and discuss conditions under which its existence can be guaranteed. Throughout the rest of the paper, an economy $\mathcal{E}$ is an atomless economy with differential information as described in Section 2.

A *private allocation* for an economy $\mathcal{E}$ is an assignment $\mathbf{x}$ such that

(3.1) for almost all $t \in T$ the function $\mathbf{x}(t, \cdot)$ is $\mathcal{F}_t$-measurable, and
(3.2) $\int_T \mathbf{x}(t, \omega) d\mu \leq \int_T \mathbf{e}(t, \omega) d\mu$ for all $\omega \in \Omega$.

A *price system* is a non-zero function $p : \Omega \to \Re^l_+$. Let $t \in T$ and let M_t be the set of all $\mathcal{F}_t$-measurable functions from Ω to $\Re^l_+$. For a price system p, define the *budget set* of t by

$$B_t(p) = \left\{ x \mid x \in M_t \text{ and } \sum_{\omega \in \Omega} p(\omega) \cdot x(\omega) \leq \sum_{\omega \in \Omega} p(\omega) \cdot \mathbf{e}(t, \omega) \right\}.$$

A *competitive equilibrium* (in the sense of Radner) for an economy $\mathcal{E}$ is a pair $(p, \mathbf{x})$ where p is a price system and $\mathbf{x}$ is private allocation such that

(3.3) for almost all $t \in T$ the function $\mathbf{x}(t, \cdot)$ maximizes h_t on $B_t(p)$, and

(3.4) $\sum_{\omega \in \Omega} p(\omega) \cdot \int_T \mathbf{x}(t, \omega) d\mu = \sum_{\omega \in \Omega} p(\omega) \cdot \int_T \mathbf{e}(t, \omega) d\mu$.

A *competitive allocation* is a private allocation $\mathbf{x}$ for which there exists a price system p such that $(p, \mathbf{x})$ is a competitive equilibrium.

In the literature, condition (3.2) in the definition of a private allocation is written usually with (strict) equality; see, e.g., Radner (1968), Krasa and Yannelis (1994), Allen (1997). Here we follow Radner (1982) who noted that the total amount to be disposed of might not be measurable with respect to the information partition of a single agent. Einy and Shitovitz (1999) provide an example of an economy with differential information which has a competitive equilibrium, but if the inequality (3.2) in the definition of a private allocation is replaced with an equality, then the economy does not have a competitive equilibrium where all prices are non-negative–see Example 2.1 in Einy and Shitovitz (1999). Condition (3.4) ensures that in a competitive equilibrium if a commodity is in excess supply its price is zero. This condition is redundant as it is implied by Walras' Law, which is satisfied in our framework. Nevertheless we include it to facilitate comparison to Radner's (1982) definition.

A function $u : \Re^l_+ \to \Re$ is *(strictly) increasing* if for all $x, y \in \Re^l_+$, $(x > y)$ $x \gg y$ implies $u(x) > u(y)$.

Throughout the paper we will often refer to the following conditions.

$(A.1)$ For every $\omega \in \Omega$ we have $\int_T \mathbf{e}(t, \omega) d\mu \gg 0$.

$(A.2)$ For almost all $t \in T$ and for every $\omega \in \Omega$, the function $u_t(\omega, \cdot)$ is continuous and increasing on $\Re^l_+$.

$(A.3)$ *Irreducibility*: for every private allocation $\mathbf{x}$ and for every two disjoint coalitions $T_1, T_2 \in \Sigma$ such that $\mu(T_1) > 0$, $\mu(T_2) > 0$, and $T_1 \cup T_2 = T$, there exists an assignment $\mathbf{y}$ such that $\mathbf{y}(t, \cdot) \in M_t$ for almost all $t \in T_2$, and such that

$(A.3.1)$ $h_t(\mathbf{y}(t, \cdot)) > h_t(\mathbf{x}(t, \cdot))$ for almost all $t \in T_2$, and

$(A.3.2)$ for all $\omega \in \Omega$: $\int_{T_1} \mathbf{e}(t, \omega) d\mu + \int_{T_2} \mathbf{x}(t, \omega) d\mu \geq \int_{T_2} \mathbf{y}(t, \omega) d\mu$.

Condition $(A.3)$, Irreducibility, was introduced in McKenzie (1959) for economies with a finite number of traders, and was extended for atomless economies by Hildenbrand (see Hildenbrand, 1974, pp. 143, 214). It expresses the idea that the endowment of every coalition is desired. Our definition is a variant of Hildenbrand's (1974).

Proposition 3.1. *Assume that an economy $\mathcal{E}$ satisfies assumption $(A.2)$. If for almost every $t \in T$ and all $\omega \in \Omega$ we have $\mathbf{e}(t, \omega) \gg 0$, then $\mathcal{E}$ satisfies Condition $(A.3)$ (Irreducibility).*

Proof Assume that $\mathbf{e}(t,\omega) \gg 0$ for almost every $t \in T$ and all $\omega \in \Omega$. Let $\mathbf{x}$ be a private allocation in $\mathcal{E}$, and let $T_1, T_2 \in \Sigma$ be two disjoint coalitions such that $\mu(T_1) > 0$ and $\mu(T_2) > 0$, and $T_1 \cup T_2 = T$. Then for all $\omega \in \Omega$ we have

$$\int_{T_1} \mathbf{e}(t,\omega)d\mu \gg 0.$$

Let $a \in \Re^l_+$ be such that $\mu(T_2)a \gg 0$, and such that for all $\omega \in \Omega$ we have

$$\int_{T_1} \mathbf{e}(t,\omega)d\mu \geq \mu(T_2)a.$$

Define $\mathbf{y} : T \times \Omega \to \Re^l_+$ by

$$\mathbf{y}(t,\omega) = \begin{cases} 0 & t \in T_1, \\ \mathbf{x}(t,\omega) + a & t \in T_2. \end{cases}$$

Then for all $t \in T_2, \mathbf{y}(t,\cdot) \in M_t$. Since for almost all $t \in T$ and all $\omega \in \Omega$, $u_t(\omega,\cdot)$ is increasing, we have

$$h_t(\mathbf{y}(t,\cdot)) > h_t(\mathbf{x}(t,\cdot)),$$

for almost all $t \in T_2$. From the choice of a it is clear that $(A.3.2)$ holds for $\mathbf{x}$ and $\mathbf{y}$. □

A *quasi equilibrium* for the economy $\mathcal{E}$ is a pair $(p,\mathbf{x})$, where p is a price system and $\mathbf{x}$ is a private allocation, such that

(3.5) for almost all $t \in T$, either $\sum_{\omega\in\Omega} p(\omega) \cdot \mathbf{e}(t,\omega) = 0$, or the function $\mathbf{x}(t,\cdot)$ maximizes h_t on $B_t(p)$, and

(3.6) $\sum_{\omega\in\Omega} p(\omega) \cdot \int_T \mathbf{x}(t,\omega)d\mu = \sum_{\omega\in\Omega} p(\omega) \cdot \int_T \mathbf{e}(t,\omega)d\mu$.

Proposition 3.2. *If an economy $\mathcal{E}$ satisfies conditions $(A.1)$– $(A.3)$, then every quasi equilibrium of $\mathcal{E}$ is a competitive equilibrium.*

Proof Proposition 3.2 is a direct consequence of Proposition 1 in Hildenbrand (1974), p. 214, when for $t \in T$ the consumption sets are M_t, the utility functions are h_t, and the production sets are $(\Re^l_-)^\Omega$. □

Theorem A. *If an economy $\mathcal{E}$ satisfies assumptions $(A.1) - (A.3)$ then it has a competitive equilibrium.*

Proof First note that our definition of quasi equilibrium is a special case of Hildenbrand's (1970, 1974) definition of quasi equilibrium for a coalition production economy where for $t \in T$ the consumption sets are M_t, the utility functions are h_t, and the production sets are $(\Re^l_-)^\Omega$ (see Hildenbrand, 1970, Sect. 2, p. 611). Therefore by Theorem 2 in Hildenbrand (1970), an economy $\mathcal{E}$ has a quasi equilibrium. Moreover, by Proposition 3.2 any quasi-equilibrium of $\mathcal{E}$ is a competitive equilibrium of $\mathcal{E}$. □

The following corollary is a direct consequence of Proposition 3.1 and Theorem A.

Corollary 3.3. *If an economy $\mathcal{E}$ satisfies* $(A.1), (A.2)$, *and in addition for every* $\omega \in \Omega$ *and almost all* $t \in T$ *we have* $\mathbf{e}(t, \omega) \gg 0$, *then* $\mathcal{E}$ *has a competitive equilibrium.*

4 The private core

In this section we extend the definition of private core introduced in Yannelis (1991) to our economy, and show that under conditions $(A.1)-(A.3)$ the set of competitive allocations of the economy coincides with the set of private core allocations.

An assignment $\mathbf{x}$ is a *private core allocation* for the economy $\mathcal{E}$ if
(4.1) $\mathbf{x}$ is a private allocation, and
(4.2) there do not exist a coalition $S \in \Sigma$ and an assignment $\mathbf{y}$ such that
(4.2.1) $\mu(S) > 0$,
(4.2.2) $\mathbf{y}(t, \cdot)$ is $\mathcal{F}_t$-measurable for all $t \in S$,
(4.2.3) $\int_S \mathbf{y}(t, \omega) d\mu \leq \int_S \mathbf{e}(t, \omega) d\mu$ for all $\omega \in \Omega$, and
(4.2.4) $h_t(\mathbf{y}(t, \cdot)) > h_t(\mathbf{x}(t, \cdot))$ for almost all $t \in S$.
The *private core* of an economy $\mathcal{E}$ is the set of all private core allocations of $\mathcal{E}$.

Proposition 4.1. *Every competitive allocation of an economy $\mathcal{E}$ is a private core allocation of $\mathcal{E}$.*

Proof Proposition 4.1 is a special case of Proposition 2 in Hildenbrand (1974), p. 216. □

Theorem B. *Under assumptions* $(A.1) - (A.3)$ *the set of competitive allocations of an economy $\mathcal{E}$ coincides with the private core of $\mathcal{E}$.*

Proof By Proposition (4.1) it suffices to show that every private core allocation in $\mathcal{E}$ is a competitive allocation. Let $\mathbf{x}$ be a private core allocation in $\mathcal{E}$. By Theorem 1 in Hildenbrand (1974), p. 216, there is a price system p such that $(p, \mathbf{x})$ is a quasi equilibrium for $\mathcal{E}$. By Proposition 3.2 we obtain that $(p, \mathbf{x})$ is a competitive equilibrium for $\mathcal{E}$. □

The following corollary is a direct consequence of Proposition 3.1 and Theorem B.

Corollary 4.2. *If an economy $\mathcal{E}$ satisfies* $(A.1), (A.2)$, *and in addition for every* $\omega \in \Omega$ and *almost all* $t \in T$ *we have* $\mathbf{e}(t, \omega) \gg 0$, *then the set of competitive allocations of $\mathcal{E}$ coincides with the private core of $\mathcal{E}$.*

We now give an example of an atomless "reducible" economy (i.e., it does not satisfy Condition $(A.3)$) with complete information satisfying $(A.1)$ and $(A.2)$ which has a non-empty core, but does not have a competitive equilibrium. Also this example shows that the private core may not satisfy the Equal Treatment Property when the economy is "reducible."

Example 4.3. Consider an atomless economy $\mathcal{E}$ in which the space of traders is $([0,2], \mathcal{B}, \lambda)$, where $\mathcal{B}$ is the σ-field of Borel subsets of $[0,2]$ and λ is the Lebesgue measure. Traders have complete information, and the commodity space is $\Re^2_+$. Every trader $t \in T_1 = [0,1]$ has an initial endowment $\mathbf{e}(t) = (1,0)$ and utility function $u_1(x,y) = x$, whereas each trader $t \in T_2 = (1,2]$ has initial endowment $\mathbf{e}(t) = (1,1)$ and utility function $u_2(x,y) = y$. The core of the economy $\mathcal{E}$ consists of all allocations $\mathbf{x}$ such that

$$u_t(\mathbf{x}(t)) = \begin{cases} \alpha(t) & t \in T_1, \\ 1 & t \in T_2, \end{cases}$$

where $\alpha : T_1 \to \Re_+$ is an integrable function such that $\alpha(t) \geq 1$ for almost all $t \in T_1$, and $\int_{T_1} \alpha(t) d\lambda \leq 2$. It is easy to see that every core allocation in $\mathcal{E}$ is a quasi equilibrium allocation with price system $p = (0,1)$. However, the economy $\mathcal{E}$ does not have a competitive equilibrium. It is worth noticing that the core of this economy contains allocations that do not have the Equal Treatment Property.

In the following example we consider a reducible economy (i.e., it does not satisfy Condition $(A.3)$) with asymmetric information in which the utility functions of the traders are strictly increasing and strictly concave. The economy does not have a competitive equilibrium, although its private core is non-empty (it consists of the initial assignment).

Example 4.4. Consider an atomless economy $\mathcal{E}$ in which the space of traders is $([0,2], \mathcal{B}, \lambda)$, where $\mathcal{B}$ is the σ-field of Borel subsets of $[0,2]$ and λ is the Lebesgue measure. The commodity space is $\Re^2_+$, and the space of states of nature is $\Omega = \{\omega_1, \omega_2\}$. The information partition of every trader t in the interval $T_1 = [0,1]$ is $\Pi_1 = \{\{\omega_1\}, \{\omega_2\}\}$, his prior belief is $q_1 = (\frac{1}{2}, \frac{1}{2})$, his initial endowments are $\mathbf{e}(t, \omega_1) = (1,0)$ and $\mathbf{e}(t, \omega_2) = (0,1)$, and his utility functions are $u_t(\omega, (x,y)) = \sqrt{x} + \sqrt{y}$ for all $\omega \in \Omega$. The information partition of every trader t in the interval $T_2 = (1,2]$ is $\Pi_2 = \{\{\omega_1, \omega_2\}\}$, his prior belief is $q_2 = (\frac{1}{2}, \frac{1}{2})$, his initial endowments are $\mathbf{e}(t, \omega_1) = \mathbf{e}(t, \omega_2) = (1,1)$, and his utility function is $u_t(\omega, (x,y)) = \sqrt{x} + \sqrt{y}$, for all $\omega \in \Omega$. It is easy to see that the economy does not have a competitive equilibrium. However, the unique private core allocation is the initial assignment $\mathbf{e}$. Note that $(p, \mathbf{e})$, where $p(\omega_1) = (0,1)$ and $p(\omega_2) = (1,0)$, is a quasi equilibrium for $\mathcal{E}$.

5 The weak fine core

In this section we extend to our model the definition of "weak fine core" introduced by Koutsougeras and Yannelis (1993) (see also Allen, 1991), and we prove an equivalence theorem for this notion of core.

We first note that since Ω is a finite set, there is a finite number of different information partitions. Let us be given an economy $\mathcal{E}$, and denote by $\Pi_1, \ldots, \Pi_n$ the n distinct information partitions of the traders. For every $1 \leq i \leq n$, let $\mathcal{F}_i$ be the field generated by Π_i, and let

$$T_i = \{t \in T \mid \mathcal{F}_t = \mathcal{F}_i\}.$$

We assume that for every $1 \leq i \leq n$ the set T_i is measurable and $\mu(T_i) > 0$. If $I \subset \{1, \ldots, n\}$ is a non-empty set, we denote by $\bigvee_{i \in I} \mathcal{F}_i$ the smallest field which contains each $\mathcal{F}_i$, $i \in I$. If $S \in \Sigma$ is a coalition with $\mu(S) > 0$, we denote

$$I(S) = \{i \mid 1 \leq i \leq n \text{ and } \mu(S \cap T_i) > 0\}.$$

An assignment $\mathbf{x}$ for the economy $\mathcal{E}$ is called a *weak fine core allocation* if
(5.1) for almost all $t \in T$ the function $\mathbf{x}(t, \cdot)$ is $\bigvee_{i=1}^{n} \mathcal{F}_i$-measurable;
(5.2) for every $\omega \in \Omega$, $\int_T \mathbf{x}(t, \omega) d\mu \leq \int_T \mathbf{e}(t, \omega) d\mu$;
(5.3) there do not exist a coalition $S \in \Sigma$ and an assignment $\mathbf{y}$ such that
 (5.3.1) $\mu(S) > 0$,
 (5.3.2) $\mathbf{y}(t, \cdot)$ is $\bigvee_{i \in I(S)} \mathcal{F}_i$-measurable for almost all $t \in S$,
 (5.3.3) $\int_S \mathbf{y}(t, \omega) d\mu \leq \int_S \mathbf{e}(t, \omega) d\mu$ for all $\omega \in \Omega$, and
 (5.3.4) $h_t(\mathbf{y}(t, \cdot)) > h_t(\mathbf{x}(t, \cdot))$ for almost all $t \in S$.

The *weak fine core* of $\mathcal{E}$ is defined as the set of all weak fine core allocations of $\mathcal{E}$.
We now introduce the following condition.

$(A.4)$ If $A \in \bigvee_{i=1}^{n} \mathcal{F}_i$ is non-empty, then $q_t(A) > 0$ for almost all $t \in T$.

We denote by $\mathcal{E}^*$ an economy identical to $\mathcal{E}$ except for the information fields of the traders, which for all $t \in T$ is taken to be $\mathcal{F}_t^* = \bigvee_{i=1}^{n} \mathcal{F}_i$. Note that the information in $\mathcal{E}^*$ is symmetric.

In the proof of the following proposition we use a result of Vind (1972), see also Proposition 7.3.2. in Mas-Colell (1985), which asserts that in atomless economy (see Aumann, 1964) if an allocation is blocked, then the blocking coalition can be chosen with a measure which is arbitrarily close to the measure of the grand coalition.

Proposition 5.1. *Assume that an economy $\mathcal{E}$ satisfies $(A.1)$, $(A.2)$ and $(A.4)$, and in addition for almost all $t \in T$ and for every $\omega \in \Omega$ the function $u_t(\omega, \cdot)$ is strictly increasing. Then the weak fine core of $\mathcal{E}$ coincides with the private core of $\mathcal{E}^*$.*

Proof It is clear that every private core allocation in $\mathcal{E}^*$ is a weak fine core allocation of $\mathcal{E}$. We prove the converse. Let $\Pi = \bigvee_{i=1}^{n} \Pi_i$ (i.e., Π is the smallest partition of Ω that refines each Π_i). Denote $\Pi = \{A_1, ..., A_k\}$, and let X be the set of all members of $(\Re_+^l)^\Omega$ which are $\bigvee_{i=1}^{n} \mathcal{F}_i$-measurable. Then every member of X is constant on every A_j, $1 \leq j \leq k$. Let the function $\alpha : X \to \Re_+^{kl}$ be defined by $\alpha(x) = \hat{x}$, where for $1 \leq j \leq k$, $\hat{x}_j = x(\omega_j)$ for some $\omega_j \in A_j$. Note that α is a one to one mapping from X onto $\Re_+^{kl}$. For every $t \in T$ we define a function $\hat{h}_t : \Re_+^{kl} \to \Re$ by $\hat{h}_t(\hat{x}) = h_t(\alpha^{-1}(\hat{x}))$. Then $\hat{h}_t$ is continuous, and by $(A.4)$ it is strictly increasing. Consider now the complete information atomless economy $\widehat{\mathcal{E}}$ in which the space of traders is (T, Σ, μ), the commodity space is $\Re_+^{kl}$, the initial assignment is $\hat{\mathbf{e}}$, where $\hat{\mathbf{e}}(t) = \alpha(\mathbf{e}(t, \cdot))$ for all $t \in T$, and the utility function of trader t is $\hat{h}_t$. Let $\mathbf{y}$ be a weak fine core allocation of $\mathcal{E}$. Assume, contrary to our claim, that $\mathbf{y}$ is not a private core allocation of $\mathcal{E}^*$. For every $t \in T$ let $\hat{\mathbf{y}}(t) = \alpha(\mathbf{y}(t, \cdot))$. Then $\hat{\mathbf{y}}$ is not in the core of the economy $\widehat{\mathcal{E}}$. Therefore by the Theorem of Vind (1972), there exists a coalition $S \in \Sigma$ and an assignment $\hat{\mathbf{z}}$ in $\widehat{\mathcal{E}}$ such that $\mu(S) > \mu(T) - \min\{\mu(T_1), ..., \mu(T_n)\}$, $\int_S \hat{\mathbf{z}}(t) d\mu \leq \int_S \hat{\mathbf{e}}(t) d\mu$, and

$\hat{h}_t(\hat{\mathbf{z}}(t)) > \hat{h}_t(\hat{\mathbf{y}}(t))$ for almost all $t \in S$. For every $t \in T$ let $\mathbf{z}(t, \cdot) = \alpha^{-1}(\hat{\mathbf{z}}(t))$. Then for every $\omega \in \Omega$ we have

$$\int_S \mathbf{z}(t,\omega)d\mu \leq \int_S \mathbf{e}(t,\omega)d\mu.$$

Since $\mu(S) > \mu(T) - \min\{\mu(T_1), ..., \mu(T_n)\}$, we have $I(S) = \{1, 2, \ldots, n\}$ and thus for all $t \in T$, $\mathbf{z}(t, \cdot)$ is $\bigvee_{i \in I(S)} \mathcal{F}_i$-measurable.

For almost all $t \in S$ we have

$$h_t(\mathbf{z}(t,\cdot)) = \hat{h}_t(\hat{\mathbf{z}}(t)) > \hat{h}_t(\hat{\mathbf{y}}(t)) = h_t(\mathbf{y}(t,\cdot)),$$

which contradicts the assumption that $\mathbf{y}$ is a weak fine core allocation of $\mathcal{E}$. □

Lemma 5.2. *Assume that an economy $\mathcal{E}$ satisfies the assumptions of Proposition 5.1, and in addition for almost every $t \in T$ there is $\omega \in \Omega$ such that* $\mathbf{e}(t, \omega) \neq 0$. *Then the economy $\mathcal{E}^*$ is irreducible, i.e., it satisfies condition* $(A.3)$.

Proof Let $\mathbf{x}$ be a private allocation in $\mathcal{E}^*$, and let T_1, T_2 be two disjoint coalitions in Σ such that $T = T_1 \cup T_2$, and $\mu(T_1) > 0$, $\mu(T_2) > 0$. For every $(t, \omega) \in T \times \Omega$ let

$$\mathbf{y}(t,\omega) = \begin{cases} 0 & t \in T_1, \\ \mathbf{x}(t,\omega) + \frac{1}{\mu(T_2)} \int_{T_1} \mathbf{e}(t,\omega)d\mu & t \in T_2. \end{cases}$$

Then for every $t \in T$, $\mathbf{y}(t, \cdot)$ is $\bigvee_{i=1}^n \mathcal{F}_i$-measurable. Since $u_t(\omega, \cdot)$ is strictly increasing for almost all $t \in T$ and all $\omega \in \Omega$, and $\frac{1}{\mu(T_2)} \int_{T_1} \mathbf{e}(t, \omega)d\mu > 0$ for some $\omega \in \Omega$, it follows from $(A.4)$ that for almost every $t \in T_2$

$$h_t(\mathbf{y}(t,\cdot)) > h_t(\mathbf{x}(t,\cdot)).$$

Moreover, for all $\omega \in \Omega$ we have

$$\int_{T_1} \mathbf{e}(t,\omega)d\mu + \int_{T_2} \mathbf{x}(t,\omega)d\mu = \int_{T_2} \mathbf{y}(t,\omega)d\mu.$$

Therefore $\mathcal{E}^*$ is irreducible. □

Theorem C. *Assume that an economy $\mathcal{E}$ satisfies the assumptions of Lemma 5.2. Then the weak fine core of $\mathcal{E}$ coincides with the set of competitive allocations of $\mathcal{E}^*$.*

Proof The proof follows directly from Proposition 5.1, Lemma 5.2 and Theorem B. □

We conclude with the following proposition.

Proposition 5.3. *If an economy $\mathcal{E}$ satisfies the assumptions* $(A.1), (A.2)$, *and in addition for every $\omega \in \Omega$ and almost all $t \in T$,* $\mathbf{e}(t, \omega) \gg 0$, *then the weak fine core of $\mathcal{E}$ coincides with the set of competitive allocations of $\mathcal{E}^*$.*

Proof The proof is the same as that of Theorem C, noticing that $\mathcal{E}^*$ is irreducible and the Theorem of Vind (1972) holds under the assumptions of Proposition 5.3. □

References

1. Allen, B.: Market games with asymmetric information and non-transferable utility: representations results, and the core. Mimeo, University of Pennsylvania (1991)
2. Aumann, R.J.: Markets with a continuum of traders. Econometrica **32**, 39-50 (1964)
3. Aumann, R.J.: Existence of competitive equilibria in markets with a continuum of traders. Econometrica **34**, 1-17 (1966)
4. Einy, E., Shitovitz, B.: Private value allocations in large economies with differential information. Games and Economic Behavior (forthcoming) (1999)
5. Einy, E., Moreno, D., Shitovitz, B.: The bargaining set of a large economy with differential information. Economic Theory (forthcoming) (1999)
6. Hildenbrand, W.: Existence of equilibria for economies with production and a measure space of economic agents. Econometrica **34**, 1–17 (1970)
7. Hildenbrand, W.: Core and equilibria in large economies. Princeton: Princeton University Press 1974
8. Koutsougeras, L.C., Yannelis, N.C.: Incentive compatibility and information superiority of the core of an economy with differential information. Economic Theory **3**, 195–216 (1993)
9. Mas-Colell, A.: The theory of general equilibrium: a differentiable approach. Cambridge, MA: Cambridge University Press 1985
10. McKenzie, L.W.: On the existence of general equilibrium for a competitive market. Econometrica **27**, 54–71 (1959)
11. Radner, R.: Competitive equilibrium under uncertainty. Econometrica **36**, 31–58 (1968)
12. Radner, R.: Equilibrium under uncertainty. In: K.J. Arrow, Intriligator, M.D. (eds.) Handbook of mathematical economics, vol. II. Amsterdam: North Holland 1982
13. Vind, K.: A third remark on the core of an atomless economy. Econometrica **40**, 585–586 (1972)
14. Wilson, R.: Information, efficiency, and the core of an economy. Econometrica **46**, 807–816 (1978)
15. Yannelis, N.C.: The core of an economy with differential information. Economic Theory **1**, 183–198 (1991)

Incentive compatible core and competitive equilibria in differential information economies*

Françoise Forges[1], **Aviad Heifetz**[2], **and Enrico Minelli**[3]

[1] THEMA, Université de Cergy-Pontoise, and Institut Universitaire de France, Paris, FRANCE
[2] School of Economics, University of Tel Aviv, Tel Aviv, ISRAEL
[3] Dipartimento di Scienze Economiche, Università di Brescia, Via San Faustino 74/B, I-25122 Brescia, ITALY (e-mail: minelli@eco.unibs.it)

Received: August 9, 1999; revised version: September 12, 1999

Summary. If the allocations of a differential information economy are defined as incentive compatible state-contingent lotteries over consumption goods, competitive equilibrium allocations exist and belong to the (ex ante incentive) core. Furthermore, any competitive equilibrium allocation can be viewed as an element of the core of the n-fold replicated economy, for every n. The converse holds under the further assumption of independent private values but not in general, as shown by a counter-example.

Keywords and Phrases: Incentive compatibility, General equilibrium, Core.

JEL Classification Numbers: D51, C71, D82.

1 Introduction

In a standard exchange economy – with complete information – competitive equilibrium allocations exist and belong to the core. Furthermore, in the replicated economy, the core has the "equal treatment" property and shrinks to the set of competitive allocations as the number of replicas gets large (Debreu and Scarf [8]). In this paper, we investigate possible extensions of these classical results in differential information economies.

When the individuals do not share the same information, various concepts of competitive equilibrium are conceivable (e.g., Walrasian equilibrium, rational expectations equilibrium) but the corresponding allocations need not be *incentive compatible*. In order to take incentives into account, we use a notion of competitive

* Financial support from European TMR network FMRX CT96 0055 is gratefully acknowledged.
Correspondence to: E. Minelli

equilibrium which is very similar to the one introduced by Prescott and Townsend [27] [28], and further analyzed by Cole [6], Minelli and Polemarchakis [25], and Kehoe, Levine and Prescott [21] among others.

We consider a finite exchange economy in which every individual's utility possibly depends on all individuals' information. Our model is thus different from the one of Prescott and Townsend, who focus on the case of "private values".

The main idea behind Prescott and Townsend's approach is that individuals trade state-contingent lotteries over the initial consumption goods. This guarantees that the consumption set (i.e. the set of incentive compatible random bundles) of every individual is convex, which is not necessarily true if one restricts attention to deterministic state-contingent allocations. Once the objects of trade are viewed as incentive compatible state-contingent lotteries over the original goods, competitive equilibria can be defined in the usual way, by understanding feasibility as expected feasibility and constructing prices of lotteries as expectations of original goods prices.

We first establish the existence of such competitive solutions (Proposition 1). This result is not an immediate consequence of Prescott and Townsend's analysis since our model is not a particular case of theirs. More importantly, our proof is direct, while they use successive discrete approximations of the underlying economy.

We then show that any competitive allocation defined above belongs to the ex ante incentive core of the economy (Proposition 2). Again, the literature contains several concepts of core for differential information economies (see, e.g., Allen [1], [2], [3], [4], Demange and Guesnerie [9], Einy, Moreno and Shitovitz [10], [11], [12], Hahn and Yannelis [17], Holmström and Myerson [19], Ichiishi and Idzik [20], Koutsougeras and Yannelis [23], Lee and Volij [24], Page [26], Vohra [33] , Volij [34], Wilson [35], Yannelis [38], ... and Forges [13] for a survey)). We are interested in *incentive compatible* core allocations, namely incentive compatible feasible allocations with the property that no coalition can propose a feasible incentive compatible allocation which improves the expected utility of all its members[1]. Allen [2], [4] and Vohra [33] defined the ex ante incentive core by relying on deterministic state-contingent allocations. Vohra [33] showed that this set might be empty. As illustrated in Forges [13] and Forges and Minelli [14], lotteries play a crucial role for the non-emptiness of the ex ante incentive core. In this paper, we use a core notion in which allocations are defined exactly as in the above competitive equilibria. Our concept is close to the "modified incentive compatible" core of Allen [2].

The natural next step is to investigate the relationship between the competitive equilibria and the ex ante incentive core in replica economies. We replicate the basic economy as in Gul and Postlewaite [16]. More precisely, we construct successive independent copies of the economy and we require that every individual's utility only depends on the information of the individuals who belong to the same copy. In this model, incentive compatibility constraints matter even if the economy is

[1] Koutsougeras and Yannelis [23] show that private core allocations are incentive compatible in a certain sense. The main difference between their approach and ours is that coalitions have a potentially larger set of objections here.

replicated many times. We prove (in Proposition 3) that the replication of any competitive allocation belongs to the core of any replicated economy, which is a standard stage toward an equivalence theorem.

The framework of replicas also allows us to make precise the fact that, thanks to the law of large numbers, average feasibility is essentially equivalent to almost sure feasibility in sufficiently large economies (Proposition 4). Hence, the issue of the possible emptiness of the ex ante incentive core disappears in sufficiently large economies.

The converse of Proposition 3, which would lead to an analog of Debreu and Scarf's theorem, does unfortunately not hold in general economies with differential information as we considered up to now. We illustrate this on a simple example in which the equal treatment property fails. In order to recover equal treatment, we strengthen our assumptions by requiring that the utility of every individual only depends on his own information ("private values"). In this case, we establish (Proposition 5) that core allocations in every replicated version of the economy are competitive allocations.

The paper follows the steps described above. The basic economy and solution concepts are described in Section 2. The next section contains the results on the finite economy (Propositions 1 and 2). Section 4 is devoted to general replicated economies (Propositions 3 and 4) while Section 5 deals with replicated economies with private values.

2 Basic economy and solution concepts

We consider an exchange economy with finitely many individuals, $i \in I$ and finitely many commodities, $l \in L$. Types of an individual are $t^i \in T^i$, a finite set. States are $t \in T = \times_i T^i$. The probability distribution on states is $q \in \Delta(T)$. For every i, the initial endowment of individual i, $e^i \in R^L_{++}$, does not depend on the realized state. The total endowment is $e = \sum_i e^i$. For every i, and every t, the utility function of individual i over consumption in state t is $u^i(t, \cdot) : R^L_+ \to R$, a continuous, weakly monotonic function[2].

Following Prescott and Townsend [27] [28], we allow individuals to trade state-contingent lotteries on consumption. Fix a compact subset $C = \{c \in R^L \mid 0 \le c \le b\}$ of R^L, with $b \gg 2e$. The commodity space is the space $\mathcal{L}(C) = (\mathcal{M}(C))^T$ of type contingent signed measures on C, endowed with the product weak topology. We denote by $\Delta(C) \subset \mathcal{M}(C)$ the space of probability measures, or lotteries, on C. A bundle for individual i is $x^i : T \to \Delta(C)$. The utility of individual i from bundle x^i is:

$$U^i(x^i) = \sum_t q(t) \int_C u^i(t, c) dx^i(c \mid t).$$

A bundle x^i is incentive compatible for individual i if, for all t^i and $r^i \in T^i$ we have:

$$\sum_{t^{-i}} q(t^{-i} \mid t^i) \int_C u^i(t, c) dx^i(c \mid t) \ge \sum_{t^{-i}} q(t^{-i} \mid t^i) \int_C u^i(t, c) dx^i(c \mid r^i, t^{-i}).$$

[2] If $x \gg y$ then $u^i(x) > u^i(y)$.

The consumption set of individual i is X^i, the set of incentive compatible bundles for individual i. The endowment of individual i is an element of X^i if we identify e^i with the corresponding (degenerate) lottery in $\mathcal{L}(C)$.

The basic economy is $E = \{I, L, (T^i, X^i, U^i, e^i)_{i\in I}, q\}$.

An allocation, $x = (x^i)_{i\in I} \in X = \times_i X^i$, specifies an incentive compatible state-contingent lottery for every individual. Feasibility is defined on average, across states and realizations of the lotteries[3]. An allocation is feasible if:

$$\sum_i \sum_t q(t) \int_C c dx^i(c \mid t) \leq \sum_i \sum_t q(t) \int_C c de^i(c \mid t) = e$$

The set of feasible allocations is $F \subset X$.

State-contingent measures over consumption are priced by the average amount of resources they use. For a given vector $p \in R^L_+$ of commodity prices, the price of $y \in \mathcal{L}(C)$ is

$$\pi_p(y) = \sum_t q(t) \int_C \sum_l p_l c_l dy(c \mid t)$$

An equilibrium is a couple (p, x) of a price and an allocation such that:

- For all i, $\pi_p(x^i) \leq \pi_p(e^i)$
- For all i, for all $y \in X^i$,

$$U^i(y) > U^i(x^i) \Rightarrow \pi_p(y) > \pi_p(e^i)$$

- $x \in F$

An allocation x is an equilibrium allocation if there exists a price p such that (p, x) is an equilibrium. The set of equilibrium allocations of E is $\mathcal{W}(E)$.

A quasi-equilibrium is a couple (p, x), with $p \neq 0$, such that:

- For all i, $\pi_p(x^i) = \pi_p(e^i)$
- For all i, for all $y \in X^i$,

$$U^i(y) \geq U^i(x^i) \Rightarrow \pi_p(y) \geq \pi_p(e^i)$$

- $x \in F$

An allocation x is a quasi-equilibrium allocation if there exists a price p such that (p, x) is a quasi-equilibrium. The set of quasi-equilibrium allocations of E is $\mathcal{Q}(E)$.

A coalition is $B \subset I \setminus \{\emptyset\}$, a nonempty subset of I. For every coalition B, let $t_B = (t^i)_{i\in B}$ and $T_B = \times_{i\in B} T^i$. The trades of individual i, when he takes part in

[3] Average feasibility is justified in large economies by the law of large numbers. See Proposition 4 for a precise result in this vein.

coalition B, depend only on the information of the members of the coalition, i.e. they belong to the set

$$X^i_B = \{x^i \in X^i \mid x^i(t) = x^i(t_B, t'_{I\backslash B}) \text{ for all } t \in T, t'_{I\backslash B} \in T_{I\backslash B}\}.$$

Notice that, with this notation, $X^i = X^i_I$. For trades in X^i_B, we write $x^i(t_B)$ as a shortcut for $x^i(t_B, t'_{I\backslash B})$, for all $t'_{I\backslash B} \in T_{I\backslash B}$. A coalition B blocks the allocation x if there exist bundles (y^i), for $i \in B$, such that :

a) For all $i \in B$, $y^i \in X^i_B$
b) For all l,

$$\sum_{i \in B} \sum_{t_B \in T_B} q(t_B) \int_C c_l dy^i(c \mid t_B) \le \sum_{i \in B} e^i_l$$

c) For all $i \in B$, $U^i(y^i) \ge U^i(x^i)$, with strict inequality for at least one $i \in B$.

An allocation (x) has the core property if it is feasible, and there does not exist a coalition that blocks it[4]. The core of E, $\mathcal{C}(E)$, is the set of allocations that have the core property.

3 Existence

Proposition 1 $\mathcal{Q}(E) \neq \emptyset$

Proof The set $F \subset X$ is non empty because it contains $(e^i)_{i \in I}$, compact because it is a closed (weak topology) subset of the compact space $\mathcal{L}(C)^I$, and convex because it is defined by linear inequalities.

For every $\lambda \in \Delta(I)$, define $W_\lambda(x) = \sum_i \lambda^i U^i(x^i)$. The problem $Max_F W_\lambda$ has a solution x_λ, because W_λ is a sum of continuous (weak topology) functions, and F is non-empty and compact.

Define the Lagrangean $L_\lambda : X \times R^L_+ \to R$ by $L_\lambda(x,p) = W_\lambda(x) - \sum_l p_l(\sum_i \sum_t q(t) \int_C c_l dx^i(c \mid t) - \sum_i e^i_l)$=$W_\lambda(x) - \sum_i[\pi_p(x^i) - \pi_p(e^i)]$. From the saddle point theorem (see for example Theorem 8.B.I in Duffie [7], p. 77), there exists p_λ such that (x_λ, p_λ) is a saddle point of L_λ. For all $x \in X$, and all $p \in R^L_+$:

$$L_\lambda(x, p_\lambda) \le L_\lambda(x_\lambda, p_\lambda) \le L_\lambda(x_\lambda, p).$$

For every individual i, if, for some $x^i \in X^i$, $U^i(x^i) \ge U^i(x^i_\lambda)$, then $\pi_{p_\lambda}(x^i) \ge \pi_{p_\lambda}(x^i_\lambda)$. Indeed, if we fix $x^j = x^j_\lambda$ for all $j \neq i$, the first inequality in the saddle point condition implies that, for all $x^i \in X^i$:

$$\lambda^i U^i(x^i_\lambda) - \lambda^i U^i(x^i) \ge \pi_{p_\lambda}(x^i_\lambda) - \pi_{p_\lambda}(x^i) \qquad (*)$$

We want to show that $p_\lambda \neq 0$. For every i, let the feasible consumption set of individual i, $F^i = Proj_{X^i} F$, be the subset of his consumption set which contains bundles that are part of feasible allocations for the economy. Let i be an individual with $\lambda^i > 0$. Such an individual exists, because $\lambda \in \Delta(I)$. At a saddle point,

[4] We shall comment on this definition in Remark 2.

individual i is not satiated. Indeed, consider the bundle $2e$ which, in all states, puts all weight on $2e \in C$. The degenerate lottery $2e$ is an element of X^i, and we have $U^i(2e) > U^i(x^i)$ for all $x^i \in F^i$, in particular for x^i_λ. If $p_\lambda = 0$, inequality $(*)$ for individual i gives $U^i(x^i_\lambda) \geq U^i(x^i)$, for all $x^i \in X^i$, a contradiction. Given that $p_\lambda \neq 0$, we can restrict prices to lie in $\Delta(L)$.

Consider the correspondence $\psi \times \phi$ of $F \times \Delta(L) \times \Delta(I)$ into itself defined component-wise by:

$$\psi(x, p) = \{\lambda \in \Delta(I) \mid \lambda^i = 0 \text{ if } \pi_p(x^i) > \pi_p(e^i)\}$$

$$\phi(\lambda) = \{(x, p) \in F \times \Delta(L) \mid (x, p) \text{ is a saddle point of } L_\lambda\}$$

The correspondence $\psi \times \phi$ is non-empty, compact, convex-valued and upper hemi-continuous, and thereby admits a fixed point, $(\hat{x}, \hat{p}, \hat{\lambda})$ by Glicksberg [15] theorem.

To show that $(\hat{x}, \hat{p})$ is a quasi-equilibrium, it is enough to prove that, for all i, $\pi_{\hat{p}}(\hat{x}^i) = \pi_{\hat{p}}(e^i)$. We first show that, for all i, $\pi_{\hat{p}}(\hat{x}^i) \leq \pi_{\hat{p}}(e^i)$. Indeed, suppose that for some i $\pi_{\hat{p}}(\hat{x}^i) > \pi_{\hat{p}}(e^i)$, then the definition of ψ implies $\lambda^i = 0$, and, from $(*)$, $\pi_{\hat{p}}(\hat{x}^i) \leq \pi_{\hat{p}}(e^i)$, a contradiction. But, from the saddle point conditions, $\sum_i \pi_{\hat{p}}(\hat{x}^i - e^i) = 0$, so that, for all i, $\pi_{\hat{p}}(\hat{x}^i) = \pi_{\hat{p}}(e^i)$. □

Lemma 1 $\mathcal{Q}(E) \subset \mathcal{W}(E)$

Proof Consider a quasi equilibrium $(\hat{x}, \hat{p})$. For all i, $\pi_{\hat{p}}(\hat{x}^i) = \pi_{\hat{p}}(e^i) > 0$. Indeed, $\pi_{\hat{p}}(e^i) = \sum_l \hat{p}_l e^i_l$, and, for all i, this is positive because $\hat{p} \in R^L_+ \setminus \{0\}$, and $e^i \in R^L_{++}$. If $(\hat{x}, \hat{p})$ was not an equilibrium, for some $i \in I$, and some $x^i \in X^i$, we would have $U^i(x^i) > U^i(\hat{x}^i)$ and $\pi_{\hat{p}}(x^i) = \pi_{\hat{p}}(e^i)$. But this would lead to a contradiction. Indeed, the lottery that puts all weight on 0 in every state is an element of X^i and individual i would prefer a convex combination between this lottery and x^i, which contradicts that $(\hat{x}, \hat{p})$ is a quasi-equilibrium. □

Lemma 2 $\mathcal{W}(E) \subset \mathcal{Q}(E)$

Proof Consider an equilibrium $(\hat{x}, \hat{p})$. For all i, $\pi_{\hat{p}}(\hat{x}^i) = \pi_{\hat{p}}(e^i)$. Indeed, agent i strictly prefers the lottery that puts all weight on $2e$ in every state, which is an element of X^i, to any lottery in his feasible consumption set F^i. If $\pi_{\hat{p}}(\hat{x}^i) < \pi_{\hat{p}}(e^i)$, individual i can take a convex combination between this lottery and $\hat{x}^i$, $z_\lambda = \lambda 2e + (1-\lambda)\hat{x}^i$; for $\lambda > 0$, $U^i(z_\lambda) > U^i(x^i)$ while, for λ small enough, $\pi_{\hat{p}}(z_\lambda) < \pi_{\hat{p}}(e^i)$, which contradicts that $(\hat{x}, \hat{p})$ is an equilibrium. The same argument shows that it cannot be the case that, for some i and some $x^i \in X^i$, $U^i(x^i) = U^i(\hat{x}^i)$ and $\pi_{\hat{p}}(x^i) < \pi_{\hat{p}}(e^i)$. □

Remark 1. The usual argument to show that an equilibrium is also a quasi-equilibrium requires local non satiation, which fails in our setting, due to the compactness of the consumption set. The argument in Lemma 2 uses the fact that, at any feasible allocation, the individual is not satiated.

Proposition 2 $\mathcal{C}(E) \neq \emptyset$

Proof We show that $\mathcal{W}(E) \subset \mathcal{C}(E)$.

Let $(\hat{x}, \hat{p})$ be an equilibrium for the economy E. If $\hat{x} \notin \mathcal{C}(E)$ there exists a coalition B, and bundles (y^i), for $i \in B$, such that conditions a, b and c are verified.

From $U^i(y^i) \geq U^i(\hat{x}^i)$, using Lemma 2 and the definition of quasi equilibrium it follows that, for all $i \in B$, $\pi_{\hat{p}}(y^i) \geq \pi_{\hat{p}}(\hat{x}^i) = \pi_{\hat{p}}(e^i)$. For the individuals for whom $U^i(y^i) > U^i(\hat{x}^i)$, the definition of equilibrium implies $\pi_{\hat{p}}(y^i) > \pi_{\hat{p}}(e^i)$. Summing over all $i \in B$ and using the definition of $\pi_{\hat{p}}$:

$$\sum_l \hat{p}_l \sum_{i \in B} \sum_t q(t) \int_C c_l dy^i(c \mid t_B) = \sum_l \hat{p}_l \sum_{i \in B} \sum_{t_B \in T_B} q(t_B) \int_C c_l dy^i(c \mid t_B)$$
$$> \sum_l \hat{p}_l \sum_{i \in B} e^i_l$$

a contradiction. □

Remark 2. We use a weak notion of blocking: it is enough for a coalition to find a deviation such that nobody is worse off and somebody strictly better off. This allows us to prove an equal treatment property (see Section 5). In the more standard definition, a deviation must lead to a strict improvement for everybody in the coalition. Under complete information, if the utility functions are continuous and monotonic, the two notions are equivalent: one can always find transfers from the individual who are made strictly better off to the others in such a way that everybody is strictly better off in the end. In our setting this is no longer true: it may be impossible to design the transfers in a way that respects incentive compatibility. The core for which we have shown non-emptiness is thus contained in the core defined with the stronger notion of blocking. Allen [2] introduced a version of the latter concept under the name "modified incentive compatible core" and showed its non-emptiness by relying on Scarf [30]'s theorem.

4 Replica economies

In the n-times replicated economy, individuals are $(i, k) \in I^n = I \times \{1, 2, \ldots, n\}$. Consumption goods are $l \in L$.

For every i, all the copies of i have the same set of types, $T^{i,k}$, a copy of T^i. The set of states is thus the product of n copies of the set of states in the basic economy, $\bar{T}_n = \times_k T_k$, with T_k a copy of T, and we assume that the probability distribution over $\bar{T}_n$ is $\bar{q}_n = \times_k q_k$, the independent product of the probability distributions over types in every replica, $q_k = q$.

For every i, all the copies of i have the same (state independent) initial endowment, $e^{i,k} = e^i \in R^L_{++}$. The total endowment is $\sum_{(i,k)} e^{i,k} = ne$. Following Gul and Postlewaite [16], we assume that the utility of each individual depends only on the types of other individuals in the same replica: for every (i, k), and every $\bar{t}_n \in \bar{T}_n$, the utility function of individual (i, k) over consumption in state $\bar{t}_n$ is:

$$u^{i,k}(\bar{t}_n, \cdot) = u^i(t_k, \cdot)$$

The objects of trade are state contingent lotteries over C. A bundle for individual (i,k) is $x^{i,k} : \bar{T}_n \to \Delta(C)$. The lotteries in replica k may thus depend on the realized types in other replicas. The commodity space is the space $\mathcal{L}_n(C) = (\Delta(C))^{\bar{T}_n}$, endowed with the (product) weak topology. The utility for individual (i,k) of bundle $x^{i,k}$ is:

$$U^{i,k}(x^{i,k}) = \sum_{\bar{t}_n} \bar{q}_n(\bar{t}_n) \int_C u^i(t_k, c) dx^{i,k}(c \mid \bar{t}_n).$$

A bundle $x^{i,k}$ is incentive compatible for individual (i,k) if, for all t_k^i and $r_k^i \in T_k^i$ we have:

$$\sum_{(t_j)_{j\neq k}} \Pi_{j\neq k} q_j(t_j) \sum_{t_k^{-i}} q_k(t_k^{-i} \mid t_k^i) \int_C u^i(t_k, c) dx^{i,k}(c \mid \bar{t}_n) \geq$$

$$\geq \sum_{(t_j)_{j\neq k}} \Pi_{j\neq k} q_j(t_j) \sum_{t_k^{-i}} q_k(t_k^{-i} \mid t_k^i) \int_C u^i(t_k, c) dx^{i,k}(c \mid (r_k^i, t_k^{-i}, (t_j)_{j\neq k}))$$

The consumption set of individual (i,k) is $X^{i,k}$, the set of incentive compatible bundles for individual (i,k). The endowment of individual (i,k) is an element of $X^{i,k}$.

The n-times replicated economy is $E^n = \{I^n, L, (T^{i,k}, X^{i,k}, U^{i,k}, e^{i,k})_{(i,k)\in I^n}, \bar{q}_n\}$

An allocation, $x = (x^{i,k})_{(i,k)\in I^n}$, specifies a state-contingent lottery for every individual. An allocation is feasible in E^n if:

$$\sum_{i,k} \sum_{\bar{t}_n} \bar{q}_n(\bar{t}_n) \int_C c dx^{i,k}(c \mid \bar{t}_n) \leq ne$$

A coalition is $B \subset I^n \setminus \{\emptyset\}$, a nonempty subset of I^n. For every coalition B, let $t_B = (t^{i,k})_{(i,k)\in B}$ and $T_B = \times_{(i,k)\in B} T^{i,k}$.

The trades of individual (i,k), when he takes part in coalition B, depend only on the information of the members of the coalition, i.e. they belong to the set

$$X_B^{i,k} = \{x^{i,k} \in X^{i,k} \mid x^{i,k}(\bar{t}_n) = x^{i,k}(t_B, t'_{I^n\setminus B}) \text{ for all } \bar{t}_n \in \bar{T}_n, t'_{I^n\setminus B} \in T_{I^n\setminus B}\}$$

For trades in $X_B^{i,k}$, we write $x^{i,k}(t_B)$ as a shortcut for $x^{i,k}(t_B, t'_{I\setminus B})$, for all $t'_{I\setminus B} \in T_{I^n\setminus B}$. A coalition B blocks the allocation x if there exist bundles $(y^{i,k})$, for $(i,k) \in B$, such that :

a') For all $(i,k) \in B$, $y^{i,k} \in X_B^{i,k}$
b') For all l,

$$\sum_{(i,k)\in B} \sum_{t_B\in T_B} \bar{q}_n(t_B) \int_C c_l dy^{i,k}(c \mid t_B) \leq \sum_{(i,k)\in B} e_l^i$$

c') For all $(i,k) \in B$, $U^{i,k}(y^{i,k}) \geq U^{i,k}(x^{i,k})$, , with strict inequality for at least one $(i,k) \in B$.

An allocation x has the core property if it is feasible, and there does not exist a coalition that blocks it. The core of E^n, $\mathcal{C}(E^n)$, is the set of allocations that have the core property.

The following lemma says that, for every incentive compatible bundle for individual (i,k), there exists a utility equivalent incentive compatible bundle that only depends on types in replica k. For all k, let $I_k = I \times \{k\}$ be the coalition of individuals who belong to replica k.

Lemma 3 *For all $x^{i,k} \in X^{i,k}$, there exists $y^{i,k} \in X^{i,k}_{I_k}$ such that $U^{i,k}(y^{i,k}) = U^{i,k}(x^{i,k})$.*

Proof For all $\bar{t}_n \in \bar{T}_n$, let $y^{i,k}(\bar{t}_n) = \sum_{(t_j)_{j\neq k}} \Pi_{j\neq k} q_j(t_j) x^{i,k}(t_k,(t_j)_{j\neq k}) = y^{i,k}(t_k)$.

The result follows by simple substitution if one notices that, for all t^i_k and $r^i_k \in T^i_k$:

$$\sum_{(t_j)_{j\neq k}} \Pi_{j\neq k} q_j(t_j) \sum_{t_k^{-i}} q_k(t_k^{-i} \mid t_k^i) \int_C u^i(t_k,c) dx^{i,k}(c \mid r^i_k, t_k^{-i}, (t_j)_{j\neq k}) =$$

$$\sum_{t_k^{-i}} q_k(t_k^{-i} \mid t_k^i) \int_C u^i(t_k,c) dy^{i,k}(c \mid r^i_k, t_k^{-i}).$$

□

An allocation x in the basic economy E can be identified with an allocation in E^n if we let $x^{i,k}(\bar{t}_n) = x^i(t_k)$, for all $(i,k) \in I^n$ and all $\bar{t}_n \in \bar{T}_n$. Lemma 3 allows us to show that:

Proposition 3 *For every n, $\mathcal{W}(E) \subset \mathcal{C}(E^n)$*

Proof Let $(\hat{x},\hat{p})$ be an equilibrium for the economy E, and suppose $\hat{x} \notin \mathcal{C}(E^n)$.

Then there exists a coalition $B \subset I^n$, and bundles $(y^{i,k})$, for $(i,k) \in B$, such that conditions a', b' and c' hold.

For all k, let $B_k = B \cap I_k$. Using Lemma 3, we can assume, without loss of generality, that $y^{i,k} \in X^{i,k}_{B_k}$, i.e. that the bundle of each individual only depends on the types of members of the coalition who also belong to his replica. Thus, with some abuse of notation, $y^{i,k} \in X^i$. But then, by the same argument as in Proposition 2, $U^{i,k}(y^{i,k}) \geq U^{i,k}(\hat{x}^{i,k})$ implies $\pi_{\hat{p}}(y^{i,k}) \geq \pi_{\hat{p}}(e^i)$, and $U^{i,k}(y^{i,k}) > U^{i,k}(\hat{x}^{i,k})$ implies $\pi_{\hat{p}}(y^{i,k}) > \pi_{\hat{p}}(e^i)$.

Summing over all $(i,k) \in B_k$ and using the definition of $\pi_{\hat{p}}$:

$$\sum_l \hat{p}_l \sum_{(i,k)\in B_k} \sum_{t_{B_k}} q_k(t_{B_k}) \int_C c_l dy^{i,k}(c \mid t_{B_k}) > \sum_l \hat{p}_l \sum_{i\in B_k} e^i_l$$

If we now sum over all $k \in \{1,2,\ldots,n\}$, we obtain a contradiction. □

In the basic economy, the equilibrium allocation $\hat{x}$ is feasible on average. By appealing to the law of large numbers, we will show that, if we let n tend to infinity, the allocation $\hat{x}$ in the replicated economy converges to an allocation which is almost surely feasible.

Let $x = (x^{i,k})_{(i,k)\in I^n}$ be a feasible allocation in E^n. Together with q, x induces a probability distribution $P_{q,x}$ over $\bar{T}_n \times C^{I^n}$. Let us denote the (random) consumption of individual (i,k) as $c^{i,k}$ and the total consumption in replica k as z^k, namely:

$$z^k = \sum_{i\in I} c^{i,k}$$

The average total consumption across replicas is then[5]:

$$\bar{z}^n = \frac{1}{n}\sum_{k=1}^{n} z^k$$

Observe that x is feasible if $E_{q,x}\bar{z}^n \leq e$, and that x is feasible almost surely if $\bar{z}^n \leq e \; P_{q,x}$ - a. s.

Let $\epsilon > 0$ and $\delta > 0$. We say that x is (ϵ, δ)-feasible in E^n iff:

$$P_{q,x}(\bar{z}^n \leq e + \epsilon) \geq 1 - \delta$$

Proposition 4 *For every $\epsilon > 0$, and every $\delta > 0$ there exists N such that for all $n \geq N$, if $x \in \mathcal{W}(\mathcal{E})$, then x is (ϵ, δ)-feasible in E^n.*

Proof Let us fix $\epsilon > 0$ and $\delta > 0$. Let $x \in \mathcal{W}(\mathcal{E})$. As above, x induces an allocation in E^n, for every n. The associated sequence $(z^k)_{k\geq 1}$ is i.i.d. and average feasibility implies that $E_{q,x}\bar{z}^n \leq e$.

By the law of large numbers, there exists N such that, for all $n \geq N$:

$$P_{q,x}(\mid \bar{z}^n - E_{q,x}\bar{z}^n \mid \leq \epsilon) \geq 1 - \delta$$

□

Let us define $\mathcal{C}_{\epsilon,\delta}(E^n)$ exactly as $\mathcal{C}(E^n)$ except that we require all feasible allocations to be (ϵ, δ)-feasible as well. Coalitions have thus less objections in $\mathcal{C}_{\epsilon,\delta}(E^n)$ than in $\mathcal{C}(E^n)$. Hence, by Proposition 3 and Proposition 4 we have:

Corollary 1 *For all $\epsilon > 0$, and all $\delta > 0$ there exists N such that for all $n \geq N$, $\mathcal{C}_{\epsilon,\delta}(E^n) \neq \emptyset$.*

---

[5] In a usual model, the consumption of every individual is deterministic and independent of the replica to which he belongs, $c^{i,k}(\bar{t}_n) = c^i(t_k)$. In this case, we recover the standard expression:

$$\bar{z}^n = \sum_t \frac{N^n(t)}{n} z(t)$$

where $N^n(t)$ is the number of replicas in which $t_k = t$, and $z(t) = \sum_i c^i(t)$.

5 Convergence

In the case of exchange economies with complete information, Debreu and Scarf [8] prove that every allocation which is in the core of all replicas must be a competitive equilibrium of the basic economy. Even with the restricted notion of replication we have chosen, this need not always be true.

Example: As our basic economy E we take the two-person two-good economy in Kreps [22]. Individuals are $I = \{1, 2\}$, and we denote the two goods by c and m respectively. There are two possible states in the economy, which coincide with the possible types of individual 1, $T = T^1 = \{s, s'\}$, with $q(s) = q(s') = 0.5$. In state $t = s, s'$, the utility function of individual 1 is $u^1(t) = a(t)lnc(t) + m(t)$, while the utility of individual 2 is $u^2(t) = b(t)lnc(t) + m(t)$, with $a(s) = b(s') = 1$ and $a(s') = b(s) = 2$. The initial endowment does not depend on the state, and is the same for both individuals, $e^i(t) = e = (1.5, 1)$. The two individuals are thus ex-ante identical, but the realized type of individual 1 determines ex-post which of the two has a higher utility from consumption of the first good. When we introduce lotteries we take the support $C = \{(c, m) \in R^2 \mid 0 \leq c \leq b_1, 0 \leq m \leq b_2\}$, with $b_1 > 6, b_2 > 4$.

Consider the allocation $\hat{x}$ defined by $(\hat{c}^1, \hat{m}^1)(s) = (\hat{c}^2, \hat{m}^2)(s') = (1, 1.5)$ and $(\hat{c}^1, \hat{m}^1)(s') = (\hat{c}^2, \hat{m}^2)(s) = (2, 0.5)$. If we restrict attention to deterministic state-contingent allocations, $\hat{x}$ is (ex ante) Pareto-optimal. Furthermore, it is easy to check that it is also incentive compatible and individually rational. In particular, each individual obtains a gain from trade equal to $U^i(\hat{x}^i) - U^i(e) = 0.085$.

To show that $\hat{x} \in \mathcal{C}(E)$, we only have to check that $\hat{x}$ is Pareto-optimal even when we allow for state-contingent lotteries.

If this were not the case, one could find a feasible allocation $y = (y^i)_{i=1,2}$, $y^i \in \Delta(C)^T$ such that $U^i(y^i) \geq U^i(\hat{x}^i), i = 1, 2$, with at least one strict inequality. But then the deterministic allocation which gives to each individual the expected value of the lottery in each state would be feasible and would dominate $\hat{x}$.

If we modify $\hat{x}$ by requiring an additional transfer of $\tau \leq 0.085$ units of good 2 from individual 2 to individual 1 in each state, we maintain incentive compatibility and individual rationality, and we obtain an allocation $\tilde{x}$ which also belong to $\mathcal{C}(E)$.

We will show that $(\hat{x}, \tilde{x})$ belongs to the core $\mathcal{C}(E^2)$ of the two fold replicated economy.

The only coalition which might possibly object is the one formed by individual 1 in the first replica and individual 2 in the second replica. But, in this coalition, if individual 1 must be guaranteed $U^1(\hat{x}^1)$, individual 2 cannot get more than his reservation utility $U^2(e)$.

To see this, consider the allocation $\bar{x}$ defined by $(\bar{c}^1, \bar{m}^1)(s) = (1, 1.5)$, $(\bar{c}^1, \bar{m}^1)(s') = (2, 0.5)$, $(\bar{c}^2, \bar{m}^2)(s) = (1.5, 0.5)$, $(\bar{c}^2, \bar{m}^2)(s') = (1.5, 1.5)$. Individual 1 obtains the same bundle as in $\hat{x}$, while individual 2, whose utility does not depend on the type of individual 1, obtains the same quantity of good 1 in both states. This allocation is incentive compatible and individually rational. Furthermore, if one restricts attention to deterministic allocations, $\bar{x}$ maximizes the sum of expected utilities in the coalition, namely it solves the problem of maximizing:

$$[lnc^1(s) + m^1(s)] + [2lnc^1(s') + m^1(s')] + \left[\frac{3}{2}lnc^2(s) + m^2(s)\right] + \left[\frac{3}{2}lnc^2(s') + m^2(s')\right]$$

under the feasibility constraints:

$$c^1(s) + c^2(s) + c^1(s') + c^2(s') \leq 6$$
$$m^1(s) + m^2(s) + m^1(s') + m^2(s') \leq 4$$

Notice that, by construction, $U^1(\bar{x}^1) = U^1(\hat{x}^1)$, so that $\bar{x}$ is also a solution of the problem of maximizing individual 2's expected utility under feasibility and the additional constraint that individual 1's obtains at least $U^1(\hat{x}^1)$. By the same argument as above, nothing changes when we allow for lotteries. Hence, individual 2's utility cannot exceed $U^2(\bar{x}^2) = U^2(e)$, as claimed. □

As the example makes clear, the dependence of the utility of a given individual on the types of other individuals in his replica creates an "informational externality", and equal treatment may fail at a core allocation.

In the special case of private values this externality is not present, and one may hope to proceed as in Debreu and Scarf [8].

Consider the basic economy $E = \{I, L, (T^i, X^i, U^i, e^i)_{i \in I}, q\}$. From now on we make the assumption of independent private values (IPV), i.e., we assume that $q = \times_i q^i$, $q^i \in \Delta(T^i)$, and $u^i(t, \cdot) = u^i(t_i, \cdot)$. The replicated economy E^n is obtained from E exactly like in Section 4, but of course now the utility of each individual only depends on his own type and the fact of belonging to one replica or another is of no consequence.

By an argument similar to the one in Lemma 3 we can prove that any level of utility achievable by individual (i, k) with an incentive compatible mechanism which depends on the information of all individuals in the replicated economy can also be achieved by an incentive compatible mechanism which depends only on his own information[6]:

Lemma 4 *Under IPV, for all $x^{i,k} \in X^{i,k}$, there exists $y^{i,k} : T^i \to \Delta(C)$ such that $y^{i,k} \in X^i$ and $U^{i,k}(y^{i,k}) = U^{i,k}(x^{i,k})$.*

The next Lemma states that at an allocation in the core of the replicated economy, all the replicas of an individual must obtain the same level of utility. Private values are crucial to the result. In general, even with our very special replication process, the lemma does not hold as shown in the previous example.

Lemma 5 *Under IPV, if $x \in C(E^n)$, then, for all (i, k) and $(i, h) \in I^n$, $U^i(x^{i,k}) = U^i(x^{i,h})$.*

---

[6] Under IPV, we thus recover, without loss of generality, the private measurability assumptions introduced by Yannelis [38]. However, observe that Lemma 4 would not hold if we did not consider random, average feasible, allocations

Proof Take $x \in \mathcal{C}(E^n)$, and assume the proposition is false. Without loss of generality, for all $(i,k) \in I^n$:

$$U^i(x^{i,1}) \leq U^i(x^{i,k})$$

and there exists $(1,h) \in I^n$ such that:

$$U^1(x^{1,1}) < U^1(x^{1,h}).$$

By Lemma 4, without loss of generality, for all (i,k), $x^{i,k} : T^i \to \Delta(C)$, and $x^{i,k} \in X^i$. Consider then the coalition formed by individuals in the first replica, I_1, and define an allocation for this coalition by $y^{i,k}(t^i) = y^i(t^i) = \frac{1}{n}\sum_{k=1}^{n} x^{i,k}(t^i)$. As a convex combination of elements of X^i, $y^i \in X^i$, for all i. Furthermore, $(y^i)_{i\in I}$ is feasible for the coalition I_1. Indeed, from feasibility of x in E^n :

$$\sum_i \sum_k \sum_{t_k^i} q_i(t_k^i) \int_C c dx^{i,k}(c \mid t_k^i) \leq ne$$

$$\sum_i \sum_{t^i} q_i(t^i) \frac{1}{n} \sum_k \int_C c dx^{i,k}(c \mid t^i) \leq e$$

By the linearity of the utility function, we have, for all i, $U^i(y^i) \geq U^i(x^{i,1})$, and for $i = 1$, $U^1(y^1) > U^1(x^{1,1})$. This contradicts the fact that $x \in \mathcal{C}(E^n)$. □

A stronger notion of equal treatment requires that all replicas of the same individual receive exactly the same bundle: for all $(i,k) \in I^n$, $x^{i,k} = x^i$. Let $\mathcal{C}^*(E^n) \subset \mathcal{C}(E^n)$ denote the set of core allocations in which all replicas of each individual i are treated equally in this stronger sense. Lemma 5 allows us to associate to every allocation in $x \in \mathcal{C}(E^n)$ an allocation $y \in \mathcal{C}^*(E^n)$ defined by $y^i(t^i) = \frac{1}{n}\sum_{k=1}^{n} x^{i,k}(t^i)$ which is indifferent to x, in utility, for every individual. From now on we thereby restrict attention to $\mathcal{C}^*(E^n)$ which we identify with a subset of $\mathcal{L}(C)^I$. From the proof of Proposition 3 we know that $\mathcal{W}(E) \subset \cap_{n\geq 1}\mathcal{C}^*(E^n)$. The next Proposition provides the opposite implication.

Proposition 5 *Under IPV,* $\cap_{n\geq 1}\mathcal{C}^*(E^n) \subset \mathcal{W}(E)$

Proof Consider an allocation $x \in \cap_{n\geq 1}\mathcal{C}^*(E^n)$. For all $i \in I$, define the set of incentive compatible net trades which are strictly preferred to x^i:

$$\psi_x^i = \{z \in \mathcal{L}(C) \mid U^i(z + e^i) > U^i(x^i),\ z + e^i \in X^i\}$$

For all i, $\psi_x^i \neq \emptyset$. It is enough to consider $z = 2e - e^i$, where we have identified $2e$ with the corresponding (incentive compatible) degenerate lottery. By monotonicity of u^i, and feasibility of x, $U^i(2e) > U^i(x^i)$ for all i.

Let $Co(\cup_i \psi_x^i)$ denote the convex hull of the set $\cup_i \psi_x^i$ and define the set A as follows:

$$A = \{a \in R^L \mid \exists z \in Co(\cup_i \psi_x^i),\ \sum_t q(t) \int_C c dz(c \mid t) = a\}$$

The set A is non-empty and convex. We now show that it does not contain any strictly negative vector: $A \cap R^L_{--} = \emptyset$.

If this was not the case, given that the sets ψ^i_x are convex (by the linearity of U^i), one could find $\hat{z}$ with $\hat{z} = \sum_i \alpha^i \hat{z}_i$, $\alpha^i \geq 0$ and $\sum_i \alpha^i = 1$, such that $U^i(\hat{z}^i + e^i) > U^i(x^i)$, $\hat{z}^i + e^i \in X^i$ and

$$\sum_t q(t) \sum_i \alpha^i \int_C cd\hat{z}^i(c \mid t) << 0$$

Given that this inequality is strict, without loss of generality α^i can be taken to be rational, $\alpha^i = \frac{\beta^i}{n}$ for some integers n and $\beta^i \leq n$. Let us then form a coalition S with β^i copies of each individual i. Consider the allocation which gives to each copy of i in the coalition the bundle $\hat{y}^i = \hat{z}^i + e^i$. By construction, for all i, $\hat{y}^i \in X^i$ and $U^i(\hat{y}^i) > U^i(x^i)$. Moreover, the preceding inequality implies feasibility:

$$\sum_t q(t) \sum_{i \in S} \int_C cd\hat{y}^i(c \mid t) =$$

$$\sum_t q(t) \sum_i \beta^i \int_C cd\hat{z}^i(c \mid t) + \sum_i \beta^i e^i =$$

$$n \sum_t q(t) \sum \alpha^i \int_C cd\hat{z}^i(c \mid t) + \sum_i \beta^i e^i << \sum_i \beta^i e^i$$

This contradicts the fact that $x \in \mathcal{C}^*(E^n)$.

By the separating hyperplane theorem, there exists $p \in R^L \setminus \{0\}$ such that $Inf\{pa \mid a \in A\} \geq Sup\{pb \mid b \in R^L_{--}\}$. The boundedness of A implies that $p \geq 0$ so that $Sup\{pb \mid b \in R^L_{--}\} = 0$.

We show that (p, x) is an equilibrium for the economy E. Consider individual i, and suppose that for some $y^i \in X^i$, $U^i(y^i) > U^i(x^i)$. Let then $z^i = y^i - e^i$. Clearly, $z^i \in \psi^i_x$, so that, by definition $a^i = \sum_t q(t) \int_C cdz^i(c \mid t) \in A$. But then, from the separation argument, $pa^i \geq 0$, i.e.,

$$\sum_t q(t) \int_C \sum_l p_l c_l dy^i(c \mid t) \geq \sum_l p_l e^i_l$$

$$\pi_p(y^i) \geq \pi_p(e^i)$$

An argument entirely analogous to the one in Lemma 1 shows that the last inequality is strict. It remains to show that, for all i, $\pi_p(x^i) \leq \pi_p(e^i)$. Consider the bundle $z_\lambda = \lambda 2e + (1 - \lambda)x^i$; for $\lambda > 0$, $U^i(z_\lambda) > U^i(x^i)$, so that, by the previous argument, $\pi_p(z_\lambda) \geq \pi_p(e^i)$. Letting λ tend to zero we obtain $\pi_p(x^i) \geq \pi_p(e^i)$ for all i, which, combined with the feasibility of x, leads to $\pi_p(x^i) = \pi_p(e^i)$. □

We can thus state the following equivalence result:

Corollary 2 *Under IPV,* $\cap_{n \geq 1} \mathcal{C}^*(E^n) = \mathcal{W}(E)$

6 Concluding remarks

We end up with a brief comment on some papers which, as the present one, study the core in large exchange economies with differential information. Allen [4] deduces the non-emptiness of approximate ex ante incentive cores in replica economies from results of Shubik and Wooders [32] and Wooders [35], but does not establish relationships between core allocations and competitive ones. The main difference between Allen's approach and ours is that, in general, unless the IPV assumption is made, Gul and Postlewaite [16]'s replicas do not satisfy one of the basic assumptions in Shubik and Wooders [32] and Wooders [36], namely that all replicas of a given agent are "substitutes" for each other. Indeed, in Gul and Postlewaite [16]'s replicated economy, the information of an agent plays a completely different role in his own replica and in the other ones.

Serrano, Volij and Vohra [31] do not consider the same replicas as in Gul and Postlewaite [16] either. They replicate a finite economy with asymmetric information in such a way that all replicas of a given agent have exactly the same information. Hence, in their model, incentive compatibility is not an issue in the replicated economy. Another difference with the present paper is that they study Wilson [35]'s coarse core, in which the coalitions make objections at the *interim* stage (see also Forges [13] and Vohra [33]). They show that the coarse core does ot converge to *any* set of price equilibrium allocations considered in the literature.

Allen [5] considers deterministic allocations in an atomless exchange economy. She imposes incentive compatibility constraints which are stronger than in the present paper. By relying on an equilibrium existence theorem of Yamazaki [37], she establishes the non-emptiness of the core.

Einy, Moreno and Shitovitz [10], [11]'s basic model is similar to Allen [5]'s one, but they further assume private values and do not require any form of incentive compatibility. In [10], they establish the equivalence between competitive allocations in the sense of Radner [29] and Yannelis [38]'s private core. In [11], they introduce the ex post core and prove its equivalence with rational expectations equilibria.

References

1. Allen, B.: Market games with asymmetric information and nontransferable utility: representation results and the core. CARESS Working paper 91-09, University of Pennsylvania (1991)
2. Allen, B.: Incentives in market games with asymmetric information: the core. CORE Discussion Paper 9221, Université Catholique de Louvain (1992)
3. Allen, B.: Market games with asymmetric information: verification and the publicly predictable core. Hitotsubashi Journal of Economics **32**, 101-122 (1993)
4. Allen, B.: Incentives in market games with asymmetric information: approximate NTU cores in large economies. In: Barnett, W., Moulin, H., Salles, M., Schofield, N. (eds.) Social choice, welfare and ethics. Cambridge: Cambridge University Press 1994
5. Allen, B.: On the existence of core allocations in a large economy with incentive compatibility constraints. In: Wooders, M. (ed), Topics in game theory and mathematical economics (Fields Institute Conference in Honor of R. Aumann). American Mathematical Society 1995
6. Cole, H.: Comment: general competitive analysis in an economy with private information. International Economic Review **30**, 249–252 (1989)

7. Duffie, D.: (1988) Security markets. New York: Academic Press 1988
8. Debreu, G., Scarf, H.: A limit theorem on the core of an economy. International Economic Review **4**, 235–246 (1963)
9. Demange, G., Guesnerie, R.: Interim design of core mechanisms. Delta, Document 97-16 (1997)
10. Einy, E., Moreno, D., Shitovitz, B.: Existence and equivalence of competitive and core allocations in large exchange economies with differential information. Universidad Carlos III de Madrid, working paper 98-71 (1998)
11. Einy, E., Moreno, D., Shitovitz, B.: Rational expectations equilibria and the ex post core of an economy with asymmetric information. Universidad Carlos III de Madrid, working paper 98-91 (1998)
12. Einy, E., Moreno, D., Shitovitz, B.: On the core of an economy with differential information. Universidad Carlos III de Madrid, working paper 98-94 (1998)
13. Forges, F.: Le cœur d'une économie d'échange en information asymétrique. Document de Travail du THEMA 9829 (1998)
14. Forges, F., Minelli, E.: A note on the incentive compatible core. Document de Travail du THEMA 9902 (1999)
15. Glicksberg, I.L.: A further generalization of the Kakutani fixed point theorem with application to Nash equilibrium points. Proceedings of the A.M.S. **3**, 170–174 (1952)
16. Gul, F., Postlewaite, A.: Asymptotic efficiency in large economies with asymmetric information. Econometrica **60**, 1273–1292 (1992)
17. Hahn, G., Yannelis, N.: Efficiency and incentive compatibility in differential information economies. Economic Theory **10**, 383-411 (1997)
18. Hildenbrand, W., Kirman, A.: Equilibrium analysis: variations on themes by Edgeworth and Walras. Amsterdam: North-Holland 1988
19. Holmström, B., Myerson, R.: Efficient and durable decision rules with incomplete information. Econometrica **51**, 1799–1819 (1983)
20. Ichiishi, T., Idzik, A.: Bayesian cooperative choice of strategies. International Journal of Game Theory **25**, 455–473 (1996)
21. Kehoe T., Levine, D., Prescott, E.: Lotteries, sunspots and incentive constraints. Mimeo, UCLA (1998)
22. Kreps D.: A note on fulfilled expectations equilibria. Journal of Economic Theory **14**, 32–43 (1977)
23. Koutsougeras, L., Yannelis, N.: Incentive compatibility and information superiority of the core of an economy with differential information. Economic Theory **3**, 195–216 (1993)
24. Lee, D., Volij, O.: Rationality and the core of an economy with differential information. Mimeo, Brown University (1996)
25. Minelli E., Polemarchakis, H.M.: Information at a competitive equilibrium. CORE DP 9583 (1995)
26. Page, F.: Market games with differential information and infinite commodity spaces: the core. Economic Theory **9**, 151–159 (1997)
27. Prescott, E., Townsend, R.: Pareto optima and competitive equilibria with adverse selection and moral hazard. Econometrica **52**, 21–45 (1984)
28. Prescott, E., Townsend, R.: General competitive analysis in an economy with private information. International Economic Review **25**, 1–20 (1984)
29. Radner, R.: Competitive equilibrium under uncertainty. Econometrica **36**, 31–58 (1968)
30. Scarf, H.: The core of an n-person game. Econometrica **35**, 50-69 (1967)
31. Serrano, R., Vohra, R., Volij, O.: (1999) On the failure of core convergence in economies with asymmetric information. Brown University, working paper 99-22 (1999)
32. Shubik, M., Wooders, M.: Approximate cores of replica games and economies. Part I: Replica games, externalities, and approximate cores. Mathematical Social Sciences **6**, 27–48 (1983)
33. Vohra, R.: Incomplete information, incentive compatibility and the core. Journal of Economic Theory **86**, 123–147 (1999)
34. Volij, O.: Communication, credible improvements and the core of an economy with asymmetric information. Brown University, Providence (1997)

35. Wilson, R.: Information, efficiency and the core of an economy. Econometrica **46**, 807–816 (1978)
36. Wooders, M.: The epsilon core of a large replica game. Journal of mathematical Economics **11**, 277–300 (1983)
37. Yamazaki, A.: An equilibrium existence theorem without convexity assumptions. Econometrica **46**, 541–555 (1978)
38. Yannelis, N.: The core of an economy with differential information. Economic Theory **1**, 183–198 (1991)

Exchange economies with asymmetric information: competitive equilibrium and core*

Stefan Maus

Department of Quantitative Economics, University of Maastricht, P.O. Box 616,
6200 MD Maastricht, THE NETHERLANDS (e-mail: s.maus@ke.unimaas.nl)

Received: February 24, 2003; revised version: July 3, 2003

Summary. A replica theorem is shown to hold for exchange economies with asymmetric information. In a replicated exchange economy with asymmetric information the set of all core elements with equal treatment is nonempty, but it is in general only a subset of the core. Nevertheless, the replica theorem and the presence of at least one core element with equal treatment suffice to show existence of a competitive quasi-equilibrium. Conditions on the initial endowments and the communication system are given to ensure that every competitive quasi-equilibrium is a competitive equilibrium.

Keywords and Phrases: Core, Competitive (quasi-)equilibrium, Equal treatment, Exchange economy with asymmetric information, Replica theorem.

JEL Classification Numbers: C70, D50, D82.

1 Introduction

The purpose of this paper is to study the relationship between the core and competitive allocations in economies with asymmetric information. This extends the analysis of uncertainty, as introduced in Chapter 7 of Debreu's 'Theory of Value' [3]. There, uncertainty is incorporated into the Arrow-Debreu theory by introducing a finite set of states of the world, and viewing the commodities as differentiated by state. Still, every agent is assumed to possess the same full information set, i.e. there are no states that cannot be distinguished. This analysis is extended by Radner [10,12], to cover the case of private information. Every agent is assigned a partition of the set of states of the world, with the interpretation that the sets in the partition contain those states which cannot be distinguished by the agent. A trade of an agent

* I thank an anonymous referee whose comments led to an improvement of the paper.

then has to be compatible with his information, that is, he cannot act differently on states that he cannot distinguish, or mathematically, his net trades have to be measurable with respect to the σ-algebra generated by his information partition. Radner [10] extends the notion of Arrow-Debreu competitive equilibrium to this model. Trades are contingent contracts before information about the realized state of nature is obtained. Information cannot be communicated and restricts the budget sets of the agents. A Radner competitive equilibrium rewards the better informed trader in the sense that his budget set is larger, thus he will in general be better off.

From the core perspective, the incorporation of information structures into the core dates back to Wilson's seminal paper [13]. However, with the core, there is more to be considered than just private information. One can devise an information structure to be parametrized not only by the agent space but also by the coalitions that an agent can be in. Private information would then be the case, where an agent's information does not change, regardless of the coalition he belongs to. But a lot of other cases are imaginable, e.g. agents pooling their information, leading to a situation in which superset coalitions have more information and the grand coalition has full information. Such a fine information system, and others, as the coarse, strong coarse and weak fine information system, have been the subject of various papers, investigating the existence of core allocations. As a starting point to this literature one can consider, apart from Wilson's work of course, the papers of Yannelis [14] and Allen [2]. The frameworks given there have to be carefully distinguished under the aspect that, in Yannelis [14], an allocation is preferred to another by an agent if its expected utility over every single set in the information partition (mathematically the conditional expected utility vector) is greater. This is often referred to as an interim concept. Allen [2] on the contrary considers an ex-ante concept, where an allocation is preferred to another if the expected utility is greater. Of course if an allocation is preferred in Yannelis [14], it will be preferred also in Allen [2], but not vice versa, making the ex-ante core a subset of the interim core. Note, however, that as pointed out yet again in the inspiring work of Debreu's Theory of Value [3], the usage of (conditional) expected utility, requiring a probability measure over the states of the world and state-dependent utilities, is not essential. Instead one can consider a utility function which maps the state-commodity space to the reals. This function could be the outcome of taking expectation over states, but this only shows that this model is able to incorporate the expected utility case. Recently, the existence of ex-ante core allocations in this general context has been adressed by Maus [9], providing a condition on the information partitions, which is equivalent to nonemptiness of the core with asymmetric information, when endowments are variables. This result will be used here to investigate the question of equal treatment in the core and the existence of an adapted version of the Radner competitive equilibrium. The core reflects the veto power of any coalition. The recent paper by Hervés-Beloso, Moreno-García and Yannelis [6] characterizes Radner Equilibria by using the veto power of the grand coalition only. They show that an allocation is a Radner equilibrium if and only if it can not be blocked by the grand coalition in every economy obtained by perturbing the original endowments in the direction of the equilibrium candidate.

The list of contributions on the relationship between Radner equilibrium and ex-ante core concept with asymmetric information comprises two other recent papers by Einy, Moreno and Shitovitz [4,5], working with a continuum of agents. In [4] they show that in an ex-ante model, if the economy is 'irreducible', then a Radner equilibrium exists and the set of Radner competitive equilibrium allocations coincides with the private core. It is also shown, that the weak fine core corresponds to competitive equilibria of an economy, where the symmetric information of the agents is their pooled private information. In the other paper [5] they show that in an interim model the fine core is a subset of the ex-post core, and consequently every fine core allocation is a selection from the equilibrium correspondence of the associated family of full information economies. Differences with the work presented here are manifold. First of all the core is defined here without referring to expected utility and prior beliefs. Secondly the concept is strictly ex-ante. Thirdly the set of agents is finite, though agents are replicated to get to the core convergence result. Fourth, the communication system is not specified. Last but not least, no free disposability of commodities is assumed. As pointed out already in a footnote by Radner [12], p.945, the disposed commodities may not be measurable with respect to the information of any single trader, making free disposability objectionable. In the presented paper, resource feasibility will be considered as denoting strict equality of the sum of the allocated commodities and the sum of the initial endowments. This complicates the existence of an equilibrium considerably, since existence results from the theory of production economies such as employed in Radner [12] or Einy, Moreno and Shitovitz [4] cannot be used. The reason for that is that the proofs of these existence results rely on the assumption that production sets have a nonempty relative interior. In a pure exchange economy that assumption translates into disposability of at least part of the endowments. Therefore, these existence results are not applicable, and a number of other, partly new results is necessary. These results are the existence result for the core from [9], a result extending equal treatment of agents of the same type in replicated economies to the case of the core with asymmetric information, and a result on the connection between quasi-competitive Radner equilibria and the cores of the replicated economies. All of these results are also of interest on their own.

The paper is organized as follows. In Section 2 the formal model is described along with the definition of core and competitive (quasi-)equilibrium. Section 3 discusses the relationship between competitive (quasi-)equilibrium allocations and core allocations in large economies. To arrive at an existence result for equilibria, further investigation of 'equal treatment in the core' is necessary, which is dealt with in Section 4. Then Section 5 establishes existence of an competitive quasi-equilibrium, and introduces some conditions under which this turns into an existence result for competitive equilibria.

2 Preliminaries

Throughout, $\mathbb{R}_+$ is the set $\{x \in \mathbb{R} | x \geq 0\}$, and $x > y$ for vectors or matrices means that every entry in x strictly exceeds the corresponding entry in y. When a constant $c \in \mathbb{R}$ is written in a place where a matrix or vector is expected, it is understood to be the matrix or vector where all entries have the value c. A product of matrices or vectors px in the same space is understood to be the sum of the products of the matching entries, i.e. the scalar product (matrices have to be interpreted as vectors for this).

2.1 Information

Let Ω be the finite set of states of the world. Let $\mathcal{P}^*$ be the set of partitions of Ω. A $\mathcal{P} \in \mathcal{P}^*$ is called an *information set*. The interpretation is that states contained in an element $S \in \mathcal{P}$ cannot be distinguished under that information set. For each $\omega \in \Omega$ denote by $\mathcal{P}(\omega)$ the element of the partition $\mathcal{P}$ that contains ω.

Let N be a finite set of agents. Each agent has an initial endowment of information, described by $\mathcal{P}_i \in \mathcal{P}$. When forming coalitions the information of agents may change, e.g. due to communication. Let $\mathcal{P}_i^S$ be the information that agent $i \in S$ has if the coalition S is formed. Throughout I assume that $\mathcal{P}_i^{\{i\}} = \mathcal{P}_i$. A collection $(\mathcal{P}_i^S)_{i \in S, S \subset N}$ is called a *communication system*.

A communication system is called *private* if $\mathcal{P}_i^S = \mathcal{P}_i$ for all $S \ni i$, i.e. the information does not change.

An information set $\mathcal{P}$ generates a σ-algebra $\sigma(\mathcal{P})$.

Information restricts the possible net trades of an agent. He cannot trade different amounts on events that he cannot distinguish. Formally this is captured by the following. Let $\mathcal{P}$ be the information the agent has. Then his trades of k goods are limited to the following set of functions

$$X_{\mathcal{P}} := \{x | x : \Omega \to \mathbb{R}^k \text{ and x is } \sigma(\mathcal{P})\text{-measurable}\}.$$

Hence, $x \in X_{\mathcal{P}}$ if and only if x is constant on elements of $\mathcal{P}$. Thus, $x : \Omega \to \mathbb{R}^k$ can be identified with $x : \mathcal{P} \to \mathbb{R}^k$, where $x(A)$ is the constant value on A. The characteristic function of any set $B \in \mathcal{P}$, denoted by

$$\begin{aligned} \mathbb{I}_B &: \Omega \to \mathbb{R}^k, \\ &: \omega \longmapsto \begin{cases} 1, & \text{if } \omega \in B \\ 0, & \text{if } \omega \notin B \end{cases}, \end{aligned}$$

is in $X_{\mathcal{P}}$ for every $\mathcal{P}$ for example. I will denote $X_{\mathcal{P}_i^S}$ by X_i^S and if $S = \{i\}$ I will write X_i. Call $X_{\mathcal{P}}$ the set of *informationally feasible trades under* $\mathcal{P}$, X_i^S the set of *informationally feasible trades of agent* i *in coalition* S and $\prod_{i \in S} X_i^S$ the set of *informationally feasible trades of the coalition* S.

Definition 1. The communication system $(\mathcal{P}_i^S)_{i,S}$ is trade bounded if for all coalitions $S \subseteq N$ and all $(z_j)_{j \in S} \in \prod_{j \in S} X_j^S$ with $\sum_{j \in S} z_j = 0$, it holds that $(z_j)_{j \in S} \in \prod_{j \in S} X_j^N$.

Informally speaking, this assures that every net trade possible in a subcoalition remains possible in the grand coalition.

2.2 Exchange economies with asymmetric information

An exchange economy with asymmetric information $\mathbb{E}$ is given by

1. a finite set of agents N,
2. a finite set Ω of states of the economy,
3. the initial endowments $e_i : \Omega \to \mathbb{R}_+^k$ for every agent $i \in N$,
4. the communication system $(\mathcal{P}_i^S)_{i\in S, S\subset N}$,
5. and utility functions $u_i : \mathbb{R}_+^{k\times\Omega} \to \mathbb{R}$ for every agent $i \in N$.

A vector of net trades $(z_i)_{i\in S}$ satisfying $e_i + z_i \geq 0$ for all $i \in S$ and $\sum_{i\in S} z_i = 0$ is called *physically feasible* for the coalition $S \subseteq N$. An *allocation* for a coalition S in an economy $\mathbb{E}$ is a function $x : S \to (\mathbb{R}^k)^\Omega$ such that the net trades $x_i - e_i$ are informationally and physically feasible for this coalition. In this paper vectors $x \in \mathbb{R}_+^{k\times\Omega}$ are identified in the natural way with the space of functions $x : \Omega \to \mathbb{R}_+^k$. Furthermore, the following spaces of utility functions will be relevant:

$$\begin{aligned}
U^{mo} &:= \{u : \mathbb{R}_+^{k\times\Omega} \to \mathbb{R}|\ u \text{ is strictly increasing,} \\
&\qquad \text{i.e. } y \geq x, y \neq x \Rightarrow u(y) > u(x)\}, \\
U_{qc} &:= \{u : \mathbb{R}_+^{k\times\Omega} \to \mathbb{R}|\ u \text{ is quasiconcave}\}, \\
U_{co,0}^{mo} &:= \{u : \mathbb{R}_+^{k\times\Omega} \to \mathbb{R}|\ u \text{ is concave and } u(0) = 0\} \cap U^{mo}, \\
U_{qc}^{mo} &:= U_{qc} \cap U^{mo} \text{ and} \\
\mathbf{C}(\mathbb{R}_+^{k\times\Omega}) &:= \{u : \mathbb{R}_+^{k\times\Omega} \to \mathbb{R}|\ u \text{ is continuous}\}.
\end{aligned}$$

2.3 The core

An NTU-game in characteristic function form is a correspondence $V : 2^N \setminus \{\emptyset\} \to \mathbb{R}^N$ satisfying

1. $V(S)$ is nonempty and closed for $S \neq \emptyset$,
2. if $x \in V(S)$ and $y \in \mathbb{R}^N$ is such that $y_i \leq x_i$ for all $i \in S$ then $y \in V(S)$,
3. for every $i \in N$ there is an $m_i \in \mathbb{R}$ with $V(\{i\}) = \{x \in \mathbb{R}^N | x_i \leq m_i\}$, and $V(N) \cap \{x \in \mathbb{R}^N | x_i \geq m_i\ \forall i \in N\}$ is nonempty and compact.

The NTU-game associated with an exchange economy $\mathbb{E}$ is defined by

$$\begin{aligned}
V_{\mathbb{E}}(S) = \{x \in \mathbb{R}^N;&\text{ there exists } (z_i)_{i\in S} \in \textstyle\prod_{i\in S} X_i^S \text{ such that} \\
&e_i + z_i \geq 0,\ \textstyle\sum_{i\in S} z_i = 0 \\
&\text{and } x_i \leq u_i(e_i + z_i)\}
\end{aligned}$$

for each coalition $S \neq \emptyset$. The core of this NTU-game is given by

$$C(V_{\mathbb{E}}) := V_{\mathbb{E}}(N) \setminus \bigcup_{\emptyset \neq S \subseteq N} int(V_{\mathbb{E}}(S)).$$

The core of the exchange economy with asymmetric information consists of all allocations for the grand coalition, resulting in a utility vector in the core of the associated NTU-game, i.e.

$$C(\mathbb{E}) := \{x \in \mathbb{R}^{N\times k\times \Omega} | x - e \text{ is an allocation for } N \text{ and } (u_i(x_i))_{i\in N} \in C(V_{\mathbb{E}})\}.$$

So $C(\mathbb{E})$ comprises all allocations x for the grand coalition to which no coalition $S \subseteq N$ has a *deviation*, i.e. an allocation y^S for that coalition such that $u_i(y_i^S) > u_i(x)$ for all $i \in S$.

The following existence result for the core from [9] will be of use.

Theorem 2. *Let $\mathbb{E}$ be an exchange economy with asymmetric information and trade bounded communication system. Assume that $u_i \in U_{qc}$ for all $i \in N$. Then the NTU-core of $V_{\mathbb{E}}$ is not empty.*

2.4 Competitive allocations

The information that an agent can use in a competitive allocation is assumed to be $\mathcal{P}_i^N$. This assumption could be viewed as implying that in a competitive equilibrium allocation the same communication takes places as in the grand coalition. Relations between core allocations and competitive allocations can in general not be expected if the measurability constraints for core allocations and competitive allocations are different.

A *price system* is a function $p : \Omega \to \mathbb{R}^k, p \neq 0$. Agents make contracts for the delivery of contingent commodities as in Chapter 7 of Debreu's Theory of Value [3], i.e. before the state of the world is revealed to them. Payments are made for contingent delivery and irrevocable even if another state of the world is realized. Additionally every agent's trades are restricted by his information.

A price system contains the information

$$\sigma(p) := \bigcap \{\mathcal{A} | \mathcal{A}\ \sigma\text{-algebra on } \Omega, p\ \mathcal{A}\text{-measurable}\}.$$

It could be argued that agents i such that $\sigma(P_i^N) \neq \sigma(p)$ can observe information from prices posted to them. This leads to rational expectations equilibria, see for example Allen [1] and Radner [11]. Here we assume that agents observe private price signals, which then reveal no information to them. This is achieved by the following construction. Let p be a price system. For an information partition $\mathcal{P}$ the price system that can be *seen* under $\mathcal{P}$ is given by $(p(A))_{A\in\mathcal{P}}$, where $p(A) := \sum_{\omega\in A} p(\omega)$. Every agent observes (possibly different) futures prices, namely $(p(A))_{A\in\mathcal{P}_i^N}$, corresponding to his possible trading activities. Another way to view this, is to consider average prices $\widetilde{p}$, which are different per agent, and are given by

$$\widetilde{p}(\omega) := \frac{1}{|\mathcal{P}(\omega)|} \sum_{\widetilde{\omega}\in\mathcal{P}(\omega)} p(\widetilde{\omega}).$$

Under both viewpoints an agent having information $\mathcal{P}$ cannot gain information from the prices seen, as these are $\sigma(\mathcal{P})$-measurable. Furthermore, any informationally feasible trading activity $z \in X^{\mathcal{P}}$ incurs the same cost under both viewpoints, as can be seen from

$$\begin{aligned} z\widetilde{p} &= \sum_{\omega \in \Omega} z(\omega)\widetilde{p}(\omega) \\ &= \sum_{A \in \mathcal{P}} \sum_{\omega \in A} z(\omega) \underbrace{\frac{1}{|\mathcal{P}(\omega)|}}_{=A} \sum_{\widetilde{\omega} \in \mathcal{P}(\omega)} p(\widetilde{\omega}) \\ &= \sum_{A \in \mathcal{P}} z(A) \sum_{\omega \in A} \frac{1}{|A|} \sum_{\widetilde{\omega} \in A} p(\widetilde{\omega}) \\ &= \sum_{A \in \mathcal{P}} z(A) \sum_{\widetilde{\omega} \in A} p(\widetilde{\omega}) \\ &= \sum z(A)p(A). \end{aligned}$$

Note that if we assume that the price system p is set by a Walrasian auctioneer who knows the information that agents have, then we can assign also the task of computing the private price signals for the agents to him. The least information that the Walrasian auctioneer has to have then is the partitions $\mathcal{P}_i^N$ for any agent $i \in N$. Thus the information $\mathcal{P}^W$ that the auctioneer has will satisfy $\sigma(\mathcal{P}^W) \supseteq \sigma(\bigcup_{i \in N} \mathcal{P}_i^N)$ and this information constraints price vectors set by the auctioneer to be $\sigma(\mathcal{P}^W)$-measurable. However, as we shall see, this constraint holds automatically for the price systems that we construct in our proofs. These correspond to separating hyperplanes of $\sigma(\bigcup_{i \in N} \mathcal{P}_i^N)$-measurable sets of allocations, so they can be chosen to be $\sigma(\bigcup_{i \in N} \mathcal{P}_i^N)$-measurable themselves. In the end $\sigma(\bigcup_{i \in N} \mathcal{P}_i^N)$-measurability of price systems is used in Lemma 17 and in Corollary 18, but anywhere in this paper it is not required. Another question, that arises if we assume that the auctioneer receives signals $\mathcal{P}_i^N$ from the agents is, whether agents, either individually or coalitionally, can manipulate the price the Walrasian auctioneer chooses by transmitting another information partition $\overline{\mathcal{P}}_i^N$ such that $\sigma(\overline{\mathcal{P}}_i^N) \subseteq \sigma(\mathcal{P}_i^N)$ to the auctioneer. This question is not adressed here. So, when thinking of prices as set by an auctioneer, it is assumed implicitly that either agents are honest or their information can be verified costlessly by the auctioneer.

Example 3. Imagine that there are three states of the world $\Omega = \{1, 2, 3\}$, which indicate the quality of a commodity (say 1 is good, 2 is medium,3 is bad). Now, compare two agents having information partitions $\{\{1, 2\}, \{3\}\}$ and $\{\{1\}, \{2, 3\}\}$ respectively (the first agent can screen for bad quality and the second agent for good quality). Assume that a price system for that commodity is given by $p = (3, 2, 1)$. Then the first agent observes the price vector $(5, 1)$, where 5 is the price for contingent delivery of a unit amount in state 1 and 2, and the second agent observers the price vector $(3, 3)$, where 3 is the price for contingent delivery of a unit amount in states 2 and 3. If both agents were interested in buying only good quality, that would be cheaper for agent 2, as agent 1 has to buy the medium quality bundled with the good quality. Instead, if the good and the medium quality were appreciated similarly by the agents, the advantage would be on the side of agent 1. The average price systems here are given by $(2\frac{1}{2}, 2\frac{1}{2}, 1)$ and $(3, 1\frac{1}{2}, 1\frac{1}{2})$ for the first and the second agent respectively.◇

For a price system p define the *budget set* of an agent $i \in N$ by

$$B_i(p) = \left\{ x \in \mathbb{R}_+^{k \times \Omega} | x - e_i \in X_i^N \text{ and } \sum_{A \in P_i^N} \sum_{\omega \in A} p(\omega)(x(\omega) - e_i(w)) \leq 0 \right\}$$

$$= \left\{ z + e_i \in \mathbb{R}_+^{k \times \Omega} | z \in X_i^N \text{ and } \sum_{A \in P_i^N} p(A) z(A) \leq 0 \right\}$$

The budget set contains all contingent allocations that the agent can afford under the given price system and that result from informationally feasible net trades. Now the definition of a competitive equilibrium can be stated.

Definition 4. A competitive equilibrium for an economy $\mathbb{E}$ is a pair (p, x), where $p \neq 0$ is a price system and x is an allocation such that x_i maximizes u_i on $B_i(p)$ for all $i \in N$. A competitive allocation is an allocation x, for which there exists a price system p such that (p, x) is a competitive equilibrium.

As mentioned in the introduction, the model considered here is able to incorporate models which use an expected utility function derived from integration of state-dependent utilities with respect to a prior on states. This can be used to argue that Definition 4 covers the notion of Radner equilibrium [10] for pure exchange economies with private information, and that Corollary 18 is also an existence result for Radner equilibrium. A standard assumption made when considering Radner equilibrium, is that initial endowments are measurable with respect to the information of the agents in the grand coalition, i.e. that $e_i \in X_i^N$ for all $i \in N$. The reason that this is not assumed here is that some of the results, e.g. about the relationships between core and equilibrium, do not need this assumption. However, in the existence result of Corollary 18, there is the condition that the initial endowments of an agent should be measurable with respect to the information that the agent uses in any coalition that he is in, i.e. $e_i \in \bigcap_{S \ni i} X_i^S$. Letting information be private, i.e. $X_i^S = X_i^N = X_i^{\{i\}}$ for all $i \in N$ and $S \ni i$, this is just the private measurability of initial endowments assumed in Radner equilibrium. Hence, Corollary 18 turns into an existence result for Radner equilibria in pure exchange economies with private information that meet the assumptions of the corollary. This demands that expected utility as derived from taking expectations in the Radner setting is quasiconcave, strictly monotone and continuous. This is met for example if the prior is strictly positive, and state-dependent utilities are concave, monotone and continuous. This is usually assumed when dealing with existence of Radner equilibrium, e.g. already in the original paper of Radner [10].

Next the weaker notion of competitive quasi-equilibrium is defined.

Definition 5. A competitive quasi-equilibrium for an economy $\mathbb{E}$ is a pair (p, x) where $p \neq 0$ is a price system and x is an allocation such that x_i maximizes u_i on $B_i(p)$ whenever $\inf pX_i^{N,\geq 0} := \inf\{px | x \in X_i^N, x \geq 0\} < pe_i$.

Obviously, every competitive equilibrium is also a competitive quasi-equilibrium. There is no difference between these two equilibrium notions if $\inf pX_i^{N,\geq 0} < pe_i$ for all $i \in N$. When utility functions are strictly increasing, these definitions imply that $(p(A))_{A \in \mathcal{P}_i^N} \geq 0$ in any competitive quasi-equilibrium and $(p(A))_{A \in \mathcal{P}_i^N} > 0$ in any competitive equilibrium. Hence, in that case one has de facto $0 = \inf pX_i^{N,\geq 0}$ and the condition $\inf pX_i^{N,\geq 0} < pe_i$ could be replaced by $0 < pe_i$ (positive income).

3 Competitive and core allocations in large economies

In this section sufficient conditions on the communication system are given such that the replica theorem holds. So first of all the replica economies $\mathbb{E}^n$, $n \in \mathbb{N}$, are defined.

Definition 6. Let $\mathbb{E} = (N, \Omega, (e_i, u_i, (\mathcal{P}_i^S)_{i \in S, S \subset N})_{i \in N})$ be an exchange economy with asymmetric information. The $n - th$ replica

$$\mathbb{E}^n = (N^n, \Omega, (\widetilde{e}_i, \widetilde{u}_i, (\widetilde{\mathcal{P}}_i^S)_{i \in S, S \subset N^n})_{i \in N^n}), n \in \mathbb{N},$$

is the exchange economy with asymmetric information, where the set of agents is $N^n := N \times \{1, \ldots, n\}$ and for an agent $(i, j) \in N^n$ one puts

1. $\widetilde{e}_{(i,j)} := e_i$
2. $\widetilde{u}_{(i,j)} := u_i$
3. $\widetilde{\mathcal{P}}_{(i,j)}^S := \mathcal{P}_i^{pr^1(S)}$, where pr^1 is the projection on the first coordinate.

So the information that agents can use in a coalition in the replica economy depends only on the type of the agents in that coalition and not on the total number of agents of a type in the coalition. Denote the agents of a given type $i \in N$, which are present in a coalition $S \subset N^n$, by

$$pr_i(S) := \{j | 1 \leq j \leq n \text{ and } (i, j) \in S\}.$$

Theorem 7. *Let $\mathbb{E}$ be an exchange economy with asymmetric information. Assume that $e_i \in X_i^N$ and $u_i \in U^{mo} \cap \mathbf{C}(\mathbb{R}_+^{k \times \Omega})$ for all $i \in N$. If $x \in \mathbb{R}^{N \times k \times \Omega}$ is such that $x^n = \prod_{j=1}^n x \in C(\mathbb{E}^n)$ for all $n \in \mathbb{N}$, then x is a competitive quasi-equilibrium allocation. Moreover the price system p decentralizing x can be chosen $\sigma(\bigcup_{i \in N} \mathcal{P}_i^N)$-measurable.*

Proof. Note that if the initial endowments e_i are $\mathcal{P}_i^N$−measurable, then not only the net trades $(z_i)_{i \in N}$ leading to a competitive equilibrium or core allocation are $\mathcal{P}_i^N$−measurable, but also the final allocation $(e_i + z_i)_{i \in N}$. Let x be such that

$x^n \in C(\mathbb{E}^n)$ for all $n \in \mathbb{N}$. It has to be shown, that there is a price system $p \in \mathbb{R}^{k\times\Omega}$ such that for any agent $i \in N$ either $\inf pX_i^{N,\geq 0} = pe_i$, or x_i maximizes u_i on $B_i(p)$.

$$H(i) := (\{y \in \mathbb{R}_+^{k\times\Omega} | u(y) > u(x_i)\} - \{e_i\}) \cap X_i^N$$

is the set of all net trades z, where agent $i \in N$ prefers $e_i + z$ to x_i, and which are informationally feasible for the agent in the grand coalition. Denote by $co(A)$ the convex hull of $A \subset \mathbb{R}^{k\times\Omega}$ and by $\mathbb{I}_B$ the characteristic function of $B \subseteq \Omega$. Let

$$X_- := \left\{ z \in \mathbb{R}^{k\times\Omega} | z = -\sum_{i\in N}\sum_{P\in\mathcal{P}_i^N} \alpha_{i,P}\mathbb{I}_P,\ \alpha_{i,P} > 0 \right\}.$$

I claim that

$$co\left(\bigcup_{i\in N} H(i)\right) \cap X_- = \emptyset.$$

Suppose on the contrary that there is some $z \in co(\bigcup_{i\in N} H(i)) \cap X_-$, then

$$z = \sum_{a\in A} \lambda_a y_a, \text{ where } A \subset N, \sum_{a\in A} \lambda_a = 1, \lambda_a > 0, y_a \in H(a),$$
$$\text{and } z = -\sum_{i\in N}\sum_{P\in\mathcal{P}_i^N} \alpha_{i,P}\mathbb{I}_P,\ \alpha_{i,P} > 0.$$

It suffices to construct a contradiction to $x^n \in C(\mathbb{E}^n)$ for all $n \in \mathbb{N}$ for the case where the λ_a are rational. In this case let n be so large that $r_a := n\lambda_a \leq n$ is a natural number for every $a \in A$. Define an allocation x^* for the coalition

$$S := \{(a,j) | 1 \leq j \leq r_a \text{ and } a \in A\} \cup \{(i, n+1) | i \in N\}$$

in the economy $\mathbb{E}^{n+1}$ by

$$x^*_{(a,i)} := e_a + y_a \ (1 \leq i \leq r_a, a \in A) \text{ and}$$
$$x^*_{(i,n+1)} := x_i + n \sum_{P\in\mathcal{P}_i^N} \alpha_{i,P}\mathbb{I}_P \ (i \in N).$$

This allocation is informationally feasible for S, as $\widetilde{\mathcal{P}}^S_{(i,j)} := \mathcal{P}_i^{pr^1(S)} = \mathcal{P}_i^N$. Every member of S prefers his bundle in x^* to that in x^{n+1}. Furthermore, x^* is also physically feasible for S, as

$$\sum_{i\in S} x_i^* = \sum_{a\in A}\sum_{1\leq i\leq r_a} x^*_{(a,i)} + \sum_{i\in N} x^*_{(i,n+1)}$$
$$= \sum_{a\in A}(r_a e_a + r_a y_a) + \sum_{i\in N}\left(x_i + n \sum_{P\in\mathcal{P}_i^N} \alpha_{i,P}\mathbb{I}_P\right)$$

$$
\begin{aligned}
&= \sum_{a \in A} r_a e_a + n \underbrace{\sum_{a \in A} \lambda_a y_a}_{=z} + \underbrace{x(N)}_{=e(N)} + \underbrace{n \sum_{i \in N} \sum_{P \in \mathcal{P}_i^N} \alpha_{i,P} \mathbb{I}_P}_{=-z} \\
&= \sum_{a \in A} r_a e_a + e(N) + nz - nz \\
&= e(S).
\end{aligned}
$$

Therefore, x^* is a deviation in $\mathbb{E}^{n+1}$ of the coalition S to the allocation $x^{n+1} \in C(\mathbb{E}^{n+1})$, a contradiction.

Now, as

$$co\left(\bigcup_{i \in N} H(i)\right) \cap X_- = \emptyset,$$

$$co\left(\bigcup_{i \in N} H(i)\right) \text{ is convex and}$$

$$X_- \text{ is open and convex,}$$

there is by a version of the separation theorem for convex sets, a $p \in \mathbb{R}^{k \times \Omega}, p \neq 0$, such that

$$pz \geq 0 \text{ if } z \in co\left(\bigcup_{i \in N} H(i)\right)$$

$$\text{and } pz \leq 0 \text{ if } z \in X_-.$$

This p can be chosen to be $\sigma(\cup_{i \in N} \mathcal{P}_i)$-measurable, as any $z \in co(\bigcup_{i \in N} H(i)) \cup X_-$ will per se, viewed as a function from Ω to $\mathbb{R}^k$, be $\sigma(\cup_{i \in N} \mathcal{P}_i)$-measurable. To show that (x, p) is a competitive quasi-equilibrium, I will need that the price system $(p(A))_{A \in \mathcal{P}_i^N}$, that can be seen by any agent under his information partition $\mathcal{P}_i^N$ in the grand coalition, is nonnegative. Assume to the contrary that there is an $A \in P_{\tilde{i}}^N$ for some $\tilde{i} \in N$ such that $p(A) < 0$. Let $z \in X_-$ be given by $\alpha_{\tilde{i},A} = K$ and $\alpha_{i,P} = 1$ otherwise. Then

$$
\begin{aligned}
pz &= p\left(-\sum_{i \in N} \sum_{P \in \mathcal{P}_i^N} \alpha_{i,P} \mathbb{I}_P\right) = -\sum_{i \in N} \sum_{P \in \mathcal{P}_i^N} \alpha_{i,P} (p\mathbb{I}_P) \\
&= -\sum_{i \in N} \sum_{P \in \mathcal{P}_i^N} \alpha_{i,P} p_P = -\left(\sum_{i \in N} \sum_{P \in \mathcal{P}_i^N, i \neq \tilde{i} \vee P \neq A} p_P\right) - Kp(A) \\
&= C - Kp(A)
\end{aligned}
$$

where the constant C is independent of K, and $p(A) < 0$. Therefore choosing K larger than $\frac{C}{p(A)}$ will make pz positive, contradicting $z \in X_-$. So $(p(A))_{A \in \mathcal{P}_i^N} \geq 0$ for any $i \in N$, implying that $\inf pX_i^{N,\geq 0} = 0$ for any $i \in N$. It remains to be shown that x_i maximizes u_i on $B_i(p)$ when $0 = \inf pX_i^{N,\geq 0} < pe_i$. So suppose there is a $y \in B_i(p)$ such that $u_i(y) > u_i(x_i)$. Then $y - e_i \in H(i) \subset co(\cup_{i \in N} H(i))$, which together with $y \in B_i(p)$ implies that $py = pe_i$. As the income of the agent is positive, y is $\mathcal{P}_i^N$–measurable, and $p(A) \geq 0$ for all $A \in \mathcal{P}_i^N$, it can be concluded from

$$0 < pe_i = py = \sum_{\omega \in \Omega} p(\omega) y_i(\omega) = \sum_{A \in \mathcal{P}_i^N} \sum_{\omega \in A} p(\omega) y_i(\omega) = \sum_{A \in \mathcal{P}_i^N} p(A) y_i(A)$$

that there must be some $A \in \mathcal{P}_i^N$ such that $y_i(A) > 0$ and $p(A) > 0$. So, by lowering consumption equally in the states contained in A, a nonnegative sequence of y_n converging to y can be obtained such that $py_n < pe_i$ for all $n \in \mathbb{N}$, i.e. $y_n - e_i \notin H(i)$. But then $u_i(y_n) \leq u_i(x_i)$, as y_n is not in $H(i)$. Since the utility functions u_i are assumed to be continuous this implies $u_i(y) \leq u_i(x_i)$, a contradiction to $u_i(y) > u_i(x_i)$. So x_i indeed maximizes u_i on $B_i(p)$ when agents have positive income. This concludes the proof that (x, p) is a quasi-equilibrium of the economy $\mathbb{E}$. □

The next example shows that the $\mathcal{P}_i^N$-measurability of the initial endowments cannot be dropped from the assumptions. The underlying reason is that an agent cannot use all of his initial endowment when trading, due to the measurability restrictions on net trades. However, when his income is calculated that is done by valuing the whole initial endowment, leading to higher demand than what can actually be achieved by trading. Thus, one could circumvent these restrictions of $e_i \in X_i^N$ by defining the income of the agents in another way, taking into account only the parts of the initial endowments which can really be used for trading. When allocations have to be nonnegative as in this paper, this would be endowments e_i' such that $e_{ij}'(\omega) := \min_{\omega' \in \mathcal{P}_i(\omega)} e_{ij}(\omega')$ for all $\omega \in \Omega$ and commodities j.

Example 8. Consider a private information economy with 3 agents $i \in N = \{1, 2, 3\}$, one commodity and three states $\Omega = \{1, 2, 3\}$. Let the initial endowments, the (private) information and the utility functions be

$$\begin{aligned} e_1 &:= (2,1,2), \mathcal{P}_1 := \{\{1,2\},\{3\}\}, u_1 := x_1 + x_2 + 3x_3, \\ e_2 &:= (2,0,2), \mathcal{P}_2 := \{\{1,3\},\{2\}\}, u_2 := x_1 + 3x_2 + x_3, \\ e_3 &:= (0,2,2), \mathcal{P}_3 := \{\{2,3\},\{1\}\}, u_3 := 3x_1 + x_2 + x_3. \end{aligned}$$

The net trades

$$z_1 = (-1, -1, 4),$$
$$z_2 = (-2, 3, -2),$$
$$z_3 = (3, -2, -2),$$

lead to the core allocation

$$x = \begin{pmatrix} 1 & 0 & 6 \\ 0 & 3 & 0 \\ 3 & 0 & 0 \end{pmatrix} \in C(\mathbb{E}).$$

Obviously, one has that $x^n \in C(\mathbb{E}^n)$ for all $n \in \mathbb{N}$. Nevertheless, x is not a competitive quasi-equilibrium allocation. To see this, consider first the case where every agent has positive income, i.e. $e_i + z_i$ maximizes u_i. Note that $p = (p_1, p_2, p_3) \geq 0$, as the prices seen by an agent have to be nonnegative and $\{3\} \in \mathcal{P}_1, \{2\} \in \mathcal{P}_2, \{1\} \in \mathcal{P}_3$. Then one calculates from the budget equations $pz_i = 0$ (these have to hold if $e_i + z_i$ maximizes u_i), that the price vector would have to be $p = (p_1, p_1, \frac{1}{2}p_1)$. As $p \neq 0$ we must have $p > 0$. The budget set of agent 1 becomes $\{(x_1, x_2, x_3) | x - e_1 \in X_1 \text{ and } px = pe_1 = 4p_1 > 0\}$. But then the unique maximizer of u_1 on this budget set is $(0, 0, 8) \neq x_1$. Hence, x cannot be a competitive quasi-equilibrium in this case, because agent 1 has positive income $4p_1$. On the other hand, if there is an agent with zero income, one can distinguish three cases.

1. $pe_1 = 0 \Rightarrow 2p_1 + p_2 + 2p_3 = 0 \overset{p \geq 0}{\Rightarrow} p = 0$, a contradiction.
2. $pe_2 = 0 \Rightarrow 2p_1 + 2p_3 = 0 \overset{p \geq 0}{\Rightarrow} p_1 = p_3 = 0 \overset{p \neq 0}{\Rightarrow} p_2 > 0 \Rightarrow pe_3 = 2p_2 > 0 \Rightarrow x_3 = (3, 0, 0)$ maximizes u_3 on $B_3(p)$, a contradiction as $p_1 = 0$.
3. $pe_3 = 0 \Rightarrow 2p_2 + 2p_3 = 0 \overset{p \geq 0}{\Rightarrow} p_2 = p_3 = 0 \overset{p \neq 0}{\Rightarrow} p_1 > 0 \Rightarrow pe_2 = 2p_1 > 0 \Rightarrow x_2 = (0, 3, 0)$ maximizes u_2 on $B_2(p)$, a contradiction as $p_2 = 0$.$\diamondsuit$

The next theorem points out that under trade boundedness of the communication system any equilibrium allocation x is also in the core. Trade boundedness here assures that all possible deviations of subcoalitions $S \subset N$ from x are also informationally feasible in the grand coalition. They are therefore excluded from the budget sets of the agents in S not for informational infeasibility, but because they cannot be afforded.

Theorem 9. *Let $\mathbb{E}$ be an exchange economy with asymmetric information and trade bounded communication system. If x is a competitive equilibrium allocation, then $x \in C(\mathbb{E})$.*

Proof. Let (x, p) be a competitive equilibrium of $\mathbb{E}$. Suppose that $x \notin C(\mathbb{E})$. As $x - e$ is informationally feasible for the grand coalition that can only be if $(u_i(x_i))_{i \in N} \in intV(S)$ for a coalition $\emptyset \neq S \subseteq N$, which means that there is an allocation y^S such that $u_i(y_i^S) > u_i(x_i)$ for all $i \in S$. As $(y_i^S - e_i)_{i \in S} \in \prod_{i \in S} X_i^S$ and $\sum_{i \in S}(y_i^S - e_i) = y^S(S) - e(S) = 0$, $(y_i^S - e_i)_{i \in S}$ is under trade boundedness also informationally feasible in the grand coalition. Thus,

$$u_i(y_i^S) > u_i(x_i) \Longrightarrow y_i^S \notin B_i(p) \Longrightarrow py_i^S > pe_i$$

for all $i \in S$. Summing up, that leads to

$$\sum_{i\in S} py_i^S > \sum_{i\in S} pe_i$$
$$\iff \quad py^S(S) > pe(S)$$
$$\implies \quad y^S(S) \neq e(S)$$

a violation of physical feasibility. Hence, $x \in C(\mathbb{E})$. □

Again trade boundedness is a vital assumption. Otherwise economies are easily constructed, where $\{e\}$ is a competitive equilibrium, but the core is empty, due to the fact that trading is only possible in strict subcoalitions of N.

If x is a competitive equilibrium allocation in $\mathbb{E}$, then it is straightforward that x^n is a competitive equilibrium allocation in the replica economy $\mathbb{E}^n$. This shows that for competitive equilibria the only if part of Theorem 7 holds in economies with a trade bounded communication system.

Corollary 10. *Let $\mathbb{E}$ be an exchange economy with asymmetric information and trade bounded communication system. If x is a competitive equilibrium allocation then $x^n \in C(\mathbb{E}^n)$ for all $n \in \mathbb{N}$.*

4 Equal treatment

One way to derive an existence result for competitive allocations in the case of full information (or only one state of the world) requires to show that agents of the same type are treated 'equally' in the core. Then a compactness argument and a version of Theorem 7 are used to show that a subsequence of the mean allocations $\overline{x}_n := \frac{1}{n}\sum_{j=1}^{n} x_{\cdot,j}$, $x_n \in C(\mathbb{E}^n)$, converges to a competitive equilibrium. Equal treatment ensures that $(\overline{x}_n)^n \in C(\mathbb{E}^n)$, and thus $\overline{x}_n \in C(\mathbb{E})$. I want to use a similar approach in Section 5. However, when trying to carry over the proof of the equal treatment property a problem arises. The enforcement of equal treatment in all core allocations relies on the possibility to redistribute parts of the mean allocation of one own's type to all agents of the other types, making them better off. Under the measurability constraints on the net trades imposed by informational feasibility, this can no longer be guaranteed for arbitrary communication systems. In fact, the following simple example shows a situation, where unequal treatment is present in the core, exactly for the described reason.

Example 11. Consider a private information economy with 3 agents $i \in N = \{1, 2, 3\}$, one commodity and three states $\Omega = \{1, 2, 3\}$. Let the initial endowments, the (private) information and the utility functions be

$$\begin{aligned} e_1 &:= (1,1,0),\, \mathcal{P}_1 := \{\{1,2\},\{3\}\},\, u_1(x) := x_1 + x_2 + 10x_3, \\ e_2 &:= (1,0,1),\, \mathcal{P}_2 := \{\{1,3\},\{2\}\},\, u_2(x) := x_1 + 10x_2 + x_3, \\ e_3 &:= (1,1,1),\, \mathcal{P}_3 := \{\{2,3\},\{1\}\},\, u_3(x) := 10x_1 + x_2 + x_3. \end{aligned}$$

The net trades

$$z_{(1,1)} = z_{(1,2)} = (-1,-1,2),$$

$$z_{(2,1)} = \left(-1, \frac{2}{10}, -1\right), z_{(2,2)} = \left(-1, \frac{38}{10}, -1\right)$$
$$z_{(3,1)} = z_{(3,2)} = (2, -1, -1),$$

lead to the allocation

$$x = \begin{pmatrix} 0 & 0 & 2 \\ 0 & 0 & 2 \\ 0 & \frac{2}{10} & 0 \\ 0 & \frac{38}{10} & 0 \\ 3 & 0 & 0 \\ 3 & 0 & 0 \end{pmatrix}.$$

The agents of type 2 are not treated equally in this allocation. Nevertheless, this allocation is in $C(\mathbb{E}^2)$. To see this, note that clearly x is informationally and physically feasible. Suppose, contrary to $x \in C(\mathbb{E}^2)$, that there is a coalition $S \subset N \times \{1,2\}, S \neq \emptyset$, and an allocation y for S such that $u_i(y_{(i,j)}) > u_i(x_{(i,j)})$ for all $(i,j) \in S$.

The first step is to show that $S \neq N \times \{1,2\}$. The allocation $\overline{x} := \frac{1}{2}(x_{i1} + x_{i2})_{i \in N}$ can easily be seen to be in $C(\mathbb{E})$, as any agent gets the complete endowment of the resource which he prefers most. Now, if $S = N \times \{1,2\}$, then $\overline{y} := \frac{1}{2}(y_{i1} + y_{i2})_{i \in N}$ is an allocation for N and the linearity of the utility functions causes $\overline{y}$ to be a deviation to $\overline{x}$ in the economy $\mathbb{E}$, contradicting $\overline{x} \in C(\mathbb{E})$.

The next step is to see that $pr^1(S) = N$. Otherwise only agents of the same type can trade with each other, which will not give them the possibility to deviate. So $S \neq N \times \{1,2\}$ and $pr_1(S) = N$.

As $S \neq N \times \{1,2\}$ the inital endowment of the coalition S satisfies $e(S) \leq (5,4,4)$. Assume for the moment that there are two agents of type 3 in the coalition S. Then $u_3(y_{(3,1)}) + u_3(y_{(3,2)}) > 30 + 30 = 60$ and, as the utility function u_3 is linear and monotone, $u_3(y_{(3,1)}) + u_3(y_{(3,2)}) = u_3(y_{(3,1)} + y_{(3,2)}) \leq u_3(e(S)) \leq u_3(5,4,4) = 58$, a contradiction. Hence, without loss of generality $S \subseteq N \times \{1,2\} - \{(3,2)\}$. This implies $e(S) \leq (5,3,3)$. By similar arguments one can now continue to show that $S \subseteq N \times \{1,2\} - \{(1,2),(2,2),(3,2)\}$, which together with $pr_1(S) = N$ implies that $S = \{(1,1),(2,1),(3,1)\}$.

The final step is now to show that there is no allocation $y = (y_{(1,1)}, y_{(2,1)}, y_{(3,1)})$ for the coalition $S = \{(1,1),(2,1),(3,1)\}$ which is a deviation, i.e. that satisfies $u_1(y_{(1,1)}) > 20$, $u_1(y_{(2,1)}) > 2$ and $u_1(y_{(3,1)}) > 30$. The problem of finding such an allocation y can be rewritten in the following way. The allocation y satisfies

$$y = \begin{pmatrix} 1+z_{11} & 1+z_{12} & z_{13} \\ 1+z_{21} & z_{22} & 1+z_{23} \\ 1+z_{31} & 1+z_{32} & 1+z_{33} \end{pmatrix}.$$

Informational feasibility requires that $z_{11} = z_{12}, z_{21} = z_{23}$ and $z_{32} = z_{33}$. Physical feasibility requires that $z_{31} = -z_{11} - z_{21}, z_{32} = -z_{12} - z_{22}$ and $z_{13} = -z_{23} - z_{33}$. Combining this one obtains

$$y = \begin{pmatrix} 1+z_{11} & 1+z_{11} & z_{11}+z_{22}-z_{21} \\ 1+z_{21} & z_{22} & 1+z_{21} \\ 1-z_{11}-z_{21} & 1-z_{11}-z_{22} & 1-z_{11}-z_{22} \end{pmatrix}.$$

So the rewritten problem is to find (z_{11}, z_{21}, z_{22}) such that $y = y(z_{11}, z_{21}, z_{22}) \geq 0$, $y(S) = e(S) = (3, 2, 2)$ and

$$\begin{aligned}\overline{u}_1 &:= u_1(y_{(1,1)}) = 2 + 12z_{11} - 10z_{21} + 10z_{22} > 20,\\ \overline{u}_2 &:= u_2(y_{(2,1)}) = 2 + 2z_{21} + 10z_{22} > 2,\\ \overline{u}_3 &:= u_3(y_{(3,1)}) = 12 - 12z_{11} - 10z_{21} - 2z_{22} > 30.\end{aligned}$$

The last three inequalities correspond to solving the linear equation system

$$\begin{pmatrix} 12 & -10 & 10 \\ & 2 & 10 \\ -12 & -10 & -2 \end{pmatrix} \begin{pmatrix} z_{11} \\ z_{21} \\ z_{22} \end{pmatrix} = \begin{pmatrix} \widehat{u}_1 \\ \widehat{u}_2 \\ \widehat{u}_3 \end{pmatrix},$$

subject to $\widehat{u}_1 > 18, \widehat{u}_2 > 0$ and $\widehat{u}_3 > 18$. Now solving this system leads to

$$\begin{aligned} z_{11} &= \frac{4\widehat{u}_1 - 5\widehat{u}_2 - 5\widehat{u}_3}{108} \\ z_{21} &= \frac{-5\widehat{u}_1 + 4\widehat{u}_2 - 5\widehat{u}_3}{108}, \\ z_{22} &= \frac{\widehat{u}_1 + 10\widehat{u}_2 + \widehat{u}_3}{108}. \end{aligned}$$

Physical feasibility implies that $1 + z_{21} \geq 0 \Leftrightarrow \widehat{u}_2 \geq \frac{5\widehat{u}_1 + 5\widehat{u}_3 - 108}{4}$, and $\widehat{u}_1 > 18, \widehat{u}_3 > 18$ implies that $\frac{5\widehat{u}_1 + 5\widehat{u}_3 - 108}{4} > \frac{180-108}{4} = 18$, so $\widehat{u}_2 > 18$. Hence, $y_{(2,1),2} = z_{22} = \frac{\widehat{u}_1 + 10\widehat{u}_2 + \widehat{u}_3}{108} > \frac{216}{108} = 2 = e_2(S)$, a contradiction, since no agent can have more of a commodity than the complete endowment. So there is no deviation y for the coalition S. This concludes the proof that $x \in C(\mathbb{E}^2)$.$\diamondsuit$

Fortunately, what is really needed for the compactness argument is not equal treatment in all core allocations, but, that there is at least one core allocation $x \in C(\mathbb{E}^n)$, in which agents are treated equally. This can be guaranteed under the conditions of the following theorem.

Theorem 12. *Let $\mathbb{E}$ be an exchange economy with asymmetric information and trade bounded communication system. Assume that $e_i \in \bigcap_{N \supseteq S \ni i} X_i^S$ and $u_i \in U_{qc}^{mo}$ for all $i \in N$. Then there is an $x \in C(\mathbb{E}^n)$, $n \in \mathbb{N}$, in which agents of the same type receive the same commodity bundle.*

Instead of proving this theorem directly, we prove Lemma 13 for concave utilities, and explain how to obtain Theorem 12 from this lemma. Since this is a standard argument, it is not given in detail. The lemma shows, that in the special case of Theorem 12 where the utility functions are concave and normalized such that $u_i(0) = 0$, even a stronger result holds, namely that the mean allocation $\overline{x}^n$ of *any* core allocation x is also in the core.

Lemma 13. *Let $\mathbb{E}$ be an exchange economy with asymmetric information and trade bounded communication system. Assume that $e_i \in \bigcap_{S \ni i} X_i^S$ and $u_i \in U_{co,0}^{mo}$ for all $i \in N$. Then $x \in C(\mathbb{E}^n)$, $n \in \mathbb{N}$, implies that $\overline{x}^n$ is also in $C(\mathbb{E}^n)$.*

The proof of the lemma relies on some other observations.

Lemma 14. *Let $\mathbb{E}$ be an exchange economy with asymmetric information and trade bounded communication system. Assume that $u_i \in U_{qc}$ for all $i \in N$. For any assignment of commodities $x : \Omega \to \mathbb{R}_+^{n\times k}$ to the agents $1, \dots, n$, $n \geq 2$, of a given type $i \in N$, and any $S \subset \{1, \dots, n\}$, $|S| \geq 2$, there is an $l \in S$ such that*

$$u_i\left(\frac{1}{|S|}\sum_{j\in S} x_j\right) \geq u_i\left(\frac{1}{|S|-1}\sum_{j\in S, j\neq l} x_j\right).$$

Proof. Suppose to the contrary that

$$u_i\left(\frac{1}{|S|-1}\sum_{j\in S, j\neq l} x_j\right) > u_i\left(\frac{1}{|S|}\sum_{j\in S} x_j\right)$$

for all $l \in S$. Then quasiconcavity yields

$$u_i\left(\frac{1}{|S|}\sum_{l\in S}\left(\frac{1}{|S|-1}\sum_{j\in S, j\neq l} x_j\right)\right) \geq \min_{l\in S} u_i\left(\frac{1}{|S|-1}\sum_{j\in S, j\neq l} x_j\right)$$
$$> u_i\left(\frac{1}{|S|}\sum_{j\in S} x_j\right).$$

But the operand on the left hand side, $\frac{1}{|S|}\sum_{l\in S}\frac{1}{|S|-1}\sum_{j\in S, j\neq l} x_j$, actually equals the operand of the right hand side, $\frac{1}{|S|}\sum_{j\in S} x_j$, a contradiction. □

This lemma allows to order $(x_1, \dots, x_n)$ in a specific way.

Lemma 15. *Let $\mathbb{E}$ be an exchange economy with asymmetric information and trade bounded communication system. Assume that $u_i \in U_{qc}$ for all $i \in N$. For any assignment of commodities $x : \Omega \to \mathbb{R}_+^{n\times k}$ to the agents $1, \dots, n$, $n \geq 2$, of a given type $i \in N$, there is a permutation π of $(1, \dots, n)$ such that for $x^\pi := (x_{\pi(1)}, \dots, x_{\pi(n)})$ it holds that*

$$u_i\left(\frac{1}{n-m+1}\sum_{j=m}^{n} x_j^\pi\right) \geq u_i\left(\frac{1}{n-m}\sum_{j=m+1}^{n} x_j^\pi\right)$$

for all $m = 1, \dots, n-1$.

Proof. π can be constructed inductively. Apply Lemma 14 to

$$S = \{1, \dots, n\}$$

and let $\pi(1) = l$. When $\pi(1), \dots, \pi(j)$ are constructed for some $2 \leq j < n$, apply Lemma 14 to

$$S = \{1, \dots, n\} \setminus \{\pi(1), \dots, \pi(j)\},$$

and let $\pi(j+1) = l$. When finally $j = n$, let $\pi(n)$ be the last remaining element in

$$S = \{1, \dots, n\} \setminus \{\pi(1), \dots, \pi(n-1)\}.$$

Note that the $\pi(j)$, $1 \leq j < n$, obtained in this way are distinct so that the process really ends with $|\{1, \dots, n\} \setminus \{\pi(1), \dots, \pi(n-1)\}| = 1$ and π is a permutation of $(1, \dots, n)$. The desired property holds by the construction via Lemma 14. □

Now the proof of Lemma 13 can be given.

Proof. Assume to the contrary that $\overline{x}^n$ is not in $C(\mathbb{E}^n)$. So there is a subcoalition S of the grand coalition $N \times \{1, \dots, n\}$ and an allocation y^S for that coalition, such that

$$u_i(y^S_{(i,j)}) > u_i(\overline{x}^n_{(i,j)}) \text{ for all } (i,j) \text{ in } S.$$

Using concavity one obtains therefore for

$$\overline{y}^S_{(i,j)} := \frac{1}{|pr_i(S)|} \sum_{j \in pr_i(S)} y^S_{(i,j)},$$

that

$$u_i\left(\overline{y}^S_{(i,j)}\right) \geq \frac{1}{|pr_i(S)|} \sum_{j \in pr_i(S)} u_i(y^S_{(i,j)}) > u_i(\overline{x}^n_{(i,j)}) \text{ for all } (i,j) \text{ in } S.$$

It is easily seen, that $\overline{y}^S$ is also physically and informationally feasible for the coalition S. As $\overline{x}^n_{(i,j)} = \overline{x}^n_{(i,k)}$ for all $k \neq j$ one can without loss of generality assume that $pr_i(S) = \{l_i, \dots, n\}$, $1 \leq l_i \leq n$, $i \in N$. Furthermore, $x \in C(\mathbb{E}^n)$ implies that for any collection of permutations $(\pi_i)_{i \in N}$ of $(1, \dots, n)$ $(x_i^{\pi_i})_{i \in N} := (x_{(i,\pi_i(j))})_{i,j}$ is also in $C(\mathbb{E}^n)$. Additionally permutations do not change the mean $\overline{x}^n_{(i,j)}$, so it can be assumed by Lemma 15 that

$$\begin{aligned}
u_i(\overline{x}^n_{(i,j)}) &= u_i\left(\frac{1}{n}\sum_{j=1}^{n} x_{(i,j)}\right) \\
&\geq u_i\left(\frac{1}{n-1}\sum_{j=2}^{n} x_{(i,j)}\right) \\
&\geq \dots \\
&\geq u_i\left(\frac{1}{n-l_i+1}\sum_{j=l_i}^{n} x_{(i,j)}\right) \\
&= u_i\left(\frac{1}{|pr_i(S)|}\sum_{j \in pr_i(S)} x_{(i,j)}\right).
\end{aligned}$$

Hence, altogether this yields

$$u_i(\overline{y}^S_{(i,j)}) > u_i\left(\frac{1}{|pr_i(S)|}\sum_{j\in pr_i(S)} x_{(i,j)}\right) \geq \frac{1}{|pr_i(S)|}\sum_{j\in pr_i(S)} u_i(x_{(i,j)}). \quad (4.1)$$

Concavity and $u_i(0) = 0$ imply that

$$u_i(\lambda x) = u_i(\lambda x + (1-\lambda)0) \geq \lambda u_i(x) + (1-\lambda)u_i(0) = \lambda u_i(x) \qquad (4.2)$$

for all $\lambda \in [0,1]$. Equation 4.1 is equivalent to

$$\sum_{j\in pr_i(S)} u_i(\overline{y}^S_{(i,j)}) > \sum_{j\in pr_i(S)} u_i(x_{(i,j)}) \geq u_i(x_{(i,\overline{j})}) \text{ for any } \overline{j} \in pr_i(S),$$

as $u_i \geq 0$ on $\mathbb{R}_+^{k\times\Omega}$ ($u_i \in U^{mo}_{co,0}$). Note that $u_i(\overline{y}^S_{(i,j)}) > 0$, so one can define

$$\lambda_{(i,j)} := \frac{u_i(x_{(i,j)})}{u_i(\overline{y}^S_{(i,j)})},$$

and using Equation 4.2 one has

$$u_i(\lambda_{(i,j)}\overline{y}^S_{(i,j)}) \geq \lambda_{(i,j)} u_i(\overline{y}^S_{(i,j)}) = \frac{u_i(x_{(i,j)})}{u_i(\overline{y}^S_{(i,j)})} u_i(\overline{y}^S_{(i,j)}) = u_i(x_{(i,j)})$$

for all $i \in pr^1(S)$, $j \in pr_i(S)$.

Furthermore,

$$\begin{aligned}
\sum_{j\in pr_i(S)} \lambda_{(i,j)}\overline{y}^S_{(i,j)} &= \left(\sum_{j\in pr_i(S)} \lambda_{(i,j)}\right)\overline{y}^S_{(i,j)} \\
&= \left(\sum_{j\in pr_i(S)} \frac{u_i(x_{(i,j)})}{u_i(\overline{y}^S_{(i,j)})}\right)\overline{y}^S_{(i,j)} \\
&= |pr_i(S)|\underbrace{\left(\frac{\sum_{j\in pr_i(S)} u_i(x_{(i,j)})}{\sum_{j\in pr_i(S)} u_i(\overline{y}^S_{(i,j)})}\right)}_{<1 \text{ by Equation 4.1}}\overline{y}^S_{(i,j)} \\
&< |pr_i(S)|\overline{y}^S_{(i,j)} = \sum_{j\in pr_i(S)} \overline{y}^S_{(i,j)}.
\end{aligned}$$

Hence, letting for all $i \in pr^1(S)$

$$y_i^r := \sum_{j\in pr_i(S)} \overline{y}^S_{(i,j)} - \sum_{j\in pr_i(S)} \lambda_{(i,j)}\overline{y}^S_{(i,j)} = \sum_{j\in pr_i(S)} (1-\lambda_{(i,j)})\overline{y}^S_{(i,j)},$$

one has that $y_i^r > 0$ and $y_i^r \in X_i^S$. Let

$$\widetilde{y}^S_{(i,j)} := \lambda_{(i,j)}\overline{y}^S_{(i,j)} + \frac{y_i^r}{|pr_i(S)|}.$$

The proof is finished, if it can be shown that $\widetilde{y}^S$ is an allocation for S, because then, as

$$u_i(\widetilde{y}^S_{(i,j)}) > u_i(x_{(i,j)}) \text{ for all } (i,j) \text{ in } S,$$

$\widetilde{y}^S$ is a deviation to x, contradicting $x \in C(\mathbb{E}^n)$. $\widetilde{y}^S$ is an allocation for S, if it is informationally and physically feasible. Informational feasibility and $\widetilde{y}^S \geq 0$ follows from

$$\begin{aligned}
\widetilde{y}^S_{(i,j)} &= \lambda_{(i,j)}\overline{y}^S_{(i,j)} + \frac{y^r_i}{|pr_i(S)|}.\\
&= \lambda_{(i,j)}\overline{y}^S_{(i,j)} + \frac{1}{|pr_i(S)|}\sum_{j\in pr_i(S)}(1-\lambda_{(i,j)})(e_{(i,j)} + \overline{y}^S_{(i,j)} - e_{(i,j)})\\
&= e_{(i,j)} + \underbrace{\lambda_{(i,j)}\overline{y}^S_{(i,j)}}_{\in X^S_i} + \frac{1}{|pr_i(S)|}\sum_{j\in pr_i(S)}\underbrace{(1-\lambda_{(i,j)})(\overline{y}^S_{(i,j)} - e_{(i,j)})}_{\in X^S_i}\\
&\quad - \underbrace{\frac{e_{(i,j)}}{|pr_i(S)|}}_{\in X^S_i}\underbrace{\sum_{j\in pr_i(S)}\lambda_{(i,j)}}_{<|pr_i(S)|}\\
&> e_{(i,j)} + \lambda_{(i,j)}\overline{y}^S_{(i,j)} + \frac{1}{|pr_i(S)|}\sum_{j\in pr_i(S)}\underbrace{(1-\lambda_{(i,j)})(\overline{y}^S_{(i,j)} - e_{(i,j)}) - e_{(i,j)}}_{\geq 0}\\
&\geq 0.
\end{aligned}$$

The remaining part of physical feasibility follows from

$$\begin{aligned}
\sum_{(i,j)\in S}\widetilde{y}^S_{(i,j)} &= \sum_{i\in pr^1(S)}\sum_{j\in pr_i(S)}\left(\lambda_{(i,j)}\overline{y}^S_{(i,j)} + \frac{y^r_i}{|pr_i(S)|}\right)\\
&= \sum_{i\in pr^1(S)}\sum_{j\in pr_i(S)}\left(\lambda_{(i,j)}\overline{y}^S_{(i,j)} + \frac{1}{|pr_i(S)|}\sum_{j\in pr_i(S)}(1-\lambda_{(i,j)})\overline{y}^S_{(i,j)}\right)\\
&= \sum_{i\in pr^1(S)}\sum_{j\in pr_i(S)}\left(\lambda_{(i,j)}\overline{y}^S_{(i,j)} + (1-\lambda_{(i,j)})\overline{y}^S_{(i,j)}\right)\\
&= \sum_{(i,j)\in S}\overline{y}^S_{(i,j)} = \sum_{(i,j)\in S} y^S_{(i,j)} = e(S).
\end{aligned}$$

□

To prove Theorem 12 from Lemma 13, use [7], p.58, Proposition A.2, to approximate $u \in U^{mo}_{qc}$ by a sequence of $(u_j)_{j\in\mathbb{N}}$ such that $u_j \in U^{mo}_{co,0}$ for all $j \in \mathbb{N}$. Then use upper hemicontinuity of the core, viewed as a compact-valued correspondence from $(N, \Omega, e, (\mathcal{P}^S_i)_{i\in S, S\subset N}) \times (U^{mo}_{qc})^N$ to $\mathbb{R}^{N\times k\times\Omega}$, i.e. when the utility functions are variables with domain U^{mo}_{qc}. Any sequence of elements from the cores of the approximating economies yields a sequence satisfying equal treatment by Lemma 13, and thus by upper hemicontinuity a core element of the quasiconcave economy.

5 Existence of competitive allocations

As a consequence of Theorem 2 one obtains in combination with Theorem 7 an existence result for the quasi-equilibrium.

Theorem 16. *Let $\mathbb{E}$ be an exchange economy with asymmetric information and trade bounded communication system. Assume that $e_i \in \bigcap_{S \ni i} X_i^S$ and $u_i \in U_{qc}^{mo} \cap \mathbf{C}(\mathbb{R}_+^{k \times \Omega})$ for all $i \in N$. Then this economy has a competitive quasi-equilibrium (p, x). Moreover the price system p can be chosen $\sigma(\bigcup_{i \in N} \mathcal{P}_i^N)$-measurable.*

Proof. By Theorem 2 $C(\mathbb{E}^n) \neq \emptyset$ for all $n \in \mathbb{N}$. By Theorem 7 it suffices to show that the set of all $x \in \mathbb{R}^{N \times k \times \Omega}$ such that $x^n \in C(\mathbb{E}^n)$ for all $n \in \mathbb{N}$ is not empty. To see this, let $x_n \in C(\mathbb{E}^n)$ be a sequence of core elements from the replica economies $\mathbb{E}^n$, $n \in \mathbb{N}$. By Theorem 12 it can be assumed that agents of the same type are treated equally, i.e. $x_n = (y_n)^n$ for all $n \in \mathbb{N}$ and allocations y_n in the economy $\mathbb{E}$. Furthermore, $y_n \in C(\mathbb{E})$ for all $n \in \mathbb{N}$, because any deviation to y_n in the economy $\mathbb{E}$ would also be a deviation to $(y_n)^n$ in the economy $\mathbb{E}^n$. As $C(\mathbb{E})$ is compact, there is a convergent subsequence $(y_{n_k})_k$, $n_k \to \infty$, of $(y_n)_n$. Let y be the limit of this subsequence. If it is shown that $y^n \in C(\mathbb{E}^n)$ for all $n \in \mathbb{N}$ the proof is finished. Assume to the contrary that y^n is not in $C(\mathbb{E}^n)$ for some $n \in \mathbb{N}$.

So there is coalition $S \subset N \times \{1, \ldots, n\}$ and an allocation x^S for that coalition such that

$$u_{(i,j)}(x_{(i,j)}^S) > u_{(i,j)}(y_{(i,j)}^n) \text{ for all } (i,j) \in S.$$

As $|S| < \infty$, there is an $\varepsilon > 0$ such that

$$u_{(i,j)}(x_{(i,j)}^S) - u_{(i,j)}(y_{(i,j)}^n) > \varepsilon \text{ for all } (i,j) \in S.$$

On the other hand, using continuity of utilities and $y_{n_k} \to y$, there is a k_0, such that for all $k \geq k_0$,

$$\varepsilon \geq u_i(y_i) - u_i((y_i)_{n_k}) \geq -\varepsilon.$$

But then, as $u_{(i,j)}(y_{(i,j)}^n) = u_i(y_i)$,

$$\begin{aligned}
&u_{(i,j)}(x_{(i,j)}^S) - u_{(i,j)}((y_{(i,j)})_{n_k}^{n_k}) \\
&= u_{(i,j)}(x_{(i,j)}^S) \underbrace{- u_{(i,j)}(y_{(i,j)}^n) + u_i(y_i)}_{=0} - u_i((y_i)_{n_k}) \\
&> \varepsilon + (-\varepsilon) \geq 0 \text{ for all } (i,j) \in S.
\end{aligned}$$

Hence, x^S is also an allocation for S in the economies $\mathbb{E}^{n_k}$, $k \geq k_0$, $n_k \geq n$, such that

$$u_{(i,j)}(x_{(i,j)}^S) > u_{(i,j)}((y_{(i,j)})_{n_k}^{n_k}) \text{ for all } (i,j) \in S.$$

However, this contradicts $(y_{n_k})^{n_k} \in C(\mathbb{E}^{n_k})$. □

Next situations are identified, where $\inf pX_i^{N,\geq 0} < pe_i$ is true for all $i \in N$, so that Theorem 16 turns into an existence result for the competitive equilibrium. As $\inf pX_i^{N,\geq 0}$ must be equal to 0 for strictly increasing utilities, it is the same to identify when every agent has positive income.

Let the *information matrix* $\mathbb{I}(S)$ of a coalition S be the matrix in $\{0,1\}^{\Omega \times \cup_{i \in S} \mathcal{P}_i}$, where the entry $\mathbb{I}(S)_{(\omega,A)}$ is equal to $\mathbb{I}_A(\omega)$, i.e.

$$\mathbb{I}(S) := (\mathbb{I}_A(\omega)_{\omega \in \Omega})_{A \in \cup_{i \in S} \mathcal{P}_i}.$$

This matrix maps

$$X_+ := \{x \in \mathbb{R}^{k \times \Omega} | x \text{ is } \sigma(\cup_{i \in N} \mathcal{P}_i)\text{-measurable and } x_A \geq 0 \text{ for all } A \in \cup_{i \in N} \mathcal{P}_i\}$$

to $\mathbb{R}^{k \times \cup_{i \in N} \mathcal{P}_i}$ in the following way. Let

$$x\mathbb{I}(S) := (x_j \mathbb{I}(S)_A)_{1 \leq j \leq k, A \in \cup_{i \in N} \mathcal{P}_i}.$$

The set of all elements in X_+ mapped to the zero matrix in $\mathbb{R}^{k \times \cup_{i \in N} \mathcal{P}_i}$ is then

$$K(\mathbb{I}(S)) := \{x \in X_+ | x\mathbb{I}(S) = 0\}.$$

Lemma 17. *Let $\mathbb{E}$ be an exchange economy with asymmetric information. Assume that $e_i \in X_i^N$, $e_i \neq 0$, $\sum_{i \in N} e_i > 0$, and $u_i \in U^{mo}$ for all $i \in N$. Each of (i) to (iv) is a sufficient condition for every agent to have positive income in a quasi-equilibrium (x, p) of this economy, where p is $\sigma(\bigcup_{i \in N} \mathcal{P}_i^N)$-measurable.*

(i) The communication system is symmetric for the grand coalition, i.e. $\mathcal{P}_i^N \equiv \mathcal{P}^N$ for all $i \in N$.
(ii) $pe_i > 0$ for some agent $i \in N$, and all other agents have positive initial endowments.
(iii) $K(\mathbb{I}(S)) = \{0\}$ for some coalition S, and all agents have positive initial endowments.
(iv) $K(\mathbb{I}(S)) = \{0\}$ for some coalition S, all agents in S have positive initial endowments, and this coalition knows collectively more than the agents in $N \setminus S$, i.e.

$$\bigcup_{i \in S} \mathcal{P}_i \supseteq \bigcup_{i \in N \setminus S} \mathcal{P}_i.$$

Proof. Let (x, p) be a quasi-equilibrium of the economy $\mathbb{E}$.

(i): In such a communication system the set of all $\sigma(\bigcup_{i \in N} \mathcal{P}_i)$-measurable price systems $p \neq 0$ such that $p(A) \geq 0$ for all $A \in \mathcal{P}_i$ is equal to the set $X_i^{N,\geq 0} \setminus \{0\}$ for any $i \in N$. So, as $\sum_{i \in N} e_i > 0$, at least one agent $\bar{i}$ has positive income. But then $p(A) > 0$ for all $A \in P_{\bar{i}}^N = \mathcal{P}_i$, $i \in N$, so every agent has positive income.

(ii): As $pe_i > 0$ leads to $p(A) > 0$ for all $A \in \mathcal{P}_i$, it can be concluded that $p(\Omega) > 0$. But then, as e_j is $\sigma(P_j^N)$-measurable and $p(A) \geq 0$ for any $A \in P_j^N$, $j \neq i$,

$$pe_j = \sum_{A \in P_j^N} p(A) e_j(A)$$

$$
\begin{aligned}
&\geq \sum_{A \in P_j^N} p(A) \min_{A \in P_j^N} e_j(A) \\
&= \left(\sum_{A \in P_j^N} p(A) \right) \min_{A \in P_j^N} e_j(A) \\
&= \underbrace{p(\Omega)}_{>0} \underbrace{\min_{A \in P_j^N} e_j(A)}_{>0} \\
&> 0,
\end{aligned}
$$

hence any agent has positive income.

(iii): I show that $K(\mathbb{I}(S)) = \{0\}$ implies that one agent in S has positive income, then (ii) can be used. So suppose to the contrary, that

$$pe_i = \sum_{A \in \mathcal{P}_i} p(A) e_i(A) = 0 \text{ for all } i \in S.$$

Then, as $p(A) \geq 0$ and $e_i(A) \geq 0$ for all $A \in \cup_{i \in N} \mathcal{P}_i$, one obtains that

$$p(A) e_i(A) = 0 \text{ for all } A \in \cup_{i \in S} \mathcal{P}_i.$$

As $e_i(A) > 0$ for all $A \in \cup_{i \in S} \mathcal{P}_i$ this leads to

$$p(A) = 0 \text{ for all } A \in \cup_{i \in S} \mathcal{P}_i,$$

or, because $p(A) \geq 0$ for all $A \in \cup_{i \in N} \mathcal{P}_i$ and p is $\sigma(\cup_{i \in N} \mathcal{P}_i)$-measurable,

$$p \in K(\mathbb{I}(S)) = \{0\}.$$

This contradicts $p \neq 0$.

(iv): As in (iii), one agent in S must have positive income. But then $p(\Omega) > 0$, so all agents in S have positive income, showing that $p(A) > 0$ for all $A \in \cup_{i \in S} \mathcal{P}_i$. As $\cup_{i \in S} \mathcal{P}_i \supseteq \cup_{i \in N \setminus S} \mathcal{P}_i$ and $e_i \geq 0$, $e_i \neq 0$ for all $i \in N \setminus S$, this means that the agents in $N \setminus S$ also have positive income. □

Note that the situations described in (i), (iii) and (iv) are independent of the given quasi-equilibrium (x, p). So the following corollary to Theorem 16 can be stated.

Corollary 18. *Let $\mathbb{E}$ be an exchange economy with asymmetric information and trade bounded communication system. Assume that $e_i \in \bigcap_{S \ni i} X_i^S$ and $u_i \in U_{qc}^{mo} \cap$ $\mathbf{C}(\mathbb{R}_+^{k \times \Omega})$ for all $i \in N$. Let the assumptions and one of the situations (i), (iii) or (iv) given in Lemma 17 hold. Then this economy has a competitive equilibrium, and any competitive quasi-equilibrium is already a competitive equilibrium.*

As stated after Definition 4, this corollary is able to cover existence of a Radner equilibrium without free disposal. Other existence results are contained in Einy, Moreno and Shitovitz [4], for the case of a non-atomic measure space of agents, and Radner [10, 12], for the case of a finite number of agents. All of these papers assume some degree of disposability, in [4] and [12] free disposal, and in [10] production sets have a nonempty relative interior, which translates into disposability of at least part of the endowments in a pure exchange economy. It is known that if one wants to dispense with free disposal one has to allow for negative prices. An example of an economy without free disposal such that equilibrium prices are negative can be found in Liu [8]. The prices as defined here are consequently also allowed to be negative. Nevertheless, in equilibrium all prices seen by agents will have to be positive, so that nobody can exploit the existence of negative prices. In contrast to the machinery of proof used in this paper the mentioned existence results are not obtained by appealing to a limit theorem for the core. Moreover, Radner has to assume concavity of state-dependent utility in order to achieve concavity of expected utility. This corresponds to concavity of the utility functions $u_i : \mathbb{R}_+^{k\times\Omega} \to \mathbb{R}$ used here. No such assumptions are necessary in [4], as they are dealing with a non-atomic measure space of agents.

References

1. Allen, B.: Generic existence of completely revealing equilibria for economies with uncertainty when prices convey information. Econometrica **49**, 1173-1199 (1984)
2. Allen, B.: Market games with asymmetric information: the core with finitely many states of the world. In: Munier, B., Machina, M.J. (eds.) Models and experiments in risk and rationality. Dordrecht: Kluwer 1994
3. Debreu, G.: Theory of value. Cowles Foundation Monograph, Vol. 17. New York: Wiley 1959
4. Einy, E., Moreno, D., Shitovitz, B.: Competitive and core allocations in large economies with differential information. Economic Theory **18**, 321-332 (2001)
5. Einy, E., Moreno, D., Shitovitz, B.: On the core of an economy with differential information. Journal of Economic Theory **94**, 262-270 (2000)
6. Hervés-Beloso, C., Moreno-García, E., Yannelis, N.: Characterizing radner equilibria. Working paper, (2003)
7. Hildenbrand, W., Kirman, A.P.: Equilibrium analysis. Variations on themes by Edgeworth and Walras. Advanced textbooks in economics, Vol. 28. Amsterdam New York Oxford: North Holland 1988
8. Liu, C.: From a theorem of Radner. Economics Letters **39**, 401-404 (1992)
9. Maus, S.: Balancedness and the core in economies with asymmetric information. Economic Theory **22**, 613-627 (2002)
10. Radner, R.: Competitive equilibrium under uncertainty. Econometrica **36**, 31-58 (1968)
11. Radner, R.: Rational expectations equilibrium: Generic existence and the information revealed by prices. Econometrica **40**, 655-678 (1979)
12. Radner, R.: Equilibrium under uncertainty. In: Arrow, K.J., Intrilligator, M.D. (eds.) Handbook of mathematical economics, Vol. II. Amsterdam New York Oxford: North Holland 1982
13. Wilson, R.: Information, efficiency and the core of an economy. Econometrica **46**, 807-816 (1978)
14. Yannelis, N.: The core of an economy with differential information. Economic Theory **1**, 183-198 (1991)

PART 3

CORE, PARETO OPTIMALITY AND INCENTIVE COMPATIBILITY

Incentive compatibility and information superiority of the core of an economy with differential information[*]

Leonidas C. Koutsougeras and Nicholas C. Yannelis

Department of Economics, University of Illinois at Urbana-Champaign, Champaign, IL 61820, USA
Received: October 31, 1991; revised version March 12, 1992

Summary. We analyze the coarse, the fine, and the private core allocation of an exchange economy with differential information. The basic questions that we address are whether the above concepts are: (i) coalitionally incentive compatible, i.e., does truthful revelation of information in each coalition occur; and (ii) taking into account the information superiority or information advantage of an agent. Moreover, the above three concepts are examined in the presence of externalities and a comparison and interpretation of all of three core notions is provided.

1 Introduction

The idea of an exchange economy with differential information [i.e., an economy consisting of a finite set of traders each of whom is characterized by a state dependent (random) utility function, a random initial endowment, a private information set, and a prior], was introduced by Radner (1968) [we will sometimes call such an economy as a Radner-type economy]. The equilibrium notion that Radner (1968) adopted to analyze trade among agents in an economy with differential information was the Walrasian equilibrium. Since the Walrasian equilibrium notion is noncooperative it precludes cooperation among groups of agents. Thus, we adopt the core, a cooperative solution concept, in order to analyze the trade among agents in a Radner-type economy. We will argue that not only does the core provide more sensible outcomes than the Walrasian equilibrium, but it is also coalitionally incentive compatible (i.e., there is truthful revelation of information in each coalition) and it takes into account explicitly the information advantage or superiority of an agent.

Throughout the paper we will denote the private information set of agent i (which is going to be a partition of a measure space) by $\mathcal{F}_i$. We will first examine

[*] We wish to thank Stefan Krasa, Frank Page, Wayne Shafer, Anne Villamil, and Myrna Wooders for several useful comments, discussions, and suggestions. The comments of two referees were also helpful and we thank them for their careful reading. Obviously, we are responsible for any remaining errors.

versions of the *coarse core* of Wilson (1978), where the blocking net trades of a coalition S are $\bigwedge_{i \in S} \mathcal{F}_i$-measurable[1] and therefore the information is common knowledge to each member of the coalition. We next examine the *fine core* concept of Wilson (1978), where the blocking net trades of a coalition S are $\bigvee_{i \in S} \mathcal{F}_i$-measurable and hence the information is pooled by the members of the coalition. Finally, we examine the *private core* of Yannelis (1991), where the blocking net trade of each member of the coalition $\mathcal{F}_i$-measurable and thus there is bargaining under differential information among the members in each coalition, contrary to the coarse and fine core.

We will show that the coarse core exists, it is coalitionally incentive compatible (i.e., there is truthful revelation of information in each coalition) and it takes into account the information superiority of an individual. However, since the coarse core always contains the private core and the latter exists and has the above properties, we will conclude that we learn nothing new from the coarse core that cannot be learned from the private core. In fact as we will show by means of an example, the coarse core is "too large", i.e., all the individually rational and Pareto optimal allocations constitute the coarse core.

Contrary to the coarse core, the basic problem with the fine core is that it is "too small" and in general it does not exist. Moreover, whenever it does exist we will show that it is not coalitionally incentive compatible and it does not take into account the information superiority of an individual. The analysis of these core concepts suggests that the private core may be the appropriate core notion in an exchange economy with differential information. This concept exists under standard continuity and concavity assumptions on the utility functions, it is coalitionally incentive compatible, and takes into account the information superiority of an agent. Moreover, we show that the private core can be used to model the idea of an intermediary [see Boyd-Prescott (1986) as well]. The intermediary is an agent with "better" information than all other agents who by using his/her superior information, executes the correct trades. The idea of an intermediary arises endogenously and naturally in our framework. Our results suggest that cooperative solution concepts may be quite useful for analyzing trade in economies with differential information and may be useful for tackling basic issues in the theory of financial markets.

The paper is organized as follows: Section 2 contains notation and the economic model. Several core notions are defined in Section 3 and some preliminary results are proved as well. Section 4 focuses mainly on the incentive compatibility of the private core. The interpretation of the different core concepts is given in Section 5. Section 6 introduces different core notions in the presence of externalities, and an existence result is proved in Section 7. Section 8 contains some concluding remarks. Finally, in Section 9 we compare the core with the value allocation of Krasa-Yannelis (1991).

---

[1] The symbol $\bigwedge_{i \in S} \mathcal{F}_i$ denotes the "meet", i.e., the maximal partition contained in all $\mathcal{F}_i$. The symbol $\bigvee_{i \in S} \mathcal{F}_i$ denotes the "join", i.e., the minimal partition containing all F_i. By an abuse of notation we will denote throughout the paper, the σ-algebra generated by the partition $\mathcal{F}_i$ also by $\mathcal{F}_i$.

2 Notation and the economic model

Before we outline our model we begin with some notation.

2.1 Notation

$\mathbb{R}^n$ denotes the n-dimensional Euclidean space.
$\mathbb{R}^N_+$ denotes the positive cone of $\mathbb{R}^n$.
$\mathbb{R}^n_{++}$ denotes the strictly positive cone of $\mathbb{R}^n$.
2^A denotes the set of all subsets of the set A.
$\varnothing$ denotes the empty set.
$\setminus$ denotes set-theoretic subtraction.
$|A|$ denotes the cardinality of the set A.

If $(\Omega, \mathcal{F}, \mu)$ is a measure space, then $\mathcal{F}_i$ will always denote a measurable partition of Ω (or a sub-σ-algebra) and $E_i(\omega)$ will denote the element of the partition $\mathcal{F}_i$ which contains $\omega \in \Omega$. If X is a linear topological space, its dual is the space X^* of all continuous linear functionals on X.

2.2 The exchange economy with differential information

Let Y denote the commodity space. For simplicity one may identify Y with the positive cone of $\mathbb{R}^{\prime}$. However, all the results in this paper remain true if Y is the positive cone of any Banach lattice with an order continuous norm.[2] Therefore, one can allow for infinitely many commodities. Denote by $(\Omega, \mathcal{F}, \mu)$ a probability measure space. An *exchange economy with differential information* $\mathcal{E}$ is given by

$\mathcal{E} = \{(\Omega, u_i, e_i, \mathcal{F}_i, \mu) : i = 1, 2, ..., n\}$ where

(1) $X_i : \Omega \to 2^Y$ is the *consumption set* of agent i,
(2) $u_i : Y \to \mathbb{R}$ is the *utility function* of agent i,
(3) $\mathcal{F}_i$ is a (measurable) partition of Ω denoting the *private information*[3] of agent i,
(4) $e_i : \Omega \to Y$ is the *initial endowment* of agent i, where e_i is $\mathcal{F}_i$-measurable, (Bochner) integrable and $e_i(\omega) \in X_i(\omega)$ $\mu - a.e.$
(5) μ is a probability measure on Ω denoting the common prior of each agent.

The *expected* utility of agent i is given by

$$\int_{\omega \in \Omega} u_i(x_i(\omega)) d\mu(\omega).$$

A possible interpretation of the above economy is the following: one may think that there are two periods where actual consumption takes place in the second period.

[2] See Section 7 for rigorous definitions.

[3] As in Kobayashi (1980) we will assume that the members of a coalition release their private information sets honestly, i.e., private information sets are common knowledge.

In period one there is uncertainty over the states of nature and in this period agents make agreements which may be contingent on the realized state of nature in the second period. It is important to note that in this setting agents have differential information with respect to the realized state of nature and know their endowment realization (i.e., the initial endowment of each agent is $\mathcal{F}_i$-measurable). Note that a common prior assumption has been adopted in this framework. However, as in Yannelis (1991), one may allow for different priors, Bayesian updating, and random utility functions. All the results of this paper remain valid in this case, but we choose not to adopt the latter modeling for simplicity of exposition and easier calculation of our examples introduced later in the paper.

3 Core notions and preliminary results

In this section we define several different core notions for an exchange economy with differential information.

3.1 The private core

The following core notion was introduced in Yannelis (1991, Definition 3.1.1, p. 187). It was subsequently used by Allen (1991) who refers to it as the private information core, or the publicly predictable information core.

Definition 3.1. An allocation $x : \Omega \to \Pi_{i=1}^n X_i$ is a *private core allocation* for $\mathcal{E}$, if the following conditions hold:

(i) each x_i is $\mathcal{F}_i$-measurable;
(ii) $\sum_{i=1}^n x_i(\omega) = \sum_{i=1}^n e_i(\omega)$ μ – *a.e.*;
(iii) it is not true that there exist $S \subset \{1, 2, ..., n\}$ and $y : \Omega \to \Pi_{i \in S} X_i$, such that each y_i is $\mathcal{F}_i$-measurable for all $i \in S, \sum_{i \in S} y_i(\omega) = \sum_{i \in S} e_i(\omega)$ μ – *a.e.* and $\int u_i(y_i(\omega))d\mu(\omega) > \int u_i(x_i(\omega))d\mu(\omega)$ for all $i \in S$.

Condition (i) implies that the net trade of each agent, i.e., $x_i - e_i$ is $\mathcal{F}_i$-measurable (recall that each e_i is $\mathcal{F}_i$-measurable) and consequently each agent knows his/her own net trade realization. Condition (ii) says that the markets are cleared for almost all states of nature. Note that since $\sum_{i=1}^n (x_i(\omega) - e_i(\omega)) = 0$ μ – *a.e.* and by (i) each $x_i - e_i$ is $\mathcal{F}_i$-measurable, it follows that $\sum_{i=1}^n (x_i(\cdot) - e_i(\cdot))$ is $\bigvee_{i=1}^n \mathcal{F}_i$-measurable and therefore, the grand coalition knows the aggregate net trade realization.

Condition (iii) says that it is not possible for a coalition of agents to get together, redistribute their resources among themselves (while each agent in the coalition use his/her own private information) and make the expected utility of each agent in the coalition better off.

It should be noted that the measurability assumptions in (i) and (iii) are *equivalent* to the fact that:

(i′) Each $x_i - e_i$ is $\mathcal{F}_i$-measurable, and

(ii′) It is not true that there exist S and $y : \Omega \to \Pi_{i \in S} X_i$, such that $y_i - e_i$ is $\mathcal{F}_i$-measurable for all $i \in S$, $\sum_{i \in S} y_i(\omega) = \sum_{i \in S} e_i(\omega)\mu$ – *a.e.* and $\int u_i(y_i) d\mu > \int u_i(x_i) d\mu$ for all $i \in S$.

Pick now arbitrarily an agent j in the coalition S. Note that since in (iii) above, $\sum_{i \in S}(y_i(\cdot) - e_i(\cdot)) = 0$, by rearranging we have that $e_j(\cdot) - y_j(\cdot) = \sum_{i \in S \setminus \{j\}}(y_i(\cdot) - e_i(\cdot))$. Since $e_j - y_j$ is $\mathcal{F}_j$-measurable and $\sum_{i \in S \setminus \{j\}}(y_i - e_i)$ is $\bigvee_{i \in S \setminus \{j\}} \mathcal{F}_i$-measurable, it is always the case that within a coalition say S the $|S - 1|$ members of the coalition can pool their private information and verify the net trade of the remaining agent.[4]

The following theorem proved in Yannelis (1991) provides sufficient conditions which guarantee the existente of a private core allocation for $\mathcal{E}$. The commodity space Y can be the positive cone of any Banach lattice with an order continuous norm.

Theorem 3.1. *Let* $\mathcal{E} = \{(X_i, u_i, e_i, \mathcal{F}_i, \mu) : i = 1, 2, ..., n)$ *be an exchange economy with differential information, satisfying the following assumptions for each* $i (i = 1, 2, ..., n)$:

(a.3.1) $X_i : \Omega \to 2^Y$ *is a convex, closed, nonempty valued correspondence;*
(a.3.2) $u_i : Y \to \mathbb{R}$ *is continuous, integrably bounded and concave.*

Then a private core allocation exists in $\mathcal{E}$.

A few technical remarks are in order. Note that in Theorem 3.1 of Yannelis (1991), the set-valued function $X_i : \Omega \to 2^Y$ is assumed to be integrably bounded and $\mathcal{F}_i$-measurable as well. The latter assumption was needed to show [see Yannelis (1991, p. 191)] that the set L_{X_i}, which is defined to be a set of all Bochner integrable and $\mathcal{F}_i$-measurable selections from the set-valued function $X_i : \Omega \to 2^Y$, is nonempty. However, since $e_i : \Omega \to Y$ is Bochner integrable and $\mathcal{F}_i$-measurable, we can conclude that $e_i \in L_{X_i}$, and hence the set L_{X_i}, is indeed nonempty. This change in assumptions allows us to relax the separability assumption on the commodity space. [Recall that the separability assumption in Yannelis (1991) was needed in one step only, in particular it was used to make the Aumann measurable selection theorem applicable and to show that L_{X_i}, is nonempty.] Finally, the utility function of each agent, $u_i : Y \to \mathbb{R}$ can only be assumed to be norm continuous instead of weakly continuous as it was assumed in Yannelis (1991). In particular, by virtue of the Lebegue Dominated Convergence Theorem one can show that $\int u_i d\mu$ is norm continuous. In view of the concavity of u_i it follows from Mazur's Theorem that $\int u_i d\mu$ is also weakly-upper semicontinuous (w-u.s.c.) and the existente proof in Yannelis (1991) remains unchanged. Note that in this case it doesn't make any difference whether we assume that each $\mathcal{F}_i$ is a partition, or a sub-σ-algebra of Ω. Moreover, the dual space of Y doesn't need to have the Radon-Nikodym Property (RNP). For more details an these issues see Balder-Yannelis (1991) or Page (1992).

[4] In view of this property of the private core, Allen (1991) refers to it as the publicly predictable information core.

3.2 The coarse core

The following definition of a coarse core allocation is taken from Yannelis (1991, p. 187). It is a variant of the coarse core concept first introduced by Wilson (1978) [see also Kobayashi (1980)].

Definition 3.2. An allocation $x : \Omega \to \Pi_{i=1}^{n} X_i$ is a coarse core allocation for $\mathcal{E}$, if the following conditions hold:

(i) Each x_i is $\mathcal{F}_i$-measurable;
(ii) $\sum_{i=1}^{n} x_i(\omega) = \sum_{i=1}^{n} e_i(\omega)$ μ – *a.e.*;
(iii) It is not true that there exist $S \subset \{1, 2, ..., n\}$ and $y : \Omega \to \Pi_{i \in S} X_i$ such that $y_i - e_i$ is $\bigwedge_{i \in S} \mathcal{F}_i$-measurable for all $i \in S$, $\sum_{i \in S} y_i(\omega) = \sum_{i \in S} e_i(\omega)$ μ – *a.e.* and $\int u_i(y_i(\omega)).\, d\mu(\omega) > \int u_i(x_i(\omega)) d\mu(\omega)$ for all $i \in S$.

Conditions (i) and (ii) are discussed above. Condition (iii) says that it is not possible for a coalition of agents by redistributing their initial endowments (based on information which is common knowledge to the coalition) to make the expected utility of each agent in the coalition better off.

We now state a result on the existence of coarse core allocations that follows directly from Theorem 3.1 in Yannelis (1991), simply by noticing that the set of all private core allocations for $\mathcal{E}$ is a strict subset of the set of all coarse core allocations for $\mathcal{E}$.

Theorem 3.2. *Let* $\mathcal{E} = \{(X_i, u_i, e_i, \mathcal{F}_i, \mu) : i = 1, 2, ..., n\}$ *be an exchange economy with differential information satisfying for each* i, $(i = 1, 2, ..., n)$ *all the assumptions of Theorem 3.1 above. Then a coarse core allocation exists in* $\mathcal{E}$.

Note that if condition (i) in Definition 3.2 is replaced by:

(i′) Each x_i is $\bigwedge_{i \in I} \mathcal{F}_i$-measurable,

then we will indicate in Section 5 that such a coarse core notion which we will call a *strong coarse core allocation* for $\mathcal{E}$, may not exist. In particular, we show in Section 5 that there exist private information exchange economies satisfying all the assumptions of Theorem 3.2, but for which strong coarse core allocations may not exist. Note that what we call here a "strong coarse core" corresponds to the "coarse core" in Allen (1991). It is exactly for this reason that Allen (1991) concludes that the strong coarse core may be empty.

3.3 The fine core

The following core notion is taken from Yannelis (1991, p. 188) and is a variant of the fine core concept introduced by Wilson (1978).

Definition 3.3. An allocation $x : \Omega \to \Pi_{i=1}^{n} X_i$ is a *fine core allocation* for $\mathcal{E}$ if the following conditions hold:

(i) Each x_i is $\mathcal{F}_i$-measurable.
(ii) $\sum_{i=1}^{n} x_i(\omega) = \sum_{i=1}^{n} e_i(\omega)$ μ – *a.e.*

(iii) It is not true that there exist $S \subset \{1, 2, ..., n\}$ and $y : \Omega \to \Pi_{i \in S} X_i$, such that $y_i - e_i$ is $\bigvee_{i \in S} \mathcal{F}_i$-measurable for all $i \in S$, $\sum_{i \in S} y_i(\omega) = \sum_{i \in S} e_i(\omega)$ μ – *a.e.* and $\int u_i(y_i(\omega))d\mu(\omega) > \int u_i(x_i(\omega))d\mu(\omega)$ for all $i \in S$.

Since conditions (i) and (ii) are the same with those in Definition 3.1 we only need to interpret condition (iii). The latter condition says that no coalition of agents can redistribute their own initial endowment using their pooled information and make every member in the coalition better off. Formally, since each net trade $y_i - e_i$ is $\bigvee_{i \in S} \mathcal{F}_i$-measurable for each $i \in S$, we can conclude that net trades are now based on the pooled information of the coalition. It was remarked in Yannelis (1991, p. 188) that the above fine core may be empty.[5] In Section 5 we show by means of an example (which satisfies all the assumptions of Theorem 3.1) that indeed the fine core may be empty.

If condition (i) in Definition 3.3 is replaced by:

(i′) Each x_i is $\bigvee_{i=1}^{n} \mathcal{F}_i$-measurable,

then we will call such a core notion a *weak fine core* allocation for $\mathcal{E}$.[6]

The theorem below indicates that a weak fine core allocation exists in $\mathcal{E}$.

Theorem 3.3. *Let $\mathcal{E} = \{(X_i, u_i, e_i, \mathcal{F}_i, \mu) : i = 1, 2, ..., n\}$ be an exchange economy with differential information satisfying for each $i (i = 1, 2, ..., n)$ all the assumptions of Theorem 3.1. Then a weak fine core allocation exists in $\mathcal{E}$.*

Proof. It follows directly from Theorem 3.1 as follows. For $S \subset \{1, 2, ..., n)$ denote $\bigvee_{i \in S} \mathcal{F}_i$ by $\mathcal{F}^S$. Define L_{X_i} as: $L_{X_i} = \{x_i : x_i : \Omega \to Y$ is Bochner integrable, $\bigvee_{S \subset I} \mathcal{F}^S$-measurable and $x_i(\omega) \in X_i(\omega)$ μ – *a.e.*$\}$. By Theorem 3.1 there exists on $x \in \Pi_{i=1}^{N} L_{X_i}$, such that:

(i) $\sum_{i=1}^{n} x_i(\omega) = \sum_{i=1}^{n} e_i(\omega)$ μ – *a.e.*

(ii) It is not true that there exist S and $(y_i)_{i \in S} \in \Pi_{i \in S} L_{x_i}$ such that $\sum_{i \in S} y_i(\omega) = \sum_{i \in S} e_i(\omega)$ μ – *a.e.* and $\int u_i(y_i)d\mu > \int \mu_i(x_i)d\mu$ for all $i \in S$.

Observe that since $x \in \Pi_{i=1}^{n} L_{X_i}$ each x_i is $\bigvee_{i=1}^{n} \mathcal{F}_i$-measurable. (ii) implies that condition (iii) of Definition 3.3 holds. To see this, suppose otherwise than there exist $S \subset I$ and $y : \Omega \to \Pi_{i \in S} X_i$ such that $y_i - e_i$ is $\bigvee_{i \in S} \mathcal{F}_i$-measurable [7] for all $i \in S$, $\sum_{i \in S} y_i(\omega) = \sum_{i \in S} e_i(\omega)$ μ – *a.e.* and $\int u_i(y_i)d\mu > \int u_i(x_i)d\mu$ for all $i \in S$. Since y_i is $\bigvee_{i \in S} \mathcal{F}_i$-measurable it is also $\bigvee_{S \subset I} \mathcal{F}^S$-measurable and therefore $y_i \in L_{X_i}$ for all $i \in S$, a contradiction to condition (ii) above. Hence, we conclude that $x : \Omega \to \Pi_{i=1}^{n} X_i$ is a weak fine core allocation for $\mathcal{E}$ and this completes the proof.

[5] Wilson (1978) has already shown by means of an example that his core notion may be empty. However, his example is not entirely consistent with the above notion. Recall that in the Wilson setting allocations and endowments are not necessarily measurable. In a public finance setting Berliant (1992) has also shown that a fine core-type notion may not exist.

[6] Clearly the set of all fine core allocations for $\mathcal{E}$ is a strict subset of the set of weak fine core allocations for $\mathcal{E}$. Also, notice that in the definitions of the fine, coarse, and weak fine core the measurability of the final allocation is equivalent to the measurability of the net trades. However, this is not the case for the strong coarse core.

[7] Note that $y_i - e_i$, $\bigvee_{i \in S} \mathcal{F}_i$-measurable for all $i \in S$ is equivalent (recall that e_i is $\mathcal{F}_i$-measurable) to the fact that y_i is $\bigvee_{i \in S} \mathcal{F}_i$- measurable for all $i \in S$.

It should be noted that our definition of a weak fine core allocation for $\mathcal{E}$ corresponds to Allen's (1991) fine core allocation for $\mathcal{E}$, who has proved a less general version of Theorem 3.3.

4 Truthful revelation of information in the core

One of the basic questions that one may ask is whether the core notions defined in the previous section are coalitionally incentive compatible. That is, is there truthful revelation of information in each coalition? We show below that indeed the private, the coarse, and the fine cores are incentive compatible. We define rigorously below a notion of incentive compatibility which was introduced in Krasa-Yannelis (1991).

Definition 4.1. A feasible allocation is said to be *coalitionally incentive compatible* if and only if the following does not hold: There exists coalition $S \subset I$ and two states a and b that members of I/S cannot distinguish (i.e., a and b are in the same event of the partition for every agent not in the coalition S) and such that members of S are better off by announcing b whenever a has actually occurred. Formally, the feasible allocation $x : \Omega \to \Pi_{i=1}^{n} X_i$ is said to be *coalitional incentive compatible* for $\mathcal{E}$ if it is not true that we can find a coalition S and states a, b with $a \in E_i(b)$ for every $i \notin S$, such that $u_i(e_i(a) + x_i(b) - e_i(b)) > u_i(x_i(a))$ for all $i \in S$.

It turns out that in the case of one commodity per state, if preferences are monotone then the $\mathcal{F}_i$-measurability of a feasible allocation implies that the allocation is also coalitionally incentive compatible. For the result below for each i, X_i is a set-valued function from Ω to $\mathbb{R}_+$, i.e., there is only one good per state.

Proposition 4.1. *Let $x : \Omega \to \Pi_{i=1}^{n} X_i$ be a feasible allocation for $\mathcal{E}$. Suppose that:*

(i) Each x_i is $\mathcal{F}_i$-measurable.
(ii) For any y, z in $\mathbb{R}_+$ and for each $i \in I$, if $y > z$ then $u_i(y) > u_i(z)$ (monotonicity). Then the allocation x is coalitionally incentive compatible.

Proof. Suppose otherwise, then there exist $S \subset I$ and $a, b, a \in E_i(b)$ for all $i \notin S$ such that

$$u_i(e_i(a) + x_i(b) - e_i(b)) > u_i(x_i(a)) \text{ for all } i \in S. \tag{1}$$

Since x is feasible it follows that

$$\sum_{i \in S}(x_i(\cdot) - e_i(\cdot)) = \sum_{i \notin S}(x_i(\cdot) - e_i(\cdot)).$$

Since by definition the initial endowment of each agent is $\mathcal{F}_i$-measurable and by assumption (i) each x_i is $\mathcal{F}_i$-measurable, it follows that $x_i - e_i$ $\mathcal{F}_i$-measurable and consequently we can conclude that for any coalition $T \subset I$, the sum $\sum_{i \in T}(x_i(\cdot) - e_i(\cdot))$ is $\bigvee_{i \in T} \mathcal{F}_i$-measurable. Since $a \in E_i(b)$ for every $i \notin S$ it follows that $a \in \bigcap_{i \notin S} E_i(b)$. Clearly $\bigcap_{i \notin S} E_i(b)$ is an element of $\bigvee_{i \notin S} \mathcal{F}_i$. By the above reasoning the sum $\sum_{i \notin S}(x_i(\cdot) - e_i(\cdot))$ is $\bigvee_{i \notin S} \mathcal{F}_i$-measurable and therefore we can conclude that

$$\sum_{i \notin S}(x_i(a) - e_i(a)) = \sum_{i \notin S}(x_i(b) - e_i(b)). \tag{2}$$

Hence,

$$\begin{aligned}\sum_{i \in S}(x_i(a) - e_i(a)) &= \sum_{i \notin S}(x_i(a) - e_i(a)) \\ &= -\sum_{i \notin S}(x_i(b) - e_i(b)) \text{ (recall (2))} \\ &= \sum_{i \in S}(x_i(b) - e_i(b)). \end{aligned} \tag{3}$$

We now show that $x_i(a) - e_i(a) = x_i(b) - e_i(b)$ for all $i \in S$. Suppose otherwise, i.e., $x_i(a) - e_i(a) \neq x_i(b) - e_i(b)$ for some $i \in S$. Without loss of generality we may assume that $x_i(b) - e_i(b) > x_i(a) - e_i(a)$ for some $i \in S$. It follows from (3) that $x_j(a) - e_j(a) > x_j(b) - e_j(b)$ for some agent $j \in S$. Since $x_j(a) > x_j(b) - e_j(b) + e_j(a)$ it follows from assumption (ii) that $u_j(x_j(a)) > u_j(x_j(b) - e_j(b) + e_j(a))$ for some $j \in S$, a contradiction to (1). Hence, we conclude that $x_i(a) - e_i(a) = x_i(b) - e_i(b)$ for all $i \in S$. But then $u_i(e_i(a) + x_i(b) - e_i(b)) = u_i(e_i(a) + x_i(a) - e_i(a)) = u_i(x_i(a))$ for all $i \in S$, a contradiction to (1). This completes the proof of the proposition.

It follows now directly from Proposition 4.1 that any private, coarse, or fine core allocation of the one commodity per state economy $\mathcal{E}$ is coalitionally incentive compatible provided that preferences are monotone. The following Corollary of Proposition 4.1 holds:

Corollary 4.1. *Let $\mathcal{E} = \{(X_i, u_i, e_i, \mathcal{F}_i \mu) : i = 1, 2, ..., n\}$ be an exchange economy with differential information satisfying assumption (ii) of Theorem 4.1. If $x : \Omega \to \Pi_{i=1}^n X_i$ is either a private, a coarse or a fine core allocation for $\mathcal{E}$, then x is coalitionally incentive compatible.*

We focus now on the private core of $\mathcal{E}$ and show that it is always coalitionally incentive compatible provided that preferences are monotone. The one good per state assumption is now dropped. Before stating the main result of this section, we modify Definition 4.1 to permit for more than one good per state.

Definition 4.2. A feasible allocation $x : \Omega \to \Pi_{i=1}^n X_i$ is said to be *weak coalitionally incentive compatible* for $\mathcal{E}$, if it is not true that there exist coalition S and states a, b in Ω such that:

(i) $E_i(a) \in \bigwedge_{i \in S} \mathcal{F}_i, \mu(E_i(a)) > 0$,
(ii) $a \in E_i(b)$ for $i \notin S$, and
(iii) $u_i(e_i(a) + x_i(b) - e_i(b)) > u_i(x_i(a))$ for all $i \in S$.

This notion of incentive compatibility states that it is not possible for any coalition S to become better off by announcing a false state, which agents not in the coalition S cannot distinguish from the true state. Conditions (ii) and (iii) are the same as in Definition 4.1. Condition (i) says that the members of the coalition S should agree on whether a state has occurred. In other words, the event containing the realized (misreported) state a, i.e., $E_i(a)$ is known to every member of the coalition. Thus, there is no "double cross" among the members of a coalition that they agree to lie. The condition $\mu(E_i(a)) > 0$ shows that there is a non-negligible possibility for misreporting.

We are now ready to state the main result of this section:

Theorem 4.1. *Let $\mathcal{E} = \{(X_i, u_i, e_i, \mathcal{F}_i, \mu) : i = 1, 2, ..., n\}$ be an exchange economy with differential information satisfying for each $i (i = 1, 2, ..., n)$, all the assumptions of Theorem 3.1. Moreover, suppose that preferences are monotone. Then any private core allocation for $\mathcal{E}$ is weak coalitionally incentive compatible.*

Proof. Let x be a private core allocation for $\mathcal{E}$. For each i, define $z_i : \Omega \to X_i$ by $z_i(\cdot) = x_i(\cdot) - e_i(\cdot)$.

Suppose that x is not weakly coalitionally incentive compatible. Then there exist $S \subset I$ and $a, b \in \Omega$ such that:

(i) $E_i(a) \in \bigwedge_{i \in S} \mathcal{F}_i, \mu(E_i(a)) > 0$,
(ii) $a \in E_i(b), i \notin S$, and
(iii) $u_i(e_i(a) + z_i(b)) > u_i(x_i(a))$ for all $i \in S$.

First notice that since $e_i(\cdot)$ and $x_i(\cdot)$ are $\mathcal{F}_i$-measurable, (iii) implies that for all $i \in S, u_i(e_i(t) + z_i(b)) > u_i(x_i(t))$ for all $t \in E_i(a)$.

Since $a \in E_i(b)$ for all $i \notin S$ we have that $a \in \bigcap_{i \notin S} E_i(b)$. Clearly $\bigcap_{i \notin s} E_i(b) \in \bigvee_{i \notin S} \mathcal{F}_i$.

We know that $\sum_{i \notin S} z_i(\cdot)$ is $\bigvee_{i \notin S} \mathcal{F}_i$-measurable and since $a \in \bigcap_{i \notin S} E_i(b)$ we conclude that $\sum_{i \notin S} z_i(a) = \sum_{i \notin S} z_i(b)$. By the feasibility of x we have that $\sum_{i \in s} z_i(a) = \sum_{i \notin S} z_i(a)$ and thus

$$\sum_{i \in S} z_i(a) = \sum_{i \in S} z_i(b). \tag{4}$$

Consider now the following net trades:[8]

$$z_i^*(t) = z_i(t)\chi_{\Omega \setminus E_i(a)} + z_i(b)\chi_{E_i(a)} \text{ for } i \in S. \tag{5}$$

The above net trades are $\mathcal{F}_i$-measurable (since each z_i is $\mathcal{F}_i$-measurable) and feasible. Indeed, since $E_i(a) \in \bigvee_{i \in S} \mathcal{F}_i$ it follows that for $t \in E_i(a), z_i(t) = z_i(a)$ for all $i \in S$. Hence, if $t \notin E_i(a), \sum_{i \in S} z_i^*(t) + \sum_{i \in S} z_i(t) = \sum_{i \in S} z_i(t) = 0$ (recall the feasibility of x).

If $t \in E_i(a), \sum_{i \in S} z_i^*(t) + \sum_{i \notin S} z_i(t) = \sum_{i \in S} z_i(b) + \sum_{i \notin S} z_i(t) = \sum_{i \in S} z_i(b) - \sum_{i \in S} z_i(t) = \sum_{i \in S} z_i(b) - \sum_{i \in S} z_i(a) = 0$ (by (3)). We can now construct the following allocation. For each i, $(i = 1, 2, ..., n)$ let

$$x_i^*(\cdot) = \begin{cases} e_i(\cdot) + z_i^*(\cdot), \, i \in S \\ e_i(\cdot) + z_i(\cdot), \, i \notin S \, . \end{cases}$$

[8] The symbol χ below denotes the characteristic function. See also Section 7.

Notice that for all $i \in S, u_i(x_i^*(t)) = u_i(e_i(t)+z_i(b)) > u_i(x_i(t))$ for $(t \in E_i(a))$. Since $\mu(E_i(a)) > 0$, we have that

$$\int u_i(x_i^*(t))d\mu(t) > \int u_i(x_i(t))d\mu(t) \quad \text{for all } \; i \in S, \; \text{ and}$$
$$\int u_i(x_i^*(t))d\mu(t) = \int u_i(x_i(t))d\mu(t) \quad \text{for all } \; i \notin S.$$

Since $\int u_i(\cdot)$ is norm-continuous [by virtue of the Lebegue Dominated Convergence Theorem], by choosing $A \in \bigwedge_{i\in I} \mathcal{F}_i$ with $\mu(A) > 0$, we can find a function $\varepsilon \cdot \chi_A$, where $\varepsilon \geqq 0$, such that for ε sufficiently small, $\|\varepsilon\chi_A\| < \delta$ so that

$$\int u_i(x_i^* - \varepsilon\chi_A)d\mu > \int u_i(x_i)d\mu \;\text{ for all } \; i \in S. \tag{6}$$

By monotonicity of preferences for $i \notin S, u_i(x_i^*(t) + \frac{1}{|I/S|}\varepsilon \cdot \chi_A) > u_i(x_i(t))$ for $t \in A$. Since $\mu(A) > 0$ we have that:

$$\int u_i\left(x_i^* + \frac{1}{|I/S|}\varepsilon\chi_A\right)d\mu > \int u_i(x_i)d\mu \quad \text{for all} \quad i \notin S. \tag{7}$$

Hence, the allocation:

$$x_i^{**} = \begin{cases} x_i^* - \varepsilon\chi_A, & i \in S \\ x_i^* + \dfrac{1}{|I/S|}\varepsilon\chi_A, & i \notin S \end{cases}$$

is $\mathcal{F}_i$-measurable (since it is x_i^* perturbed over a measurable set), it is feasible for the grand coalition and it follows from (5) and (6) that $\int u_i(x^{**})d\mu > \int u_i(x)d\mu$ for all i, $(i = 1, 2, ..., n)$, a contradiction to the fact that x has been assumed to be a private core allocation for $\mathcal{E}$. This completes the proof of the theorem.

Remark 4.1. Although the theorem says that any private core allocation is weakly coalitionally incentive compatible, the proof shows that a stronger result is true, i.e., any private Pareto optimal allocation [See Yannelis (1991), Definition 3.1.2, p. 188] will be weakly coalitionally incentive compatible as well.

5 Interpretation of the private coarse and fine core allocations

In an economy with differential information it is reasonable to expect that an agent with even a zero initial endowment but better (finer) private information than all other agents that matters to the rest of the agents, should be able to exchange his/her superior private information for actual goods. Obviously, this is not the case if we adopt as an equilibrium notion the traditional Walrasian equilibrium (i.e., any rational expectations equilibrium notion). In particular, in the Walrasian equilibrium if an agent has no initial endowment, even if his/her information is better and essential to all the other agents, he/she always ends up with zero consumption. (To see this simply note that in any Walrasian equilibrium notion this agent will have to maximize his/her expected utility conditional on his/her own private information, subject to a budget set which is zero.) We believe that for an equilibrium notion to be suitable in a differential information economy framework it should be able to reward an agent with superior information provided that the information matters to the rest of the agents (even if this agent has no endowment of physical good). The example below demonstrates that this is the case for the private core and the coarse core, but not for the fine core.

Example 5.1. Consider an economy with three agents denoted by J, K, L and three states of nature denoted by a, b, c. There is only one good in each state. The random initial endowments of the agents are given as follows: Agent J's is (10, 10, 0), agent K's is (10, 0, 10) and agent L's is (0, 0, 0). Their private information sets are: $\mathcal{F}_J = \{\{a, b\}, \{c\}\}, \mathcal{F}_K = \{\{a, c\}, \{b\}\}$ and $\mathcal{F}_L = \{\{a\}, \{b, c\}\}$. All agents have the same utility function given by $\sqrt{x}$ and each state occurs with the same probability. We first analyse the private core.

5.1 The private core

The above example satisfies all the assumptions of Theorem 3.1 and therefore a private information core allocation exists in this three-person exchange economy with differential information. We show that in a private core allocation agent L will have positive consumption. First note that since the net trade of each agent must be $\mathcal{F}_i$-measurable, J and K together cannot make any beneficial trades, i.e., the only trades possible between J and K are state independent and these trades do not make them better off. However, the participation of agent L in the economy makes everybody better off. In fact it can be easily checked that the following allocation

$$x_i^* = \begin{cases} (8, 8, 2) & \text{for} \quad i = J \\ (8, 2, 8) & \text{for} \quad i = K \\ (4, 0, 0) & \text{for} \quad i = L, \end{cases} \tag{5.1}$$

is a private core allocation, i.e., x^* is $\mathcal{F}_i$-measurable, feasible and it cannot be dominated by any coalition.

In this example, agents J and K cannot undertake any risk sharing without agent L. Since agent L has superior information he/she acts as an intermediary who executes the correct trades and as a consequence gets rewarded for this service. All three agents are better off after trade has taken place (simply note that x_i^* for $i = J, K, L$ is individually rational, i.e., $\int u_i(x^*) d\mu \gneqq \int u_i(e) d\mu$). In sharp contrast with the core notion, if we had adopted the Walrasian equilibrium, then agent L would have obtained zero consumption since he/she started with zero initial endowments. This holds no matter whether or not agent L's information is useful to the other agents.

If the private information of agent L were not useful to agents J and K then agent L would have obtained zero consumption. For instance if the private information set of agent L is $\mathcal{F}_L = \{a, b, c\}$ then the initial endowment is the unique private core allocation. This result is quite interesting because our example indicates that the private core takes into account the information advantage or superiority of an individual in an explicit way. It is exactly for this reason that we believe that the private core serves to provide more plausible outcomes than the Walrasian equilibrium.

5.2 Coarse core

Let us now examine the coarse core allocation. We know that any private core allocation is also a coarse core allocation. Hence the private core allocation (5.1) is a coarse core allocation as well.[9] To show that the coarse core allocation takes into account the superiority or information advantage of an agent, simply observe that if agent L's private information in Example 5.1 were the trivial partition, i.e., $\mathcal{F}_L = \{a, b, c\}$, then the initial endowment is the only coarse core allocation, where agent L receives zero in all states.

The *strong coarse core* in this example is empty. Simply note that the strong coarse core allocation must be $(\mathcal{F}_J \bigwedge \mathcal{F}_K \bigwedge \mathcal{F}_L)$-measurable, but the meet of these three partitions is the trivial partition, i.e., $\mathcal{F}_J \bigwedge F_K \bigwedge \mathcal{F}_L = \{a, b, c\}$. This implies that the consumption of each agent must be the same in each state. However, given the structure of the initial endowments it is easily seen that no feasible allocation can give to each agent the same consumption in all states and dominate the initial endowments. Moreover the initial endowment state is not a strong coarse allocation since it is not $(\mathcal{F}_J \bigwedge \mathcal{F}_K \bigwedge \mathcal{F}_L)$-measurable. Hence the strong coarse core is empty in this example.

5.3 Fine core

We now show that the fine core is empty in Example 5.1. To see this, note that any[10] $\mathcal{F}_i$-measurable allocation $(i = J, K, L)$ which is beneficial to agents J and K can be achieved only through agent L and this agent ends up with positive consumption (e.g., the allocation in (5.1)) in state $\{a\}$. Since improvements for agents J and K can be made with $(\mathcal{F}_J \bigvee \mathcal{F}_K)$-measurable allocations and $\{a\}$ belongs to $\mathcal{F}_J \bigvee \mathcal{F}_K = \{\{a\}, \{b\}, \{c\}\}$, it follows that all $\mathcal{F}_i$-measurable $(i = J, K, L)$ allocations are blocked by the coalition $\{J, K\}$ which in turn can share agent L's consumption in state $\{a\}$, e.g., the allocation in (5.1) can be dominated by the following allocation:

$$y_i^* = \begin{cases} (10, 8, 2) & \text{for} \quad i = J \\ (10, 2, 8) & \text{for} \quad i = K \\ (0, 0, 0) & \text{for} \quad i = L. \end{cases}$$

Hence, we can conclude that the fine core is empty.

[9] In fact it can be shown that in this example all $\mathcal{F}_i$-measurable $(i = J, K, L)$ allocations x which are individually rational (i.e., $\int u(x) d\mu \geqq \int u(e) d\mu$) constitute the coarse core.

[10] Except from the initial endowment which is dominated by the allocation in (5.1).

However, the *weak fine core* exists. It can be easily checked that the allocation

$$y_i = \begin{cases} (10,5,5) & \text{for} \quad i = J, K \\ (0,0,0) & \text{for} \quad i = L \end{cases}$$

is in the weak fine core. However, it is *not* incentive compatible. Simply note that if state a occurs then agent J can become better off by reporting state c and the latter state is not distinguishable from a by agent K. In particular, $u_J(e_J,(a) + x_J(c) - e_J(c)) = u_J(10+5) > u_J(x_J(a)) = u_J(10)$. Using the same reasoning the reader can verify that agent K can become better off by reporting state b whenever state a occurs.

In contrast with the private core, the information superiority of agent L is *not* taken into account by the fine core. Indeed, if agent L's partition is either $\mathcal{F} = \{a,b,c\}$ or $\mathcal{F}_L = \{\{a\},\{b\},\{c\}\}$ the above weak fine core allocation remains unchanged (compare with the private core). This result should not be surprising since whenever agents in a coalition pool their own information, any informational advantage that an agent may have disappears.

5.4 The unequal treatment of the private core

In Example 5.1 the agent with zero initial endowments and superior information useful to the rest of the economy facilitated the trades, i.e., he/she served as an intermediary. Obviously, by executing the correct trades he/she made all other agents better off (Pareto improvement) and was compensated for this service by consuming some of the goods.

We now provide an example with two intermediaries.

Example 5.4.1. Consider the Example 5.1 with one additional agent M whose initial endowment is zero in each state, he/she has the same utility function with the other three agents, i.e., $\sqrt{x}$ and let his/her private information set be $\mathcal{F}_M = \{\{a\},\{b\},\{c\}\}$. (Agent's J, K, L initial endowments and partitions remain the same as in Example 5.1.)

Clearly, the above four-person economy with differential information satisfies all the assumptions of Theorem 3.1 and therefore a private core allocation exists. We will show that agents L and M can serve as intermediaries. One can easily check that the allocation:

$$x_i^* = \begin{cases} (8,8,2) & \text{for} \quad i = J \\ (8,2,8) & \text{for} \quad i = K \\ \ell_1,0,0) & \text{for} \quad i = L, \ell_1 \geqq 0 \\ (m_1,0,0) & \text{for} \quad i = M, m \geqq 0 \end{cases}$$

where $\ell_1 + m_1 = 4$, is a private core allocation. Obviously either agent L or M may serve as an intermediary or both may serve simultaneously. Their final allocation in state a depends on the extent that their information was used to carry out the trades. Note that even if agent M has the same partition as agent L, i.e.,

$\mathcal{F}_M = \{\{a\}, \{b, c\}\} = \mathcal{F}_L$, the set of private core allocations remains the same. Hence, one can conclude that there is *no equal treatment*, i.e., agents with identical characteristics (utility function, initial endowment, and private information set) may receive different utility private core allocations.[11] Notice that the value allocation of an economy with differential information as defined in Krasa-Yannelis (1991) does not have the equal treatment property as well [just endow each agent in the Scafuri–Yannelis (1984) example with the full information partition].

5.5 Independence of private information sets

It should be noted that whether trade takes place or not depends crucially on the structure of the information (or the initial endowments) in a private information economy. In particular, we show below that in a differential information economy with one good per state if the private information sets of the agents are independent (a term which is defined below) then trade does not take place. Obviously, in this case there is no need for an intermediary. Hence, we can conclude that a sufficient condition for trade to take place is that the private information sets should not be independent.

We begin with a definition.

Definition 5.2. Let $\mathcal{E} = \{(X_i, u_i, e_i, \mathcal{F}_i, \mu) : i = 1, 2, ..., n\}$ be a private information economy: We say that $\mathcal{F}_i$ is *independent* of $\mathcal{F}_j, i \neq j, (i, j = 1, 2, ..., n)$ if $\mu(A \cap B) = \mu(A) \cdot \mu(B)$ for $A \in \mathcal{F}_i$ and $B \in \mathcal{F}_j$.

It can be easily shown that the above definition implies *independence of the initial endowments* (recall that each e_i is $\mathcal{F}_i$-measurable), i.e.,

$$\int [e_i(\omega) \cdot e_j(\omega)] d\mu(\omega)$$
$$= \int e_i(\omega) d\mu(\omega) \cdot \int e_j(\omega) d\mu(\omega) \text{ for } i \neq j, \quad (i, j = 1, 2, ..., n).$$

We are ready now to state the following proposition.

Proposition 5.2. *Let $\mathcal{E}$ be an exchange economy with private information satisfying all the assumptions of Theorem 3.1. Moreover, suppose that there is only one good per state, preferences are monotone, and that each $\mathcal{F}_i$ is independent of $\mathcal{F}_j, i \neq j$. Then the unique private core allocation is the initial endowment.*

The proof of this proposition follows directly from the following two lemmata. Indeed if there is a private information core allocation say $x^*(\cdot) = (x_1^*(\cdot), ..., x_n^*(\cdot))$ (other than the initial endowment), then by setting for each $i \in I, z_i(\omega) = x_i^*(\omega) - e_i(\omega)$ μ – *a.e.* and letting $I = S$ in Lemma 5.2 below we can conclude that $\int e_i(\omega) d\mu(\omega) > \int x_i^*(\omega) d\mu(\omega)$, i.e., x^* is not individually rational.

Lemma 5.1. *Let $\mathcal{E} = \{(X_i, u_i, e_i, \mathcal{F}_i, \mu) : i = 1, 2, ..., n\}$ be an exchange economy with differential information satisfying the assumptions of Proposition 5.2.*

[11] In particular, the final allocation of each intermediary depends an the volume of trades that they carry through.

Consider a coalition S whose members have independent partitions. If for each $i \in S, z_i : \Omega \to X_i$ is the net trade of agent i which is $\mathcal{F}_i$-measurable and $\sum_{i \in S} z_i(\omega) = 0$ μ – a.e., then for each $i \in S$ either $z_i(\omega) < 0$ μ – a.e., or $z_i(\omega) \geqq 0$ μ – a.e.

Proof. Choose an agent i in S. Since $\sum_{i \in S} z_i(\cdot) = 0$ it follows that

$$z_i(\cdot) = -\sum_{\substack{j \neq i \\ j \in S}} z_j(\cdot) \tag{5.4.1}$$

Since each z_j is $\mathcal{F}_j$-measurable it follows that $-\sum_{\substack{j \neq i \\ j \in S}} z_j(\cdot)$ is $\bigvee_{\substack{j \neq i \\ j \in S}} \mathcal{F}_i$-measurable and therefore by virtue of (5.4.1) we can conclude that $z_i(\cdot)$ is $\bigvee_{\substack{j \neq i \\ j \in S}} \mathcal{F}_i$-measurable. Since $z_i(\cdot)$ is $\bigvee_{\substack{j \neq i \\ j \in S}} \mathcal{F}_j$-measurable the set $z_i^{-1}([0, \infty]) = \{\omega \in \Omega : z_i(\omega) \in [0, \infty]\} = \{\omega \in \Omega : z_i(\omega) \geqq 0\}$ belongs to $\mathcal{F}_i \cap (\bigvee_{\substack{j \neq i \\ j \in S}} F_j)$. Since $\mathcal{F}_i$ is independent of $\mathcal{F}_j i \neq j, i, j$ in S, it follows that $\mathcal{F}_i$ is independent of $\bigvee_{\substack{j \neq i \\ j \in S}} \mathcal{F}_j$. Since $z_i^{-1}[0, \infty)$ is $\mathcal{F}_i \cap (\bigvee_{\substack{j \neq i \\ j \in S}} F_j)$-measurable, $z_i^{-1}[0, \infty)$ is independent to itself. Hence, $\mu(z_i^{-1}[0, \infty] \cap z_i^{-1}([0, \infty])) = \mu(z_i^{-1}[0, \infty]) - \mu(z_i^{-1}[0, \infty])$ and so $\mu(z_i^{-1}[0, \infty]) = [\mu(z_i^{-1}[0, \infty])]^2$. The latter enable us to conclude that either $\mu(z_i^{-1}[0, \infty]) = 0$ or $\mu(z_i^{-1}[0, \infty]) = 1$, i.e., either for μ – *a.e.* $z_i(\omega) < 0$ or $z_i(\omega) \geqq 0$ μ – *a.e.*

Lemma 5.2. *Let $\mathcal{E}$ be an exchange economy with differential information satisfying all the assumptions of Proposition 5.2. Then given a coalition S where net trades $z_i(\cdot)$ are $\mathcal{F}_i$-measurable for all $i \in S$ and $\sum_{i \in S} z_i(\omega) = 0$ μ – a.e.*, no trades can be beneficial to all i in S (i.e., $\int u_i(e_i + z_i)d\mu < \int u_i(e_i)d\mu$ for some i in S).

Proof. By Lemma 5.2 for each $i \in S$, either $z_i(\omega) \geqq 0$ or $z_i(\omega) < 0$ μ – *a.e.* We will show that whenever $z_i(\cdot)$ is either positive or negative for some i in S it will violate individual rationality. Since the case where $z_i(\omega) = 0$ μ – *a.e.* is trivial, we will only prove the case that for some $i \in S, z_i(\omega) > 0$ μ – *a.e.* (The case where for some $i \in S, z_i(\omega) < 0$ μ – *a.e.* can be proved along the same lines.) Since for some $i \in S, z_i(\omega) > 0$ μ – *a.e.* and $\sum_{i \in S} z_i(\omega) = 0$ μ – *a.e.*, it must be the case that for at least one $j \in S, z_j(\omega) < 0$ μ – *a.e.* Hence, $e_j(\omega) + z_j(\omega) < e_j(\omega)$, μ – *a.e.* By monotonieity we have that $u_j(e_j(\omega) + z_j(\omega)) < u_j(e_j(\omega))$ μ – *a.e.* and therefore $\int u_j(e_j(\omega) + z_j(\omega))d\mu(\omega) < \int u_j(e_j(\omega))d\mu(\omega)$. The above inequality violates individual rationality for the agent j in S, and we can conclude that for coalition S, no net trade $z_i(\cdot)$ which is $\mathcal{F}_i$-measurable and $\sum_{i \in S} z_i(\omega) = 0$ μ – *a.e.* is beneficial, to all agents i in S.

6 The α-core of an economy with differential information

In this section we will allow for interdependent preferences (i.e., externalities in consumption). In particular, the economy $\mathcal{E}$ will be identical with that described in Section 3 except that now the utility function of each agent i, is a real valued function defined on $\Pi_{j=1}^{n} Y_j$. Hence the utility function of each agent depends not only on his/her own consumption, but also on the consumption of all other agents. We will denote such an economy by Γ.

We now define the notion of a private α-core for Γ which corresponds to the private α-core strategy for a normal form game which was defined in Yannelis (1991). We will first need some notation. If $S \subset I$ then $(y^S, z^{I\backslash S})$ denotes the vector x in $\Pi_{i \in I} Y_j$ where $x_i = y_i$ if $i \in S$ and $x_i = z_i$ if $i \notin S$.

Definition 6.1. The allocation $x : \Omega \to \Pi_{i=1}^{n} X_i$ is said to be a private α-core allocation for Γ if the following conditions hold:

(i) Each $x_i : \Omega \to X_i$ is $\mathcal{F}_i$-measurable.
(ii) $\sum_{i=1}^{n} x_i(\omega) = \sum_{i=1}^{n} e_i(\omega)$ $\mu - a.e.$
(iii) It is not true that there exist $S \subset I$ and $y : \Omega \to \Pi_{i \in S} X_i$ such that each y_i is $\mathcal{F}_i$-measurable for all $i \in S, \sum_{i \in S} y_i(\omega) = \sum_{i \in S} e_i(\omega)$ $\mu - a.e.$ and $\int u_i(y^S, z^{I\backslash S} d\mu > \int u_i(x) d\mu$ for all $i \in S$ and for any $z^{I\backslash S} \in \Pi_{i \notin S} X_i$, each $z^{I\backslash S}$ is $\mathcal{F}_i$-measurable for all $i \notin S$ and $\sum_{i \notin S} z_i(\omega) = \sum_{i \notin S} e_i(\omega)$ $\mu -$ *a.e.*

Conditions (i) and (ii) have been discussed in Section 3. (iii) says that no coalition can redistribute their initial resources (while each agent in the coalition is allowed to use his/her own private information) and make the expected utility of each member better off for any feasible redistribution of the complimentary coalition.

By replacing condition (iii) of Definition 6.1 by:

(iii′) It is not true that there exist $S \subset I$ and $y : \Omega \to \Pi_{i \in S} X_i$ such that each y_i is $\bigwedge_{i \in S} \mathcal{F}_i$-measurable for all $i \in S, \sum_{i \in S} y_i(\omega) = \sum_{i \in S} e_i(\omega)$ $\mu -$ *a.e.* and $\int u_i(y^S, z^{I\backslash S}) d\mu > \int u_i(x) d\mu$ for all $i \in S$ and for any $z^{I\backslash S} \in \Pi_{i \notin S} X_i$, each $z^{I\backslash S}$ is $\bigwedge_{i \notin S} \mathcal{F}_i$-measurable for all $i \notin S$ and $\sum_{i \notin S} z_i(\omega) = \sum_{i \notin s} e_i(\omega)$ $\mu - a.e.$,

we have the notion of a coarse α-core allocation for Γ. Moreover if the measurability assumptions in (iii′) above on y_i and $z^{I\backslash S}$ are replaced by: each y_i is $\bigvee_{i \in S} \mathcal{F}_i$-measurable for all $i \in S$ and each $z_i^{I\backslash S}$ is $\bigvee_{i \notin S} \mathcal{F}_i$-measurable for all $i \notin S$, we obtain the notion of a *fine α-core allocation* for Γ.

In the next section we prove the existente of private α-core allocations for Γ.

7 The existente of private α-core allocations

We begin with some basic definitions of mathematical nature which will be needed for our existence proof.

7.1 Mathematieal preliminaries

Let $(T, \mathbf{T}, \mu)$ be a finite measure space and X be a Banach spare. Following Diestel-Uhl (1977) the function $f : T \to X$ is called *simple* if there exist $x_1, x_2, ..., x_n$

in X and $\alpha_1, \alpha_2, ..., \alpha_n$ in $\mathbf{T}$ such that $f = \sum_{i=1}^n x_i \chi_{\alpha_i}$, where $\chi_{\alpha_i}(t) = 1$ if $t \in \alpha_i$ and $\chi_{\alpha_i}(t) = 0$ if $t \notin \alpha_i$. A function $f : T \to X$ is said to be μ-measurable if there exists a sequence of simple functions $f_n : T \to X$ such that $\lim_{n\to\infty} \|f_n(t) - f(t)\| = 0$ for almost all $t \in T$. A μ-measurable function $f : T \to X$ is said to be *Bochner integrable* if there exists a sequence of simple functions $\{f_n, : n = 1, 2, ...\}$ such that

$$\lim_{n\to\infty} \int_T \|f_n(t) - f(t)\| d\mu(t) = 0.$$

In this case we define for each $E \in T$ the integral to be $\int_E (t) d\mu(t) = \lim_{n\to\infty} f_n(t) d\mu(t)$. It can be shown [sec Diestel-Uhl (1977), Theorem 2, p. 45] that if $f : T \to X$ is a μ-measurable function then, f is Bochner integrable if and only if $\int_T \|f(t)\| d\mu(t) < \infty$. It is important to note that the Lebegue Dominated Convergence Theorem holds for Bochner integrable functions. In particular, if $f_n : T \to X, (n = 1, 2, ...)$ is a sequence of Bochner integrable functions such that $\lim_{n\to\infty} f_n(t) = f(t)$ μ – *a.e.*, and $\|f_n(t)\| \leqq g(t)$ μ – *a.e.*, (where $g : T \to \mathbb{R}$ is an integrable function), then f is Bochner integrable and $\lim_{n\to\infty} \int_T \|f_n(t) - f(t)\| d\mu(t) = 0$.

For $1 \leqq p < \infty$, we denote by $L_p(\mu, X)$ the space of equivalence classes of X-valued Bochner integrable functions $x : T \to X$ normed by

$$\|x\|_p = \left(\int_T \|x(t)\|^p d\mu(t) \right)^{1\backslash p}.$$

It is a standard result that normed by the functional $\|\cdot\|_p$ above, $L_p(\mu, X)$ becomes a Banach space [see Diestel-Uhl (1977), p. 50]. Recall that a correspondence $\phi : T \to 2^x$ is said to be *integrably bounded if* there exists a map $h \in L_1(\mu, R)$ such that $\sup\{\|x\| : x \in \phi(t)\} \leqq h(t)$ μ – *a.e.*

A Banach space X has the *Radon-Nikodym Property* with respect to the measure space $(T, \mathbf{T}, \mu)$ if for each μ-continuous measure $G : T \to X$ of bounded variation there exists $g \in L_i(\mu, X)$ such that $G(E) = \int_E g(t) d\mu(t)$ for all $E \in \mathbf{T}$. A Banach space X has the Radon-Nikodym property (RNP) if X has the RNP with respect to every finite measure space. Recall now [sec Diestel-Uhl (1977, Theorem 1, p. 98)] that if $(T, \mathbf{T}, \mu)$ is a finite measure space $1 \leqq p < \infty$, and X is a Banach space, then X^* has the RNP if and only if $(L_p, (\mu, X))^* = L_q(\mu, X^*)$ where $\frac{1}{p} + \frac{1}{q} = 1$.

We now collect some basic results on Banach lattices [for an excellent treatment sec Aliprantis-Burkinshaw (1985)]. A *Banach lattice* is a Banach space L equipped with an order relation $\geq$ (i.e., $\geqq$ is a reflexive, antisymmetric and transitive relation) satisfying:

(i) $x \geqq y$ implies $x + z \geqq y + z$ for every z in L,
(ii) $x \geqq y$ implies $\lambda x \geqq \lambda y$ for all $\lambda \geqq 0$,
(iii) for all x, y in L there exists a supremum (least upper bound) $x \bigvee y$ and an infimum (greatest lower bound) $x \bigwedge y$,
(iv) $|x| \geqq |y|$ implies $\|x\| \geqq \|y\|$ for all x, y in L.

If x, y are elements of the Banach lattice L, then we define the order interval $[x, y]$ as follows:

$$[x, y] = \{z \in L : x \leqq z \leqq y\}.$$

Note that $[x, y]$ is norm closed and convex (hence weakly closed). A Banach lattice L is said to have an *order continuous norm* if, $\chi_\alpha \downarrow 0$ in L implies $\|\chi_\alpha\| \downarrow 0$. A very useful result which will play an important role in the sequel is that if L is a Banach lattice then the fact that L has an order continuous norm is equivalent to weak compactness of the order interval $[x, z] = \{y \in L : x \leqq y \leqq z\}$ for every x, z in L [see Aliprantis-Burkinshaw (1985)].

We finally note that Cartwright (1974) has shown that if X is a Banach lattice with an order continuous form (or equivalently X has weakly compact order intervals) then $L_1(\mu, X)$ has weakly compact order intervals, as well. *Cartwright's Theorem* will play a crucial role in our existence proof.

7.2 The private α-core existence proof

The following result provides sufficient conditions which guarantee the existence of a private α-core allocation for Γ where $|I| = 2$. If $|I| > 2$ then the result below is false.

Theorem 7.1. *Let Γ be an exchange economy with differential information satisfying for each i, $(i = 1, 2)$ the following assumptions.*

(a.7.1) $X_i : \Omega \to 2^Y$ is a convex, closed, nonempty valued correspondence,
(a.7.2) $u_i : \Pi_{j=1}^n Y_j \to \mathbb{R}$ is continuous, integrably bounded and concave.

Then a private α-core allocation exists in Γ.

Proof. We first state the α-core existence theorem in Yannelis (1991a) which will play a crucial role in our argument. Let $E = \{(X_i, u_i, e_i) : i = 1, 2\}$ be an exchange economy where

(1) X_i the *consumption set* of agent i is a subset of the positive cone of an ordered Hausdorff linear topological space L, which is endowed with a topology τ which is weaker than the Hausdorff topology on L, τ is a vector space topology having the property that all order intervals in L are τ-compact.
(2) The *utility function* of each agent i, $u_i : \Pi_{j=1}^n X_j \to \mathbb{R}$ is concave and τ-upper semi continuous.
(3) e_i is the *initial endowment* of agent i, where $e_i \in X_i$ for all i.

If E satisfies (1), (2), and (3) it follows from Yannelis (1991a, Theorem 4.1, p. 112) that an α-core allocation exists[12] in E.

We now construct a new economy $\bar{\Gamma}$ as follows: For each $i, i = 1, 2$ let $L_{X_i} = \{x_i \in L_1(\mu, Y) : x_1 : \Omega \to Y$ is Bochner integrable, $\mathcal{F}_i$-measurable and $x_i(\omega) \in X_i(\omega)\ \mu - a.e.\}$. For each i define $v_i : \Pi_{j=1}^n L_{X_j} \to \mathbb{R}$ by $v_i(x) = \int u_i(x(\omega))d\mu(\omega)$. Since by assumption e_i is Bochner integrable $e_i \in L_{X_i}$, (recall that $e_i(\cdot)$ is $\mathcal{F}_i$-measurable and $e_i(\omega) \in X_i(\omega)\ \mu - a.e.$) and therefore L_{X_i} is nonempty. Obviously L_{X_i} it is convex and bounded from below. It follows from Balder-Yannelis (1991) that v_i is weakly upper semicontinuous. Moreover, since u_i is concave so is v_i. We now have a new economy $\bar{\Gamma} = \{(L_{X_i}, v_i, e_i) : i = 1, 2\}$, where

(a) L_{X_i} is the consumption set of agent i,
(b) v_i is the utility function of agent i,
(c) $e_i \in L_{X_i}$, is the initial endowment of agent i.

It can be easily seen that an α-core allocation of $\bar{\Gamma}$ is a private α-core allocation for Γ. Hence all we need to show is that $\bar{\Gamma}$ satisfies the assumptions of Yannelis's (1991a) Theorem (i.e., conditions (1), (2), and (3) above).

Since Y is the positive cone of a Banach Lattice with an order continuous norm, it follows from Cartwright's theorem that order intervals are weakly compact in $L_1(\mu, Y)$. Hence, the topology τ in Yannelis's (1991a) theorem is taken here to be the weak topology, and obviously assumption (1) is satisfied. Also as noted above v_i is concave and weakly upper semicontinuous and $e_i \in L_{X_i}$ for all i. Hence (2) and (3) hold and therefore an α-core allocation exists in $\bar{\Gamma}$. The latter implies that a private α-core allocation exists in Γ. This completes the proof of the theorem.

Remark 7.1. One can easily see that the set of all private α-core allocations for Γ is contained in the set of all coarse α-core allocations for Γ. However, for $|I| = 2$ the coarse, fine and the private α-core allocations coincide. From this observation and Theorem 7.1 we can obtain the following corollary.

Corollary 7.1. *Let Γ be an exchange economy with differential information satisfying for each $i, (i = 1, 2)$ the assumptions of Theorem 7.1. Then a coarse and a fine α-core allocation exists in Γ.*

[12] To be more specific, Yannelis (1991a) allows for preferences which need not be ordered. In particular one only needs to assume that the preference correspondence of agent i, $P_i : \Pi_{j=1}^n X_j \to 2^{\Pi_{i=1}^n X_i}$ satisfies for each i the following assumptions:

(i) $X_i = Y$,
(ii) $x \notin \mathrm{con}P_i(x)$ for all $x \in \Pi_{i=1}^n X_i$ (where con denotes convex hull),
(iii) P_i has τ-open lower sections $\Big($i.e., for every $y \in \Pi_{i=1}^n X_i$, the set $P_i^{-1}(y) = \Big\{x \in \Pi_{i=1}^n X_i : y \in P_i(x)\Big\}$ is τ-open in $\Pi_{i=1}^n X_i\Big)$.

Note now that by defining the preference correspondence $P_i : \Pi_{j=1}^n X_i \to 2^{\Pi_{i=1}^n X_i}$ by $P_i(x) = \{y : u_i(y) > u_i(x)\}$, it follows from the concavity of u_i that $P_i(\cdot)$ is convex valued and clearly $x \notin \mathrm{con}P_i(x) = P(x)$ for all $x \in \Pi_{i=1}^n X_i$. Moreover the τ-upper semicontinuity of u_i implies that P_i has τ-open lower sections. Hence, Theorem 4.1 of Yannelis (1991a) applies to the above setting.

Remark 7.2. Recently Holly (1991) has shown that in an exchange economy (without incomplete information) where the set of agents is greater than two, Yannelis's (1991a) α-core existence theorem ceases to be true. It is straighforward to extend Holly's example to an exchange economy with differential information and show that if $|I| > 2$ then the coarse α-core of Γ, is empty, and therefore so is the private α-core of Γ.

Remark 7.3. If the economy Γ has one good per state, then the reader can easily verify that the private, coarse and fine α-core allocations for Γ are coalitional incentive compatible. The proof of this result is similar with that in Proposition 4.1.

8 Conclusions

The analysis of different core notions in an economy with differential information enables us to draw the following conclusions: The private core appears to be a sensible solution concept; it exists under very mild assumptions, it is coalitionally incentive compatible, and it takes into account the information superiority of an individual. Moreover, our examples indicate that it provides reasonable outcomes especially in situations where the traditional Walrasian equilibrium concept fails to do so. The coarse core appears to have the same properties as the private information core but since the latter concept is a strict subset of the former it does not provide any additional information. As our Example 5.1 indicated the coarse core is "too big". Contrary to the coarse core the fine core is "too small" and generally does not exist. However, whenever it exists (e.g., the weak fine core) it is not coalitionally incentive compatible and it does not take into account the information advantage of an agent. Nonetheless, we believe that the weak fine core may be useful for analyzing situations of adverse selection. We also showed (Sect. 7) that all the above core notions can be easily modified in order to allow for externalities in consumption. Since the private core can be used to explain intermediation, it is our belief that this concept has great potential in the theory of financial and incomplete markets. In particular, the fact that the private core rewards the agents with superior information provides interesting insights into the way that opportunities for financial intermediation or arbitrage arise in economies with differential information.

We conclude by noting that our adoption of a cooperative solution concept (e.g., the private core) to analyse economies with differential information seems to us very appealing. Indeed, in most applications agents cooperate either bilaterally or multilaterally under differential information. Although there is a non-cooperative feature in the private core notion, (i.e., private information sets are not verifiable by each member of a coalition), the resulting allocation is always coalitionally incentive compatible.

9 A comparison with the value allocation

Krasa-Yannelis (1991) examined the cardinal value allocation in an economy with differential information. Specifically, they analyzed the coarse, the fine, and the private value allocation. It was shown that the coarse and the fine value allocations are problematic (as is the case with the coarse and the line core) but the private value allocation is coalitionally incentive compatible and it takes into account the information superiority of an individual. The latter two properties are shared by the private core as shown in this paper. Despite the fact that both concepts have the same appealing properties, (i.e., they are coalitionally incentive compatible and take into account the information superiority of an agent), they redistribute the initial endowments quite differently. In particular, a private value allocation need not be a private core allocation and vice versa. Thus, since the value and the core generate different outcomes, we cannot say whether one concept is better than the other. The decision for choosing the private value over the private core (or vice versa) should be based on the economic behavior that we intend to explain or rationalize. For instance, in modeling economic behavior where the bargaining power of an individual in a private information economy plays an important role the value seems in this situation more suitable that the core.

References

Aliprantis, C.D., Burkinshaw, O.: Positive operators. New York London: Academic Press 1985

Allen, B.: Market games with asymmetric information and nontransferable utility: representation results and the core. CARESS Working Paper 91-09, University of Pennsylvania (1991)

Balder, E.J., Yannelis, N.C.: On the weak continuity of expected utility. Mimeo (1991)

Berliant, M.: On income taxation and the core. J. Econ. Theory **56**, 121–141 (1992)

Boyd, J., Prescott, E.C.: Financial intermediary coalitions. J. Econ. Theory **38**, 211–232 (1986)

Cartwright, D.: The order completeness of some spaces of vector valued functions. Bull Aust. Math. Soc. **11**, 57–61 (1974)

Diestel, J., Uhl, J.J.: Vector measures. In: Mathematical Surveys, vol. 15. Providence, Rhode Island: American Mathematical Society 1977

Holly, C.: Exchange economies can have an empty α-core. Department of Mathematics, University of Illinois at Urbana-Champaign, Working Paper (1991)

Kobayashi, T.: Equilibrium contracts for syndicates with differential information. Econometrica **48**, 1635–1665 (1980)

Krasa, S., Yannelis, N.C.: The value allocation of an economy with differential information. University of Illinois at Urbana-Champaign, Working Paper (1991)

Page, F.H.: Market games with differential information and infinite dimensional commodity spaces: the core. Mimeo (1992)

Radner, R.: Equilibrium under uncertainty. Econometrica **36**, 31–58 (1968)

Scafuri, A.J., Yannelis, N.C.: Non-symmetric cardinal value allocations. Econometrica **52**, 1365–1368 (1984)

Wilson, R.: Information, efficiency, and the core of an economy. Econometrica **46**, 807–816 (1978)

Yannelis, N.C.: The core of an economy with differential information. Econ. Theory **1**, 183–198 (1991)

Yannelis, N.C.: The core of an economy without ordered preferences. In: Khan, M.A., Yannelis, N.C. (eds.) Equilibrium theory in infinite dimensioanl spaces. Berlin Heidelberg New York: Springer 1991a

Efficiency and incentive compatibility in differential information economies★

Guangsug Hahn and Nicholas C. Yannelis
Department of Economics, University of Illinois, Champaign, IL 61820, USA

Received: March 25, 1996; revised September 5, 1996

Summary. We introduce several efficiency notions depending on what kind of expected utility is used (ex ante, interim, ex post) and on how agents share their private information, *i.e.,* whether they redistribute their initial endowments based on their own private information, or common knowledge information, or pooled information. Moreover, we introduce several Bayesian incentive compatibility notions and identify several efficiency concepts which maintain (coalitional) Bayesian incentive compatibility.

JEL Classification Numbers: 020, 226.

1 Introduction

An exchange economy with differential information consists of a finite set of agents, each of whom is characterized by a random (state dependent) utility function, random initial endowment, a private information set, and a prior. In such an economy the definition of efficiency (or Pareto optimality) is not immediate as was first alluded to in seminal papers by Wilson (1978) and by Myerson (1979). (The latter considered the Harsanyi framework rather than an exchange economy with differential information.) In particular, two main problems arise. First, if we assume that agents make agreements (contracts) before the state of nature is realized, it is important to know what kind of expected utility we adopt, *i.e.,* ex ante or interim. Moreover, how does the choice of the expected utility change the outcome? Secondly and most importantly, when all agents make a redistribution of their initial endowments, what kind of information do they use? That is, do they pool their informa-

★ This research has been supported by the Research Board of the University of Illinois at Urbana-Champaign.
Correspondence to: N.C. Yannelis

tion, or do they use common knowledge information, or do they just make redistribution based on their own information?

Before we proceed, it may be useful to accept the fact that there is no single definition of efficiency which universally works for all environments. In fact, since the information is decentralized in differential information economies, the incentive problem becomes a critical issue for a mechanism which allocates resources according to the reports of agents. This problem was first raised by Myerson (1979) and Holmström-Myerson (1983) in the Harsanyi framework. The key point is that an efficient allocation may not be Bayesian incentive compatible, *i.e.,* the set of efficient allocations is much larger than the set of Bayesian incentive compatible allocations. Our main purpose is to focus on notions of efficiency which are incentive compatible.

Several of the interim efficiency concepts that we introduce in this paper are stronger than those of Holmström-Myerson (1983), but we think they are the proper concepts to capture the efficiency idea in differential information economies. The main assumption we impose is that the net trades are private information measurable.[1] If such a condition is not satisfied, *i.e.,* a proposed net trade is not measurable with respect to private information, then it may create incentive problems and contracts may not be viable (see Example 3.1 as well as Example 6.1). Consequently, it is reasonable to impose the private information measurability condition on allocations.[2]

With the private measurability assumption, every feasible allocation turns out to be Bayesian incentive compatible. Indeed, no single agent can lie and become better off simply because if he/she becomes better off by lying, at least one other agent should be worse off by feasibility, which is impossible by the private information measurability.[3] This weak property of Bayesian incentive compatibility suggests that a stronger Bayesian incentive compatibility notion, *coalitional Bayesian incentive compatibility* may be appropriate. The idea is that no coalition can become better off by reporting false events. That is, in terms of game theory, truth-telling is a coalitional (or strong) Nash equilibrium when agents are asked to report their private information events.

Based on the "proper" efficiency notion and the private information measurability condition, we show that any "proper" efficient allocation is coalitionally Bayesian incentive compatible. This means that if we adopt

[1] The endowment is an initial signal of states and every agent has a private information generated by this signal. This means that the private information measurability of net trades is equivalent to that of allocations.

[2] It has been shown in Krasa-Yannelis (1994) that private measurability is a necessary and sufficient condition for coalitional Bayesian incentive compatibility in the one good per state differential information economy.

[3] All agents except the single liar do not distinguish the true state and the false state. The private information measurability assumption implies that their allocations in the false state are the same as in the true state. Since lying does not change the total endowment in the true state, there is no way for the single liar to become better off.

certain efficiency concepts, the incentive issue (individual or coalitional) need not be considered. It should be noted that a Holmström-Myerson type efficiency notion with the private information measurability condition does not have this property.

Finally, we consider an (interim) efficiency notion without the individual measurability assumption and propose a notion of incentive efficiency. This concept corresponds to the interim efficiency concept of Myerson (1979) and Holmström-Myerson (1983) that they have introduced for the Harsanyi framework and it is different from our other interim efficiency concepts. As in Myerson (1979), it is shown to exist whenever the utility functions are affine. This argument favors our earlier concepts of interim efficiency which exist assuming only concavity of the utility functions.

The paper is organized as follows: Section 2 outlines the basic mathematical notation and definitions. The description of the differential information economy is given in Section 3. We propose several concepts of incentive compatibility in Section 4. In Section 5, we define efficiency concepts in differential information and characterize their properties. The relationship between efficiency and incentive compatibility is examined in Section 6. There are some remarks on individual rationality in Section 7. In section 8, We show the existence of individually rational and efficient allocation. Without measurability, incentive efficiency is defined and analyzed in Section 9.

2 Notation and definitions

We begin with some notation and definitions.

2.1 Notation

$|A|$ denotes the number of elements in the set A.
2^A denotes the family of all subsets of A.
$\setminus$ denotes the set theoretic subtraction.
If A is a set, we denote by χ_A the characteristic function having the property that $\chi_A(\omega)$ is one if $\omega \in A$ and it is zero otherwise.

2.2 Definitions

Let $(\Omega, \mathscr{F}, \mu)$ be finite measure space, and X be a Banach space. Following Diestel-Uhl (1977), the function $f : \Omega \to X$ is called *simple* if there exist $x_1, x_2, \ldots, x_n$ in X and $A_1, A_2, \ldots, A_n$ in $\mathscr{F}$ such that $f = \sum_{i=1}^n x_i \chi_{A_i}$. A function $f : \Omega \to X$ is said to be *μ-measurable* if there exits a sequence of simple functions $f_n : \Omega \to X$ such that $lim_{n\to\infty} \|f_n(\omega) - f(\omega)\| = 0$ for almost all $\omega \in \Omega$. A μ-measurable function $f : \Omega \to X$ is *Bochner integrable* if there exists a sequence of simple functions $\{f_n : n = 1, 2, \ldots\}$ such that

$$\lim_{n\to\infty} \int_\Omega \|f_n(\omega) - f(\omega)\| d\mu(\omega) = 0.$$

In this case, we define for each $A \in \mathscr{F}$, the *integral* to be

$$\int_A f(\omega)d\mu(\omega) = \lim_{n\to\infty} \int_A f_n(\omega)d\mu(\omega).$$

It can be shown [see Diestel-Uhl (1977), Theorem 2, p.45] that if $f : \Omega \to X$ is a μ-measurable function, then f is Bochner integrable if and only if $\int_\Omega \|f(\omega)\| d\mu(\omega) < \infty$. It is important to note that the *Dominated Convergence Theorem* holds for Bochner integrable functions. In particular, if $\{f_n : \Omega \to X : n = 1, 2, \ldots\}$ is a sequence of Bochner integrable functions such that $\lim_{n\to\infty} f_n(\omega) = f(\omega)\mu$-*a.e.*, and $\|f_n(\omega)\| \leq g(\omega)\mu$-*a.e.*, where $g : \Omega \to \boldsymbol{R}$ is an integrable function, then f is Bochner integrable and $\lim \int_\Omega \|f_n(\omega) - f(\omega)\| d\mu(\omega) = 0$ [see Diestel and Uhl (1977), Theorem 3, p.45].

Denote by $L_p(\mu, X)$ with $1 \leq p < \infty$ the space of equivalence classes of X-valued Bochner integrable functions $x : \Omega \to X$ normed by

$$\|x\|_p = \left[\int_\Omega \|x(\omega)\|^p d\mu(\omega) \right]^{\frac{1}{p}} < \infty.$$

It is a standard result that normed by the functional $\|\cdot\|_p$ above, $L_p(\mu, X)$ becomes a Banach space [see Diestel-Uhl (1977), p.50].

We will denote by $L_\infty(\mu, X)$ the space of equivalence classes of essentially bounded Bochner integrable functions $x : \Omega \to X$ normed by

$$\|x\|_\infty = \text{ess sup} \|x\| = \inf \{\varepsilon \in \boldsymbol{R}_+ : \mu\{\omega \in \Omega : \|x(\omega)\| > \varepsilon\} = 0\}.$$

Normed by the functional $\|\cdot\|_\infty$, $L_\infty(\mu, X)$ with $1 \leq p < \infty$ becomes a Banach space [Diestel-Uhl (1977, p.50)]. It is well-known that $L_q(\mu, X^*)$ is the dual of $L_p(\mu, X)$, where $1 \leq p \leq \infty$ and $1/p + 1/q = 1$, and the value $\omega \cdot x$ of $x \in L_p(\mu, X)$ at $\omega \in L_q(\mu, X^*)$ is defined by

$$w \cdot x = \int_\Omega [w(\omega) \cdot x(\omega)] d\mu(\omega).$$

Recall that $\sigma(L_p(\mu, X), L_q(\mu, X^*))$ is defined as the weakest topology on $L_p(\mu, X)$ for which a net $x^\lambda \to x$ if and only if $w \cdot x^\lambda \to w \cdot x$ for all $w \in L_q(\mu, X^*)$. We call this topology as *weak topology* and the convergence as *weak convergence.* A function $f : X \to \boldsymbol{R}$ *is weakly upper semicontinuous* if $\lim\sup f(x^\lambda) \leq f(x)$, *weakly lower semicontinuous* if $\lim\inf f(x^\lambda) \geq f(x)$, and *weakly continuous* if it is both weakly upper semicontinous and weakly lower semicontinuous, whenever $x^\lambda \to x$ weakly.

Now we state basic results on Banach lattices [see Aliprantis-Burkinshaw (1985) for details]. A Banach space X is a *Banach lattice* if there exists an ordering $\geq$ on X with the following properties:

(1) $x \geq y$ implies $x + z \geq y + z$ for every $z \in X$,
(2) $x \geq y$ implies $\lambda x \geq \lambda y$ for every $\lambda \in \boldsymbol{R}_+$,
(3) for all $x, y \in X$, there exist a supremum $x \vee y$ and an infimum $x \wedge y$,
(4) $|x| \geq |y|$ implies $\|x\| \geq \|y\|$ for every $x, y \in X$.

For $x, y \in X$, define the *order interval* $[x, y]$ by $[x, y] = \{z \in X : x \leq z \leq y\}$. Note that $[x, y]$ is convex and norm closed, hence weakly closed (Mazur's Theorem). Cartwright (1974) has shown that if X is a Banach lattice with order continuous norm or equivalently has weakly compact order intervals, then $L_p(\mu, X)$ with $1 \leq p < \infty$ has weakly compact order intervals, as well.[4]

All the results of the paper hold true for any Banach space $L_p(\mu, X), 1 \leq p \leq \infty$. However, we will restrict ourselves to $L_1(\mu, X)$.

3 Differential information economies

Below we define the notion of an economy with differential information (or Radner-type economy). Let $(\Omega, \mathscr{F}, \mu)$ be a probability measure space denoting the states of the world and Y be an ordered Banach space denoting the commodity space.[5] An *economy with differential information* is described by $\mathscr{E} = \{(X_i, u_i, \mathscr{F}_i, \mu, e_i) : i \in I\}$, where

(1) $X_i : \Omega \to 2^{Y+}$ is the *random consumption set correspondence of agent* $i \in I$.
(2) $u_i : \Omega \times Y_+ \to \boldsymbol{R}$ is the *random utility function of agent* $i \in I$.
(3) $\mathscr{F}_i$ is a (finite) measurable partition[6] of Ω denoting the *private information of agent* $i \in I$.[7]
(4) μ is a probability measure on Ω denoting the *common prior* of each agent.
(5) $e_i : \Omega \to Y_+$ is an $\mathscr{F}_i$-measurable and Bochner integrable function denoting the *random initial endowment* of agent $i \in I$, where $e_i(\omega) \in X_i(\omega) \mu$-*a.e.*

Let us denote by L_i the set of all $\mathscr{F}_i$-measurable and Bochner integrable functions from Ω to Y, *i.e.*, $L_i = \{x_i \in L_1(\mu, Y) : x_i \text{ is } \mathscr{F}_i\text{-measurable}\}$. Denote by L_{X_i}, the set of all $\mathscr{F}_i$-measurable and Bochner integrable selections from the correspondence X_i, *i.e.*, $L_{X_i}, = \{x_i \in L_1(\mu, Y) : x_i \text{ is } \mathscr{F}_i\text{-measurable and } x_i(\omega) \in X_i(\omega) \mu\text{-}a.e.\}$. Let $L = \prod_{i \in I} L_i$ and $L_X = \prod_{i \in I} L_{X_i}$. We assume that for each $i \in I$ and each $x_i \in Y_+, u_i(\cdot, x_i)$ is integrably bounded. Denote $\bar{e} = \sum_{i \in I} e_i$.

The *ex ante expected utility function* $\overline{V}_i : L_{X_i} \to \boldsymbol{R}$ of agent i is defined by

---

4 $x_\lambda \downarrow 0$ means that x_λ is a decreasing net with $\inf x_\lambda = 0$. A Banach lattice X is said to have an *order continuous norm* if $x_\lambda \downarrow 0$ in X implies $\|x_\lambda\| \downarrow 0$. If X is a Banach lattice, X has an order continuous norm if and only if any order interval is weakly compact.

5 It is important to note that even if we assume that our commodity space $Y = \boldsymbol{R}^\ell$ (where $\boldsymbol{R}^\ell$ is the ℓ-fold Cartesian product of the reals $\boldsymbol{R}$), the space $L_p(\mu, \boldsymbol{R}^\ell), 1 \leq p \leq \infty$ is still infinite dimensional (in view of the continuum of states). Hence, to assume that $Y = \boldsymbol{R}^\ell$ does not change in any way the arguments of the main results of the paper. As a matter of fact, even if we have just one good, i.e., $Y = \boldsymbol{R}$, we will need to work with $L_p(\mu, \boldsymbol{R}), 1 \leq p \leq \infty$ which is an infinite dimensional space.

6 One may assume that $\mathscr{F}_i$ is a sub-σ-algebra of $\mathscr{F}$. The results of the paper remain unaffected.

7 Throughout our analysis, we assume that information partitions $\{\mathscr{F}_i\}_{i \in I}$ are common knowledge in the sense of Aumann.

$$\overline{V}_i(x_i) = \int_\Omega u_i(\omega, x_i(\omega))d\mu(\omega).$$

We call a set of states an event. An event E_i, which is an element of $\mathscr{F}_i$, is the maximal set of states that agent i cannot distinguish. Let $E_i(\omega)$ denote the element of $\mathscr{F}_i$ which contains $\omega \in \Omega$. This means that when a true state ω occurs, agent i knows only that $E_i(\omega)$ occurs instead. Assume that $\mu(E_i(\omega)) > 0$ for every $i \in I$ and every $\omega \in \Omega$. The *interim (conditional) expected utility function* $V_i : \Omega \times L_{X_i} \to \boldsymbol{R}$ of agent i is defined by[8]

$$V_i(\omega \cdot x_i) = \frac{1}{\mu(E_i(\omega))} \int_{E_i(\omega)} u_i(\omega', x_i(\omega'))d\mu(\omega').$$

Lemma 3.1.1: If, for every $i \in I, u_i(\omega, \cdot)$ is continuous for each $\omega \in \Omega$, then

(1) $\overline{V}_i$ is continuous,
(2) $V_i(\omega, \cdot)$ is continuous for each $\omega \in \Omega$.

Proof: Since, for every $i \in I, u_i(\omega, \cdot)$ is continuous for every $\omega \in \Omega$ and $u_i(\cdot, x_i)$ is integrably bounded for every $x_i \in Y_+$, the result follows directly from the Dominated Convergence Theorem [see Diestel-Uhl (1977), Theorem 3, p.45]. □

Lemma 3.1.2: For every $i \in I$, if $u_i(\omega, \cdot)$ is upper semicontinuous and concave for every $\omega \in \Omega, \overline{V}_i$ is weakly upper semicontinous and concave and $V_i(\omega, \cdot)$ is weakly upper semicontinuous and concave for every $\omega \in \Omega$.

Proof: See Theorem 2.8 in Balder-Yannelis (1993). □

Lemma 3.1.3: For every $i \in I, u_i(\omega, \cdot)$ is continuous and affine for every $\omega \in \Omega$ if and only if $\overline{V}_i$ is weakly continuous and $V_i(\omega, \cdot)$ is weakly continuous for every $\omega \in \Omega$.

Proof: See Corollary 2.7 and Corollary 2.9 in Balder-Yannelis (1993). □

Lemma 3.1.4: For every $i \in I$, if u_i is $\mathscr{F}_i$-measurable, then it follows that

$$V_i(\omega, x_i) = u_i(\omega, x_i(\omega)).$$

[8] One could allow agents to have different priors as follows: Let $q_i : \Omega \to \boldsymbol{R}_{++}$ be the *prior* of agent i, which is a Radon-Nikodym derivative of μ having the property that $\int_\Omega q_i(\omega)d\mu(\omega) = 1$. Then the interim expected utility function $V_i : \Omega \times L_{X_i} \to \boldsymbol{R}$ of agent i is defined by

$$V_i(\omega, x_i) = \int_{E_i(\omega)} u_i(\omega', x_i(\omega'))q_i(\omega'|E_i(\omega))d\mu(\omega'),$$

where

$$q_i(\omega'|E_i(\omega)) = \begin{cases} q_i(\omega')/\int_{E_i(\omega)} q_i(s)d\mu(s) \text{if } \omega' \in E_i(\omega) \\ 0 \qquad \text{otherwise.} \end{cases}$$

The results of the paper will remain valid under the above interim expected utility framework, but we choose not to adopt it for simplicity and convenience.

Proof: For every $i \in I$, since x_i and u_i are $\mathcal{F}_i$-measurable, $u_i(\omega', x_i(\omega'))$ is constant on $E_i(\omega)$. Therefore,

$$\int_{E_i(\omega)} u_i(\omega', x_i(\omega'))d\mu(\omega') = \mu(E_i(\omega))u_i(\omega, x_i(\omega))$$

and the conclusion follows. □

The *set of feasible allocations* is given by $\boldsymbol{A} = \{x \in L_X : \sum_{i \in I} x_i = \sum_{i \in I} e_i\}$. For each *i*, an element $z_i \in L_i$ with $z_i = x_i - e_i$ is a *net trade of agent i*. The *set of feasible net trades* is given by $\boldsymbol{Z} = \{z \in L : \sum_{i \in I} z_i = 0\}$. Let $\hat{Z} = \{\hat{z} \in \prod_{i \in I} Y_i : \sum_{i \in I} \hat{z}_i = 0\}$, where $Y_i = Y$ for every $i \in I$. Notice that the initial endowment vector denoted by $e = (e_i)_{i \in I}$ is an element of L_X. Let $L^0_{X_i} = \{x_i \in L_1(\mu, Y_+) : x_i(\omega) \in X_i(\omega) \mu\text{-}a.e.\}$ and $L^0_X = \prod_{i \in I} L^0_{X_i}$. Define the *set of ex post allocations by* $\boldsymbol{A}^0 = \{x \in L^0_X : \sum_{i \in I} x_i = \sum_{i \in I} e_i\}$. *For each partition* $\mathcal{G}$ *of* Ω, *define* $\boldsymbol{A}(\mathcal{G}) = \{x \in L^0_X : x_i$ *is* $\mathcal{G}$*-measurable for every* $i \in I$ *and* $\sum_{i \in I} x_i = \sum_{i \in I} e_i\}$ *and* $\boldsymbol{Z}(\mathcal{G}) = \{z \in \prod_{i \in I} L_1(\mu, Y) : z_i$ *is* $\mathcal{G}$*-measurable for every* $i \in I$ *and* $\sum_{i \in I} z_i = 0\}$.

We close this section by discussing the notion of private information measurability of allocations. To say that an agent's allocation is $\mathcal{F}_i$-measurable, it means that his/her consumption is the same in states that he/she cannot distinguish. Also notice that since by assumption initial endowments are $\mathcal{F}_i$-measurable, the net trade of each agents is $\mathcal{F}_i$-measurable as well. This assumption will be dropped in Section 9. However, we believe that this assumption is not only reasonable but it is also tractable from an analytical view point. The example below may be useful to bring out the importance of private measurability.

Example 3.1 Consider an economy with differential information with two agents, one good, and three states (i.e., $\Omega = \{\omega_1, \omega_2, \omega_3\}$) with equal probability (i.e., $\mu(\{\omega\}) = 1/3$ for every $\omega \in \Omega$) where utility functions, initial endowment, and private information sets are given as follows:

$$u_1(\omega, x) = \sqrt{x}, \quad e_1 = (10, 10, 0) \quad \mathcal{F}_1 = \{\{\omega_1, \omega_2\}, \{\omega_3\}\},$$
$$u_2(\omega, x) = \sqrt{x}, \quad e_2 = (10, 0, 10) \quad \mathcal{F}_2 = \{\{\omega_1, \omega_3\}, \{\omega_2\}\}.$$

In this example, we want to show that without private information measurability, a net trade may not be viable. For simplicity, we only consider the ex ante expected utility. Suppose that agent 1 proposes the net trade $z = (z_1, z_2)$ with

$$z_1 = (-2, -2, 2), \; z_2 = (2, 2, -2).$$

Note that these net trades are not private information measurable. In particular, z_2 is not $\mathcal{F}_2$-measurable. Notice that if state ω_1 is realized, agent 1 may claim that state ω_3 occurred since he/she obtains two units of the good from agent 2 at state ω_3. Observe that agent 2 cannot detect that agent 1 has misreported the state since he/she is not able to distinguish state ω_3 from state ω_1. Conversely, if state ω_1 is realized, agent 2 may claim that state ω_2

occurred since he/she obtains two units of the good from agent 1 at state ω_2 (agent 1 cannot distinguish state ω_2 from state ω_1). Consequently, the non-$\mathscr{F}_i$-measurability of the net trades has created incentive problems and the contract may not take place. In other words, trade may not be viable without private information measurability. As was mentioned in footnote 2, in this example, private measurability is necessary and sufficient for coalitional incentive compatibility. The latter concept is discussed below.

4 Coalitional Bayesian incentive compatibility

When agents have differential information, arbitrary allocations are not generally viable. In particular, arbitrary allocations might not be incentive compatible in the sense that groups of agents may misreport their information without other agents noticing it, and hence achieve different payoffs.

In Krasa-Yannelis (1994), a concept of coalitional incentive compatibility was introduced. For purposes of comparison, we modify their definition in terms of interim expected utility. An allocation $x = e + z \in A$ is *coalitionally Bayesian incentive compatible* if it is not true that there exist coalition S and states $\omega^*, \omega'(\omega^* \neq \omega')$ with $\omega' \in \bigcap_{i \notin S} E_i(\omega^*)$ such that

$$\frac{1}{\mu(E_i(\omega^*))} \int_{E_i(\omega^*)} u_i(\omega, e_i(\omega) + z_i(\omega')) d\mu(\omega) > \frac{1}{\mu(E_i(\omega^*))} \int_{E_i(\omega^*)} u_i(\omega, e_i(\omega) + z_i(\omega)) d\mu(\omega)$$

for every $i \in S$. Notice that in Krasa-Yannelis (1994), instead of the interim expected utility V_i, the ex post utility function u_i is used. In essence, this concept assures that no coalition S can make redistributions among themselves in states that the complementary coalition cannot distinguish, and become better off. In other words, if state ω^* occurs and the agents in the coalition $I \backslash S$ cannot distinguish between the state ω^* and ω', it must be the case that the agents of coalition S cannot become better off by announcing ω' instead of the actually occurred ω^*. The measurability implies that $\omega' \notin E_i(\omega^*)$ for every agent i in the coalition S.

As in Palfrey-Srivastava (1989), a *deception* for agent i is a function $\alpha_i : \mathscr{F}_i \to \mathscr{F}_i$. Let $\alpha_i^* : \mathscr{F}_i \to \mathscr{F}_i$ be the truth-telling for agent i. A deception vector $\alpha = (\alpha_i)_{i \in I}$ is *compatible* with F if $\alpha(\omega) := \cap_{i \in I} \alpha_i(E_i(\omega)) \neq \emptyset$ for every $\omega \in \Omega$. We use the following notation:[9] $\alpha_S^*(\omega) = E^s(\omega) = \cap_{i \in S} E_i(\omega)$, $\alpha_{-s}(\omega) = E^{-S}(\omega) = \cap_{i \notin S} E_i(\omega)$, $\alpha_S(\omega) = E_\alpha^S(\omega) = \cap_{i \in S} \alpha_i(E_i(\omega))$, $\alpha_{-S}(\omega) = E_\alpha^{-S}(\omega) = \cap_{i \notin S} \alpha_i(E_i(\omega))$. Let $z \in Z$ be a feasible net trade. If α is compatible with F, then $(z \circ \alpha)(\omega) = z(\alpha(\omega)) = z(\omega')$ for all $\omega' \in \alpha(\omega)$. Otherwise

[9] For example, consider the following information structure: $\mathscr{F}_1 = \{\{\omega_1, \omega_2\}, \{\omega_3\}\}, \mathscr{F}_2 = \{\{\omega_1, \omega_3\}, \{\omega_2\}\}, \mathscr{F}_3 = \{\{\omega_1\}, \{\omega_2\}, \{\omega_3\}\}$ Let us define a deception as follows: for every ω, $\alpha_i(E_i(\omega)) = \{\omega_1\}, \forall_i = 1, 2$ and $\alpha_3(E_3(\omega)) = E_3(\omega)$. Then for the coalition $S = \{1, 3\}$, $\alpha_S^*(\omega_3) = E^S(\omega_3) = \{\omega_3\}$, $\alpha_{-S}^*(\omega_3) = E^{-S}(\omega_3) = \{\omega_1, \omega_3\}$, $\alpha_S(\omega_3) = E_\alpha^S(\omega_3) = \{\omega_1\}$.

$z \circ \alpha = 0$. Note that $(z \circ [\alpha])_i = z_i \circ \alpha$ and $(z \circ [\alpha^*])(\omega) = z(\omega)$. Recall from Lemma 1 of Palfrey-Srivastava (1989, p.120) that for every $i \in I$, if $\omega' \in E_i(\omega)$, then $[\alpha](\omega') \subset E_i(\alpha(\omega))$ for every $i \in I$, where $E_i(\alpha(\omega))$ is the event that contains $[\alpha](\omega)$. In view of this Lemma, we immediately conclude that if $z \in Z$, then $z \circ \alpha \in Z$ for each α compatible with F.

In terms of the deception α_i, one can define Bayesian incentive compatibility. It captures the idea that no agent can improve his utility by using a deception, which is not detected by any other agent. Furthermore, the allocation that is generated by the deception is to be feasible. One can notice the difference between our Bayesian incentive compatibility and the standard Bayesian incentive compatibility [see for example Palfrey-Srivastava (1987)]. This property is well known and considered as a basic requirement for a desirable mechanism in differential information economies. However, it turns out that Bayesian incentive compatibility is not strong enough to play a role as a condition in our model.[10]

Definition 4.1: An allocation $x = e + z \in \boldsymbol{A}$ is said to be *Bayesian incentive compatible* (*BIC*) if for every $i \in I$, for every $\omega \in \Omega$, and for every $\alpha_i : \mathscr{F}_i \to \mathscr{F}_i$ with $(\alpha_i, \alpha^*_{-i})$ compatible with F,

$$V_i(\omega, x_i) \geq V_i(\omega, e_i + (z \circ [\alpha_i, \alpha^*_{-i}])_i),$$

where $e + z \circ [\alpha_i, \alpha^*_{-i}] \in \boldsymbol{A}$.

Definition 4.2: An allocation $x = e + z \in \boldsymbol{A}$ is said to be *coalitionally Bayesian incentive compatible* (*CBIC*) if it is not true that there exists a state $\omega \in \Omega$, a coalition $S \subset I$, and a deception $\alpha_S : \prod_{i \in S} \mathscr{F}_i \to \prod_{i \in S} \mathscr{F}_i$ such that $(\alpha_S, \alpha^*_{-S})$ is compatible with F and for every $i \in S$,

$$V_i(\omega, e_i + (z \circ [\alpha_S, \alpha^*_{-S}])_i) > V_i(\omega, x_i),$$

where $e + z \circ [\alpha_S, \alpha^*_{-S}] \in \boldsymbol{A}$.
This notion of incentive compatibility states that it is not possible for any coalition S to become better off by announcing a false event[11]. Observe that if S is a singleton, then the CBIC condition is reduced to standard BIC condition. This implies that coalitional Bayesian incentive compatibility is a stronger condition than Bayesian incentive compatibility.

Definition 4.3: An allocation $x = e + z \in \boldsymbol{A}$ is said to be *weakly coalitional Bayesian incentive compatible* (*weakly CBIC*) if it is not true that there exists a state $\omega \in \Omega$, a coalition $S \subset I$, and a deception $\alpha_S : \prod_{i \in S} \mathscr{F}_i \to \prod_{i \in S} \mathscr{F}_i$ such that for every $i \in S, \alpha_i(E_i(\omega')) = E_i(\omega') \forall \omega' \notin E_i(\omega), E_i(\omega) \in \wedge_{i \in S} \mathscr{F}_i$, and

$$V_i(\omega, e_i + (z \circ [\alpha_S, \alpha^*_{-S}])_i) > V_i(\omega, x_i),$$

[10] See Theorem 6.4.

[11] Note that whenever $E_i(\omega)$ and $\alpha_i(E_i(\omega))$ are singletons for every $i \in S$, our notion coincides with that of Krasa-Yannelis (1994), provided that ex post utility functions are used.

where $e + z \circ [\alpha_S, \alpha^*_{-S}] \in \boldsymbol{A}$.

This is obtained from adding $E_i(\omega) \in \bigwedge_{i\in S} \mathscr{F}_i, \forall_i \in S$ to the deceiving conditions in CBIC. In particular, the true event, which is misreported, is common knowledge to the deceiving coalition.

We can now define a much stronger notion of incentive compatibility.

Definition 4.4: An allocation $x = e + z \in \boldsymbol{A}$ is said to be *strongly coalitional Bayesian incentive compatible* (*SCBIC*) if it is not true that there exist a state $\omega \in \Omega$, a coalition $S \subset I$, and a deception $\alpha_S : \prod_{i\in S} \mathscr{F}_i \to \prod_{i\in S} \mathscr{F}_i$ such that $(\alpha_S, \alpha^*_{-S})$ is compatible with F and for every $i \in S$,

$$V_i(\omega, e_i + (z \circ [\alpha_S, \alpha^*_{-S}])_i) \geq V_i(\omega, x_i)$$

with strict inequality for some $i \in S$, where $e + z \circ [\alpha_S, \alpha^*_{-S}] \in \boldsymbol{A}$.

Definition 4.5: An allocation $x = e + z \in \boldsymbol{A}$ is said to be *t-coalitional Bayesian incentive compatible* (*TCBIC*)[12] if it is not true that there exist a state $\omega \in \Omega$, a coalition $S \subset I$, and a deception $\alpha_S : \prod_{i\in S} \mathscr{F}_i \to \prod_{i\in S} \mathscr{F}_i$, and a transfer $(t_i)_{i\in S} \in \prod_{i\in S} L_i$ with $\sum\limits_{i\in S} t_i = 0$, *each* t_i *is* $\bigwedge\limits_{i\in S} \mathscr{F}_i$-measurable such that $(\alpha_S, \alpha^*_{-S})$ is compatible with F and for every $i \in S$,

$$V_i(\omega, e_i + (z \circ [\alpha_S, \alpha^*_{-S}])_i + t_i) > V_i(\omega, x_i),$$

where $e + z \circ [\alpha_S, \alpha^*_{-S}] \in \boldsymbol{A}$.

The t-coalitional Bayesian incentive compatibility models the idea that it is impossible for any coalition to cheat the complementary coalition by misreporting the event and making side payments to each other which cannot be observed by agents who are not members of this coalition. When $u_i(\omega, \cdot)$ is monotone and continuous for every $i \in I$ and $\omega \in \Omega$, one can easily show that the notion of SCBIC is equivalent to the TCBIC. In particular, if an allocation is not SCBIC, the agent who became strictly better off can make side payments to every agent in the deceiving coalition and make them strictly better off.

By observing the definitions, one can easily check that the following relationship between these concepts of incentive compatibility holds:

$$TCBIC \Rightarrow SCBIC \Rightarrow CBIC \Rightarrow weakly\ CBIC \Rightarrow BIC.$$

5 Efficiency

5.1 Efficiency concepts in differential information economies

The notions of informational efficiency discussed below are distinguished depending the degree of private information. The ex ante efficiency is defined

---

[12] This is an interim version of the *strong coalitional incentive compatibility* of Krasa-Yannelis (1994).

at the stage where every agent has private information but no state is yet realized. The interim efficiency is defined at the stage where every agent knows his/her private information event which contains the realized state. The ex post efficiency is defined at the stage where every agent has complete information. Because the interim stage and the ex post stage depend on states, it is more difficult to define the notions of efficiency. In particular, for the definition of interim efficiency, the possibility of communication between all the agents when they block the proposed allocation make the problem even harder. In order to address the possibility of communication among agents, we will introduce more notation.

Denote by $\bigwedge_{i\in I}\mathscr{F}_i$ the *finest common coarsening* of $\{\mathscr{F}_i : i \in I\}$, *i.e.*, the finest partition of Ω which is coarser than $\mathscr{F}_i$ for every $i \in I$. An event E is said to be *common knowledge at* ω if $(\bigwedge_{i\in I}\mathscr{F}_i)(\omega) \subset E$ where $(\bigwedge_{i\in I}\mathscr{F}_i)(\omega)$ is the event of $\bigwedge_{i\in I}\mathscr{F}_i$ containing ω. Notice that $(\bigwedge_{i\in I}\mathscr{F}_i)(\omega)$ itself is common knowledge at ω. We also call $\bigwedge_{i\in I}\mathscr{F}_i$ the *common knowledge partitions* of Ω. Denote by $\bigvee_{i\in I}\mathscr{F}_i$ coarsest common refinement of $\{\mathscr{F}_i : i \in I\}$, *i.e.*, the coarsest partition of Ω which is finer than $\mathscr{F}_i$ for every $i \in I$. Denote by $(\bigvee_{i\in I}\mathscr{F}_i)(\omega)$ the event of $\bigvee_{i\in I}\mathscr{F}_i$ containing ω. We also call $\bigvee_{i\in I}\mathscr{F}_i$ the *pooled information partition.*[13]

Several notions of efficiency will be defined below. The main differences of these concepts are basically two. Firstly, the degree of information sharing of the grand coalition, *i.e.*, do agents make redistribution of their initial endowment based on their own private information, common knowledge information, or pooled information? Secondly, what kind of expected utility is used, i.e., interim, ex ante, or ex post?

5.2 Ex ante efficiency

The notion of ex ante efficiency is defined in terms of the ex ante expected utility. If the grand coalition of agents is allowed to redistribute their resources among themselves to become better off by using the common knowledge information, the ex ante coarse efficiency is a natural concept of efficiency.

Definition 5.2.1: An allocation $x \in \boldsymbol{A}$ is *ex ante coarse efficient* if there is no $x' \in \boldsymbol{A}$ such that $x' - e \in Z(\bigwedge_{i\in I}\mathscr{F}_i)$ and $\overline{V}_i(x'_i) > \overline{V}_i(x_i)$ for every $i \in I$.

If it is possible for the grand coalition of agents to redistribute their initial endowments among themselves to become better off by using their own private information, the ex ante private efficiency can be adopted.

[13] In the context of σ-algebra, $\wedge_{i\in I}\mathscr{F}_i$ denotes the *meet, i.e.* the maximal (finest) σ-algebra contained in every σ-algebra $\mathscr{F}_i$ and $\vee_{i\in I}\mathscr{F}_i$ denotes the *join, i.e.*, the minimal (coarsest) σ-algebra containing every σ-algebra $\mathscr{F}_i$.

Definition 5.2.2: All allocation $x \in \boldsymbol{A}$ is *ex ante private efficient*[14] there is no $x' \in \boldsymbol{A}$ such that $\overline{V}_i(x_i') > \overline{V}_i(x_i)$ for every $i \in I$.

If it is possible for the grand coalition of agents to redistribute their initial endowments among themselves to become better off by pooling and sharing their private information, the ex ante fine efficiency can be defined as follows.

Definition 5.2.3: An allocation $x \in \boldsymbol{A}$ is *ex ante fine efficient* if there is no $x' \in \boldsymbol{A}(\bigvee_{i \in I} \mathscr{F}_i)$ such that $\overline{V}_i(x_i') > \overline{V}_i(x_i)$ for every $i \in I$.

In addition, if a feasible allocation is allowed to be measurable with respect to the pooled information, then a weaker concept can be defined.

Definition 5.2.4: An allocation $x \in \boldsymbol{A}(\bigvee_{i \in I} \mathscr{F}_i)$ is *ex ante weak fine efficient* if there is no $x' \in \boldsymbol{A}(\bigvee_{i \in I} \mathscr{F}_i)$ such that $\overline{V}_i(x_i') > \overline{V}_i(x_i)$for every $i \in I$.

5.3 Interim efficiency

The interim efficiency notions below will be defined in terms of the interim expected utility. If the grand coalitian of agents can redistribute their resources among themselves to become better off by using the common knowledge information, the interim coarse efficiency can be defined as follows.

Definition 5.3.1: An allocation $x \in \boldsymbol{A}$ is *interim coarse efficient* if there is no $x' \in \boldsymbol{A}$ such that $x' - e \in Z(\bigwedge_{i \in I} \mathscr{F}_i)$ and for some $\omega \in \Omega$, $V_i(\omega, x_i') > V_i(\omega, x_i)$ for every $i \in I$.

If it is possible for the grand coalition of agents to redistribute their initial endowments among themselves to become better off by using their own private information, the interim private efficiency can be defined as follows.

Definition 5.3.2: An allocation $x \in \boldsymbol{A}$ is *interim private efficient*[15] if there is no $x' \in \boldsymbol{A}$ such that for some $\omega \in \Omega$, $V_i(\omega, x_i') > V_i(\omega, x_i)$ for every $i \in I$.

If it is possible for the grand coalition of agents to redistribute their initial endowments among themselves to become better off by pooling and sharing their information, the interim fine efficiency can be defined as follows.

Definition 5.3.3: An allocation $x \in \boldsymbol{A}$ is *interim fine efficient* if there is no $x' \in \boldsymbol{A}(\bigvee_{i \in I} \mathscr{F}_i)$ such that for some $\omega \in \Omega$, $V_i(\omega, x_i') > V_i(\omega, x_i)$ for every $i \in I$.[16]

---

[14] Notice that if $u_i(\omega, \cdot)$ is continuous and monotone, this definition is equivalent to: An allocation $x \in \boldsymbol{A}$ is *strongly ex ante private efficient* if there is no $x' \in \boldsymbol{A}$ such that $\overline{V}_i(x_i') \geq \overline{V}_i(x_i)$ for every $i \in I$ with strict *inequality for some* **i** $\in I$.

[15] An allocation $x \in \boldsymbol{A}$ is *strongly interim private efficient* if there is no $x' \in \boldsymbol{A}$ such that for some $\omega \in \Omega$, $V_i(\omega, x_i') \geq V_i(\omega, x_i)$ for every $i \in I$ with strict inequality for some $i \in I$.

[16] One may consider an interim expected utility which takes into account the pooled information. But throughout the paper, we ignore this effect.

If the feasible allocation is allowed to be measurable w.r.t to the pooled information, then a weaker concept can be defined as follows.

Definition 5.3.4: An allocation $x \in \boldsymbol{A}$ $(\bigvee_{i \in I} \mathscr{F}_i)$ is *interim weak fine efficient* if there is no $x' \in \boldsymbol{A}(\bigvee_{i \in I} \mathscr{F}_i)$ such that for some $\omega \in \Omega$, $V_i(\omega, x'_i) > V_i(\omega, x_i)$ for every $i \in I$.

Moreover, if the event where every agent becomes better off is common knowledge to the grand coalition, then the notion of weakly interim efficiency can be defined as follows.

Definition 5.3.5: An allocation $x \in \boldsymbol{A}$ is *weakly interim efficient* if there is no $x' \in \boldsymbol{A}$ such that for some $E \in \bigwedge_{i \in I} \mathscr{F}_i$, $V_i(\omega, x'_i) > V_i(\omega, x_i)$ for every $\omega \in E$ and for every $i \in I$.

Since interim efficiency does depend on states, we have one more notion of interim efficiency, *HM interim efficiency,* which is widely used as interim efficiency notion in economics literature [for example, see Holmström-Myerson (1983, p.1805)].

Definition 5.3.6: An allocation $x \in \boldsymbol{A}$ is *HM interim efficient*[17] if there is no $x' \in \boldsymbol{A}$ such that $V_i(\omega, x'_i) \geq V_i(\omega, x_i)$ for every $\omega \in \Omega$ and for every $i \in I$ with strict inequality for some $\omega \in \Omega$ and for some $i \in I$.

In the same way as in interim efficiency, one can define *strongly ex post efficiency, ex post efficiency,* and *HM ex post efficiency* by using ex post utility u_i and ex post feasible set A^0.

5.4 Relationship of the efficiency concepts

In economies with certainty, it is known that if the preferences are monotone and continuous, strong efficiency and efficiency are equivalent. In the same way, one could get corresponding equivalence[18] for differential information economies. Furthermore, one can easily prove that efficiency concepts are stronger if the information sharing of the grand coalition is finer in either ex ante or interim case as the following proposition indicate:

Proposition 5.4.1: The following statements hold.

(a) Every ex ante fine efficient allocation in $\mathscr{E}$ is also ex ante private efficient.
(b) Every ex ante private efficient allocation in $\mathscr{E}$ is also ex ante coarse efficient.
(c) Every ex ante fine efficient allocation in $\mathscr{E}$ is also ex ante weak fine efficient.

[17] This is different from that of Homström-Myerson (1983) in that they do not impose the private information measurability.

[18] Assume that $u_i(\omega,)$ is monotone and continuous for every $i \in I$ and $\omega \in \Omega$. By simply observing the definitions, one can easily check that an allocation is strongly interim private (strongly ex ante private, strongly ex post, resp.) efficient if and only if it is interim private (ex ante private, ex post, resp.) efficient.

Proof: (a) Let x be an ex ante fine efficient allocation. Suppose that it is not ex ante private efficient. Then there exist $x' \in \boldsymbol{A}$ such that $\overline{V}_i(x_i') > \overline{V}_i(x_i)$ for every $i \in I$. Since $\boldsymbol{A} \subset \boldsymbol{A}(\bigvee_{i \in I} \mathscr{F}_i)$, we have $x' \in \boldsymbol{A}(\bigvee_{i \in I} \mathscr{F}_i)$ such that $\overline{V}_i(x_i') > \overline{V}_i(x_i)$ for every $i \in I$, a contradiction.

(b) Let x be an ex ante private efficient allocation. Suppose that it is not ex ante coarse efficient. Then there exists $x' \in \boldsymbol{A}$ such that $x' - e \in Z(\bigwedge_{i \in I} \mathscr{F}_i)$ and $\overline{V}_i(x_i') > \overline{V}_i(x_i)$ for every $i \in I$. Since $Z(\bigwedge_{i \in I} \mathscr{F}_i) \subset Z$,we have $x' \in \boldsymbol{A}$ such that $\overline{V}(x_i') > \overline{V}_i(x_i)$ for every $i \in I$, a contradiction.

(c) Let $x \in \boldsymbol{A}$ be an ex ante fine efficient allocation. Suppose that it is not ex ante weak fine efficient. Then there exists $x' \in \boldsymbol{A}(\bigvee_{i \in I} \mathscr{F}_i)$ such that $\overline{V}_i(x_i') > \overline{V}_i(x_i)$ for every $i \in I$. Since $\boldsymbol{A} \subset \boldsymbol{A}(\bigvee_{i \in I} \mathscr{F}_i)$, we have $x \in \boldsymbol{A}(\bigvee_{i \in I} \mathscr{F}_i)$ such that there exists $x' \in \boldsymbol{A}(\bigvee_{i \in I} \mathscr{F}_i)$ such that $\overline{V}(x_i') > \overline{V}_i(x_i)$ for every $i \in I$, a contradiction. □

Applying the same arguments about the information sharing, we get the same results for the concepts of interim efficiency.

Proposition 5.4.2: The following statements hold.

(a) Every interim fine efficient allocation in $\mathscr{E}$ is also interim private efficient.
(b) Every interim private efficient allocation in $\mathscr{E}$ is also interim coarse efficient.
(c) Every interim fine efficient allocation in $\mathscr{E}$ is also interim weak fine efficient.

Proof: Follow the argument adopted for the proof of Proposition 5.4.1. □

Proposition 5.4.3: The following statements hold.

(a) Every interim private efficient allocation in $\mathscr{E}$ is also weakly interim efficient.
(b) Every HM interim efficient allocation in $\mathscr{E}$ is also weakly interim efficient.
(c) If $u_i(\omega, \cdot)$ is monotone and continuous, then every interim private efficient allocation in $\mathscr{E}$ is also HM interim efficient.

Proof: (a) Let x be an interim efficient allocation. Suppose that x is not weakly interim efficient. Then there is a feasible allocation x' such that for some common knowledge event $E \in \bigwedge_{i \in I} \mathscr{F}_i$, $V_i(\omega, x_i') > V_i(\omega, x_i)$ for every $\omega \in E$ and for every $i \in I$. This implies that for some $\omega \in \Omega$, $V_i(\omega, x_i')$ $> V_i(\omega, x_i)$ for every $i \in I$. Hence, x is not interim efficient, a contradiction.

(b) Let x be a HM interim efficient allocation. Suppose that x is not weakly interim efficient. Then there is a feasible allocation x' such that for some common knowledge event $E \in \bigwedge_{i \in I} \mathscr{F}_i$, $V_i(\omega, x_i') > V_i(\omega, x_i)$ for every $\omega \in E$ and for every $i \in I$. Consider a new allocation $x^* = (x_i^*)_{i \in I} \in \boldsymbol{A}$, where

$$x_i^*(\omega') = \begin{cases} x_i'(\omega') & \text{if } \omega' \in E, \\ x_i(\omega') & \text{otherwise.} \end{cases}$$

It follows that $V_i(\omega, x_i^*) \geq V_i(\omega, x_i)$ for every $\omega \in \Omega$ and for every $i \in I$. Moreover, $V_i(\omega, x_i^*) > V_i(\omega, x_i)$ for some $\omega \in \Omega$ and for some $i \in I$. Hence, x is not HM interim efficient, a contradiction.

(c) Let x be an interim private efficient allocation. Suppose that x is not HM interim efficient. Then there is a feasible allocation x' such that, $V_i(\omega, x_i') \geq V_i(\omega, x_i)$ for every $\omega \in \Omega$ and for every $i \in I$ with strict inequality for some $\omega^* \in \Omega$ and for some $k \in I$. Since for each $i \in I$ and for each fixed $\omega \in \Omega$, $V_i(\omega, \cdot)$ is continuous by Lemma 3.1.1, there is an $\varepsilon > 0$ such that that $V_k(\omega^*, x_k' - \varepsilon \cdot 1 > V_k(\omega^*, x_k)$. Consider a new allocation $x^* = (x_i^*)_{i \in I} \in \boldsymbol{A}$ with

$$x_i^* = \begin{cases} x_i' - \varepsilon \cdot \mathbf{1} & \text{if } i = k, \\ x_i' + \frac{1}{|I|-1} \varepsilon \cdot \mathbf{1} & \text{otherwise.} \end{cases}$$

Since $V_i(\omega, \cdot)$ is monotone, it follows that $V_i(\omega^*, x_i^*) > V_i(\omega^*, x_i)$ for every $i \in I$. Hence, x is not interim private efficient, a contradiction. □

Recall that the ex ante expected utility and the interim expected utility are related in the following way:

$$\overline{V}_i(x_i) = \sum_{E_i(\omega) \in \mathscr{F}_i} \mu(E_i(\omega)) V_i(\omega, x_i). \tag{5.4.1}$$

This gives the relationship between ex ante private efficiency and HM interim efficiency.

Proposition 5.4.4: Assume that $u_i(\omega, \cdot)$ is monotone and continuous for every $i \in I$ and $\omega \in \Omega$. Every ex ante efficient allocation in $\mathscr{E}$ is also HM interim efficient.[19]

Proof: Let x be an ex ante private efficient allocation. Suppose that x is not HM interim efficient. Then there is an feasible allocation x' such that $V_i(\omega, x_i') \geq V_i(\omega, x_i)$ for every $\omega \in \Omega$ and for every $i \in I$ with strict inequality for some $\omega^* \in \Omega$ and some $k \in I$. It follows from (5.4.1) that $\overline{V}_i(x_i') \geq \overline{V}_i(x_i)$ for every $i \in I$ with strict inequality for $k \in I$. Since $\overline{V}_i$ is continous (recall Lemma 3.1.1), there is an $\varepsilon > 0$ such that $\overline{V}_k(x_k' - \varepsilon \cdot \mathbf{1}) > \overline{V}_k(x_k)$. Consider a new allocation $x^* = (x_i^*)_{i \in I} \in \boldsymbol{A}$ where

$$x_i^* = \begin{cases} x_i' - \varepsilon \cdot \mathbf{1} & \text{if } i = k, \\ x_i' + \frac{1}{|I|-1} \varepsilon \cdot \mathbf{1} & \text{otherwise.} \end{cases}$$

Since $\overline{V}_i$ is monotone, $\overline{V}_i(x_i^*) > \overline{V}_i(x_i)$ for every $i \in I$. This implies that x is not ex ante private efficient, a contradiction. □

Corollary 5.4.5: Assume that $u_i(\omega, \cdot)$ is monotone and continuous for every $i \in I$ and $\omega \in \Omega$. Every ex ante private efficient allocation in $\mathscr{E}$ is weakly interim efficient.

---

[19] When the private information measurability is not imposed, one can show that the strong ex ante efficiency implies the HM interim efficiency, which in turn implies the HM ex post efficiency. This is a well-known fact [see Holmström-Myerson (1983)]. For comparison, let T_i be the type space of agent i. Then $E_i(t) = \{t_i\} \times T_{-i}$ with $t = (t_i, t_{-i})$. Thus, in the context of type representation of private information, the private information measurability is described by $x_i(t_i, t_{-i}) = x_i(t_i, t'_{-i})$ for every t_{-i} and t'_{-i} in T_{-i} since $E_i(t_i, t_{-i}) = E_i(t_i, t'_{-i})$.

Proof: It follows from Proposition 5.4.3 (b) and Proposition 5.4.4. □

Unlike Holmström-Myerson (1983), it turns out that there is no direct implication between the ex ante private efficiency and the interim private efficiency, as the proposition below indicates.

Proposition 5.4.6: An ex ante private efficient (weakly interim efficient, or HM interim efficient) allocation in $\mathscr{E}$ may not be interim private efficient.

Proof: Consider an economy with differential information with three agents, two goods, and three equally probable states, where utility functions, random initial endowments, and private information sets are given as follows:

$$u_1(\omega, x^1, x^2) = \sqrt{x^1 x^2}, e_1 = ((10,0),(10,0),(10,0)), \mathscr{F}_1 = \{\{\omega_1, \omega_2, \omega_3\}\},$$
$$u_2(\omega, x^1, x^2) = \sqrt{x^1 x^2}, e_2 = ((4,4),(1,5),(1,5)), \quad \mathscr{F}_2 = \{\{\omega_1\}, \{\omega_2, \omega_3\}\},$$
$$u_3(\omega, x^1, x^2) = \sqrt{x^1 x^2}, e_3 = ((0,1),(1,3),(3,4)), \quad \mathscr{F}_3 = \{\{\omega_1\}, \{\omega_2\}, \{\omega_3\}\}.$$

The allocation $x = (x_1, x_2, x_3)$ with

$$x_1 = ((6,2),(6,2),(6,2)),$$
$$x_2 = ((7,3),(5,3),(5,3)),$$
$$x_3 = ((1,0),(1,3),(3,4))$$

is an ex ante private (weakly interim efficient, or HM interim efficient) but is not interim private efficient, since the allocation $x' = (x'_1, x'_2, x'_3)$ with

$$x'_1 = ((6.1,2),(6.1,2),(6.1,2)),$$
$$x'_2 = ((7,3),(4,4),(4,4)),$$
$$x'_3 = ((0.9,0),(1.9,2),(0.9,3))$$

results in $V_i(\omega_2, x'_i) > V_i(\omega_2, x_i)$ for every $i \in I$. □

Denote by $\overline{\mathscr{E}}$ an economy as defined in Section 3, with the only difference that now $Y_+ = \boldsymbol{R}_+$, *i.e.*, we have one good per state. In this case, the set of feasible allocations lies in the infinite dimensional space $L_1(\mu, \boldsymbol{R}_+)$. In a one good economy, the set of feasible allocations is equivalent to the set of interim efficient allocations. It is obvious that every interim efficient allocation is feasible. The other direction is clear too. Indeed, from a given feasible allocation, a change to any other feasible allocation makes at least one agent become worse off at some state because there is only one good. It can be proved formally as follows.

Proposition 5.4.7: Every feasible allocation in $\overline{\mathscr{E}}$ is interim coarse efficient.

Proof: Suppose that a feasible allocation $x \in \boldsymbol{A}$ is not interim coarse efficient. Then there exist a state $\omega \in \Omega$, a agent $i \in I$, an allocation $x' \in \boldsymbol{A}$ such that $x' - e \in Z(\bigwedge_{i \in I} \mathscr{F}_i)$ and for every $i \in I$,

$$V_i(\omega, x_i') > V_i(\omega, x_i). \tag{5.4.2}$$

Since there is only one good, (5.4.2) implies that for every $i \in I$, $x_i'(\omega) > x_i(\omega)$ by monotonicity and measurability. Hence, $\bar{e}(\omega) = \sum_{i\in I} x_i'(\omega) > \sum_{i\in I} x_i(\omega) = \bar{e}(\omega)$, a contradiction. □

Corollary 5.4.8: Every feasible allocation in $\bar{\mathscr{E}}$ is interim private efficient.

Proof: Since every interim coarse efficient allocation is interim private efficient [Proposition 5.4.2(b)], the conclusion follows from Proposition 5.4.7. □

6 Relationship of efficiency with incentive compatibility

It is well-known that the Bayesian incentive compatibility condition is too restrictive for achieving socially desirable allocations. In particular, Myerson (1979) recognized that most interim efficient allocations may not be Bayesian incentive compatible. However, if a simple condition is assumed (that is, the private information measurability), the BIC condition turns out to be so weak that every feasible allocation is BIC. However, the CBIC condition seems more appropriate and one can show that several efficiency concepts defined in Section 5 are always coalitionally Bayesian incentive compatible.

Proposition 6.1: Every interim coarse efficient allocation in $\mathscr{E}$ is TCBIC.

Proof: Suppose that $x = e + z \in \boldsymbol{A}$ is interim coarse efficient but it is not TCBIC. Then there exists a state $\omega^* \in \Omega$, a coalition S, a deception $\alpha_S : \prod_{i\in S} \mathscr{F}_i \to \prod_{i\in S} \mathscr{F}_i$, and a transfer $(t_i)_{i\in S} \in \prod_{i\in S} L_i$ with $\sum_{i\in S} t_i = 0$, each t_i is $\bigwedge_{i\in S} \mathscr{F}_i$-measurable and such that for every $i \in S$,

$$V_i(\omega^*, e_i + (z \circ [\alpha_S, \alpha^*_{-S}])_i + t_i) > V_i(\omega^*, x_i),$$

where $e + z \circ [\alpha_S, \alpha^*_{-S}] \in \boldsymbol{A}$. Since for every $\omega' \in E^S_\alpha(\omega^*) \cap E^{-S}(\omega^*)$ it holds that $z_i(\omega') = z_i(\omega^*)$, *i.e.*, $(z \circ [\alpha_S, \alpha^*_{-S}])_i(\omega^*) = z_i(\omega^*)$ for every $i \notin S$, it must be the case that for every $i \notin S$,

$$V_i(\omega^*, e_i + (z \circ [\alpha_S, \alpha^*_{-S}])_i) = V_i(\omega^*, x_i). \tag{6.1}$$

Since for each $i \in I$ and for each fixed $\omega \in \Omega$, $V_i(\omega, \cdot)$ is continuous by Lemma 3.1.1, there exists an $\varepsilon > 0$ such that for every $i \in S$,

$$V_i(\omega^*, e_i + z \circ [\alpha_S, \alpha^*_{-S}]_i + t_i - \varepsilon \cdot \mathbf{1}) > V_i(\omega^*, x_i). \tag{6.2}$$

Let us define $z' = (z_i')_{i\in I} : \Omega \to \hat{Z}$ by $z_i'(\omega) = (z_i \circ [\alpha_S, \alpha^*_{-S}])(\omega^*) + t_i(\omega^*)$ for every $\omega \in \Omega$, where $t_i = 0$ for every $i \notin S$. Define $x' = (x_i')_{i\in I}$ by

$$x_i' = \begin{cases} e_i + z_i' - \varepsilon \cdot \mathbf{1} & \text{if } i \in S, \\ e_i + z_i' + \frac{|S|}{|I \setminus S|} \varepsilon \cdot \mathbf{1} & \text{if } i \notin S. \end{cases}$$

Note that $x_i' - e_i$ is $\bigwedge_{i\in I} \mathscr{F}_i$-measurable and x' is a feasible allocation since $\sum_{i\in I} (z \circ [\alpha_S, \alpha^*_{-S}])_i = 0$. However, (6.2) implies that $V_i(\omega^*, x_i') > V_i(\omega^*, x_i)$ for every $i \in S$. Because $V_i(\omega, \cdot)$ is monotone for every $i \in I$, (6.1) implies that

$V_i(\omega^*, x_i') > V_i(\omega^*, x_i)$ for every $i \notin S$. Hence we have a contradiction to the fact that x is interim coarse efficient. □

Corollary 6.2: Every interim coarse efficient allocation in $\mathscr{E}$ is CBIC.

Proof: Since TCBIC implies CBIC, the conclusion follows from Proposition 6.1. □

Corollary 6.3: Every interim private efficient allocation in $\mathscr{E}$ is CBIC.

Proof: Since every interim private efficient allocation is interim coarse efficient [Proposition 5.4.2 (b)], Corollary 6.2 leads to the assertion. □

Since CBIC implies BIC, we can therefore obtain the following from Corollary 6.3.

Corollary 6.4: Every interim private efficient allocation in $\mathscr{E}$ is BIC.

Proposition 6.5: Assume that $u_i(\omega, \cdot)$ is monotone and continuous for every $\omega \in \Omega$ and for every $i \in I$. Then every weakly interim efficient allocation in ε is weakly CBIC.

Proof: Suppose $x = e + z \in \boldsymbol{A}$ is weakly interim efficient but it is not weakly CBIC. Then there exist a state $\omega^* \in \Omega$, a coalition S, and a deception $\alpha_S : \prod_{i \in S} \mathscr{F}_i \to \prod_{i \in S} \mathscr{F}_i$ such that for every $i \in S, \alpha_i(E_i(\omega')) = E_i(\omega')$ $\forall \omega' \notin E_i(\omega^*)$, $E_i(\omega^*) \in \bigwedge_{i \in S} \mathscr{F}_i$, and

$$V_i(\omega^*, e_i + (z \circ [\alpha_S, \alpha^*_{-S}])_i) > V_i(\omega^*, x_i),$$

where $e + z \circ [\alpha_S, \alpha^*_{-S}] \in \boldsymbol{A}$. Since for every $\omega' \in E^S_\alpha(\omega^*) \cap E^{-S}(\omega^*)$ it holds that $z_i(\omega') = z_i(\omega^*), i.e., (z \circ ([\alpha_S, \alpha^*_{-S}])_i(\omega^*) = z_i(\omega^*)$ for every $i \notin S$, it must be the case that for every $i \notin S$,

$$V_i(\omega', e_i + (z \circ [\alpha_S, \alpha^*_{-S}])_i) = V_i(\omega^*, x_i). \tag{6.3}$$

Since for each $\omega \in \Omega$, $V_i(\omega, \cdot)$ is continuous for every $i \in I$ by Lemma 3.1.1, there exists an $\varepsilon > 0$ such that for every $\omega \in \Omega$ and for every $i \in S$

$$V_i(\omega, e_i + (z \circ [\alpha_S, \alpha^*_{-S}])_i - \varepsilon \cdot \mathbf{1}) > \mathbf{V_i}(\omega, \mathbf{x_i}). \tag{6.4}$$

Let us define $x' = (x_i')_{i \in I}$ by

$$x_i'(\omega') = \begin{cases} e_i + (z \circ [\alpha_S, \alpha^*_{-S}])_i - \varepsilon \cdot \mathbf{1} & \text{if } i \in S, \\ x_i + \frac{|S|}{|I \backslash S|} \varepsilon \cdot \mathbf{1} & \text{otherwise.} \end{cases}$$

Note that $x' \in \boldsymbol{A}$. (6.4) implies that $V(\omega, x_i') > V_i(\omega, x_i)$ for every $\omega \in E$ and for every $i \in S$. Because $V_i(\omega, .)$ is monotone for every $\omega \in \Omega$ and for every $i \in I$, (6.3) means that $V_i(\omega, x_i') > V_i(\omega, x_i)$ for every $\omega \in \Omega$ and for every $i \notin S$. Hence x is not weakly interim efficient, a contradiction. □

Corollary 6.6: Assume that $u_i(\omega,\cdot)$ is monotone and continuous for every $\omega \in \Omega$ and for every $i \in I$. Then every HM interim efficient allocation in $\mathscr{E}$ is weakly CBIC.

Proof: It follows from Proposition 5.4.3 (b) and Proposition 6.5. □

Corollary 6.7: Assume that $u_i(\omega,\cdot)$ is monotone and continuous for every $\omega \in \Omega$ and for every $i \in I$. Then every ex ante private efficient allocation in $\mathscr{E}$ is weakly CBIC.

Proof: It follows from Corollary 5.4.5 and Proposition 6.5. □

Proposition 6.8: A weakly interim efficient allocation in $\mathscr{E}$ may not be CBIC.

Proof: Consider the same economy as in Proposition 5.4.5. The allocation $x = (x_1, x_2, x_3)$ with

$$x_1 = ((6,2),(6,2),(6,2)),$$
$$x_2 = ((7,3),(5,3),(5,3)),$$
$$x_1 = ((1,0),(1,3),(3,4))$$

is weakly interim efficient allocation but is not interim private efficient allocation. However, the allocation x is not coalitional Bayesian incentive compatible, since, at ω_2, coalition $S = \{2,3\}$ with a deception $\alpha_i(E_i(\omega)) = \{\omega_1\}$ for every $\omega \in \Omega$ and $i \in S$ will make its members better off, *i.e.*,

$$V_2(\omega_2, e_2 + (z \circ [\alpha_S, \alpha_1^*])_2) > V_2(\omega_2, x_2),$$
$$V_3(\omega_2, e_3 + (z \circ [\alpha_S, \alpha_1^*])_3) > V_3(\omega_2, x_3).$$

□

Corollary 6.9: A HM interim efficient allocation in $\mathscr{E}$ may not be CBIC.

In fact, we can show that any feasible allocation is Bayesian incentive compatible. This means that the Bayesian incentive compatibility is too weak to play a role as a condition.

Proposition 6.10: Every feasible allocation in $\mathscr{E}$ is BIC.

Proof: Suppose a feasible allocation $x \in \boldsymbol{A}$ is not BIC. Then there exist a state $\omega \in \Omega$, an agent $i \in I$, and a deception $\alpha_i : \mathscr{F}_i \to \mathscr{F}_i$ such that

$$V_i(\omega, e_i + (z \circ [\alpha_i, \alpha_{-i}^*])_i) > V_i(\omega, x_i), \tag{6.5}$$

where $e + z \circ [\alpha_i, \alpha_{-i}^*] \in \boldsymbol{A}$. Since for every $\omega' \in E_\alpha^i(\omega) \cap E^{-i}(\omega)$ it holds that $z_k(\omega') = z_k(\omega)$, *i.e.*, $(z \circ [\alpha_i, \alpha_{-i}^*])_k(\omega) = z_k(\omega)$ for every $k \neq i$, it follows from the feasibility that

$$(z \circ [\alpha_i, \alpha_{-i}^*])_i(\omega) = z_i(\omega).$$

By measurability, we obtain

$$V_i(\omega, e_i + (z \circ [\alpha_i, \alpha_{-i}^*])_i) = V_i(\omega, x_i),$$

a contradiction to (6.5). □

It is worth noting that in an economy $\overline{\mathscr{E}}$ with one good per state, the set of interim private efficient allocations coincides with the set of Bayesian incentive compatible allocations. This can be shown by combining Corollary 5.4.8 with Proposition 6.10.

Corollary 6.11: The set of interim private efficient allocations in $\overline{\mathscr{E}}$, the set of Bayesian incentive compatible allocations $\overline{\mathscr{E}}$, and the set of feasible allocations $\overline{\mathscr{E}}$, are all equivalent.

Since every BIC allocation is feasible, Proposition 6.10 implies that the set of feasible allocations is equivalent to the set of BIC allocations. Apparently, our result looks contradicting that of Myerson (1979), *i.e.*, an interim efficient allocation may not be Bayesian incentive compatible. Note that the interim efficiency of Myerson (1979) is equivalent to our HM interim efficiency except the private information measurability assumption on allocations. In view of Proposition 5.4.3 (c), our interim efficiency is stronger than that of Myerson (1979). As it will be shown with an example below, Myerson's argument is robust without the imposition of private information measurability (*i.e.*, an interim efficient allocation may not be Bayesian incentive compatible). This is still true even when our interim efficiency notion is adopted. However, when allocations are private information measurable, the adoption of our notion of interim private efficiency (Definition 5.3.2) guarantees that indeed any interim private efficient allocation is always CBIC (BIC). One may think of Corollary 6.2 as an improvement of that of Myerson (1979), in the sense that a stronger notion of interim efficiency with a simple condition (private information measurability) makes any interim private efficient allocation CBIC (BIC).

Example 6.1: Consider an economy with differential information with two agents, two goods, and two equally probable states, where utility functions, random initial endowments, and private information sets are given as follows:

$$u_1(\omega, x^1, x^2) = \sqrt{x^1 x^2}, \quad e_1 = ((10,0),(10,0)), \quad \mathscr{F}_1 = \{\{\omega_1, \omega_2\}\},$$
$$u_2(\omega, x^1, x^2) = \sqrt{x^1 x^2}, \quad e_2 = ((0,8),(0,10)), \quad \mathscr{F}_2 = \{\{\omega_1\},\{\omega_2\}\}.$$

The allocation $x = (x_1, x_2)$ with

$$x_1 = ((5,4),(5,5)),$$
$$x_2 = ((5,4),(5,5))$$

is a strongly ex post efficient (ex post efficient, or HM ex post efficient) allocation. But it is neither ex ante nor interim efficient allocation because x_i is not $\mathscr{F}_i$-measurable for $i = 1,2$. However, if we do not impose private information measurability on the allocations as in Myerson (1979), Holmström-Myerson (1983), and Palfrey-Srivastava (1987), this allocation is ex ante efficient, interim efficient, and ex post efficient. But, observe that it is not CBIC (BIC). Suppose that ω_2 is realized. Since

$$V_2(\omega_2, e_2 + (z \circ [\alpha_2, \alpha_1^*])_2) > V_2(\omega_2, x_2)$$

with $\alpha_2(E_2(\omega)) = \{\omega_1\}$ for every $\omega \in \Omega$, it is not CBIC (BIC). Therefore, this example shows that *an interim efficient allocation without private information measurability may not be* CBIC (BIC). This also illustrates that Bayesian incentive compatibility is incompatible with the ex post efficiency.

Proposition 6.12: An interim weak fine efficient allocation in $\mathscr{E}$ may not be CBIC.

Proof: Observe that the allocation x in Example 6.1 is also a weak fine efficient allocation. □

7 Are efficient and incentive compatible allocations individually rational?

Even though a mechanism is efficient, it cannot be achieved unless it is individually rational, otherwise someone may not be willing to trade. Therefore the individual rationality condition is a fundamental requirement for a mechanism. As with the efficiency notions, the individual rationality can be defined according to ex ante, interim, and ex post utility functions. In this section, we show that efficient allocations may not be individually rational.

Definition 7.1: An allocation $x \in \boldsymbol{A}$ is *interim individually rational* if for every $\omega \in \Omega$, $V_i(\omega, x_i) \geq V_i(\omega, e_i)$ holds for every $i \in I$.

An allocation $x \in \boldsymbol{A}$ is *ex ante individually rational* if the same condition holds for ex ante expected utility $\overline{V}_i$. An allocation $x \in \boldsymbol{A}^0$ is *ex post individually rational* if the same condition holds for ex post utility u_i and ex post feasible set $\boldsymbol{A}^0$.

We begin with a simple result for an economy with one good per state.

Proposition 7.1: The initial endowment is the unique interim individually rational allocation in $\bar{\mathscr{E}}$.

Proof: First of all, note that the initial endowment is interim individually rational. Suppose that a feasible allocation $x \neq e$ is individually rational. Then for every $\omega \in \Omega$ and every $i \in I$,

$$V_i(\omega, x_i) \geq V_i(\omega, e_i),$$

Since there is only one good and $x \neq e$, this implies that $x_i(\omega) \geq e_i(\omega)$ for every $\omega \in \Omega$ and $i \in I$, and $x_i(\omega^*) > e_i(\omega^*)$ for some $\omega^* \in \Omega$ and for some $i \in I$ by measurability. Thus, $\overline{e}(\omega^*) = \sum_{i \in I} x_i(\omega^*) > \sum_{i \in I} e_i(\omega^*) = \sum_{i \in I} \overline{e}(\omega^*)$, a contradiction. □

Proposition 7.2: An ex ante private efficient allocation in $\mathscr{E}$ may not be interim individually rational.

Proof: Consider an economy with differential information with three agents, one good, and three states (*i.e.*, $\Omega = \{\omega_1, \omega_2, \omega_3\}$) with equal probability

($i.e., \mu(\{\omega\}) = 1/3$ for every $\omega \in \Omega$) where utility functions, initial endowment, and private information sets are given as follows:

$$\begin{array}{lll} u_1(\omega,x) = \sqrt{x} & e_1 = (9,9,1) & \mathscr{F}_1 = \{\{\omega_1,\omega_2\},\{\omega_3\}\} \\ u_2(\omega,x) = \sqrt{x} & e_2 = (9,1,9) & \mathscr{F}_2 = \{\{\omega_1,\omega_3\},\{\omega_2\}\} \\ u_3(\omega,x) = \sqrt{x} & e_3 = (0,0,0) & \mathscr{F}_3 = \{\{\omega_1\},\{\omega_2\},\{\omega_3\}\}. \end{array}$$

It can be shown that the allocation $x = (x_1, x_2, x_3)$ is ex ante private efficient and ex ante individually rational where

$$x_1 = (8,8,2),\ x_2 = (8,2,8),\ x_3 = (2,0,0).$$

However, the initial endowment is the *unique* and interim individually rational allocation. □

Proposition 7.3: A CBIC (BIC) allocation in $\mathscr{E}$ may not be interim individually rational.

Proof: Consider an economy with differential information with two agents, one good, and two equally probable states, where utility functions, random initial endowments, and private information sets are given as follows:

$$\begin{array}{lll} u_1(\omega,x) = \sqrt{x}, & e_1 = (8,8), & \mathscr{F}_1 = \{\{\omega_1,\omega_2\}\}, \\ u_2(\omega,x) = \sqrt{x}, & e_2 = (1,1), & \mathscr{F}_2 = \{\{\omega_1\},\{\omega_2\}\}. \end{array}$$

The allocation $x = (x_1, x_2)$ with

$$x_1 = (9,9),$$
$$x_2 = (0,0)$$

is a CBIC (BIC) allocation but is not interim individually rational.

8 On the existence of individually rational and efficient allocations

8.1 Existence of individually rational and efficient allocations

Before we state the results for the individually rational and efficient allocations, we need two preliminary lemmata. These concern the properties of the upper contour set and a selection theorem.

Lemma 8.1.1: Suppose that $u_i(\omega,\cdot)$ is upper semicontinuous and concave for every $\omega \in \Omega$. Define the correspondence $P_i : \Omega \times L_{X_i} \to 2^{L_{X_i}}$ by

$$P_i(\omega, x_i) = \{x_i' \in L_{X_i} : V_i(\omega, x_i') > V_i(\omega, x_i)\}.$$

Then for every $\omega \in \Omega$, $P_i(\omega,\cdot)$ is

(a) irreflexive, convex-valued, and
(b) it has weakly open lower sections[20].

[20] Let X, Y be linear topological spaces. A correspondence $\Psi : X \to 2^X$ is said to be irreflexive if $x \notin \Psi(x)$ for every $x \in X$. A correspondence $\Psi : X \to 2^Y$ is said to have (weakly) open lower sections if for every $y \in Y$, the set $\Psi^{-1}(y) := \{x \in X : y \in \Psi(x)\}$ is (weakly) open in X.

Proof: (a) It follows from the concavity of $u_i(\omega,\cdot)$ that $V_i(\omega,\cdot)$ is concave as well and therefore the correspondence $P_i(\omega,\cdot)$ is convex-valued. It can be easily checked that $P_i(\omega,\cdot)$ is irreflexive, for if $x_i \in P_i(\omega,x_i)$ for some x_i, then $V_i(\omega,x_i) > V_i(\omega,x_i)$, a contradiction.
(b) Fix $\omega \in \Omega$. To show that $P_i(\omega,\cdot)$ has weakly open lower sections in L_{X_i}, define the correspondence $R_i : \Omega \times L_{X_i} \to 2^{L_{X_i}}$ by

$$R_i(\omega,x_i) = L_{X_i}\backslash P_i^{-1}(\omega,x_i) = \{x_i' \in L_{X_i} : V_i(\omega,x_i') \geq V_i(\omega,x_i)\}.$$

It suffices to show that $R_i(\omega,x_i)$ is weakly closed for every x_i. Fix x_i and take a net $\{x_i^\lambda\}$ such that x_i^λ converges weakly to x_i^* in L_{X_i} and $x_i^\lambda \in R_i(\omega,x_i)$. Since $x_i^\lambda \in R_i(\omega,x_i)$, it follows that $V_i(\omega,x_i^\lambda) \geq V_i(\omega,x_i)$. By Lemma 3.1.2, $V_i(\omega,\cdot)$ is weakly upper semicontinuous, *i.e.*, if x^λ converges weakly to x^*, we have $V_i(\omega,x_i^*) \geq \lim\sup V_i(\omega,x_i^\lambda)$.Notice that $\lim\sup V_i(\omega,x_i^\lambda) \geq V_i(\omega,x_i^\lambda)$. Therefore, $V_i(\omega,x_i^*) \geq V_i(\omega,x_i)$, *i.e.*, $x_i^* \in R_i(\omega,x_i)$. Hence $R_i(\omega,x_i)$ is weakly closed and we can conclude that $P_i(\omega,\cdot)$ has weakly open lower sections in L_{X_i}. □

Lemma 8.1.2: If X be a paracompact Hausdorff space and Y be a topological space. Suppose that a correspondence $\Psi : X \to 2^Y$ is non-empty-valued, convex-valued, and having open lower sections. Then there exists a continuous function $f : X \to Y$ such that $f(x) \in \Psi(x)$ for every $x \in X$.

Proof: See Theorem 3.1 in Yannelis-Prabhakar (1983). □

For the theorem below we will assume that Ω in finite. This will simplify the proof.

Theorem 8.1.3: If $u_i(\omega,\cdot)$ is upper semicontinuous and concave for every $i \in I$ and every $\omega \in \Omega$, then an interim individually rational and weakly interim efficient allocation exists in $\mathscr{E}$.

Proof: Let $\boldsymbol{B}$ be the set of all interim individually rational allocations:

$$\boldsymbol{B} = \{x \in L_X : \forall\omega \in \Omega, V_i(\omega,x_i) \geq V_i(\omega,e_i), \forall i \in I\}.$$

Since $e \in \boldsymbol{B}$, $\boldsymbol{B}$ is nonempty. Since $V_i(\omega,\cdot)$ is weakly upper semicontinuous, $\boldsymbol{B}$ is a weakly closed subset of the order interval $[0,\bar{e}]^{|I|} = [0,\bar{e}] \times \cdots \times [0,\bar{e}]$, which is weakly compact (Cartwright's Theorem). This implies that $\boldsymbol{B}$ is also weakly compact.

Let $\bigwedge_{i\in I}\mathscr{F}_i = \{E^1,E^2,\cdots,E^k,\cdots,E^n\}$ be the common knowledge partition. Fix $E^k \in \bigwedge_{i\in I}\mathscr{F}_i$. Let us define

$$\begin{aligned}\boldsymbol{B}^k &= \{x \cdot \chi_{E^k} : x \in \boldsymbol{B}\},\\ L_{X_i}^k &= \{x_i \cdot \chi_{E^k} : x_i \in L_{X_i}\},\\ L_X^k &= \{x \cdot \chi_{E^k} : x \in L_X\},\end{aligned}$$

Note that $\boldsymbol{B}^k$ is weakly compact. Define the correspondence $P_i^k : E^k \times L_{X_i}^k \to 2^{L_{X_i}^k}$ by

$$P_i^k(\omega,x_i \cdot \chi_{E^k}) = \{x_i' \cdot \chi_{E^k} \in L_{X_i}^k : V_i(\omega,x_i') > V_i(\omega,x_i)\}.$$

and define the correspondence $P^k : \boldsymbol{B}^k \to 2^{\boldsymbol{B}^k}$ by

$$P^k(x\cdot\chi_{E^k}) = \bigcap_{\omega\in E^k}\left[\prod_{i\in I} P_i^k(\omega, x_i\cdot\chi_{E^k})\bigcap \boldsymbol{B}^k\right].$$

It follows from Lemma 8.1.1 that P^k is irreflexive, convex, and has weakly open lower sections in $\boldsymbol{B}^k$.

Now let $x\cdot\chi_{E^k}\in\boldsymbol{B}^k$ and suppose that there is $x'\cdot\chi_{E^k}\in L_X^k$ such that $x_i'\cdot\chi_{E^k}\in P_i(\omega, x_i\cdot\chi_{E^k})$ for every $\omega\in E^k$. Since $x\cdot\chi_{E^k}$ belongs to $\boldsymbol{B}^k$, so does $x'\cdot\chi_{E^k}$. Therefore, P^k is nonempty-valued. Hence, there is a weakly continuous function $f: B^k\to\boldsymbol{B}^k$ such that $f(x\cdot\chi_{E^k})\in P^k(x\cdot\chi_{E^k})$ for every $x\cdot\chi_{E^k}\in\boldsymbol{B}^k$. By the Brouwer-Schauder-Tychonotf fixed point theorem, there exists a fixed point, i.e., $\bar{x}\cdot\chi_{E^k}\in f(\bar{x}\cdot\chi_{E^k})\in P(\bar{x}\cdot\chi_{E^k})$, a contradiction to the irreflexivity of P^k. Therefore, there exists $x^k\cdot\chi_{E^k}\in\boldsymbol{B}^k$ such that $P^k(x^k\cdot\chi_{E^k})=\emptyset$ for every $k=1,\cdots,n$. Construct $x^*=\sum_{k=1}^n x^k\cdot\chi_{E^k}$. It is clear that x^* is interim individually rational. To show that it is weakly interim efficient, suppose otherwise. Then there is $x'\in\boldsymbol{A}$ such that for some common knowledge event $E^k\in\bigwedge_{i\in I}\mathscr{F}_i$, $V_i(\omega,x')>V_i(\omega,x^*)$ for every $\omega\in E^k$ and for every $i\in I$. This means $x_i'\cdot\chi_{E^k}\in P_i^k(\omega,x_i^k\cdot\chi_{E^k})$ for every $\omega\in E^k$ and for every $i\in I$. Since $x^k\cdot\chi_{E^k}\in\boldsymbol{B}^k$, it follows that $x'\cdot\chi_{E^k}\in\boldsymbol{B}^k$. This contradicts that $P^k(x^k\cdot\chi_{E^k})=\emptyset$. □

Theorem 8.1.4: If the $u_i(\omega,\cdot)$ is upper semicontinuous and concave for every $i\in I$ and every $\omega\in\Omega$, then the set of ex ante individually rational and ex ante private efficient allocations of ε is nonempty.

Proof: Let $\boldsymbol{H}$ be the set of all ex ante individually rational allocations:

$$\boldsymbol{H}=\{x\in L_X:\overline{V}_i(x_i)\geq\overline{V}_i(e_i),\forall i\in I\}.$$

Since $e\in\boldsymbol{H}$, $\boldsymbol{H}$ is nonempty. Since $\overline{V}$ is weakly upper semicontinuous, $\boldsymbol{H}$ is a weakly closed subset of the order interval $[0,\bar{e}]^{|I|}=[0,\bar{e}]\times\cdots\times[0,\bar{e}]$, which is weakly compact (Cartwright's Theorem). This implies that $\boldsymbol{H}$ is also weakly compact. Define the correspondence $\bar{P}_i: L_{X_i}\to 2^{L_{X_i}}$ by

$$\bar{P}_i(x_i)=\{x_i'\in L_{X_i}:\overline{V}_i(x_i')>\overline{V}_i(x_i)\}.$$

and define the correspondence $\bar{P}:\boldsymbol{H}\to 2^{\boldsymbol{H}}$ by

$$\bar{P}(x)=\prod_{i\in I}\bar{P}_i(x_i)\bigcap\boldsymbol{H}.$$

In the same way as in Lemma 8.1.3, we can show that $\bar{P}$ is irreflexive, convex-valued, and it has weakly open lower sections in $\boldsymbol{H}$.

Now let x be an ex ante individually rational allocation. Suppose that it is not ex ante private efficient. Then there is an allocation $x'\in\boldsymbol{A}$ such that $x_i'\in\bar{P}_i(x_i)$ for every $i\in I$. Note that $x\in\boldsymbol{H}$ implies $x'\in\boldsymbol{H}$. It follows that $x'\in\bar{P}(x)$ and therefore, $\bar{P}$ is nonempty-valued. By Lemma 8.1.2, there is a weakly continuous function $f:\boldsymbol{H}\to\boldsymbol{H}$ such that $f(x)\in\bar{P}(x)$ for every $x\in\boldsymbol{H}$. By the Brouwer-Schauder-Tychonoff Theorem, there exists a fixed point $x^*=f(x^*)\in\bar{P}(x^*)$, a contradiction to the irreflexivity of $\bar{P}$. Hence we conclude that there exists an ex ante individually rational and ex ante efficient allocation. □

8.2 Nonexistence of individually rational and efficient allocations

Below we show that in well-behaved differential information economies, that is, where agents' utility functions are monotone, continuous, and concave, an interim fine efficient allocation may not exist.

Proposition 8.2.1: An interim fine efficient allocation may not exist in $\mathscr{E}$.

Proof: Consider an economy with differential information with two agents, two goods, and three equally probable states, where utility functions, random initial endowments, and private information sets are given as follows:

$$u_1(\omega, x^1, x^2) = \sqrt{x^1 x^2},\ e_1 = ((10,0),(10,0),(10,0)),\ \mathscr{F}_1 = \{\{\omega_1, \omega_2\}, \{\omega_3\}\},$$
$$u_2(\omega, x^1, x^2) = \sqrt{x^1 x^2},\ e_2 = ((10,0),(0,10),(0,10)),\ \mathscr{F}_2 = \{\{\omega_1\}, \{\omega_2, \omega_3\}\}.$$

Suppose that ω_1 is realized. Then there is no trade in that state. It follows that agent 1 will not trade at ω_2 and agent 2 will not trade ω_3, which implies that there is no trade at every state. The allocation $x = e$ is the unique feasible allocation. Consider a new allocation $x' \in \boldsymbol{A}(\bigvee_{i \in I} \mathscr{F}_i)$:

$$x_1' = ((10,0),(5,5),(5,5))$$
$$x_2' = ((10,0),(5,5),(5,5)).$$

Since $V_i(\omega_2, x_i') > V_i(\omega_2, e_i)$ for $i = 1, 2$, the initial endowment is not interim fine efficient. Hence there is no interim fine efficient allocation. □

Proposition 8.2.2: An interim individually rational and interim coarse efficient allocation need not exist in $\mathscr{E}$.

Proof: Consider an economy with differential information with two agents, two goods, and two equally probable states, where utility functions, random initial endowments, and private information sets are given as follows:

$$u_1(\omega, x) = \sqrt{x^1 x^2},\quad e_1 = ((10,2),(10,2)),\quad \mathscr{F}_1 = \{\{\omega_1, \omega_2\}\},$$
$$u_2(\omega, x) = \sqrt{x^1 x^2},\quad e_2 = ((2,10),(2,6)),\quad \mathscr{F}_2 = \{\{\omega_1\}, \{\omega_2\}\}.$$

The set of all interim coarse efficient allocation is

$$\{[((12,8),(12,8)),((0,4),(0,0))]; [((0,0),(0,0)),((12,12),(12,8))]\}.$$

Hence, no interim coarse efficient allocation is interim individually rational.

Corollary 8.2.3: An interim individually rational and interim private efficient allocation need not exist in $\mathscr{E}$.

Proof: Since the interim private efficiency implies the interim coarse efficiency [Proposition 5.4.2 (b)], the claim follows from Proposition 8.2.2. □

8.3 Compactness of the set of individually rational and efficient allocations

In this section, we show that the set of interim individually rational and interim private efficient allocations is weakly compact.

Theorem 8.3.1: If $u_i(\omega,\cdot)$ is upper semicontinuous for every $i \in I$ and every $\omega \in \Omega$, the set of interim individually rational and interim private efficient allocations of $\mathscr{E}$ is weakly compact.

Proof: Define the correspondence $P_i : \Omega \times L_{X_i} \to 2^{L_{X_i}}$ by

$$P_i(\omega, x_i) = \{x_i' \in L_{X_i} : V_i(\omega, x_i') > V_i(\omega, x_i)\}.$$

and define the correspondence $P : \Omega \times \boldsymbol{B} \to 2^{L_X}$ by

$$P(\omega, x) = \prod_{i \in I} P_i(\omega, x_i). \tag{8.3.1}$$

It follows from Lemma 8.1.1 that $P(\omega,\cdot)$ is irreflexive, and has weakly open lower sections in $\boldsymbol{B}$ for every $\omega \in \Omega$. Let M be the set of interim individually rational and interim private efficient allocations. Formally,

$$M = \{x \in \boldsymbol{B} : \forall \omega, P(\omega, x) = \emptyset\}$$

Then it follows that

$$\begin{aligned}
\boldsymbol{B} \backslash M &= \{x \in \boldsymbol{B} : \exists \omega \in \Omega, P(\omega, x) \neq \emptyset\} \\
&= \{x \in \boldsymbol{B} : \exists \omega \in \Omega \text{ and } \exists y \in P(\omega, x)\} \\
&= \{x \in \boldsymbol{B} : \exists \omega \in \Omega \text{ and } \exists y \in L_X \text{ such that } x \in P^{-1}(\omega, y)\} \\
&= \bigcup_{\omega \in \Omega} \bigcup_{y \in L_X} P^{-1}(\omega, y)
\end{aligned}$$

Since $P(\omega,\cdot)$ has weakly open lower sections, $\boldsymbol{B} \backslash M$ is weakly open. Hence M is a weakly closed subset of the weakly compact set B and therefore we can conclude that M is weakly compact. □

Notice that if $u_i(\omega,\cdot)$ is affine (a rather strong assumption which rules out risk aversion), the set M can be shown to be convex. This is parallel to the results of Myerson (1979) and Holmström-Myerson (1983) who show that if $u_i(\omega,\cdot)$ is linear, the set M is convex.

Theorem 8.3.2: If the $u_i(\omega,\cdot)$ is upper semicontinuous for every $i \in I$ and every $\omega \in \Omega$, then the set of ex ante individually rational and ex ante private efficient allocations of $\mathscr{E}$ is weakly compact.

Proof: Define the correspondence $\bar{P}_i : L_{X_i} \to 2^{L_{X_i}}$ by

$$\bar{P}_i(x_i) = \{x_i' \in L_{X_i} : \overline{V}_i(x_i') > \overline{V}_i(x_i)\}.$$

and define the correspondence $\bar{P} : \boldsymbol{H} \to 2^{L_X}$ by

$$\bar{P}(x) = \prod_{i \in I} \bar{P}_i(x_i). \tag{8.3.2}$$

Define $\boldsymbol{H}$ as in Theorem 8.1.4. Then $\boldsymbol{H}$ is nonempty and weakly compact, and $\bar{P}(\cdot)$ is irreflexive and has weakly open lower sections in $\boldsymbol{H}$. Let M^a be the set of ex ante individually rational and ex ante private efficient allocations. Formally,

$$M^a = \{x \in \boldsymbol{H} : \bar{P}(x) = \emptyset\}$$

Then it follows that

$$\begin{aligned} \boldsymbol{H} \setminus M^a &= \{x \in \boldsymbol{H} : \bar{P}(x) \neq \emptyset\} \\ &= \{x \in \boldsymbol{H} : \exists y \in \bar{P}(x)\} \\ &= \{x \in \boldsymbol{H} : \exists y \in L_X \text{ such that } x \in \bar{P}^{-1}(\omega, y)\} \\ &= \bigcup_{y \in L_X} \bar{P}^{-1}(y) \end{aligned}$$

Since $\bar{P}$ has weakly open lower sections, $\boldsymbol{H} \setminus M^a$ is weakly open. Hence M^a is a weakly closed subset of the weakly compact set $\boldsymbol{H}$ and therefore we can conclude that M^a is weakly compact. □

Theorem 8.3.3: If the $u_i(\omega, \cdot)$ is upper semicontinuous for every $i \in I$ and every $\omega \in \Omega$, then the set of ex post individually rational and ex post efficient allocations of $\mathscr{E}$ is nonempty and weakly compact.

Proof: One can proceed in a similar way as in Theorem 8.3.1. □

9 Incentive efficiency

As we saw in Example 6.1, the private information measurability of allocations is the key assumption to obtain our results. That is, without it, an interim efficient allocation may be not Bayesian incentive compatible. We now propose the concept of incentive efficiency. This notion is defined as before but no measurability conditions are imposed. Hence, we disregard the private information measurability assumption and define the concept of incentive efficiency. We show that an incentive efficient allocation exists and that the set of incentive efficient allocations is weakly compact. This kind of approach is not the first one in the literature. Holmström-Myerson (1983) show that an efficient (HM interim efficient) allocation may be not Bayesian incentive compatible and propose the concept of incentive efficiency as an appropriate efficiency concept in incomplete information environment. However, we have a different setting, *i.e.*, a differential information economy, rather than the Harsanyi model. Moreover, we allow for a continuum of states and commodities as well. We define below our incentive efficiency notion. Note that we keep all the definitions of efficiency and incentive compatibility as before except the imposition of private information measurability on allocations. An allocation $x = e + z \in \boldsymbol{A}^0$ is *incentive compatible* if for every $\omega \in \Omega$, every $i \in I$, and every deception $\alpha_i : \mathscr{F}_i \to \mathscr{F}_i$ such that

$(\alpha_i, \alpha^*_{-i})$ is compatible with F and $V_i(\omega, x_i) \geq V_i(\omega, e_i + (z \circ [\alpha_i, \alpha^*_{-i}])_i)$ where $e + z \circ [\alpha_i, \alpha^*_{-i}] \in \boldsymbol{A}^0$. Let us define

$$D_i = \{\alpha_i | \alpha_i : \mathscr{F}_i \to \mathscr{F}_i\}$$

and let

$$C_i(\omega, \alpha_i) = \{x \in \boldsymbol{A}^0 : V_i(\omega, x_i) \geq V_i(\omega, e_i + (z \circ [\alpha_i, \alpha^*_{-i}])_i) \text{ with } e + z \circ [\alpha_i, \alpha^*_{-i}] \in \boldsymbol{A}^0\}.$$

Then the set of all incentive compatible allocations is given by

$$\boldsymbol{C} = \bigcap_{\omega \in \Omega} \bigcap_{i \in I} \bigcap_{\alpha_i \in D_i} C_i(\omega, \alpha_i).$$

Definition 9.1: An allocation $x = e + z \in \boldsymbol{C}$ is *incentive efficient* if there is no $x' \in \boldsymbol{C}$ such that for some $\omega \in \Omega$, $V_i(\omega, x'_i) > V_i(\omega, x_i)$ for every $i \in I$.

It should be noted that our interim efficiency (Definition 5.3.2) is not comparable to the incentive efficiency without private information measurability because the set of feasible allocations with the private information measurability assumption is smaller but it is more difficult for the grand coalition to block them with the private information measurability assumption. Hence, neither set contains the other one. As in Myerson (1979), for the existence of incentive efficient allocations, risk neutrality is required as the theorem below indicates (compare this result with the existence results of Section 8 where risk aversion is allowed).

Theorem 9.2: If $u_i(\omega, \cdot)$ is continuous and affine $\omega \in \Omega$ and $i \in I$, there exists an incentive efficient allocation in $\mathscr{E}$.

Proof: Since $e \in \boldsymbol{C}, \boldsymbol{C}$ is nonempty. Since $V_i(\omega, \cdot)$ is weakly continuous by Lemma 3.1.3, it follows that $\boldsymbol{C}$ is a weakly closed subset of the weakly compact set $[0, \bar{e}]^{|I|} = [0, \bar{e}] \times [0, \bar{e}] \times \cdots \times [0, \bar{e}]$. This implies that $\boldsymbol{C}$ is weakly compact.

Fix $\omega \in \Omega$ and consider the maximization problem:

$$\max_{x \in \boldsymbol{C}} \sum_{i \in I} \lambda_i(\omega) u_i(\omega, x_i(\omega)),$$

where for all $i \in I$, $\lambda_i : \Omega \to \boldsymbol{R}_+$ and $\lambda = (\lambda_i)_{i \in I} \neq 0$. Since C is nonempty and weakly compact, and the maximand is weakly continuous in x, there is a solution x^ω. Then x^* with $x^*(\omega) = x^\omega(\omega)$ is an incentive efficient allocation. □

Theorem 9.3: If $u_i(\omega, \cdot)$ is continuous and affine for every $\omega \in \Omega$ and for every $i \in I$, the set of incentive efficient allocations of $\mathscr{E}$ is weakly compact.

Proof: One can proceed in a similar way as in Theorem 8.3.1. □

References

Aliprantis, C. D., O. Burkinshaw: Positive Operators. New York: Academic Press 1985

Balder, E. J., N. C. Yannelis: On the Continuity of Expected Utility. Economic Theory **3**, 625–643 (1993)

Cartwright, D.: The Order Completeness of Some Spaces of Vector Valued Functions. Bulletin of the Australian Mathematical Society **11**, 57–61 (1974)

Diestel, J., J. J. Uhl: Vector Measures, Mathematical Surveys, No. 15. Providence: American Mathematical Society 1977

Hahn, G., Nicholas C. Yannelis: Coalitional Bayesian Nash Implementation in Differential Information Economies. Working Paper, University of Illinois at Urbana-Champaign (1995)

Holmström, B., R. Myerson: Efficient and Durable Decision Rules with Incomplete Information. Econometrica **51**, 1799–1819 (1983)

Kobayashi, T.: Equilibrium Contracts for Syndicates with Differential Information, Econometrica **48**, 1635–1665 (1980)

Koutsougeras, L., N. C. Yannelis: Incentive Compatibility and Information Superiority of the Core of an Economy with Differential Information. Economic Theory **3**, 195–216 (1993)

Krasa, S., N. C. Yannelis: The Value Allocation of an Economy with Differential Information Econometrica **62**, 881–900 (1994)

Myerson, R.: Incentive Compatibility and the Bargaining Problem. Econometrica **47**, 61–74 (1979)

Palfrey, T., S. Srivastava: Implementation with Incomplete Information in Exchange Economies. Econometrica **57**, 115–134 (1986)

Radner, R.: Equilibrium under Uncertainty. Econometrica **36**, 31–58 (1968)

Wilson, R.: Information, Efficiency, and the Core of an Economy. Econometrica **46**, 807–816 (1978)

Yannelis, N. C.: The Core of an Economy with Differential Information. Economic Theory **1**, 183–198 (1991)

Yannelis, N. C., N. D. Prabhakar: Existence of Maximal Elements and Equilibria in Linear Topological Spaces. Journal of Mathematical Economics **12**, 233–245 (1983)

Incentives in market games with asymmetric information: the core*

Beth Allen**

Department of Economics, University of Minnesota, 1035 Heller Hall, 271 – 19th Avenue South, Minneapolis, MN 55455, USA (e-mail: assist@econ.umn.edu)

Received: December 26, 2001; revised version: June 11, 2002

Summary. This paper examines the *ex ante* core of a pure exchange economy with asymmetric information in which state-dependent allocations are required to satisfy incentive compatibility. This restriction on players' strategies in the cooperative game can be interpreted as incomplete contracts or partial commitment. An example is provided in which the incentive compatible core with nontransferable utility is empty; the game fails to be balanced because convex combinations of incentive compatible net trades can violate incentive compatibility. However, randomization of such strategies leads to *ex post* allocations which satisfy incentive compatibility and are feasible on average. Hence, convexity is preserved in such a model and the resulting cooperative games are balanced. In this framework, an incentive compatible core concept is defined for NTU games derived from economies with asymmetric information. The main result is nonemptiness of the incentive compatible core.

Keywords and Phrases: Incentive compatability, Core, NTU cooperative games, General equilibrium with asymmetric information, Balancedness.

JEL Classification Numbers: D82, D71, D51, D79.

---

* This work was financed, in part, by contract No 26 of the programme "Pôle d'attraction interuniversitaire" of the Belgian government, and, in part, by research grant SBR93-09854 from the U.S. National Science Foundation. Much of my thinking about this topic was developed during a wonderful visit to CORE for the 1991–1992 academic year (on sabbatical from the University of Pennsylvania). This paper was originally circulated in December 1991 as CARESS Working Paper #91-38, Center for Analytic Research in Economics and the Social Sciences, Department of Economics, University of Pennsylvania and in February 1992 as CORE Discussion Paper 9221, Center for Operations Research and Econometrics, Université Catholique de Louvain, Louvain-la-Neuve, Belgium.

** At the very start of my research, Jean-François Mertens was almost a co-author. François Forges provided detailed comments at a later stage, during my visit to THEMA, Université Cergy-Pontoise, in Spring 1997. They are entitled to the customary disclaimer.

1 Introduction

Incentive considerations are interesting in cooperative games because they provide an argument that agreements made within coalitions are self-enforcing. For example, if contracts are incomplete,[1] a fully credible verification and enforcement arrangement may be lacking. More generally, commitments may be only partially enforceable. Games with partial commitment power define an interesting territory between cooperative and noncooperative game theory.

The introduction of information into cooperative game theory was initiated by Wilson (1978) and extended, in the standard Harsanyi framework of games with incomplete information, by Myerson (1984) and Rosenmüller (1990). A series of more recent papers – i.e., as initiated by Yannelis (1991) and Allen (1991) – has examined the (NTU) core with asymmetric information, but without incentive compatibility requirements.

On the other hand, Myerson (1984) and Rosenmüller (1990) do discuss incentives, but do not analyze any core concept. Allen (1993) examines verification – the notion that a player's or coalition's strategy must be measurable with respect to the information available to other players – but this is very different from incentive compatibility. Krasa and Yannelis (1994) define the (NTU) value with incentive compatibility for a simple class of economies with private information sharing and (essentially) finitely many states of the world. Koutsougeras and Yannelis (1995) find core allocations (with private, fine, and the balanced cover of coarse information sharing) and check that private information sharing allocations satisfy incentive compatibility. However, they do not require incentive compatibility for blocking allocations. [Allen (1995 and 1999, respectively) demonstrates positive existence results for the (NTU) incentive-compatible core in large economies under either a relaxation to an approximate core concept or a dispersion hypothesis.]

Recent attention has focused on a variety of definitions of coalitional incentive compatibility. For an overview of part of this large and growing literature, see Allen and Yannelis (2001) and the references cited there.

For a class of taxation problems having a very special structure, Berliant (1992) defines and analyzes the incentive compatible core; he argues that it may, in general, be empty. In the context of financial intermediation, Boyd and Prescott (1986) and Kahn (1987) find incentive compatible core allocations directly. Marimon (1989) examines an asymmetric information core with adverse selection.

However, all of this work has failed to define the incentive compatible core in general economies with asymmetric information and to analyze when the incentive compatible core is nonempty. The work of Berliant (1992) suggests that problems can be expected to arise in general, but his model is sufficiently different from (for instance) the general exchange economy paradigm that such a conclusion is not fully justified.

In this paper, I formulate the incentive compatible core for exchange economies in which agents with asymmetric information have state-dependent utilities and initial endowments. Incentive compatibility is taken to mean that players' strategies (in

[1] See, for example, Section 3 of Hart and Holmstrom (1977) and the article by Hart and Moore (1988).

the induced cooperative game with nontransferable utility) satisfy a self-selection constraint that the state-dependent net trade must give rise to allocations that are at least as good as those for the state-dependent net trade of another state of the world. Here "at least as good" refers to the true state-dependent utility function if the player is perfectly informed and to conditional expected utility given his information otherwise. I focus on games with nontransferable utility because this is the appropriate setting for the study of incentives. Incentive compatibility or truthful revelation does not pose a problem when the goal of all coalition members is to maximize a single objective function.

My notion of incentive compatibility should be interpreted as a restriction on players' strategies rather than a specification of what happens if a lie occurs. Indeed, I do not emphasize mechanisms to define allocations when one or more players lie about their information. Instead, one performs the thought experiment in which the player receives the net trade corresponding to whatever information he announces, even though all of these net trades may not be feasible for the whole economy (unless all players lie in an internally consistent way). Any state-dependent net trade commitments that give anyone a strict incentive to lie are eliminated from the set of strategies that players may use. For example, a contract may detail outcomes based on the information provided by a party to the contract but may be independent of the information given by others in similar contracts. Any strategy that is not self-enforceable in the sense described above is simply not believed to be a credible commitment by other members of a coalition.

Existence of allocations in the incentive compatible core requires a weakening of the usual solution concept. Convex combinations of incentive compatible net trades may violate incentive compatibility, so that the usual argument to show balancedness of cooperative games derived from exchange economies fails, even though traders are assumed to have concave utilities. To bypass this obstacle, I introduce a randomization.[2] Incentive compatible core allocations thus consist of probabilities over incentive compatible state-dependent net trades such that the actual state-dependent allocation is, for almost all states of the world, feasible on average with respect to the randomization.

The next section furnishes an example of a pure exchange economy with three traders, five commodities, and five states of the world in which incentive compatibility leads to nonconvex strategy sets. As a result, the game is not balanced and its (NTU) incentive compatible core is empty. This demonstrates that the randomization is indeed necessary. The example further shows that resource feasibility must be expressed in terms of averages over the randomization within every state rather than with probability one with respect to the randomization.

The necessity for weakening the resource feasibility requirement suggests that perhaps the phenomenon should be viewed as a negative result. While this argument indeed has its rationale, asset markets, inventories and the law of large numbers could be used to justify my average feasibility condition. Nevertheless, the existence

[2] This is reminiscent of the work of Prescott and Townsend (1984a,b), who analyze competitive equilibria and Pareto optimality in large economies with asymmetric information and lotteries or many independent risks.

of such incentive compatible core allocations should be interpreted with caution. On the other hand, my analysis has the virtue of providing a cooperative game-theoretic explanation of the endogenous and strategic choice of a mechanism in the presence of asymmetric information. The choice of a cooperative solution point can be interpreted as the choice of a mechanism. While the core cannot be expected to contain, in general, a unique incentive-compatible state-dependent allocation (and hence to correspond to a unique equivalence class of mechanisms), allocations or mechanisms outside of my core can be blocked with incentive compatible outcomes so that points not in the core can be eliminated from consideration when strategic agents can cooperate within coalitions.

The remainder of this paper is organized as follows: Section 2 gives the example illustrating that incentives can destroy convexity, balancedness, and nonemptiness of the set of incentive compatible core allocations. Section 3 presents the model. Incentive constraints and randomizations are discussed in Sections 4 and 5 respectively, while Section 6 presents and proves the main result. Section 7 concludes with some remarks.

2 The example

This section presents an example which demonstrates the main ideas of the paper. The example illustrates a pure exchange economy under uncertainty in which the induced strategy sets of the cooperative game fail to be convex due to incentive compatibility constraints. The set of utility vectors attainable by the grand coalition is also nonconvex, although the nonconvexities in payoff space occur where they are inessential for the existence of points in the core. The nonconvexity of incentive-compatible strategy sets leads to a game with nontransferable utility which is not balanced and has an empty (NTU) core.

However, the introduction of randomizations over allocations restores convexity, balancedness, and nonemptiness of the core at the expense of weakening feasibility requirements. Almost sure resource feasibility is replaced by feasibility, in each state of the world, of (state-dependent) allocations on average with respect to the randomization. While randomization over almost surely feasible allocations would convexify the sets of attainable payoffs, it does not guarantee either balancedness or the existence of core allocations. Hence the original game with incentive constraints derived from my example fails to have convex feasible payoff sets, has convex strategy sets, is not balanced, and has an empty core. Randomization with almost sure feasibility is sufficient to ensure convexity of feasible payoffs, but does not solve any of the other problems. Randomization with mean feasibility enlarges the attainable utility sets further and suffices to convexify the underlying strategy sets so as to give a balanced game which therefore has a nonempty core. Yet removal of the incentive compatibility constraints would alter the game still more and lead to larger core payoffs so that my randomization operation is definitely not equivalent to the elimination of incentive compatibility.

The example features three traders, five states of the world, and five commodities, of which the first two matter in the first two states while each of the remaining goods is of consequence in exactly one state. All states are equally likely, so that total utility is proportional to expected utility (which slightly simplifies the calculations). Let $\Omega = \{a, b, c, d, f\}$ denote the set of states of the world. Use subscripts $i = 1, 2, 3$ to distinguish players, and let commodities be indicated by x, y, z, r, and t. Nonnegative amounts of all goods can be consumed in every state $s \in \Omega$.

Traders' initial endowment vectors do not depend on the state of the world in this pure exchange example. Write $e_i(s) \in \mathbb{R}^5_+$ for player i's initial endowment in state $s \in \Omega$ and suppose that $e_1(s) = (1, 1, 3, 0, 0)$, $e_2(s) = (0, 4, 0, 1, 0)$, and $e_3(s) = (4, 0, 0, 0, 1)$ for all $s = a, b, c, d, f$.

The second and third economic agents have almost identical state-dependent cardinal utilities. They are given by $u_2(x, y, z, r, t; a) = \sqrt{x}$, $u_3(x, y, z, r, t; a) = \sqrt{y}$, $u_2(x, y, z, r, t; b) = -\infty I(y < 4)$, $u_3(x, y, z, r, t; b) = -\infty I(x < 4)$, $u_2(x, y, z, r, t; c) = u_3(x, y, z, r, t; c) = z$, $u_2(x, y, z, r, t; d) = r$, $u_3(x, y, z, r, t; f) = t$ and $u_3(x, y, z, r, t; d) = u_2(x, y, z, r, t; f) = 0$.[3]

Trader 1 has rather complicated utility functions in states a and b, but straightforward utilities in the remaining three states. Let $u_1(x, y, z, r, t; c) = z/3$, $u_1(x, y, z, r, t; d) = r/2$, $u_1(x, y, z, r, t; f) = t/2$, and $u_1(x, y, z, r, t; b) = (x + y)/2$ if $x + y \leq 2$, $u_1(x, y, z, r, t; b) = 1$ if $x = 0$ and $y \geq 2$ or if $y = 0$ and $x \geq 2$, and set $u_1(x, y, z, r, t; b) > 1$ if $x > 0$, $y > 0$, and $x + y > 2$ so as to be concave and continuous. Finally, for state a, define[4] $u_1(x, y, z, r, t; a) = \frac{1}{4}\min(x + y - 1 + \sqrt{(x + y - 1)^2 + 8x}, x + y - 1 + \sqrt{(x + y - 1)^2 + 8y})$. Note that, in particular, utility along each axis in the plane is given by $(y - 1)/2$ if $x = 0$ or $(x - 1)/2$ if $y = 0$. This utility function is more easily described by its utility level m indifference surfaces as the union of the line segment joining $(0, 2m + 1)$ and (m, m) and the line segment joining (m, m) and $(2m + 1, 0)$ in the $x - y$ plane, so that the points $(3, 0)$, $(0, 3)$ and $(1, 1)$ are associated with a utility level of 1, $(4, 0)$, $(0, 4)$ and $(3/2, 3/2)$ have utility $3/2$, $(5, 0)$, $(0, 5)$ and $(2, 2)$ give utility 2, and $(7, 0)$, $(0, 7)$, and $(3, 3)$ give utility 3 regardless of trader 1's allocation of commodities z, r, and t. Note that this utility function is concave and, moreover, is the least concave representation for these indifference curves because it is linear along the diagonal.

Writing payoffs as total utilities (summed over the five states) gives 3 as the individually rational utility level for player 1 and 1 as the individual rationality constraint for players 2 and 3; these are the total utilities associated with each player's initial endowment. By interchanging their initial endowments in state a, the coalition composed of players 2 and 3 can achieve utilities of 3 each; this presumes that they can distinguish state a from state b.

With complete information and in the absence of any incentive compatibility considerations, the (total utility) imputation $(5, 3.5, 3.5)$ belongs to the core. It can be achieved with the state-dependent allocations $((4, 4, 3, 0, 0), (1, 1, 3, 0, 0), (1, 1, 0, 0, 0), (1, 1, 3, 0, 0), (1, 1, 3, 0, 0))$ for player 1, $((1, 0, 0, 1, 0), (0, 4, 0, 1, 0),$

---

[3] Define $\infty \cdot 0 = 0$ and $\infty \cdot 1 = \infty$, so that $-\infty \cdot I(y < 4) = -\infty$ if $y < 4$, $-\infty \cdot I(y < 4) = 0$ if $y \geq 4$, $-\infty \cdot I(x < 4) = -\infty$ if $x < 4$ and $-\infty \cdot I(x < 4) = 0$ if $x \geq 4$, where $I(\cdot)$ denotes the indicator function of a set.

[4] A discussion with Heraklis Polemarchakis helped me to transform the desired indifference curves into this utility function.

$(0,4,1.5,1,0)$, $(0,4,0,1,0)$, $(0,4,0,1,0))$ for player 2, and $((0,1,0,0,1)$, $(4,0,0,0,1)$, $(4,0,1.5,0,1)$, $(4,0,0,0,1)$, $(4,0,0,0,1))$ for player 3. These outcomes are individually rational and cannot be blocked by the coalition $\{2,3\}$. Clearly the allocation is Pareto optimal and hence cannot be blocked by the grand coalition. Finally, notice that it cannot be blocked by $\{1,2\}$ or $\{1,3\}$, as these coalitions can, at best, achieve utility imputations such as $(3.94, 3.5)$, $(5.19, 0)$ or $(4.95, 0.7)$.

Incentive compatibility implies that the above core allocation (as well as the latter two imputations for coalitions $\{1,2\}$ or $\{1,3\}$) cannot be sustained. Indeed, player 1 prefers his allocation in state a to his allocation in state b when the true state of the world is b. The role of the particular indifference curve specified for player 1 in state b is to demonstrate that incentives destroy convexity of sets of feasible net trades. Note that $(5,0,0,0,0)$ or $(0,5,0,0,0)$ in state a with $(1,1,0,0,0)$ in state b is an incentive compatible allocation for trader 1, but the convex combination of $(2.5,2.5,0,0,0)$ in state a violates incentive compatibility if trader 1 receives $(1,1,0,0,0)$ in state b, as he would then always claim that the state is a even when b is true. In this way, incentive considerations change the cooperative game in characteristic function form that is derived from this economy.

An easy way to see why this is true in general is to observe that if there were only one commodity and if all agents' state-dependent utilities were strictly increasing in that commodity, then the only incentive compatible allocations are those that give, to each trader, identical amounts of the good in each state that occurs with positive probability. Clearly this reduces the possibilities for risk sharing and for gains from trade based on different preferences or subjective probabilities.

To construct the cooperative game $V\colon 2^I \to \mathbb{R}^3$ in characteristic function form (where $I = \{1,2,3\}$ is the player set) derived from my economic example with incentive compatibility constraints, examine efficient, incentive compatible and feasible state-dependent commodity allocations and their associated total utility payoff vectors for each coalition. In order to do so, I must first specify the information of each player. To set notation, let $v\colon 2^I \to \mathbb{R}$ be the associated game with transferable utility, so that $v(S)$ is the total worth of coalition S [set $v(\emptyset) = 0$], and if T is any subset of some Euclidean space, let $\mathrm{comp}(T)$ denote its comprehensive hull, where $\mathrm{comp}(T) = T - \mathbb{R}^n_+$ if T is considered as a subset of $\mathbb{R}^n$.

Information will be defined for each agent so as to make player 2 or player 3 need another player in order to be able to distinguish state a from state b. Player 1 knows the precise state of the world – i.e., his information is specified by the partition $\{\{a\},\{b\},\{c\},\{d\},\{f\}\}$ of S. Players 2 and 3 each receive a signal drawn randomly, with equal probabilities, from the set $\{0,1,c,d,f\}$. These signals are correlated with states of the world as follows: If the signal received by a player is c, d, or f, then the same signal is received by the other player and that (common) signal equals the state of the world. Conditional on the state (and signal) not being equal to c, d, or f, the combinations $00, 01, 10$ and 11 are equally likely. If both players receive the same signal, the true state is a, while if they receive different signals, the state is b. Thus, player 2 or player 3 has the information partition $\{\{a,b\},\{c\},\{d\},\{f\}\}$ while all other coalitions, including $\{2,3\}$, have the complete information partition $\{\{a\},\{b\},\{c\},\{d\},\{f\}\}$.

Since incentive compatibility does not affect initial endowments – indeed, agents' endowments in the example do not depend on the state of the world – the worth of any singleton equals its worth in the game without incentive constraints. Hence $v(\{1\}) = \sum_{s \in \Omega} u_1(1,1,3,0,0;s) = 1+1+1+0+0 = 3$, $v(\{2\}) = \sum_{s \in \Omega} u_2(0,4,0,1,0;s) = 0+0+0+1+0 = 1$, and $v(\{3\}) = \sum_{s \in \Omega} u_3(4,0,0,0,1;s) = 0+0+0+0+1 = 1$ in the TU game while $V(\{1\}) = \text{comp}(v(\{1\})) \times \mathbb{R}^2$, $V(\{2\}) = \mathbb{R} \times \text{comp}(v(\{2\})) \times \mathbb{R}$, and $V(\{3\}) = \mathbb{R}^2 \times \text{comp}(v(\{3\}))$ in the NTU game. Alternatively, $V(\{1\}) = \{p \in \mathbb{R}^3 \mid p_1 \leq 3\}$, $V(\{2\}) = \{p \in \mathbb{R}^3 \mid p_2 \leq 1\}$, and $V(\{3\}) = \{p \in \mathbb{R}^3 \mid p_3 \leq 1\}$. Clearly these are nonempty closed convex comprehensive cylinder sets.

The efficient feasible allocations for $\{2,3\}$ are defined by giving all of good x to player 2 in state a (giving him a utility of $\sqrt{4} = 2$ in state a) and all of good y to player 3 in state a (for a utility of $\sqrt{4} = 2$ in state a). Goods x and y are not traded in state b, nor are r and t traded in states d and f. The resulting allocations are incentive compatible and assign three units of total utility to each player in the coalition $\{2,3\}$. Therefore, $v(\{2,3\}) = 6$ and $V(\{2,3\}) = \mathbb{R} \times \text{comp}((3,3)) = \{p \in \mathbb{R}^3 \mid p_2 \leq 3$ and $p_3 \leq 3\}$. Note that this coalition uses its combined (complete) information and neither player can gain from cheating in information revelation.

Now consider the remaining two-player coalitions, $\{1,2\}$ and $\{1,3\}$. As players 2 and 3 play symmetric roles here, it suffices to examine the incentive compatible utility imputations of one of these coalitions. Individual rationality precludes trades in state b. Therefore the (unique) efficient and incentive compatible allocation in state a gives all five units of y to player 1 and one unit of x to player 2. In state c, the three units of commodity z can be divided arbitrarily between the two players and similarly for the one unit of good r in state d. Satiation in utilities implies that state-dependent allocations of all other goods do not matter providing that they do not cause violations of incentive compatibility. Hence $V(\{1,2\}) = \{p \in \mathbb{R}^3 \mid p_1 \leq 4.5 - z/3$ and $p_2 \leq 1 + z$ for $z \in [0,3]$ or $p_1 \leq 3.5 - r/2$ and $p_2 \leq 4 + r$ for $r \in [0,1]\}$. Similarly, $V(\{1,3\}) = \{p \in \mathbb{R}^3 \mid p_1 \leq 4.5 - z/3$ and $p_3 \leq 1 + z$ for $z \in [0,3]$ or $p_1 \leq 3.5 - t/2$ and $p_3 \leq 4 + t$ for $t \in [0,1]\}$ Therefore, in the game with transferable utility, $v(\{1,2\}) = \max\{p_1 + p_2 | p \in V(\{1,2\})\} = 8$ and $v(\{1,3\}) = \max\{p_1 + p_3 | p \in V(\{1,3\})\} = 8$. Note that both $V(\{1,2\})$ and $V(\{1,3\})$, like $V(\{2,3\})$, are convex sets.

Finally, the grand coalition $\{1,2,3\} = I$ is also restricted by incentive compatibility. Individual rationality for players 2 and 3 implies that no trades occur in state b. This forces player 1 in state a to receive either 1 unit each of x and y or to receive an allocation consisting of zero units of either x or y and a quantity between three units and five units (inclusive) of the other good. If player 1 keeps his one unit of x and one unit of y while players 2 and 3 interchange their initial endowments of x and y in state a, the sum total utilities received by players is maximized. This defines worth $v(I) = 11$ of the grand coalition in the game with transferable utility. The corresponding nontransferable utility imputations are specified by $(2 + z_1/3 + r_1/2 + t_1/2, 2 + z_2 + r_2, 2 + z_3 + t_3)$ where $z_1 + z_2 + z_3 = 3$, $r_1 + r_2 = 1$, $t_1 + t_3 = 1$, and z_1, z_2, z_3, r_1, r_2, t_1, t_3 are all nonnegative. However, the imputations of this form that are individually rational for player 1 $[v(\{1\}) = 3]$ are dominated by those based on giving all five units of x or y to player 2 or 3 respectively and the remainder to player 1 in state a. This yields imputations of the form $(3 + z_1/3 + r_1/2 + t_1/2, 2.236 + z_2 + r_2, 0 + z_3 + t_3)$ and $(3 + z_1/3 + r_1/2 + t_1/2, 0 + z_2 + r_2, 2.236 + z_3 + t_3)$ where $z_1 + z_2 + z_3 = 3$, $r_1 + r_2 = 1, t_1 + t_3 = 1$, and $z_1, z_2, z_3, r_1, r_2, t_1, t_3$ are all nonnegative. Inspection of the imputations $(5, 2.236, 0)$ and $(5, 0, 2.236)$ shows that $V(I)$ is not convex; the utility vector $(5, \frac{1}{2}\sqrt{5}, \frac{1}{2}\sqrt{5})$ cannot be realized by incentive compatible allocations.

Recapitulation of the game with transferable utility gives $v(\{1\}) = 3, v(\{2\}) = v(\{3\}) = 1$, $v(\{1,2\}) = v(\{1,3\}) = 8$, $v(\{2,3\}) = 6$, and $v(\{1,2,3\}) = 11$. This game is balanced and (equivalently, by the theorem of Bondareva (1962) and Shapley (1967)) has a nonempty core. The easiest way to see this is to observe that $(5, 3, 3)$ is the unique core imputation for the TU game derived from my example with incentive compatibility constraints. However, notice that $(5, 3, 3) \notin V(I)$; this utility vector cannot be attained in the game with nontransferable utility unless incentive compatibility is violated (in which case $(5, 3.5, 3.5)$ can be achieved); the total utility of 11 in the NTU game uses the imputation $(2, 4.5, 4.5)$, which fails individual rationality for player 1.

A sufficient (see Scarf, 1967) but not necessary condition for nonemptiness of the (NTU) core of a cooperative game with nontransferable utility is balancedness. This means that for any balanced collection $\mathcal{B}$ of subsets S of I with balancing weights $\gamma_S \geq 0$ [where $\sum_{\substack{S \in \mathcal{B} \\ S \ni i}} \gamma_S = 1$ for all $i \in I$], $V(I) \supseteq \sum_{T \subseteq I} \gamma_T V(T)_T$, where $V(T)_T = \{w \in V(T) \mid w_i = 0 \text{ if } i \notin T\}$. A weaker condition which suffices for nonemptiness of the NTU core is quasibalancedness, or $\bigcap_{S \in \mathcal{B}} V(S) \subseteq V(I)$ for every balanced collection $\mathcal{B}$ of coalitions with nonnegative balancing weights. To see that my example does not generate a balanced game,[5] consider the collection $\{\{1,2\}, \{1,3\}, \{2,3\}\}$ of two-player coalitions with balancing weights $\frac{1}{2}$ each. If one takes $(3, 5, 0) \in V(\{1,2\})_{\{1,2\}}$, $(3, 0, 5) \in V(\{1,3\})_{\{1,3\}}$ and $(0, 3, 3) \in V(\{2,3\})_{\{2,3\}}$, the sum $\frac{1}{2}(3,5,0) + \frac{1}{2}(3,0,5) + \frac{1}{2}(0,3,3) = (3,4,4) \notin V(I)$, so that the game is not balanced. However, these vectors do not show that the game fails to be quasibalanced, as $(3, 3, 3) \in V(I)$. Alternatively, checking $(3\frac{5}{6}, 3, 0) \in V(\{1,2\})_{\{1,2\}}$, $(3\frac{5}{6}, 0, 3) \in V(\{1,3\})_{\{1,3\}}$, and $(0, 3, 3) \in V(\{2,3\})_{\{2,3\}}$ gives $\frac{1}{2}(3\frac{5}{6}, 3, 0) + \frac{1}{2}(3\frac{5}{6}, 0, 3) + \frac{1}{2}(0, 3, 3) = (3\frac{5}{6}, 3, 3) \notin V(I)$ but $(3\frac{5}{6}, 3, 3) \in V(\{1,2\}) \cap V(\{1,3\}) \cap V(\{2,3\})$ so that the game is neither balanced nor quasibalanced.[6] Unfortunately, this need not prove that the NTU core is empty.

Perhaps the easiest way to verify that the core actually is empty is to examine the maximum payoffs that players can obtain in $V(I)$ subject to certain other coalitions

---

[5] Contrast this to the standard case of a pure exchange economy without uncertainty, which necessarily generates a balanced game, as demonstrated by Scarf (1971).

[6] Almost by definition, the (NTU) game is superadditive: $V(T) \cap V(T') \subseteq V(T \cup T')$ whenever $T \cap T' = \emptyset$. This is true because resources are additive in an exchange economy and incentive compatibility is a restriction on an individual's allocation which is independent of the coalition, so that the union of disjoint coalitions can always choose any allocations that were permitted in the smaller coalitions.

being unable to block. The logic here is to observe that if the core is nonempty, then there are strictly Pareto optimal core allocations, where strict optimality means that no single agent can be made better off without making someone else worse off. (Contrast this to the definition of blocking by the grand coalition, which requires strict improvements for every player.) Requiring individual rationality for players 1 and 3 leads to the problem of maximizing w_2 subject to $(w_1, w_2, w_3) \in V(I)$ with $w_1 \geq 3$ and $w_3 \geq 1$. Its solution is $(3, 6.236, 1) \in V(I)$. However, this imputation can be blocked by the coalition $\{1, 3\}$, which can attain 4 for player 1 and 2 for player 3. The argument works the same way if the roles of players 2 and 3 are reversed. Alternatively, the condition that coalition $\{2, 3\}$ cannot block combined with individual rationality for players 2 and 3 yields the problem of maximizing w_1 subject to $(w_1, w_2, w_3) \in V(I)$ and $w_2 \geq 3$, $w_3 \geq 1$ (or, equivalently, $w_2 \geq 1$ and $w_3 \geq 3$). The solution is $(4.413, 3, 1) \in V(I)$, which can be blocked by $\{1, 3\}$ with the imputation of 4.45 to player 1 and 1.15 to player 3. This proves that all strictly Pareto optimal feasible allocations for the grand coalition can be blocked by some smaller coalition. Therefore, my example has an empty core.

Randomization over almost surely feasible state-dependent allocations satisfying incentive compatibility would convexify the sets $V(S)$ for all $S \subseteq I$. In my example, this randomization operation is thus equivalent to replacing $V(I)$ by its (closed) convex hull in the definition of the NTU game with incentive constraints. However, this does not suffice to give a balanced game. In fact, the solutions to the two optimization problems examined above would be the same if $V(I)$ were replaced by its convex hull. Hence, almost surely feasible randomization does not insure nonemptiness of the core.

On the other hand, randomization with feasibility on average in every state of the world does lead to the existence of an incentive compatible core. Indeed, for my example, giving probability $\frac{1}{2}$ to the allocation $((0, 5, 0, 1, 1)$, $(1, 1, 0, 1, 1)$, $(1, 1, 0, 1, 1)$, $(1, 1, 0, 1, 1)$, $(1, 1, 0, 1, 1))$ to player 1, $((4, 0, 1, 0, 0)$, $(0, 4, 1, 0, 0)$, $(0, 4, 1, 0, 0)$, $(0, 4, 1, 0, 0)$, $(0, 4, 1, 0, 0))$ to player 2 and $((0, 1, 2, 0, 0)$, $(4, 0, 2, 0, 0)$, $(4, 0, 2, 0, 0)$, $(4, 0, 2, 0, 0)$, $(4, 0, 2, 0, 0))$ to player 3 (so that total utilities are $4, 3$ and 3 while the total resource allocation is $(4, 6, 3, 1, 1)$ in state a and $(5, 5, 3, 1, 1)$ in all other states) and probability $\frac{1}{2}$ to the allocation $((5, 0, 0, 1, 1)$, $(1, 1, 0, 1, 1)$, $(1, 1, 0, 1, 1)$, $(1, 1, 0, 1, 1)$, $(1, 1, 0, 1, 1))$ to player 1, $((1, 0, 2, 0, 0)$, $(0, 4, 2, 0, 0)$, $(0, 4, 2, 0, 0)$, $(0, 4, 2, 0, 0)$, $(0, 4, 2, 0, 0))$ to player 2 and $((0, 4, 1, 0, 0)$, $(4, 0, 1, 0, 0)$, $(4, 0, 1, 0, 0)$, $(4, 0, 1, 0, 0)$, $(4, 0, 1, 0, 0))$ to player 3 (for total utilities of $4, 3$, and 3 again with a total allocation of $(6, 4, 3, 1, 1)$ in state a and $(5, 5, 3, 1, 1)$ in all other states) is an incentive compatible core allocation.

Note that this randomization is not equivalent to removal of incentive compatibility constraints. Without incentive compatibility, the above randomized core allocation could be blocked (and strictly Pareto improved) by the nonrandomized state-dependent allocation of $((2.5, 2.5, 0, 1, 1)$, $(1, 1, 0, 1, 1)$, $(1, 1, 0, 1, 1)$, $(1, 1, 0, 1, 1)$, $(1, 1, 0, 1, 1))$ to player 1, $((2.5, 0, 1.5, 0, 0)$, $(0, 4, 1.5, 0, 0)$, $(0, 4, 1.5, 0, 0)$, $(0, 4, 1.5, 0, 0)$, $(0, 4, 1.5, 0, 0))$ to player 2 and $((0, 2.5, 1.5, 0, 0)$, $(4, 0, 1.5, 0, 0)$, $(4, 0, 1.5, 0, 0)$, $(4, 0, 1.5, 0, 0)$, $(4, 0, 1.5, 0, 0))$ to player 3 which yields total utilities of 4.5, 3.08, and 3.08 respectively. Of course, this allocation violates incentive compatibility for player 1, who prefers his state a allocation to his state b

allocation when the true state of the world is b; the problem is clearly nonconvexity of player 1's set of incentive compatible allocations for state a given the allocation which is forced by the utilities of players 2 and 3 in state b.

It is relatively easy to find examples in which incentive compatibility violates convexity of strategy sets and balancedness, but the economy nevertheless admits nonrandom incentive compatible core allocations. Indeed, if one were to delete the last two states and two commodities from the example analyzed above, the resulting game is not balanced but has a nonempty core. In this case, $V(\{1\}) = (-\infty, 3] \times \mathbb{R}^2$, $V(\{2\}) = \mathbb{R} \times (-\infty, 0] \times \mathbb{R}$, $V(\{3\}) = \mathbb{R}^2 \times (-\infty, 0]$, and $V(\{2,3\}) = \mathbb{R} \times \text{comp}((2,2))$. Coalition $\{1,2\}$ can trade to the incentive compatible allocation $((0,5,3),(1,1,3),(1,1,3))$ for player 1 and $((1,0,0),(0,4,0),(0,4,0))$ for player 2 (with total utilities of 4 and 1 respectively). Similarly, $\{1,3\}$ can attain 4 and 1 using the allocations $((5,0,3),(1,1,3),(1,1,3))$ and $((0,1,0),(4,0,0),(4,0,0))$. However, placing balancing weights of one-half each on the balanced collections of two-player coalitions requires $\frac{1}{2}(4,1,0) + \frac{1}{2}(4,0,1) + \frac{1}{2}(0,2,2) = (4, 1.5, 1.5) \in V(\{1,2,3\})$, for balancenedness, which is false. Nor is the game quasi-balanced $[(4,1,1) \notin V(\{1,2,3\})]$, although it is superadditive. The efficient and individually rational points in $V(\{1,2,3\})$ are of the form $(3 + z_1/3, \sqrt{5} + z_2, z_3)$ or $(3 + z_1/3, z_2, \sqrt{5} + z_3)$ where $z_i \geq 0$ for all i and $z_1 + z_2 + z_3 = 3$. (This also shows that $V(\{1,2,3\})$ is not convex. Convexification of this set via extending strategy sets to include almost surely feasible random allocations does not lead to balanced game.) The imputations $(4, 2.236, 0)$ and $(4, 0, 2.236)$ belong to the incentive compatible core. They can be achieved by the incentive compatible (nonrandom) allocations $((0,5,3),(1,1,3),(1,1,3))$ to player 1, $((5,0,0),(0,4,0),(0,4,0))$ to player 2, and $((0,0,0),(4,0,0),(4,0,0))$ to player 3 or $((5,0,3),(1,1,3),(1,1,3))$, $((0,0,0),(0,4,0),(0,4,0))$, and $((0,5,0),(4,0,0),(4,0,0))$ respectively. Recall that $\{1,2\}$ or $\{1,3\}$ can obtain exactly 4 and 1. These two core points cannot be blocked in the convexified game with almost surely feasible random allocations, but its core also includes closed line segments with these endpoints; more precisely, the core contains all total utility vectors of the form $(4, \lambda\sqrt{5}, (1-\lambda)\sqrt{5})$ for $\lambda \in [0, 2/\sqrt{5}] \cup [1 - 2/\sqrt{5}, 1]$. If one allows randomizations that are only feasible on average, then the imputation $(3,3,3)$ belongs to the core. It can be obtained from the allocation $((0,5,0),(1,1,0),(1,1,0))$, $((4,0,0),(0,4,0),(0,4,0))$, and $((0,1,3),(4,0,3),(4,0,3))$ with probability $\frac{1}{2}$ (for a total allocation of $(4,6,3)$ in state a and $(5,5,3)$ in states b and c) and $((5,0,0),(1,1,0),(1,1,0))$, $((1,0,3),(0,4,3),(0,4,3))$, and $((0,4,0),(4,0,0),(4,0,0))$ with probability $\frac{1}{2}$ (for a total allocation of $(6,4,3)$ in state a and $(5,5,3)$ in states b and c). The (nonrandomized) transferable utility version of this example with three states has $v(\{1\}) = 3$, $v(\{2\}) = v(\{3\}) = 0$, $v(\{1,2\}) = v(\{1,3\}) = 7$, $v(\{2,3\}) = 4$, and $v(\{1,2,3\}) = 9$. Its core equals the imputation $(5,2,2)$ which cannot be achieved in the nontransferable utility game with incentive compatibility constraints.

For a somewhat more dramatic example of the difference between balancedness and the existence of core allocations with incentive compatibility, remove the third state and third commodity from the preceding example – i.e., take only the first two states and two good in the original example. Then $V(\{1\}) = (-\infty, 2] \times \mathbb{R}^2$,

$V(\{2\}) = \mathbb{R} \times \mathbb{R}_- \times \mathbb{R}$, $V(\{3\}) = \mathbb{R}^2 \times \mathbb{R}_-$, $V(\{2,3\}) = \mathbb{R} \times (-\infty, 2] \times (-\infty, 2]$, $V(\{1,2\}) = (-\infty, 3] \times (-\infty, 1] \times \mathbb{R}$, and $V(\{1,3\}) = (-\infty, 3] \times \mathbb{R} \times (-\infty, 1]$. The game is not balanced because $\frac{1}{2}(3,1,0) + \frac{1}{2}(3,0,1) + \frac{1}{2}(0,2,2) = (3, 1.5, 1.5) \notin V(\{1,2,3\})$, nor is it quasibalanced as $(3,1,1) \notin V(\{1,2,3\})$. On the other hand, it is superadditive; to check this directly verify that the points $(2,2,2)$, $(3,1,0)$, $(3,0,1)$, and $(2,0,0)$ all belong to $V(\{1,2,3\})$. The transferable utility version of the game has $v(\{1\}) = 2$, $v(\{2\}) = v(\{3\}) = 0$, $v(\{1,2\}) = v(\{1,3\}) = v(\{2,3\}) = 4$, and $v(\{1,2,3\}) = 6$. The TU core consists of the (unique) point $(2,2,2)$, which can be obtained under an incentive compatible allocation $[((1,1),(1,1)),((4,0),(0,4)),((0,4)(4,0))]$ which belongs to the NTU core. However, every Pareto optimal and individually rational imputation of the NTU game with incentive compatibility constraints belongs to its core, including also the imputations $(2.5, \sqrt{5}, 1)$, $(2.5, 1, \sqrt{5})$, $(2, \sqrt{5}, \sqrt{2})$, $(2, \sqrt{2}, \sqrt{5})$, $(3, \sqrt{5}, 0)$, and $(3, 0, \sqrt{5})$.

Finally, a further remark regarding the original example is in order. It can be perturbed slightly so as to give a pure exchange economy having strictly positive initial endowments (which are constant across the world) and strictly concave and continuous – or even smooth – utility functions for which the incentive compatible core is empty.

3 The model

To begin, let Ω be a finite set of states of the world (with typical element ω) and let μ be a probability on $(\Omega, \mathcal{F})$ where $\mathcal{F} = 2^{\Omega}$. For convenience, assume that the (subjective) probability measure μ is the same for all agents and that $\mu(\omega) > 0$ for all $\omega \in \Omega$ (otherwise reduce Ω by a μ-null set). Interpret Ω as a description of all of the relevant uncertainty in the economy, where points in Ω represent systematic risk or states of the world common to all agents. Alternatively, think of Ω as the set of all possible profiles of agents' types. Note that if Ω is infinite but each agent's information consists of a finite partition of Ω, one could redefine a finite set of states of the world by events in the pooled information partition.

Assume that there is a finite number, ℓ, of commodities potentially available in each state of the world. Take $\mathbb{R}^{\ell}_{+}$ to be the consumption set of each consumer in each state of the world.

Finitely many economic agents are present in my pure exchange economy. Let I be the set of traders (or players in the induced games), write #I for its cardinality, and use subscript i $(i \in I)$ to signify a typical individual agent.

Each trader is endowed with a nonnegative vector $e_i \in \mathbb{R}^{\ell}_{+}$ of commodities. For simplicity, these initial allocations are assumed to be constant as states of the world vary. Thus, an agent's initial endowment is always incentive compatible and does not contain any information. Nor does the economy's total resource endowment vector contain any information. If incentive compatibility were violated for state-dependent endowments, the worth of a singleton might not be well defined as the utility level that an agent can guarantee itself should be attainable with an incentive compatible allocation if incentive compatibility is required for all other coalitions.

Preferences are described by state-dependent cardinal utility functions. Expected utilities define payoffs. For $i \in I$, write $u_i \colon \mathbb{R}^{\ell}_{+} \times \Omega \to \mathbb{R}$ and assume that, for all $\omega \in \Omega$, $u_i(\cdot;\omega)$ is a continuous and concave function on $\mathbb{R}^{\ell}_{+}$. (For the generalization to infinitely many basic states of the world, $\mathcal{F}$-measurability in ω and continuity on $\mathbb{R}^{\ell}_{+}$ imply joint measurability of this mapping.)

The data of the pure exchange economy under uncertainty generates a cooperative game. The correspondence $V \colon 2^I \to \mathbb{R}^{\#I}$ defines a cooperative game with nontransferable utility if $V(\emptyset) = \{0\}$ and for all $S \subseteq I, S \neq \emptyset, V(S)$ is a nonempty closed comprehensive cylinder set. In the absence of incentive compatibility constraints, my economic model generates a balanced game with a nonempty core.[7]

4 Information and incentives

With asymmetric information, the incentive compatible core of a pure exchange economy under uncertainty should consist of exactly those feasible incentive compatible state-dependent allocations that cannot be blocked by any coalition. Here blocking requires strict improvement in expected utility using state-dependent allocations that are incentive compatible and feasible for the coalition. Equivalently, one could define the NTU game generated by the economy with incentive compatibility constraints and examine the core of the induced game. Analysis of either concept requires stating the information that each trader possesses and formulating the appropriate incentive compatibility restriction.

To specify traders' information, for each $i \in I$, let S_i be a finite set and let $\mathrm{s}_i \colon \Omega \to S_i$ be a function. Then s_i generates a finite partition P_i of Ω and a finite sub-σ-field $\mathcal{F}_i$ of $\mathcal{F}$. Interpret S_i as the set of signals that i can receive about the state of the world and P_i or $\mathcal{F}_i$ as i's initial information. Think of the sets S_i and the maps $\mathrm{s}_i \colon \Omega \to S_i$ for all $i \in I$ as common knowledge for all agents (and to the planner or mechanism designer). Take S_i also to be the set of messages that agents can implicitly communicate. In other words, an agent can convey a (true or false) subset of his actual information partition. Note that the realizations $\mathrm{s}_i(\omega)$ of i's signal are not observable to agents other than i (or to the planner or mechanism designer). Note also that "random signals" are allowed in this model in that otherwise identical "copies" of $\omega \in \Omega$ could be mapped into different elements of S_i, so that this is equivalent to expanding Ω to a larger finite set. Let $S = S_1 \times \cdots \times S_{\#I}$ with typical element $s = (s_1, \ldots, s_{\#I})$ and write $(s_i', s_{-i}) = (s_1, \ldots, s_{i-1}, s_i', s_{i-1}, \ldots \mathrm{s}_{\#I}) \in S$ and $s_{-i} \in S_{-i} = \prod_{j \neq i} S_j$. Let $\mathrm{s} \colon \Omega \to S$ be defined by $\mathrm{s}(\omega) = (\mathrm{s}_1(\omega), \ldots, \mathrm{s}_{\#I}(\omega))$ and define s_{-i} in the obvious way. Write $\mu(\cdot \mid s_i)$ and $\mu(\cdot \mid s)$ for the conditional probabilities on Ω given $s_i \in S_i$ or $s \in S$ respectively. The distribution μ on Ω and the map $\mathrm{s} \colon \Omega \to S$ induce a distribution ν on S. For $i \in I$, denote its conditional distribution on S_{-i} given $s_i \in S_i$ by $\nu_i(\cdot \mid s_i)$.

A state-dependent allocation $x_i \colon \Omega \to \mathbb{R}^{\ell}_{+}$ for trader $i \in I$ is *strongly incentive compatible* if $x_i(\cdot)$ is $\sigma(\bigcup_{i \in I} \mathcal{F}_i)$-measurable [i.e., if $x_i(\omega) = x_i(\omega')$ whenever $\mathrm{s}(\omega) = \mathrm{s}(\omega')$] and if for all $\omega \in \Omega$, $u_i(x_i(\omega);\omega) \geq u_i(x_i(\omega');\omega)$ for all $\omega' \in \Omega$.

[7] Hildenbrand and Kirman (1976) is a useful reference.

It is *Bayesian incentive compatible* if $x_i(\cdot)$ is $\sigma(\bigcup_{i\in I}\mathcal{F}_i)$-measurable and if, for all $s_i \in S_i$ and all $s'_i \in S_i$, one has

$$\sum_{\omega\in\Omega} u_i(x_i(\mathbf{s}(\omega));\omega)\mu_i(\omega \mid s_i) \geq \sum_{\omega\in\Omega} u_i(x_i(s'_i, \mathbf{s}_{-i}(\omega));\omega)\mu(\omega \mid s_i)$$

where $x_i(\mathbf{s}(\omega)) = x_i(\omega)$, $x_i(s'_i, \mathbf{s}_{-i}(\omega)) = x_i(\omega')$ if $\mathbf{s}(\omega') = (s'_i, \mathbf{s}_{-i}(\omega))$, and $x_i(s'_i, \mathbf{s}_{-i}(\omega)) = 0$ if there is no $\omega' \in \Omega$ for which $\mathbf{s}(\omega') = (s'_i, \mathbf{s}_{-i}(\omega))$. Both definitions require that allocations depend only on signals; each agent's state-dependent allocation must be measurable with respect to the joint information received – or reported – by all agents. Both formulations also capture the notion that agent i can do no better by reporting s'_i rather than s_i when his true signal is $s_i \in S_i$. Strong incentive compatibility clearly implies Bayes incentive compatibility, but not conversely. The difference is that strong incentive compatibility applies to each possible realization of $\omega \in \Omega$ separately, while the Bayes incentive compatibility requirement is stated in terms of trader i's expected utility given the signal that he has received. Strong incentive compatibility is more appropriate if players do not know the probabilities affecting their opponents. Either definition can be applied (consistently) in the remainder of this paper.

Say that an allocation belongs to the (strong or Bayes) *incentive compatible core* if it is feasible and (strongly or Bayesian) incentive compatible and if it cannot be blocked by any coalition using an allocation that is (strongly or Bayesian) incentive compatible and feasible for the coalition. More formally, the *strongly incentive compatible core* (respectively, *Bayesian incentive compatible core*) of a pure exchange economy with asymmetric information consists of $((x_i(\cdot), \ldots, x_{\#I}(\cdot))\colon \Omega \to \mathbb{R}_+^{\#I\ell}$ such that $\sum_{i\in I} x_i(\omega) = \sum_{i\in I} e_i(\omega)$ for (almost) all $\omega \in \Omega$, $x_i(\cdot)$ is strongly incentive compatible (Bayesian incentive compatible) for all $i \in I$, and there does not exist $T \subseteq I$, $T \neq \emptyset$, and $x'_i(\cdot)\colon \Omega \to \mathbb{R}_+^\ell$ for $i \in T$ such that $\sum_{i\in T} x'_i(\omega) = \sum_{i\in T} x_i(\omega)$ for (almost) all $\omega \in \Omega$ with $x'_i(\cdot)$ strongly incentive compatible (Bayesian incentive compatible) for all $i \in T$ and $\int_\Omega u_i(x'_i(\omega);\omega)\,d\mu(\omega) > \int_\Omega u_i(x_i(\omega);\omega)\,d\mu(\omega)$ for every $i \in T$. Allocations in the strongly incentive compatible core (Bayesian incentive compatible core) of an economy correspond to imputations in the (NTU) core of the strongly incentive compatible game (Bayesian incentive compatible game) $V^S\colon 2^I \to \mathbb{R}^{\#I}(V^B\colon 2^I \to \mathbb{R}^{\#I})$ with nontransferable utility defined by $V(\emptyset) = \{0\}$ and for $T \subseteq I$, $T \neq \emptyset$, $V(T) = \{(w_1, \ldots, w_{\#I}) \in \mathbb{R}^{\#I} \mid$ there exist strongly incentive compatible (Bayesian incentive compatible) $x_i\colon \Omega \to \mathbb{R}_+^\ell$ for $i \in T$ with $\sum_{i\in T} x_i(\omega) = \sum_{i\in T} e_i(\omega)$ for (almost) all $\omega \in \Omega$ and $w_i \leq \int_\Omega u_i(x_i(\omega);\omega)\,d\mu(\omega)\}$.

Any Bayesian incentive compatible allocation (and hence any strongly incentive compatible allocation) can be achieved as a Nash equilibrium of a mechanism in which each player's message space is S_i. However, to do so may require the mechanism to waste resources outside of (this) equilibrium. The reason is that initial endowment vectors may dominate the desired incentive compatible allocation for some signal realization, so that the usual device of forcing all agents to keep their endowments whenever the messages are inconsistent may not cause truthful revelation. This requires, of course, the extremely weak monotonicity assumption on state-dependent preferences that for all $i \in I$ and all $\omega \in \Omega$, $u_i(0;\omega) \leq u_i(x_i;\omega)$ for all $x_i \in \mathbb{R}^\ell_+$.

5 Randomization

To demonstrate existence of incentive compatible core allocations, one might enlarge players' strategy spaces to as to ensure balancedness of the induced game. Since balancedness is a type of convexity condition – albeit involving convex combinations of objects taken from different sets being required to belong to yet another set – extension to "mixed strategies" in the cooperative game intuitively appears to be a promising approach.

However, as the example of Section 2 proves, randomization over state-dependent allocations which are feasible with probability one for the economy serves to convexify $V(I)$ [and also $V(S)$ for $S \subseteq I$ if needed] but does not guarantee that this modified game has a nonempty core. Balancedness relates to convexification over allocations that are feasible for different coalitions in the balanced family. Because different players can belong to these coalitions, that total resource requirement may therefore also be random, although feasibility can be preserved on average in each state of the world.

My story is as follows: Agents reveal their messages, perhaps strategically, and a commonly observed and verifiable random device selects an allocation according to a known, verifiable, and agreed upon probability over allocations that are state-dependent and incentive compatible. The *ex ante* random allocation is incentive compatible (in terms of its expected utility, not the utility of its expectation) as is its *ex post* realization. In this sense, the sequencing of communication and the randomization does not matter. Messages can be sent before, during, or after the random drawing occurs in my model. When averaged over the randomization, the (state-dependent) allocations satisfy mean resource feasibility in each state of the world, although the given realization of the allocation need not be feasible.

The law of large numbers could perhaps justify mean feasibility. For instance, the deviation from feasibility vanishes if the state-dependent resource allocation problem with incentives is repeated many times or if there are many independent "copies" (i.e., "islands") of the economy or the game. Inventories could also play a role.

However, perhaps a better interpretation is to appeal to asset markets.[8] Imagine that the grand coalition (or planner or mechanism designer) buys and sells risky commodity contracts so as to offset the discrepancy between the total incentive compatible state-dependent resource allocation and the group's total initial endowment. The randomization could even be defined to depend on the outcome in the risky asset market so that after the outside commodity market's random addition to or subtraction from total resources, feasibility is exactly satisfied.

To summarize this discussion more formally, first consider the case of almost surely feasible randomizations. Define the *convexified strongly incentive compatible game* $V^{CS}: 2^I \to \mathbb{R}^{\#I}$ by $V^{CS}(T) = \text{conv}(V^S(T))$ for $T \subseteq I$. Sim-

[8] It's a pleasure to acknowledge a helpful discussion with Jacques Drèze on this point.

ilarly, define the *convexified Bayesian incentive compatible game* $V^{CB}: 2^I \to \mathbb{R}^{\#I}$ by $V^{CB}(T) = \operatorname{conv}(V^B(T))$ for $T \subseteq I$. [If S is a closed subset of $\mathbb{R}^n$, $\operatorname{conv}(S)$ denotes its (closed) convex hull.] The *convexified strongly incentive compatible core* is the core of the convexified strongly incentive compatible game and the *convexified Bayesian incentive compatible core* is the core of the convexified Bayesian incentive compatible game. Not surprisingly, core allocations in the respective incentive-constrained economies give rise to core imputations in the corresponding games. Hence, the convexified strongly (Bayesian) incentive compatible core of an economy includes precisely those randomizations over strongly (Bayesian) incentive compatible allocations that are resource feasible with probability one and that cannot be blocked by any coalition using an almost surely feasible randomization over strongly (Bayesian) incentive compatible allocations for the coalition. Somewhat more formally, these core concepts are defined by the statement that the convexified strongly (Bayesian) incentive compatible core of a pure exchange economy with asymmetric information consists of probability measures α on state-dependent allocations with $\alpha(\{(x_1(\cdot), \ldots, x_{\#I}(\cdot)) : \Omega \to \mathbb{R}_+^{\#I\ell} | \sum_{i\in I} x_i(\omega) = \sum_{i\in I} e_i(\omega)$ for (almost) all $\omega \in \Omega$ and for all $i \in I$, $x_i(\cdot)$ is strongly (Bayesian) incentive compatible$\}) = 1$ such that there does not exist $T \subseteq I, T \neq \emptyset$, and a probability measure β on state-dependent allocations for T with $\beta(\{(x'_i(\cdot))_{i\in T} : \Omega \to \mathbb{R}_+^{\#T\ell} | \sum_{i\in T} x_i(\omega) = \sum_{i\in T} e_i(\omega)$ for (almost) all $\omega \in \Omega$ and for all $i \in T$, $x'_i(\cdot)$ is strongly (Bayesian) incentive compatible$\}) = 1$ such that $\int\int u_i(x_i(\omega); \omega)\, d\mu(\omega)\, d\alpha(x_i(\cdot)) < \int\int u_i(x'_i(\omega); \omega)\, d\mu(\omega)\, d\beta(x'_i(\cdot))$ for all $i \in T$.

It remains to define these concepts when the randomization is only required to satisfy resource feasibility on average. The term "modified" signifies this distinction. The *modified strongly (Bayesian) incentive compatible game* $V^{MS}: 2^I \to \mathbb{R}^{\#I}$ $(V^{MB}: 2^I \to \mathbb{R}^{\#I})$ is defined by $V^{MS}(\emptyset) = V^{MB}(\emptyset) = \{0\}$ and, for $T \subseteq I, T \neq \emptyset$, $V^{MS}(T)$ (respectively $V^{MB}(T)$) equals the set $\{(w_1, \ldots, w_{\#I}) \in \mathbb{R}^{\#I} |$ there exists a probability measure α on state-dependent allocations for coalition T, with $\alpha(\{(x_i(\cdot))_{i\in T} : \Omega \to \mathbb{R}_+^{\#T\ell} | x_i(\cdot)$ is strongly (Bayesian) incentive compatible$\}) = 1$ and $\sum_{i\in T} \int x_i(\omega)\, d\alpha(x_i(\cdot)) = \sum_{i\in T} e_i(\omega)$ for (almost) every $\omega \in \Omega$, for which $w_i \leq \int\int u_i(x_i(\omega); \omega)\, d\mu(\omega)\, d\alpha(x_i(\cdot))$ for all $i \in T\}$. Then the modified strongly (Bayesian) incentive compatible core of the economy equals the set of randomized strongly (Bayesian) incentive compatible allocations, with resource feasibility on average, that yield imputations in the (NTU) core of the modified strongly (Bayesian) incentive compatible game. To state this more explicitly, the *modified strongly (Bayesian) incentive compatible core* of a pure exchange economy with asymmetric information consists of those probability measures α on state-dependent allocations such that $\alpha(\{(x_1(\cdot), \ldots, x_{\#I}(\cdot)) : \Omega \to \mathbb{R}_+^{\#I\ell} | x_i$ is strongly (Bayesian) incentive compatible for all $i \in I\}) = 1$, $\sum_{i\in I} \int x_i(\omega)\, d\alpha(x_i(\cdot)) = \sum_{i\in I} e_i(\omega)$ for (almost) all $\omega \in \Omega$, and there is no coalition $T \subseteq I$ $(T \neq \emptyset)$ with probability measure β on state-dependent allocations for T such that $\beta(\{(x'_i(\cdot))_{i\in T} : \Omega \to \mathbb{R}_+^{\#T\ell} | x'_i(\cdot)$ is strongly (Bayesian) incentive compatible for each $i \in T\}) = 1$, $\sum_{i\in T} \int x'_i(\omega)\, d\beta(x'_i(\cdot)) = \sum_{i\in T} e_i(\omega)$ for (almost) all $\omega \in \Omega$, and $\int\int u_i(x_i(\omega);\omega)d\mu(\omega)\,d\alpha(x_i(\cdot)) < \int\int u_i(x_i'(\omega);\omega)\,d\mu(\omega)\,d\beta(x_i'(\cdot))$ for all $i \in T$.

Since my main result does not depend on the difference between strong incentive compatibility and Bayes incentive compatibility, call either concept the modified game and the modified incentive compatible core. To simplify notation, write $\tilde{V}: 2^I \to \mathbb{R}^{\#I}$ for either V^{MS} or V^{MB}. Call $\tilde{V}$ the modified NTU game with incentive compatibility constraints and randomization with feasibility on average.

6 Nonemptiness of the modified incentive compatible core

The purpose of this section is to state and prove the main result that there are (state-dependent) allocations in the modified incentive compatible core (as defined in Section 5), or, equivalently, that the modified game has a nonempty core. As discussed above, this holds because balancedness follows from the extension of strategy sets to permit randomization over state-dependent and incentive compatible allocations that are feasible on average for every state of the world.

Theorem. *The modified NTU game $\tilde{V}: 2^I \to \mathbb{R}^{\#I}$ with incentive compatibility constraints (and randomization with feasibility on average) is balanced and its (modified incentive compatible) core is nonempty.*

Proof. Let $\mathcal{B}$ be a balanced family of coalitions with (nonnegative) balancing weights $\gamma_T \geq 0$ for $T \in \mathcal{B}$. I need to show that $\sum_{T\in\mathcal{B}} \gamma_T \tilde{V}(T)_T \subseteq \tilde{V}(I)$. First choose $w^T \in V(T)_T$ for $T \in \mathcal{B}$. Recall that this means $w^T \in V(T) \subseteq \mathbb{R}^{\#I}$ and $w_i^T = 0$ if $i \notin T$. By definition, there are (nonrandom) incentive compatible (state-dependent) allocations $x_i^T: \Omega \to \mathbb{R}_+^\ell$ such that, for (almost) all $\omega \in \Omega$,

$$\sum_{i\in T} x_i^T(\omega) = \sum_{i\in T} e_i(\omega)$$

and for all $i \in T$

$$\sum_{\omega\in\Omega} u_i(x_i^T(\omega);\omega)\mu(\omega) \geq w_i^T.$$

Assign $i \in I$ the random allocation which equals $x_i^T(\cdot)$ with probability γ_T if $i \in T$. Since

$$\sum_{\substack{T\in\mathcal{B}\\ T\ni I}} \gamma_T = 1 \qquad \text{for all} \qquad i \in I,$$

this defines a probability measure λ on state-dependent allocations for the grand coalition. Feasibility holds on average because

$$\begin{aligned}
&\sum_{i\in I}\sum_{T\ni i} x_i^T(\omega)\lambda(\{x_1'(\cdot),\ldots,x_{\#I}'(\cdot))|x_i'(\omega) = x_i^T(\omega)\})\\
&= \sum_{T\in\mathcal{B}}\sum_{i\in T} \gamma_T x_i^T(\omega) = \sum_{T\in\mathcal{B}} \gamma_T \sum_{i\in T} x_i^T(\omega)\\
&= \sum_{T\in\mathcal{B}} \gamma_T \sum_{i\in T} e_i(\omega) = \sum_{i\in I} e_i(\omega).
\end{aligned}$$

If, in place of $w_T \in V(T)_T$, we had $w_T \in \tilde{V}(T)_T$, then there would exist finitely many (nonrandom) incentive compatible allocations $\tilde{x}_i^T : \Omega \to \mathbb{R}_+^\ell$ such that the resulting random allocation is feasible on average and always generates expected utility w_i^T. Repetition of the above argument completes the proof that $\tilde{V}: 2^I \to \mathbb{R}^{\#I}$ is balanced. By Scarf's (1967) theorem, $\tilde{V}$ therefore has a nonempty core. Hence there exist (random) allocations in the modified incentive compatible core. □

*Remark.*The same argument shows that $\tilde{V}$ is totally balanced and hence that all subgames have nonempty modified incentive compatible cores.

7 Remarks

1. If for all agents $i \in I$ and all pairs $\omega, \omega' \in \Omega$ of states of the world, the state-dependent preferences represented by the utilities $u_i(\cdot;\omega)$ and $u_i(\cdot;\omega')$ on $\mathbb{R}_+^\ell$ satisfy the "single crossing" property, then the randomization is not necessary for the existence of state-dependent allocations in the incentive compatible core. In this case, incentive compatibility may require the disposal of some resources. As one would expect, even without randomization the core and the incentive compatible core are (generally) distinct.
2. If there are infinitely many states of the world (and especially if Ω is an uncountable set and the support of μ fails to be at most countable), closedness of the $V(S)$ sets may prove problematic since incentive compatibility destroys the convexity of strategy sets. Technically, the problem is that the proof in Allen (1991) that the NTU game $V: 2^I \to \mathbb{R}^{\#I}$ is well defined (based on the Theorem of Dunford and Pettis) relies on the fact that strongly closed *convex* subsets of an $\mathcal{L}^1$ space are also weakly closed or, equivalently, that strongly continuous and concave utilities must be weakly upper semicontinuous functions.
3. Communication requirements or communication restrictions could be modeled as exogenous information sharing rules. Such limits to the endogeneity of information revelation constraints should satisfy the boundedness condition for information sharing rules to avoid destroying balancedness of the resulting NTU game. See Allen (1991).
4. The incentive compatibility constraints in the definition of the (modified) incentive compatible core could be altered to reflect the specific asymmetric information situation. For instance, particular restrictions may not be desirable if they involve states of the world that agents know. However, such modifications move one away from the mechanism story.

References

Allen, B.: Market games with asymmetric information and nontransferable utility: Representation results and the core. CARESS Working Paper #91–09 (1991)

Allen, B.: Market games with asymmetric information: Verification and the publicly predictable information core. Hitotsubashi Journal of Economics (Special issue: Proceedings of Tokyo Conference in Honor of Lionel McKenzie) **34**, 101–122 (1993)

Allen, B.: Incentives in market games with asymmetric information: Approximate NTU cores in large economies. In: Barnett, W.A., Moulin, H., Salles, M., Schofield, N.J. (eds) Social choice, welfare and ethics: Proceedings of the Eighth International Symposium on Economic Theory and Econometrics, Caen, France, (1992). Cambridge: Cambridge University Press 1995

Allen, B.: On the existence of core allocations in a large economy with incentive compatibility constraints. In: Wooders, M.H. (ed) Topics in game theory and mathematical economics in honor of Robert Aumann. Fields Institute Communications, American Mathematical Society (1999)

Allen, B., Yannelis, N.C.: Differential information economies: introduction. Economic Theory **18**, 263–273 (2001)

Berliant, M.: On income taxation and the core. Journal of Economic Theory **56**, 121–141 (1992)

Bondareva, O.N.: Theory of core in an n-person game. Vestnik Leningradskogo Universiteta, Seriia Matematika, Mekhaniki i Astronomii **13**, 141–142 (1962)

Boyd, J.H., Prescott, E.C.: Financial intermediary-coalitions. Journal of Economic Theory **38**, 211–232 (1986)

Hart, O., Holmstrom, B.: The theory of contracts. In: Bewley, T. (ed) Advances in economic theory. Cambridge: Cambridge University Press 1985

Hart, O., Moore, J.: Incomplete contracts and renegotiation. Econometrica **56**, 755–785 (1988)

Hildenbrand, W., Kirman, A.P.: Introduction to equilibrium analysis. Amsterdam: North Holland 1976

Kahn, J.A.: Endogenous financial structure in an economy with private information. Rochester Center for Economic Research Working Paper No. 96 (1987)

Koutsourgeras, L., Yannelis, N.C.: Incentive compatibility and information superiority of the core of an economy with differential information. Economic Theory **3**, 195–216 (1995)

Krasa, S., Yannelis, N.C.: The value allocation of an economy with differential information. Econometrica **62**, 881–900 (1994)

Marimon, R.: The core of private information economies. UAB/IAE Discussion Papers W.P. 131.90, Universitat Autónoma de Barcelona (1989)

Myerson, R.B.: Cooperative games with imcomplete [sic] information. International Journal of Game Theory **13**, 69–96 (1984)

Prescott, E.C., Townsend, R.M.: Pareto optima and competitive equilibria with adverse selection and moral hazard. Econometrica **52**, 21–45 (1984a)

Prescott, E.C., Townsend, R.M.: General competitive analysis in an economy with private information. International Economic Review **25**, 1–20 (1984b)

Rosenmüller, J.: Fee games: (N)TU games with incomplete information. Institut für Mathematische Wirtschaftsforschung Working Paper No. 190, Universität Bielefeld (1990)

Scarf, H.E.: On the existence of a cooperative solution for a general class of n-person games. Journal of Economic Theory **3**, 169–181 (1971)

Scarf, H.E.: The core of an n-person game. Econometrica **35**, 50–69 (1967)

Shapley, L.S.: On balanced sets and cores. Naval Research Logistics Quarterly **14**, 453–460 (1967)

Wilson, R.: Information, efficiency and the core of an economy. Econometrica **46**, 807–816 (1978)

Yannelis, N.C.: The core of an economy with differential information. Economic Theory **1**, 183–198 (1991)

Cooperative *interim* contract and re-contract: Chandler's M-form firm★

Tatsuro Ichiishi[1,2] **and Murat R. Sertel**[2,3]

[1] Department of Economics, Ohio State University, Columbus, OH 43210-1172, USA
[2] Center for Economic Design, Boğaziçi University, TURKEY
[3] Department of Economics, Faculty of Economic and Administrative Sciences, Boğaziçi University, 80815 Bebek – Istanbul, TURKEY

Received: May 17, 1996; revised version: January 14, 1997

Summary. At an *interim* stage players possessing only their private information freely communicate with each other to coordinate their strategies. This results in a core strategy, which is interpreted as an equilibrium set of players' alternative type-contingent contract offers to their fellows. From this set of offers each player then chooses an optimal one and engages in some subsequent action, thus possibly revealing some private information to the others. Now with new information thus obtained from each other, the players play a new game to re-write their contract. In all of the optimization and gaming just described, Bayesian incentive compatibility plays a central role. These ideas are formulated within a model of a profit-center game with incomplete information which formally describes interaction of the asymmetrically informed profit-centers in Chandler's multidivisional firm.

JEL Classification Numbers: D2, D8, L2.

1 Introduction

The present paper proposes new concepts and a new scenario with which to analyze cooperative *interim* contracting and re-contracting in an organization. Cooperation is understood here simply as coordination of strategies. Individuals have diverse (and most likely conflicting) interests, yet they talk to each other and coordinate their strategies, because by doing so they better serve their diverse interests. Since a firm is created as an organization within which these individuals can take advantage of coordination of strategies,

★ Many thanks to Nicholas Yannelis and an anonymous referee for their thoughtful suggestions, which significantly improved the paper. We owe the entire idea of Remark 3.5 of this paper, including its incisive example, to the referee, whom we hereby thank for this contribution.

Correspondence to: T. Ichiishi (USA address)

analysis of cooperative choice of strategies is essential to the theory of the firm.

Yannelis [18] opened up a lively revival of research in strategic cooperative game theory in the presence of incomplete information. Most works in this literature study *ex ante* cooperative choice of strategies (type-contingent actions); see, e.g., the literature discussions by Ichiishi and Idzik [6, Section 6]. When applied to the theory of contracting, therefore, they are concerned with contracts signed at a time prior to players' receiving their respective private information. In the absence of information to be handled at this stage of the game, information processing is also missing during the signing of contracts.[1] Information-processing becomes an issue only after a contract is established and the players receive their individual private information.

On the other hand, noncooperative principal-agent theory offers a successful analysis of contracts signed at the *interim* stage (when each agent is in possession of his private information). Here, the very act of signing a contract transmits and processes information, for an agent's signing will generally reveal some of his private information, making it public.[2]

The purpose of the present paper is to propose a conceptual framework within which to analyze a cooperatively determined set of *interim* contracts. Such a framework would be appropriate for analyzing problems of several privately informed principals, who freely communicate with each other in order to reach a coordinated decision about contract offers, and who also function as agents. The scenario arises in decision-making within an organization or across organizations, rather than in a *market* (e.g., insurance market) where customer or sellers are typically anonymous and so their coordination of strategies is precluded.

Basic to the proposed framework are two key ideas derived from the above two strands (the strategic cooperative game theory with incomplete information, and the principal-agent theory) in the theory of information processing. First, the *ex ante* solution of the cooperative theory of games with incomplete information is suitably strengthened here, and its strengthened version is interpreted as a set of type-contingent contract offers. Second, as in the theory of principal and agent, private information is revealed through the act of choosing a contract from this set.

An organization is formed in order to perform tasks in accordance with an initially agreed contract. It is essential, therefore, to study the members' subsequent actions. By this time, however, private information of the members has been made public, at least partially. So, given this updated information, the players can re-evaluate their committed strategies and possible alternatives. Re-contracting may or may not occur.

---

[1] Possible exceptions are the pioneering work by Wilson [17] and the more recent contributions of Berliant [1] and Demange and Guesnerie [3], all of which set out to analyze a cooperative *interim* contract.

[2] Recall the pioneering work of Rothschild and Stiglitz [12], proposing the concept of separating equilibrium in insurance markets.

The proposed framework for *interim* contracting and re-contracting thus consists of three consecutive games: (1) an *interim* game which determines a set of type-contingent contract offers; (2) choice of an *interim* contract from among the offered contracts; and (3) an *interim* game which determines a re-contract. In all three games, Bayesian incentive compatibility will play a central role. Relationships between the first game and Wilson's [17] *interim* cooperative games will be discussed in Remark 3.4. The second game is actually not a genuine game but merely a collection of individual optimization problems, each division i choosing from the offered contracts one that is best for i. The third game will turn out to be essentially a classical static game.

In parallel with Rothschild and Stiglitz's [12] equilibrium, which they emphasized to be *ex post* inefficient, so also does the cooperatively determined *interim* contract prove to be *ex post* inefficient. Even a re-contract, agreed upon at a later stage, will turn out to be unable to remove this *ex post* inefficiency.

The present paper formulates the above ideas on *interim* contracting and re-contracting, and on *ex post* welfare loss, by using Ichiishi and Radner's [8] recent model of profit-center game with incomplete information, a formal model to describe interaction of the asymmetrically informed profit-centers (divisions) in Chandler's [2] firm in multidivisional form (M-form firm). There are several reasons why this model is chosen here: First, the significance of M-form firms in the present-day economy has increasingly been recognized. Second, in spite of their significance, the neoclassical paradigm and noncooperative game theory have failed to analyze the workings of such firms. Third, strategy-choice and information-processing in an M-form firm (indeed, in any organization) are intrinsically cooperative, so what need to be applied here is strategic cooperative game theory with incomplete information. Fourth, by simplicity of the model, one can easily obtain clear-cut conclusions.

The first simple result of the present paper (Proposition 4.1) says that every *ex ante* core contract of the profit-center game is *interim* equilibrium type-contingent contract offers of the first game. (Sufficient conditions for the existence of an *ex ante* core contract were established by Ichiishi and Radner [8].) The second simple result (Proposition 4.2) is the existence of a core re-contract of the third game, both for the convex case and for the case characterized by a structural relationship among the divisions (the [complementary supplier]-customer relationship). This structural relationship was proposed first by Radner [11], and covers non-convex as well as convex cases. A re-contract improves the welfare of each division. The example at the end of Section 4 (Example 4.4) illustrates in the full-information-revealing case how a re-contract removes the inefficiency that was caused by Bayesian incentive compatibility. Nevertheless, *ex post* efficiency can still not be achieved, due to the constraint still imposed by the information-pooling rule.

Section 2 reviews the model of a profit-center game with incomplete information and the *ex ante* core contract concept. Section 3 develops a scenario of *interim* contracting and re-contracting. Section 4 presents simple existence results and an example to illustrate the present *interim* concepts.

2 *Ex Ante* contracting in a profit-center game

A model of a profit-center game with incomplete information was introduced by Ichiishi and Radner [8], as a Bayesian extension of Radner's [11] formulation of Chandler's [2] firm in multidivisional form. This section briefly reviews Ichiishi and Radner's model.

The firm is divided into finitely many *divisions*, indexed by $i \in I$. Denote[3] by $\mathscr{I}$ the family of all nonempty subsets of I; $\mathscr{I} := 2^I \setminus \{\emptyset\}$. A member of $\mathscr{I}$ is called a *coalition* of divisions. The goods, services, and resources that are used, produced, bought or sold by the divisions are called *commodities*. The commodities here are classified into two categories: those k_m commodities, called *market commodities*, that are bought or sold by the divisions, and those k_n commodities, called *nonmarket commodities*, that are owned or produced by the firm and used internally. Nonmarket commodities can be multilaterally exchanged among the divisions. Denote by K_m the set of k_m market commodities, and by K_n the set of k_n nonmarket commodities. Let $k := k_m + k_n$, $K := K_m \cup K_n$.

Division i's possible types are described by a finite set T_i. Define $T_S := \prod_{i \in S} T_i$ for each $S \in \mathscr{I}$, and write $T := T_I$ for simplicity. Formally, division i's private information structure is defined as the algebra $\mathscr{T}_i$ on T generated by the cylinder-sets, $\{\{t_i\} \times T_{I \setminus \{i\}} \mid t_i \in T_i\}$. Each division i knows all the objective *ex ante* probabilities π_j on T_j, $j \in I$. Define $\pi(t) := \prod_{i \in I} \pi_i(t_i)$. *It is postulated*[4] *that* $\pi \gg \mathbf{0}$.

A *net output plan* of division i is a function $y_i : T \to \mathbf{R}^k$, which assigns to each type-profile t a net output contingent upon t, $y_i(t) := (m_i(t), n_i(t))$; the subvectors $m_i(t) \in \mathbf{R}^{k_m}$ and $n_i(t) \in \mathbf{R}^{k_n}$ correspond to the market and nonmarket commodities, respectively. The usual sign convention is adopted here, so a coordinate of $y_i(t)$ is positive or negative according to whether the corresponding commodity is a net output or a net input of division i. The *production set* $Y_i \subset \mathbf{R}^{k|T|}$ is the set of all technologically feasible net output plans of i, where $|T|$ is the cardinality of T. Define $Y_S := \prod_{i \in S} Y_i$, and $y_S := (y_i)_{i \in S}$. Function y_S is a *net output plan* of coalition S.

Given a type-profile $t \in T$, the vector $r_i(t) \in \mathbf{R}^{k_n}$ of *initial resources* describes the quantities of nonmarket commodities that are assigned to (or owned by) division i contingent upon t before the production process.

---

[3] The notation "$a := b$" means "a is equal to b by definition."

[4] For two vectors π and π', $\pi \geq \pi'$ means $\pi(t) \geq \pi'(t)$ for all t, $\pi > \pi'$ means $[\pi \geq \pi'$ and $\pi \neq \pi']$, and $\pi \gg \pi'$ means $\pi(t) > \pi'(t)$ for all t.

A net output will be sold in the market at a price vector $p \in \mathbf{R}^{k_m}$. The profit[5] that division i brings in to the firm by selling its net output of market commodities $m_i(t)$ is the inner product $p \cdot m_i(t)$. In the present partial equilibrium analysis, the price vector p is exogenously given. *It is postulated that* $p \gg \mathbf{0}$.

Definition 2.1 A *profit-center game with incomplete information* is a list of exogenously given data, $\mathscr{D} := (\{T_i,\ \pi_i\}_{i \in I},\ \{Y_i,\ r_i(\cdot)\}_{i \in I},\ p)$, such that $\pi_i \gg \mathbf{0}$ for all $i \in I$, and $p \gg \mathbf{0}$.

Given a type-profile $t \in T$, a *profit imputation* of coalition S is a vector $x_S(t) := (x_i(t))_{i \in S}$ whose ith coordinate is to be interpreted as the accounting profit attributed to division i. The incentive of division i is to claim as high a profit as possible for each t. A *profit imputation plan* of coalition S is a function $x_S : T \to \mathbf{R}^S$, $t \mapsto x_S(t)$. A pair of a profit imputation plan and a net output plan, (x_S, y_S), is called a *plan* of coalition S. A net output plan involves transfer of nonmarket commodities from a division to another, but their monetary values cannot be computed in the market (since nonmarket commodities do not have prices). Nevertheless, a profit imputation plan takes into account how much payment should be made for such transfer.

The *ex ante* stage is defined as the stage in which no division knows the true type of any division. The *interim* stage is defined as the stage in which each division in coalition S knows its own true type and possibly has some information about the true types of some other divisions in S. The *ex post* stage is the stage in which each division knows the true type-profile in T.

Ichiishi and Radner [8] presented a scenario of how a production process in coalition S takes place throughout the *interim* stage, in accordance with a contract which was agreed upon at the preceding *ex ante* stage. Here, an *ex ante* contract is identified with a plan (x_S, y_S). There are two periods of the *interim* stage; the first period for the divisions' simultaneous supply/demand decisions about some commodities, and the second period for supply/demand decisions about the other commodities and for imputation of the profit that is made by sale/purchase of the market commodities.

With K_1 (K_2, resp.) denoting the index set for the commodities that are produced/used in the first period (in the second period, resp.), the family $\{K_1, K_2\}$ is a partition of K, possibly different from $\{K_n, K_m\}$. Set $k_1 := |K_1|$, $k_2 := |K_2|$. A net output plan y_i may be written as $y_i =: (y_{1i}, y_{2i})$; the functions $y_{1i} : T \to \mathbf{R}^{k_1}$, $t \mapsto y_{1i}(t)$, and $y_{2i} : T \to \mathbf{R}^{k_2}$, $t \mapsto y_{2i}(t)$, correspond to K_1 and K_2, respectively. Define $y_{1S} := (y_{1i})_{i \in S}$, $y_{2S} := (y_{2i})_{i \in S}$.

Let $\mathscr{T}_S$ be the algebra on T generated by $\{\mathscr{T}_i\}_{i \in S}$, $\mathscr{T}_S := \vee_{i \in S} \mathscr{T}_i$. This is an upper bound for the information structures that are available to members in S during the *interim* stage. A function on T is $\mathscr{T}_S$-measurable, iff it depends only upon t_S. Denote by F_S the set of all *technologically attainable* plans of a

[5] The present definition of profit is different from the neoclassical definition of entrepreneurial profit, in that it need not reflect the cost of resources, such as capital, that are not under the control of the divisions.

coalition S, that is, the set of all $\mathscr{T}_S$-measurable functions $(x_S, y_S) : T \to \mathbf{R}^{(1+k)|S|}$ such that y_S is technologically feasible, i.e.,

$$y_S \in Y_S,$$

and such that the total resource constraint is satisfied within S, i.e.,

$$\forall\, t \in T : \sum_{i \in S} \begin{pmatrix} x_i(t) \\ \mathbf{0} \end{pmatrix} \leq \sum_{i \in S} \begin{pmatrix} p \cdot m_i(t) \\ n_i(t) + r_i(t) \end{pmatrix} .$$

Not all plans in F_S are available to S as candidates for *ex ante* contracts. The two postulates (see Postulates 2.2, 2.3) below impose further economically meaningful conditions.

First, each division i's plan should reflect the information available to i at the time of its execution. In order to present this constraint, definition of some notation is in order: Let W be any vector space, and let $f : T \to W$ be any function. Denote by $\mathscr{A}(f)$ the algebra generated by f, that is, the smallest algebra on T that contains the family of sets $\{f^{-1}(w) \mid w \in W\}$. Let $\mathscr{B}$ be any algebra on T. Denote by $E(f \mid \mathscr{B}) : T \to W$ the conditional expectation of f given algebra $\mathscr{B}$, defined as

$$E(f \mid \mathscr{B})(t) := \frac{1}{\pi(B(t))} \sum_{s \in B(t)} f(s)\pi(s) ,$$

where $B(t)$ is the minimal element of $\mathscr{B}$ that contains t. When $\mathscr{B} = \mathscr{T}_S$, one may write $E(f \mid \bar{t}_S) := E(f \mid \mathscr{T}_S)(\bar{t})$.

For a division i's strategy to be feasible, it has to prescribe the same action for two type-profiles that i cannot discern. The private information structure that division i holds at the beginning of the first *interim* period is $\mathscr{T}_i$, so y_{1i} has to be $\mathscr{T}_i$-measurable. Assuming that choice c_{1i} is made according to the true type, the other divisions infer that i's true type must be in the subset $(y_{1i})^{-1}(c_{1i})$ of T_i. In other words, the information structure $\vee_{j \in S} \mathscr{A}(y_{1j})$ becomes common knowledge in coalition S at the beginning of the second period. Define

$$\hat{\mathscr{T}}_i(y_{1S}) := \mathscr{T}_i \vee (\vee_{j \in S} \mathscr{A}(y_{1j})) .$$

The members can design y_S at the *ex ante* stage, so that each (x_i, y_{2i}) is $\hat{\mathscr{T}}_i(y_{1S})$-measurable. Thus, one can make the following postulate:

Postulate 2.2 (Information-Pooling Rule) The members of coalition S can design only those plans (x_S, y_S) such that for each $i \in S$ it follows that

(i) y_{1i} is $\mathscr{T}_i$-measurable,
(ii) (x_i, y_{2i}) is $\hat{\mathscr{T}}_i(y_{1S})$-measurable.

Second, the members of coalition S need to have the common knowledge that each division has the incentive to make a choice in the first period according to its true type. After all, the agreement will not be enforced, if some member has the incentive to make a choice with false pretensions about its true type either during the first period or during the second period. If the members of S foresee at the time of designing a plan that a particular plan (x_S, y_S) may later induce such false pretensions, they do not agree on the plan (x_S, y_S). They will consider only those plans that are Bayesian incentive-compatible. Bayesian incentive compatibility is defined in the next two paragraphs.

A *pretension function* of division i is a function $\sigma : T_i \to T_i$, which says that when its true type is t_i, it acts (makes a choice) as though its type were $\sigma(t_i)$. In the first period, division i's information structure is given as $\mathscr{T}_i$, which is private. So, division i is not caught in the first period no matter which choice it makes from $\{y_{1i}(t_i) \mid t_i \in T_i\}$; that is, it can make a choice according to any pretension function σ, so that when division i's true type is $\bar{t}_i$, it makes the choice $y_{1i}(\sigma(\bar{t}_i))$. By acting according to the function $y_{1i} \circ \sigma$, division i with its true type $\bar{t}_i$ passes on to all the other members of S the information that event $A := y_{1i}^{-1}(y_{1i} \circ \sigma(\bar{t}_i))$ has occurred. This information may be false, that is, $\bar{t}_i$ may not be a member of A, but the other members take it as i's testimony about itself and expect that i will act according to this information in the second period, that is, i will have to make a choice from $(x_i, y_{2i})(A)$ in the second period. Therefore, i's pretension function in the second period has to be of the form $\tau \circ \sigma$ for some pretension function τ which maps each minimal set of $\mathscr{A}(y_{1i})$ into itself. The idea of pretension here follows the *pretend-but-perform principle* of Sertel [15, Ch. 6, pp. 77–80], later developed by Koray and Sertel (see, e.g., Koray and Sertel [9, 10].) Accordingly, when this principle is made operant, agents are allowed to *pretend* to have their chosen types, *but* must thereafter *perform* so as not to belie them.

The present concept of Bayesian incentive compatibility says that division i cannot benefit from any pair of pretension functions that are not caught. Formally, a plan (x_S, y_S) is called *Bayesian incentive-compatible*, if for all $i \in S$, all pretension functions $\sigma : T_i \to T_i$, and all pretension functions $\tau : T_i \to T_i$ which map each minimal set of $\mathscr{A}(y_{1i})$ into itself,

$$\forall\, t \in T:\ E\Big(x_i \,\Big|\, \hat{\mathscr{T}}_i(y_{1S})\Big)(t) \geq E\Big(x_i \circ (\tau \circ \sigma, \mathrm{id}) \Big| \hat{\mathscr{T}}_i(y_{1S})\Big)(t)\ ,$$

where id is the identity map on $T_{S\setminus\{i\}}$.

Postulate 2.3 (Bayesian Incentive Compatibility) The members of coalition S can design only Bayesian incentive-compatible plans.

The following fact, taken from Ichiishi and Radner [8, Fact 6.2 (i)], characterizes the Bayesian incentive compatibility.

Fact 2.4 *Let $(x_S, y_S) \in F_S$ be a plan which satisfies the information-pooling rule. Then, it is Bayesian incentive-compatible, iff*

$$\forall\, i \in S:\ \forall\, t_i, t_i' \in T_i:\ \forall\, \bar{t}_{I\setminus\{i\}} \in T_{I\setminus\{i\}}:$$
$$E\Big(x_i \,\Big|\, \hat{\mathscr{T}}_i(y_{1S})\Big)(t_i, \bar{t}_{I\setminus\{i\}}). \;=\; E\Big(x_i \,\Big|\, \hat{\mathscr{T}}_i(y_{1S})\Big)(t_i', \bar{t}_{I\setminus\{i\}})\ .$$

Let $\hat{F}_S$ be the set of all plans in F_S that satisfy Postulates 2.2 and 2.3. Plan (x_S, y_S) is a candidate for coalition S's agreement iff $(x_S, y_S) \in \hat{F}_S$.

Suppose the grand coalition I is considering a possible agreement $(x_I^*, y_I^*) \in \hat{F}_I$ at the *ex ante* stage. Coalition S will improve upon (x_I^*, y_I^*) if it finds a plan $(x_S, y_S) \in \hat{F}_S$ such that, for each member i, the expectation Ex_i of x_i is greater than the expectation Ex_i^* of x_i^*. If plan (x_I^*, y_I^*) is not improved upon by any coalition, it becomes a self-enforcing agreement. An *ex ante core plan* is a plan of the grand coalition which is feasible and which cannot be improved upon by any coalition; this is signed by all members of I as an *ex ante contract*.

Definition 2.5 An *ex ante core plan* of profit-center game with incomplete information $(\{T_i, \pi_i\}_{i \in I}, \{Y_i, r_i(\cdot)\}_{i \in I}, p)$ is a plan (x_I^*, y_I^*) of the grand coalition I such that

(i) $(x_I^*, y_I^*) \in \hat{F}_I$; and
(ii) it is not true that there exist $S \in \mathscr{I}$ and $(x_S, y_S) \in \hat{F}_S$ such that for all $i \in S$, $Ex_i > Ex_i^*$.

In the present section on an *ex ante* contract, it is explicitly interpreted that the game is over as soon as a contract (strategy bundle) is signed. The rest of the time sequence (that is, the *interim* stage) is spent for contract-execution. A contract, once agreed upon, is binding in the subsequent periods.

Remark 2.6 Ichiishi and Radner [8] proposed a weaker notion of *ex ante* core plan than the present notion (Definition 2.5). They addressed the existence issue within the framework of a modified profit-center game by further postulating the headquarters' insurability. Actually, they established existence theorems for the present (stronger) notion of *ex ante* plan in the modified game. The present paper studies the unmodified model, because the headquarters' insurability postulate turns out to be irrelevant to the results (Propositions 4.1 and 4.2) and the welfare analysis (Example 4.4) given here. □

3 A story of *interim* contracting and re-contracting

The divisions of the firm may wish to make a contract at the *interim* stage, when the true type of each division i is i's private information. By signing such an *interim* contract, a division typically passes on to the other divisions (a part of) its private information. Furthermore, having obtained possibly parts of others' private information through their contracts, some members may wish to re-write their contracts. The purpose of this section is to present a scenario for *interim* contracting and re-contracting. On the one hand, while the presentation here is within the framework of the profit-center game $\mathscr{D} := (\{T_i, \pi_i\}_{i \in I}, \{Y_i, r_i(\cdot)\}_{i \in I}, p)$, it is given in such a way that the scenario is readily extended to a more general framework of a two-*interim*-period Bayesian society. On the other hand, the present section covers only the specific instances of the profit-center games given by

$$\forall\, i \in I: \; Y_i = \prod_{t_i \in T_i} Y_i(t_i)$$

(see Remarks 3.3, 4.5 and 5.3 in Ichiishi and Radner [8]). Only these instances are meaningful in the present study of the profit-center game, because each division already has its private information at the time of signing a contract. It is also postulated that each r_i depends only upon t_i.

For signing a contract to pass on information, there have to be several alternative contract offers. By choosing one offer, division i reveals to the other divisions the information that, of all the available offers, i's chosen contract gives him the highest conditional expected imputation given his true type (assuming the Bayesian incentive compatibility), hence the others' deduction about i's true type. This setup is certainly true for the principal-agent model of a labor contract in a setting where the principal offers several alternative contracts to laborers and each laborer chooses one of them. It turns out that mere offers of alternative contracts do not work in the present context of profit-center game, due to the simplicity of each division's preference relation (division i's utility level is identified with i's profit imputation). Therefore, an additional structure among the offered *interim* contracts will be introduced.

As in the previous section, there are two *interim* periods. An *interim contract* $(\eta_{1I}, (\eta_{2I}, \xi_I))$ specifies for each division i: (1) its initial activity η_{1i} $(\in \mathbf{R}^{k_1})$ chosen in the first period, (2) its subsequent activity $\eta_{2i}(t_{I\setminus\{i\}})$ $(\in \mathbf{R}^{k_2})$ chosen in the second period contingent upon each type-profile $t_{I\setminus\{i\}}$ of the other members, which is identified with the function $\eta_{2i} : T_{I\setminus\{i\}} \to \mathbf{R}^{k_2}$, and (3) a level of profit imputed to i in the second period contingent upon each type-profile of the other members, $\xi_i : T_{I\setminus\{i\}} \to \mathbf{R}$.

A division i's *ex ante* plan (x_i, y_i) in the profit-center game is interpreted as type-contingent *interim* contracts,

$$t_i \mapsto (y_{1i}(t_i), (y_{2i}(t_i, \cdot), x_i(t_i, \cdot))) \; .$$

This type-contingency is the additional structure needed for the set of alternative contracts offered to i, $\{(y_{1i}(t_i), (y_{2i}(t_i, \cdot), x_i(t_i, \cdot))) \mid t_i \in T_i\}$.

The scenario consists of a sequence of several games. In the first stage, the divisions play a cooperative game whose outcome is a set of type-contingent *interim* contract offers. As yet there is no action of taking an offer, so division i's true type $\bar{t}_i$ remains its private information throughout this stage. For a detailed discussion of why information is not processed during this step, see Remark 3.5 below.

The game in the first stage is played much in the same way as in the preceding section; in particular, a feasible strategy bundle for coalition S is a member (x_S, y_S) of $\hat{F}_S$. However, the equilibrium concept here is stronger, because the game is played at an *interim* stage. Suppose that the grand coalition I is considering a possible strategy bundle $(x_I^*, y_I^*) \in \hat{F}_I$. If there exists a coalition S and its strategy $(x_S, y_S) \in \hat{F}_S$ such that each member $i \in S$

feels better off with (x_S, y_S) than with (x_I^*, y_I^*) when evaluating them based on its private information, that is,

$$\forall\, i \in S:\ E(x_i \mid \bar{t}_i) > E(x_i^* \mid \bar{t}_i)\ ,$$

then the members S will defect from the grand coalition I. One might, therefore, propose a coalitional stability condition on (x_I^*, y_I^*) as:

$$\neg\, \exists\, S \in \mathscr{I} :\ \exists\, (x_S, y_S) \in \hat{F}_S :\ \forall\, i \in S:\ E(x_i \mid \bar{t}_i) > E(x_i^* \mid \bar{t}_i)\ .$$

This condition is not operational, however, since nobody in $I \setminus \{i\}$ knows i's true type. A stronger, but operational coalitional stability condition says that no coalition S can improve upon (x_I^*, y_I^*) regardless of whichever type-profile is true; formally,

$$\neg\, \exists\, S \in \mathscr{I} :\ \exists\, t_S \in T_S : \exists\, (x_S, y_S) \in \hat{F}_S :\ \forall\, i \in S:\ E(x_i \mid t_i) > E(x_i^* \mid t_i)\ .$$

This is a coalitional strengthening of Harsanyi's [5] Bayesian equilibrium concept. A strategy bundle $(x_I^*, y_I^*) \in \hat{F}_I$ satisfying this (stronger) condition is defined as equilibrium type-contingent contract offers; sometimes it is also called an *interim core strategy*.

Definition 3.1 *Equilibrium type-contingent contract offers* of profit-center game with incomplete information $\mathscr{D}$ are a plan (x_I^*, y_I^*) of the grand coalition I such that

(i) $(x_I^*, y_I^*) \in \hat{F}_I$; and
(ii) it is not true that there exist a coalition $S \in \mathscr{I}$, its type-profile $t_S \in T_S$ and a plan $(x_S, y_S) \in \hat{F}_S$ such that for all $i \in S$, $E(x_i \mid t_i) > E(x_i^* \mid t_i)$.

Remark 3.2 A *mechanism* is defined as a function $f : M \to Z$ from a message-profile space M to an outcome space Z; typically a message is a type. In the standard principal-agent theory (see, e.g., Rothschild and Stiglitz [12]), the principal designs a mechanism f and offers to the agents the family of *interim* contracts, $\{f(m) \mid m \in M\}$. In the present profit-center game, type-contingent contract offers $f_I := (x_I, y_I) : T \to \mathbf{R}^{(1+k)|I|}$ may be considered a mechanism, endogenously determined by the divisions I. The family of *interim* contracts offered to division i is then defined as the set $\{f_i(t_i, \cdot) \mid t_i \in T_i\}$. □

In the second stage, each division i, having the private information of its true type $\bar{t}_i$, signs one of the contracts available to i,

$$(y_{1i}^*(t_i), (y_{2i}^*(t_i, \cdot), x_i^*(t_i, \cdot))),\ \ t_i \in T_i\ .$$

By the Bayesian incentive compatibility, it chooses the contract

$$(\bar{\eta}_{1i}, (\bar{\eta}_{2i}(\cdot), \bar{\xi}_i(\cdot))) := (y_{1i}^*(\bar{t}_i), (y_{2i}^*(\bar{t}_i, \cdot), x_i^*(\bar{t}_i, \cdot)))\ .$$

This is not a genuine multiperson game but an individual optimization problem.

As division i chooses $(\bar{\eta}_{1i}, \bar{\eta}_{2i}(\cdot), \bar{\xi}_i(\cdot))$ and is engaged in its initial activity $\bar{\eta}_{1i}$ during the first *interim* period, $i \in I$, the information is passed on to the other divisions that i's true type should be in the set

$$\bar{T}_i := (x_i^*, y_i^*)^{-1}(\bar{\xi}_i(\cdot), \bar{\eta}_{1i}, \bar{\eta}_{2i}(\cdot)) \ .$$

(Here, the strong structure of type-contingency for the set of contract offers is used.) Let $\bar{T}_S := \prod_{i \in S} \bar{T}_i$, and $\bar{T} := \bar{T}_I$, and let $\bar{\pi}_i$ be the conditional probability of π_i given $\bar{T}_i$. Let $\bar{Y}_{2i}$ be the set of type-profile-contingent second-period activities $y_{2i} : \bar{T} \to \mathbf{R}^{k_2}$ that are technologically feasible, given choice $\bar{\eta}_{1i}$ of the first period; formally,

$$\bar{Y}_{2i} := \{y_{2i} : \bar{T} \to \mathbf{R}^{k_2} \mid \forall\, t \in \bar{T} : (\bar{\eta}_{1i}, y_{2i}(t)) \in Y_i(t_i)\} \ .$$

Notice that $\bar{\xi}_i(\cdot)$, $\bar{\eta}_{2i}(\cdot)$ are constant functions on $\bar{T}_{I\setminus\{i\}}$ due to the measurability requirement on x_i^*, y_{2i}^*, so one may denote their values by $\bar{\xi}_i$, $\bar{\eta}_{2i}$. While the divisions $I \setminus \{i\}$ still cannot identify i's true type in the set $\bar{T}_i$, they know i's contract, $(\bar{\xi}_i, \bar{\eta}_{2i})$ $(\in \mathbf{R} \times \mathbf{R}^{k_2})$. The third stage is for the latter part of the *interim* period when every division has already observed the event $\bar{T}$. At this stage, division i can choose activity $\bar{\eta}_{2i}$ and receive imputation $\bar{\xi}_i$ as specified in the contract, or in the case of re-contracting it faces the one-period profit-center game,

$$\bar{\mathscr{D}} := (\{\bar{T}_j, \bar{\pi}_j\}_{j \in I}, \bar{Y}_{2j}, r_{2j}(\cdot)\}_{j \in I},\ p) \ . \tag{1}$$

Since the profit imputation $\bar{\xi}_i$ is guaranteed to i as long as i insists on legal enforcement of the contract, value $\bar{\xi}_i$ serves as i's *disagreement point* in game $\bar{\mathscr{D}}$. Division i's *plan* in game $\bar{\mathscr{D}}$ is a pair of a net output plan $y_{2i} : \bar{T} \to \mathbf{R}^{k_2}$ and a profit imputation plan $x_i : \bar{T} \to \mathbf{R}$ for the second *interim* period. Denote by $\bar{\mathscr{T}}_i$ the algebra on $\bar{T}$ generated by the cylinders $\{\{t_i\} \times \bar{T}_{I\setminus\{i\}} \mid t_i \in \bar{T}_i\}$. No further information about the others' types is available to i during the re-contracting period, so i's plan $(y_{2i}, x_i) : \bar{T} \to \mathbf{R}^{k_2+1}$ has to be *allowable* in the sense that it is $\bar{\mathscr{T}}_i$-measurable, that is, it depends only upon t_i. (Yannelis [18] called this measurability requirement the private information case, and Ichiishi and Idzik [6] called it the I-NP case (information non-pooling case).)

Suppose divisions S are to defect from the grand coalition by signing a re-contract among themselves. Such a defection would be *approvable* by the remaining divisions i, if they were to receive from S their disagreement points $\bar{\xi}_i$, $i \in I \setminus S$; see Sertel [16] for the role of *approved exit*, albeit in a different economic context. Denote by G'_S the set of all *allowable*, *approvable*, *technologically attainable*, and *individually rational* plans of coalition S in game $\bar{\mathscr{D}}$, that is, the set of plans $(x_S, y_{2S}) : \bar{T} \to \mathbf{R}^{(1+k_2)|S|}$ such that each (x_i, y_{2i}) is $\bar{\mathscr{T}}_i$-measurable, $i \in S$, such that y_{2S} is technologically feasible, i.e.,

$$y_{2S} \in \bar{Y}_{2S} \ ,$$

such that the total resource constraint is satisfied within S, i.e.,

$$\forall\, t \in \bar{T} : \sum_{i \in I\setminus S} \begin{pmatrix} \bar{\xi}_i \\ \mathbf{0} \end{pmatrix} + \sum_{i \in S} \begin{pmatrix} x_i(t_i) \\ \mathbf{0} \end{pmatrix} \leq \sum_{i \in S} \begin{pmatrix} p \cdot m_i(t_i) \\ n_{2i}(t_i) + r_{2i}(t_i) \end{pmatrix} ,$$

where $m_{1i}(t_i)$ is given by the corresponding subvector of $\bar{\eta}_{1i}$, and such that the profit imputation is no less than the disagreement point for some type-profile, i.e.,

$$\forall\, i \in S:\ \exists\, t_i \in \bar{T}_i:\ x_i(t_i) \geq \bar{\xi}_i\ .$$

(The last condition is weaker than the meaningful condition,

$$\forall\, i \in S:\ x_i(\bar{t}_i) \geq \bar{\xi}_i\ ,$$

but it is operational.)

The Bayesian incentive compatibility of plan (x_S, y_{2S}) for the I-NP case is defined by

$$\forall\, i \in S:\ \forall\, t_i',\ t_i'' \in \bar{T}_i : \bar{E}(x_i \mid t_i') \geq \bar{E}(x_i(t_i'') \mid t_i')$$

where the expectation $\bar{E}$ is taken over $\bar{T}$ with respect to probability $\bar{\pi}$, and $x_i(t_i'')$ is identified with the constant function $t_i \mapsto x_i(t_i'')$. Denote by $\hat{G}_S$ the set of all Bayesian incentive-compatible plans in G_S'. (The present condition of Bayesian incentive compatibility was used in the I-P case by Ichiishi and Idzik [6] in a more general setup.)

A *core re-contract* is a Bayesian incentive-compatible plan of the grand coalition in game $\bar{\mathscr{D}}$ which cannot be improved upon by any coalition:

Definition 3.3 A *core re-contract* of the one-period profit-center game $\bar{\mathscr{D}}$ is a plan (x_I^{**}, y_{2I}^{**}) of the grand coalition such that

(i) $(x_I^{**}, y_{2I}^{**}) \in \hat{G}_I$; and
(ii) it is not true that there exist $S \in \mathscr{I}$, $t_S \in \bar{T}_S$ and $(x_S, y_{2S}) \in \hat{G}_S$ such that for all $i \in S$, $\bar{E}(x_i \mid t_i) > \bar{E}(x_i^{**} \mid t_i)$.

If there is a core re-contract, the divisions will agree on it at the beginning of the second *interim* period. Otherwise, they will choose $(\bar{\xi}_I, \bar{\eta}_{2I})$ as specified in the original contract.

The possibility of re-contracting arises, because given the updated information (summarized by the event $\bar{T}$) in the third stage, the divisions realize that the initial contracts they drew (in the first stage) and signed (in the second stage) are inefficient. This does not mean that the contracts were made irrationally. Under imperfect information, only those contracts that obey the information-pooling rule are feasible, and only the Bayesian incentive-compatible contracts can truthfully be executed. The divisions institute the Bayesian incentive compatibility in order to level off the difference in imputations (with respect to types), in fact in order to bring the imputations down to the lowest level, so that there be no incentive for misrepresentation. By engaging in re-contract upon update of information, the divisions are taking back the removed excess of the imputations over the lowest level.

Remark 3.4 Wilson [17] extended the core of a static pure exchange economy and the α-core of a game in normal form to the situations in which players differ in their endowments of information structure. Here, player i's information structure is defined as an algebra $\mathcal{T}_i$ on an arbitrarily given finite state space.[6] Let u_i be player i's von Neumann-Morgenstern utility function defined[7] on his pure-strategy space X_i. Consider coalition S which is trying to see whether its joint strategy bundle x_S improves upon a prevailing strategy x^* of the grand coalition. Wilson focused on two cases, based on the information structure $\mathscr{C}_S$ defined as the members' common information, $\mathscr{C}_S := \wedge_{i \in S} \mathcal{T}_i$, and the information structure $\mathscr{F}_S$ defined as the members' fully pooled information, $\mathscr{F}_S := \vee_{i \in S} \mathcal{T}_i$. (1) In the first case, the members conclude that x_S improves upon x^* (hence form a defecting coalition S), if there exists a nonnull event $A \in \mathscr{C}_S$ such that $Eu_i(x_i \mid \mathcal{T}_i) > Eu_i(x_i^* \mid \mathcal{T}_i)$ on A. (2) In the second case, the members conclude that x_S improves upon x^* (hence form a defecting coalition S), if (i) by using information structure $\mathscr{F}_S$, there exists a nonnull event $A \in \mathscr{F}_S$ such that $Eu_i(x_i \mid \mathscr{F}_S) > Eu_i(x_i^* \mid \mathscr{F}_S)$ on A, or (ii) by using information structure $\mathscr{C}_S$, there exists a nonnull event $A \in \mathscr{C}_S$ such that $Eu_i(x_i \mid \mathscr{C}_S) > Eu_i(x_i^* \mid \mathscr{C}_S)$ on A. He called the core and α-core concepts *coarse* or *fine*, accordingly as they correspond to cases (1) or (2), respectively.[8]

Wilson's strategy concept is the same as the strategy concept of the present first game which determines type-contingent *interim* contract offers, in the sense that both are a function from the type-profile space T to an action space. The two differ in that, while Wilson is not explicit about the measurability requirement on strategy bundles,[9] the present paper imposes a specific measurability on feasible strategies. Also Bayesian incentive compatibility is postulated in the present paper as a part of the feasibility condition, and its implication is explored.

There is an important difference between Wilson [17] and the present paper: In the present paper, endogenous information-processing through signing a contract is explicitly modelled. Indeed, the outcome of the first game is interpreted as a *family* of *interim* contract offers, and each division chooses and signs a contract, hence reveals its true type at least partially, in the optimization problem of the second stage (Remark 3.2).

Another difference between the two works is the role that various information structures play. Wilson's coarse core and coarse α-core concepts are not considered here because, due to the extreme asymmetry of infor-

---

[6] The framework of the present paper is more specific than his in that the state space here is given as a type-profile space.

[7] The extension of this definition to the normal-form game is straightforward.

[8] The above definition of "blocking" in the second case is precisely Wilson's general definition [17, p. 813, the third paragraph] applied to the situation where two communication structures are available to coalition S: one in which every member of S has access to information structure $\mathscr{F}_S$ as a communication structure, and the other in which every member of S has access to information structure $\mathscr{C}_S$. It seems, however, that in his demonstration of the non-existence of a fine core allocation (Wilson [17, p. 814, the first paragraph]) he replaces condition (ii) by the following more plausible condition: By using information structure $\mathscr{C}_S$, there exists a nonnull event $A \in \mathscr{C}_S$ such that $Eu_i(x_i \mid \mathcal{T}_i) > Eu_i(x_i^* \mid \mathcal{T}_i)$ on A.

[9] In one situation he defines it as the measurability with respect to $\mathscr{F}_S$ for case (1); see the measurability of d_i^M in the second-from-the-last paragraph on page 815 in Wilson [17].

mation modelled here (independence of the algebras held by any two players), the common information $\mathscr{C}_S$ is the trivial algebra $\{\emptyset, T\}$ for a non-singleton coalition S, so the coalitional stability condition of a coarse core strategy bundle x^* against S's strategy bundle x_S becomes

$$\neg \, \forall \, i \in S: \; E(x_i \mid \mathscr{T}_i) > E(x_i^* \mid \mathscr{T}_i) \text{ on } T \ ,$$

which is always satisfied by an *ex ante* core strategy bundle; that is, an *ex ante* core strategy bundle which satisfies *interim* individual rationality is *by definition* a coarse *interim* core. Wilson's fine core and fine α-core concepts are not considered here either, because the issue here is how the members of a coalition endogenously pool their private information (Postulate 2.2), so that member i will eventually have an information structure that is finer than $\mathscr{T}_i$ and is coarser than $\mathscr{F}_S$. □

Remark 3.5 In general, private information may be revealed in the process of coalition formation, which in turn influences whether or not the coalition can eventually be formed. In the present specific context of a profit-center game, however, this information revelation at the coalition-formation stage does not occur. To illustrate information revelation in coalition-formation, consider the following example. Let $S = \{1, 2\}$, $T_i = \{H_i, L_i\}$, $\pi_i(H_i) = \pi_i(L_i) = 1/2$, $i \in S$, and consider two strategies $x_S^*, x_S : T \to \mathbf{R}^S$ defined by

$$x_i^*(t) := 1, \text{ for all } i \text{ and } t \ ;$$

$$x_1(t) := \begin{cases} 4, & \text{if } t = (H_1, L_2) \ , \\ 0, & \text{otherwise} \ ; \end{cases}$$

$$x_2(t) := \begin{cases} 4, & \text{if } t = (L_1, H_2) \ , \\ 0, & \text{otherwise} \ ; \end{cases}$$

Suppose S contemplates whether to use x_S to improve upon x_S^*. Notice that

$$E(x_i \mid H_i) = 2 > 1 = E(x_i^* \mid H_i), \text{ for every } i \in S \ ,$$

so S can improve upon x_S^* when the true type-profile is $\bar{t} = (H_1, H_2)$. According to Definition 3.1 (ii), therefore, x_S^* could not be an *interim* core strategy. However, player 1 knows that player 2 agrees to the joint strategy x_S only when 2's true type is H_2, since

$$E(x_2 \mid L_2) = 0 < 1 = E(x_2^* \mid L_2) \ .$$

Then player 2's agreement to x_S reveals the information to player 1 that 2's true type is H_2. Given this information, player 1 does not agree to x_S since

$$x_1(t_1, H_2) = 0 < 1 = x_1^*(t_1, H_2), \text{ for } t_1 = H_1, L_1 \ .$$

Thus, strategy x_S cannot serve as a "blocking" strategy against x_S^*.

The fact that information revelation does not occur in the present profit-center game is clear from the observation that any strategy $(x_S, y_S) \in \hat{F}_S$ is Bayesian incentive-compatible, so $E(x_i \mid \mathscr{T}_i)$ is constant on T for every $i \in S$ (Fact 2.4). For a "blocking" strategy would then have to make every $i \in S$

better off (with respect to the *interim* expected imputation) for all possible type-profiles. Thus, the fact that division i joins a particular coalition does not reveal any information to the other divisions of the coalition. □

4 The story specific to the Bayesian profit-center game

The first simple result of this paper is the following characterization of *interim* equilibrium type-contingent contract offers as an *ex ante* core plan. Existence results for the latter were established by Ichiishi and Radner [8], so they also serve as existence results for the former in view of this characterization result.

Proposition 4.1 *A plan of a profit-center game with incomplete information is an ex ante core plan if and only if it is equilibrium type-contingent contract offers.*

Proof. Every strategy bundle (x_S, y_S) in $\hat{F}_S$ is Bayesian incentive-compatible, so by Fact 2.4

$$\forall\, i \in S:\ \forall\, t_i, t_i' \in T_i : \forall\, \bar{t}_{I\setminus\{i\}} \in T_{I\setminus\{i\}}:$$
$$E\Big(x_i \;\Big|\; \hat{\mathscr{T}}_i(y_{1S})\Big)(t_i, \bar{t}_{I\setminus\{i\}}). = E\Big(x_i \;\Big|\; \hat{\mathscr{T}}_i(y_{1S})\Big)(t_i', \bar{t}_{I\setminus\{i\}}) \ .$$

Therefore,

$$Ex_i = E(x_i \mid t_i), \text{ for every } t_i \in T_i \ .$$

The coalitional stability condition of the *ex ante* core plan is then equivalent to the stability condition of the equilibrium type-contingent contract offers. □

In the light of the Bayesian incentive compatibility, no division has any incentive to misrepresent its type, so it signs its contract according to its true type. The next simple result guarantees the existence of a core re-contract in the third game. Radner [8, Subsection 7.2] paid a particular attention to the following structural relationship among the divisions. The divisions I are partitioned into two types, the suppliers and the customers: The *suppliers* produce and supply to the customers nonmarket intermediate commodities, and the *customers* use these nonmarket intermediate commodities, produce market commodities and bring in profit to the firm. If the products of all suppliers are needed for each customer's production activities, the suppliers are called *complementary*.

Proposition 4.2 *Let (x_I^*, y_I^*) be equilibrium type-contingent contract offers of profit-center game $\mathscr{D}$, $\bar{t}$ be the true type-profile, and $\bar{\mathscr{D}}$ be the one-period profit-center game given by (1). Assume either that each Y_i is convex or that there exists the [complementary supplier]-customer relationship among the divisions. Then there exists a core re-contract of the game $\bar{\mathscr{D}}$.*

Proof. Write $\bar{\xi}_i$ and $\bar{\eta}_{2i}$ for the values of the (constant) functions $x_i^*(\bar{t}_i, \cdot)$ and $y_{2i}^*(\bar{t}_i, \cdot)$ on $\bar{T}_{I \setminus \{i\}}$, respectively. The allowability requirement of plan $(x_S, y_{2S}) \in G'_S$ of $\bar{\mathscr{D}}$ says that for each $i \in S$, (x_i, y_{2i}) is a function of t_i only, so one may safely write $(x_i(t_i), y_{2i}(t_i))$ instead of $(x_i(t), y_{2i}(t))$. By the same argument as in the proof of Ichiishi and Radner [8, Fact 6.2], one can easily show that plan $(x_S, y_{2S}) \in G'_S$ of $\bar{\mathscr{D}}$ is Bayesian incentive-compatible iff, for each $i \in S$, $\bar{E}(x_i \mid \mathscr{T}_i)$ is a constant function on $\bar{T}$ (hence, x_i is a constant function on $\bar{T}$). By slight abuse of notation, let x_i also denote the value of the constant function.

Define the non-side-payment game $\bar{V} : \mathscr{I} \to \mathbf{R}^I$ by

$$\bar{V}(S) := \{u \mid \exists\ (x_S, y_{2S}) \in \hat{G}_S : \forall\ i \in S : u_i \le x_i\} \ .$$

Proof for the convex technology case is now given. Choose any balanced subfamily $\mathscr{B}$ of $\mathscr{I}$ with associated balancing coefficients $\{\lambda_S\}_{S \in \mathscr{B}}$, and take any $u \in \cap_{S \in \mathscr{B}} \bar{V}(S)$. Then, for each $S \in \mathscr{B}$, there exists $(x_S^{(S)}, y_{2S}^{(S)}) : \bar{T} \to \mathbf{R} \times \mathbf{R}^{k_2}$ such that $x_S^{(S)}$ is constant, each $y_{2i}^{(S)}$ is $\bar{\mathscr{T}}_i$-measurable, $y_{2S} \in \bar{Y}_{2S}$,

$$\forall\ t \in \bar{T} : \sum_{i \in S} \begin{pmatrix} x_i^{(S)} \\ \mathbf{0} \end{pmatrix} \le \sum_{i \in I \setminus S} \begin{pmatrix} \bar{\xi}_i \\ \mathbf{0} \end{pmatrix} + \sum_{i \in S} \begin{pmatrix} x_i^{(S)} \\ \mathbf{0} \end{pmatrix}$$
$$\le \sum_{i \in S} \begin{pmatrix} p \cdot m_i(t_i) \\ n_{2i}(t_i) + r_{2i}(t_i) \end{pmatrix} ,$$

$$\text{and } \forall\ i \in S : \quad x_i^{(S)} \ge \max\ \{\bar{\xi}_i, u_i\} \ .$$

Define (x_S, y_{2S}) by

$$(x_i, y_{2i}) := \sum_{S \in \mathscr{B} : S \ni i} \lambda_S (x_i^{(S)}, y_{2i}^{(S)}) \ .$$

Then, x_S is a constant function, each y_{2i} depends only on t_i, $y_{2S} \in \bar{Y}_{2S}$ (by convexity of Y_i),

$$\forall\ t \in \bar{T} : \sum_{i \in I} \begin{pmatrix} x_i \\ \mathbf{0} \end{pmatrix} \le \sum_{i \in I} \begin{pmatrix} p \cdot m_i(t_i) \\ n_{2i}(t_i) + r_{2i}(t_i) \end{pmatrix} ,$$

$$\text{and } \forall\ i \in I : \quad x_i \ge \max\ \{\bar{\xi}_i, u_i\} \ .$$

Therefore, $u \in \bar{V}(I)$. Game $\bar{V}$ is balanced, so it has a nonempty core. A core strategy is a core re-contract of S.

Proof for the [complementary supplier]-customer relationship case is the same as the proof of Theorem 5.2 of Ichiishi and Radner [8], so is left to the reader. □

Remark 4.3 In the above proof of Theorem 4.2 for the convex case, it is the convexity of $\bar{Y}_{2S}$ that is used in an essential way. This assumption is consistent with several cases in which the original set Y_S is nonconvex and exhibits increasing returns to scale. On the other hand, if all the second-period nonmarket commodities are only used as inputs (so they cannot be intermediate commodities), and if the set $\bar{Y}_{2i}(t_i)$ is distributive (see Scarf [14, pp. 410–411] for a definition; this condition implies increasing returns to scale), then by the same argument as in Scarf [14, Theorems 1 and 6], one can establish the existence of a core re-contract of game $\bar{\mathscr{D}}$. However, distributiveness of $\bar{Y}_{2i}(t_i)$ is a questionable assumption, even when the original set $Y_i(t_i)$ is distributive. □

The following example of the [one supplier]-[one customer] relationship illustrates full-information-revealing equilibrium type-contingent contract offers in the first game, and shows how a core re-contract in the third game removes the inefficiency that was caused by the Bayesian incentive compatibility in the first game. It will be pointed out, however, that *ex post* inefficiency may persist, because the divisions may have made some of the choices at the time when each had no information about the others' types, that is, because they had to follow the information-pooling rule.

Example 4.4 Consider a firm with two divisions, $I = \{1,2\}$, in which division 1 is a *supplier* (of an intermediate nonmarket commodity) and division 2 is a *customer* (of an intermediate nonmarket commodity). There are three goods; a market commodity whose quantity is denoted by m, an intermediate nonmarket commodity whose quantity is denoted by n_s, and a primary nonmarket commodity whose quantity is denoted by n_z. Assume that the primary commodity is used in the first *interim* period, the intermediate nonmarket commodity is produced and used in the second *interim* period, and the market commodity is produced in the second *interim* period. The market price of the output, p, is normalized to be equal to 1. Each division has two types, $T_i = \{H_i, L_i\}$, and the *ex ante* probability on T_i is given as the equal probability $\pi_i(\{H_i\}) = \pi_i(\{L_i\}) = 1/2$, $i = 1,2$. The supplier produces only the intermediate commodity, and its production set is given by: $y_1(t) := (m_1(t), n_{s1}(t), n_{z1}(t_1)) \in Y_1(t_1)$ iff

$$
\begin{aligned}
m_1(t) &\le 0, \text{ for } t_1 = H_1, L_1 \ , \\
n_{s1}(t) &\le \begin{cases} |n_{z1}(H_1)|^2, & \text{if } t_1 = H_1 \ , \\ |n_{z1}(L_1)|, & \text{if } t_1 = L_1 \ , \end{cases} \\
n_{z1}(t_1) &\le 0, \text{ for } t_1 = H_1, L_1 \ .
\end{aligned}
$$

The customer uses both the primary commodity and the intermediate commodity to produce the market commodity, and its production set is given by: $y_2(t) := (m_2(t), n_{s2}(t), n_{z2}(t_2)) \in Y_2(t_2)$ iff

$$
\begin{aligned}
m_2(t) &\le \begin{cases} |n_{s2}(t_1, H_2) \cdot n_{z2}(H_2)|, & \text{if } t_2 = H_2 \ , \\ \sqrt{|n_{s2}(t_1, L_2) \cdot n_{z2}(L_2)|}, & \text{if } t_2 = L_2 \ , \end{cases} \\
n_{s2}(t) &\le 0, \text{ for } t_2 = H_2, L_2 \ , \\
n_{z2}(t_2) &\le 0, \text{ for } t_2 = H_2, L_2 \ .
\end{aligned}
$$

The initial nonmarket resources are given by

$$r_{si}(t_i) = 0, \quad \text{for } t_i = H_i, L_i \ ,$$
$$r_{zi}(t_i) = \begin{cases} 5, & \text{if } t_i = H_i \ , \\ 1, & \text{if } t_i = L_i \ , \end{cases}$$

for each division $i \in I$.

Clearly, the maximal profit that each singleton can make given any of the two types is 0.

For the grand coalition, the full-information-revealing net output plans which are technologically attainable and satisfy the information-pooling rule (Postulate 2.2),

$$y_I = \left\{ \begin{pmatrix} m_1(t) \\ n_{s1}(t_1) \\ n_{z1}(t_1) \end{pmatrix}, \begin{pmatrix} m_2(t) \\ n_{s2}(t_1) \\ n_{z2}(t_2) \end{pmatrix} \right\}_{t \in T} ,$$

are computed now. Here, n_{zi} depends only on t_i while m_i and n_{si} depend in general on t, due to the information-pooling rule and the present assumption that n_{zi} is 1-1 on T_i. Notice the dependence of n_{s1} only on t_1, in view of the fact that the intermediate commodity is produced from $|n_{z1}(t_1)|$-units of the primary commodity. Without loss of generality, one may assume that $n_{s2} = -n_{s1}$, so that n_{s2} is a function only of t_1, because division 2 can use the entire amount of the intermediate commodity that division 1 produces. By Lemma 6.3 of Ichiishi and Radner [8], for each i there exists a function $\bar{\bar{n}}_{zi} : T_i \to \mathbf{R}$ such that $\bar{\bar{n}}_{zi} \le n_{zi}$ and

$$\forall\, t \in T : \ -\sum_{i \in I} \bar{\bar{n}}_{zi}(t_i) = \sum_{i \in I} r_{zi}(t_i) \ .$$

Without loss of generality, therefore, one may also assume that

$$\forall\, t \in T : \ -\sum_{i \in I} n_{zi}(t_i) = \sum_{i \in I} r_{zi}(t_i) \ .$$

Let $a := |n_{z1}(L_1)|$. Clearly, $0 \le a \le r_{z1}(L_1) + r_{z2}(L_2) = 2$, and $|n_{z2}(L_2)| = 2 - a$. For type-profile (L_1, H_2), $|n_{z1}(L_1)| + |n_{z2}(H_2)| = r_{z1}(L_1) + r_{z2}(H_2) = 6$, so $|n_{z2}(H_2)| = 6 - a$. Similarly, $|n_{z1}(H_1)| = 4 + a$. Each n_{zi} is indeed 1-1 on T_i, $i \in I$. By using the production function of division 1,

$$n_{s1}(t_1) = \begin{cases} (4+a)^2, & \text{if } t_1 = H_1 \ , \\ a, & \text{if } t_1 = L_1 \ . \end{cases}$$

By using the identity $n_{s2} = -n_{s1}$ and the production function of division 2, the total profit at each type-profile is computed as

$$m_2(t) = \begin{cases} \sqrt{a(2-a)}, & \text{if } t = (L_1, L_2) \ , \\ a(6-a), & \text{if } t = (L_1, H_2) \ , \\ \sqrt{(4+a)^2(2-a)}, & \text{if } t = (H_1, L_2) \ , \\ (4+a)^2(6-a), & \text{if } t = (H_1, H_2) \ . \end{cases} \tag{2}$$

In view of $0 \le a \le 2$, the information-pooling rule thus imposes the following bound on m_2:

$$\begin{array}{rcccl} 0 & \le & m_2(L_1,L_2) & \le & 1 \ , \\ 0 & \le & m_2(L_1,H_2) & \le & 8 \ , \\ 0 & \le & m_2(H_1,L_2) & \le & \sqrt{32} \ , \\ 96 & \le & m_2(H_1,H_2) & \le & 144 \ . \end{array}$$

The above net-output plan y_I constrains nonnegative profit imputation plan x_I by

$$\forall\, t \in T : \ x_1(t) + x_2(t) \ \le \ m_2(t) \ .$$

Since y_I fully reveals information, plan (x_I, y_I) is Bayesian incentive compatible iff

$$\begin{array}{ll} x_1(L_1,L_2) = x_1(H_1,L_2), & x_1(L_1,H_2) = x_1(H_1,H_2) \ , \\ x_2(L_1,L_2) = x_2(L_1,H_2), & x_2(H_1,L_2) = x_2(H_1,H_2) \ . \end{array}$$

(Fact 2.4). A little calculation using (2) yields

$$\begin{aligned} x_1(H_1,H_2) + x_2(H_1,H_2) &\le a(6-a) - \sqrt{a(2-a)} + \sqrt{(4+a)^2(2-a)} \\ &\le a(6-a) + \sqrt{(4+a)^2(2-a)} \\ &\le 8 + \sqrt{32} \ . \end{aligned} \tag{3}$$

Let (x_I^*, y_I^*) be equilibrium type-contingent contract offers, with the associated value $a^* := |n_{z1}^*(L_1)|$. Suppose (H_1,H_2) is the true type-profile. Then division i chooses the offer $(n_{zi}^*(H_i), n_{si}^*(H_i,\cdot), m_i^*(H_i,\cdot), x_i^*(H_i,\cdot))$.

In the third stage (for a re-contract), the divisions have already used primary commodities, $(n_{z1}^*(H_1), n_{z2}^*(H_2)) = -(4+a^*, 6-a^*)$, and each division i has the disagreement point $x_i^*(H_1,H_2)$. But from the primary-commodity allocation $(n_{z1}^*(H_1), n_{z2}^*(H_2))$ the grand coalition can produce a total profit no less than 96, given the true type-profile $((H_1,H_2))$. Re-contracting thus occurs, improving upon the disagreement-point allocation $(x_1^*(H_1,H_2), x_2^*(H_1,H_2))$ which is bounded by $x_1^*(H_1,H_2) + x_2^*(H_1,H_2) \le 8 + \sqrt{32}$ (inequality (3)).

The present example confirms the widely understood claim that Bayesian incentive compatibility is a major cause of (Pareto-)inefficiency. Re-contracting is a frequent practice, instituted in order to decrease the degree of the inefficiency.[10]

Green and Laffont [4] emphasized the Bayesian incentive compatibility as a cause of *ex post* Pareto non-optimality. In the present example of the full-information-revealing case, re-contracting removes the difficulty about Bayesian incentive compatibility, yet *ex post* Pareto optimality cannot be

[10] In the employment contract (which is beyond the present framework), one sometimes observes *merit raise* after the initial contract. This is a form of a re-contract when the employee's ability, his private information, is truthfully made public.

achieved, as the computation in the next paragraph shows. The *ex post* inefficiency in the present setup is due to the information-pooling rule, which stipulates in particular that the primary commodity used in the first *interim* period should depend only upon private information. Thus, in state (H_1, H_2) the divisions can use the primary commodity subject to

$$\begin{array}{ccccccc} 4 & \leq & |n_{z1}(H_1)| & = & 4+a & \leq & 6 \ , \\ 4 & \leq & |n_{z2}(H_2)| & = & 6-a & \leq & 6 \ , \end{array}$$

whereas there is no such constraint for computation of the *ex post* optimality given (H_1, H_2).

It remains to show that (x_I^*, y_I^*) is not *ex post* efficient given (H_1, H_2). For the grand coalition, the maximal total profit subject to technological attainability given (H_1, H_2) is provided by the activities,

$$\begin{pmatrix} m_1(H_1, H_2) \\ n_{s1}(H_1, H_2) \\ n_{z1}(H_1, H_2) \end{pmatrix} = \begin{pmatrix} 0 \\ 400/9 \\ -20/3 \end{pmatrix}, \begin{pmatrix} m_2(H_1, H_2) \\ n_{s2}(H_1, H_2) \\ n_{z2}(H_1, H_2) \end{pmatrix} = \begin{pmatrix} 4000/27 \\ -400/9 \\ -10/3 \end{pmatrix}.$$

But

$$m_2(H_1, H_2) = 4000/27 > 144 \geq m_2^*(H_1, H_2) \ ,$$

so (x_I^*, y_I^*) cannot be *ex post* efficient. □

References

1. Berliant, M.: On income taxation and the core. Journal of Economic Theory **56**, 121–141 (1992)
2. Chandler, Jr., A. D.: Strategy and structure. Cambridge, MA: MIT Press 1962
3. Demange, G., Guesnerie, R.: Interim design of core mechanisms. Mimeo (1995)
4. Green, J. R., Laffont, J.-J.: Posterior implementability in a two-person decision problem. Econometrica **55**, 69–94 (1987)
5. Harsanyi, J. C.: Games with incomplete information played by 'Bayesian' players. Management Science: Theory **14**, 159–182 (Part I), 320–334 (Part II), 486–502 (Part III) (1967/1968)
6. Ichiishi, T., Idzik, A.: Bayesian cooperative choice of strategies. International Journal of Game Theory **25**, 455–473 (1996)
7. Ichiishi, T., Idzik, A., Zhao, J.: Cooperative processing of information via choice at an information set. International Journal of Game Theory **23**, 145–165 (1994)
8. Ichiishi, T., Radner, R.: A profit-center game with incomplete information. Cahiers eco & maths 95.10, Université de Paris I – Panthéon-Sorbonne (1994); revised: September 1996
9. Koray, S., Sertel, M. R.: The welfaristic characterization of two-person revelation equilibria under imputational government. Social Choice and Welfare **9**, 49–56 (1992)
10. Koray, S., Sertel, M. R.: A multiplication of solution concepts, and pretend-but-perform implementation. Paper presented at the International Conference on Game Theory and Economic Applications, Bangalore, India, January 2–6, 1996
11. Radner, R.: Transfer payments and the core of a profit-center game. In: Dasgupta, P., et al. (eds.) Economic analysis of markets and games (Essays in honor of Frank Hahn), pp. 316–339. Cambridge, MA: MIT Press 1992
12. Rothschild, M., Stiglitz, J.: Equilibrium in competitive insurance market: An essay on the economics of imperfect information. Quarterly Journal of Economics **90**, 629–649 (1976)
13. Scarf, H.: The computation of economic equilibria. New Haven, CT: Yale Univ. Press 1973

14. Scarf, H.: Notes on the core of a productive economy. In: Hildenbrand, W., et al. (eds.) Contributions to mathematical economics (in honor of Gerard Debreu), pp. 401–429. Amsterdam New York: North-Holland 1986
15. Sertel, M. R.: Workers and incentives. Amsterdam: North-Holland 1982
16. Sertel, M. R.: Lecture notes in microeconomic theory. Unpublished, Boğaziçi University, Istanbul (1990)
17. Wilson, R.: Information efficiency, and the core of an economy. Econometrica **46**, 807–816 (1978)
18. Yannelis, N. C.: The core of an economy with differential information. Economic Theory **1**, 183–198 (1991)

Optimal multilateral contracts*

Stefan Krasa and Anne P. Villamil

Department of Economics, University of Illinois, 1206 South Sixth Street, Champaign, IL 61820, USA

Received: March 27, 1992

Summary. The purpose of this paper is to derive the structure of optimal multilateral contracts in a costly state verification model with multiple agents who may be risk averse and need not be identical. We consider two different verification technology specifications. When the verification technology is deterministic, we show that the optimal contract is a multilateral debt contract in the sense that the monitoring set is a lower interval. When the verification technology is stochastic, we show that transfers and monitoring probabilities are decreasing functions of wealth. The key economic problem in this environment is that optimal contracts are *interdependent*. We are able to resolve this interdependency problem by using abstract measure theoretic tools.

1. Introduction

In the Arrow-Debreu model complete insurance markets exist and agents are able to attain unconstrained Pareto efficient consumption allocations. In addition, the structure of the set of financial securities that support these allocations is indeterminate (i.e., the Modigliani-Miller Theorem states that a firm's debt-equity ratio is irrelevant when there are no market imperfections). Clearly, firms have determinate debt-equity ratios and insurance markets are incomplete. The costly state verification model, proposed by Townsend (1979), provides one plausible explanation of these outcomes that is consistent with (constrained) Pareto efficient behavior. The model imbeds an information friction in an Arrow-Debreu'economy and has two essential elements. First, agents have asymmetric information since they know the (common) distribution of a random variable (i.e., endowments) but the realization of a particular agent's random variable is costlessly observed only by the agent.

* We wish to thank Mark Feldman, Wayne Shafer and Nicholas Yannelis for useful comments. We also gratefully acknowledge financial support from the National Science Foundation (SES 89-09242).

Second, a technology exists that can be used to publicly announce the realization to all agents ex post, but it is costly to use. The model has proved useful for analyzing many economic problems (cf. Sect. 2). In this paper we generalize the model to permit its application to a broader class of economic problems - environments with multiple heterogeneous agents where contracts (and hence consumption allocations) may be interdependent.

We characterize the nature of contracts that support information and resource constrained Pareto efficient consumption allocations when trade is not restricted a priori to be symmetric, bilateral, or separable in endowments and agents may be risk averse. We use abstract measure theoretic arguments to show that "debt-like" securities are optimal when the monitoring technology is deterministic (i.e., the optimal multilateral contract has a lower interval), and transfer functions and monitoring probabilities are monotonically decreasing functions of wealth when the monitoring technology is stochastic. The key problem when agents are risk averse and contracts are explicitly multilateral is that non-trivial interdependencies among agents exist. Measure theoretic tools such as the Isomorphism Theorem and Lusin's Theorem appear to be necessary to solve this interdependency problem since they allow us to change contracts so that the expected utility of one agent is affected while the expected utility of all others remains the same. We then show that given any arbitrary initial contract, unless one starts with a lower interval monitoring set (for deterministic monitoring) or monotonically decreasing transfer and monitoring functions (for stochastic monitoring) at least one agent can be made better off ceteris paribus, which contradicts the Pareto optimality of the arbitrary initial contract. These results are stated formally in Theorem 1, and Theorem 2 and Corollary 1, respectively.

2. Discussion of the literature

Townsend (1979) proved that when costly state verification is deterministic (i.e., monitoring occurs with either probability one or zero) the optimal contract that supports information and resource constrained Pareto efficient consumption allocations resembles debt because the monitoring set is a "lower interval" (i.e., it is optimal to monitor only announcements below a cut-off point). This result is important because it is consistent with many stylized facts observed in actual markets (e.g., debt payment characteristics and institutional features of bankruptcy).[1] Previous lower interval results have been established only under several restrictive assumptions: Agents are either assumed to be risk neutral or their trades are exogenously restricted to be symmetric, separable in endowments, and bilateral.[2] Townsend (1979, p. 281) notes that these restrictions are "unpleasant" because

[1] This result is also obtained by Gale and Hellwig (1985) when only the agent who pays for monitoring gets the information. The distinction between public and private monitoring is irrelevant in a two agent economy but is important with multiple agents. See Williamson (1986) and Krasa and Villamil (1992) for multiple agent costly state verification environments wich private monitoring reports.

[2] Recently, Winton (1991) considers a costly state verircation model wich multiple risk averse agents and derives conditions under which subordinated debt is optimal. Boyd and Smith (1993) study credit rationing in a similar model with multiple risk neutral agents with a particular form of heterogeneity.

they are motivated by technical, rather than economic considerations. Further, they may preclude optimal risk sharing arrangements even in two-agent contracting problems.[3] More fundamentally, however, multilateral structures which permit interdependencies among heterogeneous agents are important because these features are inherent in many economic problems. For example, Krasa and Villamil (1992) consider a costly state verification model with deterministic monitoring and risk neutral agents where a financial intermediary arises endogenously. They show that when the intermediary contracts with both depositors and entrepreneurs and is subject to non-trivial default risk, its deposit and loan contracts are interdependent. As is common in the literature, their results depend crucially on risk neutrality. In contrast, the multilateral model that we present permits many types of agent heterogeneity: multiple types of risk aversion, (endowment) distribution functions, transfer functions, and monitoring cost functions.[4]

Border and Sobel (1987) prove that when the verification technology is stochastic (i.e., monitoring need not occur with probability one) the optimal contract that supports information and resource constrained Pareto efficient consumption allocations specifies transfer and monitoring procedures that resemble those commonly used by insurance companies and tax collection agencies: transfers and monitoring probabilities are monotonically decreasing functions of an agent's reported wealth. This result is important because it is consistent with the following stylized facts. In insurance markets a large loss can be viewed as a low wealth realization, thus a monotonic contract implies that policy holders receive higher transfers when they claim larger losses and the probability of being audited is correspondingly higher for such reports. In a public finance context laxes can be viewed as negative transfers and low wealth reports can be viewed as high itemized deduction claims, so the monotonicity result implies that larger (total) tax payments are associated with larger wealth reports and the probability of a tax audit is decreasing in reported wealth (where low wealth claims not low wealth make an audit more likely). Border and Sobel's model has two risk neutral agents (one having a random endowment of wealth), and information conditions that are identical to those in Townsend's model. They note (p. 533) that risk neutrality is essential for their argument and that "it is not known if the monotonicity result ... extends to the risk averse case." This open question is particularly important for insurance applications of the model as risk aversion is typically a driving force behind most insurance arrangements.

3. The model

Consider a two period exchange economy with finitely many individuals indexed by $i = 1, ..., n$. Traders are described by von Neumann-Morgenstern utility functions,

[3] Townsend (1979, p. 281) provides an example where two agents have utility functions of the form $u(c) = c^{\alpha+1}/(\alpha+1)$, where $-1 < \alpha < 0$. The optimal symmetric transfer function implied by this common utility specification is not separable in endowments as required by the exogeneous restriction.

[4] Using the core, Boyd and Prescott (1986) and Wilson (1968) also argue that group (syndicate) structures are important in finance and insurance problems. Höwever, Boyd and Prescott note: "An extension which is not so easy [in their model] is to allow for more than two agent or project types." Such an extension is straightforward in our framework.

u_i, defined over second period consumption, c_i, and random endowments, X_i. Let u_i be concave and monotonically increasing in consumption. Assume that the X_i are independent random variables. Denote a particular realization of X_i by x_i, let F_i be the distribution of X_i, let F^n be the joint distribution of $X_1, ..., X_n$, and assume that all distributions are non-atomic.[5] To ensure non-negative consumption, assume that the support of F_i, is contained in $[m, \infty)$, where $m > 0$. The information conditions are as follows. Each agent i privately observes his/her endowment realization, x_i, ex-post, but all agents have access to a costly state verification technology that can be used to publicly announce the realization to other agents. Let $\phi_i(\cdot)$ be the cost incurred by agent i from using the verification technology.[6]

Denote by $t_i(x_1, ..., x_n,)$ the net transfer function of agent i, which describes the payment between the coalition and each agent i. This payment may be positive (indicating a state-contingent payment from the coalition to the agent), negative (indicating a payment by the agent to the coalition), or zero. Assume that agents' verification costs are an arbitrary positive function of the transfer payments, $\phi_i(t_i(\cdot))$. Because transfers need not be identical across agents, verification costs may differ as well.[7] At time zero agents have the opportunity to write binding contracts to provide for consumption next period. The structure of optimal contracts will depend on the specification of agents' preferences, the distributions of random variables, the verification technology, and the nature of information in the economy.

4. The case of deterministic verification

In this section we study the form of Pareto efficient multilateral contracts under deterministic monitoring. Transfers, $t_i(\cdot)$, can be contingent only an endowment realizations of agent i which are publicly verified. In private information states, all transfers must be non-contingent. Let S_i denote the set of all announced realizations of X_i for which verification occurs, and let S_i^c denote the complement of S_i. Define a multilateral contract as follows.

Definition 1. *A multilateral contract with deterministic verification for each agent $i = 1, ..., n$ is a pair (t_i, S_i), where $t_i(x_1, , ..., x_n)$ is a net-transfer function for agent i from $\mathbb{R}^n$ into $\mathbb{R}$ and S_i is a set of endowment realizations announced by agent i for which monitoring occurs (with probability one). If agent i is verified, the endowment becomes public information.*

We restrict the analysis to the class of incentive compatible contracts:

Definition 2. *A collection of multilateral contracts (t_i, S_i) with deterministic verification is incentive compatible if $S_i = \bar{S}_i$ and $t_i(\cdot) = \bar{t}_i(\cdot)$ for every $i = 1, ..., n$,*

[5] A distribution is non-atomic if every single point has probability zero. This follows automatically if the distribution has a density.

[6] When monitoring occurs the true endowment is publicly reported without error.

[7] Townsend (1979, p. 269) considers two verification cost specifications, and our cost function includes both as special cases. In his first case the verification cost is a fixed constant, and hence independent of the actual realization. In the second case the verification cost of agent i depends on the transfer t_i, where the costs are strictly monotonic.

where (t_i, S_i) *denotes the pre-state contractual commitment and* $(\bar{t}_i, \bar{S}_i)$ *denotes the post-state outcome.*

Definition 2 indicates that under an incentive compatible contract, agents do not misrepresent their private information (i.e., pre-state commitments are fulfilled ex post). Townsend (1988, pp. 416–418) uses a revelation principle argument to prove that incentive compatibility can be imposed without loss of generalizy. The following conditions generalize the incentive compatibility specification of Lemma 5.1 in Townsend (1979):

(IC 1) $x_i \mapsto t_i(x_1, ..., x_i, ..., x_n)$ is constant an S_i^c, for a.e. $x_j, j \neq i$.
(IC2) $t_i(x_1, ..., x_i, ..., x_n) - \phi_i(t_i(\cdot)) \geq t_i(x_1, ..., y, ..., x_n)$; for a.e. $x_i \in S_i$, for every $y \in S_i^c$, and for a.e. $x_j, j \in i$.

IC1 says that when agent i's endowment announcement is not verified (ceteris paribus), his/her net-transfer is constant (because transfers cannot depend on private information). IC2 says that it is (at least weakly) optimal for agent i to request verification when the endowment realization is in the verification set. Thus, it ensures that agent i requests verification when $x_i \in S_i$. We assume that the incentive constraints are satisfied a.e. Thus, there exists a set of realizations of the agent's endowment which has measure zero in which it might be optimal to misreport. See Section 6 for a discussion of implementation and alternative specifications of the incentive constraints.

We now state an information constrained optimization problem whose solutions characterize optimal multilateral contracts. The objective is to Coose Pareto efficient net transfer functions, $t_i(x_t, ..., x_n)$, and sets of endowment realizations for which verification occurs, S_i, to maximize a weighted average of agents' utilities, subject to feasibility and information constraints. The λ_i denote weights on agents' utility functions.

Problem 3.1. *Choose* t_i *and* S_i *for* $i = 1, ..., n$ *to maximize*

$$\sum_{i=1}^{n} \lambda_i \int u_i[c_i(x_1, ..., x_n)] dF^n(x_1, ..., x_n), \tag{3.1}$$

subject to

$$0 \leq c_i \leq x_i + t_i(x_1, ..., x_n) - \phi_i(t_i(\cdot)) \text{ a.e. for all } i, \tag{4.2}$$

$$\sum_{i=1}^{n} t_i \leq 0 \text{ a.e.}, \tag{4.3}$$

$$t_i \text{ is incentive compatible for every } i \tag{4.4}$$

$$S_i \text{ is a measurable set for every } i. \tag{4.5}$$

The optimal multilateral contract maximizes the expected utility of all agents (3.1), subject to: (3.2) a budget constraint for each agent which holds almost everywhere; (3.3) an aggregate feasibility constraint which holds almost everywhere;

(3.4) incentive-compatibility conditions IC1 and IC2; and (3.5) a standard measurability condition. The problem's solution characterizes the nature of optimal contracts when verification is deterministic. (constrained) Pareto efficient multilateral contracts have lower interval monitoring sets, except for nullsets. Thus, we show that there exists a y_i such that $S_i = [m, \gamma_i)$ for all i, except for a set of measure zero, where the lower interval may be trivial. Because monitoring is deterministic, it follows immediately from this result that the transfer function is constant for all $x_i \in S_i^c$ (for fixed $x_j, j \neq i$).

Townsend (1979, p. 283) proves a related lower interval result under several exogenous restrictions which he describes as "unpleasant" because they are necessary for technical reasons, but are not motivated by economic considerations. He assumes:

(i) all transfers and verification costs are symmetric;
(ii) all trades are bilateral; and further
(iii) when both agents are verified, the transfer function is separable in endowment realizations (i.e., in our notation $t(x_1, x_2) = \hat{t}_1(x_1) + \hat{t}_2(x_2)$).[8]

Before beginning our formal analysis we describe the relationship between our result and Townsend's, and give an overview of the proof of Theorem 1. Townsend specifies an optimization problem which involves the maximization of a weighted average of utilities, subject to information and resource constraints. However, instead of characterizing t_i and S_i directly as in our Problem 3.1, Townsend reformulates an analog of Problem 3.1 as a standard constrained maximization problem. The key difference between our approaches is that the maximizer in his reformulated problem is a function of only one variable. This follows from restrictions (i) and (iii), as they immediately impiy that the transfer function is of the form $t(x_1, x_2) = \hat{t}(x_1) + \hat{t}(x_2)$. Under these restrictions it is only necessary to choose a one-dimensional transfer function, $\hat{t}$. Townsend considers the multilateral case (pp. 278–283) but reduces it to a similar one-dimensional problem by using (ii). This approach has two limitations. First, it precludes certain types of agent heterogeneity (i.e., (i) rules out transfer and cost function differences). Second, even when agents' transfer and cost functions are identical, restrictions (ii) and (iii) preclude some economically plausible risk-sharing arrangements.

In contrast, we characterize the solutions to Problem 3.1 directly. The maximizers are explicitly multi-dimensional transfer functions and verification sets, where transfer and verification cost functions need not be symmetric. We use abstract measure theoretic arguments to obtain our results; and these mathematical tools appear to be essential in our more general setting. We proceed as follows: Our main result in this section is Theorem 1, which establishes that in a multi-agent economy with deterministic costly state verification, *all* solutions to Problem 3.1 have lower interval verification sets (except for sets of measure zero). We prove the Theorem indirectly by assuming that there exists some arbitrary initial contract $(t_i(\cdot), S_i)$ which is optimal but is not a lower interval. We then define a measure preserving mapping, g, which allows us to transform the transfer functions, monitoring sets,

[8] Separability is equivalent to the slope of the net-transfer function of agent i depending only on agent i's realization. This precludes most interdependencies among agents.

and monitoring costs associated with the initial contract into an alternative contract (t_i', S_i') such that the new contracts are feasible, incentive compatible, strictly increase the expected utility of at least one agent, and leave the expected utility of all other agents unaffected. This contradicts the optimality of the original (non-lower monitoring interval) contract, hence it establishes the optimality of contracts with lower monitoring intervals.

Roughly speaking, we contradict the optimality of non-lower intervals in the following way. We move a part of the original (non-lower interval) monitoring set of one agent (say agent one) to the left, mapping it into a set where there was previously no state verification. Such sets (with positive measure) always exist if the initial contract was not a lower monitoring interval, and we construct these sets to be compact. The existence of a measure preserving one-to-one mapping, g, between these two sets follows from the Isomorphism Theorem which says that measure preserving one-to-one mappings exist between all separable and complete measure spaces (where both spaces have the same measure). Since compact subsets of $\mathbb{R}$ are separable and complete (in the induced topology) the Theorem can be applied.

Feasibility and incentive compatibility of the alternative contract are straightforward to show because g is measure preserving and one-to-one. It is also reasonably straightforward to show that the expected utility of agent one increases by a Rothschild and Stiglitz increasing risk argument. Townsend (1979, p. 288) uses a similar argument in the proof of Proposition 3.2, which is his lower-interval result for two-agents, one risk neutral, with *fixed* monitoring costs. Thus, the reader may wonder why we use abstract measure theory to obtain our results. The remaining and key step in the proof is to show that the utility of all other agents does not decrease under the alternative contract. In Townsend's proof, this follows immediately from risk neutrality and fixed verification costs.[9] In our setting with multiple risk-averse agents and arbitrary verification cost functions his argument breaks down exactly at this step because all contracts are *interdependent*. Without an additional argument, it is not possible to avoid affecting other agents' expected utility nor to see in which direction their utilities change. Measure preserving mappings impose the necessary structure to overcome this problem.

We begin our analysis by defining a measure preserving mapping. As indicated above, this concept is crucial for the arguments that follow.

Definition 3. *Let* $(Y_i, \beta_i, \mu_i), i = 1, 2$ *be two measure spaces and let* $g : Y_1 \to Y_2$ *be a measurable function. For every* $A \in \beta_2$ *define* $gA = \{ga : a \in A\}$. *Then* g *is measure preserving iff* $\mu_1(g^{-1}A) = \mu_2(A)$.

[9] Townsend (1979, Proposition 3.1) proves a second lower interval result for a *bilateral* contracting problem where agents may be risk averse and the monitoring tost function is convex with $\phi_i(0) < 1$. His Euler equation argument depends crucially on restrictions (i), (ii), and (iii). It does not appear that this approach can be readily extended to the multilateral case because of the interdependency problem.

The following Remark is an immediate consequence of Definition 3.[10]

Remark 1. Let f be an integrable function on Y_2, and let g be a measure preserving transformation as defined above. Then $f \circ g$[11] is integrable and the following holds:

$$\int_{Y_2} f(x) d\mu_2(x) = f \int_{Y_1} f(g(x)) d\mu_1(x).$$

Remark 1 corresponds to Theorem 1.6.12 of Ash (1972) or Remark 28.14 of Parthasarathy (1977). For completeness we give the proof in the Appendix. This Remark is essential for the proofs of our main results as it establishes that whenever we change the payoffs to one agent in a measure preserving way (i.e., choose a measure preserving function g), then the expected utility from an arbitrary initial contract $t_i(x_1, ..., x_n)$ and a transformed alternative contract $t_i(g(x_1), x2, ..., x_n)$ is the same for all other agents.

To construct measure preserving mappings we use the Isomorphism Theorem from measure theory (cf., Parthasarathy (1977) Proposition 26.6).

Isomorphism Theorem. *Let $Y_i, i = 1, 2$, be complete and separable metric spaces, and let μ_i, be non-atomic Borel measures on Y_i such that $\mu(Y_1) = \mu(Y_2) > 0$. Then the two measure spaces are isomorphic, i.e., there exist two sets of measure zero, $N_i, i = 1, 2$, and there exists a measure preserving transformation, $g : Y_1 \backslash N_1 \mapsto Y_2 \backslash N_2$, whose inverse exists and is also measure preserving.*[12]

We now state our main result concerning the nature of optimal contracts in a multi-agent economy with deterministic costly state verification.

Theorem 1. *Assume that the utility functions of all agents are twice continuously differentiable and that $u'' < 0$. Furthermore assume that $\phi_i(t) > 0$ for every agent i and for every $t \in \mathbb{R}$. Let the endowments of the agents be described by independent random variables X_i for all $i = 1, ..., n$. Then all solutions to Problem 3.1 have lower interval verification sets, except for sets of measure zero (i.e., there exists a γ_i such that $S_i \Delta \{x : x < \gamma_i\}$ has measure zero).*[13]

Proof. We proceed indirectly. Without loss of generality, assume that the monitoring set of agent one is not a lower interval. Let μ be the distribution of the endowment of agent one. Then there exist compact sets $K_i, i = 1, 2$ with positive measure, and such that $k_1 < k_2$ for all $k_i \in K_i$ and such that $K_1 \subset \mathbb{R} \backslash S_1$ and

[10] Consider the following example of a measure preserving mapping. Let $Y_1 = [0, 1] \cup \{2\}$ and $Y_2 = [1, 2]$. On both sets consider the standard Lebesgue measure. Then the function

$$g(x) = \begin{cases} x + 1 & \text{if } x \in [0, 1]; \\ 1 & \text{if } x = 2; \end{cases}$$

is measure preserving in this example (though not a one-to-one mapping). ‘

[11] $f \circ g$ is the composition of f and g, i.e. $f \circ g(x) = f(g(x))$.

[12] "$\backslash$" denotes set theoretic subtraction.

[13] "Δ" denotes the symmetric difference: $A \Delta B = (A \backslash B) \cup (B \backslash A)$, for arbitrary sets A and B.

$K_2 \subset S_1$. By regularity[14] and non-atomicity of the measure, we can assume that $\mu(K_1) = \mu(K_2)$.

Note that the K_i are reparable and complete because they are compact. Thus, by the Isomorphism Theorem there exists a measure preserving mapping $h : K_1 \backslash N_1 \to K_2 \backslash N_2$ such that h^{-1} exists and is also measure preserving, where $N_i, i = 1, 2$ are sets of measure zero. Note that h can be extended to $\mathbb{R}$ by

$$g(x) = \begin{cases} h(x) & \text{if } x \in K_1 \backslash N_1; \\ h^{-1}(x) & \text{if } x \in K_2 \backslash N_2; \\ x & \text{otherwise.} \end{cases}$$

Clearly, g is again measure preserving.

Recall that $t_i(x_1, ..., x_n)$ are transfer functions associated with some arbitrary initial contract, where the monitoring set of agent one is not a lower interval. Thus for every agent i, now define new transfers t_i' by

$$t_i'(x_1, ..., x_n) = t_i(g(x_1), x_2, ..., x_n).$$

Further, define the new monitoring set of agent one by $S_1' = g^{-1}(S_1)$ and $S_i' = S_i$ for $i = 2, ..., n$. The strategy of the proof is to show the following: (i) The transfer functions associated with the new contracts $(t_i'(\cdot), S_i')$ are feasible; (ii) the new contracts are incentive compatible; (iii) the utility of all other agents $i \neq 1$ does not change, and (iv) the utility of agent one strictly increases. This gives the contradiction to the assumed optimality of a non-lower interval contract. (i)–(iv) are proved as follows:

(i) Let $A = \{(x_1, ..., x_n) : \sum_{i=1}^n t_i(x_1, ..., x_n) > 0\}$. Define $\tilde{g}$ on $\mathbb{R}^n$ by $(x_1, ..., x_n) \mapsto (g(x_1), x_2, ..., x_n)$. Clearly, $\tilde{g}$ is measure preserving with respect to the joint distribution of the X_i. Then, $\tilde{g}^{-1}A = \{(y_1, ..., y_n) : g(y_1) = x_1; y_i = x_i$ for all $i > 1$, and $\sum_{i=1}^n t_i(x_1, ..., x_n) > 0\} = \{(y_1, ..., y_n) : \sum_{i=1}^n = t_i(g(y_1), y_2, ..., y_n) > 0\}$. Since $\tilde{g}$ is measure preserving, (3.3) implies that $\tilde{g}^{-1}A$ has measure zero. Hence,

$$\sum_{i=1}^{n} t_i(g(x_1), x_2, ..., x_n) \leq 0 \text{ a.e.}$$

which proves feasibility.

(ii) Incentive compatibility requires IC1 and IC2 to be fulfilled. IC1 is obvious. Let $\bar{t}_i(x_1, ..., x_{i-1}, x_{i+1}, ..., x_n)$ denote the constant payment to agent i in non-monitoring states. We first show that IC2 is satisfied for $i \geq 2$ (the argument is similar to that given for (i)). Define $\tilde{g}$ as above, but now let $A = \{(x_1, ..., x_n) : t_i(x_1, ..., x_n) - \phi_i(t(x_1, ..., x_n)\}$. Then it follows that $\tilde{g}^{-1}A = \{(x_1, ..., x_n) : t_i(g(x_1), ..., x_n) - \phi_i(t(g(x_1), ...x_n)) < \bar{t}_i(g(x_1), ...x_n)\}$. Since $\tilde{g}$ is measure preserving, IC2 implies that $\tilde{g}^{-1}A$ has measure zero. Hence IC2 holds for the new

[14] Regularity means that $\mu(A) = \inf\{\mu(O) : O \subset A, O \text{ open}\} = \sup\{\mu(F) : F \subset A, F \text{ closed}\}$. Our measure μ is regular, since every probability measure on a metric space is regular (cf., Parthasarathy (1977) Proposition 19.13).

contract for all agents $i \geq 2$. It remains to give the proof for $i = 1$. This, however, follows immediately from the argument for $i \geq 2$ and the fact that

$$\begin{aligned} \bar{t}'_1(x_2, ..., x_n) &= \sup_{y_1 \in S_1'^c} t_1(g(y_1), ..., x_n) \\ &= \sup_{y_1 \in S_1'^c} t_1(y_1, ..., x_n) = \bar{t}_1(x_2, ..., x_n), \end{aligned}$$

because g is one-to-one. This proves (ii).

(iii) Apply Remark 1 and Fubini's Theorem (cf., Ash (1972), Theorem 2.6.4). Let c_i' denote consumption under the new contract, and let c_i denote consumption under the original contract for agent i. Note that for every $i \neq 1$ we have $c_i(g(x_1, x_2, ..., x_n) = c_i'(x_1, ..., x_n)$. We must show that $\int u_i(c_i)dF^n = \int u_i(c_i')dF^n$, which means that the expected utilities are the same. This follows from Fubini's Theorem since

$$\begin{aligned} &\int\int ... \int u_i(c_i(g(x_1), x_2, ..., x_n))dF_1(x_1), dF_2(x_2), ..., dF_n(x_n) \\ &= \int\int ... \int u_i(c_i(x_1, x_2, ..., x_n))dF_1(x_1)dF_2(x_2)...dF_n(x_n). \end{aligned}$$

Equality follows from Remark 1, i.e., the fact that g is measure preserving. This proves (iii).

(iv) For given $(x_2, ..., x_n)$ define

$$f(x_1) = t_1(x_1, ..., x_n) - \phi_1(t_1(x_1, ...x_n)).$$

Because of IC1 and IC2, transfers (net of monitoring costs) in monitoring states are always higher than transfers in non-monitoring states. g moves these high transfers to the left (i.e., to low income states) and vice versa.[15] By Lemma 2 in the Appendix, agent one is strictly better off under the new contract. This contradicts the assumed optimality of the original contract, proving the Theorem.

5. The case of stochastic verification

In this section we study the form of Pareto efficient multilateral contracts that arise among agents under stochastic monitoring. We begin by defining a multilateral contract for this economy.

Definition 4. *A multilateral contract with stochastic verification for each agent $i = 1, ..., n$ is a pair (t_i, p_i), where $t_i(x_1, ..., x_n)$ is a net-transfer function for agent i from $\mathbb{R}^n$ into $\mathbb{R}$, and $p_i : [m, \infty]^n \to [0, 1]$ is a function which indicates the probability that agent i's endowment announcement is verified. If agent i is verified, the endowment becomes public information.*

[15] That is $f(k_1) < f(k_2)$ for every $k_1 \in K_1$, and for every $k_2 \in K_2$. This is exactly the condition under which Lemma 2 holds.

Alternative formulations of the stochastic monitoring problem have been studied by other authors. Townsend (1988, p. 424) reports the results of systematic numerical analyses of costly state verification economies with stochastic monitoring and gives examples of non-monotonic monitoring probabilities. His results stem from the fact that his monitoring probability function, p_i, is defined on $[m, \infty]$, so whether or not an agent is verified depends only on the agent's own announcement (and is independent of all other agents' announcements).[16] In contrast, in our model monitoring depends on the agents own endowment announcement *and* on the announcements of all other agents (i.e., p_i is defined on $[m, \infty]^n$ in Definition 4). This specification seems reasonable for many stochastic auditing applications of the model. For example, the probability of a tax audit is related not only to an individual's own income tax return, but also to the returns filed by all other individuals in the economy.[17] Border and Sobel also prove a monotonicity result, but their arguments depend crucially on risk neutrality (cf. Sect. 2).

Our main goal in this section is to characterize the solutions to an information constrained optimization problem with stochastic monitoring. Before beginning our formal analysis we first discuss an inherent difficulty that emerges in economies with stochastic monitoring *and* risk averse agents. The problem stems from the fact that stochastic monitoring generates additional uncertainty into expected consumption allocations, and this additional uncertainty decreases the expected utility of risk averse agents.[18] The key problem is that states with low endowment realizations are the same states where the probability of monitoring is the highest. These high variance states are precisely the states of most concern to risk averse agents. In general it is difficult to characterize the marginal loss of utility to an agent from the additional uncertainty caused by stochastic monitoring. Transfers which are contingent not only on all agents' endowment realizations (as they are in our model), but also on whether or not monitoring is actually performed (which does not occur in our model) might ameliorate the negative utility effects associated with stochastic monitoring somewhat. However, it is unlikely that such transfers would eliminate these effects entirely.

We consider two polar cases which are designed to address the "marginal utility loss" problem experienced by risk averse agents. We first consider the Gase where monitoring costs are borne by each individual agent, but restrict agents' utility functions to be separable in consumption and monitoring cost. This approach is often employed in the literature (e.g., Moohkerjee and Png (1989)), hence we use it in the statement of Problem 4.1 below. However, our proofs also apply to an alternative specification where agents are abie to diversify their individual specific monitoring cost risk (e.g., if monitoring occurs, the monitoring costs of agent i are

[16] Townsend's example is for a discrete (hence atomic) distribution. However, because it is an equal distribution our proof immediately goes through (but breaks down for discrete distributions which are not equal distributions). We are not aware of an example where the "discreteness" is solely responsible for the non-monotonicity.

[17] That is, an individual with a university salary is more likely to be audited in a small college town (Urbana, IL) than in the Silicon Valley (Palo, Alto, CA).

[18] Note that stochastic monitoring also has the countervailing beneficial effect of reducing expected monitoring costs (relative to deterministic monitoring).

borne by all other agents $i \neq j$). We defer discussion of this second specification until after we have proved our main results (Theorem 2 and Corollary 1).

We now state the optimization problem for this economy:

Problem 4.1. *Choose $t_i(\cdot)$ and $p_i(\cdot)$ for $i = 1, ..., n$ to maximize:*

$$\sum_{i=1}^{n} \lambda_i \int [v_i(x_i + t_i(\cdot)) - p_i(\cdot)\phi_i(t_i(\cdot))] dF^n(x_1, ...x_n), \tag{4.1}$$

subject to

$$0 \leq c_i \leq x_i + t_i(x_1, ..., x_n) \text{ a.a. for all } i, \tag{4.2}$$

$$\sum_{i=1}^{n} t_i \leq 0, \text{ a.e.} \tag{4.3}$$

$$\begin{aligned} & v_i(x_i + t_i)(x_1, ..., x_i, ..., x_n)) - p_i(x_i, ..., x_n)\phi_i(t_i(\cdot)) \\ & \quad \geq (1 - p_i(x_1, ..., y, ..., x_n))v_i(x_i + t(x_1, ..., y, ..., x_n)) \\ & \qquad + p_i(x_1, ..., y, ..., x_n)[v_i(0) - \phi_i(t_i(\cdot))], \quad \text{for all } i. \end{aligned}$$

$$\text{for all } y, \quad \text{and for a.e. } x_i; \quad \text{and} \tag{4.4}$$

$$0 \leq p_i(x_i, ..., x_n) \leq 1, \text{ for every } x_i. \tag{4.5}$$

(4.1) reflects the consumption and monitoring cost separability restriction described above. Separability implies that each agent's utility from consumption is independent of the non-pecuniary (effort cost) imposed on the agent by the monitoring procedure. Loosely speaking, the idea is that the monitoring process causes no additional utility or disutility other than the direct costs. (4.3) is the same as in Problem 3.1. (4.4) is the incentive compatibility constraint under stochastic monitoring.[19] The left-hand side of (4.4) is the expected utility of agent i from truthfully reporting endowment realization x_i; and the right-hand side is the expected utility of agent i from announcing any other realization $y \neq x_i$. When agent i misreports and is verified, he/she receives a zero transfer and the entire endowment is confiscated, so utility is $v_i(0) - \phi_i(t_i(\cdot))$. We implicitly assume that it is optimal to punish an agent as much as possible (by seizing the entire endowment) for misreporting, but this is straightforward to show since maximizing the penalty minimizes the propensity to cheat. See Section 6 for further discussion of incentive compatibility. Finally, (4.5) states that the p_i are probabilities.

We now give an overview of the proof of Theorem 2. This Theorem shows that the transfer function associated with the optimal contract is a decreasing function of wealth when monitoring is stochastic. As in Theorem 1, we proceed indirectly: Assume that the transfer function of agent one is not a monotonically decreasing function of wealth over the entire support of the distribution. We again wish to use the Isomorphism Theorem to find a measure preserving one-to-one function g which

[19] Townsend (1988, pp. 416–418) uses a relevation princple argument to prove that this restriction can be imposed without loss of generality.

maps arbitrary initial contracts into an alternative contract which is "more monotonic."[20] We show that this "more monotonic" alternative contract: (i) is feasible; (ii) is incentive compatible; (iii) does not decrease the expected utility of all other agents; and (iv) strictly increases the expected utility of agent one. This establishes the optimality of contracts with monotonically decreasing transfer functions.

The first step of the proof, since the argument is indirect, is to establish a uniform violation of (decreasing) monotonicity of an arbitrary initial (non-monotonic) transfer function. We begin by showing that it is possible to find two compact sets with positive measure, denoted $\mathcal{U}$ and $\mathcal{V}$, where $\mathcal{U}$ is strictly to the left of $\mathcal{V}$, and such that all values of the transfer function in $\mathcal{U}$ are strictly below the values which the transfer function assumes in $\mathcal{V}$.[21] To construct such sets, we use Lusin's Theorem (cf., Parthasarathy (1977) Proposition 24.21 and Corollary 24.22), which says that for any integrable function (on a complete and separable metric space) there exist arbitrary large compact subsets of the domain such that the restriction of a function on this compact subset is continuous. We use this continuity to establish the desired (uniform) violation of monotonicity of the transfer function on $\mathcal{U}$ and $\mathcal{V}$. The main insight in this part of the proof is that it is not sufficient to establish a violation of monotonicity of the transfer function for single points as the analysis necessarily excludes sets of measure zero. Hence, starting with two points z_1, z_2 for which monotonicity is violated, we must establish a violation which also holds on a set of positive measure contained in neighborhoods of there two points. For continuous functions this is obviously always the case. Fortunately, Lusin's Theorem implies that this is also true almost everywhere for arbitrary measurable functions (by continuity of such a function on compact subsets).

The remainder of the proof is similar in structure to Theorem 1: We apply a version of the Isomorphism Theorem (proved in Lemma 3) to get a measure preserving mapping h between the arbitrary initial (non-monotonic) contract and a (more monotonic) alternative contract, on the two compact sets $\mathcal{U}$ and $\mathcal{V}$. We then show that (i)–(iv) hold. However, unlike in Theorem 1 with deterministic monitoring, when we apply the Isomorphism Theorem in the stochastic case, we must apply it "slice-wise."[22] The basic problem is that the sets $\mathcal{U}$ and $\mathcal{V}$ do not necessarily have a product structure, i.e., we cannot represent $\mathcal{U}$ or $\mathcal{V}$ in the form $A \times C$ where $A \subset \mathbb{R}$ and $C \subset \mathbb{R}^{n-1}$ and $\mathcal{V}$ as $B \times C$ where $B \subset \mathbb{R}$. If the sets had a product structure, then we could apply the Isomorphism Theorem to construct a

[20] In general it is not possible (even for very simple cases) to construct a monotonic contract directly with a measure preserving transformation. Consider the following example: Choose the interval $[0, 1]$ with the standard Lebesgue measure. Let $f(x) = x(1-x)$. Now assume (indirectly) that there exists a measure preserving transformation g on $[0, 1]$ such that $f \circ g(x) = g(x)(1 - g(x))$ is monotonic. The function is quadratic, so there are two solutions $x_i, i = 1, 2$ to any equation $x(1-x) = z$. Hence, there exist $x_1 \neq x_2$ such that $f \circ g(x_1) = f \circ g(x_2)$. Assume that $x_1 < x_2$. Since $f \circ g$ is monotonic, f is constant on the image of the interval $[x_1, x_2]$ under g. This, however, means that $g([x_1, x_2])$ contains at most two points. This is a contradiction to g being measure preserving.

[21] For technical reasons we prove an even stronger violation of monotonicity. We show that if the endowment realization x lies in $\mathcal{U}$ and a transfer corresponding to an arbitrary state in $\mathcal{V}$ is used instead of the transfer $t_1(x)$, then the consumption of agent one strictly increases. A similar condition holds if the realized state is an element of $\mathcal{V}$.

[22] A slice of a set $A \subset \mathbb{R}$ is given by $A_{(x_2,\ldots,x_n)} = \{(x_1, x_2, \ldots, x_n) : (x_1, x_2, \ldots, x_n) \in A\}$.

measure preserving mapping between A and B for agent one when the realizations of all other agents are fixed. In Lemma 3 we generalize the Isomorphism Theorem so that for fixed realizations $(x_2, ..., x_n)$ of all other agents we can still establish an isomorphism between respective "slices" of the sets A and B. We define the mapping on every slice in Lemma 3 in a way which ensures that we get a measure preserving "slice-wise" mapping, and then use Fubini's Theorem to get a measure preserving mapping h between the sets A and B. The technical problem in applying the argument is to ensure measurability of h, but this follows from a measurable selection Theorem (also contained in Lemma 3). The strategies of the arguments for (i), (ii), (iii), and (iv) remain similar to those used in Theorem 1.

We now state our main result concerning the nature of optimal contracts in a multi-agent economy with stochastic verification.

Theorem 2. *Let (t_i, p_i) for $i = n$ be a collection of Pareto optimal contracts. Then there exists a set of measure zero N such that for every agent i and for every $z_1 = (x_1, ..., x_i, ..., x_n)$, and $z_2 = (x_1, ..., y_i, ..., x_n)$ with $z_1, z_2 \in \mathbb{R}^n \backslash N$ it follows that $t_i(z_1) \geq t_i(z_2)$ if $x_i \leq y_i$, i.e., the transfers are monotonically decreasing a.e.*

Proof. We proceed indirectly. Without loss of generality we can assume that the transfer function of agent one is not monotonic a.e. Let $\mathcal{O}$ be the union of all open sets with measure zero. Then $\mathcal{O}$ itself is open and has measure zero. By Lusin's Theorem (cf. Ash (1972), Corollary 4.3.17(b)) there exists for every $\varepsilon > 0$, a compact subset $K \subset \mathbb{R}^n$ with $\mu(\mathbb{R}^n \backslash K) < \varepsilon$ and such that t_1 is continuous on K. Without loss of generality we can assume that $\mathcal{O} \cap \emptyset$ (otherwise take $K \backslash \mathcal{O}$). Hence, we can construct an increasing sequence of compact sets K_i such that t_i is continuous on each of the K_i and such that $\mathbb{R}^n \backslash \cup_{i=1}^{\infty} K_i$ has measure zero. Since t_i is not monotonic a.e. there must exist $z^1 = (x_1, x_2, ..., x_n)$, and $z^2 = (y_1, x_2, ..., x_n)$ such that $x_1 < y_1$, and $t_1(z^1) < t_1(z^2)$, and such that $z^1, z^2 \in \cup_{n=1}^{\infty} K_n$. For a sufficiently large n we can assure that $z^1, z^2 \in K_n$. Thus, t_1 is continuous on K_n.[23] Choose γ_1 and γ_2 such that $x_1 + t_1, z^1) < \gamma_1 < x_1 + t_1(z^2))$ and $y_1 + t_1(z^1) < \gamma_2 < \gamma_1 + t_1(z^2)$. Then there exist compact neighborhoods $\mathcal{U}$ of z^1 and $\mathcal{V}$ of z^2 such that

(a) $u_1 + t_1(u) < \gamma_1 < u_1 + t, (v)$; and
(b) $v_1 + t_1(u) < \gamma_2 < v_1 + t_1(v)$,

for every $u \in \mathcal{U}$ and $v \in \mathcal{V}$, where u_1 and v_1 are the first coordinates of u and v, respectively. Furthermore, we can assume that $\mathcal{U}$ is to the left of $\mathcal{V}$, i.e., for every $u \in \mathcal{U}$ and for every $v \in \mathcal{V}$ we have $u_1 < v_1$. Since $\mathcal{U}$ and $\mathcal{V}$ are neighborhoods, they must have positive measure (since their intersection with $\mathcal{O}$ is empty). By the "generalized Isomorphism Theorem" (Lemma 3) there exist subsets $A \subset \mathcal{U}$ and $B \subset \mathcal{V}$ and measure preserving mappings $h^1 : A \to B$ and $h^2 : B \to A$ such that for fixed $(x_2, ..., x_n)$ the mappings $x \mapsto h^i(x_1, ..., x_n)$, $i = 1, 2$ are measure preserving on $A_{(x_2,...,x_n)}$ and $B_{(x_2,.-.,x_n)}$, respectively.[24]

[23] Note that all neighborhoods are in the *induced* topology on K_n and not in the original topology of $\mathbb{R}^n$, i.e., $\mathcal{U}$ is a neighborhood of $x \in K_n$ if there exists a neighborhood $\mathcal{W}$, of x in $\mathbb{R}^n$ such that $\mathcal{U} = K_n \cap \mathcal{W}$.

[24] $A_{(x_2,...,x_n)} = \{(x_1, x_2, ..., x_n) : (x_1, ..., x_n) \in A\}$ and similar for B.

Now define

$$f(x_1, ..., x_n) = \begin{cases} h^1(x_1, ..., x_n) & \text{if } x \in A; \\ h^2(x_1, ..., x_n) & \text{if } x \in B; \\ (x_1, x_2, ..., x_n) & \text{otherwise.} \end{cases}$$

Then for fixed $(x_2, ..., x_n)$ the mapping $x_1 \mapsto f(x_1, ..., x_n)$ is a measure preserving transformation on $\mathbb{R}^n_{(x_2,...,x_n)}$, where $\mathbb{R}^n_{(x_2,...,x_n)}$ is given by the set $\{(x_1, x_2, ..., x_n) : x_1 \in \mathbb{R}\}$. Let g denote the first coordinate of $f(x) = (f_1(x), ..., f_n(x))$. Then $x_1 \mapsto g(x_1, ..., x_n)$ is a measure preserving transformation on $\mathbb{R}$ for fixed $(x_2, ..., x_n)$.

Now define new transfers denoted by $t_i(g(x_1, ..., x_n), x_2, ..., x_n)$ and new monitoring probabilities denoted by $p_i(g(x_1, ..., x_n), x_2, ..., x_n)$. We show that these new contracts are: (i) feasible, (ii) incentive compatible, (iii) preserve the utility of all agents $i \neq 1$, (iv) increase the utility of agent one.

(i) Feasibility follows as in the proof of Theorem 1.

(ii) Incentive compatibility requires (4.4) to be satisfied. There are three possible cases. First, assume the true realization $(x_1, ...x_n)$ lies in B. If it is profitable to cheat in this situation under the alternative contract, then it must also have been profitable with the initial contract in state $g^{-1}(x_1, x_2, ..., x_n)$, because the transfers are the same under the two contracts but under the initial contract the endowment of agent one was lower (hence the penalty if detected cheating was less severe). This contradicts incentive compatibility of the initial contract. Second, assume the realization lies in A. If it is profitable to cheat in this situation under the alternative contract, then it must have been even more profitable under the initial contract as ihe transfer was lower. This again contradicts optimality of the initial contract. Finally, for all other realizations the two contracts are the same. This proves (ii). However, also note that the monitoring probabilities can be reduced slightly without violating incentive compatibility.

(iii) The expected utility of all other agents is unchanged, cf. Theorem 1.[25]

(iv) Since (a) and (b) are fulfilled, and since A is to the left of B we can apply Lemma 1 for agent one for fixed $x_2, ..., x_n$. Agent one is therefore strictly better off with the alternative contract (we exchange high transfers to low income states and vice versa). Since the monitoring probabilities can be reduced slightly without violating incentive compatibility because of(ii), the utility of agent one can be strictly increased. Thus, Fubini's Theorem implies that the expected utility of agent one is vtrictly greater under the alternative contract. Thus, contracts which are non-monotonic cannot be optimal. This proves the Theorem.

The following Corollary follows immediately from Theorem 2.

[25] The utility of an agent depends on $x_1, ..., x_n$. Using Fubini's theorem, we can first integrate over the realization x_1 in order to compute the expected utility. However, for fixed $x_2, ..., x_n$, the mapping $x_1 \mapsto g(x_1, ..., x_n)$ is measure preserving. Thus, g drops out of the integral when integrating over x_1 (cf. Remark 1).

Corollary 1. *Under the assumptions of Theorem 2 it follows that $p_i(z^1) \geq p_i(z^2)$, i.e., the probabilities of verification are monotonically decreasing a.e. in endowments.*

Proof. The Corollary follows immediately from the fact that the transfers are monotonically decreasing: Let $x_1 \leq y_1$. Consider two endowments $z^1 = (x_1, ..., x_n)$, and $z^2 = (y_1, x_2, ..., x_n)$, and assume that monotonicity of the probabilities is violated for agent one, i.e., $p_1(z_1) < p_1(z_2)$. By Theorem 2, $t_1(z^1) \geq t_1(z^2)$. Now choose $p(z_1)$ as the monitoring probability for z^2. Using this construction we could lower the probabilities of verification if the contracts are not monotonic and thus increase the payoff of the agent. This is a contradiction to the optimality of the original contracts. It remains to prove that the contracts with the lower probabilities of verification are still incentive compatible. This, however, follows immediately. Suppose it were profitable for the agent to cheat in some other state and announce z^1. Then it would be at least as profitable to announce zt since the transfer is at least as high and the probability that cheating is detected is lower which contradicts incentive compatibility of the original contract. This proves the Corollary.

We conclude this section by diseussing the alternative monitoring cost specification described before the statement of Problem 4.1. That is, instead of assuming that each risk averse agent i privately bears the entire “utility loss” stemming from stochastic monitoring, Theorem 2 and Corollary 1 continue to hold if we assume that a mechanism exists whereby the monitoring costs of agent i are borne by all agents $j \neq i$ (when monitoring occurs). This follows from the fact that steps (i), (ii) and (iii) from the proof of Theorem 2 remain valid under either specification of the model because the transfers and monitoring probabilities have the same expected value and the same distribution (although we did not use this fach in the proof of Theorem 2 because of the assumed separability of the utility function). Examples of mechanisms in actual economies which appear to be qualitatively similar to this second (publicly borne) cost specification are tax surcharges (levied by a government) or a reduction in the “dividend credits” commonly rebated to policy holders by insurance companies (e.g., TIAA-CREF and many other insurance companies follow this practice).

6. Concluding remarks

This paper generalizes the costly state verification model to allow risk averse agents who need not be identical ex ante to write multilateral contracts. Following standard practice in the literature we impose incentive compatibility constraints. Townsend (1988) notes that in order to justify this restriction in costly state verification models one can formulate the underlying revelation game as follows: Contracts are written before uncertainty is revealed. Uncertainty is then privately revealed, and each agent sends a message (i.e., reports a state). Thus, agents play a Nash game in messages where each agent has beliefs over whether all other agents teil the truth. When the analysis is restricted to truth-telling equilibria, it follows that each agent expects all other agents to tell the truth. In such a framework the point-wise incentive constraints used in our model follow. This formulation implicitly contains a great deal of communication among agents in the sense that decisions are made based on the expected announcements of all other agents.

An alternative formulation is to consider is a game with no communication among agents. This suggests two issues. First, what are the implications of such an environment for the form of the incentive constraints? Second, which environment (one with or without communication) is most plausible? We begin with the first issue. In a game with no communication among agents, each agent makes an announcement with no knowledge of other agents' announcements. This corresponds to a Harsanyi (1967) type Bayesian Nash game, where the incentive constraints need not hold pointwise but only in expected value.[26] Theorem 2 and Corollary 1 immediately go through under this alternative formulation of the constraint because we do not use incentive compatibility in any essential way in the proof. Rather, we need only check that it remains satisfied.[27] From a technical point of view, this step of the proof requires us to show that our construction does not move us out of the set of all incentive compatible contracts – and this is of course easier to show if the constraint set is bigger. Thus, the expected value form of the incentive constraint does not change the structure of the optimal contract with stochastic monitoring. In fact, it facilitates the technical arguments necessary to prove the result.

In contrast, in Theorem 1 we again check that incentive compatibility conditions IC1 and IC2 are satisfied in step (ii) of the proof, but we also use there conditions in step (iv) in an essential way. In particular, we use them in (iv) to show that the transfer in every non-monitoring state is always higher than the transfer in every monitoring state. Thus, the final step in the proof of Theorem 1 does not go through with an incentive constraint which holds only in expected value. In fact, it turns out that under the mathematically weaker expected value constraint, the transfers associated with the optimal contract need no longer be constant on the nonmonitoring set. We first show this in a simple (but not pathological) example and then provide an economic interpretation of the result.

Example 1. Consider a discrete distribution and two agents: agent one is risk neutral and agent two is very risk averse. The same kind of example also goes through for continuous distributions and if one agent is (slightly) risk averse. Assume that there are four states which occur with equal probability. The endowment of agent one is given by $(7, 7, 3, 3)$ and of agent two by $(7, 3, 7, 3)$. Clearly, the two endowments are independent. Let ϕ be a constant monitoring cost. Choose $S_1 = \emptyset$ and $S_2 = \{3\}$, i.e., agent one is *never* monitored and agent two is monitored in the low state. Pareto optimal contracts are given by $t_1 = -t_2 = (2 + c, -2, 2 + c, -2)$, since under this contract agent two is completely insured, i.e., consumption is state-independent (net of monitoring costs). However, agent one's net-transfer is not constant even though the agent is never monitored. Incentive compatibility for

[26] IC1 is then: $x_i \mapsto \int t_i(x_1, ..., x_i, ..., x_n) dF(x_1, ..., x_{i-1}, x_{i+1}, ..., x_n)$ is constant on S_i^c; while IC2 is: $\int t_i(x_1, ..., x_i, ..., x_n) - \phi(\cdot) dF(x_1, ..., x_{i-1}, x_{i+1}, ..., x_n) \geq \int t_i(x, ..., y, ..., x_n) dF(x_1, ..., x_{i-1}, x_{i+1}, ..., x_n)$ for all $x_i \in S_i$ and for every $y \in S_i^c$. In both cases $dF(\cdot)$ denotes integration with respect to the joint distribution of the random variables $X_j, j \neq i$. Unlike in the pointwise specirication, these constraints need only hold on average.

[27] Use the argument in step (ii) of the proof of Theorem 2 and take the expected value.

agent two is straightforward. Incentive compatibility for agent one is fulfilled in expected value: Assume that agent one gets the high realization. The expected net transfer is $c/2$, the same expected net-transfer the agent would get in the low state. The argument goes through even if agent one is slightly risk averse because this arrangement economizes on monitoring costs: Choosing $S_1 = \{3\}$ increases monitoring costs by a discrete amount since monitoring is deterministic.

Some readers may be tempted to construe Example 1 as refuting the optimality of debt even under deterministic verifcation. We regard this interpretation as misguided. As Townsend (1987, p. 382) notes, the motivation for an analysis such as our Theorem 1 is to "begin with some striking arrangement [e.g., debt] in an actual economy and ask whether any theoretical environment might yield such an arrangement ... without making the [model] too complicated or implausible." We view the question – is a model with an expected value incentive constraint better than a model with a point-wise constraint? – to be methodologically equivalent to the question – is a model with stochastic monitoring better than a model with deterministic monitoring? In our opinion the answer is clearly no. Mathematical generality is not the desideratum per se, rather it is the consistency of the *structure and results* of alternative models with those observed in actual economic environments which determines which model is more appropriate for the problem at hand.

7. Appendix

Proof of Remark 1. Let t be a simple function on Y_2, i.e., there exist $A \in \beta_2$, and $\lambda_i \in \mathbb{R}$ such that $t = \sum_{i=1}^{n} \lambda_i \mathbf{1}_{A_i}$, where

$$\mathbf{1}_{A_i}(x) = \begin{cases} 1 \; x \in A_i; \\ 0 \text{ otherwise.} \end{cases}$$

Then $t(g(x)) = \sum_{i=1}^{n} \lambda_i \mathbf{1}_{g^{-1}A_i}(x)$. Hence,

$$\int_{Y_1} t(g(x))d\mu_1(x) = \sum_{i=1}^{n} \lambda_i \int_{Y_1} \mathbf{1}_{g^{-1}A_i}(x)d\mu_1(x) = \sum_{i=1}^{n} \lambda_i \mu_1(g^{-1}A_i)$$
$$= \sum_{i=1}^{n} \lambda_i \mu_2(A) = \sum_{i=1}^{n} \lambda_i \int_{Y_2} \mathbf{1}_{A_i}(x)d\mu_2(x) = \int_{Y_2} t(x)d\mu_2(x).$$

The third equality follows because g is measure preserving. Since the Remark holds for all simple functions, it also holds for all integrable functions.[28]

Lemma 1. *Let μ be a measure on $\mathbb{R}$ and let A, B be two subsets of $\mathbb{R}$ with the same measure. Let f be an integrable function on $\mathbb{R}$. Assume that $a < b$ for every $a \in A$ and for every $b \in B$. Assume that f is bounded on $A \cup B$. Let g be a measure*

[28] This is a standard approximation argument in measure theory: All integrable functions can be approximated by simple functions.

preserving isomorphism on $\mathbb{R}$ *such that* $g(x) = x$ *for every* $x \in \mathbb{R}\backslash A \cup B$. *Assume that there exist* $\gamma_A, \gamma_B \in \mathbb{R}$ *such that* $x + f(X) \leq \gamma_A \leq x + f(g(x))$, *for every* $x \in A$; *and* $x + f(g(x)) \leq \gamma_B \leq x + f(x)$, *for every* $x \in B$. *Then* $x + f(g(x))$ *is less risky than* $x + f(x)$ *in the Rothschild and Stiglitz sense (i.e., every risk averse agent prefers* $x + f(g(x))$ *over* $x + f(x)$*).*

Proof. Here we need only check that the integral condition of Rothschild and Stiglitz (1970) holds. Let F be the distribution of $x + f(g(x))$ and let G be the distribution of $x + f(x)$. Then

(i) $G(t) - F(t) \geq 0$ for every $t < \gamma_A$.
(ii) $G(t) - F(t)$ is monotonically decreasing for $\gamma_A \leq t < \gamma_B$; and
(iii) $G(t) - F(t) \leq 0$ for every $t \geq \gamma_B$.

(i) follows from the fact that g is measure preserving. In particular:

$$\mu(\{x \in A : x + f(x) \leq t\}) = \mu(\{x \in B : g(x) + f(g(x)) \leq t\}).$$

Note that $\mu(\{x \in B : x+f(x) \leq t\}) = 0$, and $\mu(\{x \in A : x+f(g(x)) \leq t\}) = 0$ for every $t < \gamma_A$. Since $g(x) \leq x$ for every $x \in B$ this proves (i). (ii) follows immediately from the definition of γ_A and γ_B. Finally, the argument for (iii) is similar to the argument for (i).

Since f is bounded and since A and B are bounded there exists an $M > 0$ such that $G(t) - F(t) = 0$ for every $t \notin [-M, M]$.[29] Let $T(y) = \int_{-M}^{y} G(t) - F(t)dt$. By Rothschild and Stiglitz (1970, Theorem 2) it is sufficient to prove that the following two conditions are satisfied.

(a) $T(M) = \int_{-M}^{M} [G_i(x) - F_i(x)]dx = 0$;
(b) $T(y) \geq O$ for $-M \leq y \leq M$.

(a) follows immediately from integration by parts and from the fact that g is measure preserving.[30] (b) follows immediately from (a) and from conditions (i), (ii) and (iii). This concludes the proof.

Lemma 2. *Let* u *be a utility function which is twice continuously differentiable. Assume that* $u''(x) < 0$ *for every* x. *Let* A, B *be two subsets of* $\mathbb{R}$ *with the same measure. Let* f *be integrable and let* g *be a measure preserving transformation such that* $g(g(x)) = x$ *for every* x, *such that* $g(A) = B$, *and* $f(x) = x$ *for every* $x \notin A \cup B$. *Assume that* $f(a) < f(b)$ *for every* $a \in A$ *and for every* $b \in B$. *Then Lemma 1 holds with a strict inequality, i.e., the agent strictly prefers the contract* $x + f(g(x))$ *to* $x + f(x)$.

---

[29] In order to be able to apply the Theorem we need that the points of increase of the distribution functions lie in a compact interval.

[30] Measure preservingness implies $\int_{-M}^{M} tdF(t) = \int_{-M}^{M} tdG(t)$. Partial integration therefore yields

$$\int_{-M}^{M} G(t) - F(t)dt = t(G(t) - F(t))|_{-M}^{M} - \left(\int_{-M}^{M} tdG(t) - \int_{-M}^{M} tdF(t)\right) = 0.$$

Proof. Here we need only check that the integral condition of Rothschild and Stiglitz holds with a stritt inequality, and then use partial integration to show that the agent strictly prefers $x + f(g(x))$ (cf., Rothschild and Stiglitz (1970), footnote 10).

Let $\varepsilon > 0$. Then there exists a $\delta > 0$ such that $f(b) - f(a) > \delta$ except on sets $S_A \subset A$ and $S_B \subset B$ with $\mu(S_A) = \mu(S_B) < \varepsilon$. Since $g(g(x)) = x$ for every $x \in \mathbb{R}$, we can construct a finite partition $A_i, i = 1, ..., n$ of $A \backslash S_A$ and a finite partition $B_i, i = 1, ..., n$ of $B \backslash S_B$ such that $g(A_i) = B_i$ and such that the condition of Lemma 1 is fulfilled for each A_i and B_i.[31] Thus, since we can subsequently exchange the transfers of A_i with the transfers of B_i and since ε was chosen arbitrarily, Lemma 1 implies that $x + f(g(x))$ is less risky than $x + f(x)$. Thus, the integral conditions of Rothschild and Stiglitz (1970) hold. Note that $T(y) > 0$ on a set of positive measure.

Integration by parts yields

$$\begin{aligned}\int_{-M}^{M} u(x)dS(x) &= u(x)S(x)|_{-M}^{M} - \int_{-M}^{M} u'(x)S(x)dx \\ &= -u'(x)T(x)|_{-M}^{M} + \int_{-M}^{M} u''(x)T(x)dx, \qquad (A.1)\end{aligned}$$

since $u(x)S(x)|_{-M}^{M} = 0$ by (a). Further, since T is strictly positive on a set of positive measure, and since $u'' < 0$ it follows that

$$\int_{-M}^{M} u''(x) : T(x)dx < 0. \qquad (A.2)$$

(A.1), (A.2) and $u'(x)T(x)|_{-M}^{M} = 0$ immediately imply that the agent's utilityis strictly greater under contract $x + f(g(x))$. This proves the Lemma.

Next we state a "generalized" version of the Isomorphism Theorem. The problem we face in the proof of Theorem 2 is that the sets $\mathcal{U}$ and $\mathcal{V}$ are not necessarily representable as the product of lower dimensional spaces. However, for the proof of the Theorem we need an isomorphism between subsets $A \subset \mathcal{U}$ and $B \subset \mathcal{V}$ which is also an isomorphism between the corresponding "slices" of A and B. The existence of such an isomorphism and of the subsets is provided by the following Lemma. The central step of the argument is the use of a theorem on measurable selections.

Lemma 3. Let K^i, $i = 1, 2$ be two compact subsets of $\mathbb{R} \times \mathbb{R}^n$. Let μ_1, and μ_n be probability measures on $\mathbb{R}$ and $\mathbb{R}^n$, respectively, and let μ denote the product measure. Then there exist measurable subsets $A^i \subset K^i$ and measure preserving mappings $h^1 : A^1 \to A^2$ and $h_2 : A^2 \to A^1$ such that $x \mapsto h_i(x, y), i = 1, 2$ are measure preserving mappings from A^1_y to A^2_y and from A^2_y to A^1_y, respectively, for every $y \in \mathbb{R}^n$ where $A^i_y = \{(x, y) : (x, y) \in K^i\}$.[32]

[31] That means that there exist γ_A, γ_B, for $i = t, ..., n$ such that $x + f(x) \leq \gamma_{A_i} \leq x + f(g(x))$, for every $x \in A_i$; and $x + f(g(x)) \leq \gamma_{B_i} \leq x + f(x)$, for every $x \in B_i$.

[32] Let B be a subset of A^i_y. Then $B = B' \times \{y\}$. By slight abuse of notation, we define $\mu_1(B)$ to be $\mu_1(B')$.

Proof. Define a function $f : \mathbb{R} \times \mathbb{R}^n \times \mathbb{R} \to \mathbb{R}_+$ by

$$f(x, y, t) = |\mu_1((-\infty, x) \times \{y\} \cap K^1_y) - \mu_1((-\infty, x + t) \times \{y\} \cap K^2_y)|.$$

Note that f is jointly measurable in x and y.[33] Furthermore, for fixed x and y, the function $t \mapsto f(x, y, t)$ is continuous on K_1. By compactness of K_1 and K_2 there always exists a $\bar{t}$ such that $f(x, y, \bar{t}) = \inf_t f(x, y, t)$. Thus, Landers' Theorem [cf. 6.10 of Strasser (1985)][34] implies that there exists a Borel measurable function $\phi : \mathbb{R} \times \mathbb{R}^n \to \mathbb{R}$ such that

$$f(x, y, \phi(x, y)) = \inf_t f(x, y, t).$$

Note that ϕ is part of the measure preserving transformation defined below, however, we still must construct the sets A^i, i.e., the sets where the measure of the slices coincide. We do this as follows. Let

$$A^1 = \{(x, y) : f(x, y, \phi(x, y)) = 0\}.$$

Then A^1 is measurable. Define h_1 on A^1 by $(x, y) \mapsto (x + \phi(x, y), y)$. In a similar way we can construct a measurable subset A^2 of K^2 and a mapping h_2 on A_2. It immediately follows from the above construction that $h_1(A^1_y) \subset A^2_y$ and $h_2(A^2_y) \subset A^1_y$. Thus, $h_1(A^1) \subset A^2$ and $h_2(A^2) \subset A^1$. It now remains to show that the h_i are measure preserving. This can be established as follows:

Let $y \in \mathbb{R}^n$. By construction, h_i preserves the μ_1-measure of all sets of the form $(-\infty, a) \times \{y\} \cap A^i_y$ for $i = 1, 2$. These sets generate the σ-algebra of all measurable sets. Thus h_i is a measure preserving transformation on K^i_y for $i = 1, 2$. Finally, note that Fubini's Theorem proves that the mappings h_i are measure preserving on K^i. This concludes the proof of the Lemma.

---

[33] This can be established as follows: Let $g : K \to \mathbb{R}$ be defined by $(x, y) \mapsto \mu_1((-\infty, x)x\{y\} \cap K^1_y)$. For fixed x, the mapping $y \mapsto g(x, y)$ is measurable by Fubini's Theorem. Furthermore, note that for fixed y the mapping $x \mapsto g(x, y)$ is continuous. Thus, g is jointly measurable [See Castaing and Valadier (1977)]. A similar argument shows that $(x, y, t) \mapsto \mu_1, ((-\infty, x + t) \times \{y\} \cap K^2_y)$ is also jointly measurable in x and y. This proves the measurability of f in x and y.

[34] Landers' Theorem: Let $(\Omega, \mathcal{A}, \mu)$ be a measurable spare and (X, d) a σ-compact metric space. Let $h : \Omega \times X to \mathbb{R}$ be a function such that $\omega \mapsto h(\omega)x)$ is measurable for every $x \in S$ and $t \mapsto rightarrow(\omega, t)$ is continuous for every $\omega \in \Omega$. If

$$B(\omega) = \left\{ x \in X : h(\omega, x) = \inf_{y \in X} h(\omega, y) \right\} \neq \emptyset$$

for a.e. ω then there is a measurable function $\phi : \Omega \to X$ such that

$$h(\omega, \phi(\omega)) = \inf\{h, (\omega, y) : y \in X\}, \quad \mu - \text{a.e.}$$

References

Ash, R.: Real analysis and probability. New York: Academic Press 1972

Border, K., Sobel, J.: Samurai accountant: a theory of auditing and plunder. Rev. Econ. Stud. **54**, 525–540 (1987)

Boyd, J., Prescott, E.C.: Financial intermediary coalitions. J. Econ. Theory **38**, 211–232 (1986)

Boyd, J., Smith, B.: The equilibrium allocation of investment capital in the presence of adverse selection and costly state verification. Econ. Theory (1993) (forthcoming)

Castaing, C., Valadier, M.: Convex analysis and measurable multifunctions. Lect. Notes Math. Berlin Heidelberg New York: Springer 1977

Gate, D., Hellwig, M.: Incentive-compatible debt contracts: the one period problem. Rev. Econ. Stud. **52**, 647–663 (1985)

Harsanyi, J.: Games with incomplete information played by Bayesian players. Manag. Sci. **14**, 159–182 (1967)

Krasa, S., Villamil, A.: Monitoring the monitor An incentive structure for a Financial intermediary. J. Econ. Theory **57**, 197–221 (1992)

Mookherjee, D., Png, I.: Optimal auditing, insurance and redistribution. Quart. J. Econ. **104**, 399–415 (1989)

Parthasarathy, K.: Introduction to probability and measure.
London: McMillan Press 1977

Rothschild, M., Stiglitz, J.: Increasing risk: 1. A definition. J. Econ. Theory **2**, 225–243 (1970)

Strasser, H.: Mathematical theory of statistics. Berlin New York: de Gruyter 1985

Townsend, R.M.: Optimal contracts and competitive markets with costly state verification. J. Econ. Theory **21**, 1–29 (1979)

Townsend, R.M.: Arrow-Debreu programs as microfoundations of macro-economics. In: Bewley, T. (ed.) Advances in economic theory. Econometric Society Monograph. Cambridge: Cambridge University Press 1987

Townsend, R.M.: Information constrained insurance: the revelation principle extended. J. Monet. Econ. **21**, 411–450 (1988)

Williamson, S.D.: Costly monitoring, Financial intermediation, and equilibrium credit rationing. J. Monet. Econ. **18**, 159–179 (1986)

Wilson, R.: On the theory of syndicates. Econometrica **36**, 119–132 (1968)

Winton, A.: Costly state verification and multiple investors: the rote of seniority. Mimeo, Northwestern University (1991)

Risk aversion and incentive compatibility with *ex post* information asymmetry*

Martin F. Hellwig

Fakultät für Volkswirtschaftslehre, Universität Mannheim, Seminargebäude A5, 68131 Mannheim, GERMANY (e-mail: hellwig@@pool.uni-mannheim.de)

Received: December 7, 1998; revised version: June 9, 1999

Summary. The paper extends Diamond's (1984) analysis of financial contracting with information asymmetry *ex post* and endogenous "bankruptcy penalties" to allow for risk aversion of the borrower. The optimality of *debt contracts*, which Diamond obtained for the case of risk neutrality, is shown to be nonrobust to the introduction of risk aversion. This contrasts with the costly state verification literature, in which debt contracts are optimal for risk averse as well as risk neutral borrowers.

Keywords and Phrases: Debt contracts, Risk sharing under asymmetric information.

JEL Classification Numbers: D82, G32.

1 Introduction

Under a *standard debt contract*, a borrower's obligation to his financiers is independent of his actual returns or his ability to pay. If he cannot fulfil the obligation, he goes bankrupt, and the financiers confiscate his remaining assets. The use of such contracts is commonly explained by differences in information of the borrower and his financiers about outcomes. If the borrower's obligation is independent of his own returns, it is easy for financiers to determine whether the obligation is being fulfilled or not. If the borrower's obligation depends on his returns, financiers have to ascertain what these returns actually are. This may be difficult or costly; the use

* I am grateful for helpful comments from Paul Povel and an anonymous referee. I am also grateful for research support from the Deutsche Forschungsgemeinschaft.

of debt, with a non-contingent payment obligation, avoids this difficulty, at least in normal circumstances when the obligation is met.

The confiscation of assets in the event of bankruptcy, i.e., when the obligation cannot be fulfilled, provides the borrower with an incentive to avoid bankruptcy; moreover by maximizing the financiers' returns in bankruptcy, it makes it possible to keep the nominal debt service obligation, i.e., the financiers' return outside of bankruptcy, low and therefore to minimize the incidence of bankruptcy with all the costs and difficulties that it may entail.

The literature contains two distinct formalizations of this argument. The costly state verification approach of Townsend (1979) or Gale and Hellwig (1985) assumes that the information asymmetry can be lifted if resources are spent to provide the financier with information about the borrower's true ability to pay. If all participants are risk neutral, an optimal contract provides for such costly state verification if and only if the borrower cannot pay the prescribed amount; if this occurs, the borrower's remaining assets are confiscated. This is interpreted as "bankruptcy".[1]

In contrast, Diamond (1984) assumes that the information asymmetry cannot be lifted at all. It is however possible to use nonpecuniary "bankruptcy penalties" to discourage the borrower from claiming that he cannot repay the financiers. These nonpecuniary penalties need not be hours in debtor's prison; they may represent a loss of future opportunities that can be imposed on the borrower (without a corresponding gain to his financiers). Imposition of these penalties is endogenous, in accordance with the initial contract; their magnitude, or equivalently the probability of their being imposed, are made to depend on the amount by which the borrower's payment falls short of his debt service obligation. If all participants are risk neutral, an optimal contract in this setting also takes the form of a standard debt contract (Diamond, 1984; see also Povel and Raith, 1999).

The apparent similarity of the two approaches disappears if the assumption of risk neutrality of the borrower is dropped. If the borrower is risk averse and, as usual in models of financial contracting, financiers are risk neutral and non-wealth-constrained, the costly state verification approach still yields a modified version of a standard debt contract. In this modified version, the debtor has a state-independent debt service obligation, and state verification occurs if and only if he fails to fulfil this obligation. In the event of "bankruptcy", i.e., when state verification occurs, the borrower's assets are not always confiscated: He may be left with a positive living allowance. This living allowance is the same in all "bankruptcy" states, providing an element of insurance against return risk across bankruptcy states (Townsend, 1979; Gale and Hellwig, 1985)[2]. Under the additional incentive constraint that the borrower must not want to destroy returns before the financiers get to verify anything

[1] For details see Gale and Hellwig (1985). As emphasized by a referee, Townsend (1979) does not actually consider standard debt contracts. His notion of "debt" allows for arbitrary return sharing arrangements in bankruptcy states and requires only that (i) payments to financiers be non-state contingent for all non-bankruptcy states and (ii) non-bankruptcy states are exactly those states in which the borrower's returns do not suffice for this non-bankruptcy debt service.

[2] As emphasized by a referee, this result is not valid if financiers as well as the borrower are wealth-constrained and, moreover, the total wealth of financiers and the borrower is insufficient to finance the optimal living allowance as well as the desired level of investment.

(Innes, 1990), the optimal level of this living allowance in the event of bankruptcy is actually equal to the lowest nonbankruptcy consumption of the borrower, i.e., the consumption he has if he can barely fulfil his debt service obligation.[3]

In contrast, the present paper shows that in Diamond's (1984) approach the optimality of standard debt contracts is not generally robust to the introduction of risk aversion. The underlying incentive considerations are significantly more complex, and an optimal incentive compatible contract should not be expected to have a simple mathematical form. The nonlinearity of the borrower's utility function implies that the nonpecuniary "bankruptcy penalty" that is required to discourage the borrower from underreporting his ability to pay will itself be given by a *nonlinear* function of the amount of underreporting. Moreover, an optimal contract will involve an element of risk sharing as well as finance. These two considerations interact in such a way that *an optimal incentive-compatible contract will typically not take the form of a standard debt contract, even one with a minimum living allowance.*

The difference is illustrated in Figures 1 and 2. The heavy line in Figure 1 exhibits the relation between the return realization y of the borrower and his consumption $c(y)$ under a standard debt contract, with $c(y) = 0$ for y below the repayment obligation $\hat{y}$ and $c(y) = y - \hat{y}$ when y exceeds the repayment obligation. The dashed line in Figure 1 exhibits the same relation under a standard debt contract with a living allowance $\varepsilon > 0$. In contrast, Figure 2 exhibits the relation between y and $c(y)$ under an optimal contract à la Diamond (1984) when (i) financiers are risk neutral, (ii) the borrower exhibits constant relative risk aversion, and (iii) the ex ante distribution of returns is uniform over some interval $[0, Y]$. For high values of y, the dependence of $c(y)$ on y looks similar in all three cases, but for low values of y, contracting à la Diamond (1984) with risk aversion looks quite different from standard debt[4] - with or without a minimum living allowance. This suggests that Diamond's model of incentive contracting with endogenous "bankruptcy penalties" is rather less closely related to the costly state verification literature than the parallel results on the optimality of standard debt contracts under risk neutrality would seem to indicate.

I came across these findings when I wanted to extend Diamond's (1984) analysis of financial intermediation to allow for risk aversion of the potential financial intermediaries. Diamond (1984) had used his result on the optimality of standard debt contracts as an ingredient in the analysis of the conditions under which financial in-

[3] As pointed out by Garino and Simmons (1998), optimal contracting in the simple Townsend-Gale-Hellwig model with risk aversion of the borrower requires that the bankruptcy living allowance have a marginal utility equal to the expected marginal utility of the borrower's consumption in nonbankruptcy states. This implies that the bankruptcy living allowance exceeds the lowest nonbankruptcy consumption level of the borrower. The Innes incentive condition eliminates this possibility, because in "bad nonbankruptcy states" the borrower must not want to destroy output in order to get into bankruptcy and avail himself of the bankruptcy living allowance.

[4] As emphasized by the referee, the contract corresponding to Figure 2 would still be called "debt" under the weaker definition of Townsend (1979), with $y_2 - c(y_2)$ as the borrower's debt service obligation and $[0, y_2)$ as the set of bankruptcy states. If we forget about the semantics of "debt" and "standard debt", the point of Figure 2 is that the different approaches give qualitatively different results for optimal risk sharing in low-return states.

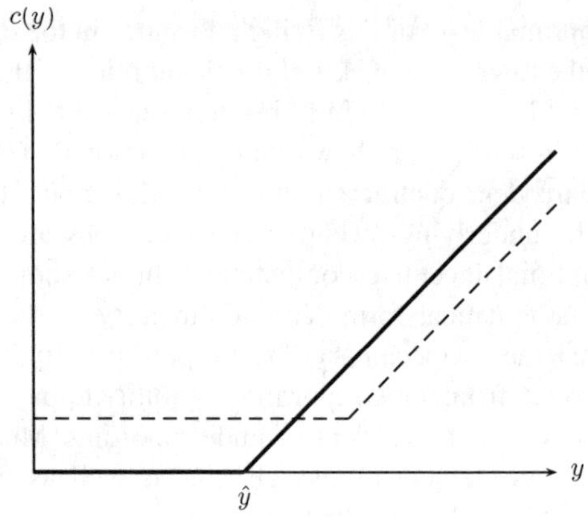

Figure 1

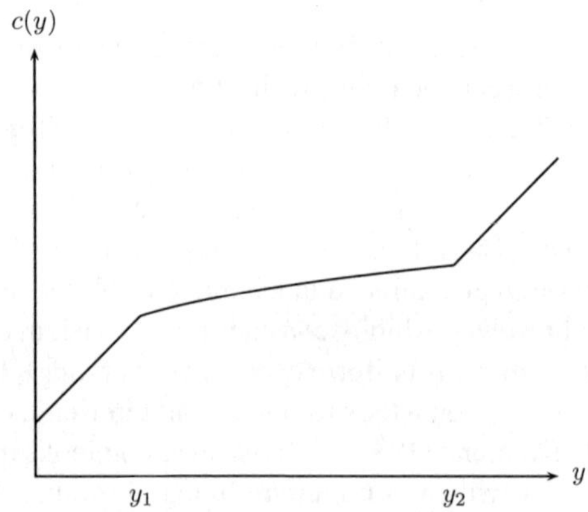

Figure 2

termediation is efficient in the sense that the overall agency costs of intermediated finance are less than the agency costs of direct finance even though intermediation lengthens the chain of transactions. This analysis involves a diversification argument, which makes essential use of the assumption that intermediaries are risk neutral and raises the question of robustness to the introduction of risk aversion. On the way to answering this question, I found that risk aversion complicates not only the diversification argument for financial intermediation, but also the underlying model of incentive contracting. This latter complication is studied here; on the basis of this analysis, the viability of financial intermediation with risk aversion is studied in a companion paper (Hellwig, 1998b). That paper shows that the central results of Diamond (1984) on diversification across borrowers as a basis for intermediation are indeed robust to the introduction of risk aversion.

In the following, Section 2 develops the basic model of incentive contracting with *ex post* information asymmetry and endogenous bankruptcy penalties for a risk averse borrower. Section 3 discusses optimal contracts and explains the economics underlying the contract exhibited in Figure 2. Proofs are presented in the Appendix.

2 *Ex post* information asymmetry and incentive contracting with nonlinear utility

Like Diamond (1984), I consider the financing of a venture that requires a fixed investment $I > 0$ and bears a random return $\tilde{y}$. The random variable $\tilde{y}$ has a probability distribution G with a density g, which is continuous and strictly positive on the interval $[0, Y]$. The expected return of the venture is strictly greater than the cost I, i.e.,

$$\int_0^Y y\, dG(y) > I. \tag{1}$$

The owner/manager of the venture, with own funds $w \geq 0$, wants to raise external finance, either because his funds are less than the investment outlay, or because he wants to share the risk of his venture with others.

Outside financiers know the return distribution G, but – in contrast to the entrepreneur – they are unable to observe the realizations of the return random variable $\tilde{y}$. The agency problems caused by this information asymmetry can be reduced through the use of nonpecuniary penalties as a device to discourage misreporting of return realizations. These penalties are determined endogenously as part of the finance contract. Moreover they can be made to depend on the entrepreneur's report about his return realization and his actual payment to his financiers. As mentioned above, these nonpecuniary penalties need not literally correspond to hours in debtor's prison or something like that; they can also be interpreted as expectations of losses of subsequent opportunities that are imposed on the borrower by the intervention of the lenders (see, e.g., Povel and Raith, 1999).

A finance contract is represented by a number L indicating the funds provided by outside financiers and by two functions $r(.)$ and $p(.)$ such that for any $z \in [0, Y]$, $r(z)$ is the payment to financiers and $p(z) \geq 0$ is the nonpecuniary penalty the entrepreneur suffers when he reports that his return realization is equal to z. With outside funds L, his own financial contribution to his project is $E = I - L \leq w$. Any excess of w over E is invested in an alternative asset, which is safe and has a gross rate of return equal to one.

Given a finance contract $(L, r(.), p(.))$, the entrepreneur's consumption is $w + L - I + y - r(z)$ if the true return realization is y and the reported return realization is z; the corresponding payoff realization is $u(w + L - I + y - r(z)) - p(z)$. A contract $(L, r(.), p(.))$ is said to be *feasible* if $L \geq I - w$ and moreover,

$$w + L - I + y - r(y) \geq 0 \tag{2}$$

for all $y \in [0, Y]$, so the entrepreneur's consumption is never negative. A contract $(L, r(.), p(.))$ is said to be *incentive compatible* if it is feasible and moreover

$$u(w + L - I + y - r(y)) - p(y) \geq u(w + L - I + y - r(z)) - p(z) \tag{3}$$

for all $y \in [0, Y]$ and all $z \in [0, Y]$ such that $w + L - I + y \geq r(z)$, so he has no incentive to misreport his return realization.

The utility function $u(.)$ is assumed to be strictly increasing and strictly concave as well as twice continuously differentiable on $\Re_{++}$; moreover, $u(0) = lim_{c\to 0}u(c)$, with the usual conventions when $lim_{c\to 0}u(c) = -\infty$, e.g., when $u(.) = ln(.)$. Given these assumptions, standard arguments from incentive theory yield:

Proposition 1 *A finance contract* $(L, r(.), p(.))$ *satisfying (2) for all* $y \in [0, Y]$ *is incentive compatible if and only if:*
(i) the function $r(.)$ *is nondecreasing on* $[0, Y]$ *and*
(ii) for all $y \in [0, Y]$,

$$p(y) = p(Y) + \int_y^Y u'(w + L - I + x - r(x))\, dr(x). \tag{4}$$

Condition (4) shows that for a given loan size L and repayment function $r(.)$, incentive compatibility determines the penalty function $p(.)$ up to a constant of integration, $p(Y)$. If $r(.)$ is differentiable, this condition is actually equivalent to the differential equation

$$\frac{dp}{dy} = -u'(w + L - I + y - r(y))\, \frac{dr}{dy}, \tag{5}$$

showing that as the return realization y goes down, the penalty $p(y)$ goes up at a rate which depends on the rate $\frac{dr}{dy}$ at which the payment $r(y)$ goes down as y goes down.

As an illustration, consider the class of finance contracts $(L, r(.), p(.))$ such that

$$r(y) = w + L - I + \min(y, \hat{y}) - \varepsilon \tag{6}$$

for some fixed $\varepsilon \geq 0$, $\hat{y} \in (0, Y)$, and all $y \in [0, Y]$. Such contracts can be interpreted as *standard debt contracts with a minimum living allowance* ε. The amount $w + L - I + \hat{y} - \varepsilon$ represents a return-independent debt service obligation. If the entrepreneur can meet this obligation he does so and retains the excess of his actual return y over $\hat{y}$ as well as ε. If he cannot meet the obligation $w + L - I + \hat{y} - \varepsilon$, he defaults and retains just the minimum living allowance ε. If the minimum living allowance is zero, (6) is the repayment function for a *standard debt contract* as studied by Diamond (1984) or Gale and Hellwig (1985).

By (6), a standard debt contract has $\frac{dr}{dy} = 0$ if $y > \hat{y}$ and the obligation $w + L - I + \hat{y} - \varepsilon$ is met, but $\frac{dr}{dy} = 1$ if $y < \hat{y}$ and the entrepreneur defaults on his obligation. Thus condition (5) entails $\frac{dp}{dy} = 0$ if $y > \hat{y}$, and $\frac{dp}{dy} = -u'(\varepsilon)$ if $y < \hat{y}$; condition (4) reduces to:

$$p(y) = p(Y) + \max(\hat{y} - y, 0)\, u'(\varepsilon). \tag{7}$$

If $\varepsilon = 0$ and $u'(0) = 1$, (7) is exactly the condition that Diamond (1984) gives for the incentive compatibility of a standard debt contract for the case of risk neutrality[5], requiring that the difference $p(y) - p(Y)$ be just equal to the amount of money that the entrepreneur saves by paying $r(y)$ rather than $r(Y)$. For $y \in [\hat{y}, Y]$ of course, (6) implies $r(y) = r(Y)$ and hence $p(y) = p(Y)$. If $u'(\varepsilon) \neq 1$, the money gain $r(\hat{y}) - r(y) = \hat{y} - y$ from reporting $y < \hat{y}$ rather than $\hat{y}$ under the repayment function (6) has to be weighted by $u'(\varepsilon)$ so as to as to make the penalty $p(y)$ commensurate with the utility gain from reporting y rather than $\hat{y}$ and paying $r(y)$ rather than $r(\hat{y})$. (If $u'(0)$ is very large, this militates against the use of a standard debt contract as opposed to one with a minimum living allowance $\varepsilon > 0$.)

Turning to the choice between contracts, I note that the entrepreneur's expected payoff from an incentive-compatible contract $(L, r(.), p(.))$ is equal to:

$$\int_0^Y u(w + L - I + y - r(y))\, dG(y) - \int_0^Y p(y)\, dG(y). \tag{8}$$

Upon using (4) to substitute for $p(y)$ and integrating the resulting double integral by parts, one finds that this is equal to

$$\begin{aligned} &\int_0^Y u(w + L - I + y - r(y))\, dG(y) \\ &- \int_0^Y u'(w + L - I + y - r(y))\, G(y)\, dr(y) - p(Y). \end{aligned} \tag{9}$$

As for the financiers, I assume that there are enough of them dividing the uncertain return $r(\tilde{y})$ among each other so that they assess the contract $(L, r(.), p(.))$ as if they were risk neutral. Moreover on aggregate, they are not wealth-constrained. They are only concerned as to whether the expected gross return $\int_0^Y r(y)\, dG(y)$ is enough to cover the opportunity cost of their putting up the funds L. From their perspective, an incentive-compatible finance contract $(L, r(.), p(.))$ is *acceptable*, if and only if

$$\int_0^Y r(y)\, dG(y) \geq L. \tag{10}$$

Condition (1) ensures that the set of acceptable contracts is nonempty. An acceptable incentive-compatible finance contract $(L, r(.), p(.))$ is called *optimal* if it maximizes the entrepreneur's expected payoff (8), respectively (9), over the set of all acceptable incentive-compatible contracts.

In the remainder of the paper, I study the properties of optimal incentive-compatible contracts. I begin with the observation that, as shown in (9), the entrepreneur wants $p(Y)$, the penalty he suffers when he reports the maximum possible return, to be as small as possible. As for the financiers, (10) shows that their

[5] Risk neutrality is not compatible with the assumption that $u(.)$ is strictly concave. A careful analysis of the proof of Proposition 1 shows that this assumption is used *only* to establish that incentive compatibility entails weak monotonicity of $r(.)$. The other parts of Proposition 1, i.e., the sufficiency of (i) and (ii) and the necessity of (ii) for incentive compatibility, go through even if $u(.)$ is merely weakly concave.

payoff is independent of $p(Y)$; moreover (4) shows that incentive compatibility hinges on the difference $p(y) - p(Y)$ rather than the level of $p(Y)$. Trivially then one obtains:

Remark 1 Any optimal incentive-compatible contract satisfies $p(Y) = 0$.

3 Optimal incentive-compatible contracts

In view of Proposition 1, the problem of finding an optimal incentive-compatible contract is equivalent to the problem of finding a loan size $L \geq I - w$ and a nondecreasing repayment function $r(.)$ so as to maximize (9), with $p(Y) = 0$, subject to the feasibility constraint (2) and the acceptability condition (10). It is convenient to rewrite this problem in terms of the entrepreneur's consumption pattern $c(.)$, where for any $y \in [0, Y]$,

$$c(y) := w + L - I + y - r(y). \tag{11}$$

Since (11) implies $dr(y) = dy - dc(y)$, the objective function (9) with $p(Y) = 0$ can be rewritten as

$$\int_0^Y u(c(y))\, g(y)\, dy - \int_0^Y u'(c(y))\, G(y)\, dy + \int_0^Y u'(c(y))\, G(y)\, dc(y). \tag{12}$$

Upon combining the first and the third term and integrating, one can further rewrite this as

$$u(c(Y)) - \int_0^Y u'(c(y))\, G(y)\, dy. \tag{13}$$

The financiers' participation constraint (10) is similarly rewritten as:

$$\int_0^Y c(y)\, dG(y) \leq w - I + \int_0^Y y\, dG(y). \tag{14}$$

Finally, with (4) subsumed in (9), respectively (13), feasibility and incentive compatibility reduce to the requirements that

$$c(y) \geq 0 \tag{15}$$

and that $r(.)$ be nondecreasing or, equivalently, that

$$c(y) - c(z) \leq y - z \tag{16}$$

for all $y, z \in [0, Y]$ such that $y \geq z$. The optimal-contracting problem has thus been reduced to the problem of choosing a function $c(.)$ on $[0, Y]$ so as to maximize (13) under the constraints (14)-(16).

Proposition 2 *Under the maintained assumptions, in particular (1), an optimal incentive-compatible contract exists.*

Proposition 3 *If $c(.)$ corresponds to an optimal incentive-compatible contract, then $c(.)$ is continuous on $(0, Y]$. Moreover $c(Y) > 0$ and $c(y) = c(Y) + y - Y$ for y sufficiently close to Y.*

Given that the range of the random return $\tilde{y}$ is bounded, an optimal contract *always* exhibits the feature of a debt contract whereby for high realizations of the borrower's return a further increase in his return leaves his payment to the financiers unaffected, i.e., all of this increase serves to raise his consumption. This reflects the prominence of $u(c(Y))$ in (13): For return levels close to Y, it is important to have $c(y)$ increase as much as possible with y so as to make $c(Y)$ and hence $u(c(Y))$ large. Accordingly the consumption patterns in Figures 1 and 2 all have a slope $\frac{dc}{dy}$ equal to one when the return level y is close to the upper bound Y.[6]

In contrast, for low realizations of the borrower's return, an optimal contract in the presence of risk aversion does *not always* exhibit the feature of a debt contract that lenders confiscate everything "in the event of bankruptcy". Indeed Propositions 5 - 7 below show that for many specifications of the borrower's utility function the repayment owed to lenders is insensitive to the borrower's return realization when the latter is *low* as well as when it is high, and the borrower's consumption $c(y)$ is bounded away from zero.

The analysis uses control-theoretic methods. If $c(.)$ was known to be absolutely continuous, the problem of maximizing (13) under the constraints (14) - (16) could be formulated as a standard optimum-control problem with control $v(y) := \frac{dc}{dy}(y)$ and (16) equivalent to the requirement that $v(y) \leq 1$ for all y. The assumptions here do not actually guarantee absolute continuity of $c(.)$. Even so, the consumption pattern induced by an optimal contract must satisfy a suitable analogue of Pontryagin's conditions. This is the point of:

Proposition 4 *Let $c(.)$ correspond to an optimal incentive-compatible contract. Then there exist a scalar $\mu > 0$ and a continuously differentiable real-valued function $\psi(.)$ on $[0, Y]$ such that for all $y \in [0, Y]$,*

$$\frac{d\psi}{dy} \leq u''(c(y))\, G(y) + \mu\, g(y), \text{ with equality if } c(y) > 0, \tag{17}$$

$$\psi(y) \geq 0, \text{ with equality unless in a neighbourhood of } y \tag{18}$$
$$c(.) \text{ is continuously differentiable with } \frac{dc}{dy} = 1,$$

[6] If the range of the random return $\tilde{y}$ is unbounded, this statement is not generally true any more. Suppose for instance that $\tilde{y}$ has an exponential distribution with density $\lambda\, e^{-\lambda y}$ for some $\lambda > 0$ and that the borrower has constant absolute risk aversion $\delta > 0$. If $\lambda < \delta$, the same arguments as in the proofs of Propositions 4 and 5 below show that for this specification the optimal incentive-compatible contract is unique and involves the consumption pattern $c(.)$ such that for some $y_1 > 0$ and all $y \geq y_1$,

$$c(y) = c(y_1) + \left[\ln(e^{\lambda y} - 1) - \ln(e^{-\lambda y_1} - 1)\right] / \delta$$

and $\frac{dc}{dy} = \lambda/\delta(1 - e^{-\lambda y})$, with $\lim_{y \to \infty} \frac{dc}{dy} = \lambda/\delta \in (0, 1)$.

$$\psi(Y) = u'(c(Y)), \tag{19}$$

$$\psi(0) = 0. \tag{20}$$

If $u''(.)$ is a strictly increasing function, these conditions are sufficient as well as necessary for $c(.)$ to maximize (13) under the constraints (14)-(16); in this case, the optimal contract is unique in the sense that consumption patterns corresponding to different optimal contracts all coincide on $(0, Y]$.

It is instructive to consider the case of *constant absolute risk aversion*. In this case, as in the more general case of nonincreasing absolute risk aversion, $u''(.)$ is automatically a strictly increasing function, so the maximand (13) is strictly concave in $c(.)$, and the last part of Proposition 4 applies, i.e., the consumption pattern corresponding to an optimal contract is completely characterized by the Pontryagin conditions (17)-(20). This yields:

Proposition 5 *Assume that $u(.)$ exhibits constant absolute risk aversion, i.e., that $u(c) \equiv -e^{-\delta c}$ for some $\delta > 0$. Assume further that the distribution $G(.)$ is uniform, i.e., that $g(y) \equiv 1/Y$, and let $c(., \delta)$ be the consumption pattern corresponding to an optimal incentive-compatible contract.*

(a) If δ is sufficiently close to zero, then $c(., \delta)$ has the form shown in Figure 1, i.e.,

$$c(y, \delta) = \max(w - I, 0) + \max(0, y - \hat{y}), \tag{21}$$

where $\hat{y} \in [0, Y)$ is chosen so that (14) holds with equality.

(b) If δ is sufficiently large, then $c(., \delta)$ has the form shown in Figure 2, i.e., $c(0, \delta) > 0$, and there exist $y_1(\delta), y_2(\delta) \in (0, Y)$ such that

$$\frac{dc}{dy} = 1 \text{ if } y < y_1(\delta) \text{ or } y > y_2(\delta) \tag{22}$$

and

$$c(y, \delta) = c(y_1, \delta) + \delta^{-1}[\ln y - \ln y_1(\delta)] \text{ if } y_1(\delta) \le y \le y_2(\delta). \tag{23}$$

Moreover as δ goes out of bounds, $y_1(\delta)$ converges to zero, $y_2(\delta)$ converges to Y, and $c(., \delta)$ converges to the constant function with value $w - I + \int y \, dG(y)$, uniformly on $[0, Y]$.

With constant absolute risk aversion, a standard debt contract is optimal if risk aversion is close to zero, but *not* if risk aversion is large. To understand the economics behind this result, go back to the borrower's objective function as specified in (12) and rewrite this in the form

$$\int_0^Y u(c(y))\, g(y)\, dy - \int_0^Y u'(c(y))\, G(y)\, (1 - v(y))\, dy \tag{24}$$

where, for any y, $v(y) := \frac{dc}{dy}$. (In the constellation of Proposition 5 this is actually legitimate.) In the case of risk neutrality, with $u(c) \equiv c$ and $u'(c) \equiv 1$, (24) simplifies to

$$\int_0^Y c(y)\, g(y)\, dy - \int_0^Y G(y)\, (1 - v(y))\, dy. \tag{25}$$

An optimal contract must obviously satisfy (14) with equality. In the case of risk neutrality, this fixes the first term in (25) as $w - I + \int y dG$, regardless of any other aspect of the consumption pattern $c(.)$. The *shape* of $c(.)$ is then chosen solely with a view to minimizing the expected value $\int_0^Y G(y)\,(1 - v(y))\, dy$ of nonpecuniary penalties. If $w \geq I$, this requires $v(y) = 1$ for all y; if $w < I$, it requires that $v(y) = 1$ if $G(y)$ is large and $v(y) = c(y) = 0$ if $G(y)$ is small. This explains Diamond's (1984) result on the optimality of standard debt under risk neutrality.[7]

The appearance of the weights $G(y)$ in the expressions for expected nonpecuniary penalties in (25) reflects the fact that the incentive compatibility condition (4) relates *changes* in $p(.)$ to *changes* in $r(.)$. If $\frac{dr}{dy} = 1 - v(y)$ is positive over some interval $(y - \Delta,\ y]$, the increase in penalties as one goes from y to $y - \Delta$ affects the level of penalties not just at $y - \Delta$, but at *all* return levels $y' < y$ (to discourage the entrepreneur from misreporting y' instead of $y - \Delta$). This explains why the "increase" $dr = (1 - v(y))\, dy$ enters (25) with the weight $G(y)$ of the set of *all* return levels less than y. Given this appearance of the weights $G(y)$ in (25), under risk neutrality it is desirable to concentrate the deviations of $v(y)$ from one at low levels of y.

A simple comparison of (24) and (25) shows that this argument for the optimality of debt contracts is heavily dependent on the assumption of risk neutrality. If the von Neumann-Morgenstern utility function $u(.)$ is strictly concave, two additional considerations must be taken into account: First, when the borrower is risk averse, the first term in (24) depends on the riskiness as well as the mean of the random variable $c(\tilde{y}) = w + L - I + \tilde{y} - r(\tilde{y})$. If $u'(c)$ is large when c is close to zero, this militates against $c(\tilde{y})$ being zero with positive probability. Secondly, the weight with which "the increase" dr enters the expected value of the nonpecuniary penalties in the last term in (24) depends on $u'(c(y))$ as well as $G(y)$, the point being that the nonpecuniary penalties have to compensate for *utility* gains from false reporting, not just the money gains. If marginal utility is large, a given money gain from false reporting may translate into a large utility gain, requiring a large penalty to keep the borrower honest. Whereas under risk neutrality, expected nonpecuniary penalties are minimized by concentrating the increases of $r(.)$ at those return levels where $G(y)$ is small, with risk aversion, they are minimized by concentrating them at those return levels where $u'(c(y))G(y)$ is small. This need not be where y is small.

The consumption pattern in Figure 2 reflects these considerations. In Figure 2, in contrast to Figure 1, the slope $v(y) = \frac{dc}{dy}$ is equal to one for very *low* as well as very high values of y; this reflects the possibility that the weight $u'(c(y))G(y)$ of

[7] Alternatively, the reader may observe that under risk neutrality, (13) takes the form $c(Y) - \int G(y)dy$, which is maximal if $c(Y)$ is maximal, which in turn is the case when $c(.)$ takes the form (21); for details see Lemma 9 in Hellwig (1998a).

the term $(1 - v(y))$ in (24) may be large if $c(y)$ is small and $u'(c(y))$ is large. In an intermediate range in Figure 2, $v(y) = \frac{dc}{dy}$ lies strictly between zero and one, reflecting a *tradeoff at the margin* between considerations of risk sharing (calling for a low value of $v(y)$), the need to repay the financiers (again calling for a low value of $v(y)$) and the desire to keep nonpecuniary penalties low (calling for a high value of $v(y)$).

The important point is that in the presence of risk aversion the finance contract provides for risk sharing as well as finance. Even if $w \geq I$, i.e., if the entrepreneur is able to finance his project on his own, he may still want to bring in an external investor as this enables him to maintain his consumption when project returns are low. He has to pay for this insurance in terms of nonpecuniary penalties, but depending on his risk preferences and on the distribution of returns, he may well find this worthwhile. This is, e.g., always the case in the constellation of Proposition 5 when δ is large; in this case, regardless of the relation of w and I, an optimal incentive-compatible contract will provide the entrepreneur with a consumption pattern close to the nonrandom constant $w - I + \int y dG(y)$.

More generally, for the case of constant absolute risk aversion, Proposition 5 shows that risk sharing considerations play no role if risk aversion is low, but entail the nonoptimality of debt contracts if risk aversion is high. For other utility functions, risk sharing considerations *always* preclude the optimality of debt contracts if $u'(c)$ becomes large as c becomes small, e.g., if the von Neumann-Morgenstern utility function exhibits *constant relative risk aversion*. This is shown in:

Proposition 6 *Let $c(.)$ correspond to an optimal incentive-compatible contract. If $\lim_{c \to 0} u'(c) = \infty$, then $c(y) > 0$ for all $y \in (0, Y]$. If $\lim_{c \to \infty} u''(c)c = -\infty$, then $c(0) > 0$ and $v(y) = 1$ for any y that is sufficiently close to 0.*

Proposition 7 *Assume that $u(.)$ exhibits constant relative risk aversion, i.e., that $u'(y) \equiv c^{\delta - 1}$ for some $\delta < 1$. Assume further that the distribution function $G(.)$ is uniform, i.e., that $g(y) \equiv 1/Y$, and let $c(.)$ be the consumption pattern corresponding to an optimal incentive-compatible contract. If $w < I$, then $c(.)$ has the form shown in Figure 2, i.e., $c(0) > 0$, and there exist $y_1, y_2 \in (0, Y)$ such that*

$$\frac{dc}{dy} = 1 \text{ if } y < y_1 \text{ or } y > y_2 \tag{26}$$

and

$$c(y) = c(y_1)(y/y_1)^{1/(2-\delta)} \text{ if } y_1 \leq y \leq y_2. \tag{27}$$

For other specifications of utility and distribution functions, yet more complicated finance contracts may be optimal. To see this, note that if $v(y) = \frac{dc}{dy}$ is not equal to one, then, by (18), the costate variable $\psi(.)$ must have a local minimum at y. If in addition $c(y) > 0$, then by (17), one must have

$$u''(c(y))\, G(y) + \mu\, g(y) = 0. \tag{28}$$

This equation represents the tradeoff at the margin mentioned above that underlies an interior choice of $v(y) = \frac{dc}{dy}$. If $u''(.)$ is a strictly increasing function, (28) can be rewritten as

$$c(y) = u''^{-1}(-\mu \frac{g(y)}{G(y)}). \tag{29}$$

Using (29), one easily verifies that if $u''(.)$ is strictly increasing and $g(.)/G(.)$ is nondecreasing, then the consumption pattern $c(.)$ that corresponds to an optimal incentive-compatible contract is nondecreasing. Otherwise, e.g., if $u''(.)$ is increasing and the hazard rate function $g(.)/G(.)$ is *not* everywhere nondecreasing, $c(.)$ may be *decreasing* somewhere.[8] Moreover, depending on the slope of the function $u''^{-1}(-\mu g(.)/G(.))$, the number of switches back and forth between intervals where $\frac{dc}{dy}$ takes an interior value and intervals where $\frac{dc}{dy} = 1$ may be arbitrarily large. Optimal incentive-compatible contracts are thus very sensitive to the specification of the functions $u''(.)$ and $g(.)/G(.)$.

A Appendix

Proof of Proposition 1. Suppose first that a finance contract satisfies conditions (i) and (ii) of the proposition. For any $y \in [0, Y]$ and any $z \in [0, Y]$ such that $r(z) \in w + L - I + y$, one then has, from (4)

$$u(w + L - I + y - r(y)) - p(y) - [u(w + L - I + y - r(z)) - p(z)] \tag{A.1}$$
$$= -\int_{r(z)}^{r(y)} u'(w + L - I + y - r)\, dr + \int_{z}^{y} u'(w + L - I + x - r(x))\, dr(x).$$

If $z < y$, concavity of $u(.)$ implies $u'(w + L - I + x - r(x)) \geq u'(w + L - I + y - r(x))$ for all $x \in [z, y]$. By the monotonicity of $r(.)$, one then has

$$\int_{z}^{y} u'(w + L - I + x - r(x))\, dr(x) \geq \int_{z}^{y} u'(w + L - I + y - r(x))\, dr(x)$$
$$= \int_{r(z)}^{r(y)} u'(w + L - I + y - r)\, dr,$$

which means that (A.1) implies (3). Alternatively, if $z > y$, concavity of $u(.)$ implies $u'(w + L - I + x - r(x)) \leq u'(w + L - I + y - r(x))$ for all $x \in [y, z]$. By the monotonicity of $r(.)$, one then has

$$\int_{y}^{z} u'(w + L - I + x - r(x))\, dr(x) \leq -\int_{y}^{z} u'(w + L - I + y - r(x))\, dr(x)$$
$$= \int_{r(y)}^{r(z)} u'(w + L - I + y - r)\, dr,$$

[8] This is not incompatible with Innes's (1990) condition whereby the borrower must not have an incentive to destroy output. The Innes condition requires monotonicity of the total payoff $u(c(y)) - p(y)$ rather than monotonicity of just $u(c(y))$. Monotonicity of $u(c(y)) - p(y)$ is implied by (4).

and again (A.1) implies (3). This shows that any finance contract which satisfies assertions (i) and (ii) in the proposition is incentive-compatible.

Conversely, suppose that a contract $(L, r(.), p(.))$ is incentive-compatible. Let $0 \leq y_1 < y_2 \leq Y$. Apply the incentive compatibility condition (1) once with $y = y_2$ and $z = y_1$, and, assuming that $r(y_2) \leq w + L - I + y_1$,[9] once with $y = y_1$ and $z = y_2$, and add the resulting inequalities.

This yields

$$u(w + L - I + y_2 - r(y_2)) + u(w + L - I + y_1 - r(y_1)) \\ \geq u(w + L - I + y_2 - r(y_1)) + u(w + L - I + y_1 - r(y_2))$$

or, after a rearrangement of terms,

$$\int_{y_1}^{y_2} u'(w + L - I + x - r(y_2))\, dx \geq \int_{y_1}^{y_2} u'(w + L - I + x - r(y_1))\, dx.$$

Given that $u(.)$ is strictly concave, this inequality implies $r(y_2) \geq r(y_1)$, proving that $r(.)$ is nondecreasing on $[0, Y]$.

To prove that the contract also satisfies (4), note that for any $y \in [0, Y]$ and $x < y$, (3) implies

$$\begin{aligned} p(y) - p(x) &\leq u(w + L - I + y - r(y)) - u(w + L - I + y - r(x)) \\ &= - \int_{r(x)}^{r(y)} u'(w + L - I + y - r)\, dr \\ &\leq - \int_{r(x)}^{r(y)} u'(w + L - I + x - r)\, dr. \end{aligned} \tag{A.2}$$

For any $y^* \in [0, Y]$ and any sequence $\{y_i\}_{i=1}^n$ with $y_1 = y^* < y_2 < ... < y_n = Y$, a repeated application of (A.2) with $y = y_i, x = y_{i-1}, i = 2, ..., n$, yields

$$p(Y) - p(y^*) \leq - \sum_{i=2}^{n} \int_{r(y_{i-1})}^{r(y_i)} u'(w + L - I + y_{i-1} - r)\, dr. \tag{A.3}$$

Further, a precisely parallel argument, based on incentive compatibility relative to upward deviations in reports, yields

$$p(Y) - p(y^*) \geq - \sum_{i=1}^{n-1} \int_{r(y_{i-1})}^{r(y_i)} u'(w + L - I + y_i - r)\, dr. \tag{A.4}$$

Given that the right-hand sides of (A.3) and (A.4) are just the approximating sums for the Stieltjes integral in (4), the validity of (4) follows immediately. This completes the proof of Proposition 1.

[9] If $r(x_2) > w + L - I + x_1$, the desired result, namely $r(x_2) \geq r(x_1)$, is trivial.

For the proof of Proposition 2 the reader is referred to Hellwig (1998a). The argument is completely standard and does not contribute anything to the understanding of optimal contracts in the present context.

Proof of Proposition 3. Suppose first that $c(.)$ is not continuous on $(0, Y]$. If $c(.)$ is not continuous at Y, then, by (16), one must have $\lim_{y' \nearrow Y} c(y') > c(Y)$, and an increase in $c(Y)$ will raise the value of (13) without affecting the validity of (14) - (16). If $c(.)$ is not continuous on $(0, Y)$, then, again by (16), there exists $y \in (0, Y)$ such that $\lim_{y' \nearrow y} c(y') > \lim_{y' \searrow y} c(y')$, and, by the strict monotonicity of $u'(.)$, $\lim_{y' \nearrow y} u'(c(y')) < \lim_{y' \searrow y} u'(c(y'))$. But then a small reduction in $c(y')$ for y' belonging to a small interval to the left of y, combined with a suitably chosen small increase in $c(y')$ for y' belonging to a small interval to the right of y, will raise the value of (13) without affecting the validity of (14) - (16). The assumption that $c(.)$ is not continuous on $(0, Y]$ thus leads to a contradiction and must be false.

To prove that $c(Y) > 0$, I note that, by standard arguments, there exists a Lagrange multiplier μ such that if $c(.)$ maximizes (13) under the constraints (14)-(16), then $c(.)$ also maximizes

$$u(c(Y)) - \int_0^Y u'(c(y))\, G(y)\, dy + \mu\, (K - \int_0^Y c(y)\, g(y)\, dy) \tag{A.5}$$

under the constraints (15) and (16), with $K > 0$ defined by

$$K := w - I + \int y dG(y) > 0. \tag{A.6}$$

Since both $u(.)$ and $-u'(.)$ are strictly increasing functions, μ must be strictly positive. For any $\varepsilon > 0$, consider the consumption pattern $\hat{c}_\varepsilon(.)$ such that $\hat{c}_\varepsilon(y) = c(y)$ if $y \leq Y - \varepsilon$ and $\hat{c}_\varepsilon(y) = c(Y - \varepsilon) + y - Y + \varepsilon$ if $y \geq Y - \varepsilon$. Clearly, $\hat{c}_\varepsilon(.)$ satisfies (15) and (16), and so does $\hat{c}^\lambda_\varepsilon(.) = (1 - \lambda)c(.) + \lambda\hat{c}_\varepsilon(.)$ for any $\lambda \in [0, 1]$. It follows that for any $\varepsilon > 0$ the derivative

$$\frac{d}{d\lambda}[u(\hat{c}^\lambda_\varepsilon(Y)) - \int_0^Y (u'(\hat{c}^\lambda_\varepsilon(y))\, G(y) + \mu\, \hat{c}^\lambda_\varepsilon(y)\, g(y))dy]$$

must be nonpositive at $\lambda = 0$, and one must have

$$\begin{aligned} &u'(c(Y))(\varepsilon + c(Y - \varepsilon) - c(Y)) \\ &- \int_{Y-\varepsilon}^Y (u''(c(y))\, G(y) + \mu\, g(y))(y - Y + \varepsilon + c(Y - \varepsilon) - c(y))dy \leq 0. \end{aligned}$$

Since $u''(c(y)) < 0$ for all y and, by (16), $0 \leq y - Y + \varepsilon + c(Y - \varepsilon) - c(y) \leq \varepsilon + c(Y - \varepsilon) - c(Y)$ for all $y \in [Y - \varepsilon, Y]$, it follows that

$$[u'(c(Y)) - \mu(1 - G(Y - \varepsilon))](\varepsilon + c(Y - \varepsilon) - c(Y)) \leq 0 \tag{A.7}$$

for any $\varepsilon > 0$. However if ε is close to zero, $[u'(c(Y)) - \mu(1 - G(Y - \varepsilon))] > 0$, so (A.7) must imply $c(Y) = \varepsilon + c(Y - \varepsilon) \geq \varepsilon > 0$. For $y = Y - \varepsilon$ close to Y, this in turn yields $c(y) = c(Y) + y - Y$. The second statement of the proposition is thereby proved.

Proof of Proposition 4. To prove Proposition 4, I note that the consumption pattern $c(.)$ has a Lebesgue decomposition

$$c(.) = c_A(.) + c_S(.) + c_D(.)$$

into an absolutely continuous function $c_A(.)$, a singular, continuous function $c_S(.)$, and a jump function $c_D(.)$. This follows from the observation that the repayment function $r(.)$ that corresponds to $c(.)$ is nondecreasing and therefore has a Lebesgue decomposition

$$r(.) = r_A(.) + r_S(.) + r_D(.)$$

into an absolutely continuous function $r_A(.)$, a singular, continuous function $r_S(.)$, and a jump function $r_D(.)$. These three functions are all nondecreasing; moreover there is no loss of generality in assuming that $r_S(0) = r_D(0) = 0$. The corresponding decomposition of $c(.)$ is given by setting $c_A(y) \equiv y - r_A(y)$, $c_S(y) \equiv -r_S(y)$, and $c_D(y) \equiv -r_D(y)$; the continuity of $c(.)$ established in Proposition 3 implies $c_D(y) \equiv -r_D(y) \equiv 0$. As for $c_S(.)$ and $c_A(.)$, the monotonicity of $r_S(.)$ and $r_A(.)$ implies that $c_S(.)$ is nonincreasing and that the (Radon-Nikodym) derivative $v(.)$ of $c_A(.)$ satisfies

$$v(y) \leq 1 \text{ for all } y \in [0, Y]. \tag{A.8}$$

Now let $\mu > 0$ again be the Lagrange multiplier in (A.5) and consider the control problem

$$\max_{F(.),f(.)} \left[u(c(Y) + F(Y)) - \int_0^Y (u'(c(y) + F(y))G(y) + \mu(c(y) + F(y))g(y))\,dy\right] \tag{A.9}$$

with the constraints

$$c(y) + F(y) \geq 0 \text{ for all } y \in [0, Y], \tag{A.10}$$

$$\frac{dF}{dy} = f(y) \text{ for all } y \in [0, Y], \tag{A.11}$$

and

$$f(y) \leq 1 - v(y) \text{ for all } y \in [0, Y]. \tag{A.12}$$

I claim that the pair $(F_0(.),\ f_0(.))$ with $F_0(y) \equiv f_0(y) \equiv 0$ solves this control problem. Given (A.8), clearly $(F_0(.),\ f_0(.))$ satisfies the constraints (A.10)-(A.12). Moreover for *any* pair $(F(.),\ f(.))$ that satisfies (A.10)-(A.12) one easily finds that the consumption pattern $\hat{c}_F(.) := c(.)+F(.)$ satisfies (15), and, for any $y, z \in [0, Y]$ such that $y \geq z$,

$$\begin{aligned}\hat{c}_F(y) - \hat{c}_F(z) &= c(y) - c(z) + F(y) - F(z) \\ &= \int_z^y v(x)\, dx + c_S(y) - c_S(z) + \int_z^y f(x)\, dx \\ &\leq \int_z^y v(x)\, dx + \int_z^y (1 - v(x))\, dx \\ &\leq y - z,\end{aligned}$$

i.e., $\hat{c}_F(.)$ also satisfies (16). Given that $c(.)$ maximizes (A.5) under the constraints (15)-(16), it follows that

$$\begin{aligned}&u(c(Y)) - \int_0^Y (u'(c(y))\, G(y) + \mu\, c(y)\, g(y))\, dy \\ &\geq u(\hat{c}_F(Y)) - \int_0^Y (u'(\hat{c}_F(y))\, G(y) + \mu\, \hat{c}_F(y)\, g(y))\, dy \\ &= u(c(Y) + F(Y)) - \int_0^Y (u'(c(y) + F(y))\, G(y) + \mu(c(y) + F(y))\, g(y))\, dy\end{aligned}$$

for any pair $(F(.),\ f(.))$ satisfying (A.10)-(A.12), and hence that $(F_0(.),\ f_0(.))$ maximizes (A.9) under the constraints (A.10)-(A.12).

Pontryagin's maximum principle now implies the existence of a continuously differentiable real-valued function $\psi(.)$ such that for all $y \in [0, Y]$,

$$\frac{d\psi}{dy}(y) \leq u''(c(y))\, G(y) + \mu\, g(y), \text{ with equality unless } c(y) = 0, \tag{A.13}$$

$$f_0(y) = 0 \in \arg \max_{f \leq 1 - v(y)} \psi(y)\, f; \tag{A.14}$$

moreover, since $c(Y) > 0$, one has the transversality conditions

$$\psi(Y) = u'(c(Y)) \tag{A.15}$$

and

$$\psi(0) \leq 0, \text{ with equality unless } c(0) = 0. \tag{A.16}$$

Note that (A.13) and (A.15) are the same as (17) and (19). Further, (A.14) implies

$$\psi(y) \geq 0, \text{ with equality unless } v(y) = 1. \tag{A.17}$$

To establish (18), it is therefore necessary and sufficient to show that if $\psi(y) > 0$, then in a neighbourhood of y the singular component $c_S(.)$ of $c(.)$ is constant. For this purpose, note that $c_A(.)$, the absolutely continuous component of $c(.)$, satisfies

(15) and (16), and so does $\hat{c}_\lambda(.) = (1-\lambda)\, c(.) + \lambda\, c_A(.)$, for any $\lambda \in [0,1]$. Given that $c(.)$ maximizes (A.5) subject to (15) and (16), it follows that the derivative

$$\frac{d}{d\lambda}\left[u(\hat{c}_\lambda(Y)) - \int_0^Y (u'(\hat{c}_\lambda(y))\, G(y) + \mu\, \hat{c}_\lambda(y)\, g(y))\, dy\right]$$

must be nonpositive at $\lambda = 0$, or that

$$u'(c(Y))(-c_S(Y)) - \int_0^Y (u''(c(y))\, G(y) + \mu\, g(y))(-c_S(y))\, dy \leq 0. \quad \text{(A.18)}$$

Upon adding $\int_0^Y \frac{d\psi}{dy}(-c_S(y))dy - \psi(Y)(-c_S(Y)) + \int_0^Y \psi(y)d(-c_S(y)) = 0$ to the left-hand side and rearranging terms, using (A.15) and the fact that $c_S(0) = 0$, one can rewrite (A.18) as

$$\int_0^Y \psi(y)\, d(-c_S(y)) \quad \text{(A.19)}$$

$$-\int_0^Y [u''(c(y))\, G(y) + \mu\, g(y) - \frac{d\psi}{dy}(y)](-c_S(y))dy \leq 0.$$

Given that $c_S(.)$ is nonincreasing and $c_S(0) = 0$, (A.13) implies that the last term on the left-hand side is nonnegative, hence

$$\int_0^Y \psi(y)\, d(-c_S(y)) \leq 0. \quad \text{(A.20)}$$

Since $c_S(.)$ is nonincreasing, (A.20) can only hold if $c_S(.)$ is constant in the neighbourhood of any point y satisfying $\psi(y) > 0$. (18) is thereby proved. As for (20), this follows trivially from (A.14) and (A.17).

To complete the proof of Proposition 4, assume that $u''(.)$ is a strictly increasing function. Then $-u'(.)$ is a strictly concave function, and (13) defines a strictly concave functional on the set of consumption plans $c(.)$. Since the set of plans satisfying (14)-(16) is convex, the optimal $c(.)$ is unique up to a set of measure zero; given the continuity property established in Proposition 3, the optimal $c(.)$ in fact is unique up to the possible discontinuity at $y = 0$. Sufficiency of Pontryagin's conditions for characterizing this optimal $c(.)$ follows as in Theorem 1, p. 141, of Mangasarian (1966).

Proof of Proposition 5. (a) Given the consumption pattern specified in (21), set $\mu = \hat{\mu}(\delta)$ and $\psi(.) = \hat{\psi}(., \delta)$ where

$$\hat{\mu}(\delta) = \frac{1 + \delta\hat{y} - e^{-\delta(Y-\hat{y}+c(0))}}{Y-\hat{y}} \quad \text{(A.21)}$$

and for any $y \in [0, Y]$,

$$\hat{\psi}(y,\delta) = 0 \quad \text{if } y < \hat{y}, \quad \text{(A.22)}$$

$$\hat{\psi}(y,\delta) = [(1+\delta y)e^{-\delta(y-\hat{y}+c(0))} - (1+\delta\hat{y})e^{-\delta c(0)} + \mu(y-\hat{y})]/Y \quad \text{(A.23)}$$
$$\text{if } y \geq \hat{y}.$$

For $y < \hat{y}$, (A.22) implies

$$\frac{d\psi}{dy}(y) = \frac{\partial\hat{\psi}}{\partial y}(y,\delta) = 0 \leq -\delta^2\frac{y}{Y} + \hat{\mu}(\delta)\frac{1}{Y}$$

$$= u''(0)G(y) + \mu g(y)$$

if δ is sufficiently small so that $\delta^2\hat{y} \leq \hat{\mu}(\delta)$. For $y > \hat{y}$, $(A.23)$ implies

$$\frac{d\psi}{dy}(y) = \frac{\partial\hat{\psi}}{\partial y}(y,\delta) = -\delta^2 e^{-d(y-\hat{y}+c(0))}\frac{y}{Y} + \hat{\mu}(\delta)\frac{1}{Y}$$

$$= u''(c(y))G(y) + \mu g(y).$$

In either case, (17) is verified. For $y > \hat{y}$, (A.23) and (A.21), also imply $\lim_{\delta\to 0}\frac{\partial\hat{\psi}}{\partial y}(y,\delta) = \lim_{\delta\to 0}\hat{\mu}(\delta)/Y = 1/(Y-\hat{y})Y$, uniformly in y. Hence, if δ is sufficiently small, one has $\psi(y) = \hat{\psi}(y,\delta) > 0$ for all $y > \hat{y}$. In combination with (A.22), this shows that the given $c(.)$, μ, and $\psi(.)$ also satisfy (18) if δ is sufficiently small. As for (19) and (20), these are obviously implied by (A.21) and (A.23). Part (a) of the proposition therefore follows from the last part of Proposition 4 and the observation that $u''(c) = -\delta^2 e^{-\delta c}$ is strictly increasing in c.

(b) To prove part (b), I begin by specifying the critical values $y_1(\delta), y_2(\delta)$ for the kinks in the consumption pattern $c(.,\delta)$. By standard calculus arguments, there exists a unique $z_1 > 1$ such that

$$e^{z_1} = 1 + z_1 + z_1^2. \tag{A.24}$$

For any $X \geq 3$, there exists $z_2(X) \in (1, X)$ such that

$$z_2(X) = \frac{1 - e^{-(X - z_2(X))}}{X - z_2(X) - 1}. \tag{A.25}$$

(The function $z_3 \to f(z_3) := z_3 + (1 - e^{-z_3})/(z_3 - 1)$ ranges between $+\infty$ and $3 - e^{-2}$ as z_3 ranges from 1 to 2, so put $z_2(X) = \max_{z_3\in f^{-1}(X)\cap[1,2]}(1 - e^{-z_3})/(z_3 - 1)$.) For $\delta \geq 3/Y$, put

$$y_1(\delta) = z_1/\delta \tag{A.26}$$

and

$$y_2(\delta) = z_2(\delta Y)/\delta. \tag{A.27}$$

From (A.26), one has $\lim_{\delta\to\infty} y_1(\delta) = 0$. From (A.25) and the fact that $z_2(X) > 1$ for $X \geq 3$, one also has $\lim_{X\to\infty}(X - z_2(X)) = 1$, so (A.27) implies $\lim_{\delta\to\infty}\delta(Y - y_2(\delta)) = 1$, hence $\lim_{\delta\to\infty} y_2(\delta) = Y$. In particular, one has $0 < y_1(\delta) < y_2(\delta) < Y$ if δ is sufficiently large.

Let δ be such that $0 < y_1(\delta) < y_2(\delta) < Y$, and, for any parametrically given $\mu > 0$, consider the consumption pattern $\hat{c}(.,\delta,\mu)$ such that

$$\hat{c}(y,\delta,\mu) = \frac{1}{\delta}\ln(\delta y_1(\delta)) + y - y_1(\delta) - \frac{1}{\delta}\ln(\mu/\delta) \text{ if } y < y_1(\delta), \quad \text{(A.28)}$$

$$\hat{c}(y,\delta,\mu) = \frac{1}{\delta}\ln(\delta y) - \frac{1}{\delta}\ln(\mu/\delta) \qquad \text{if } y \in (y_1(\delta), y_2(\delta)), \quad \text{(A.29)}$$

$$\hat{c}(y,\delta,\mu) = \frac{1}{\delta}\ln(\delta y_2(\delta)) + y - y_2(\delta) - \frac{1}{\delta}\ln(\mu/\delta) \text{ if } y > y_2(\delta). \quad \text{(A.30)}$$

Given the monotonicity of $\hat{c}(.,\delta,\mu)$ in μ, there exists a unique $\mu(\delta)$ such that

$$\int \hat{c}(y,\delta,\mu(\delta))\, dG(y) = K \quad \text{(A.31)}$$

where $K > 0$ is again given by (A.6). I claim that for any sufficiently large δ the consumption pattern $c(.,\delta) := \hat{c}(.,\delta,\mu(\delta))$ corresponds to an optimal incentive-compatible contract.

To establish this claim, I first show that if δ is sufficiently large, then $c(.,\delta)$ satisfies the constraints (14)-(16). (14) holds trivially, by the definition of $\mu(\delta)$. As for (15), I note that (A.26)-(A.30) imply

$$-y_1(\delta) \;\leq\; c(y,\delta) + \frac{1}{\delta}\ln(\mu(\delta)/\delta) \;\leq\; \frac{1}{\delta}\ln(\delta Y) + Y - y_2(\delta)$$

for all y and δ, hence

$$\lim_{\delta\to\infty}\left[c(y,\delta) + \frac{1}{\delta}\ln(\mu(\delta)/\delta)\right] = 0,$$

uniformly in y. From (A.31) this implies $\lim_{\delta\to\infty}\frac{1}{\delta}\ln(\mu(\delta)/\delta) = K$ and hence $\lim_{\delta\to\infty} c(y,\delta) = K$, uniformly in y. Since $K > 0$, it follows that $c(y,\delta) > 0$, confirming (15) for all y if δ is sufficiently large. Finally, (A.26)-(A.30) imply that $\frac{dc}{dy} = 1$ if $y < y_1(\delta)$ or $y > y_2(\delta)$, and $\frac{dc}{dy} = \frac{1}{\delta y} \leq \frac{1}{\delta y_1(\delta)} = \frac{1}{z_1} < 1$ if $y \in (y_1(\delta), y_2(\delta))$, so (16) holds as well.

To show that $c(.,\delta) := \hat{c}(.,\delta,\mu(\delta))$ is actually optimal, I specify the costate variable $\psi(.) = \hat{\psi}(.,\delta)$ where

$$\hat{\psi}(y,\delta) = \left[(1+\delta y)e^{-\delta(y-y_1(\delta))} - e^{\delta y_1(\delta)} + \delta^2 y_1(\delta) y\right]\frac{\mu(\delta)}{\delta^2 y_1(\delta) Y} \quad \text{(A.32)}$$
$$\text{if } y < y_1(\delta),$$

$$\hat{\psi}(y,\delta) = 0 \qquad \text{if } y \in (y_1(\delta), y_2(\delta)), \quad \text{(A.33)}$$

$$\hat{\psi}(y,\delta) = \Big[(1+\delta y)e^{-\delta(y-y_2(\delta))} - 1 \quad \text{(A.34)}$$
$$-\delta y_2(\delta) + \delta^2 y_2(\delta)(Y - y_2(\delta))\Big]\frac{\mu(\delta)}{\delta^2 y_2(\delta) Y}$$
$$\text{if } y > y_2(\delta).$$

Given the conditions (A.24)-(A.27) for $y_1(\delta)$ and $y_2(\delta)$, it is straightforward to check that the consumption pattern $c(.,\delta)$, the Lagrange multiplier $\mu(\delta)$, and the costate variable $\psi(.) = \hat{\psi}(.,\delta)$ defined by (A.32)-(A.34) satisfy conditions (17)-(20). As in part (a) of the proposition, the optimality of the consumption pattern $c(.,\delta)$ for any sufficiently high δ now follows from the last part of Proposition 4.

Proof of Proposition 6. Let $\lim_{c\to 0} u'(c) = \infty$. If $c(y) = 0$ for a nonnull set of return levels $y \in [0, Y]$, then the integral in (13) is undefined and one cannot be at a maximum of (13) under the constraints (14)-(16). Therefore one must have $c(y) > 0$ for all but a null set of return levels $y \in [0, Y]$. Suppose that, contrary to the first statement of the proposition, $c(y') = 0$ for some $y' > 0$. Then there exists a sequence $\{y^k\}$ converging to y' from below such that the associated sequence $\{c(y^k)\}$ converges to $c(y') = 0$ monotonically from above and moreover $v(y^k) < 1$ for all k. By (18), it follows that $\psi(y^k) = 0$ for all k and hence that $\frac{d\psi}{dy}(y') = 0$, which is incompatible with (17). This proves the first statement of the proposition.

Next impose the stronger assumption that $\lim_{c\to\infty} u''(c)c = -\infty$. To prove the second statement of the proposition, I first show that this assumption implies $\psi(y) > 0$ and hence, by (18), $\frac{dc}{dy} = 1$ for any y that is sufficiently close to zero. For suppose that this claim is false. Then there exists a sequence $\{y^k\}$ converging to zero such that $\psi(y^k) = 0$ for all k. Then (18) implies that for each k, $\psi(.)$ has a local minimum at y^k, i.e., one must have

$$\frac{d\psi}{dy}(y^k) = 0 \tag{A.35}$$

for all k. Since $\lim_{c\to\infty} u''(c)c = -\infty$ implies $\lim_{c\to 0} u'(c) = \infty$, for each k one must also have $c(y^k) > 0$. By (17), it follows that (A.35) entails

$$u''(c(y^k))\, G(y^k) + \mu g(y^k) = 0 \tag{A.36}$$

for all k. Since $g(.)$ is continuous and strictly positive on $[0, Y]$, and moreover $\lim_{k\to\infty} G(y^k) = 0$, it follows that $\lim_{k\to\infty} u''(c(y^k)) = -\infty$ and hence that $c(0) = \lim_{k\to\infty} c(y^k) = 0$. From (16), this implies $c(y^k) \le y^k$ for all k. However upon rewriting (A.36) in the form

$$u''(c(y^k))c(y^k)\frac{y^k}{c(y^k)}\frac{G(y^k)}{y^k} + \mu g(y^k) = 0 \tag{A.37}$$

and using l'Hospital's rule, one also finds that $\lim_{k\to\infty} u''(c(y^k))c(y^k) = -\infty$ implies $\lim_{k\to\infty}(y^k/c(y^k)) = 0$ and hence that $y^k < c(y^k)$ for any sufficiently large k. The assumption that there exists a sequence $\{y^k\}$ converging to zero such that $\psi(y^k) = 0$ for all k has thus led to a contradiction and must be false. This proves that $\psi(y) > 0$ and $\frac{dc}{dy}(y) = 1$ for any y that is sufficiently close to zero.

By Proposition 4, this latter conclusion in turn implies that

$$0 < \psi(y) = \psi(y) - \psi(0) \le \int_0^y [u''(c(0) + x)\, G(x) + \mu\, g(x)]\, dx \tag{A.38}$$

for any y that is sufficiently close to zero. Now (A.38) implies

$$-\underline{g} \int_0^y u''(c(0)+x)\, x\, dx \leq \mu\, \bar{g}\, y \tag{A.39}$$

where $\underline{g}$ and $\bar{g}$ are the minimum and the maximum, respectively, of the density $g(.)$ on $[0, \bar{Y}]$. If (A.39) is to hold for any y that is sufficiently close to zero, it follows that $-u''(c(0)+x)x$ must be uniformly bounded even as x is close to zero. Under the assumption that $\lim_{c\to 0} u''(c)c = -\infty$, this is possible only if $c(0) > 0$. This completes the proof of Proposition 6.

Proof of Proposition 7. Since $\lim_{c\to 0} c^{\delta-1} = \infty$ and $\lim_{c\to 0} u''(c)c = (\delta - 1)c^{\delta-1} = -\infty$, Proposition 6 implies $c(0) > 0$ and $\frac{dc}{dy}(y) = 1$ for y close to zero. Since $w < I$, (14) implies that $c(y') < y'$ for some $y' \in (0, Y]$. By (18) it follows that the set $I = \{y \in (0, Y) |\ \psi(y) = 0\}$ has positive measure. Define $y_1 = \inf I$ and $y_2 = \sup I$. I claim that in fact $I = [y_1, y_2]$, i.e., I is a closed interval. For suppose not. Then there exists $y' \in [y_1, y_2]$ such that $\psi(y') > 0$. By continuity, $\psi(y)$ must actually be positive on some neighbourhood of y'; this implies $y' \neq y_1$, $y' \neq y_2$ and hence $y' \in (y_1, y_2)$. One may therefore define $\bar{y}_1 = \sup I \cap [0, y']$ and $\bar{y}_2 = \inf I \cap [y', Y]$, and one has $\bar{y}_1 < \bar{y}_2$. For $y \in (\bar{y}_1, \bar{y}_2)$, one has $\psi(y) > 0$, hence, by (18),

$$c(\bar{y}_2) - c(\bar{y}_1) + \bar{y}_2 - \bar{y}_1. \tag{A.40}$$

Next I note that, again by (18), at any $y \in I$, the function $\psi(.)$ has a minimum and satisfies $\frac{d\psi}{dy} = 0$. By (17) in conjunction with the fact that $c(y) > 0$ for all y, this implies that

$$u''(c(y))G(y) + \mu g(y) = 0 \tag{A.41}$$

for all $y \in I$. For the given utility function and distribution function, it follows that

$$c(y) = \hat{c}(y, \mu) \text{ where} \tag{A.42}$$

$$\hat{c}(y, \mu) = (1-\delta)\left(\frac{y}{\mu}\right)^{\frac{1}{2-\delta}} \tag{A.43}$$

for all $y \in I$. By continuity, one has $\psi(\bar{y}_i) = 0$, hence $\bar{y}_i \in I$ for $i = 1, 2$. From (A.42), one then has

$$c(\bar{y}_i) = \hat{c}(\bar{y}_i, \mu) \tag{A.44}$$

for $i = 1, 2$. From (A.40) and (A.44), it follows that there is some $y'' \in (\bar{y}_1, \bar{y}_2)$ such that

$$\frac{\partial \hat{c}}{\partial y}(y'', \mu) = 1. \tag{A.45}$$

Given that the function $\hat{c}(., \mu)$ is strictly concave, (A.45) implies

$$\frac{\partial \hat{c}}{\partial y}(y, \mu) > 1 \tag{A.46}$$

for all $y \leq \bar{y}_1$. From (16), (A.44), and (A.45) one then obtains

$$\begin{aligned} c(y) &\geq c(\bar{y}_1) - \bar{y}_1 + y \\ &> \hat{c}(\bar{y}_1, \mu) - \int_y^{\bar{y}_1} \frac{\partial \hat{c}}{\partial y}(y, \mu) dy \\ &= \hat{c}(y, \mu) \end{aligned} \tag{A.47}$$

for all $y < \bar{y}_1$. In view of (A.43), (A.47) implies $u''(c(y))G(y) + \mu g(y) \neq 0$ and hence, by (17), that $\frac{d\psi}{dy} \neq 0$ for all $y < \bar{y}_1$. Given that $\psi(\bar{y}_1) = 0$, this is incompatible with the transversality condition (20) requiring that $\psi(0) = 0$. The assumption that the set $I = \{y \in (0, Y) | \ \psi(y) = 0\}$ is *not* a closed interval $[y_1, y_2]$ has thus led to a contradiction and must be false.

Given that $I = [y_1, y_2]$, the argument just given shows that $c(.)$ satisfies (A.42) and (A.43) and hence (27) on $[y_1, y_2]$. Since $\lim_{y \to 0} \frac{\partial \hat{c}}{\partial y}(y, \mu) = \infty$, it follows that $y_1 > 0$. For $y \in (0, y_1)$, $\psi(y) > 0$ and hence by (18), $\frac{dc}{dy} = 1$. Finally, Proposition 3 implies that $y_2 < Y$. For $y \in (y_2, Y]$, $\psi(y) > 0$ and hence, again by (18), $\frac{dc}{dy} = 1$. This shows that $c(.)$ also satisfies (26) and completes the proof of Proposition 7.

References

1. Billingsley, P.: Convergence of probability measures. New York: Wiley 1968
2. Diamond, D.: Financial intermediation as delegated monitoring. Review of Economic Studies **51**, 393–414 (1984)
3. Gale, D., Hellwig, M.: Incentive-compatible debt contracts: the one-period problem. Review of Economic Studies **52**, 647–663 (1985)
4. Garino, G., Simmons, P.J.: A note on costly state verification contracts with risk aversion. Mimeo, University of York, UK (1998)
5. Hellwig, M.: Risk aversion and incentive compatibility with ex post information asymmetry. Discussion Paper No. 98-38, SFB 504, University of Mannheim (1998a)
6. Hellwig, M.: Financial intermediation with risk aversion, Discussion Paper No. 98-39, SFB 504, University of Mannheim (1998b)
7. Innes, R.D.: Limited liability and incentive contracting with ex-ante action choices. Journal of Economic Theory **52**, 45–67 (1990)
8. Mangasarian, O.L.: Sufficient conditions for the optimal control of nonlinear systems. SIAM Journal of Control **4**, 139–152 (1966)
9. Povel, P., Raith, M.: Endogenous debt contracts with undistorted incentives. Discussion Paper No. 99-61, SFB 504, University of Mannheim (1999)
10. Townsend, R.M: Optimal contracts and competitive markets with costly state verification. Journal of Economic Theory **21**, 1–29 (1979)

Ex ante contracting with endogenously determined communication plans*

Jülide Yazar

Department of Economics, Ohio Wesleyan University, Delaware, OH 43015, USA
(e-mail: jjyazar@cc.owu.edu)

Received: 4 March 1998; revised version: 17 September 1999

Summary. In this paper we introduce a new model of *ex ante* contracting for economies with asymmetric information to examine endogenously determined communication plans for information sharing in the *interim* stage. In contrast to the models used in previous research, in the present model agents negotiate not only on a contract of state contingent allocations but also on a communication plan, a set of rules describing how agents will reveal part of their private information at the *interim* stage to execute the trade contracts. We prove a result about the nested structure of the set of allocations implementable by various communication plans and establish the existence of core strategies for this cooperative game under various regularity conditions.

Keywords and Phrases: Exchange economies with asymmetric information, Cooperative games, Information processing.

JEL Classification Numbers: D51, C71, D82.

1 Introduction

As first introduced by Radner [7], an exchange economy with asymmetric information consists of a finite set of agents contracting to trade their allocations in an uncertain economic environment. The uncertainty in the economy is formalized by a set of possible states of nature and agents differ in their private information about the prevailing state of nature. Each economic agent is characterized by a state dependent utility function, a state dependent initial endowment, a prior and an information partition.

* I wish to thank Tatsuro Ichiishi and the referee for several valuable comments and suggestions.

Wilson [11] initiated research in strategic cooperative game theory in the presence of asymmetric information by introducing core and efficiency concepts for asymmetric information economies. He considered several scenarios differing in the amount of information that the agents are allowed to rely on to execute their trades. In particular the *fine core* concept introduced by Wilson allowed information sharing between agents, however ignored the issue of incentive compatibility.

Yannelis [10] opened up a lively revival of research in this area by formalizing information processing between agents by means of appropriate measurability restrictions on the trade contracts. Indeed, Yannelis [10], Koutsougeras-Yannelis [6], Allen [1], Ichiishi-Idzik [5] and Hahn-Yannelis [3] introduced new cooperative equilibrium concepts with explicit treatment of information processing and incentive compatibility. Koutsougeras-Yannelis [6] remarked that the *fine core* concept suggested by Wilson is flawed in the sense that whenever Wilson's core notion exist it is not incentive compatible and whenever it is incentive compatible it does not exists. They showed that exogenously determined measurability restrictions on trade contracts can resolve these difficulties. In particular, private information measurable trade contracts provide a rich set of contracts satisfying Bayesian incentive compatibility (BIC) conditions which facilitate truthful implementation of these contracts by agents with asymmetric information.

In this paper, we present a new model of *ex ante* contracting for economies with asymmetric information. Our starting point is the measurability approach to asymmetric information economies. We seek to endogenize the choice of the measurability restrictions or equivalently, the choice of the information sharing rule by introducing an explicit communication stage before trade execution into the model, while adopting appropriate individual incentive compatibility conditions to ensure truthful implementation. In contrast to the models used in previous research, in the present model agents negotiate not only on a contract of state contingent allocations but also on a communication plan, a set of rules describing how agents will reveal part of their private information at the *interim* stage. The contract execution relies only on the pooled information revealed by agents as described in the communication plan and therefore, the trade vector for each agent becomes common knowledge after the communication stage. Thus, the set of allocations implementable with a given communication plan are restricted by appropriate measurability conditions. The presence of asymmetric information further restricts the set of implementable communication plan-allocation pairs. Specifically, only communication plan-allocation pairs satisfying certain individual Bayesian incentive compatibility conditions are considered to ensure truthful execution of the information revealing process. These incentive compatibility conditions are appropriate generalizations of the individual BIC conditions introduced by Hahn and Yannelis in [3].

We define and analyze core strategies for this new *ex ante* contracting model with endogenized communication plans. We prove that the set of allocations implementable by various communication plans has a nested structure. As a corollary to this result we obtain an information revealing property for the core strategies. Specifically, we show that if the core is nonempty, then there exists always an element of the core that is associated with the full information revealing communication plan. Using this result the existence of core strategies is established under certain regularity conditions.

2 Model and definitions

We begin by defining the concept of economy with asymmetric information [7]. The set Ω denotes the set of possible states of the nature. We assume Ω is finite. μ is a probability measure on Ω denoting the common prior of all agents.

Definition 2.1. A pure exchange economy with a finite set of agents N, with asymmetric information is specified with a list of data, $\mathcal{E}_{pe} = (N, \{u^i, e^i, \mathcal{F}^i\}_{i\in N}, \mu)$.

(i) At each state of the nature, there are l goods available for trade. For every agent, $\mathbb{R}^l_+$ is the consumption set.

(ii) Preferences of agent i is given by state dependent utility function $u^i : \Omega \times \mathbb{R}^l_+ \to \mathbb{R}$.

(iii) Information structure will be described by a collection of fields, $\mathcal{F} = \{\mathcal{F}^i\}_{i\in N}$. The private information of agent i ($i \in N$) is given by the field $\mathcal{F}^i$ which forms a partition $\mathcal{PF}^i$ on Ω. Specifically, if the true state of the world is ω then agent i is informed of the unique element $\mathcal{PF}^i(\omega)$ of $\mathcal{PF}^i$ that contains ω.

(iv) The function $e^i : \Omega \to \mathbb{R}^l$ specifies the endowment vector $e^i(\omega) \in \mathbb{R}^l$ of agent i, for every state $\omega \in \Omega$. The function e^i is $\mathcal{F}^i$ measurable.

We formalize information sharing between agents using explicitly designed communication plans. A communication plan $\mathcal{C}$ is a collection of fields $\mathcal{C}^i$, ($\mathcal{C} = \{\mathcal{C}^i\}_{i\in N}$) such that for every agent i, $\mathcal{C}^i$ induces a partition $\mathcal{PC}^i$ on Ω which is coarser than $\mathcal{PF}^i$,(alternatively for all $i \in N, \mathcal{F}^i \supset \mathcal{C}^i$). Specifically, when the true state of nature is ω, agent i is informed that the event $\mathcal{PF}^i(\omega)$ has occurred and reveals to other agents that the event $\mathcal{PC}^i(\omega)$ has occurred, as agreed in communication plan $\mathcal{C}$.

Every state dependent allocation $x^i : \Omega \to \mathbb{R}^l_+$, naturally defines a state dependent net trade vector $z^i : \Omega \to \mathbb{R}^l$, by $z^i(\omega) = x^i(\omega) - e^i(\omega)$. Also for an allocation x^i the expected utility of agent i, conditioned on event $A \subset \Omega$ is defined as:

$$EU^i\left(x^i(\omega)|A\right) = \sum_{\omega\in A} u^i\left(\omega, x^i(\omega)\right)\mu(\omega|A)$$

For simplicity the unconditional expected utility of agent i for a state dependent allocation $x^i : \Omega \to \mathbb{R}^l_+$ is defined as: $EU^i\left(x^i\right) = EU^i\left(x^i(\omega)|\Omega\right)$.

The timing of the model consists of two stages. In the first stage there is uncertainty over the states of nature and in this stage agents negotiate on a coalitionally stable, state contingent trades and an incentive compatible communication plan $\mathcal{C}$, which will pool the necessary information to execute the trades. In the second (*interim*) stage, the state of the nature is realized and agents reveal part of their private information according to the previously agreed communication plan. Then trades are executed contingent on the pooled information. The presence of asymmetric information restricts the set of communication plans over which truthful cooperation is possible. Therefore, during negotiations agents consider only the contracts, which satisfy Bayesian incentive compatibility conditions which ensure truthful revelation of the information in the communication stage. These incentive compatibility conditions are appropriate generalizations of the individual BIC conditions introduced by Hahn and Yannelis in [3].

First, consider agent i in coalition S participating in the communication plan $\mathcal{C}^S = \{\mathcal{C}^j\}_{j \in S}$ for the execution of contract x^S. Assume that the true state of nature is ω^*. In the *interim* stage agent i will learn that the event $E = \mathcal{PF}^i(\omega^*)$ has occurred.

Agent i knows that, for any $\omega \in E$, if everyone in S truthfully participates in the communication plan, each agent j will signal that the event $\mathcal{PC}^j(\omega)$ has occurred. Therefore, after information pooling, it will be common knowledge that the event $\bigcap_S \mathcal{PC}^j(\omega)$ has occurred. Agent i will execute the net trade $z^i(\bigcap_S \mathcal{PC}^j(\omega)) = z^i(\omega)$ and his final utility will be given by $u^i(\omega, x^i(\omega))$.

On the other hand, agent i can signal a wrong message $E' \in \mathcal{PC}^i$, provided that it will result in a meaningful outcome of the communication plan for all possible prevailing states of nature if :

$$\text{For every } \omega \in E,\ A(\omega) = E' \cap \left(\bigcap_{j \neq i} \mathcal{PC}^j(\omega) \right) \neq \emptyset \tag{1}$$

(For some cases of information structures it may be impossible for the agents participating in the communication plan to infer which agent is causing the conflict. However we assume that all agents participating in the communication plan are strongly penalized in the case of conflicting messages. As a consequence each agent only considers cheating with a deceptive message E', only if he is certain that it would not conflict with the truthful messages of the others.)

If the true state is ω, this false signal will result in a utility level $u^i(\omega, z^i(A(\omega)) + e^i(\omega))$. Since any of the $\omega \in E$ can be the prevailing state, agent i will benefit from signaling E' if:

$$EU^i\left(z^i(A(\omega)) + e^i(\omega) | E\right) > EU^i\left(x^i(\omega) | E\right) \tag{2}$$

Definition 2.3. A state dependent allocation, communication plan pair $(x^S, \mathcal{C}^S)$ is incentive compatible with the communication system $\mathcal{C}^S$, if there exists no agent $i \in S$ and a $E \in \mathcal{PF}^i$, with a deceptive message $E' \in \mathcal{PC}^i$ such that (1) and (2) holds true.

Remark 2.4. Alternatively the incentive compatibility definition in 2.3 can be formulated using the deception function framework adopted by Hahn and Yannelis [3]. A deception for agent i is a function $\alpha^i : \mathcal{PF}^i \to \mathcal{PC}^i$ and truth telling α^{*i} is characterized by $\alpha^{*i}(\mathcal{PF}^i(\omega)) \supset \mathcal{PF}^i(\omega)$. A deception vector $\alpha^S = \{\alpha^i\}_{i\in S}$ is compatible with $\mathcal{C}^S$ if $\forall\omega \in \Omega : \alpha^S(\omega) = \cap_{i\in S}\alpha^i(\mathcal{PF}^i(\omega)) \neq \emptyset$. If α^S is a compatible deception vector with $\mathcal{C}^S$, then $[z^i \circ \alpha^S](\omega) = z^i\left(\alpha^S(\omega)\right)$.

Then, the state dependent allocation, communication plan pair $(x^S, \mathcal{C}^S)$ is incentive compatible, if for every $i \in S$, there exists no deception $\alpha^i : \mathcal{PF}^i \to \mathcal{PC}^i$ and an $\overline{\omega} \in \Omega$ such that:

(i) $(\alpha^i, \alpha^{*S/\{i\}})$ is compatible with $\mathcal{C}^S$ and

(ii) $EU^i(e^i(\omega) + [z^i \circ (\alpha^i, \alpha^{*S/\{i\}})](\omega)|\mathcal{PF}^i(\overline{\omega})) > EU^i(x^i(\omega)|\mathcal{PF}^i(\overline{\omega}))$

For a communication plan $\mathcal{C}^S$ we define the set of *implementable* allocations $X^S(\mathcal{C}^S)$ as the set of $\bigvee_S \mathcal{C}^j$ measurable allocations, which form an incentive compatible pair with $\mathcal{C}^S$.

Definition 2.5. The set of contracts implementable by a coalition S, with a communication plan $\mathcal{C}^S = \{\mathcal{C}^j\}_{j\in S}$ is defined as:

$$X^S\left(\mathcal{C}^S\right) = \left\{ \{x^i\}_{i\in S} \;\middle|\; \begin{array}{l} \forall i \in S : z^i \text{ is } \bigvee_S \mathcal{C}^j \text{ measurable and there does} \\ \text{not exist } E \in \mathcal{PF}^i \text{ and } E' \in \mathcal{PC}^i : E \not\subset E', \\ (i)\ \forall\omega \in E, A(\omega) = E' \cap \left(\bigcap_{j\neq i} \mathcal{PC}^j(\omega)\right) \neq \emptyset \\ (ii)\ EU^i\left(z^i(A(\omega)) + e^i(\omega)|E\right) > EU^i\left(x^i(\omega)|E\right) \end{array} \right\}$$

It is relatively easy to show that two classes of exogenous measurability restrictions adopted in previous literature: common information measurable and private measurable contracts can be obtained as special cases of the implementable contracts considered in this research by choosing an appropriate communication plan [12].

Next we define an *ex ante* core concept for the pure exchange economy with asymmetric information.

Definition 2.6. The *ex ante* core of the pure exchange economy with endogenous communication plan (EC-core) consists of all pairs $(x, \mathcal{C})$ of state dependent allocation $x = \{x^i : \Omega \to \mathbb{R}^l_+\}_{i\in N}$ and communication plan $\mathcal{C} = \{\mathcal{C}^i\}_{i\in N}$ where $\mathcal{C}^i \subset \mathcal{F}^i$ for every $i \in N$ such that:

1. $x \in X(\mathcal{C})$ and for all $\omega \in \Omega$, $\sum_N x^j(\omega) = \sum_N e^j(\omega)$
2. there does not exist a coalition S with a communication plan $\overline{\mathcal{C}}^S = \{\{\overline{\mathcal{C}}^i\}_{i\in S}|\overline{\mathcal{C}}^i \subset \mathcal{F}^i \text{ for every } i \in S\}$ such that there exists a state dependent allocation $\overline{x}^S = \{\overline{x}^i\}_{i\in S} \in X^S(\overline{\mathcal{C}}^S)$, such that $\forall j \in S : \sum_S \overline{x}^j(\omega) = \sum_S e^j(\omega)$ and $EU^j(\overline{x}^j) > EU^j(x^j)$

3 Properties and existence of the EC-core

We first prove a lemma about the nested structure of the set of allocations which can be implemented with various communication plans. As a corollary to this result we note that there exists always an element of the core, that is associated with the full information revealing communication plan.

Lemma 3.1. (*Nested Structure*) *For a coalition $S \in \mathcal{N}$, let $\mathcal{C}^S$ and $\overline{\mathcal{C}}^S$ be two communication plans. If $\mathcal{PC}^j$ is a refinement of $\mathcal{P}\overline{\mathcal{C}}^j$ for all $j \in S$, then $X^S(\overline{\mathcal{C}}^S) \subset X^S(\mathcal{C}^S)$.*

Proof. Let $x^S = \{x^i(\omega)\}_{i \in S} \in X^S(\overline{\mathcal{C}}^S)$. First, it is clear that for all $i \in S$, $z^i(\omega)$ is $\bigvee_S \mathcal{C}^j$ measurable, since $\mathcal{P}(\bigvee_S \mathcal{C}^j)$ is a refinement of $\mathcal{P}(\bigvee_S \overline{\mathcal{C}}^j)$. Therefore, to prove that $x^S \in X^S(\mathcal{C}^S)$ we just need to show that x^S forms an incentive compatible pair with $\mathcal{C}^S$: Assume on the contrary that there exists $i \in S$ such that $x^i(\omega)$ is not incentive compatible with $\mathcal{C}^S$. Then, there exists $E \in \mathcal{PF}^i$ and $E' \in \mathcal{PC}^i : E \not\subset E'$ such that:

$$\forall \omega \in E : A(\omega) = E' \cap \left(\bigcap_{\substack{j \in S \\ j \neq i}} \mathcal{PC}^j(\omega) \right) \neq \emptyset \tag{3}$$

$$EU^i\left(z^i(A(\omega)) + e^i(\omega)|E\right) > EU^i\left(x^i(\omega)|E\right) \tag{4}$$

Next, we will show that agent i would also have the opportunity and the incentive to cheat under the communication plan $\overline{\mathcal{C}}^S$, with a deceptive message $\overline{E}'$ defined as $\overline{E}' = \mathcal{P}\overline{\mathcal{C}}^i(E')$:

First we note that $E \not\subset \overline{E}'$, because otherwise we would have

$$\forall \omega \in E : A(\omega) = E' \cap \left(\bigcap_{j \neq i} \mathcal{PC}^j(\omega) \right) \subset E' \cap \left(\bigcap_{j \neq i} \mathcal{P}\overline{\mathcal{C}}^j(\omega) \right) \subset \bigcap_{j \in S} \mathcal{P}\overline{\mathcal{C}}^j(\omega)$$

which implies $z^i(A(\omega)) = z^i(\omega)$, (because $z^i(\omega)$ is $\bigvee_S \overline{\mathcal{C}}^j$ measurable) a contradiction to (4). Similarly, we can show :

$$\forall \omega \in E : \overline{A}(\omega) = \overline{E}' \cap \left(\bigcap_{j \neq i} \mathcal{P}\overline{\mathcal{C}}^j(\omega) \right) \supset E' \cap \left(\bigcap_{j \neq i} \mathcal{PC}^j(\omega) \right) = A(\omega) \neq \emptyset \tag{5}$$

Then, note (5) and $\bigvee_S \overline{\mathcal{C}}^j$ measurability of $z^i(\omega)$ implies $z^i(\overline{A}(\omega)) = z^i(A(\omega))$, and we have

$$EU^i\left(z^i(\overline{A}(\omega)) + e^i(\omega)|E\right) > EU^i\left(x^i(\omega)|E\right) \tag{6}$$

(5) and (6) implies that $(x^S, \overline{\mathcal{C}}^S)$ is not incentive compatible, a contradiction.

Lemma 3.1 implies the following theorem about the EC-core strategies:

Theorem 3.2. *If $\mathcal{C}$ and $\overline{\mathcal{C}}$ are two communication plans such that $\mathcal{PC}^j$ is a refinement of $\mathcal{P}\overline{\mathcal{C}}^j$ for all $j \in N$, then: if $(x, \overline{\mathcal{C}})$ is an EC-core strategy pair then $(x, \mathcal{C})$ is also an EC-core strategy pair.*

In particular, if an allocation x forms a core strategy pair with a communication plan $\mathcal{C}$, then x also forms a core strategy pair with the full information revealing communication plan $\mathcal{F}$. This shows that if the EC-core is nonempty, then there exists always an element of the core associated with the full information revealing communication plan $\mathcal{F}$.

Remark 3.3. We also note that the proof of Lemma 3.1 and Theorem 3.2 can be extended easily to the case where the individual BIC conditions are replaced by the stronger coalitional BIC conditions [12]. Following Hahn and Yannelis [3] we define a state dependent allocation, communication plan pair $(x^S, \mathcal{C}^S)$ as coalitionally incentive compatible if it is not true that there exists a state $\overline{\omega} \in \Omega$, a subcoalition $\overline{S} \subset S$, and a deception vector $\alpha^{\overline{S}} : \prod_{\overline{S}} \mathcal{PF}^i \to \prod_{\overline{S}} \mathcal{PC}^i$ for coalition $\overline{S}$, such that:

(i) $(\alpha^{\overline{S}}, \alpha^{*S/\{\overline{S}\}})$ is compatible with $\mathcal{C}^S$ and
(ii) For all $i \in \overline{S}$, $EU^i(e^i(\omega) + [z^i \circ (\alpha^{\overline{S}}, \alpha^{*S/\{\overline{S}\}})](\omega)|\mathcal{PF}^i(\overline{\omega}))$
$> EU^i(x^i(\omega)|\mathcal{PF}^i(\overline{\omega}))$

To establish conditions for the existence of core strategies, it is more convenient to work in the abstract framework of NTU games. The associated NTU game is defined in the characteristic function $V : \mathcal{N} \to \mathbb{R}^n$ form by

$$V(S) = \left\{ u \in \mathbb{R}^N \;\middle|\; \begin{array}{l} \exists \mathcal{C}^S = \{\mathcal{C}^i \subset \mathcal{F}^i\}_{i \in S} \text{ and } \exists x^S \in X^S(\mathcal{C}^S): \\ (i)\ \forall \omega \in \Omega, \sum_S x^j(\omega) = \sum_S e^j(\omega) \\ (ii)\ \forall i \in S : EU^i(x^i) \geq u_i \end{array} \right\} \tag{7}$$

From Lemma 3.1 we know that $X^S(\mathcal{C}^S) \subset X^S(\mathcal{F}^S)$ and therefore $V(S)$ defined by (7) can be characterized as:

$$V(S) = \left\{ u \in \mathbb{R}^N \;\middle|\; \begin{array}{l} \exists x^S \in X^S(\mathcal{F}^S): \\ (i)\ \forall \omega \in \Omega, \sum_S x^j(\omega) = \sum_S e^j(\omega) \\ (ii)\ \forall i \in S : EU^i(x^i) \geq u_i \end{array} \right\}$$

For the special case of affine linear utility functions, we prove the balancedness of the NTU game given in (7) and therefore the nonemptiness of the EC-core [9].

Theorem 3.4. *If for all $i \in N$ and $\omega \in \Omega$, $u^i(\omega, x) = a^i(\omega) \cdot x + b^i(\omega)$ for some $a^i(\omega) \in \mathbb{R}^l_+$ and $b^i(\omega) \in \mathbb{R}$, then the EC-core is nonempty.*

Proof. Let $\mathcal{B}$ be a balanced family, with the associated balancing coefficients $\{\lambda_S\}_{S \in \mathcal{B}}$ and $u \in \bigcap_{S \in \mathcal{B}} V(S)$. Then for all $S \in \mathcal{B}$ there exists an allocation $x^{(S)} \in X^S(\mathcal{F}^S)$ such that $\sum_S x^{(S),i}(\omega) = \sum_S e^i(\omega)$ for all states $\omega \in \Omega$ and for every agent $i \in S$, $EU^j(x^{(S),i}) \geq u_i$. Now for each state $\omega \in \Omega$ define the allocation $x = \{x^j(\omega)\}_{j \in N}$ by:

$$x^j(\omega) = \sum_{\substack{S \in \mathcal{B} \\ S \ni j}} \lambda_S x^{(S),j}(\omega)$$

Then the allocation x is feasible at each state $\omega \in \Omega$.

$$\begin{aligned}
\forall \omega \in \Omega : \sum_{j \in N} x^j(\omega) &= \sum_{j \in N} \sum_{\substack{S \in B \\ S \ni j}} \lambda_S x^{(S),j}(\omega) \\
&= \sum_{S \in B} \lambda_S \sum_{j \in S} x^{(S),j}(\omega) \\
&= \sum_{S \in B} \lambda_S \sum_{j \in S} e^j(\omega) \\
&= \sum_{j \in N} e^j(\omega) \sum_{\substack{S \in B \\ S \ni j}} \lambda_S = \sum_{j \in N} e^j(\omega)
\end{aligned}$$

$$\Rightarrow \forall \omega \in \Omega : \sum_{j \in N} x^j(\omega) = \sum_{j \in N} e^j(\omega) \tag{8}$$

Next we will show that x attains the utility level u by:

$$\begin{aligned}
\forall j \in N : EU^j\left(x^j\right) &= \sum_{\omega \in \Omega} u^j(\omega, \sum_{\substack{S \in B \\ S \ni j}} \lambda_S x^{(S),j}(\omega))\mu(\omega) \\
&= \sum_{\omega \in \Omega} \sum_{\substack{S \in B \\ S \ni j}} \lambda_S u^j\left(\omega, x^{(S),j}(\omega)\right)\mu(\omega) \\
&= \sum_{\substack{S \in B \\ S \ni j}} \lambda_S \sum_{\omega \in \Omega} u^j\left(\omega, x^{(S),j}(\omega)\right)\mu(\omega) \\
&\geq \sum_{\substack{S \in B \\ S \ni j}} \lambda_S u_j = u_j
\end{aligned}$$

$$\Rightarrow \forall j \in N : EU^j\left(x^j\right) \geq u_j \tag{9}$$

Finally we will show that, $x \in X(\mathcal{F})$. First, for every agent $j \in N$, $z^j(\omega)$ is $\bigvee_N \mathcal{F}^i$ measurable since for each S, $z^{(S),j}(\omega)$ is $\bigvee_S \mathcal{F}^i$ ($\subset \bigvee_N \mathcal{F}^i$) measurable. To prove that the net trades x^i is incentive compatible with the communication plan $\mathcal{F}$ for each agent $i \in N$, assume on the contrary that there exists an agent j such that x^j is not incentive compatible with full information revealing communication plan $\mathcal{F}$. Then we have:

$$\exists E \in \mathcal{PF}^j, \exists E' \in \mathcal{PF}^j : E \neq E' :$$
$$(i)\ \forall \omega \in E : A(\omega) = E' \cap \left(\bigcap_{i \neq j} \mathcal{PF}^i \right) \neq \emptyset,$$
$$(ii)\ EU^j\left(z^j\left(A(\omega)\right) + e^j(\omega)|E\right) > EU^j\left(x^j(\omega)|E\right)$$

$$\Rightarrow \sum_{\omega \in E} \left[a(\omega) \cdot \left(z^j\left(A(\omega)\right) + e^j(\omega)\right) + b(\omega)\right] \mu(\omega|E)$$
$$> \sum_{\omega \in E} \left[a(\omega) \cdot x^j(\omega) + b(\omega)\right] \mu(\omega|E)$$
$$\Rightarrow \sum_{\omega \in E} a(\omega) \cdot \sum_{\substack{S \in B \\ S \ni j}} \lambda_S z^{(S),j}\left(A(\omega)\right) \mu(\omega|E)$$
$$> \sum_{\omega \in E} a(\omega) \cdot \sum_{\substack{S \in B \\ S \ni j}} \lambda_S z^{(S),j}(\omega) \mu(\omega|E)$$
$$\Rightarrow \sum_{\substack{S \in B \\ S \ni j}} \lambda_S \sum_{\omega \in E} a(\omega) \cdot z^{(S),j}\left(A(\omega)\right) \mu(\omega|E)$$
$$> \sum_{\substack{S \in B \\ S \ni j}} \lambda_S \sum_{\omega \in E} a(\omega) \cdot z^{(S),j}(\omega) \mu(\omega|E)$$
$$\Rightarrow \exists S \ni j : \sum_{\omega \in E} a(\omega) \cdot z^{(S),j}\left(A(\omega)\right) \mu(\omega|E) > \sum_{\omega \in E} a(\omega) \cdot z^{(S),j}(\omega) \mu(\omega|E)$$
$$\Rightarrow \exists S \ni j : EU^j\left(z^{(S),j}\left(A(\omega)\right) + e^j(\omega)|E\right) > EU^j\left(x^{(S),j}(\omega)|E\right)$$

The last equation implies that there exists a coalition $S \in \mathcal{B}$ and $S \ni j$ such that agent j in coalition S with the allocation $x^{(S),j}$ has an incentive and an opportunity to use the wrong signal E', if he is informed that the event E has occurred, because:

$$\tilde{A}(\omega) = E' \cap \left(\bigcap_{\substack{i \neq j \\ i \in S}} \mathcal{PF}^i(\omega) \right) \supset A(\omega) \neq \emptyset$$

and

$$EU^j\left(z^{(S),j}\left(\tilde{A}(\omega)\right) + e^j(\omega)|E\right) = EU^j\left(z^{(S),j}\left(A(\omega)\right) + e^j(\omega)|E\right)$$
$$> EU^j\left(x^{(S),j}(\omega)|E\right)$$

(because $\tilde{A}(\omega) \in \mathcal{P}(\bigvee_S \mathcal{F}^i)$ and therefore $z^{(S),j}(\omega)$ is constant on $\tilde{A}(\omega)$).

This contradiction with the measurability proves that $x \in X(\mathcal{F})$. Combining this last result with (8) and (9) we have $u \in V(N)$.

The restriction of affine linear utility function is a strong condition of risk neutrality on the player's preference relations if the allocations $x^S(\omega)$ are interpreted as pure choices in X^S. As pointed out by Ichiishi *et al* [4], if the allocations are interpreted as correlated strategies over pure choices in X^S then the assumption of linearity is automatically satisfied because the expected utility is linear in probabilities.

Remark 3.5. The affine linear utility assumption can be relaxed in the case of two person economy with continuous utility functions. For the two person economy, EC-core allocations are precisely the feasible, individually rational allocations in $X(\mathcal{F})$, which are (*ex ante*) pareto optimal among the set of feasible allocations in $X(\mathcal{F})$. The individually rational and feasible allocations in $X(\mathcal{F})$ form a compact set in $\mathbb{R}^l$. Since the initial random endowments satisfy these constraints, this compact set is nonempty. Then, the existence of the EC-core strategies follows from the continuity of the *ex ante* utility functions of both agents.

4 Extensions

In this section we extend the results of Section 3, to coalitional production economies and discuss an example of such an economy where affine linear utility functions are commonly adopted.

Definition 4.1. A coalition production economy with asymmetric information with a finite set of agents N, is a list of data $\mathcal{E}_{cpe} := (N, \{u^i, e^i, \mathcal{F}^i\}_{i \in N}, \mu, Y)$.

The production possibilities for coalition S in state ω is given by the production set $Y(\omega, S) \subset \mathbb{R}^l$. The allocation schemes $X^S(\mathcal{C}^S)$ which can be implemented by coalition S using a communication plan $\mathcal{C}^S = \{\{\mathcal{C}^i\}_{i \in S} | \mathcal{C}^i \subset \mathcal{F}^i \text{ for every } i \in S\}$ are given by:

$$X^S(\mathcal{C}^S) := \left\{ \{x^i\}_{i \in S} \;\middle|\; \begin{array}{l} \forall i \in S : z^i \text{ is } \bigvee_S \mathcal{C}^j \text{ measurable and there does not} \\ \text{exist } E \in \mathcal{PF}^i \text{ and } E' \in \mathcal{PC}^i : E \not\subset E', \\ \\ (i)\ \ \forall \omega \in E, A(\omega) = E' \cap \left(\bigcap_{j \neq i} \mathcal{PC}^j(\omega) \right) \neq \emptyset \\ (ii)\ EU^i\left(z^i(A(\omega)) + e^i(\omega) | E\right) > EU^i\left(x^i(\omega) | E\right) \end{array} \right\}$$

then the associated NTU game V is defined by:

$$V(S) := \left\{ u \in \mathbb{R}^N \;\middle|\; \begin{array}{l} \exists y : \Omega \to \mathbb{R}^l : \forall \omega \in \Omega : y(\omega) \in Y(\omega, S), \\ \\ \exists \mathcal{C}^S = \{\mathcal{C}^j \subset \mathcal{F}^j\}_{j \in S} \text{ and } \exists x^S \in X^S(\mathcal{C}^S) : \\ \\ (i)\ \ \forall \omega \in \Omega : \sum_S x^j(\omega) = y(\omega) + \sum_S e^j(\omega), \\ (ii)\ \forall i \in S : EU^i\left(x^i\right) \geq u_i \end{array} \right\} \tag{10}$$

Based on this NTU game, EC-core strategies can be defined for the coalitional production economy.

Definition 4.2. $(x, y, \mathcal{C})$ is an endogenous communication core (EC-core) strategy if it results in a utility level in the core of the NTU game V given in (10). The following lemma is still valid in the CPU framework.

Lemma 4.3. *For a coalition $S \in \mathcal{N}$, let $\mathcal{C}^S$ and $\overline{\mathcal{C}}^S$ be two communication plans. If $\mathcal{PC}^j$ is a refinement of $\mathcal{P}\overline{\mathcal{C}}^j$ for all $j \in S$ then $X^S(\overline{\mathcal{C}}^S) \subset X^S(\mathcal{C}^S)$.*

This lemma implies that we can concentrate only on full information revealing communication plans to find the attainable utility levels for coalition S. Therefore $V(S)$ consists of precisely the utility vectors using an allocation and production plan implementable with the full information revealing communication plan.

The NTU game given in (10) can be characterized as:

$$V(S) = \left\{ u \in \mathbb{R}^N \;\middle|\; \begin{array}{l} \exists y : \Omega \to \mathbb{R}^l : \forall \omega \in \Omega : y(\omega) \in Y(\omega, S), \\ \text{and } \exists x^S \in X^S(\mathcal{F}^S) : \\ (i)\ \forall \omega \in \Omega, \sum_S x^j(\omega) = y(\omega) + \sum_S e^j(\omega) \\ (ii)\ \forall i \in S : EU^i\left(x^i\right) \geq u_i \end{array} \right\}$$

The following results are analogous to the theorems given in the previous section. The proofs are very similar and therefore omitted.

Theorem 4.4. *If $\mathcal{C}$ and $\overline{\mathcal{C}}$ are two communication plans such that $\mathcal{PC}^j$ is a refinement of $\mathcal{P}\overline{\mathcal{C}}^j$ for all $j \in S$ then: if $(x, y, \overline{\mathcal{C}})$ is an EC-core strategy pair then $(x, y, \mathcal{C})$ is also an EC-core strategy pair.*

Theorem 4.5. *The NTU game associated with CPE is balanced if:*

(i) $\forall \omega \in \Omega : \sum_{S \in B} \lambda_S Y(\omega, S) \subset Y(\omega, N)$

(ii) $u^i(\omega, x) = a^i(\omega) \cdot x(\omega) + b^i(\omega)$, $a^i(\omega) \in \mathbb{R}^l_+$, $b^i(\omega) \in \mathbb{R}$.

One important model in the cooperative production economy framework is the model of the firm in multidivisional form (M-form firm, in short) studied by Chandler [2] in 1962. An M-form firm is a corporation in which several divisions (or profit centers) are operated semiautonomously. Although each division is an independent decision-maker, they share information with each other and coordinate their production activities. Finally a share of the total profit is allocated to each division. Therefore divisions should agree on specific plans for coordinated activities such as production and information sharing and a plan to determine final profit imputations. The divisions can only settle at self enforcing agreements, where no coalition of divisions can improve upon it by its own effort. Radner [8] recently formulated the internal organization of an M-form firm as a static model of a profit-center game, defined the core as the set of self enforcing agreements and studied its properties. In his cooperative game theoretic model, the utility of agents (divisions) are given by their final share of profit, which satisfies linearity condition. Therefore the results of this chapter are readily applicable to M-form firm models. In particular the main result of this chapter indicates that the full information revealing principle holds true for M-form firm models with *ex ante* contracting.

References

1. Allen, B.: Market games with asymmetric information and nontransferable utility: Representation results and the core. CARESS Working Paper, No. 91–09 (1991)
2. Chandler, A.D.: Strategy and structure. Cambridge, MA: MIT Press 1962
3. Hahn, G., Yannelis, N.C.: Efficiency and incentive compatibility in differential information economies. Economic Theory **10**, 383–411 (1997)
4. Ichiishi, T., Idzik, A., Zhao, J.: Cooperative processing of information via choice at an information set. International Journal of Game Theory **23**, 145–165 (1994)
5. Ichiishi, T., Idzik, A.: Bayesian cooperative choice of strategies. International Journal of Game Theory **25**, 455–473 (1996)
6. Koutsougeras, L., Yannelis, N.C.: Incentive compatibility and informational superiority of the core of an economy with differential information. Economic Theory **3**, 195–216 (1993)
7. Radner, R.: Equilibrium under uncertainty. Econometrica **36**, 31–58 (1968)
8. Radner, R.: Transfer payments and the core of a profit-center game. In: Dasgupta, P., et al. (eds.) Economic analysis of markets and games (Essays in Honor of Frank Hahn), pp 316–339. Cambridge, MA: MIT Press 1992
9. Scarf, H.: The core of an N-person game. Econometrica **35**, 50–69 (1967)
10. Yannelis, N. C.: The core of an economy with differential information. Economic Theory **1**, 183–198 (1991)
11. Wilson, R.: Information, efficiency and the core of an economy. Econometrica **46**, 807–816 (1978)
12. Yazar, J.: Communication in economies with asymmetric information. Working Paper, The Ohio State University (1998)

Incentive compatible contractible information*

Philip Bond

Kellogg School of Management, Northwestern University, Evanston IL 60208-2001, USA
(e-mail: p-bond@northwestern.edu)

Received: August 7, 2000; revised version: December 21, 2001

Summary. The paper seeks to characterize what information is always available for contracting, independent of the form of the contract and the probabilities of different states of nature. The paper denotes such information as *contractible*. It is established that it is possible to speak uniquely of *maximal contractible information*. Several characterizations are exhibited. In particular, it is shown that if either (a) punishments are bounded everywhere, or (b) deviations from truth-telling are either always or never detected, then maximum contractible information coincides with $\bigwedge_i \bigvee_{j \neq i} \mathcal{E}_j$ where $\mathcal{E}_j$ is the information partition of agent j. An argument is given for why (b) may be expected to hold.

Keywords and Phrases: Contractible information, Incentive compatibility, Information partition, Common knowledge, Cross-reporting.

JEL Classification Numbers: D78, D82.

1 Introduction

Incentive constraints play a central role in understanding the restrictions imposed by private information (i.e. information uniquely possessed by single agents). Dual to this, much of the contract theory literature has assumed that information possessed by at least two agents can be freely contracted upon, regardless of the nature of the contract and the probability of different states.[1] For instance, in principal-agent models[2] effort is assumed private to the agent, but output is assumed to be observed by both the principal and agent and hence to be contractible.

* I thank Michael Chwe, Douglas Diamond, Lars Stole, Robert Townsend, Nicholas Yannelis and an anonymous referee for helpful comments.

[1] The exceptions are the implementation theory and incomplete contracting literatures.

[2] See, for instance, Mirrlees (1999).

This paper seeks to extend our understanding of what information is contractible in three ways. First, what information can be said to be contractible in more complex informational settings? For example, suppose that in a principal-agent setting the agent either has malaria, the flu, a cold, or is entirely healthy, but is unable to distinguish between flu and a cold. Suppose (for the sake of example) that malaria and the flu produce identical symptoms, and that the principal (but not the agent, who in this example lacks medical training) is able to identify these symptoms. So the principal knows when the agent is suffering from malaria or the flu, but not which, and knows when the agent either has a cold or is healthy, but not which. Is any aspect of the agent's health always contractible here? The answer turns out to depend on whether or not the size of punishments available is bounded.

Second, when there are more than two agents, is information private unless it is entirely shared by two agents? For instance, suppose effort has two components – duration and intensity. If the principal observes hours worked, and a coworker intensity, is the agent's effort contractible?

Third, is it possible to speak uniquely of maximal *contractible information*, in the sense of a uniquely defined maximal amount of information that is freely contractible? If so, how is such information characterized?

Before proceeding to the analysis, it is worth pausing to note how this paper differs from the work initiated by Crémer and McLean (1985)[3] establishing circumstances in which incentive compatibility places no constraint on the ability of an uninformed outsider to extract surplus. This body of work is concerned with incentive compatibility's effect on what welfare levels are attainable, and depends on the assumptions of risk neutral agents and correlated information. In contrast, the present paper considers what *contracts* are unaffected by incentive concerns, and makes only weak assumptions about preferences, and none about stochastic structure.

I conclude the introduction with a note on methodology. This paper is a *mechanism design* paper, in the sense of being concerned with whether outcomes can be supported as the equilibria of some pre-specified game. It neglects the concerns of the parallel *implementation theory* literature about whether there also exist other (undesirable) equilibria of these games. It is well known (see Maskin, 1999) that the multiple equilibrium problem is important in complete information settings (i.e. those in which agents share all information) when the equilibrium concept employed is Nash. However, papers by – among others – Matsushima (1993), Arya et al. (1995), and Duggan (1997) suggest that the multiple equilibrium problem is of limited importance when information is incomplete. Moreover, in complete information settings Moore and Repullo (1988) find that multiple equilibria concerns can be avoided if one looks instead at implementation in subgame perfect equilibria. In this paper I appeal to this broad class of results for justification in considering the mechanism design problem in isolation.

The paper proceeds as follows. Section 2 gives a general specification of the problem to be analyzed. Section 3 establishes that maximal contractible information is uniquely defined. Section 4 presents several examples. Section 5 shows that

[3] See also Crémer and McLean (1988), together with the related work of McAfee and Reny (1992).

if punishments are unlimited, maximum contractible information has an easy characterization. Section 6 establishes that the same characterization holds with limited punishments if information satisfies a certain restriction, and gives an argument for why that restriction is reasonable. Section 7 discusses the case in which preferences as well as probabilities are allowed to vary. Section 8 gives an alternative characterization of maximum contractible information. Section 9 relates the results of this paper to previous research.

2 Preliminaries

An economy is a quintuple $(\Omega, N, A, \{u_i\}, \{\mathcal{E}_i\})$ where:

- Ω is the state space, with $\omega \in \Omega$ a typical member. Let M be the (finite) cardinality of Ω.
- N is the (finite) set of agents.
- A is the (possibly infinite) set of outcomes.
- $u_i : A \times \Omega \to \Re$ is the utility mapping of agent $i \in N$, giving the utility of each agent given an outcome and state of the world. Agents are assumed to maximize expected utility.
- $\mathcal{E}_i$ is a partition of Ω giving the information[4] possessed by agent $i \in N$.

Also, let P be the set of probability mappings $p : \Omega \to (0,1)$, where $p(\omega)$ is the probability of state ω. Thus $P \subset (0,1)^M$.

The economy $(\Omega, N, A, \{u_i\}, \{\mathcal{E}_i\})$ and probability mapping $p \in P$ are assumed to be common knowledge to the agents N. We will characterize what information is always available for contracting, regardless of the probability mapping p. That is, if an outside observer sees the preferences and information of each agent in the economy, what information can that observer infer is *always* available for contracting, no matter what the probabilities are? In Section 7 we will also briefly consider the related question of what information is contractible independent of agents' preferences.

Throughout the paper, agents' preferences will be assumed to satisfy the following two properties:

Condition SP (Strict Preferences): *For all states $\omega \in \Omega$ and agents $i \in N$, there exist outcomes a_1 and a_2 such that $u_i(a_2;\omega) > u_i(a_1;\omega)$.*

Condition FD (Free Disposal): *For any finite subset $A_k = \{a_1, \dots, a_k\}$ of A, there exists an outcome $a_0 \in A$ such that $u_i(a_0,\omega) \leq u_i(a,\omega)$ for all agents $i \in N$, states $\omega \in \Omega$, and outcomes $a \in A_k$.*

Condition SP ensures that agents always have *strict preferences* over at least two outcomes, and serves to rule out degenerate cases in which information is contractible simply because agents face no meaningful choices. Condition FD represents a generalization of *free disposal of goods* in the following sense. Suppose

[4] See, e.g., Osborne and Rubinstein (1994) for a description of how information can be represented by partitions of the state space.

that outcomes relate to the provision of (public or private) goods. Then provided all agents weakly prefer more goods to less in all states, the outcome a_0 just relates to taking the minimum provision of each good from the outcomes in A_k. Condition FD ensures that a central planner does not need to have too much information in order to effectively punish agents. Note that the condition says nothing about the *size* of punishments available, a point that will be returned to below.

Throughout the paper, $\mathcal{E}_S$ will be used to denote the collective information $\bigvee_{i \in S} \mathcal{E}_i$ of a coalition $S \subset N$, and $\mathcal{E}_{-i}$ will denote the collective information of all agents other that agent i, $\mathcal{E}_{N \setminus \{i\}}$. Also, $E_i(\omega)$ (respectively $E_S(\omega), E_{-i}(\omega)$) will denote the element of the partition $\mathcal{E}_i$ (respectively $\mathcal{E}_S, \mathcal{E}_{-i}$) that contains the state ω.

Given a particular probability map p, incentive compatibility is defined as normal:

Definition 1 *A mapping $f : \Omega \to A$ is said to be p-incentive compatible*[5] *if when probabilities are given by $p \in P$, there exists a mechanism $(\{\mathcal{M}_i\}, F : \times_i \mathcal{M}_i \to A)$ such that for all $\omega \in \Omega$ there is a Bayesian equilibrium $m(\omega)$ of the mechanism with $F(m(\omega)) = f(\omega)$.*

We are interested in what information can always be contracted upon, independent both of the desired outcomes and the particular probabilities of different states. Such information will be said to be *contractible*. Informally, information is contractible if an outside observer who is ignorant of the stochastic structure of the economy can nonetheless infer that the information is available for contracting. Formally,

Definition 2 *A partition $\mathcal{G}$ of Ω is said to be contractible information if whenever $f : \Omega \to A$ is $\mathcal{G}$-measurable, then f is p-incentive compatible for all $p \in P$.*

Clearly it is never possible to contract on information not possessed by *any* agent. That is,

Lemma 1 *A partition $\mathcal{G}$ is contractible information only if $\mathcal{G} \preccurlyeq \mathcal{E}_N$.*

Proof. Suppose not. Then for any $p \in P$ there exists a mapping $f : \Omega \to A$ and states $\omega_1, \omega_2 \in \Omega$ such that f is p-incentive compatible, $f(\omega_1) \neq f(\omega_2)$ and $E_N(\omega_1) = E_N(\omega_2)$. The latter implies that $E_i(\omega_1) = E_i(\omega_2)$ for all $i \in N$. But the revelation principle implies that there exists a direct mechanism – so

$$F(E_1(\omega_1), \ldots, E_N(\omega_1)) = f(\omega_1)$$
$$F(E_1(\omega_2), \ldots, E_N(\omega_2)) = f(\omega_2)$$

for some function $F : \times_i \mathcal{E}_i \to A$. But since $f(\omega_1) \neq f(\omega_2)$ this gives a contradiction, completing the proof.

---

[5] Note that incentive compatibility is sometimes referred to elsewhere as weak implementability.

3 Maximal contractible information

3.1 Punishments

Consider the example economies of Figures 1 and 2. The connected points depict the information partitions – so in Figure 1, $\mathcal{E}_1 = \{\{\omega_1\}, \{\omega_2, \omega_3\}, \{\omega_4\}\}$ and $\mathcal{E}_2 = \{\{\omega_1, \omega_2\}, \{\omega_3, \omega_4\}\}$. The economy of Figure 1 corresponds to the health example given in the introduction. That is, the states $\omega_1, \ldots, \omega_4$ correspond respectively to"malaria", "flu", "cold" and "healthy". Agent 1 is the worker, and cannot distinguish between the flu and a cold. Agent 2 is the employer, who can spot the identical symptoms produced by malaria and the flu.

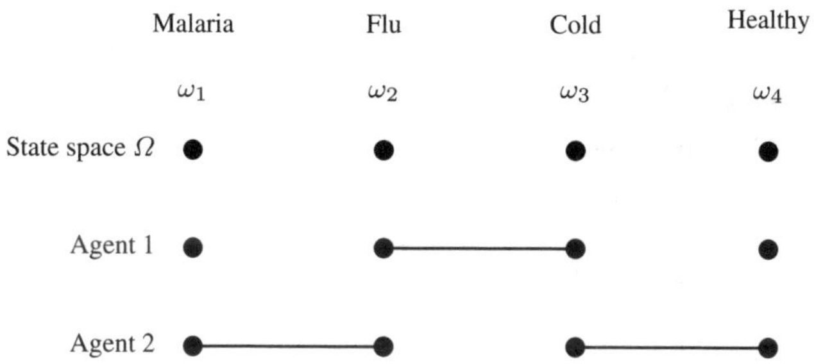

Figure 1. Information about health

Is the health of the worker available for contracting, independent of the form of the contract? The answer depends on the size of punishments available. To see this, consider the situation faced by the employer when she observes the symptoms associated with malaria and the flu (i.e. when her information is $\{\omega_1, \omega_2\}$). If she reports that the worker has no symptoms, the worker will know she is lying if he in fact has malaria (state ω_1), but not if he has the flu or a cold (states ω_2, ω_3), since in this case the worker is aware his illness might only be a cold. If it is possible to heavily punish the employer, then the threat of detection will be enough to keep her honest, and she will always tell the truth. In this case, contracting upon at least some aspects of the worker's health (here, symptoms/no symptoms) is always possible. But if punishments are more limited, she may (depending on what is at stake) take the risk of detection and misreport the worker's medical condition – so arbitrary contracts cannot be written on any aspect of the worker's health.

The size of punishments is important here because of the availability of misreports that are detected sometimes, but not always. No such reporting deviations are available in the economy of Figure 2, and so – as we will see – the size of punishment available does not play a critical role.

With the above discussion in mind, at each state $\omega \in \Omega$ the agents N are partitioned into two groups $N_\infty(\omega)$ and $N_0(\omega)$ as follows: Agent i is in $N_\infty(\omega)$ if it holds that for all $x \in \Re$ there exists $a \in A$ such that $u_i(a, \omega) < x$. Otherwise

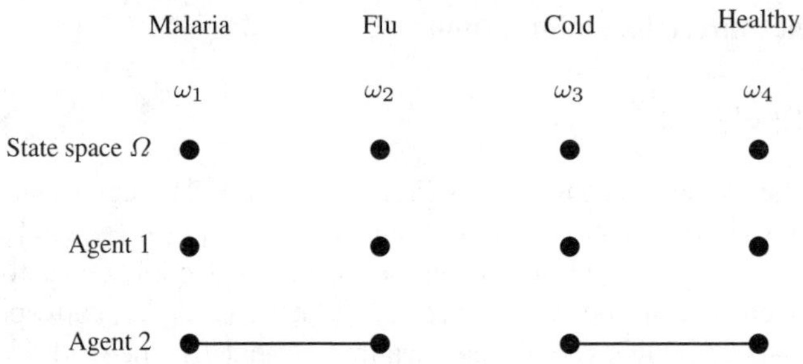

Figure 2. Alternative information about health

agent i is in $N_0(\omega)$. That is, $N_\infty(\omega)$ consists of those agents who at state ω can be punished *arbitrarily* harshly. For those agents in $N_0(\omega)$ for whom punishments are limited, define the maximum level of punishment $z_i(\omega)$ by

$$z_i(\omega) = \inf_{a \in A} u_i(a, \omega)$$

3.2 A candidate partition

Based on the above considerations about types of deviation and size of punishment available, I start by constructing a candidate partition for the maximum contractible information.

The revelation principle tells us that we can think in terms of agents truthfully reporting their information (i.e. elements of $\mathcal{E}_i$), and focuses attention on what incentives agents must be given to do so. Denote the report of each agent by $\tilde{E}_i \in \mathcal{E}_i$. If an agent's report is *always* inconsistent with the truthful reports of others (i.e. if $\tilde{E}_i \cap E_{-i}(\omega) = \emptyset$ for all $\omega \in E_i$, where E_i is agent i's actual information), then it will be easy to deter untruthful reports by punishing the agent at these report combinations. Note that the punishments will not have to very large, since the deviating agent knows with certainty that he will be detected.

Report deviations that are only sometimes undetected will be harder to deter, and indeed may be impossible to do so for arbitrary outcome functions $f : \Omega \to A$. With this in mind, for each $i \in N$ define a binary relation $\rightharpoonup_i$ on Ω by:

$$\omega_1 \rightharpoonup_i \omega_2 \text{ if and only if } \omega_2 \in E_{-i}(\omega_1)$$

So the relations $\rightharpoonup_i$ define states which may be contractibly indistinguishable, since agent i can report $E_i(\omega_2)$ when her information is $E_i(\omega_1)$, and escape detection at least sometimes.

Whether they are or not depends, as we saw above, on the level of punishments available. With heavy punishments, occasional detection is a powerful deterrent – but with limited punishments this is not the case. Define a further set of binary relations $\omega_1 \rightarrow_i \omega_2$ by

$$\omega_1 \rightarrow_i \omega_2 \text{ if and only if } \omega_1 \rightharpoonup_i \omega_2 \text{ and}$$
$$i \in N_0(\omega) \text{ whenever } \omega \in E_i(\omega_1) \text{ and } E_i(\omega_2) \cap E_{-i}(\omega) = \emptyset$$

So if $\omega_1 \rightarrow_i \omega_2$, agent i can report $E_i(\omega_2)$ at $E_i(\omega_1)$ knowing that this deviation will not always be detected, and in those states where it is punishments are limited.

Finally define the graph $(\Omega, \rightarrow)$ by

$$\omega_1 \rightarrow \omega_2 \text{ if and only if } \omega_1 \rightarrow_i \omega_2 \text{ for some } i \in N$$

Define $\mathcal{E}^*$ as the partition induced by the components of the graph $(\Omega, \rightarrow)$. Below it is shown that $\mathcal{E}^*$ is the unique maximal contractible information partition.

First, note that the partition $\mathcal{E}^*$ is coarser than the collective knowledge of all agents, $\mathcal{E}_N$.

Lemma 2 $\mathcal{E}^* \preceq \mathcal{E}_N$.

Proof. Take $\omega_1, \omega_2 \in \Omega$ such that $E_N(\omega_1) = E_N(\omega_2)$. Take any $i \in N$. Then $E_{-i}(\omega_1) = E_{-i}(\omega_2)$ and $E_i(\omega_1) = E_i(\omega_2)$. So certainly $\omega_1 \rightharpoonup_i \omega_2$. Moreover, since $E_i(\omega_1) \cap E_{-i}(\omega) \neq \emptyset$ for all $\omega \in E_i(\omega_1)$, it must also hold that $E_i(\omega_2) \cap E_{-i}(\omega) \neq \emptyset$ for all $\omega \in E_i(\omega_1)$. So $\omega_1 \rightarrow_i \omega_2$, completing the proof.

Moreover, $\mathcal{E}^*$ is indeed contractible information:

Lemma 3 *$\mathcal{E}^*$ is contractible information.*

Proof. Take an arbitrary $\mathcal{E}^*$-measurable function $f : \Omega \rightarrow A$ and probability mapping $p \in P$. We must show that f is p-incentive compatible. The proof is by construction, and consists of exhibiting a direct mechanism $F : \times_{i \in N} \mathcal{E}_i \rightarrow A$ such that

$$F\left((E_i(\omega))_{i \in N}\right) = f(\omega) \tag{1}$$

and such that for all $i \in N$ and $E_i, \tilde{E}_i \in \mathcal{E}_i$ the incentive constraint

$$\sum_{\omega \in E_i} p(\omega) u_i(f(\omega), \omega) \geq \sum_{\omega \in E_i} p(\omega) u_i\left(F\left(\tilde{E}_i, E_{-i}(\omega)\right), \omega\right) \tag{2}$$

holds.

First, choose $a_f \in A$ such that for $i \in N$ and $\omega, \omega' \in \Omega$, $u_i(a_f, \omega) \leq u_i(f(\omega'), \omega)$. Such a choice is always possible by Condition FD.

Next, construct a set of functions $F^i : (\times_{j \in N} \mathcal{E}_j) \times \mathcal{E}_i \rightarrow A$ as follows. First, if $\left(\left(\tilde{E}_j\right)_{j \in N}, E_i\right) \in (\times_{j \in N} \mathcal{E}_j) \times \mathcal{E}_i$ is such that $\bigcap_{j \in N} \tilde{E}_j \neq \emptyset$, then set $F^i\left(\left(\tilde{E}_j\right)_{j \in N}, E_i\right) = f\left(\bigcap_{j \in N} \tilde{E}_j\right)$. Note that this is a well-defined choice, since by assumption f is $\mathcal{E}^*$-measurable and hence $\mathcal{E}_N$-measurable by Lemma 2.

Over the remainder of $(\times_{j\in N}\mathcal{E}_j)\times\mathcal{E}_i$, choose $F^i(\cdot)$ so that the incentive constraints

$$\sum_{\omega\in E_i} p(\omega)u_i(f(\omega),\omega) \geq \sum_{\omega\in E_i \text{ s.t. } \tilde{E}_i\cap E_{-i}(\omega)\neq\emptyset} p(\omega)u_i\left(f\left(\tilde{E}_i\cap E_{-i}(\omega)\right),\omega\right) + \sum_{\omega\in E_i \text{ s.t. } \tilde{E}_i\cap E_{-i}(\omega)=\emptyset} p(\omega)u_i\left(F^i\left(\tilde{E}_i,(E_j(\omega))_{j\in N}\right),\omega\right) \tag{3}$$

are satisfied. That is, $F^i\left(\tilde{E}_i,(E_j(\omega))_{j\in N}\right)$ gives the punishment for agent i when she says $\tilde{E}_i$ at $E_i(\omega)$, and the remainder of agents say $(E_j(\omega))_{j\neq i}$. Any choice of $F^i(\cdot)$ such that these punishments are sufficiently great to make (3) hold is OK. To see that such a choice is always possible, fix the agent i, her information E_i and deviation $\tilde{E}_i$ and note that one of the following three cases must hold:

1. $\tilde{E}_i\cap E_{-i}(\omega)=\emptyset$ for all $\omega\in E_i$: In this case, $F^i\left(\tilde{E}_i,(E_j(\omega))_{j\in N}\right)=a_f$ will satisfy constraint (3).
2. $\tilde{E}_i\cap E_{-i}(\hat{\omega})\neq\emptyset$ for some $\hat{\omega}\in E_i$, and $i\in N_0(\omega)$ whenever $\omega\in E_i$ and $\tilde{E}_i\cap E_{-i}(\omega)=\emptyset$: So if $\omega\in E_i$ is such that $\tilde{E}_i\cap E_{-i}(\omega)\neq\emptyset$, and if $\omega'\in\tilde{E}_i\cap E_{-i}(\omega)$, then it follows that $\omega\rightarrow_i\omega'$. Thus whenever $\tilde{E}_i\cap E_{-i}(\omega)\neq\emptyset$, it holds that $f(\omega)=f\left(\tilde{E}_i\cap E_{-i}(\omega)\right)$. So setting $F\left(\tilde{E}_i,(E_j(\omega))_{j\in N}\right)=a_f$ whenever $\tilde{E}_i\cap E_{-i}(\omega)=\emptyset$ will satisfy the constraint.
3. $\tilde{E}_i\cap E_{-i}(\hat{\omega})\neq\emptyset$ for some $\hat{\omega}\in E_i$, and there exists ω^* such that $\tilde{E}_i\cap E_{-i}(\omega^*)=\emptyset$ and $i\in N_\infty(\omega^*)$: Just choose $F^i\left(\tilde{E}_i,E_{-i}(\omega^*),E_i\right)$ such that $u_i\left(F^i\left(\tilde{E}_i,E_{-i}(\omega^*),E_i\right),\omega^*\right)$ is sufficiently low for (3) to hold. Such a choice is always possible since $i\in N_\infty(\omega^*)$.

Finally, construct $F:\times_{i\in N}\mathcal{E}_i\rightarrow A$ itself as follows. If $e=\left(\tilde{E}_i\right)_{i\in N}\in\times_{i\in N}\mathcal{E}_i$ is such that $\bigcap_{i\in N}\tilde{E}_i\neq\emptyset$, set $F(e)=f\left(\bigcap_{i\in N}\tilde{E}_i\right)$. (As above, this is a well defined construction). For all other points in $\times_{i\in N}\mathcal{E}_i$, define

$$A_e=\left\{F^i(e,E_i):i\in N,E_i\in\mathcal{E}_i\right\}$$

Then choose $F(e)=a_0$ where a_0 satisfies $u_i(a_0,\omega)\leq u_i(a,\omega)$ for all $i\in N,\omega\in\Omega$ and $a\in A_e$. Such a choice is always possible by Condition FD. By construction F satisfies (1) and (2), establishing that f is p-incentive compatible. Since $p\in P$ and f $\mathcal{E}^*$-measurable were chosen arbitrarily, this completes the proof.

Since $\mathcal{E}^*$ is contractible information, it is clear that any coarser information (i.e. $\mathcal{G}\preceq\mathcal{E}^*$) is also contractible. I next prove the converse, namely that information is contractible *only if* it is a coarsening of $\mathcal{E}^*$. That is, no information finer than $\mathcal{E}^*$ is contractible. Moreover, $\mathcal{E}^*$ is the unique partition with this property. Thus $\mathcal{E}^*$ represents the *maximum contractible information.*

Lemma 4 *A partition $\mathcal{G}$ is contractible information only if $\mathcal{G} \preceq \mathcal{E}^*$.*

Proof. The proof is by contradiction. Suppose to the contrary that there exists some partition $\mathcal{G}$ that is contractible information and for which $\mathcal{G} \npreceq \mathcal{E}^*$. Now, $\mathcal{E}^*$ not finer than $\mathcal{G}$ is equivalent to[6]

$$\neg\left(\forall E \in \mathcal{E}^* \exists G \in \mathcal{G} \text{ s.t. } E \subset G\right)$$

or equivalently

$$\exists \hat{E} \in \mathcal{E}^* \text{ s.t. } \forall G \in \mathcal{G}, \neg\left(\hat{E} \subset G\right)$$

Take any such $\hat{E}$, along with $\omega_1, \omega_2 \in \hat{E}$ such that $\omega_1 \to \omega_2$ but $G(\omega_1) \neq G(\omega_2)$. (If no such pair of elements existed, we would have $G \in \mathcal{G}$ such that $\hat{E} \subset G$). So $\omega_1 \to_i \omega_2$ for some $i \in N$.

Define the function $f : \Omega \to A$ by

$$f(\omega) = \begin{cases} a_1 \text{ if } \omega \in G(\omega_1) \\ a_2 \text{ otherwise} \end{cases}$$

where $a_1, a_2 \in A$ are such that $u_i(a_1, \omega_1) < u_i(a_2, \omega_1)$. Such a choice is always possible by Condition SP. Since f is $\mathcal{G}$-measurable, by hypothesis it is p-incentive compatible for all $p \in P$.

By the revelation principle, there exists a direct mechanism F implementing f. So it must be the case that F is such that agent i prefers reporting $E_i(\omega_1)$ to $E_i(\omega_2)$ at $E_i(\omega_1)$:

$$\begin{aligned} \sum_{\omega \in E_i(\omega_1)} p(\omega) u_i(f(\omega), \omega) \geq & \sum_{\omega \in B} p(\omega) u_i\left(f\left(E_i(\omega_2) \cap E_{-i}(\omega)\right), \omega\right) \\ & + \sum_{\omega \in E_i(\omega_1) \backslash B} p(\omega) u_i\left(F\left(E_i(\omega_2), (E_j(\omega))_{j \neq i}\right), \omega\right) \end{aligned} \tag{4}$$

where $B = \{\omega \in E_i(\omega_1) : E_i(\omega_2) \cap E_{-i}(\omega) \neq \emptyset\}$. Since $\omega_1 \to_i \omega_2$, $E_i(\omega_2) \cap E_{-i}(\omega_1) = E_N(\omega_2)$ so that (4) can be rewritten

$$\begin{aligned} & p(\omega_1)\left(u_i\left(f(\omega_2), \omega_1\right) - u_i\left(f(\omega_1), \omega_1\right)\right) \leq \\ & \quad \sum_{\omega \in B \backslash \{\omega_1\}} p(\omega)\left(u_i\left(f(\omega), \omega\right) - u_i\left(f\left(E_i(\omega_2) \cap E_{-i}(\omega)\right), \omega\right)\right) \\ & \quad + \sum_{\omega \in E_i(\omega_1) \backslash B} p(\omega)\left(u_i\left(f(\omega), \omega\right) - u_i\left(F\left(E_i(\omega_2), E_{-i}(\omega)\right), \omega\right)\right) \end{aligned} \tag{5}$$

Since $\omega_1 \to_i \omega_2$, it follows that $i \in N_0(\omega)$ for all $\omega \in E_i(\omega_1) \backslash B$. So define ε_1 and ε_2 by

$$\varepsilon_1 = \max_{\omega \in B \backslash \{\omega_1\}} \{|u_i(a_1, \omega) - u_i(a_2, \omega)|\}$$

$$\varepsilon_2 = \max_{\omega \in E_i(\omega_1) \backslash B} \{u_i(a_1, \omega) - z_i(\omega), u_i(a_1, \omega) - z_i(\omega)\}$$

[6] Here, $\neg$ denotes the logical operator “not”.

The left hand side of (5) is strictly positive by construction, and is increasing in $p(\omega_1)$. On the other hand, the right hand side of (5) is bounded above by $(1 - p(\omega_1))\max(\varepsilon_1, \varepsilon_2)$. So for all $p \in P$ with $p(\omega_1)$ sufficiently large, inequality (5) fails to hold, in contradiction to the hypothesis that f is p-incentive compatible. This completes the proof.

Thus we have:

Proposition 1 *The partition $\mathcal{E}^*$ is the unique maximal contractible information partition.*

4 Examples

In the economy of Figure 1, we have $\omega_1 \rightharpoonup_1 \omega_2$ and also $\omega_1 \rightarrow_1 \omega_2$, and symmetrically $\omega_4 \rightarrow_1 \omega_3$. That is, agent 1 is always able to dishonestly claim to have the flu when he in fact has malaria, and to claim that he has a cold when in fact he is healthy. Also, $\omega_2 \rightharpoonup_1 \omega_3$ does not hold, since agent 2 would always recognize a dishonest claim by agent 1 to have a cold when in fact he has the flu. It follows that neither $\omega_2 \rightarrow_1 \omega_3$ or $\omega_3 \rightarrow_1 \omega_2$ holds.

To complete the analysis, we need to determine whether or not one of $\omega_2 \rightarrow_2 \omega_3$ or $\omega_3 \rightarrow_2 \omega_2$ holds. Certainly $\omega_2 \rightharpoonup_2 \omega_3$ and $\omega_3 \rightharpoonup_2 \omega_2$. Also, $\omega_2 \rightarrow_2 \omega_3$ only if $2 \in N_0(\omega_1)$ and $\omega_3 \rightarrow_2 \omega_2$ only if $2 \in N_0(\omega_4)$. That is, for agent 2 to be able to lie about whether she has observed any symptoms (i.e. whether she has observed $\{\omega_1, \omega_2\}$ or $\{\omega_3, \omega_4\}$) then it must be the case that she cannot be punished arbitrarily heavily when she lies and is found out. This happens if she denies agent 1 has symptoms when he in fact has malaria, of when she claims agent 1 has symptoms when in fact he is healthy.

Combining the above statements, we have $\mathcal{E}^* = \{\{\omega_1, \omega_2\}, \{\omega_3, \omega_4\}\}$ if agent 2 can be punished arbitrarily heavily in both of the states ω_1 and ω_4, and $\mathcal{E}^* = \{\Omega\}$ otherwise.

For the economy of Figure 2, it is straightforward to show that $\mathcal{E}^* = \{\{\omega_1, \omega_2\}, \{\omega_3, \omega_4\}\}$ independent of the level of punishments available. This is an instance of the familiar result that since agent 1 and 2 can both fully distinguish $\{\omega_1, \omega_2\}$ from $\{\omega_3, \omega_4\}$, this information is contractible.

For a slightly more complex example, consider an economy with state space $\Omega = \{\omega_1, \omega_2, \dots, \omega_M\}$, with $M \geq N$. Each agent $i \in N$ is a specialist in identifying state ω_i, in the sense that $\mathcal{E}_i = \{\{\omega_i\}, \{\omega_j : j \neq i\}\}$. That is, agent i knows when state ω_i occurs, but otherwise cannot distinguish any of the other states. In such an economy, is the realization of the state available for contracting?

For any agent i, the information collectively possessed by all other agents is

$$\mathcal{E}_{-i} = \{\{\omega_j\} : j \in N \backslash \{i\}\} \cup \{\{\omega_j : j \in (M \backslash N) \cup \{i\}\}\}$$

Now, $\omega \rightharpoonup_i \omega'$ if and only if $\{\omega, \omega'\} \subset \{\omega_j : j \in (M \backslash N) \cup \{i\}\}$ or $\omega = \omega'$. Moreover, if $M > N$ then $\omega_i \not\rightarrow_i \omega'$ for any $\omega' \in \{\omega_j : j \in (M \backslash N)\}$, since $\omega_i \rightharpoonup_i \omega'$ and $E_i(\omega') \cap E_{-i}(\omega_i) = \{\omega_j : j \in (M \backslash N)\} \neq \emptyset$. So $\mathcal{E}^* = \{\Omega\}$ if $M > N$, and $\mathcal{E}^* = \{\{\omega_i\} : i \in N\}$ if $M = N$.

That is, if there is a specialist for *every* state in this economy ($M = N$), then collectively all *other* agents also know the information, and the specialist cannot lie. So in such a case all information is available for contracting.

However, introducing even one state that no-one is able to recognize (i.e. $M = N + 1$) radically alters this conclusion. In this case, for any state ω_i the coalition of agents $N \setminus \{i\}$ is never able to distinguish between ω_i and ω_M, which enables agent i to deny knowledge of state ω_i. It then follows that *no* information is available for contracting. This conclusion holds independently of the size of punishments possible.

5 Common knowledge and limited punishments

Contrary to what might have been expected, even in the two-person case maximum contractible information does not coincide with common knowledge.

Consider again the example of Figure 1. There, if punishments are unlimited in all states for both agents (i.e. if $N_\infty(\omega) = N$ for all $\omega \in \Omega$), then $\mathcal{E}^* = \{\{\omega_1, \omega_2\}, \{\omega_3, \omega_4\}\}$ whereas $\mathcal{E}_1 \wedge \mathcal{E}_2 = \{\Omega\}$. That is, even though the agents have no non-trivial common knowledge, it is still possible to contract upon the employer's information about the worker's health, since the employer cannot misreport her information without a positive probability of detection.

On the other hand, if punishments are everywhere bounded, common knowledge and maximum contractible information do coincide in this example. This property is in fact general to two-agent economies, and can be generalized to the N-agent case as:

Proposition 2 *If $N_0(\omega) = N$ for all $\omega \in \Omega$, then $\mathcal{E}^* = \bigwedge_{i \in N} \mathcal{E}_{-i}$.*

Proof. The partition $\bigwedge_{i \in N} \mathcal{E}_{-i}$ is that induced by components of the graph $(\Omega, \leftrightarrow)$ where

$$\omega_1 \leftrightarrow \omega_2 \text{ if and only if } E_{-i}(\omega_1) = E_{-i}(\omega_2) \text{ for some } i \in N$$

If $\omega_1 \leftrightarrow \omega_2$ then $\omega_1 \rightharpoonup_i \omega_2$, and then $\omega_1 \rightarrow_i \omega_2$ since punishments are limited everywhere. Conversely, if $\omega_1 \rightarrow \omega_2$ then $\omega_1 \rightarrow_i \omega_2$ and thus $\omega_1 \rightharpoonup_i \omega_2$ for some agent i, so $\omega_2 \in E_{-i}(\omega_1)$ and hence $\omega_1 \leftrightarrow \omega_2$. Thus the graphs $(\Omega, \leftrightarrow)$ and $(\Omega, \rightarrow)$ have the same components, and so $\mathcal{E}^* = \bigwedge_{i \in N} \mathcal{E}_{-i}$, completing the proof.

The following corollary, which holds independently of the boundedness of punishments, is worth stating separately. The proof is simply the second half of the proof of Proposition 2.

Corollary 1 $\bigwedge_{i \in N} \mathcal{E}_{-i} \preccurlyeq \mathcal{E}^*$.

When applicable, this characterization greatly eases the application of Proposition 1. For instance, consider a three-agent economy in which agent 1 is a worker, agent 2 is a coworker and agent 3 is an employer. Suppose we are interested solely in agent 1's effort, which has two components – hours worked, which can either be long (L) or short (S), and intensity, which can either be (H) or low (0). So the state space is $\Omega = \{\omega_{LH}, \omega_{L0}, \omega_{SH}, \omega_{S0}\}$. Agent 1 knows exactly how hard he has worked, so $\mathcal{E}_1 = \{\{\omega_{LH}\}, \{\omega_{L0}\}, \{\omega_{SH}\}, \{\omega_{S0}\}\}$. The coworker, agent 2, observes only the intensity, so $\mathcal{E}_2 = \{\{\omega_{LH}, \omega_{SH}\}, \{\omega_{L0}, \omega_{S0}\}\}$, and the employer observes only hours worked, so $\mathcal{E}_3 = \{\{\omega_{LH}, \omega_{L0}\}, \{\omega_{SH}, \omega_{S0}\}\}$. What aspects of agent 1's effort are contractible here? The answer follows easily from Corollary 1. Trivially $\mathcal{E}_{-2} = \mathcal{E}_{-3} = \mathcal{E}_1$. Moreover, $\mathcal{E}_{-1} = \mathcal{E}_1$. Thus $\bigwedge_{i\in N} \mathcal{E}_{-i} = \mathcal{E}_1$ – so *all* information is contractible. The fact that agents 2 and 3 separately see different and incomplete aspects of agent 1's information does not prevent their joint information being useful.

6 Acquiring information

From the discussion so far, it should be clear that the availability or otherwise of unbounded punishments only affects the shape of the maximum contractible information in cases where deviations from truth-telling are sometimes but not always detected. Whenever the economy is such that deviations are either *always* or *never* detected, unbounded punishments play no role. In such cases, maximum contractible information should be again expected to coincide with the partition $\bigwedge_{i\in N} \mathcal{E}_{-i}$, and to reduce to common knowledge in the two-agent case. In this section I confirm this result, and then give an argument for why the case of always-or-never detected deviations can be expected to hold.

We start with a couple of definitions which formalize the property that deviations from truth-telling are either always or never detected. First,

Definition 3 *A pair of partitions $(\mathcal{G}_1, \mathcal{G}_2)$ is said to have the property of pairwise intersection if for any element of the common knowledge partition $\overline{G} \in \mathcal{G}_1 \wedge \mathcal{G}_2$ and any pair of partition elements $G_1 \in \mathcal{G}_1$ and $G_2 \in \mathcal{G}_2$ satisfying $G_1 \cup G_2 \subset \overline{G}$ it holds that $G_1 \cap G_2 \neq \emptyset$.*

Next,

Definition 4 *A set of partitions $\{\mathcal{G}_i\}_{i\in N}$ is said to satisfy global pairwise intersection (GPI) if $\mathcal{G}_i$ and $\mathcal{G}_{-i}$ satisfy pairwise intersection for all $i \in N$.*

When applied to information partitions, global pairwise intersection is exactly the property of single-agent deviations from truth-telling either being always or never detected when all others tell the truth. We then have

Proposition 3 *If $\{\mathcal{E}_i\}_{i\in N}$ satisfy GPI then the maximum contractible information $\mathcal{E}^*$ is equal to $\bigwedge_{i\in N} \mathcal{E}_{-i}$.*

Proof. Observe that under GPI, $\omega_1 \rightarrow_i \omega_2$ holds if and only if $\omega_1 \rightharpoonup_i \omega_2$. "Only if" is immediate from the definitions. For the converse, suppose $\omega_1 \rightharpoonup_i \omega_2$. So $E_i(\omega_2) \cap E_{-i}(\omega_1) \neq \emptyset$, and so there exists $\overline{E} \in \mathcal{E}_i \wedge \mathcal{E}_{-i}$ such that $E_i(\omega_2) \cup E_{-i}(\omega_1) \subset \overline{E}$. So $E_i(\omega_1) \subset \overline{E}$, and thus $E_{-i}(\omega) \subset \overline{E}$ for all $\omega \in E_i(\omega_1)$. By GPI, $E_i(\omega_2) \cap E_{-i}(\omega) \neq \emptyset$ for all $\omega \in E_i(\omega_1)$, and so $\omega_1 \rightarrow_i \omega_2$.

The result then follows exactly as in the proof of Proposition 2.

Up to this point, the information $\mathcal{E}_i$ of agents has simply been taken as given. Where does this information come from? Suppose that each agent i starts life with information given by $\mathcal{E}_i^0$. The set of information partitions $\{\mathcal{E}_i^0\}$ satisfy GPI if agent j, based on her own information $E_j^0 \in \mathcal{E}_j^0$, is unable to rule out any combination of other agents information $\{E_i^0\}_{i \neq j}$.

GPI would seem a reasonable enough property for these "primal" information partitions $\{\mathcal{E}_i^0\}$, but over time agents may acquire new information. Without any loss of generality, this process can be thought of as occurring in a number of sequential rounds, $r = 1, \ldots, R$. At each round agents may refine their information – regardless of the process by which this occurs, it is clear that the refinement must still be some coarsening of the total information possessed by all other agents. Thus if $\mathcal{E}_i^r$ denotes the information of agent i at round r, that agent's information at round $r+1$ must be of the form

$$\mathcal{E}_i^{r+1} = \mathcal{E}_i^r \vee \mathcal{G}_i^r$$

where $\mathcal{G}_i^r$ is some coarsening of the combined information of all other agents, $\mathcal{E}_{-i}^r$. Under this very general description of information acquisition, it can be shown that information after round $r+1$ satisfies GPI whenever the information at round r did.

To establish this result, it is convenient to start by noting that:

Lemma 5 *Let $\mathcal{E}_1, \mathcal{E}_2, \mathcal{F}_1, \mathcal{F}_2$ be partitions of Ω. Then $\mathcal{E}_1 \wedge \mathcal{E}_2 \preceq (\mathcal{E}_1 \vee \mathcal{F}_1) \wedge (\mathcal{E}_2 \vee \mathcal{F}_2)$.*

Proof. $\mathcal{E}_l \preceq (\mathcal{E}_l \vee \mathcal{F}_l)$ for $l = 1, 2$. Thus

$$\mathcal{E}_1 \wedge \mathcal{E}_2 \preceq (\mathcal{E}_1 \vee \mathcal{F}_1) \wedge \mathcal{E}_2 \preceq (\mathcal{E}_1 \vee \mathcal{F}_1) \wedge (\mathcal{E}_2 \vee \mathcal{F}_2)$$

The key result is then:

Lemma 6 *Suppose that $\mathcal{E}_1$ and $\mathcal{E}_2$ are partitions of a finite set Ω and satisfy pairwise intersection. Then if $\mathcal{F}_1$ and $\mathcal{F}_2$ are coarsenings of $\mathcal{E}_1$ and $\mathcal{E}_2$ respectively, the pair of partitions $(\mathcal{E}_1 \vee \mathcal{F}_2, \mathcal{E}_2 \vee \mathcal{F}_1)$ also satisfies pairwise intersection.*

Proof. Suppose to the contrary that the pair of partitions $(\mathcal{E}_1 \vee \mathcal{F}_2, \mathcal{E}_2 \vee \mathcal{F}_1)$ does not satisfy pairwise intersection. So there exist $E_1 \in \mathcal{E}_1, E_2 \in \mathcal{E}_2, F_1 \in \mathcal{F}_1, F_2 \in \mathcal{F}_2$ and $\bar{F} \in (\mathcal{E}_1 \vee \mathcal{F}_2) \wedge (\mathcal{E}_2 \vee \mathcal{F}_1)$ such that $E_1 \cap F_2 \subset \bar{F}$ and $E_2 \cap F_1 \subset \bar{F}$ but $(E_1 \cap F_2) \cap (E_2 \cap F_1) = \emptyset$.

By Lemma 5 $\exists \bar{E} \in \mathcal{E}_1 \wedge \mathcal{E}_2$ such that $\bar{F} \subset \bar{E}$. Then certainly $E_1, E_2 \subset \bar{E}$, and so $E_1 \cap E_2 \neq \emptyset$ by the pairwise intersection property of $\mathcal{E}_1$ and $\mathcal{E}_2$. Next, for $i = 1, 2$ write $F_i = \bigcup_{j=1}^{n_i} F_{ij}$ with $F_{ij} \in \mathcal{E}_i$ – this is always possible since $\mathcal{F}_i$ is a coarsening of $\mathcal{E}_i$ – and with $F_{1j} \cap E_2 \neq \emptyset$ and $F_{2j} \cap E_1 \neq \emptyset$. The elements F_{ij} are all contained in $\bar{E}$, and so by pairwise intersection again, $F_{1j} \cap F_2 = \bigcup_{k=1}^{n_2} (F_{1j} \cap F_{2k}) \neq \emptyset$, and thus $F_1 \cap F_2 = \bigcup_{j=1}^{n_1} (F_{1j} \cap F_2) \neq \emptyset$.

By hypothesis $(E_1 \cap E_2) \cap (F_1 \cap F_2) = \emptyset$. Define $\hat{F} = \bar{F} \cap (\bar{E} \setminus (F_1 \cap F_2))$. Now, $\hat{F} \neq \bar{F}$ since $F_1 \cap F_2 \neq \emptyset$ by above, and $\hat{F} \neq \emptyset$ since $E_1 \cap E_2$ is non-empty, has non-empty intersection with $\bar{F} \cap \bar{E}$, and by hypothesis is disjoint to $F_1 \cap F_2$.

To complete the proof, it is sufficient to establish that there exists a subset of $\hat{F}$ that is a member of $(\mathcal{E}_1 \vee \mathcal{F}_2) \wedge (\mathcal{E}_2 \vee \mathcal{F}_1)$, which since $\hat{F}$ is a *proper* subset of $\bar{F}$ gives the required contradiction. Take any element G of $\mathcal{E}_1 \vee \mathcal{F}_2$. Then $G \cap (F_1 \cap F_2) \in \{\emptyset, G\}$ since $\mathcal{F}_1 \vee \mathcal{F}_2 \preceq \mathcal{E}_1 \vee \mathcal{F}_2$ and $G \cap (\bar{F} \cap \bar{E}) \in \{\emptyset, G\}$ since $(\mathcal{E}_1 \vee \mathcal{F}_2) \wedge (\mathcal{E}_2 \vee \mathcal{F}_1) \preceq \mathcal{E}_1 \vee \mathcal{F}_2$. Thus $G \cap \hat{F} \in \{\emptyset, G\}$. An identical argument establishes that for any $G \in \mathcal{E}_2 \vee \mathcal{F}_1$, $G \cap \hat{F} \in \{\emptyset, G\}$. Thus there must be some subset of $\hat{F}$ that is a member of $(\mathcal{E}_1 \vee \mathcal{F}_2) \wedge (\mathcal{E}_2 \vee \mathcal{F}_1)$.

Applied to any agent i and coalition $N \backslash \{i\}$, Lemma 6 ensures that the pair of partitions $(\mathcal{E}_i^{r+1}, \mathcal{E}_{-i}^{r+1})$ satisfies pairwise intersection whenever $(\mathcal{E}_i^r, \mathcal{E}_{-i}^r)$ does. This gives:

Proposition 4 *The partitions $\{\mathcal{E}_i^{r+1}\}$ satisfy GPI provided $\{\mathcal{E}_i^r\}$ satisfies GPI. Thus information at round r satisfies GPI if the information at round 0 satisfies GPI.*

7 Unknown preferences

Up to this point, we have been concerned with characterizing what information an outside observer can conclude is always available for contracting, given that the observer knows all features of the economy *other than* the actual probabilities of different states (i.e. p). A closely related question is that of what information is known to be contractible by an observer who does not observe the *preferences* of agents,[7] $\{u_i\}$ (though it is assumed that he does know the preferences satisfy Conditions SP and FD).

Now, if an outside observer knows *neither* the preferences $\{u_i\}$ *nor* the probability mapping p, then the maximal information that he can conclude is available for contracting is again $\bigwedge_{i \in N} \mathcal{E}_{-i}$. This observation is immediate from Proposition 2 and Corollary 1: If preferences are not known, then it is possible that punishments are limited for all agents in all states, and so the maximal contractible information is $\bigwedge_{i \in N} \mathcal{E}_{-i}$ – which we know is weakly coarser than the contractible information under any other assumption about the boundedness of punishments.

It is worth noting in passing that any $\bigwedge_{i \in N} \mathcal{E}_{-i}$-measurable function f can be supported with a mechanism that is independent of the probability mapping p, and which depends on preferences only to the extent of making use of a punishment a_0 which is worse than any outcome in the range of f for all agents in all states (see the proof of Lemma 3).[8] Thus not only can an outside observer conclude that $\bigwedge_{i \in N} \mathcal{E}_{-i}$ is contractible information even if he cannot observe preferences or probabilities, but a planner charged with designing a mechanism to support an

[7] Arguably, preferences are easier for an outsider to observe than probabilities. Preferences are usually believed to be reasonably stable over time, allowing an observer to infer them from agents' past actions. On the other hand, if the world is non-stationary then historical data will be of little use in inferring the probabilities of future events.

[8] This is of course just the familiar forcing contract, which punishes agents for inconsistent reports by imposing an outcome that all agents dislike.

$\bigwedge_{i \in N} \mathcal{E}_{-i}$-measurable function f can do so without any knowledge of the probabilities, and with only very limited knowledge of the preferences of agents.

What information could an outside observer who did observe probabilities *but not preferences* conclude is contractible? Certainly $\bigwedge_{i \in N} \mathcal{E}_{-i}$ will be available for contracting. Whether or not any additional information is available for contracting depends on the class from which preferences are being drawn.

On the one hand, if *no* restrictions are placed on the class of possible preferences (beyond conditions SP and FD) then one can show that $\bigwedge_{i \in N} \mathcal{E}_{-i}$ is the finest information partition that an outsider could conclude is definitely available for contracting. The proof is similar to that of Lemma 4.

On the other hand, a restriction on the preference class as mild as the requirement that $u_i(a, \cdot)$ be $\mathcal{E}_i$-measurable for all $a \in A$ is enough to lead to probability mappings p for which strictly more information than $\bigwedge_{i \in N} \mathcal{E}_{-i}$ is available for contracting even when preferences are arbitrary. To see this, consider the following simple example. The state space is $\Omega = \{\omega_1, \omega_2, \omega_3, \omega_4\}$, there are two agents $N = \{1, 2\}$, and their information is given by $\mathcal{E}_1 = \{\{\omega_1, \omega_2\}, \{\omega_3, \omega_4\}\}$ and $\mathcal{E}_2 = \{\{\omega_2, \omega_3\}, \{\omega_4, \omega_1\}\}$. So

$$\bigwedge_{i \in N} \mathcal{E}_{-i} = \mathcal{E}_1 \wedge \mathcal{E}_2 = \{\Omega\}$$

However, for the probability mapping p defined by $p(\omega) = 1/4$ for all $\omega \in \Omega$, the information $\mathcal{F} \equiv \{\{\omega_1, \omega_3\}, \{\omega_2, \omega_4\}\}$ is available for contracting. To see this, let $f : \Omega \to A$ be any $\mathcal{F}$-measurable function. We will show that f is p-incentive compatible.

Without loss, $f(\omega_1) = f(\omega_3) = a$ and $f(\omega_2) = f(\omega_4) = b$ for some $a, b \in A$. Take the truth-telling mechanism $\mathcal{M}_i = \mathcal{E}_i$ for $i = 1, 2$ and $F(E_1, E_2) = f(E_1 \cap E_2)$. Then agent 1's truth-telling constraint upon observing $\{\omega_1, \omega_2\}$ is

$$\frac{1}{2}\left(u_1(a, \{\omega_1, \omega_2\}) + u_1(b, \{\omega_1, \omega_2\})\right) \geq \frac{1}{2}\left(u_1(b, \{\omega_1, \omega_2\}) + u_1(a, \{\omega_1, \omega_2\})\right)$$

which is satisfied for any choice of a and b and any specification of preferences. A similar argument applies to all the other truth-telling constraints. Thus f is p-incentive compatible as claimed.

To summarize, when an outside observer does not know either the probability mapping p or the preferences of agents $\{u_i\}$, then the most information that he can conclude is definitely available for contracting is $\bigwedge_{i \in N} \mathcal{E}_{-i}$. However, if the outside observer knows *either* probabilities *or* preferences, then there are circumstances under which he can conclude that strictly more information than $\bigwedge_{i \in N} \mathcal{E}_{-i}$ is available for contracting.

8 Cross reporting

Thus far it has been established that maximum contractible information is a well-defined concept, and that for two important classes (bounded punishments, and information satisfying GPI) of economies is equal to the partition $\bigwedge_{i\in N}\mathcal{E}_{-i}$. This finding may seem more intuitive under the following lattice theory result, the proof of which is in the appendix.

Lemma 7 $\bigwedge_{i\in N}\mathcal{E}_{-i} = \bigvee_{i\in N}(\mathcal{E}_i \wedge \mathcal{E}_{-i})$

Lemma 7 says that when punishments are bounded or information satisfies GPI, the maximum contractible information can be thought of as follows. Each agent i can misreport only that information that is not known by the coalition of all other agents. Thus agent i can only deviate from truth-telling when deviations respect the partition $\mathcal{E}_i \wedge \mathcal{E}_{-i}$. The maximum contractible information is then that obtained by combining these restrictions for all agents, resulting in the partition $\bigvee_{i\in N}(\mathcal{E}_i \wedge \mathcal{E}_{-i})$. By Lemma 7 and Propositions 2 and 3, this partition is precisely the maximum contractible information.

9 Relationship to literature

This is not the first paper to consider what information is easy to contract upon. In this section, I show how some previous results can be seen as special cases of the analysis conducted here.

9.1 Harris and Townsend (1981)

An early version of the result that any information shared by at least two agents is contractible can be found in Harris and Townsend (1981).[9] In their framework, the state space is of the form

$$\Omega = \times_{m\in M}\Omega_m$$

with typical element $\omega = (\omega_1, \ldots, \omega_M)$. Each agent i observes a subset M_i of the M random variables $\omega_1, \ldots \omega_M$. Thus the information of agent i is given by

$$\mathcal{E}_i = \{\{\omega : \omega_{M_i} = \theta_{M_i}\} : \theta_{M_i} \in \times_{m\in M_i}\Omega_m\}$$

It follows that if M_{-i} is defined by $M_{-i} = \cup_{j\neq i}M_j$, then the combined information of all agents other than i is given by

$$\mathcal{E}_{-i} = \left\{\{\omega : \omega_{M_{-i}} = \theta_{M_{-i}}\} : \theta_{M_{-i}} \in \times_{m\in M_{-i}}\Omega_m\right\}$$

The authors define the private information of agent i as those variables ω_m that are observed only by that agent. Denote these privately observed variables by M_i^P. Harris and Townsend's Theorem 2 establishes that any allocation satisfying the incentive constraints

$$\sum_{\tilde{\omega}:\tilde{\omega}_{M_i}=\omega_{M_i}} p(\tilde{\omega})u_i\left(f(\tilde{\omega}),\tilde{\omega}\right) \geq \sum_{\tilde{\omega}:\tilde{\omega}_{M_i}=\omega_{M_i}} p(\tilde{\omega})u_i\left(f(\delta_{M_i^P},\omega_{-M_i^P}),\tilde{\omega}\right)$$

[9] See Theorem 2 and also Section 10.

$\forall i \in N, \omega \in \Omega, \delta_{M_i^P} \in \times_{m \in M_i^P} \Omega_m$ is incentive compatible i.e. the only incentive constraints that matter are those related to individuals' private information. An immediate implication is the result that any variable ω_m observed by at least two agents is freely contractible, since there are then *no* incentive constraints to satisfy. Formally, the set of "contractible" random variables is defined by $M^* = M \backslash \bigcup_{i \in N} M_i^P$, and any allocation that is measurable with respect to

$$\{\{\omega : \omega_{M^*} = \theta_{M^*}\} : \theta_{M^*} \in \times_{m \in M^*} \Omega_m\}$$

is incentive compatible. In light of Lemma 3, this implication is equivalent to simply noting that under the stochastic restriction imposed by Harris and Townsend, global pairwise intersection holds and thus

$$\mathcal{E}^* = \bigwedge_{i \in N} \mathcal{E}_{-i} = \{\{\omega : \omega_{M^*} = \theta_{M^*}\} : \theta_{M^*} \in \times_{m \in M^*} \Omega_m\}$$

9.2 Postlewaite and Schmeidler (1986)

In an implementation theory context, Postlewaite and Schmeidler (1986) define *non-exclusivity of information (NEI)* as holding if $E_{-i}(\omega) = E_N(\omega)$ for all states ω and all agents $i \in N$. They establish that under NEI incentive constraints place no restrictions on implementability. In terms of this paper, their result can be expressed as:

$$\bigwedge_{i \in N} \mathcal{E}_{-i} = \mathcal{E}_N \text{ if } E_{-i}(\omega) = E_N(\omega)$$

In fact,

Lemma 8 *$\bigwedge_{i \in N} \mathcal{E}_{-i} = \mathcal{E}_N$ if and only if $E_{-i}(\omega) = E_N(\omega)$ for all $\omega \in \Omega$ and $i \in N$.*

Proof. Suppose first that $E_{-i}(\omega) = E_N(\omega)$ for all ω and $i \in N$. So $\mathcal{E}_{-i} = \mathcal{E}_N$, and the result follows immediately.

Next suppose that $\bigwedge_{i \in N} \mathcal{E}_{-i} = \mathcal{E}_N$. Suppose that contrary to the hypothesis, there exists $\omega \in \Omega$ and $j \in N$ such that $E_{-j}(\omega) \neq E_N(\omega)$. Then

$$\bigwedge_{i \in N} \mathcal{E}_{-i} \preceq \mathcal{E}_{-j} \prec \mathcal{E}_N$$

since $E_{-j}(\omega) \neq E_N(\omega)$. But this contradicts $\mathcal{E}^* = \mathcal{E}_N$, and so completes the proof.

From Propositions 2 and 3 it then follows that if either punishments are limited for all agents (i.e. $N_0(\omega) = N$) or if $\{\mathcal{E}_i\}_{i \in N}$ satisfy GPI, then Postlewaite and Schmeidler's NEI condition is both sufficient *and* necessary for incentive constraints to place no restrictions on implementability.

9.3 The private core of a differential information economy

Yannelis (1991) defines the *private core* of a differential information economy[10] as the class of resource-feasible allocations satisfying *private measurability* and the condition that no coalition of agents can redistribute its combined allocation in a privately measurable way while increasing the utility of all coalition members. Koutsougeras and Yannelis (1993) have shown that any allocation satisfying private measurability will satisfy coalitional incentive compatibility, and hence individual incentive compatibility. Thus any allocation in the private core is incentive compatible.

In the current paper, this property of the private core can be seen as follows. In the differential information economies considered in the above papers, the outcome set A is effectively a set of net-trade vectors, i.e. commodity transfers that sum to zero. For any mapping $f : \Omega \to A$, let f_j denote the net-trade made by agent j and f_{-j} the net-trade made by the coalition $N \setminus \{j\}$. The mapping f is said to be privately measurable if f_j is $\mathcal{E}_j$-measurable for every agent $j \in N$. Private measurability clearly implies that f_{-j} is $\mathcal{E}_{-j}$-measurable, and since $f_j + f_{-j} \equiv 0$ it also implies that f_j is $\mathcal{E}_{-j}$-measurable. Hence f_j is $\mathcal{E}_{-i}$ measurable for all $i \in N$, and so is $\bigwedge_{i \in N} \mathcal{E}_{-i}$-measurable. As this is true for any $j \in N$, the mapping f is also $\bigwedge_{i \in N} \mathcal{E}_{-i}$-measurable, and hence incentive compatible by Lemma 3 and Corollary 1.

Although every allocation of the private core is $\bigwedge_{i \in N} \mathcal{E}_{-i}$-measurable, the converse is clearly not true. For instance, the final example of Section 4 (with $N = M$) is an instance of a case in which $\bigwedge_{i \in N} \mathcal{E}_{-i} = \{\{\omega\} : \omega \in \Omega\}$ and so any mapping $f : \Omega \to A$ is $\bigwedge_{i \in N} \mathcal{E}_{-i}$-measurable, but in which $\mathcal{E}_j$ is strictly coarser than $\bigwedge_{i \in N} \mathcal{E}_{-i}$ for all agents j. Thus there exist $\bigwedge_{i \in N} \mathcal{E}_{-i}$-measurable mappings f that cannot possibly lie in the private core.

Appendix: Omitted proofs

Proof of Lemma 7. First note that the order relations $\preccurlyeq$ and $\succcurlyeq$ induce a lattice over the set of all possible partitions of Ω. This lattice is isomorphic to that induced by set inclusion over the corresponding set of σ-algebras. It is a standard result that any lattice of sets is distributive (and hence modular as well).[11]

The proof is by induction over N. The base case $N = 2$ is immediate.

[10] See also Allen and Yannelis (2001) for a review of work on differential information economies.

[11] See, e.g., Davey and Priestley (1990). If a lattice is distributive then

$$a \wedge (b \vee c) = (a \wedge b) \vee (a \wedge c)$$
$$a \vee (b \wedge c) = (a \vee b) \wedge (a \vee c)$$

for all lattice elements a, b, c. If a lattice is modular then whenever $a \geq c$

$$a \wedge (b \vee c) = (a \wedge b) \vee c$$

We are required to show that

$$\bigwedge_{i\in N}\bigvee_{j\neq i}\mathcal{E}_j=\bigvee_{i\in N}\left(\mathcal{E}_i\wedge\bigvee_{j\neq i}\mathcal{E}_j\right)$$

Now, take any $i_0 \in N$. Then

$$\begin{aligned}\bigwedge_{i\in N}\bigvee_{j\neq i}\mathcal{E}_j&=\bigvee_{j\neq i_0}\mathcal{E}_j\wedge\left(\bigwedge_{i\neq i_0}\bigvee_{j\neq i}\mathcal{E}_j\right)\\&=\bigvee_{j\neq i_0}\mathcal{E}_j\wedge\left(\mathcal{E}_{i_0}\vee\bigwedge_{i\neq i_0}\bigvee_{j\neq i,i_0}\mathcal{E}_j\right)\end{aligned}$$

by distributivity. Then by the inductive step

$$\bigwedge_{i\in N}\bigvee_{j\neq i}\mathcal{E}_j=\bigvee_{j\neq i_0}\mathcal{E}_j\wedge\left(\mathcal{E}_{i_0}\vee\bigvee_{i\neq i_0}\left(\mathcal{E}_i\wedge\bigvee_{j\neq i,i_0}\mathcal{E}_j\right)\right)$$

Next note that

$$\bigvee_{j\neq i_0}\mathcal{E}_j\succcurlyeq\bigvee_{i\neq i_0}\left(\mathcal{E}_i\wedge\bigvee_{j\neq i,i_0}\mathcal{E}_j\right)$$

so that by modularity

$$\begin{aligned}\bigwedge_{i\in N}\bigvee_{j\neq i}\mathcal{E}_j&=\left(\bigvee_{j\neq i_0}\mathcal{E}_j\wedge\mathcal{E}_{i_0}\right)\vee\bigvee_{i\neq i_0}\left(\mathcal{E}_i\wedge\bigvee_{j\neq i,i_0}\mathcal{E}_j\right)\\&=(\mathcal{E}_{i_0}\wedge\mathcal{E}_{-i_0})\vee\bigvee_{i\neq i_0}\left(\mathcal{E}_i\wedge\bigvee_{j\neq i,i_0}\mathcal{E}_j\right)\end{aligned}\tag{6}$$

Now

$$\mathcal{E}_i\wedge\bigvee_{j\neq i,i_0}\mathcal{E}_j\preccurlyeq\mathcal{E}_i\wedge\bigvee_{j\neq i}\mathcal{E}_j=\mathcal{E}_i\wedge\mathcal{E}_{-i}$$

Thus since (6) is true for arbitrary $i_0 \in N$ then

$$\begin{aligned}\bigwedge_{i\in N}\bigvee_{j\neq i}\mathcal{E}_j&=\bigvee_{i_0\in N}\left\{(\mathcal{E}_{i_0}\wedge\mathcal{E}_{-i_0})\vee\bigvee_{i\neq i_0}\left(\mathcal{E}_i\wedge\bigvee_{j\neq i,i_0}\mathcal{E}_j\right)\right\}\\&=\bigvee_{i\in N}(\mathcal{E}_i\wedge\mathcal{E}_{-i})\end{aligned}$$

□

References

[1] Allen, B., Yannelis, N.C.: Differential information economies: Introduction. Economic Theory **18**, 263–273 (2001)

[2] Arya, A., Glover, J., Young, R.: Virtual implementation in separable Bayesian environments using simple mechanisms. Games and Economic Behavior **9**, 127–138 (1995)

[3] Crémer, J., McLean, R.P.: Optimal selling strategies under uncertainty for a discriminating monopolist when demands are interdependent. Econometrica **53**, 345–362 (1985)

[4] Crémer, J., McLean, R.P.: Full extraction of the surplus in Bayesian and dominant strategy auctions. Econometrica **56**, 1247–1257 (1988)

[5] Davey, B.A., Priestley, H.A.: Introduction to lattices and order. Cambridge, UK: Cambridge University Press 1990

[6] Duggan, J.: Virtual Bayesian implementation. Econometrica **65**, 1175–1199 (1997)

[7] Harris, M., Townsend, R.: Resource allocation under asymmetric information. Econometrica **49**, 33–64 (1981)

[8] Koutsougeras, L.C., Yannelis, N.C.: Incentive compatibility and information superiority of the core of an economy with differential information. Economic Theory **3**, 195–216 (1993)

[9] McAfee, R.P., Reny, P.J.: Correlated information and mechanism design. Econometrica **60**, 395–421 (1992)

[10] Maskin, E.: Nash equilibrium and welfare optimality. Review of Economic Studies **66**, 23–38 (1999)

[11] Matsushima, H.: Bayesian monotonicity with side payments. Journal of Economic Theory **59**, 107–121 (1993)

[12] Mirlees, J.A.: The theory of moral hazard and unobservable behaviour: Part I. Review of Economic Studies **66**, 3–21 (1999)

[13] Moore, J., Repullo, R.: Subgame perfect implementation. Econometrica **56**, 1191–1220 (1988)

[14] Osborne, M.J., Rubinstein, A.: A course in game theory. Cambridge, MA: The MIT Press 1994

[15] Postlewaite, A., Schmeidler, D.: Implementation in differential information economies. Journal of Economic Theory **39**, 14–33 (1986)

[16] Yannelis, N.C.: The core of an economy with differential information. Economic Theory **1**, 183–198 (1991)

PART 4

CONTINUITY AND STABILITY

Core concepts in economies where information is almost complete★

Stefan Krasa and Wayne Shafer

Department of Economics, University of Illinois, 1206 South Sixth Street, Champaign, IL 61820, USA (e-mail: skrasa@uiuc.edu)

Received: March 15, 2000; revised version: August 24, 2000

Summary. The paper analyzes the properties of cores with differential information, as economies converge to complete information. Two core concepts are investigated: the private core, in which agents' net trades are measurable with respect to agents' private information, and the incentive compatible core, in which coalitions of agents are restricted to incentive compatible allocations.

Keywords and Phrases: Differential information, Core.

JEL Classification Numbers: C71, D51, D82.

1 Introduction

The main focus of this paper can be seen in two ways. First, is the complete information core a good predictor in environments with "almost complete" information? Second, are existing notions of a core with differential information close to the complete information core when informational asymmetries are small?

We consider two alternative core concepts with differential information that represent the main approaches in the literature. The first concept is that of the private core of Yannelis (1991). In his concept coalitions of agents are restricted to allocations that are measurable with respect to each agent's private information. The second concept that we consider is the incentive compatible core of Allen (1994), Ichiishi and Idzik (1996), and Vohra (1999). In this concept coalitions of agents are restricted to those allocations that are Bayesian incentive compatible. These two approaches differ substantially. However, because there is no clear benchmark against which the predictions of these two very different concepts can be compared

★ We wish to thank an anonymous referee for very helpful comments.

Correspondence to: S. Krasa

when there are significant asymmetries of information, one of the main motivations of this paper is to compare them when informational asymmetries are small and when the standard (complete information) core can be used as such a benchmark.

Consider a pure exchange economy with differential information in which each agent receives a possibly noisy signal about the true state. Thus, agents' information can be specified by means of a prior over the signals and the true states. There is complete information if the prior assigns probability 1 to each agent receiving the correct signal. In order to describe what it means to be "close to complete information," we use the priors to parameterize economies. Behavior close to complete information is analyzed by considering sequences of priors that converge to the complete information prior.

Our first Theorem provides a generic result on the convergence behavior of the private core. We show that the private core does not converge to the standard complete information core for all sequences of priors, for which information is asymmetric before the limit. More precisely, we prove that generically the set of limit points of private core allocations has empty intersection with the standard (complete information) core, as the noise in the agents' signals converges to zero. Thus, the complete information core cannot be seen as an approximation of private cores of economies with almost complete information. The intuition for this result is that the private core models the difficulty of information sharing by assuming that agents base trades only on their private information. Therefore even "small" informational asymmetries lead to very different outcomes when compared to the core with complete information.

Our second and third Theorem analyze the incentive compatible core. In contrast to the private core, the incentive compatible core need not exist in general (Allen, 1994; Vohra, 1999). However, in Theorem 2 we show that it does exist close to complete information. Moreover, Theorem 2 also shows that almost every standard core allocation is the limit point of incentive compatible core allocations. Does this imply that the incentive compatible core behaves more like the standard core close to complete information? It turns out that this is not the case. Theorem 3 shows that there is a robust class of economies for which the set of limit points of incentive compatible cores is strictly larger than the standard core.

As mentioned above, the two core notions analyzed in this paper are representative of two tracks of research. Specifically, in the literature on core concepts with differential information authors either impose restrictions on how information is shared by coalitions of agents (see Wilson, 1978; Yannelis, 1991; Allen, 1992; Berliant, 1992; Koutsougeras and Yannelis, 1993; Koutsougeras, 1998), or they impose incentive compatibility restrictions on the allocations a coalition of agents can obtain (Boyd, Prescott and Smith, 1988; Allen, 1994; Ichiishi and Idzik, 1996; Vohra, 1999; Ichiishi and Sertel, 1998).

In addition to the private core, the first group of papers also investigates other core concepts, most notably the coarse core and the fine core. In the coarse core, coalitions of agent are restricted to trades that are measurable with respect to common knowledge information. In contrast, in the fine core a coalition can use the pooled information of its members (c.f., Yannelis, 1991; or Koutsougeras and Yannelis, 1993). In this paper, we investigate the private core because it has been shown to have desirable properties (c.f., Koutsougeras and Yannelis, 1993). That is, the private core exists in general, it takes informational asymmetries into account, and it is incentive compatible.[1] Moreover, our main result for the private core implies that the same result holds for the coarse core.

In the second group of papers, where authors impose incentive compatibility restrictions, the concepts differ with respect to the participation constrained used, i.e., whether the blocking notion is ex-ante or interim. In this paper we use the ex-ante notion, because it avoids information leakage problems that arise when coalitions can block in the interim period (c.f., Krasa, 1999).

2 The model

Consider an exchange economy with n agents, indexed by $i \in I = \{1, \dots, n\}$. There is uncertainty over the state of nature $\omega \in \Omega$, where Ω is finite. Each agent i receives a possibly noisy signal $\phi_i \in \Phi_i$ about ω. For simplicity assume that $\Phi_i = \Omega$ for all agents i. Let $\Phi = \prod_{i=1}^{n} \Phi_i$. Any $\phi = (\phi_1, \dots, \phi_n) \in \Phi$ will be also denoted by (ϕ_{-i}, ϕ_i). Let π be a probability on $\Omega \times \Phi$ which is the common prior of all agents over states and signals. Let μ be the marginal probability on Ω.

Assume there are ℓ goods, and let $X_i = \mathbb{R}_+^\ell$ be the consumption space of agent i. Each agent i's preference ordering is given by a state dependent von Neumann-Morgenstern utility function $u_i \colon \Omega \times X_i \to \mathbb{R}$. Note that an agent's utility depends directly only on consumption and the true state ω. Consumption itself, however, will depend on the signals. A consumption bundle for agent i is therefore given by $x_i \colon \Omega \times \Phi \to \mathbb{R}_+^\ell$. An allocation x is a collection of consumption bundles x_i, $i \in I$ for all agents. Agent i's ex-ante expected utility is then given by

$$V_i(x_i) = \int_{\Omega \times \Phi} u_i\big(\omega, x_i(\omega, \phi)\big)\, d\pi(\omega, \phi).$$

Agent i's endowment is given by $e_i \colon \Omega \times \Phi \to \mathbb{R}_+^\ell$. We assume that the endowment e_i only depends on the true state ω. Thus, with a slight abuse of notation, we will often write $e_i(\omega)$ to denote agent i's endowment in state ω.

In a complete information economy, each agent i observes the true state ω. Thus, the signal ϕ is given by $\phi = \delta(\omega) = (\omega, \dots, \omega)$. Let $\Delta = \{(\omega, \delta(\omega)) | \omega \in \Omega\}$. Then $\pi(\Delta) = 1$ in a complete information economy, and only the consumption in $(\omega, \delta(\omega))$ matters. As a consequence, for complete information economies we will often denote agent i's consumption in $(\omega, \delta(\omega))$ by $x_i(\omega)$.

Finally, we describe our notion of convergence of allocations of the incomplete information economies to an allocation in a complete information economy.

For each $k \in \mathbb{N}$, let x_i^k, $i \in I$ be an allocation of the incomplete information economy with prior π_k such that $\lim_{k \to \infty} \pi_k(\Delta) = 1$. Then x_i^k, $i \in I$ converges to x_i, $i \in I$ if $\lim_{k \to \infty} x_i^k(\omega, \delta(\omega)) = x_i(\omega, \delta(\omega))$ for all $\omega \in \Omega$ and all agents i.

[1] Krasa and Yannelis (1994) show that if the grand coalition cannot block, then coalitional incentive compatible notions are fulfilled. These incentive notions are stronger than Bayesian incentive compatibility, and therefore imply Bayesian incentive compatibility.

3 The core concepts

3.1 Complete information economies

Consider an economy, in which the signals perfectly reveal the true state. Thus, there is uncertainty at $t = 0$ about the state ω, but ω becomes known to all agents at $t = 1$. As mentioned above, a consumption bundle can then be written as a function of ω alone, i.e., $x_i \colon \Omega \to \mathbb{R}^\ell_+$. Agent i's ex-ante expected utility is then given by $V_i(x_i) = \int_\Omega u_i(\omega, x_i(\omega))\, d\mu(\omega)$. Thus, we can use the standard definition of the core of an exchange economy.

Definition 1 *An allocation x is in the* **core** *of the complete information economy if and only if*

(i) $\sum_{i\in I} x_i = \sum_{i\in I} e_i$, *$\mu$-a.e. (feasibility);*
(ii) The following does not hold:
There exists a coalition $S \subset I$ and $y_i \colon \Omega \to \mathbb{R}^\ell_+$, $i \in S$ with
(ii.i) $\sum_{i\in S} y_i = \sum_{i\in S} e_i$, *$\mu$-a.e.;*
(ii.ii) $V_i(y_i) \geq V_i(x_i)$ *for all $i \in S$, where at least one inequality is strict.*

An allocation x is a **strict core** *allocation if x is a core allocation and if the same utilities cannot be obtained by any strict subcoalition, i.e., there do not exist $S \subsetneqq I$ and $y_i \colon \Omega \to \mathbb{R}^\ell_+$ with $\sum_{i\in S} y_i = \sum_{i\in S} e_i$, μ-a.e., and $V_i(y_i) = V_i(x_i)$ for all $i \in S$.*

For example, it is easy to show that if there exists a competitive equilibrium (x, p) with the property that $\sum_{i\in S} x_i \neq \sum_{i\in S} e_i$ for all coalitions $S \subsetneqq I$ then x is also a strict core allocation. Note that under the above assumption, p is no longer a competitive equilibrium price vector if the economy is decomposed into two parts. Thus, the existence of a strict core allocation (which we require in Theorem 2 below) can be viewed as an indecomposibility assumption on the economy.

3.2 Economies with differential information

If the signals are noisy, then agents are differentially informed. We provide two different core notions for differential information economies.

3.2.1 Definition of the private core In the private core of Yannelis (1991), each agent i is restricted to consumption bundles that are measurable with respect to his private information $\mathcal{F}_i$. We first provide the definition of the private core, and then describe in (1) how $\mathcal{F}_i$ is derived from the signal ϕ_i and the observed endowment realization $e_i(\omega)$. Also note that consumption bundles and endowments are now written as functions of ω and ϕ.

Definition 2 *An allocation x is in the* **private core** *of the differential information economy if and only if*

(i) $\sum_{i\in I} x_i = \sum_{i\in I} e_i$, *$\pi$-a.e. (feasibility);*

(ii) x_i *is* $\mathcal{F}_i$*-measurable for all agents* i.
(iii) *The following does not hold:*
There exist a coalition $S \subset I$ *and* $y_i : \Omega \times \Phi \to \mathbb{R}^{\ell}_{+}$, $i \in S$ *with*
(iii.i) y^i *is* $\mathcal{F}_i$*-measurable for all* $i \in S$;
(iii.ii) $\sum_{i \in S} y_i = \sum_{i \in S} e_i$, π*-a.e.;*
(iii.iii) $V_i(y_i) \geq V_i(x_i)$ *for all* $i \in S$, *where at least one inequality is strict.*

Note that the main difference between the private core and the complete information core is the measurability restriction imposed on the core allocation and on any blocking allocation. Yannelis (1991) provides a very general existence result for the private core.

It now remains to describe how each agent i's private information $\mathcal{F}_i$ is derived. Let $\sigma(e_i)$ be the information generated by e_i, which can be interpreted as a partition of $\Omega \times \Phi$.[2] In addition, agent i knows the signal ϕ_i. Thus,

$$\mathcal{F}_i = \sigma(e_i) \vee \left\{ \Omega \times \Phi_{-i} \times \{\phi_i\} \;\middle|\; \phi_i \in \Phi_i \right\}. \tag{1}$$

For example, consider the case where all agents' signals are accurate. We now show that any complete information core allocation x corresponds to a private core allocation $\hat{x}$. Define $\hat{x}_i(\omega, \phi_{-i}, \phi_i) = x_i(\phi_i)$, where (with a slight abuse of notation) $x_i(\phi_i)$ corresponds to agent i's consumption in the complete information core allocation if the state is $\omega = \phi_i$. Then each $\hat{x}_i$ is $\mathcal{F}_i$ measurable. Because the signal is accurate, $\sum_{i \in I} \hat{x}_i = \sum_{i \in I} e_i$, π-a.e., i.e., the allocation is feasible. Finally, $\hat{x}$ cannot be dominated by another $\mathcal{F}_i$-measurable allocation for any coalition S. Thus, $\hat{x}$ is in the private core.

3.2.2 Definition of the incentive compatible core We first provide the standard definition of incentive compatibility.

Definition 3 *A consumption bundle* x_i *is incentive compatible for agent* i *if and only if*

$$\int_{\Omega \times \Phi_{-i}} u_i(\omega, x_i(\omega, \phi))\, d\pi\left(\omega, \phi_{-i} \mid \phi_i, e_i\right)$$
$$\geq \int_{\Omega \times \Phi_{-i}} u_i(\omega, x_i(\omega, \phi_{-i}, \phi_i'))\, d\pi\left(\omega, \phi_{-i} \mid \phi_i, e_i\right),$$

for all $\phi_i, \phi_i' \in \Phi_i$.

We now provide the definition of the incentive compatible core. The main difference between this core and the core of a complete information economy is that the core allocation itself and any allocation used by a blocking coalition are required to be incentive compatible. The trades of members of coalition S must be measurable with respect to the pooled information of all of its members. Otherwise, the coalition could not execute the trades. Again, $\mathcal{F}_i$ is given by (1).

[2] Thus, $\sigma(e_i)$ is the information generated by the sets of the form $\{\omega | e_i(\omega) = \bar{e}_i\} \times \Phi$, where $\bar{e}_i \in \mathbb{R}^{\ell}_{+}$.

Definition 4 *An allocation x is in the* **incentive compatible core** *of the differential information economy if and only if*

(i) $\sum_{i\in I} x_i = \sum_{i\in I} e_i$, *$\pi$-a.e. (feasibility);*
(ii) x_i *is incentive compatible and* $\bigvee_{i\in I} \mathcal{F}_i$*-measurable for all agents* $i \in I$.
(iii) The following does not hold:
There exist a coalition $S \subset I$ *and* $y_i : \Omega \times \Phi \to \mathbb{R}^{\ell}_{+}$, $i \in S$ *with*
(iii.i) y^i *is incentive compatible and measurable with respect to* $\bigvee_{i\in S} \mathcal{F}_i$ *for all* $i \in S$*;*
(iii.ii) $\sum_{i\in S} y_i = \sum_{i\in S} e_i$, *$\pi$-a.e.;*
(iii.iii) $V_i(y_i) \geq V_i(x_i)$ *for all* $i \in S$*, where at least one inequality is strict.*

As indicated in Allen (1994) and Vohra (1999), the incentive compatible core does not exist in general. However, in Theorem 2 we show that it exists for economies that are sufficiently close to a complete information economy.

4 The convergence results

In this section we analyze whether or not the private core and the incentive compatible core are close to the complete information core if the economy is close to a complete information economy. Our economies could be parameterized by endowments, preferences, and priors. When we characterize properties of the core with "almost" complete information, we fix endowments and preferences. An economy is then close to complete information if $\pi(\Delta)$ is close to 1, i.e., if the probability that all agents receive the correct signal is close to 1. In all of our Theorems we therefore consider sequences of priors $\lim_{k\to\infty} \pi_k = \pi$, where $\pi(\Delta) = 1$, and investigate whether or not the limit points of sequences of incomplete information core allocations coincide with the complete information core. The Theorems investigate for what type of sequences π_k, $k \in \mathbb{N}$ convergence can be obtained.

4.1 Convergence of the private core

We now show that private core allocations of economies with almost complete information will in general differ substantially from complete information core allocations. In particular, Theorem 1 below shows that generically the set of limit points of private core allocations has an empty intersection with the complete information core. We first illustrate the main intuition of Theorem 1 by means of an example.

Example 1 Assume there are two agents $i = 1, 2$ and three states ω_i, $i = 1, 2, 3$, each of which occurs with positive probability. There is one good in each state. The agents' preferences are given by

$$u_1(\omega, x) = \begin{cases} \sqrt{x} & \text{if } \omega = \omega_1; \\ x & \text{otherwise;} \end{cases} \qquad u_2(\omega, x) = x, \text{ for all } \omega.$$

Each agent has a state independent endowment of 1 unit of the consumption good.

Assume that agent 1 can perfectly observe ω_1, but that his signals about states ω_2 and ω_3 are incorrect with probability $\varepsilon > 0$. In contrast, agent 2 correctly observes state 3, but his signals about states ω_1 and ω_2 are also incorrect with probability ε. Because the endowments are state independent, agent i's consumption in the private core can only depend on ϕ_i, and can therefore be denoted by $x_i(\phi_i)$. Feasibility requires that $x_1(\phi_1) + x_2(\phi_2) = e_1 + e_2$ for π_k a.e. $\phi = (\phi_1, \phi_2)$. Given the noise in the signals described above, all $\phi \in \Phi$ occur with positive probability. Thus, feasibility implies that x_1 and x_2 are independent of the signals. Hence, the private core consists only of the agents' endowments.

In contrast, it easy to see that in a complete information core allocation, trade will always occur. Agent 1 will give up a strictly positive quantity of the good in state ω_1 in exchange for an increased consumption in states ω_2, ω_3.

The result illustrated in Example 1 will hold for a generic economy, with genericity over agents' preferences. It is easy to see that we can only get a generic result. For example, consider an economy, in which the endowment is Pareto efficient in the complete information economy. Then no trade is also the only private core allocation, and both core notions will therefore coincide.

In order to provide a generic result, we parameterize each agent i's utility function in state $\omega \in \Omega$ by $\theta_i(\omega) \in \Theta_{i,\omega}$ where $\Theta_{i,\omega}$ is an open subset of $\mathbb{R}^\ell$. Agent i's utility is therefore given by $u^i(\omega, x, \theta_i(\omega))$. As we allow agent specific perturbations of utility functions in different states ω, the entire parameter space, Θ, has dimension $\ell|\Omega|n$. We say that a result holds for a generic set of economies, if there exists a set $\tilde{\Theta}$ which is closed in Θ and has Lebesgue measure 0, such that the result holds for all economies except possibly those in $\tilde{\Theta}$.

In the following, let $\theta \in \Theta_{i,\omega}$, and $x \in \mathbb{R}^\ell_{++}$. For the genericity argument, the following standard assumptions must be fulfilled.

Assumption A1

(1) Each $u_i(\omega, x, \theta)$ is smooth, has strictly positive first derivatives with respect to x, and has a negative definite matrix of second derivatives $D^2_{xx} u_i(\omega, x, \theta)$.
(2) $D^2_{x\theta} u_i(\omega, x, \theta)$ is non-singular.
(3) For all $\omega \in \Omega$, and for all sequence x^k, $k \in \mathbb{N}$, with $x^k \in \mathbb{R}^\ell_{++}$ and $\lim_{k\to\infty} x^k_l = 0$ for some good l, it follows that $\lim_{k\to\infty} \|D_x u_i(\omega, x^k, \theta)\| = \infty$.
(4) Each agent's endowment is strictly positive.

The main result of this section, Theorem 1, shows that the private core does not converge to the complete information core as long as for each agent i, one of the signals is noisy in a state that agent i does not learn about from the endowment realization.

Theorem 1 *Assume that the economy fulfills Assumption A1 and that there are at least two goods in each state. Then for a generic set of economies the following holds:*

Let $\pi_k \to \pi$ be an arbitrary sequence of priors that fulfills

1. $\pi(\Delta) = 1$;

2. *for every $k \in \mathbb{N}$ and for each agent i there exist states $\omega_i \neq \omega_i'$ with $e^i(\omega_i) = e^i(\omega_i')$ and $\pi_k(\phi_i = \omega_i' \mid \omega_i) > 0$.*

Let $\mathcal{E}_k$ be the economy with prior π_k. For each $k \in \mathbb{N}$, let x^k be a private core allocation of $\mathcal{E}_k$. Then none of the limit points of x^k, $k \in \mathbb{N}$ is a complete information core allocation.

Before proving the Theorem, we need Lemma 1 below. Lemma 1 shows that generically Pareto efficient allocations of the complete information economy will provide agents a different level of consumption in different states. Because there are n agents, we add more than n independent restrictions on Pareto efficient allocations if we require each agent i's consumption to be the same in two different states for all goods. Lemma 1 therefore follows from the fact that the Pareto set itself has only dimension $n-1$. The proof of Lemma 1 is in the Appendix.

Lemma 1 *For all agents i, let $\omega_i \neq \omega_i'$. Let $\mathcal{P}_\theta$ be the set of all Pareto efficient, (ex-ante) individually rational allocations with $x_i(\omega_i) = x_i(\omega_i')$, for all agents i. Then $\mathcal{P}_\theta = \emptyset$ for generic θ.*

We now prove Theorem 1.

Proof of Theorem 1. Let $\omega_i \neq \omega_i'$, $i \in I$ be arbitrary. Then Lemma 1 implies that there exists a generic set of economies such that no Pareto efficient allocation fulfills $x_i(\omega_i) = x_i(\omega_i')$. We next show that the limit of private core allocations must always fulfill such restrictions.

Let x^k be a private core allocation for the economy with prior π_k. Then because each agent i has a noisy signal, there exist $\phi_i' = \omega_i' \neq \omega_i$, such that $\pi_k(\phi_i'|\omega_i) > 0$. Then

$$x_i^k(\omega_i, \delta_{-i}(\omega_i), \phi_i') = x_i^k(\omega_i', \delta_{-i}(\omega_i'), \phi_i'), \tag{2}$$

because agent i's consumption must be measurable with respect to his information $\mathcal{F}_i$ (he can neither distinguish the states from observing his signal, nor from learning the endowment realization). Now note that in state ω_i, all agents other than agent i cannot determine whether agent i received signal ϕ_i or signal ϕ_i', because the signal is private information to agent i. Thus $x_j^k(\omega_i, \delta(\omega_i)) = x_j^k(\omega_i, \delta_{-i}(\omega_i), \phi_i')$ in all private core allocations for all agents $j \neq i$. Because the aggregate endowment only depends on ω and not on the signals, and because both ϕ_i and ϕ_i' can occur with positive probability in state ω_i, feasibility implies

$$x_i^k(\omega_i, \delta(\omega_i)) = x_i^k(\omega_i, \delta_{-i}(\omega_i), \phi_i'). \tag{3}$$

Thus, (2) and (3), and the fact that $\omega_i' = \phi_i'$ imply

$$x_i^k(\omega_i, \delta(\omega_i)) = x_i^k(\omega_i', \delta(\omega_i')). \tag{4}$$

Now consider a limit point $x(\omega)$ of the sequence $x^k(\omega, \delta(\omega))$, $k \in \mathbb{N}$ of private core allocations. We can assume without loss of generality that (4) holds for the same states ω_i, ω_i', for all elements in the subsequence of x^k, $k \in \mathbb{N}$ that converges to x. Therefore $x_i(\omega_i, \delta(\omega_i)) = x_i(\omega_i', \delta(\omega_i'))$. Thus, for a generic economy the limit points of private core allocations are not Pareto efficient in the complete information economy, and therefore not in the complete information core. This proves the Theorem. □

4.2 *Convergence of the incentive compatible core*

We now investigate the convergence of the incentive compatible core. First, we require that there are at least three agents. If there are only 2 agents, we cannot expect to get convergence as Example 2 indicates.

Example 2 Assume there are two agents $i = 1, 2$ and two states ω_1, ω_2. Each state occurs with probability $1/2$. There is one good in each state. Agents' preferences are given by

$$u_1(\omega, x) = \begin{cases} \sqrt{x} & \text{if } \omega = \omega_1; \\ x & \text{if } \omega = \omega_2; \end{cases} \qquad u_2(\omega, x) = \begin{cases} x & \text{if } \omega = \omega_1; \\ \sqrt{x} & \text{if } \omega = \omega_2. \end{cases}$$

Each agent has a state independent endowment of one unit of the consumption good.

First, consider the complete information economy, where each of the agents learns the state ω when it is realized. It is easy to see that any ex-ante individually rational and feasible allocation x must fulfill $x_1(\omega_1) \leq 1; x_1(\omega_2) \geq 1; x_2(\omega_1) \geq 1$; and $x_2(\omega_2) \leq 1$. Moreover, at least one of the inequalities must be strict for complete information core allocations. We now show that no such allocation is the limit of a sequence of incentive compatible core allocations.

Assume that each agent receives a noisy signal about ω. The signal is correct with probability $1-\varepsilon$, and incorrect with probability ε. Let x^ε be an incentive compatible core allocations for the economy with noise ε. Without loss of generality assume that $x^\varepsilon(\omega, \delta(\omega))$ converges. We denote the limit by $x(\omega)$.

Because agents' endowments are state independent, we can write their consumption as a function of the reported signals only. Thus, $x_i^\varepsilon(\omega, \omega')$ denotes agent i's consumption if agent 1 reports signal $\phi_1 = \omega$ and agent 2 reports signal $\phi_2 = \omega'$. Incentive compatibility of x^ε implies

$$E_{\Omega\times\Phi_2}\Big(u_1(\cdot, x_1^\varepsilon(\omega_1, \cdot)) \;\Big|\; \phi_1 = \omega_1\Big) \geq E_{\Omega\times\Phi_2}\Big(u_1(\cdot, x_1^\varepsilon(\omega_2, \cdot)) \;\Big|\; \phi_1 = \omega_1\Big) \quad (5)$$

$$E_{\Omega\times\Phi_1}\Big(u_2(\cdot, x_2^\varepsilon(\cdot, \omega_2)) \;\Big|\; \phi_2 = \omega_2\Big) \geq E_{\Omega\times\Phi_1}\Big(u_2(\cdot, x_2^\varepsilon(\cdot, \omega_1)) \;\Big|\; \phi_2 = \omega_2\Big) \quad (6)$$

where (5) is the incentive constraint for agent 1 if he observes $\phi_1 = \omega_1$, and (6) is the incentive constraint for agent 2 if he observes $\phi_2 = \omega_2$.[3]

If we take the limit on both sides of (5) and (6) for $\varepsilon \to 0$ we get $\sqrt{x_1(\omega_1)} \geq \limsup_{\varepsilon\to 0} \sqrt{x_1^\varepsilon(\omega_2, \omega_1)}$ and $\sqrt{x_2(\omega_2)} \geq \limsup_{\varepsilon\to 0} \sqrt{x_2^\varepsilon(\omega_2, \omega_1)}$, which implies

$$x_1(\omega_1) + x_2(\omega_2) \geq \limsup_{\varepsilon\to 0} x_1^\varepsilon(\omega_2, \omega_1) + x_2^\varepsilon(\omega_2, \omega_1). \quad (7)$$

Now assume by way of contradiction that x is a complete information core allocation. Then as noted above $x_1(\omega_1) \leq 1$ and $x_2(\omega_2) \leq 1$, where at least one inequality is strict. Thus, (7) implies $x_1^\varepsilon(\omega_2, \omega_1) + x_2^\varepsilon(\omega_2, \omega_1) < 2$ for sufficiently

[3] Note that the expectation operator itself depends on the prior over $\Omega \times \Phi$, and therefore depends on ε.

small ε, a contradiction to feasibility.[4] Therefore any limit of incentive compatible core allocations is not in the complete information core.

Example 2 demonstrates that if there are only two agents, incentive compatible core allocations do not necessarily converge to complete information core allocations. The reason for this result is that incentive compatibility would require that both agents are penalized by a low level of consumption when reports $\phi_1 = \omega_2$ and $\phi_2 = \omega_1$ are made. If there are more than two agents, penalties can be executed by transferring the consumption good to other agents. In fact, Theorem 2 below shows that with three or more agents convergence can be obtained. That is, we show that almost every complete information core allocation is the limit of incentive compatible core allocations. In particular, this result also implies existence of incentive compatible core allocations close to complete information (see the first statement in Theorem 2). This is a useful result, because as mentioned earlier, the incentive compatible core may be empty. In an interesting recent paper, McLean and Postlewaite (2000) provide alternative conditions under which such core allocations exist.

In Theorem 2 we use the following regularity assumptions.

Assumption A2

(1) Each $u_i(\omega, x)$ is smooth, has strictly positive first derivatives with respect to $x \in \mathbb{R}^\ell_{++}$, and has a negative definite matrix of second derivatives $D^2_{xx} u_i(\omega, x)$.
(2) For all $\omega \in \Omega$, and for all sequence x^k, $k \in \mathbb{N}$, with $x^k \in \mathbb{R}^\ell_{++}$ and $\lim_{k\to\infty} x^k_l = 0$ for some good l, it follows that $\lim_{k\to\infty} \|D_x u_i(\omega, x^k)\| = \infty$.
(3) Each agent's endowment is strictly positive.

Theorem 2 *Consider an economy where*

(i) $|I| \geq 3$, *(i.e., at least three agents);*
(ii) π *is a prior over* $\Omega \times \Phi$ *with* $\pi(\Delta) = 1$ *(i.e., signals are not noisy under* π*);*
(iii) assumption A2 holds;
(iv) there exists a strict core allocation x *in the complete information economy.*

Let C *be the set of core allocations of a complete information economy. Then there exists a closed set* N *of lower dimension than* C*, such that for all sequences of priors* π_k, $k \in \mathbb{N}$ *that converge to the complete information prior* π:

1. *Incentive compatible core allocations exist in the economy with prior* π_k *for all sufficiently large* k.
2. *Every core allocation* $x \in C \setminus N$ *is the limit of a sequence of incentive compatible core allocations of the incomplete information economies with priors* π_k.

The proof of Theorem 2 is in the Appendix. We now explain the intuition.

Lemma 3 below demonstrates that any allocation of the complete information economy is the limit of incentive compatible allocations. Thus, in order to prove Theorem 2, one must show that the approximating sequence can be chosen to be in the incentive compatible core.

[4] Note that throughout this paper we assume that there is not free disposal.

The existence of a strict core allocation ensures that the core has full dimension (i.e., dimension $n-1$, where n is the number of agents). Using Lemma 2 below, one can then show that all complete information core allocations except those in a set N of lower dimension than $n-1$ can be approximated by strict core allocations. Thus, it is sufficient to prove the result for all strict core allocations x.

The proof proceeds by way of contradiction. Assume we have a sequence of incentive compatible allocations x^k, $k \in \mathbb{N}$ that converges to a core allocation x of the complete information economy, but that x^k is not in the incentive compatible core for all k. One can show that each x^k can be selected such that the grand coalition cannot improve upon x^k by choosing another incentive compatible allocation. Thus, if x^k is not in the incentive compatible core, there must exist a coalition $S \subsetneqq I$, which can block it. Taking the limit as $k \to \infty$ implies that there exists a coalition $S \subsetneqq I$ which can obtain for its members the same utilities as in the strict core allocation x, a contradiction that proves the Theorem.

Finally, we state Lemma 2 and Lemma 3. The proofs are in the Appendix.

In the following let $U(S)$ be the set of attainable utilities of coalition S. Thus, $U(S) = \{w \in \mathbb{R}^n \mid \text{there exists } x \text{ with } \sum_{i \in S} x_i = \sum_{i \in S} e_i \text{ such that } w_i \leq V_i(x_i),$ for all $i \in S\}$. Let $\operatorname{bd} U(S)$ be the boundary of this set.

Lemma 2 *Assume that A2 holds. Then* $\operatorname{bd} U(S) \cap \operatorname{bd} U(T)$ *has dimension* $n-2$ *for all coalitions* $S \neq T$ *with* $\emptyset \subsetneqq S, T \subsetneqq I$.

Lemma 3 *Assume that:*

1. *There are at least three agents;*
2. x *is an allocation with* $\sum_{i \in I} x_i(\omega, \delta(\omega)) = \sum_{i \in I} e_i(\omega)$ *(feasibility if information is complete);*
3. $x_i(\omega, \delta(\omega)) \in \mathbb{R}^{\ell}_{++}$, *for all* $i \in I$, $\omega \in \Omega$;
4. π_k, $k \in \mathbb{N}$ *is an arbitrary sequence of priors with* $\pi_k \to \pi$, *and* $\pi(\Delta) = 1$.

Then there exists a sequence x^k, $k \in \mathbb{N}$, *with* $\lim_{k \to \infty} \int u_i(\cdot, x_i^k(\cdot))\, d\pi_k = V_i(x_i)$ *for all* $i \in I$, *where each* x^k *is a Bayesian incentive compatible allocation for the economy with prior* π_k.

Let $\tilde{C}$ be the set of limit points of all sequence of incentive compatible core allocations, for a given sequence of priors π_k, $k \in \mathbb{N}$. Let C denote the set of core allocations of the complete information economy. Theorem 2 shows that $\tilde{C}$ contains C, except possibly for a negligible set. Are there cases where $\tilde{C}$ is strictly larger than C? Theorem 3 below shows that this is the case. The intuition for this result is as follows.

In the incentive compatible core, blocking can be difficult for two agent coalitions. We have already pointed this difficulty in Example 2. Thus, in order to find economies where $\tilde{C}$ is strictly larger than C, it is sufficient to construct economies in which blocking by two agent coalitions matters. Apart from constructing an economy that has these required properties, Theorem 3 uses an argument similar to that of Theorem 2 to show that allocations that can only be blocked by a particular two agent coalition are limit points of incentive compatible core allocations.

Finally, we state Theorem 3. The proof is in the Appendix.

Theorem 3 *Let $|\Omega| \geq 4$ and $|I| \geq 3$. There exist economies that fulfill all conditions of Theorem 2, but for which the set of limit points of incentive compatible core allocations $\tilde{C}$ is strictly larger than the complete information core C. The set $\tilde{C} \setminus C$ is not negligible, and the economies are robust with respect to perturbations of endowments and preferences.*

5 Appendix

Proof of Lemma 1. In this proof, let $\Omega = \{\omega_1, \dots, \omega_{|\Omega|}\}$. For every agent i let $\omega_{k_i}, \omega_{k'_i}$ be the two states in which consumption should be the same. We now define agent i's expected utility by $V_i(x_i, \theta_i) = \sum_\omega \mu(\omega) u_i(\omega, x_i(\omega), \theta_i(\omega))$. Then let $\hat{\mathcal{P}}_\theta$ be the set of all $(x_1, \dots, x_n, p, \lambda_2, \dots, \lambda_n)$ which solve

(E1) $D_{x_1} V_1(x_1, \theta_1) - p = 0$;
(E2) $D_{x_i} V_i(x_i, \theta_i) - \lambda_i p = 0$, $i = 2, \dots, n$;
(E3) $e - \sum_{i=1}^n x_i = 0$;
(E4) $x_{i1}(\omega_{k_i}) - x_{i1}(\omega_{k'_i}) = 0$, for $i < n$; and $x_{n2}(\omega_{k_n}) - x_{n2}(\omega_{k'_n}) = 0$.

Clearly, $\hat{\mathcal{P}}_\theta$ is homeomorphic to the set of all Pareto efficient allocations for which (E4) holds, a set that contains $\mathcal{P}_\theta$. Therefore, it is sufficient to prove that $\hat{\mathcal{P}}_\theta = \emptyset$ for generic θ.

The matrix of derivatives of this system of equations is given by

$$E = \begin{pmatrix} \tilde{C} & \tilde{B} \\ \tilde{A} & 0 \end{pmatrix},$$

where $\tilde{C}$, the matrix of derivatives of (E1)–(E3) with respect to $x_1, \dots, x_n, p, \lambda_2, \dots, \lambda_n$, is given by

$$\begin{pmatrix}
D^2_{x_1x_1} V_1(x_1,\theta_1) & \cdots & 0 & \cdots & 0 & -I & \cdots & 0 & \cdots & 0 \\
\vdots & \ddots & \vdots & \ddots & \vdots & \vdots & \ddots & \vdots & \ddots & \vdots \\
0 & \cdots & D^2_{x_ix_i} V_i(x_i,\theta_i) & \cdots & 0 & -\lambda_i I & \cdots & -p & \cdots & 0 \\
\vdots & \ddots & \vdots & \ddots & \vdots & \vdots & \ddots & \vdots & \ddots & \vdots \\
0 & \cdots & 0 & \cdots & D^2_{x_nx_n} V_n(x_n,\theta_n) & -\lambda_n I & \cdots & 0 & \cdots & -p \\
-I & \cdots & -I & \cdots & -I & 0 & \cdots & 0 & \cdots & 0
\end{pmatrix},$$

and the matrix of derivatives of (E1)–(E3) with respect to $\theta_1, \dots, \theta_n$ is

$$\tilde{B} = \begin{pmatrix}
D^2_{x_1\theta_1} V_1(x_1, \theta_1) & \cdots & 0 & \cdots & 0 \\
\vdots & \ddots & \vdots & \ddots & \vdots \\
0 & \cdots & D^2_{x_i\theta_i} V_i(x_i, \theta_i) & \cdots & 0 \\
\vdots & \ddots & \vdots & \ddots & \vdots \\
0 & \cdots & 0 & \cdots & D^2_{x_n,\theta_n} V_n(x_n \theta_n) \\
0 & \cdots & 0 & \cdots & 0
\end{pmatrix}.$$

Finally, $\tilde{A} = (A_1, \cdots, A_n)$, where A_i is the derivative of (E4) with respect to x_i, λ, and p. The only non-zero entries in matrix A_i correspond to the derivatives with respect to $x_{i1}(\omega_{k_i})$ and $x_{i1}(\omega_{k'_i})$ if $i < n$, and $x_{n2}(\omega_{k_n})$ and $x_{n2}(\omega_{k'_n})$, otherwise.

We now show that matrix E has full rank. Thus, consider a linear combination of the rows of E which is equal to 0. The vector of scalars for the rows are denoted by w_1 for (E1), $w_2, \ldots, w_n$ for (E2); a for (E3); and b for (E4). Matrix $\tilde{B}$ implies that $w_i D^2_{x_i\theta} V_i(x_i) = 0$, $i = 1, \ldots, n$. Because $D^2_{x_i\theta} V_i(x_i)$ has full rank, it follows that $w_i = 0$ for $i = 1, \ldots, n$. Let $a_{1,\omega,k}$ be the scalar multiplier corresponding to the part of (E3) that ensures feasibility for good 1 in state ω_k. Because $w_i = 0$, the linear combination of the column elements of E corresponding to the derivative with respect to $x_{n1}(\omega_{k_i})$ yields $a_{1,\omega,k_i} = 0$. Similarly, it follows that $a_{1,\omega,k'_i} = 0$. Now let $i < n$ and consider the linear combination of column elements of E corresponding to the derivatives with respect to $x_{i1}(\omega_{k_i})$. Then since $w_i = 0$ and $a_{1,\omega,k_i} = 0$ we get $b_i = 0$. Similarly, we can show that $b_n = 0$. This immediately implies that $a = 0$. Hence all scalars are equal to 0 and E has therefore full rank. Because there are more equations than unknowns, the transversality theorem therefore implies that $\hat{\mathcal{P}}_\theta = \emptyset$ except for a set $\tilde{\Theta} \subset \mathbb{R}^\ell$ that has measure 0.

We now show that $\tilde{\Theta}$ is closed. Let θ_k, $k \in \mathbb{N}$ be a sequence in $\tilde{\Theta}$ with $\lim_{k\to\infty} \theta_k = \theta$. Let (x^k, p^k, λ^k) be a solution of (E1)–(E4) given θ^k. Then the associated matrix E will not have full rank, i.e. some of the rows of E will be collinear. Without loss of generality we can assume that the same rows are collinear for all $k \in \mathbb{N}$. Because feasible allocations are bounded we can assume without loss of generality that x^k converges to x as $k \to \infty$. Since all x^k are individually rational and because of assumption A1 it follows that each x_i is not on the boundary of agent i's consumption set. (E1) therefore implies that p^k converges to p, where $p > 0$. Thus, (E2) implies that λ^k also converges. Therefore, the rows of matrix E are collinear for (x, p, λ) given θ. Thus, $\theta \in \tilde{\Theta}$. □

Proof of Lemma 2. Let e_S and e_T be the aggregate endowments of coalitions S and T, respectively. First, note that $u \in \operatorname{bd} U(S) \cap \operatorname{bd} U(T)$ if and only if there exist allocations x, y with the following properties:

x and y are feasible for coalitions S and T, respectively; x cannot be improved upon by another allocation x' with $\sum_{i\in S} x'_i = e_S$, and similar for y; $u_i = V_i(x_i)$ for $i \in S$ and $u_i = V_i(y_i)$, for $i \in T$.

Without loss of generality we renumber the agents such that $S = \{1, \ldots, k+j\}$ and $T = \{k, \ldots, k+j, \ldots, m\}$, where $m \leq n$. Then x and y must fulfill the following equations:

(E1) $D_{x_i} V_i(x_i) - \lambda_i p = 0$, $i \in S$;
(E2) $D_{y_i} V_i(y_i) - \mu_i q = 0$, $i \in T$;
(E3) $V_i(x_i) - V_i(y_i) = 0$, $i \in S \cap T$;
(E4) $\sum_{i\in S}(x_i - e_i) = 0$;
(E5) $\sum_{i\in T}(y_i - e_i) = 0$;

where $\lambda_i, \mu_i > 0$, $\lambda_1 = \mu_k = 1$, and $p, q > 0$.

We now show that the matrix of derivatives (A B C) of (E1)–(E5) with respect to x, y, p, λ, q, has full rank. The matrix A of derivatives with respect to x is given by

$$
\begin{array}{r}
\\
\\
(E1)\\
\\
\\
\\
(E2)\\
\\
\\
(E3)\\
\\
(E4)\\
(E5)
\end{array}
\begin{array}{c}
\begin{array}{ccccc}
x_1 & \dots & x_k & \dots & x_{k+j}
\end{array}\\
\left(\begin{array}{ccccc}
D^2_{x_1x_1}V_1(x_1) & \dots & 0 & \dots & 0\\
\vdots & \ddots & \vdots & \ddots & \vdots\\
0 & \dots & D^2_{x_kx_k}V_k(x_k) & \dots & 0\\
\vdots & \ddots & \vdots & \ddots & \vdots\\
0 & \dots & 0 & \dots & D^2_{x_{k+j}x_{k+j}}V_{k+j}(x_{k+j})\\
0 & \dots & 0 & \dots & 0\\
\vdots & \ddots & \vdots & \ddots & \vdots\\
0 & \dots & 0 & \dots & 0\\
0 & \dots & D_{x_k}V_k(x_k) & \dots & 0\\
\vdots & \ddots & \vdots & \ddots & \vdots\\
0 & \dots & 0 & \dots & D_{x_{k+j}}V_{k+j}(x_{k+j})\\
I & \dots & I & \dots & I\\
0 & \dots & 0 & \dots & 0
\end{array}\right)
\end{array}
$$

The matrix B of derivatives with respect to y is

$$
\begin{array}{r}
\\
(E1)\\
\\
\\
\\
(E2)\\
\\
\\
\\
(E3)\\
\\
(E4)\\
(E5)
\end{array}
\begin{array}{c}
\begin{array}{ccccc}
y_k & \dots & y_{k+j} & \dots & y_m
\end{array}\\
\left(\begin{array}{ccccc}
0 & \dots & 0 & \dots & 0\\
\vdots & \ddots & \vdots & \ddots & \vdots\\
0 & \dots & 0 & \dots & 0\\
D^2_{y_ky_k}V_k(y_k) & \dots & 0 & \dots & 0\\
\vdots & \ddots & \vdots & \ddots & \vdots\\
0 & \dots & D^2_{y_{k+j}y_{k+j}}V_{k+j}(y_{k+j}) & \dots & 0\\
\vdots & \ddots & \vdots & \ddots & \vdots\\
0 & \dots & 0 & \dots & D^2_{y_my_m}V_m(y_m)\\
-D_{y_k}V_k(y_k) & \dots & 0 & \dots & 0\\
\vdots & \ddots & \vdots & \ddots & \vdots\\
0 & \dots & -D_{y_{k+j}}V_{k+j}(y_{k+j}) & \dots & 0\\
0 & \dots & 0 & \dots & 0\\
I & \dots & I & \dots & I
\end{array}\right)
\end{array}
$$

Finally, the matrix C of derivatives with respect to the remaining variables is

$$
\begin{array}{r}
\\
\\
\\
(E1)\\
\\
\\
\\
\\
\\
(E2)\\
\\
\\
\\
(E3)\\
\\
(E4)\\
(E5)
\end{array}
\begin{array}{c}
\begin{array}{cccccccc}
p & \dots & \lambda_k & \dots & \lambda_{k+j} & q & \dots & \mu_m
\end{array}\\
\left(\begin{array}{cccccccc}
-I & \dots & 0 & \dots & 0 & 0 & \dots & 0\\
\vdots & \ddots & \vdots & \ddots & \vdots & \vdots & \ddots & \vdots\\
-\lambda_k I & \dots & -p & \dots & 0 & 0 & \dots & 0\\
\vdots & \ddots & \vdots & \ddots & \vdots & \vdots & \ddots & \vdots\\
-\lambda_{k+j} I & \dots & 0 & \dots & -p & 0 & \dots & 0\\
0 & \dots & 0 & \dots & 0 & -I & \dots & 0\\
\vdots & \ddots & \vdots & \ddots & \vdots & \vdots & \ddots & \vdots\\
0 & \dots & 0 & \dots & 0 & -\mu_{k+j} I & \dots & 0\\
\vdots & \ddots & \vdots & \ddots & \vdots & \vdots & \ddots & \vdots\\
0 & \dots & 0 & \dots & 0 & -\mu_m & \dots & -q\\
0 & \dots & 0 & \dots & 0 & 0 & \dots & 0\\
\vdots & \ddots & \vdots & \ddots & \vdots & \vdots & \ddots & \vdots\\
0 & \dots & 0 & \dots & 0 & 0 & \dots & 0\\
0 & \dots & 0 & \dots & 0 & 0 & \dots & 0\\
0 & \dots & 0 & \dots & 0 & 0 & \dots & 0
\end{array}\right)
\end{array}
$$

We now show that the rows of the matrix $(A\,B\,C)$ are linearly independent. That is, consider an arbitrary linear combination of the rows which is equal to 0. Denote the vectors of scalar multipliers corresponding to (E1)–(E5) by w_i, z_i, α_i, a, and b, respectively. We must show that all multipliers are zero.

From the columns corresponding to the derivatives with respect to x_i we get

$$w_i D^2_{x_i x_i} V_i(x_i) + aI = 0, \text{ for } i < k; \tag{8}$$

$$w_i D^2_{x_i x_i} V_i(x_i) + \alpha_i \lambda_i p + aI = 0, \text{ for } i \geq k. \tag{9}$$

As $\lambda_1 = 1$, the column corresponding to the derivative with respect to p yield

$$\sum_{i=1}^{k+j} \lambda_i w_i = 0. \tag{10}$$

Finally, from the derivatives with respect to λ_i we get $pw_i = 0$ for $i \geq 2$. This, and (10) implies

$$pw_i = 0, \quad \text{for all } i \in I. \tag{11}$$

Now multiply from the right both sides of (8) and (9) by w_i, and use (11). This yields $w_i D^2_{x_i x_i} V_i(x_i) w_i + a w_i = 0$. Then

$$\sum_{i=1}^{k+j} \Big[\lambda_i w_i D^2_{x_i x_i} V_i(x_i) w_i + \lambda_i a w_i \Big] = 0. \tag{12}$$

Now (10) and (12) imply that $\sum_{i=1}^{k+j} \lambda_i w_i D^2_{x_i x_i} V_i(x_i) w_i = 0$. However, since $D^2_{x_i x_i} V_i(x_i)$ is negative definite it follows that $w_i D^2_{x_i x_i} V_i(x_i) w_i < 0$ if $w_i \neq 0$.

Thus, $w_i = 0$ for $i = 1, \ldots, k+j$. Now equation (8) immediately implies that $a = 0$. Equation (9) therefore implies

$$w_i = -\alpha_i \lambda_i p \big(D^2_{x_i x_i} V_i(x_i)\big)^{-1}. \tag{13}$$

Thus, we get $-\alpha_i \lambda_i p \big(D^2_{x_i x_i} V_i(x_i)\big)^{-1} p = 0$, when we multiply both sides of (13) from the right with p, and use (11). Note that $p\big(D^2_{x_i x_i} V_i(x_i)\big)^{-1} p < 0$ because $p \neq 0$. Thus, $\alpha_i \lambda_i = 0$. Because $\lambda_i \neq 0$, we therefore get $\alpha_i = 0$.

Similarly, we can prove that z_i and b are zero. Thus, the matrix of derivatives of (E1)–(E5) has full rank. Because there are $m - 2$ more equations than unknowns, the set of solutions is therefore a $m - 2$ dimensional manifold. Thus $\operatorname{bd} U(S) \cap \operatorname{bd} U(T) \cap \mathbb{R}^m$ has dimension $m - 2$. Consequently, $\operatorname{bd} U(S) \cap \operatorname{bd} U(T)$ has dimension $n - 2$. □

Proof of Lemma 3. First, note that we can assume without loss of generality that the information an agent receives from observing the endowment realization is also contained in the signal. Formally, let $\omega_i, \omega_i' \in \Omega$ with $e_i(\omega) \neq e_i(\omega')$. Then $\pi(\phi_i = \omega | \omega') = 0$. We can therefore assume that allocations in the incomplete information economy depend only on all signals $\phi = (\phi_1, \ldots, \phi_n)$ but not on ω.

Now let x be a feasible allocation of the complete information economy. We define an allocation $\hat{x}$ of the incomplete information economy as follows.

If $\phi = (\omega, \ldots, \omega)$ then $\hat{x}_i(\phi) = x_i(\omega)$. If $\phi = (\delta_{-i}(\omega), \omega')$, for $\omega' \neq \omega$ then $\hat{x}_i(\phi) = 0$ and $\hat{x}_j(\phi) = x_j(\omega) + (1/(I-1)) x_i(\omega)$. Finally, for all other signal profiles let $\hat{x}_i(\phi)$ be the agent's endowment $e_i(\omega)$.

We now show that $\hat{x}(\phi)$ is incentive compatible given prior π_k for all sufficiently large k. That is, we must show that

$$\int_{\Omega \times \Phi_{-i}} u_i\big(\omega, \hat{x}_i(\phi_{-i}, \phi_i)\big)\, d\pi_k(\omega, \phi_{-i} | \phi_i) \geq \int_{\Omega \times \Phi_{-i}} u_i\big(\omega, \hat{x}_i(\phi_{-i}, \phi_i')\big)\, d\pi_k(\omega, \phi_{-i} | \phi_i). \tag{14}$$

If $k \to \infty$ then $\pi_k\big(\omega, \phi_{-i} \mid \phi_i\big)$ converges to 1 if $\phi_{-i} = \delta_{-i}(\phi_i)$, $\phi_i = \omega$, and to 0 otherwise. Let $\omega_i = \phi_i$. Then the lefthand side of (14) converges to $u_i(\omega_i, \hat{x}(\delta(\phi_i))) = u_i(\omega_i, x_i(\omega_i))$. The righthand side of (14) converges to $u_i(\omega_i, \hat{x}(\delta_{-i}(\phi_i), \phi_i')) = u_i(\omega_i, 0)$. Because $x_i(\omega_i) > 0$ and u_i is strictly monotone, it follows that

$$\lim_{k\to\infty} \int_{\Omega \times \Phi_{-i}} u_i\big(\omega, \hat{x}_i(\phi_{-i}, \phi_i)\big)\, d\pi_k(\omega, \phi_{-i} | \phi_i) = u_i(\omega_i, x_i(\omega_i)) > u_i(\omega_i, 0) = \lim_{k\to\infty} \int_{\Omega \times \Phi_{-i}} u_i\big(\omega, \hat{x}_i(\phi_{-i}, \phi_i')\big)\, d\pi_k(\omega, \phi_{-i} | \phi_i).$$

Thus, for each $i \in I$ and $\phi_i \in I$ there exists $K_{i\phi_i} > 0$ such that (14) holds for all $k \geq K_{i\phi_i}$. Because the number of states and agents is finite, we can find $K > 0$ such that (14), and hence incentive compatibility holds for all $k \geq K$ and all ϕ_i.

Because $\lim_{k\to\infty} \pi_k = \pi$ and $\hat{x}(\delta(\omega)) = x(\omega)$ we get

$$\lim_{k\to\infty} \int u_i(\omega, \hat{x}_i(\omega,\phi))\, d\pi_k(\omega,\phi) = \int u_i(\omega, x_i(\omega,\phi))\, d\pi(\omega,\phi) = V_i(x_i).$$

Thus, we can define the sequence of allocations as follows: Let $x^k = \hat{x}$ for all $k \geq K$; and let x^k be an arbitrary incentive compatible allocation for $k < K$. This proves the Lemma. □

Proof of Theorem 2. We first show that the core has dimension $n-1$. More precisely, the core contains a set which is homeomorphic to an open subset of $\mathbb{R}^{n-1}$.

Because utility is strictly concave, the function $\psi(x) = (V_1(x_1), \ldots, V_n(x_n))$ is a homeomorphism between Pareto efficient allocations and $U(I)$. By assumption, there exists a strict core allocation x. Let $\bar{u} = \psi(x)$ be the corresponding vector of utilities. Then by the definition of a strict core allocation $\bar{u} \notin U(S)$ for all $S \subsetneqq I$.

Now recall that $U(I)$ has dimension $n-1$, i.e., it is homeomorphic to a set containing an nonempty open subset of $\mathbb{R}^{n-1}$. Since $\bar{u} \notin U(S)$ for all $S \neq I$ and since the sets $U(S)$ are closed it follows that there exists a neighborhood $W(\bar{u})$ of $\bar{u}$ in $U(I)$ with $W(\bar{u}) \cap U(S) = \emptyset$ for all $S \neq I$. Thus, $W(\bar{u})$ is homeomorphic to a subset of the set of core allocations. Because $W(\bar{u})$ has dimension $n-1$, the core has dimension $n-1$.

Now define $\hat{N} = \bigcup_{S\neq T; S,T\neq I} \operatorname{bd} U(S) \cap \operatorname{bd} U(T)$. By Lemma 2, $\hat{N}$ is a closed set of dimension at most $n-2$. Let $N = \psi^{-1}(\hat{N})$. Then the intersection of N with the set of core allocations has at most dimension $n-2$.

Let $\tilde{C}$ be the set of all core allocations y with $\psi(y) \notin U(S)$ for all $S \neq I$. Let x be a core allocation with $x \notin N$. We now show that x is in the closure of $\tilde{C}$.

Assume that $x \notin \tilde{C}$. Then $\bar{u} = \psi(x) \in U(S)$ for a coalition $S \neq I$. We now construct a sequence of allocations x^k, $k \in \mathbb{N}$ in $\tilde{C}$ that converges to x.

Let x^k, $k \in \mathbb{N}$ be a sequence of Pareto efficient allocations that fulfill $\lim_{k\to\infty} x^k = x$ and $V_i(x_i^k) > V_i(x_i)$ for all $i \in S$. Then $\bar{u}_k = \psi(x^k) \notin U(S)$. In order to show that $x^k \in \tilde{C}$ for sufficiently large k, it is remains to prove that $\bar{u}_k \notin U(T)$ for all $T \neq I$.

Assume by way of contradiction that $\bar{u}_k \in U(T)$ for a coalition $T \neq I$ for all large k. By construction $T \neq S$. Because $U(T)$ is closed, $\bar{u} \in U(T)$. Because x is a core allocation, it follows that $\bar{u} \in \operatorname{bd} U(T)$. Moreover, by assumption $\bar{u} \in U(S)$ and hence $\bar{u} \in \operatorname{bd} U(S)$. This, however, is a contradiction to $x \notin N$.

It now remains to prove that every core allocation $x \notin N$ is the limit of incentive compatible core allocations of incomplete information economies. Because the set of limit points of sequences is closed, it is sufficient to provide a proof for all $x \in \tilde{C}$.

For any consumption bundle y, let $V_i^k(y_i) = \int u_i(\omega, y_i(\omega,\phi))\, d\pi_k$. Let $x \in \tilde{C}$. By Lemma 3 there exists a sequence of Bayesian incentive compatible allocations x^k, $k \in \mathbb{N}$ for the incomplete information economies π_k, $k \in \mathbb{N}$ with $\lim_{k\to\infty} V_i^k(x_i^k) = V_i(x_i)$ for all agents $i \in I$. We show that one can assume x^k to be constrained Pareto efficient[5] for all k.

[5] That is, there does not exist another incentive compatible allocation that makes all agents weakly and at least some agents strictly better off.

If x^k is not constrained Pareto efficient, choose a constrained Pareto efficient allocation $\tilde{x}^k$ with $V_i^k(\tilde{x}_i^k) \geq V_i^k(x_i^k)$ for all $i \in I$. Then because of compactness, $\tilde{x}_i^k$, $k \in \mathbb{N}$ has a subsequence that converges. By slight abuse of notation we denote this subsequence again by $\tilde{x}_i^k$, $k \in \mathbb{N}$. Let $\tilde{x}$ be the limit. Clearly, $V_i(\tilde{x}_i) = \lim_{k\to\infty} V_i^k(\tilde{x}_i^k) \geq V_i(x_i)$. However, since x is Pareto efficient and because utility is strictly concave it therefore follows that $\tilde{x} = x$. Hence, $\tilde{x}^k$ converges to x.

It now remains to prove that x^k is in the core for sufficiently large k. We proceed by way of contradiction. Without loss of generality we can assume that there exists a coalition S that blocks x^k for all k. Thus, for every k there exist y^k with $V_i^k(y_i^k) \geq V_i^k(x_i^k)$ and $\sum_{i\in S} y_i^k = \sum_{i\in S} e_i$. By compactness we can assume without loss of generality that y^k converges to an allocation y.[6] Then $\sum_{i\in S} y_i = \sum_{i\in S} e_i$. Moreover, $V_i(y_i) = \lim_{k\to\infty} V_i^k(y_i^k) \geq \lim_{k\to\infty} V_i^k(x_i^k) = V_i(x_i)$. Thus, $\psi(x) \in U(S)$, a contradiction to the assumption that $x \in \tilde{C}$. □

Proof of Theorem 3. The proof proceeds as follows. First, we construct an economy in which blocking by the two agent coalition $\{2,3\}$ matters. We denote by E the set of allocations that are blocked only by $\{2,3\}$ but not by any other coalition. In the economy that we construct, E has the same dimension as the core. The economy has also strict core allocations. Then we show that these properties are robust if we perturb agents' utility functions. The perturbed economies have utility functions that fulfill assumption A2. Finally, we use an argument similar to that of Theorem 2 to show that the set of limits of incentive compatible core allocations contains E, and is therefore larger than the core.

To simplify notation in the proof, we will consider the core in the set of attainable utilities rather than in the set of allocations. In particular, if $U(S)$ denotes the set of attainable utilities of coalition S, then v is in the core if $v \in U(I)$ but not in the interior of any $U(S)$.

We start by constructing the example economy.

There are $n \geq 3$ agents. Assume there are four states, ω_i, $i = 1, \ldots, 4$. The argument immediately generalizes to any number of states greater than four. There is one consumption good in each state. Agents' utility functions are given by

$$u_1(\omega, x) = \begin{cases} \sqrt{x} \text{ if } \omega = \omega_1; \\ x \quad \text{otherwise;} \end{cases} \qquad u_2(\omega, x) = \begin{cases} \sqrt{x} \text{ if } \omega = \omega_2, \omega_3; \\ x \quad \text{otherwise;} \end{cases}$$

$$u_i(\omega, x) = \begin{cases} \sqrt{x} \text{ if } \omega = \omega_2; \\ x \quad \text{otherwise.} \end{cases} \quad \text{for } i \geq 3$$

Each agent's endowment in the four states is (a, a, a, b), where $b \geq (n+1)a$. Agents therefore know at $t = 1$ whether state 4 has occurred. However, their information about states ω_1, ω_2, and ω_3 is noisy. Note that state ω_4 is included to make utility functions quasilinear and to avoid boundary problems. Thus, for the case of complete information, the economy can be transformed into a game with transferable utility. In particular, the set of allocations for a coalition S where marginal rates

[6] More precisely, define $y_i^k = y_i$ for all $i \notin S$. The resulting sequence is bounded by the feasibility restriction for coalition S.

of substitution are equated have the property that agent i's consumption is $1/4$ in state ω if $u_i(\omega, x) = \sqrt{x}$ and there exists $j \in S$ with $u_j(\omega, x) = x$.

Now normalize each agent's utility function such that $E[u_i(e_i)] = 0$. Let $z = a - \sqrt{a} + 1/4$. Let m be the number of members of a coalition S. The payoff of any coalition S with at least two members is then given by

(i) $V(S) = (m+1)z$ if $1, 2 \subset S$;
(ii) $V(S) = mz$ if $1 \in S$ but $2 \notin S$;
(iii) $V(S) = z$ if $2 \in S$ but $1 \notin S$.

The payoff of any single agent coalition is 0. Let $v = (v_1, \ldots, v_n)$ be an allocation of agents' expected utilities. Let D be the set of all v that fulfill the following conditions.

$v_1 > (n-1)z$, $v_2 > z$, $v_j > 0$, for all j, and $\sum_{i=1}^{n} v_i = (n+1)z$. Then D is a subset of the core. In fact, D contains only strict core allocations, because $\sum_{i \in S} v_i > v(S)$ for all $v \in D$, and for all $S \subsetneqq I$. Moreover, because D has dimension $n-1$, the core has full dimension $n-1$.

Let E be the set of all v that fulfill the following conditions.

$v_1 = 4z - t_1$, $v_2 = t_2$, $v_3 = t_3$, $v_j = z - t_j$, for $j \geq 4$, $t_i > 0$ for all i, $t_1 < z$, $\sum_{j \neq 2,3} t_j < z$, and $t_2 + t_3 = \sum_{j \neq 2,3} t_j$.

Note that E has dimension $n-1$. Moreover, none of the allocations in E is in the core, because they can be blocked by coalition $S = \{2, 3\}$. In particular, $v(\{2,3\}) = z$. However, since $\sum_{j \neq 2,3}^{n} t_j < z$ it follows that $v_2 + v_3 < z$. Thus, $\{2, 3\}$ can block. Also note that $\{2, 3\}$ is the only coalition that can block. In fact, $\sum_{i \in S} v_i > v(S)$, for all coalitions $S \neq I, \{2, 3\}$. That is, all allocations of utilities in E are strict in the sense that the same utilities cannot be obtained for its member by a coalition $S \neq I, \{2, 3\}$.

Because agents' consumption is strictly greater than 0 in all states, we can modify agents utility function such that all agents' marginal utility at 0 is infinite in all states. We now perturb the utility functions.

Let $\varepsilon > 0$ be arbitrary. Consider the set $\mathcal{U}_\varepsilon$ of all utility functions for the n agents $\tilde{u}_i$, $i = 1, \ldots, n$ with $\left|u_i(\omega, x) - \tilde{u}_i(\omega, x)\right| < \varepsilon$ for all $\omega \in \Omega$, $i \in I$ and for all $0 \leq x \leq \sum_{i \in I} e_i$. Clearly, $\mathcal{U}_\varepsilon$ contains preferences which are strictly concave. Thus, the conditions of Theorem 2 are fulfilled for such preferences. Let $U(S)$ and $\tilde{U}(S)$ be the set of attainable utilities generated by u_i and by $\tilde{u}_i$, respectively. Then $U(S)$ and $\tilde{U}(S)$ will differ by less than ε. That is, let $v \in U(S)$ be arbitrary. Then there exists $\tilde{v} \in \tilde{U}(S)$ with $\|v - \tilde{v}\| < \varepsilon$. Similarly, for all $\tilde{v} \in \tilde{U}(S)$ there exists $v \in U(S)$ with $\|v - \tilde{v}\| < \varepsilon$.

Let $v \in D$. Then as shown above $v \notin U(S)$ for all $S \subsetneqq I$. Now choose $\varepsilon < (1/2)\,\mathrm{dist}(x, U(S))$. Let $W_\varepsilon(v)$ be an ε-neighborhood of v. Then $v' \notin \tilde{U}(S)$ for all $v' \in W_\varepsilon(v)$, where $\tilde{U}(S)$ is generated by utility functions $\tilde{u} \in \mathcal{U}_\varepsilon$. The core of the economy with utility functions $\tilde{u}_i$ therefore contains all $\tilde{v} \in \mathrm{bd}\,\tilde{U}(I)$ with $\tilde{v} \geq v'$ for some $v' \in W_\varepsilon(v)$. Thus, $\tilde{v} \notin \tilde{U}(S)$ for all $S \subsetneqq I$. Thus, there exist strict core allocations in the perturbed economy. As a consequence, the core has full dimension $n-1$.

Similarly, we can pick $v \in E$ and prove that there exists a neighborhood $W_\varepsilon(v)$ such that $\tilde{v} \in \mathrm{bd}\,\tilde{U}(I)$ and $\tilde{v} \notin U(S)$ for all $S \neq I, \{2, 3\}$.

Now pick utility functions in $\mathcal{U}_\varepsilon$ that fulfill the assumptions of Theorem 2. Let $v \in E$, and $\tilde{v}$ be a vector of utilities generated by a Pareto efficient allocation x with $\tilde{v} \geq v'$ for some $v' \in W_\varepsilon(v)$. Note that the set of all such allocations x has dimension $n-1$. Moreover, x is not in the core as it can be blocked by coalition $S = \{2,3\}$. It thus remains to show that x is nevertheless the limit of utilities of incentive compatible core allocations.

Let $\pi_k \to \pi$ such that agents 2 and 3 are not completely informed about states ω_1, ω_2 and ω_3, i.e., $\pi_k\big(\phi_k = \omega_i \mid \omega_j\big) > 0$, for $i,j = 1,2,3$ and $k = 2,3$. We now proceed as in the last part of the proof of Theorem 2.

Lemma 3 implies that there exists a sequence of Bayesian incentive compatible allocations x^k, $k \in \mathbb{N}$ for the economies with priors π_k, $k \in \mathbb{N}$ such that $\lim_{k\to\infty} \int u_i(\omega, x_i^k(\omega,\phi))\, d\pi_k = V_i(x_i)$. Again, one can assume that x^k is constrained Pareto efficient.

Now suppose there exists a coalition S that can block x^k. Than as in Theorem 2 one can conclude that S can block x. Thus, $S = \{2,3\}$. In order to prove that x^k is an incentive compatible core allocation, it therefore remains to prove that $\{2,3\}$ cannot block x^k for all sufficiently large k.

In order for agent 2 and 3 to improve through trade and receive an allocation close to x, agent 2 must make a transfer to agent 3 in state ω_3 which is strictly larger than the transfer in the other states. Thus, similar to Example 2, the trades are not incentive compatible. Agent 2 is better off reporting $s = \omega_1$ if $s = \omega_2$ or ω_3 has occurred. Similarly, agent 3 is better off reporting ω_3 when ω_2 has occurred. Thus, the resulting allocation is not incentive compatible and agents 2 and 3 can therefore not block. Hence x is a limit of incentive compatible core allocations. □

References

Allen, B.: Market games with asymmetric information: the private information core. Working Paper, University of Pennsylvania (1992)

Allen, B.: Incentives in market games with asymmetric information: approximate NTU cores in large economies. In: Barnett, W., Moulin, H., Salles, M., Schofield, N. (eds.) Social choice, welfare and ethics. Cambridge: Cambridge University Press 1994

Berliant, M.: On income taxation and the core. Journal of Economic Theory **56**, 121–141 (1992)

Boyd, J., Prescott, E., Smith, B.: Organizations in economic analysis. Canadian Journal of Economics **21**, 477–491 (1988)

Ichiishi, T., Idzik, A.: Bayesian cooperative choice of strategies. International Journal of Game Theory **25**, 455–473 (1996)

Ichiishi, T., Sertel, M.: Cooperative interim contract and re-contract: Chandler's M-form firm. Economic Theory **11**, 523–543 (1998)

Koutsougeras, L.: A two stage core with applications to asset market and differential information economies. Economic Theory **11**, 563–584 (1998)

Koutsougeras, L., Yannelis; N.C.: Incentive compatibility and information superiority of the core of an economy with differential information. Economic Theory **3**, 195–216 (1993)

Krasa, S.: Unimprovable allocations in economies with incomplete information. Journal of Economic Theory **87**, 144–168 (1999)

Krasa, S., Yannelis, N.: The value allocation of an economy with differential information. Econometrica **62**, 881–900 (1994)

McLean, R., Postlewaite, A.: Informational size, incentive compatibility and the core of a game with incomplete information. Working Paper (2000)

Vohra, R.: Incomplete information, incentive compatibility and the core. Journal of Economic Theory **86**, 123–147 (1999)
Wilson, R.: Information, efficiency, and the core of an economy. Econometrica **46**, 807–816 (1978)
Yannelis, N.: The core of an economy with differential information. Economic Theory **1**, 183–198 (1991)

On coalitional stability of anonymous interim mechanisms★

Gabrielle Demange and Roger Guesnerie

DELTA, 48 Boulevard Jourdan, 75014 Paris 14, FRANCE

Received: September 6, 1999; revised version: December 3, 1999

Summary. In a situation where agents have private information, we investigate the stability of mechanisms with respect to coalitional deviations. In the cooperative tradition, we first extend the notion of Core, taking into account the information a coalition may have when it forms and the conjectures of outsiders. This leads us to propose a family of Cores rather than a single one. Secondly, we study the stability of Core mechanisms to secession proposals in simple noncooperative games. The two different stability analyses, normative and strategic, tend to give support to the more natural extension of the Core, called Statistical Core, only in situations where some strong form of increasing returns to coalition is met. Without this property, arguments for a concept of Core that is non empty in a reasonably large class of problems are less compelling. Applications to taxation and insurance are given.

Keywords and Phrases: Core, Incentive mechanism.

JEL Classification Numbers: C71, D82.

1 Introduction

We consider the stability of arrangements made in a given Society (the Grand Coalition) when the arrangements may be challenged by subgroups of the Society (coalitions). However, because information is incomplete and asymmetric, social choice alternatives, for any coalition, are incentive mechanisms that simultaneously allow to extract information and to allocate resources. The set of such incentives devices that are available to the Society, or to its coalitions, depends upon the exact

★ The paper has benefited from the remarks of N. Yannelis and an anonymous referee.

Correspondence to: G. Demange

nature of the informational gaps that constrain mechanism design. In the cooperative tradition, we take a viewpoint associated with the concept of Core. But the context leads to propose a family of plausible Cores rather than a single Core. All make related assumptions on the information that is available to coalitions but they reflect different conjectures of outsiders to the coalition. First, attention is focused on the relationship between these concepts. Second, we study the connections between the cooperative viewpoint, which the different concepts that are introduced reflect, and the assessment of secession threats in simple non cooperative games.

A number of recent studies have been concerned with asymmetric information in cooperative games. It is useful to stress, at the outset, two main possible sources of differences with ours.

First, we consider problems in which both the mechanism *design* and the analysis of *objections* by coalitions are made at the *interim stage*, i.e. when agents have learnt the part of the information that is privately held. This feature can be contrasted with the options taken in previous studies. Wilson (1978) in his pioneering work, or more recently Yannelis (1991), Koutsougeras-Yannelis (1993) are concerned with *ex ante* arrangements (in an exchange economy) that are challenged, as we do here, at the *interim* stage. They do not use however the mechanism design approach to tackle the revelation problem that appears in case of asymmetric information. Instead they consider various degrees of pooling of information among the members of a coalition when it forms, each one leading to a different Core, such as the coarse, fine or private Core. Another line of research (Allen, 1992; Vohra, 1997; Forges-Heifetz-Minelli, 1999) studies the *ex ante* coalitional stability of such mechanisms (which are implemented at the *interim* stage).

Second, attention is focused on a framework that is special from the viewpoint of mechanism design theory. We assume that the distribution of the characteristics on which there is incomplete information is known at the level of the whole Society. In line with this assumption, we further assume that the incentives mechanisms used to extract information are anonymous, in the sense that what an agent receives only depends on his announcement and on the distribution of announced characteristics.[1]

These two first options make clear that the present work incorporates preoccupations that are similar to those of the literature on the coalitional stability of taxation rules or on insurance under adverse selection. Although these two pieces of literature concentrate on formal models that have substantial differences, in both insurance and taxation models agents have private information, respectively on "ability" and "risk". Mechanism design as well as stability threats have to be envisaged at the *interim* stage. Earlier studies that take a related viewpoint include Berliant (1992) who deals with taxation problems, Kahn-Mookherjee (1995) who are mainly concerned with insurance issues, and Hammond (1989) who, however, adopts alternative views of the blocking process.[2]

[1] The justifications and properties of such anonymous mechanisms have been the focus of a subset of the incentives literature that includes Hammond (1979), Guesnerie (1981), Mas Colell-Vives (1993), Guesnerie (1995) (Chap.1).

[2] The paper by Boyd-Prescott (1986) and more recently the work by Vohra (1998) in an exchange economy also take an interim viewpoint.

The informational hypothesis underlying the blocking procedure that we are considering in the present paper deserves special comments at the outset. We assume, throughout the paper, that the information that may be public knowledge within a coalition is comparable to the information available to the Grand Coalition: in other words, the coalitions have information on the distribution of their characteristics. Such an assumption is not absent from the theoretical literature,[3] although it is debatable on a priori grounds. Furthermore, as we shall argue later, the assumption is overwhelmingly adopted within the vivid and rapidly growing literature on the coalitional stability of tax rules in Nations or in Federations. Partly, our paper may be viewed as an attempt of assessing the validity of the just evoked most popular informational assumption from basic premises.

As suggested at the beginning of this introduction, we first define a family of Cores that are immune to blocking threats by coalitions that would be able to discover the distribution of the characteristics of their members. With the Statistical Core, the distribution is assumed to be known, whereas in the whole family of Beliefs-Based Cores such a knowledge is compatible with guesses of the coalitions' members. Such guesses are triggered by considerations of attractiveness of blocking offers, that derive themselves from parameterized beliefs of the outsiders to the coalition. The Status quo Core, one of our Beliefs Based Cores, corresponds to the conservative beliefs, à la Rothschild-Stiglitz, that outsiders can keep their standing utility levels. We show how, in general, all these Cores can be ordered by inclusion. We then introduce notions (effectivity of monotonic blocking) that reflect the existence of different forms of "increasing returns to size", or "increasing returns to coalitions". Under the corresponding assumptions, the general inclusion order previously alluded to is simplified: in particular, the Statistical Core is identical to the Status quo Core.

The final part of the paper (Section 4) comes back on the problems just evoked - the evaluation of the adequacy of the different Core concepts - but now with a non cooperative perspective. Our concern, in contrast with some recent literature, is not the "implementation" of the various Cores - i.e. exhibiting games whose equilibria lead to the Core[4] - but rather a robustness analysis: assuming that a Core outcome has been implemented, we consider games that are intended to test the stability of the outcome to outside offers. Our setting and terminology are reminiscent of the theory of local public goods where the stability of arrangements is challenged within the framework of "developer" games.[5] The conclusions of this section reinforce the conclusions of the previous "cooperative" analysis: when there are increasing returns to size that imply the identity of the Statistical and the Status quo Cores, then the corresponding Core mechanisms are the only ones that are not "destabilized" – in a sense that refers to robust equilibria associated with iterative elimination of dominated strategies – in the developer game.

---

[3] In particular, in the just two quoted articles, the assumption is explicitly made in Berliant (1992), or is an implicit consequence of the approach taken in Kahn-Mokherjee (1995).

[4] There is a significant literature on the so-called implementation problem (see e.g. Perry-Reny, 1994), that mainly focuses attention on the complete information case.

[5] Such a viewpoint also evokes the durability concept of Holmstrom-Myerson (1983).

The plan is as follows. Feasible incentives mechanisms are defined for the Grand Coalition and the coalitions in Section 2. Examples on taxation and insurance illustrate the framework. Several notions of Core are introduced, and their relationships are examined in Section 3. Section 4 defines developer games, studies their equilibria and their relationships with the Core outcomes. Conclusions are drawn in Section 5, and some proofs are gathered in Section 6.

2 The model

We consider a society A, also called the Grand Coalition. Its population consists of a continuum of infinitesimal agents but definitions are always chosen so that, in case, they remain congruent with those that would be taken if the number of agents were finite. The population is endowed with a probability structure (A, λ). Each individual is characterized by his type θ, his preferences being represented by a utility function $u(., \theta)$. He will be called a θ-agent. There is a finite number of possible types, also called characteristics, $T = \{\theta_1, ..., \theta_p\}$. The mapping $\tilde{\theta} : a \in A \to \tilde{\theta}(a) \in T$ describes the agents' profile of characteristics.

As is well known, the decision or allocation rules that can be used by the society crucially depend on the nature of public information and on the repartition of private information. We assume that:

– *each agent knows his own type, and this privately held information is neither publicly verifiable nor observable,*

– *the individual characteristics are independent draws from a probability distribution $\overline{\mu}$ on T,*
– *the distribution $\overline{\mu}$ is public knowledge.*

In the usual terminology, we are at an *interim stage* where the state of the world, i.e the exact profile of the individual types, is unknown but agents have private information on their own type. Each agent only observes the others' names and these names are not correlated with types. By the law of large numbers, the distribution is exactly $\overline{\mu}$: the measure of agents of type θ_j is equal to $\overline{\mu}(j)$ for any $j = 1, ..., p$.[6]

A *distribution of characteristics* is a positive vector μ, the j-th coordinate being interpreted as the measure of type j. We shall have to consider the set of all possible distributions:

$$M = \left\{ \mu / \sum_j \mu(j) \leq 1, \mu(j) \geq 0,\ j = 1...p \right\}.$$

[6] For recent discussions on the law of large numbers when the set A is a continuum, see Al-Najjar (1995) and Khan-Sun (1999).

2.1 Feasible and incentive compatible mechanisms

In this set up, the society has to decide upon some allocation mechanism which both allocates resources and extracts information, i.e. which must satisfy feasibility and incentive compatibility constraints.

A *mechanism* specifies the outcome for all members of the society as a function of the announcements. Feasibility is assumed to depend on the distribution of characteristics only, as it often occurs in economic examples: $X(\mu)$ will denote the set of feasible outcomes when the distribution of characteristics is μ. Concerning incentives, we require mechanisms to be incentive compatible in *dominant strategy* so that, by the revelation principle, we may restrict attention to direct and truthful mechanisms.

Finally we consider mechanisms that are *anonymous* in two ways: first an agent's allocation does not depend upon his name " a", and second it is influenced only by the distribution of the characteristics announced by the others and not by their precise profile. To avoid tedious repetition when we consider coalitions, the mechanisms are assumed to be defined for all distributions on T of size smaller than 1, i.e. for any μ in M.

Definition 1 *A (direct anonymous) mechanism f is a function defined over $T \times M$ which assigns to any (θ, μ) an element $f(\theta, \mu)$, interpreted as the outcome for a θ-agent belonging to a group whose distribution is μ.*

The Grand Coalition. A priori, the set of possible announcements profiles for the Grand Coalition is:

$$M(1) = \left\{\mu \in M / \sum_j \mu(j) = 1\right\}$$

Definition 2 *A mechanism is* **incentive compatible and feasible (ICF)** *if*

$$u(f(\theta, \mu), \theta) \geq u(f(\theta', \mu), \theta), \forall \theta, \theta' \in T, \forall \mu \in M(1) \tag{1}$$

and

$$(f(., \mu)) \in X(\mu), \ \forall \mu \in M(1) \tag{2}$$

Condition (1) says that truth telling is a dominant strategy. It uses the fact that, each agent being infinitesimal, his announcement has no effect on the distribution of announced characteristics. Condition (2) requires more comments. In general, whether an allocation rule $(f(., \mu))$ is feasible depends not only upon the distribution of announced characteristics, but also on the true distribution. Hence, except in some special cases such as the taxation game we shall introduce later on, condition (2) has to be interpreted as involving feasibility at equilibrium, i.e. when agents tell the truth, rather than out of equilibrium.[7]

[7] Note that feasibility could have been required only for the true distribution $\overline{\mu}$ since it is exactly known. However, in the finite population case, the exact distribution is unknown since the profile is a $N-$sample of $\overline{\mu}$. Then, the present formulation introduces no discontinuity between large but finite populations and infinite ones. Note also that if the distribution of a finite population were known, then incentive compatibility constraints would not bind, since high punishment mechanisms would implement first best allocations. However this result is not robust to some uncertainty on the exact distribution, whatever small, that occurs in an N-sample where a single lie cannot be detected for sure.

Coalitions. In our framework, a coalition, as the whole society, takes decisions on the basis of mechanisms that should be feasible and incentive compatible with its members' information. A **coalition** C is a subset of A. It is characterized by the distribution μ^C of its members' types, where $\mu^C(j)$ gives the measure of agents with type θ_j in C. Its size s, equal to $s = \sum_j \mu^C(j)$ is observable. Relevant coalitions are of positive size. The set of types who have non zero measure in the coalition is called its support. Its feasibility set is denoted $X(\mu^C)$. The set of distributions that are now considered is:

$$M(s) = \left\{ \mu \in M / \sum_j \mu(j) = s \right\}.$$

Definition 3 *A mechanism* f *is* **incentive compatible and feasible for a coalition of size** s *if*

$$u(f(\theta, \mu), \theta) \geq u(f(\theta', \mu), \theta), \forall \theta, \theta' \in T, \forall \mu \in M(s) \tag{3}$$

and

$$(f(., \mu)) \in X(\mu), \ \forall \mu \in M(s). \tag{4}$$

Definition 3 when applied to the Grand Coalition coincides with Definition 2. If, in addition, the coalition knows more than its size, for example if it knows its distribution as we consider further on, we stick to the same definition in line with the argument that we developed in footnote 7 for the Grand Coalition.[8] Also previous comments on the meaning of feasibility still apply. Note that, as is standard in cooperative game theory, we view objections as "minmax threats". Accordingly, what is feasible for a coalition is what it would get on its own without the help of agents outside the coalition.

Finally it will be convenient to consider mechanisms that are incentive compatible and feasible for all coalitions.

Definition 4 *A mechanism is* **universally ICF** *or, for short,* **universal**, *if it is incentive compatible and feasible for all coalitions of positive size.*

We shall illustrate this general framework with two examples, a public good model à la Diamond-Mirrlees (1971), and the insurance model with adverse selection of Rothschild–Stiglitz (1976).

[8] Recall that, in an infinite agents framework, adding a finite number of agents does not affect the distribution μ, hence the outcome. Therefore, the utility level $u(f(\theta, \mu), \theta)$ is well defined even for an agent with a type θ outside the support of μ and incentive compatibility is required for any type, even if the support is known. One may object that if the coalition knows its distribution it knows a fortiori its support S, and could use such an information. Accordingly, the last part of the incentive compatibility condition (3) $\forall \mu \in M(s)$ could be modified into (3)': $\forall \mu \in M(s)$ with support S. However, a mechanism that satisfies (3)' is always the restriction of a mechanism that satisfies (3): here, the incentive compatibility constraints for different μ can be analyzed separately. The domain on which the mechanism is designed introduces significant differences only if restrictions on the outcome, when the distribution varies, are introduced (such as the continuity restrictions in Mas-Colell-Vives, 1993).

2.2 Examples

1. A pure public good model There are two private goods, one consumption good and labor, and a public good. Both consumption and public goods are produced from labor through technologies with constant returns to scale. Preferences over private and public goods are separable, and represented for a θ-agent by:

$$u(x, l, y, \theta) = v(x, l, \theta) + w(y)$$

where x, l, y are respectively the amounts of consumption good, labor and public good. Standard regularity, monotonicity, and concavity assumptions are made.

Some anonymous feasible incentive compatible mechanisms can be associated with tax systems. More precisely, take labor as the numeraire and consider a specific tax t on the consumption good together with a uniform lump sum tax (possibly negative) r. Given such taxes (t, r), let us define an equilibrium.[9] Due to the separability of preferences, one may proceed as follows. The competitive production price for the private good is equal to its constant average cost c, and the consumption price to $c + t$. The consumer's demand at that price, $x(t, r, \theta), l(t, r, \theta)$, solves:

$$\text{maximize } v(x, l, \theta) \text{ over } (x, l), \text{ such that } (c + t)x \leq l - r.$$

If all collected taxes are used to produce $y(t, r, \mu)$ units of public good, of unit cost 1, then, with straightforward notation:

$$y(t, r, \mu) = r + tE_{\mu}x(t, r, \theta).$$

Consumption good market clearing obtains through an appropriate choice of the production scale. From that, together with Government Budget balance, the labor market also clears.

Consider now the mechanism that assigns the equilibrium allocation associated with the distribution of announced preferences μ: produce $Y_{tr}(\theta, \mu) = y(t, r, \mu)$ that has just been computed, (and that is independent of θ), and assign to an agent who has announced θ: $X_{tr}(\theta, \mu) = x(t, r, \theta)$, and $L_{tr}(\theta, \mu) = l(t, r, \theta)$ where $x()$ and $l()$ are as just defined (and are actually independent of μ). The mechanism (Y_{tr}, X_{tr}, L_{tr}) associated with tax system (t, r) is incentive compatible and feasible, and even universal in the sense of definition 4.[10]

Conversely, if labor revenue is not observable and trade on consumption good is costless and non verifiable, nothing in our setting is lost in welfare terms when restricting attention to mechanisms associated with tax systems. The just sketched identification of tax systems with (anonymous) incentives mechanisms can be extended to models à la Diamond-Mirrlees (1972).[11]

The reader should note an important property of the mechanisms just described, when taxes and lump sum contributions are positive: if new agents enter a coalition,

[9] Some conditions on (t, r) have to be imposed, for example r should not be too high.

[10] Actually, it is feasible, even "out of equilibrium" (cf the comment on definition 2).

[11] These assertions are substantiated, for example, in Guesnerie (1995), where a key insight is Theorem 4 in chapter 1, and where bibliographical references, starting from Hammond (1979) and Guesnerie (1981), are given.

whatever their types, the incentives system is such that the incumbents, the initial members of the coalition, are necessarily better off. This property, labelled later full-monotonicity, reflects the existence of sufficiently strong increasing returns to size in the public good model; such property cannot be expected, as we shall see now, in an insurance context.

2. Insurance models with adverse selection Individuals are alike except for their probability θ of having an accident, which costs d. Their income is w. A mechanism assigns an income $w_1(\theta, .)$ in case of no accident and $w_2(\theta, .)$ in case of accident yielding a utility level of:

$$(1-\theta)u(w_1(\theta, .)) + \theta u(w_2(\theta, .)).$$

Assuming the law of large numbers, feasibility requires if μ is the actual distribution of characteristics:[12]

$$E_\mu[(1-\theta)w_1(\theta, .) + \theta w_2(\theta, .)] \leq w - dE_\mu[\theta].$$

The mechanism f of the general model identifies with the vector-function $(w_1(.,.), w_2(.,.))$. Besides feasibility, incentive compatibility, referred to as self-selection in the insurance literature, is defined as above.

3 Cores

3.1 Information of coalitions

Now the methodology of our investigation is the following. We are at the interim stage: private information has been revealed to the agents. *What are the objections that can be made, at the interim stage, at some mechanism f proposed by the Grand Coalition ?* In our framework, a coalition, as the whole society, has to use mechanisms which should be incentive compatible (see Definition 3). A coalition blocks (*interim*) if it is common knowledge for its members that there exists an ICF mechanism which makes all of them better off. Of course blocking opportunities crucially depend on any specific information a coalition may acquire at the *interim* stage. We shall consider two cases: in the first one, interim information can be seen as exogenous, in the second case, endogenous.

– *statistical information*: some coalitions will learn the distribution of their members' characteristics but it is not known which coalition(s) will be informed at the interim stage. In such circumstances, a mechanism can be viewed as "stable" if it is common knowledge that no informed coalition will block. Another possible justification for considering statistical information is *normative*. Even if coalitions are unable to strategically block, one still may be concerned by what they would achieve if they had access to the same type of information as the Grand Coalition.

12 The next equation makes clear that, as asserted precedently, feasibility depends upon the actual and the announced distribution of characteristics. Then, contrary to the previous case, the mechanism is not necessarily feasible if agents lie.

– *beliefs-based information* : coalition members infer information from the attractiveness, for the different types, of seceding proposals. In other words, self-selection arguments provide information to potentially blocking coalitions.

In our context, the latter approach may a priori look more sensible than the former one. That statistical information matters either for blocking or for evaluating the power of coalitions, is however explicitly accepted by a large body of the existing literature which fits the present framework. It is the case with the contributions concerned with the coalitional stability of taxation rules, when considering either the Core or the Shapley value (see for example respectively the survey by Greenberg, 1994; Aumann-Kurz, 1977). Similarly, in the more recent literature concerned with fiscal federalism, almost all contributions assume that coalitions are sufficiently informed so that they can predict the level of tax receipts. The fact that this standard approach is *a priori* debatable is one of the starting point of the present paper. For a critical assessment, we undertake here the comparison of the stability of mechanisms under alternative informational assumptions.

3.2 Blocking and cores

Statistical blocking, as defined below, is the standard blocking condition for a coalition who knows its distribution.

Definition 5 *Let f be an ICF mechanism for the Grand Coalition. A coalition C, with distribution μ^C, size s and support S, statistically blocks if there is a mechanism g that is feasible and incentive compatible for it such that:*

$$u(g(\theta,\mu^C),\theta) > u(f(\theta,\overline{\mu}),\theta), \forall\theta\in S. \tag{5}$$

f is then said to be statistically blocked by coalition C with the mechanism g.

In words any member of C, who knows the distribution μ^C, correctly expects a higher utility level from secession with the blocking mechanism than from the standing mechanism.

"Beliefs-based blocking", that will be now defined, relies on the idea that a seceding coalition is able to guess its distribution from selection arguments: the proposed mechanism is not only preferred by the agents inside the coalition to the standing mechanism, as above, but, also, is constrained not to be chosen by the agents outside the coalition. The selection crucially depends on the utility levels that the latter expect in not joining a successful secession, and on the fact that these levels are common knowledge.

As a simple benchmark, we shall first assume that an agent expects some exogenous reservation utility level which only depends on his type. Denote by $u^* = (u^*(\theta_1), ..., u^*(\theta_p))$ such a vector of reservation levels. In such a context, agents of a given type cannot be distinguished and, hence, face the same dilemna: namely either to join or not to join. Therefore the coalitions that may block are necessarily *type-full* in the following sense:

Given a subset S of types, $S \subset T$, the *S-full coalition* is the coalition composed with all agents whose type belongs to S, the distribution of which is denoted $\overline{\mu}_S$ (note that $\overline{\mu}_A = \overline{\mu}$).

Definition 6 *An incentive compatible and feasible mechanism f for the Grand Coalition is u^*-**beliefs blocked** if for some S, $S \subset T$, the S-full coalition statistically blocks with an ICF mechanism g i.e.:*

$$u(g(\theta, \overline{\mu}_S), \theta) > u(f(\theta, \overline{\mu}), \theta) \text{ for all } \theta \text{ in } S \tag{6}$$

and moreover

$$u(g(\theta, \overline{\mu}_S), \theta) \leq u^*(\theta) \text{ for all } \theta \text{ not in } S. \tag{7}$$

Equation (6) expresses the blocking argument for the S-full coalition. (7) states the self-selection argument introduced above: no single agent with type outside S wants to join the coalition, although he could without modifying the coalition distribution, given his exogenous reservation utility level. This allows, assuming that an agent blocks only if he has a strict incentive to do so, the coalition to be sure of its composition and to "learn" $\overline{\mu}_S$.

Beliefs-based blocking is more difficult than statistical blocking: not only agents who block benefit from secession, but also other agents should not be interested in joining. The lower the exogenous utility reference levels, the more difficult the latter condition and the larger the corresponding Core (to be defined) is.

The reservation utility levels can be made endogenous. There are a number of different ways to do so, which may depend upon the context. Here we focus on the *status quo reservation levels* which are the utility levels that are obtained at the standing mechanism when applied to the whole society:[13]

$$u_f^*(\theta) = u(f(\theta, \overline{\mu}), \theta)$$

We now define in the usual way the Cores that correspond to the different blocking concepts.

Definition 7 *The* **Statistical Core** *is the set of ICF mechanisms that are not statistically blocked. The u^*-**Beliefs-Based Core** (u^*-BB Core) is the set of ICF mechanisms that are not u^*-Beliefs-Based blocked. Status quo blocking is defined accordingly, and the* **Status quo Core**[14] *is the set of ICF mechanisms that are not status quo blocked.*

As usual, an ICF mechanism is said to be (second best) *Pareto optimal* if there is no other ICF mechanism under which every type is strictly better off. Second best comparisons that take into account incentive compatibility constraints are the meaningful ones. Obviously an ICF mechanism is Pareto optimal if and only if it is

[13] In an insurance context, the view adopted here is in line with the Rothschild-Stiglitz approach as explained below. Other endogenous reservation levels are relevant, and might reflect stories that would be reminiscent of Riley (1978), or Wilson (1977). For example, if the standing mechanism is universal, the reservation payoffs could be derived from the standing mechanism applied to the non-deviating population.

[14] In an income taxation model, Berliant (1992) introduces a similar notion that he calls the IC-Core. Also, in a context of adverse selection, Kahn-Mookherjee (1995) define a concept of coalition-proof equilibrium that implicitly refers to status quo beliefs. Connections between coalition-proof and strong equilibria are examined in Konishi-Le Breton-Weber (1999).

not blocked by the Grand Coalition. As a consequence all Cores are included in the set of Pareto optima. We gather some simple results on the relationships between Cores in the next proposition.[15]

Proposition 1 *The following inclusions hold true:*

1. *The set of Pareto optimal mechanisms coincides with the* $-\infty$-BB *Core and contains any Beliefs-Based Core.*
2. *if* $u^* > v^*$
   $+\infty$-BB *Core* $\subset$ u^*-BB *Core* $\subset$ v^*-BB *Core* $\subset$ $-\infty$-BB *Core*
3. *Statistical Core* $\subset$ $+\infty$-BB *Core*

Proof The fact that any Pareto optimum is not blocked for u^* equal to $-\infty$ follows from the definition: every agent wants to join the seceding coalition because otherwise he gets $-\infty$. Therefore no coalition, except the Grand Coalition, can block because of (7). The other assertions follow from earlier remarks. □

Before going further it is worth noting that in a full information private goods economy, the Statistical Core, Status quo Core, and u^*-Beliefs-Based Cores would all be identical for reservation values u^* above the individual rationality level. However, in the adverse selection context under consideration, the distinction just introduced makes sense, as it will be next illustrated.

First, consider the insurance model with two types. Second best optimality requires full insurance for the high risk agents. The Statistical Core is empty: the low risk agents, if they could recognize themselves and separate, could fully insure themselves with a lower premium. Now the conditions of blocking in the status quo sense closely relate with the conditions of the competition between contracts considered by Rothschild and Stiglitz: if a coalition of a given type blocks, there is a contract which, if it is taken only by the agents of this type, (1) is preferred by them to the standing contracts, (2) is not preferred by the other types (implicitly assuming that they do not anticipate that, in case of secession, their payoffs might be modified). It follows that *the Status quo Core coincides with the Rothschild-Stiglitz equilibrium*[16] *if the latter exists and is optimal;* otherwise the Status quo Core is empty.

In simple versions of the public good model, the Statistical Core is not empty under conditions that have an easy economic interpretation (Guesnerie-Oddou, 1981) and that are generalizable to some extent.[17] The Status quo Core is a priori less easy to analyze: however, the results we present now will show that it (often) coincides with the Statistical Core.

[15] Although, in many applications, it looks sensible to consider exogenous utility levels that are below the endogenous utility levels obtained in the Grand Coalition, no restriction is put here on their values.

[16] The candidate for the Rothschild-Stiglitz equilibrium is the mechanism which, without subsidy between the two types, fully ensures high risk agents. If it is not dominated by a pooling mechanism, it is an equilibrium in the sense of Rothschild-Stiglitz where destabilizing offers can consist only of single contracts. It may be nevertheless dominated by a pair of contracts involving cross subsidization, in which case it is not in our Core because all coalitions, including the Grand Coalition can propose a pair of contracts.

[17] See Demange-Henriet (1992) and Demange-Guesnerie (1997).

3.3 Further relationships between the different cores

At the level of generality adopted until now, Proposition 1 exhausts what can be said about the relationships between the different Cores. We now introduce additional properties of the game that imply a more specific inclusion relationship. The properties of Effectivity of monotonic mechanisms say that the game has "enough" universal mechanisms that display some form of increasing returns to distribution.

3.3.1 Monotonic mechanisms We first define a concept of monotonicity with respect to a subset of types.

Definition 8 *Let R be a nonempty subset of types, $R \subset T$. A universal mechanism g is* **R-monotonic** *if for any coalitions C, D, with distributions μ^C and μ^D such that*

$$\mu^D(j) \geq \mu^C(j) \text{ for any } \theta_j \in R \text{ and } \mu^D(j) \leq \mu^C(j) \text{ for any } \theta_j \notin R \text{ then}$$

$$u(g(\theta, \mu^D), \theta) \geq u(g(\theta, \mu^C), \theta) \text{ for all } \theta \text{ in } R.$$

A mechanism that is T-monotonic is called **full-monotonic**.

The universal mechanism under consideration is called R-monotonic because the welfare of all type-R agents increases when their number increases (increasing returns to size hold so far as the subgroup R is concerned). In addition, if R differs from T, the presence of agents of an other type (in $T - R$) is not desired by the R-agents: there is a conflictual dimension in the mechanism that only disappears if $T = R$. Indeed T-monotonicity, or equivalently full-monotonicity, says that for any two coalitions C, D, where D contains C

$$u(g(\theta, \mu^D), \theta) \geq u(g(\theta, \mu^C), \theta) \text{ for all } \theta.$$

In words the mechanism g has the property that incumbents (members of C) are never hurt by newcomers (members of $D - C$).

Two key points of the analysis are stated in the following lemma.

Lemma 1 *Let a mechanism f be statistically blocked by coalition C with a universal mechanism g. Let S be the support of C.*

1. *if g is R-monotonic and $R \cap S \neq \emptyset$, then f is also statistically blocked by the $R \cap S$-full coalition. As a consequence f does not belong to the $+\infty$-Beliefs Based Core.*
2. *if g is full-monotonic then f is also statistically blocked by the S-full coalition and status quo blocked by some S'-full coalition with $S \subset S'$. As a consequence f does not belong to the Status quo Core.*

Proof Let f be an ICF mechanism that is statistically blocked by coalition C of support S with a mechanism g:

$$u(g(\theta, \mu^C), \theta) > u(f(\theta, \overline{\mu}), \theta) \text{ for all } \theta \text{ in } S.$$

1) If g is R-monotonic and $R \cap S \neq \emptyset$, let D be the $R \cap S$-full coalition. By construction D contains at least as many agents with type in R than C and no agent with type in $T - R$. Hence R-monotonicity implies:

$$u(g(\theta, \mu^D), \theta) \geq u(g(\theta, \mu^C), \theta) \text{ for all } \theta \text{ in } R,$$

and finally

$$u(g(\theta, \mu^D), \theta) > u(f(\theta, \overline{\mu}), \theta) \text{ for all } \theta \text{ in } R \cap S$$

2) If g is full-monotonic, because $\overline{\mu}_S \geq \mu^C$ we have:

$$u(g(\theta, \overline{\mu}_S), \theta) \geq u(g(\theta, \mu^C), \theta) > u(f(\theta, \overline{\mu}), \theta) \text{ for all } \theta \text{ in } S,$$

so that the S-full coalition (statistically) blocks. Now consider all the S'-full coalitions that block f with g and such that $S \subset S'$. Take a maximal S'. Maximality implies that (7) is also met i.e.:

$$u(f(\theta, \overline{\mu}), \theta) \geq u(g(\theta, \overline{\mu}_{S'}), \theta), \text{ for any } \theta \text{ not in } S'.$$

If not, $u(g(\theta', \overline{\mu}_{S'}), \theta') > u(f(\theta', \overline{\mu}), \theta')$ for some θ' not in S' so that by full-monotonicity the $S' \cup \{\theta'\}$-full coalition would block f with g as well, in contradiction with the maximality of S'. Hence the S'-full coalition status quo blocks f. □

3.3.2 Effectivity of monotonic mechanisms Per se, the existence of universal mechanisms being R-monotonic, for some R, is a rather weak[18] restriction. Effectivity properties, that are now stated, bear on the whole set of mechanisms of the game. They assert, in which sense and to which extent, standard mechanisms can be extended to monotonic mechanisms.

Definition 9 Effectivity of monotonic mechanisms *(EMM) holds if:*
For any mechanism g ICF for a coalition C, there exist R, a nonempty subset of the support S of C, and a (universal) R-monotonic mechanism g' such that

$$u(g'(\theta, \mu^C), \theta) \geq u(g(\theta, \mu^C), \theta) \text{ for all } \theta \text{ in } R \quad (8)$$

In words, given any ICF mechanism applied to a coalition, there always exists a subgroup of types R which may be at least as well off under some R-monotonic mechanism. As we shall see from the examples, the property is not too demanding. One may also require, in the same spirit, that the mechanism g' is full-monotonic, a much stronger property indeed.

Definition 10 Effectivity of full-monotonic mechanisms *(EFMM) holds if:*
For any mechanism g ICF for a coalition C, there exist a nonempty subset D of C and a universal full-monotonic mechanism g' such that

$$u(g'(\theta, \mu^D), \theta) \geq u(g(\theta, \mu^C), \theta) \text{ for all } \theta \text{ in the support of } D \quad (9)$$

If, in the above definition, D may always be taken equal to C we say that **strong** *Effectivity of full-monotonic mechanisms holds.*

[18] It is for example implied by the fact that a type acting as a dictator in a coalition can take advantage of an increasing size, a fact that holds in most models where congestion effects are ruled out.

Strong Effectivity of full-monotonic mechanisms implies that, as far as statistical blocking is concerned, nothing is lost by considering full-monotonic blocking mechanisms. This is no longer the case if Effectivity is not strong. EFMM however is enough for our purpose and is a much weaker property as we shall see below in a public finance context.

Let us discuss these properties in the models introduced in Section 2.

In the insurance model, Effectivity of monotonic mechanisms holds. If for example the support of a coalition consists of both types, a mechanism can be extended to a "low-risk"-monotonic mechanism, that would give more to the low-risk agents, when there are more low-risk agents and less high-risk agents.[19] However, it cannot be extended to a full-monotonic mechanism: with more high risk agents and the same number of low-risk agents, some group is necessarily losing. Therefore EFMM does not hold.

Consider now the public good model with non observable income. A mechanism ICF for a coalition is payoff-equivalent to a tax system associated with a consumption tax and a poll tax. If both taxes are positive, the mechanism can straightforwardly be extended to a full-monotonic mechanism, the one that mimics the same tax levels, whatever the coalition. Accordingly, if mechanisms are required to involve no subsidy, strong Effectivity of full-monotonic mechanisms holds.

However, if a poll tax may be negative, i.e. if it is a subsidy, the mechanism can only be extended to a R-monotonic mechanism, where R is the set of types the overall fiscal contributions of which are positive. Accordingly, if redistribution is allowed, without further specification, we can only say that Effectivity of monotonic mechanisms holds.

Nevertheless, the difficulty partly relates here with the linearity of the income tax system, which is only justified if revenue is not observable. Let us now consider that the public good is financed through a (possibly) non-linear income tax system. Let $g(w)$ be the after tax revenue for an agent who earns w. Consider a mechanism that statistically blocks through g in C. If g entails no subsidies, i.e. if $g(w) \leq w$ for any possible w, then g is (can be extended to) a full-monotonic mechanism. Assume now that the previous condition does not hold for all w. We claim that the subcoalition composed of all agents who are not subsidized may be at least as well off by using a full-monotonic mechanism. For that it suffices to change g into g' defined as $g'(w) = \min(g(w), w)$. Effectivity of full-monotonic mechanisms holds although strong Effectivity does not. To sum up, in a public finance context, EFMM is always satisfied if the tax system is fully non linear (which implicitly assumes full observability of consumption and revenue) or if subsidies go through the non linear part of the tax systems.

Theorem 1

1. *Under Effectivity of monotonic mechanisms, the Statistical Core is identical to the $+\infty$-Beliefs Based Core.*
2. *Under Effectivity of full-monotonic mechanisms, the Statistical Core, the $+\infty$-Beliefs Based Core and the Status quo Core are identical.*

[19] This property extends to a model with more than two types.

Proof Let f be statistically blocked by a coalition C through an ICF mechanism g. Under Effectivity of (resp. full) monotonic mechanisms, it is blocked by a mechanism that satisfies condition 1), (resp. 2), of Lemma 1: hence f does not belong to the $+\infty$-beliefs-based Core (resp. Status quo). The conclusion follows from the inclusions in Proposition 1. □

As a consequence of the theorem, in many versions of the public good model, the three just mentioned Cores do coincide. This provides an argument in favour of the stability analysis often adopted in the taxation literature that more or less explicitly refers to the Statistical Core. The statement also suggests an entry into the understanding of the cases where the Statistical Core and the Status quo Core differ: in view of the previous example, a mechanism that statistically blocks may not block in the sense of the status quo because it has some redistributive features too much attractive or generous.

3.4 Relationships with the coarse core

Statistical information is the maximal information a coalition can reasonably be supposed to have at the *interim stage*. One may also consider the opposite situation in which a coalition gets no information. This leads to ideas that are reminiscent of those associated with the concept of *coarse* Core introduced by Wilson[20] (1978). In our framework this concept is typically rather weak but provides a good benchmark. Indeed consider a coalition that is formed. Agents recognize themselves from their names a, and only observe the size s of the coalition. No information is inferred from the fact that the coalition is formed. Therefore, under our assumption that the names are not correlated with the characteristics, μ^C is perceived as a random variable, the distribution of which is conditioned by the size s of the coalition; and this is the only fact that is common knowledge. Under the assumption of an infinite number of agents, the law of large numbers suggests that μ^C is almost surely $s\overline{\mu}$: the coalition is a smaller replica of the grand coalition. Therefore, under some weak form of increasing returns to size,[21], a quite innocuous assumption in the absence of congestion effects, a mechanism that is blocked by a coalition is blocked by the whole society as well. As a consequence the coarse Core is non empty and coincides with the set of second best Pareto optima[22].

[20] Wilson's set up is slightly different: in a private economy, agents choose ex ante contracts contingent on a state of a world which will be observed ex post but on which agents learn some private information interim

[21] More precisely if $[u(f(\theta, s.\overline{\mu}), \theta), \theta \in T]$ is increasing with s for any mechanism. The effect of an increase in s is truly a "size" effect, without the "composition" -change in the distribution- effect that has been introduced before. This idea is developed in a previous version of this paper where the corresponding concept of Core was labelled *generic* rather than coarse. In a context such similar to ours, i.e. an adverse selection economy with a continuum of agents, Hammond (1989) also considers that coalitions do not get information at the interim stage. If one restricts attention to anonymous mechanisms, one would fall back to the so called "generic" viewpoint.

[22] In an exchange economy, Koutsougeras-Yannelis (1993) also note that the coarse Core is "too large", i.e. is constituted with all the individually rational and Pareto optimal allocations. The private Core, introduced by Yannelis (1991), which is nonempty in their framework, is more sensible.

4 Extensive form stability of the cores

Under complete information, the Core concept, besides of normative justification, can be viewed as capturing strategic stability - secession is not plausible. In our complete information framework, we introduce some extensive form games, hereafter called developer games, that provide further evaluation of the strategic stability of the different Core concepts that have just been defined.

4.1 Developer games

As a reference starting point, let us consider a complete information framework, in which the agents' characteristics are known and verifiable.

4.1.1 Complete information The following simplistic game helps to assess the "stability" associated with the standard Core concept.

Developer game: Let f be the standing allocation. First the "developer" makes a public announcement of a feasible proposal to some coalition C of positive measure. Second, the members of C are asked to say *yes* or *no*. The outcome is the following:

- If almost all the members of C say *yes*, secession succeeds: the developer implements the proposal and receives $\epsilon > 0$ and the agents get the secession utility minus ϵ/s.[23].
- If a subset of non-zero measure of C says *no*, the Grand Coalition A implements f.

The developer's behaviour is not fully defined. We only assume that he reacts to the incentives of a positive reward ϵ, even very small, by making a successful proposal whenever it is possible to do so ; however when there are many such successful proposals, we say nothing on which one is chosen.

Let f be the standing allocation. Consider first the Nash equilibria of the developer game. If secession succeeds, the seceding coalition C is a blocking coalition: f does not belong to the Core. The converse is not entirely true: if f does not belong to the Core and the developer makes a proposal to a blocking coalition, there is a trivial Nash equilibrium in which every player says *no*, because he expects the other members of the coalition to say *no*. But such strategies are weakly dominated: if they are deleted, secession succeeds. Therefore, if one sticks to strategies that satisfy some robustness criterium, *the developer may "destabilize" the standing allocation if and only if it does not belong to the Core*.

4.1.2 Incomplete information We now adapt the developer game in order to take into account the incomplete information aspects of our problem. Within the complete information framework, the outcome of the developer game does not depend upon what happens to the agents who are not invited to secede. With incomplete information, the developer is unable to check the precise characteristics of the

[23] ϵ can be viewed as the equivalent of a monetary payment made to the developer by seceding agents.

seceding group. Any agent may join and his decision depends not only on the proposal itself but also on his outside options. We define here the developer game, by assuming that the outside option gives everybody the status quo utility levels $u_f^*(\theta) = u(f(\theta, \overline{\mu}), \theta)$ defined above. Such an outside option may reflect, for example, the fact that, in the outside world, there are similar communities where the standing mechanism f or an equivalent mechanism prevails.

The developer game. Let f be the standing mechanism. A developer proposes a minimal size $s, 0 < s \leq 1$ for the coalition to be formed and a mechanism g, feasible and incentive compatible for all coalitions of size larger than s.

- *Stage 1*: Each agent is asked to say *yes* or *no*. All agents play simultaneously. Two cases occur depending on the size of the group of agents who choose *yes*: If the size is strictly smaller than s, the game is over, and the Grand Coalition A implements f; if it is at least s, secession succeeds and a second stage is played:
- *Stage 2*: Agents who said *no* at the first stage are asked whether they *join* or *not join*.
  Call C the coalition formed with the agents who said either *yes* at the first stage or *join* at the second stage. The developer implements g on C and receives $\epsilon > 0$ from the members of C. Agents outside C get the standing level under f applied to the whole population, $u(f(\theta, \overline{\mu}), \theta)$.

Note that the game introduced by Holmstrom and Myerson (1983) in their study of durability is a special case of the game considered here in which the developer was required to ask for unanimity ($s = 1$), a case in which, in particular, the second stage is irrelevant.

4.2 Nash equilibria

We first concentrate on Nash equilibria.

Proposition 2 *Let f be a feasible and incentive compatible mechanism for A and consider the developer game.*

1. *If f belongs to the Statistical Core, then whatever the proposal and whatever $\epsilon > 0$, secession never succeeds at a Nash equilibrium.*
2. *If f does not belong to the Status quo Core, let C be a blocking S-full coalition. Then, there is some proposal such that C secedes at a Nash equilibrium for $\epsilon > 0$ small enough.*

Proof 1. At a Nash equilibrium, an agent correctly expects the others' strategies. As a consequence, if secession succeeds the distribution μ^C of the seceding coalition is correctly expected. Therefore:

$$u(g(\theta, \mu^C), \theta) > u(f(\theta, \overline{\mu}), \theta) + \epsilon/s, \forall \theta \in S$$

so that C statistically blocks, a contradiction.

2. Consider a S-full coalition C that blocks with g. Choose $\epsilon > 0$ small enough so that the members of C still block even if they pay ϵ/s. Let the developer propose

g, s equal to the size of C, and ϵ. Consider the following strategies: the members of C say *yes* at the first stage, agents outside C say *no* and *not join* at the first and second stages. They form a Nash equilibrium at which C secedes. □

4.3 Destabilization

As in the complete information case, assertion 2 of Proposition 2 is not entirely satisfactory. There are always trivial Nash equilibria without successful secession. Our aim now is to get rid of them and to find conditions under which a proposal is successful at all "robust" equilibria. In order to define robustness some care is needed. Since the population is infinite, an agent has no influence on whether secession occurs or not: he is never "pivotal" as might be the case with a finite population. As a consequence some strategies are payoff equivalent: for example the strategy *yes* at the first stage is equivalent to the strategy: *no* at the first stage and *join* at the second stage, if any. However it is sensible to assume that an agent favors the outcome he most prefers in cases where there is no ambiguity.

Positive influence. *Consider a step in the process of elimination of dominated strategies. A player eliminates no and says yes at the first stage of the game if*

$$u(g(\theta,\mu),\theta) > u(f(\theta,\overline{\mu}),\theta)$$

for any possible distribution μ of a seceding coalition, that may occur when taking into account the eliminated strategies at this step.

Definition 11 *A mechanism is said to be destabilized if there is a proposal and $\epsilon > 0$ such that secession occurs at all equilibria obtained by successive elimination of dominated strategies under the assumption of positive influence.*

Proposition 3 *Let f be a feasible and incentive compatible mechanism for A. If it is strongly Pareto dominated, it is destabilized.*

Proof Assume that f is strongly Pareto dominated by some g. Let the developer propose g and a minimal size equal to 1. If secession occurs, any agent is strictly better off than under the standing mechanism. Therefore, from the assumption of positive influence,[24] all say *yes*, at the first stage. □

If positive influence is not assumed, secession may fail at an equilibrium even if a strictly Pareto improving mechanism is proposed: the strategy *no join* is payoff equivalent to *yes*. Contrary to what would happen in the finite population case, weak dominance arguments do not help here to eliminate such a trivial equilibrium.

Theorem 2 *Suppose that Effectivity of full-monotonic mechanisms holds true. Let f be a feasible and incentive compatible mechanism for A that does not belong to the Statistical Core. Then it is destabilized in the developer game.*

[24] Note that, in view of the above remark on the Holmstrom-Myerson game,the statement shows that in our framework (second best) Pareto optimality and durability coincide. This could have been checked in another way: interim information does not modify the vector of utility payoffs that can be obtained by the agents contingent on their types.

The Proof of Theorem 2 is more intricate than the preceding ones and is postponed to the appendix. An idea of the proof can be given. Under EFMM, a mechanism that is not in the Statistical Core is blocked by a type-full coalition with a maximal support and the blocking mechanism can be chosen full-monotonic (see Lemma 1). Let the developer propose this mechanism and a minimal size equal to the size of this blocking coalition. The process of elimination of dominated strategies is as follows. We construct recursively sets of types included in $T - S$: T_1 is the set of types of the agents who are always worse off by being in a secession than under the initial situation: they have *no* and *not join* as a dominant strategy. Knowing that no agent with a type in T_1 will ever secede, a new set of agents, who will never join the secession, is found, with types in T_2, and so on until exhausting all types outside S. Then the agents with types in S know that if secession occurs they will be all together and they say *yes*. Therefore secession succeeds and the mechanism is destabilized.

Combining the above Theorem 2 together with Proposition 1 and Theorem 1, we get one of the main results of the paper:

Theorem 3 *Assume Effectivity of full-monotonic mechanisms. Let f be a feasible and incentive compatible mechanism for A. The three properties are equivalent:*

1. *f is not destabilized*
2. *f belongs to the Statistical Core*
3. *f belongs to the Status quo Core.*

This result hopefully provides a convincing illustration of the duality between the "normal form" approach, that suggests the definition of the Statistical Core, and the "extensive form" approach that has been chosen to assess robustness to secession proposals.

We stress again that the result provides support for the consideration of the concept of Statistical Core whenever Effectivity of full-monotonic mechanisms holds.

A final remark can be made. Different after-secession stories would lead to different developers games, as in Hellwig (1987). The beliefs would then be endogenous and would depend upon the outcome of a continuation game taking place after stage 2. Although the analysis would be different being possibly more intricate, the understanding obtained here on the mechanics of destabilization would likely remain precious.

5 Conclusion

This paper attempts at providing reasonable criteria for assessing the stability of a society that faces interim asymmetries of information. Our analysis is conducted within an abstract framework and follows lines of reflection associated with the Core concept.

The first lesson of our investigation is that both concepts, blocking and secession, are significantly more complex and ambiguous under asymmetric information than under symmetric information. This is of no surprise in view of the role of beliefs, adverse selection, acquisition of information that previous literature has already emphasized. However, the precise nature of increased complexities and ambiguities is enlightening: beliefs of outsiders and information of insiders are crucial to trigger blocking so that a family of concepts is relevant rather than a single one. Also, the inclusion relationship that orders this family of Cores is significantly simplified when some increasing returns to coalitions, such as those trigerred by the assumption Effectivity of full-monotonic mechanisms, are present.

A second lesson is both more specific and positive. In taxation like situations, where some enough strong form of increasing returns prevail, then our analysis gives a lot of support to the concept of Statistical Core: on the normative side, this Core coincides with the Status quo Core that refers to conservative but well defined beliefs of the agents, on the stability side this Core basically coincides with the mechanisms that are not destabilized in our developer games. We have then provided a cluster of arguments in order to identify the stability of an arrangement with the fact that it belongs to the Statistical Core.

At the opposite, and this is our third lesson, in games in which such a form of increasing returns do not hold, the compelling arguments supporting the statistical Core are weakened when at the same time the emptiness of this Core is becoming likely. There is furthermore no Core concept that would provide a fully convincing alternative to the previous one, even if the Status quo Core keeps some merits in view of our previous results.

6 Appendix

Proof of Theorem 2. Let f be not in the Statistical Core. As in the Proof of Lemma 1, because of Effectivity of full-monotonic mechanisms, there is a blocking coalition C that satisfies: it blocks through a full-monotonic mechanism g, it is S-full, and it has a maximal support among the coalitions that block f with g. Therefore since C blocks with g, there is $\varepsilon > 0$ such that:

$$u(g(\theta, \mu^C), \theta) > u(f(\theta, \overline{\mu}), \theta) + \varepsilon/s \text{ any } \theta \text{ in } S$$

and since g is a full-monotonic mechanism g:

$$u(g(\theta, \mu^D), \theta) > u(f(\theta, \overline{\mu}), \theta) + \varepsilon/s \text{ any } \theta \text{ in } S, \text{ any } D \supseteq C \tag{P1}$$

Secondly, by maximality of S, for any S', $S \subset S'$, $S \neq S'$ there exists θ in S' (but not in S) such that

$$u(f(\theta, \overline{\mu}), \theta) \geq u(g(\theta, \overline{\mu}_{S'}), \theta). \tag{P2}$$

Let the developer propose g, a minimal size equal to the size of C, and fee ε.

The process of elimination of dominated strategies is as follows. We construct recursively sets of types included in $T - S$: T_1 are the types of the agents who have *no* at all stages as a dominant strategy. Knowing that the agents whose type belongs to T_1 never secede, a new set of agents who will never join the secession is T_2 of types, and so on until exhausting all the types that are not in S. Then the agents with types in S know that if secession accurs they will be all together and they say *yes*.

Step 1

* *case a*: C is the whole set A. We are done: g strongly Pareto dominates f and we fall back on Proposition 7.

* *case b*: C is not A, and therefore S is a strict subset of T. Applying $(P2)$ to $S' = T$ we know that there is some θ not in S

$$u(f(\theta, \overline{\mu}), \theta) \geq u(g(\theta, \overline{\mu}), \theta). \tag{P3}$$

Call T_1 the set of types such that (P3) is true. By monotonicity of g

$$u(f(\theta, \overline{\mu}), \theta) \geq u(g(\theta, \overline{\mu}), \theta) \geq u(g(\theta, \mu^D), \theta) \text{ any } D \subset A, \theta \in T_1$$

Therefore, because of the fee, A player with type in T_1 is strictly worse off by joining the secession when it occurs than under the standing mechanism. It means that (*no,not join*) is a dominant strategy for him.

Step 2

**case a*: $T_1 = T - S$. We are done: agents with types in S know that they are the only ones to say *yes*. Because of the size requirement, it implies that secession occurs only if almost of all of them say *yes*. Therefore the final seceding coalition surely contains C. From (P1) and the positive influence assumption, members of C eliminate *no*. Secession succeeds and f is destabilized.

**case b*: T_1 is a strict subset of $T - S$. Apply $(P2)$ to $S' = T - T_1$. There are surely some types not in T_1 (and not in S) such that

$$u(f(\theta, \overline{\mu}), \theta) \geq u(g(\theta, \overline{\mu}^{S'}), \theta) \tag{P4}$$

Call T_2 the set of types for which (P4) is true. By monotonicity of g, if θ is in T_2:

$$u(f(\theta, \overline{\mu}), \theta) \geq u(g(\theta, \mu^D), \theta) \text{ for any } D \text{ with support } \subset T - T_1 \tag{P5}$$

Everybody knows that only the agents with types outside T_1 may join the secession either at the first or at the second stage of the game. Therefore, by (P5), agents with characteristics in T_2 have (*no, not join*) as a dominant strategy.

One goes to Step 3, etc... and proceeds as follows.

Step k

At the beginning of step k, there are sets $T_j, j = 1, .., k-1$. Everybody knows that the agents belonging to them will never join a secession. Two cases occur.

* *case a*:: any type not in S belongs to one of the set $T_j, j = 1, .., k-1$; the process stops as in case a of step 1: S secedes.

* *case b*: one constructs a non empty subset T_k, not intersecting any T_j nor S such that all agents with types in T_k says (*no not join*). The process surely stops since there are a finite number of types. □

References

Allen, B.: Incentives in market games with asymmetric information: the Core. Core discussion paper, 9221

Al-Najjar, N.: Decomposition and characterization of risk with a continuum of random variables. Econometrica **63**, 1195–1224 (1995)

Aumann, R., Kurz, M.: Power and taxes. Econometrica **45**, 1137–1161 (1977).

Berliant, M.: On income taxation and the core. Journal of Economic Theory **56**, 121–141 (1992)

Boyd, J., Prescott, E.: Financial intermediary coalitions. Journal of Economic Theory **38**, 221–232 (1986)

Demange, G., Guesnerie, R.: Nonemptiness of the core: low dimensional decisions spaces and one-dimensional preferences. Ricerche Economice **51**, 7–17 (1997)

Demange, G., Henriet, D.: Sustainable Oligopolies. Journal of Economic Theory **54**, 417–428 (1991)

Diamond, P.A, Mirrless, J.: Optimal taxation and public production. American Economic Review **61**(8–27), 261–78 (1971)

Forges, F., Heifetz, A., Minelli, E.: Incentive compatible core and competitive equilibria in differential information economies. Working paper 99–06 (1999). Economic Theory (forthcoming)

Greenberg, J.: Coalition structures. Handbook of game theory, vol. II, chap. 36. Amsterdam: North Holland 1994

Guesnerie, R.: On taxation and incentives; further remarks on the limits to redistribution. Bonn Discussion Paper 89 (1981)

Guesnerie, R.: A contribution to the pure theory of taxation. Cambridge: Cambridge University Press 1995

Guesnerie, R., Oddou, C.: Second best taxation as a game. Journal of Economic Theory **25**, 67–91 (1981)

Hammond, P.: Straightforward incentive compatibility in large economies. Review of Economic Studies **46**, 263–282 (1979)

Hammond, P.: Perfected option markets in economies with adverse selection. EUI Working paper 89–426 (1989)

Hellwig, M.: Some recent developments in the theory of competition in markets with adverse selection. European Economic Review **31**, 319–325 (1987)

Holmstrom, B., Myerson, R.: Efficient and durable decision rules with incomplete information. Econometrica **51**, 1799–1820 (1983)

Kahn, C. M., Mookherjee, D.: Coalition proof equilibrium in an adverse selection insurance economy. Journal of Economic Theory **66**(1), 113–138 (1995)

Khan, M.A., Sun, Y.: Weak measurability and characterizations of risk. Economic Theory **13**(3), 541–560 (1999)

Konishi, H., Le Breton, M., Weber, S.: On coalition–proof Nash equilibria in common agency games. Economic Theory **85**, 122–139 (1999)

Koutsougeras, L., Yannelis, N.C.: Incentive compatibility and information superiority of the Core of an economy with differential information. Economic Theory **3**, 195–216 (1993)

Mas Colell, A., Vives, X.: Implementation in economies with a continuum of agents. Review of Economic Studies **60**(3), 613–630 (1993)

Perry, M., Reny, P.: A noncooperative view of coalition formation and the core. Econometrica **62**(4), 795–817 (1994)

Rothschild, M., Stiglitz, J.: Equilibrium in competitive Insurance Markets: An essay on the economics of imperfect information. Quarterly Journal of Economics **90**, 629–649 (1976)

Riley, J.: Informational equilibrium. Econometrica **47**, 331–359 (1979)

Vohra, R.: Incomplete information, incentive incompatibility and the core. Journal of Economic Theory **86**, 123–147 (1999)

Wilson, R.: Information, efficiency, and the core of an economy. Econometrica **52**, 1365–1368 (1978)

Wilson, C.: A model of insurance markets with asymmetric information. Journal of Economic Theory **16**, 167–207 (1977)

Yannelis, N.C.: The core of an economy with differential information. Economic Theory **1**, 183–198 (1991)

Signaling in markets with two-sided adverse selection*

Douglas Gale

Department of Economics, New York University, 269 Mercer Street, New York, NY 10003, USA
(e-mail: douglas.gale@nyu.edu)

Received: September 3, 1999; revised version: December 3, 1999

Summary. The paper analyzes an economy with two-sided adverse selection, focusing on equilibria that satisfy a refinement based on the notion of strategic stability. In the familiar case of one-sided adverse selection, agents reveal all of their private information as long as the contract space is rich enough. However, with two-sided adverse selection, the sufficient conditions for separation are much stronger.

Keywords and Phrases: Signaling, Adverse selection, Markets, Rationing, Types, Equilibrium, Refinement, Strategic stability, Walrasian.

JEL Classification Numbers: D5, D82.

1 Introduction

One of the classical problems in the analysis of markets with incomplete information is to discover conditions under which agents reveal their private information. There are several reasons why this problem is important. The market has long been recognized as a mechanism for facilitating the transfer of information. If agents have an incentive to withhold information, the resulting equilibrium may be informationally inefficient. The vast empirical and theoretical literature on the informational efficiency of financial markets attests to the interest of this issue. At the same time,

* This paper is an outgrowth of work that has been presented at Carnegie Mellon University; the Studia della Economia/*Workshop on Economic Theory*, Venice, Italy; the *Conference on Endogenous Incompleteness* at the Federal Reserve Bank of Minneapolis; *SITE* at Stanford University. I would like to thank the participants for their comments, Nicholas Yannelis and an anonymous referee for numerous helpful suggestions, and Piero Gottardi and Alberto Bisin for stimulating conversations over a number of years. The financial support of the National Science Foundation and the C.V. Starr Center for Applied Economics is gratefully acknowledged.

it is well known that the attempt to signal private information can cause distortions that make the market allocation inefficient. For example, in markets with adverse selection, an agent's choice of education, insurance, or borrowing may reveal his private information about his productivity, probability of having an accident, or the riskiness of his project. The early papers of Spence (1973) on market signaling and Rothschild and Stiglitz (1976) on screening provide sufficient conditions for the existence of equilibria in which agents reveal their private information. These models are further refined by Wilson (1977) and Riley (1979). In order to signal their private information, agents have to incur a private cost (otherwise their signal could be imitated). The cost of signaling is a deadweight loss from society's point of view and for this reason equilibria in which agents can signal their private information are typically inefficient. An analysis of the efficiency properties of signaling equilibria is contained in Gale (1996).

This paper focuses on the conditions that ensure agents signal their private information. The theoretical results in the literature are mixed. While screening models sometimes have no equilibrium, signaling models have a multiplicity of equilibria in which the amount of information revealed varies from full revelation to no revelation. If we want to reduce the set of equilibria and make tighter predictions about information revelation, some refinement of equilibrium is needed.

Refinements of the Nash equilibrium were developed for games of incomplete information in the nineteen-eighties (Kohlberg and Mertens (1986), Banks and Sobel (1987), Grossman and Perry (1986) and Mailath, Okuno-Fujiwara, and Postlewaite (1993)). Applications to signaling games by Cho and Kreps (1984) and Cho and Sobel (1990) select a separating equilibrium, in which agents' actions reveal their type, as the only outcome that satisfies a (strong) refinement. There are exceptions, however. Hellwig (1987) adds a third stage to the canonical signaling game, to capture the flavor of the Wilson (1977) equilibrium, and finds that for some parameter values the pooling equilibrium is the unique equilibrium satisfying the Kohlberg-Mertens refinement.

The classical signaling game is special in a number of respects.

- There are only two agents, an informed agent and an uninformed agent.
- The informed agent has private information (his type) and moves first.
- The uninformed agent observes the informed agent's action before choosing his own action.

The signaling game can also be interpreted as a market game. Instead of a single informed agent with a probability distribution of types, the market-game interpretation assumes a continuum of informed agents with a known cross-sectional distribution of types. Under certain conditions (e.g., the uninformed agents are risk-neutral firms) the reactions of a continuum of uninformed agents mimics the response of a single uninformed agent to a single informed agent with a random type.

The signaling game provides a simple and tractable framework in which to study the informational properties of equilibrium, but it has limitations. There are many ways in which the signaling game could be extended. One alternative framework, which accommodates a richer set of environments and has proved to be very tractable, is developed in Gale (1991, 1992, 1996). A complete set of contracts represents all the possible forms of interactions between agents on the two sides of the markets. Agents choose the contracts they most prefer, taking into account the information revealed by the contract choices of the other agents. Thus, information is symmetrically and simultaneously revealed by equilibrium contract choices and used by agents in making those choices.

In contrast to the signaling game, this framework allows for adverse selection on one or both sides of the market.

- There is a continuum of 'buyers' and 'sellers', each of whom may have private information.
- All agents simultaneously choose the contracts they would like to exchange.
- The 'buyers' and 'sellers' who have chosen a given contract are randomly matched.

Because agents move simultaneously, they do not observe any information prior to their choice of contract. However, in equilibrium they know the strategies of the other agents and correctly assess the probability of being matched with a given type of 'buyer' or 'seller' conditional on the contract they choose. Because there is a large number (non-atomic continuum) of agents, the market is competitive. This is reflected in the fact that agents take as given the matching probabilities conditional on each choice of contract. The matching probabilities play the role of prices in the classical theory of competitive equilibrium, that is, they determine which contracts can be traded and which cannot.

The refinements of Nash equilibrium developed in the game theory literature have natural counterparts here. Gale (1992) uses a refinement derived from the Kohlberg-Mertens concept of strategic stability to select a unique separating equilibrium allocation in which each agent chooses to reveal his private information. The selection theorem in Gale (1992) is restrictive in one important respect, however. While it allows for heterogeneous types on each side of the market, there is only adverse selection on one side. More precisely, in a market consisting of buyers and sellers, there are several types of buyers but the sellers do not care which type of buyer they trade with. This model allows for assortative matching, which cannot occur in the standard signaling game, but it avoids the difficulties of analysing two-sided adverse selection.

The present paper has two objectives. First, it is argued that when there is one-sided uncertainty (OSU), separation will always occur if the space of contracts traded in the economy is sufficiently rich relative to the type space. Secondly, it is argued that the case of two-sided uncertainty (TSU) is much more difficult to analyze and requires stronger conditions to ensure full revelation of information.

The reason for the greater difficulty of analyzing TSU is that, roughly speaking, with OSU an informed agent has preferences over contracts but not over types of agents on the other side of the market. For example, if the sellers are workers with different productivities and the buyers are identical, risk neutral firms, then the sellers have preferences over different contracts (specifications of wages, hours of work, and education) but do not care what kind of firm they work for. Thus, beliefs (about the buyers' types) do not affect the sellers' behavior in a significant way. By contrast, in the case of TSU, an informed agent has preferences over contracts *and* the types of agents on the other side of the market. For example, if there are different types of firms and workers care about the type of firm they work for, then workers have to take into account both the nature of the contract they choose and the type of firm that is likely to offer that particular contract.

Refinements of equilibrium work by restricting 'plausible' out-of-equilibrium beliefs. With TSU it becomes much more difficult to say whether an out-of-equilibrium belief is 'plausible' or not. An agent's choice of contract depends both on the physical characteristics of the contract and the probability distribution of types with which an agent will be matched. This means that in order to discuss what is a reasonable or plausible belief, one has to look at both sides of the market simultaneously. For example, a seller may have the belief that a particular contract is traded by a bad type of buyer. This belief may discourage the seller from choosing that contract. In order to determine whether the buyer's belief is plausible, we have to ask whether it is plausible to expect a bad buyer to choose that contract. That in turn will depend not just on the bad buyer's preferences over contracts, but also on his beliefs about the type of seller he will be matched with if he chooses that contract. But then we have to ask whether the buyer's belief is plausible and that in turn depends on the sellers' behavior, that is, on the preferences and beliefs of the sellers.

The theory developed in this paper focuses on decentralized and uncoordinated decision making under incomplete information. Other theories focus on the optimality properties of competitive markets (Harris and Townsend, 1982; Prescott and Townsend, 1984; Yannelis, 1991; Myerson, 1993; Koutsougeras and Yannelis, 1993; Krasa and Yannelis, 1994; Jerez 1999). The two approaches should be thought of as complementary. There is no single paradigm of competitive markets with incomplete information.

The rest of the paper is organized as follows. The basic model of an economy with incomplete information is presented in Section 2, where equilibrium is defined and a refinement of equilibrium is introduced. This refinement is shown to impose restrictions on beliefs about matching probabilities. In Section 3 these restrictions are applied to an economy with OSU. Under conditions that are weaker than the standard Spence-Mirrlees monotonicity conditions, the only kind of equilibrium that satisfies the refinement is a separating equilibrium. An economy with TSU is analyzed in Section 4, where two sets of sufficient conditions for separation are discussed. Proofs are gathered in Section 5.

2 An economy with incomplete information

The economy consists of two mutually exclusive classes of agents who are conventionally referred to as buyers and sellers. There is a finite number of different types of agents. Let $K = S \cup T$ denote the set of types, where S denotes the set of seller types and T denotes the set of buyer types. Each type $k \in K$ consists of a non-atomic continuum of identical agents whose measure is $N(k) > 0$.

The objects traded in this economy are *contracts*. The set of all possible contracts is denoted by Θ. Initially, Θ is assumed to be finite. Later the theory is extended to an infinite set of contracts.

Contracts are exchanged between buyers and sellers, with one buyer and one seller for each contract. If a seller of type s and a buyer of type t exchange a contract of type θ, the seller's payoff is $u(\theta, s, t)$ and the buyer's payoff is $v(\theta, s, t)$. Each agent is assumed to have an outside option that determines his utility if he chooses not to trade a contract. The payoff functions are normalized so that the value of each agent's outside option is equal to 0.

Agents are allowed to trade at most one contract. Under this assumption, the equilibrium choices of all the agents can be described by an *allocation* $f : \Theta \times K \to \mathbf{R}_+$, where $f(\theta, k)$ is the measure of type-k agents who choose a contract θ. An allocation is *attainable* if it satisfies the adding-up condition

$$\sum_\theta f(\theta, k) \leq N(k),$$

for every k. The number of agents of type k who choose not to trade is $N(k) - \sum_\theta f(\theta, k) \geq 0$. Let F denote the set of attainable allocations.

To avoid some pathological cases, it is assumed that there is a small disutility of participating in a market. Let $c(k) > 0$ denote the participation cost for agents of type k. Note that an agent has to pay the participation cost $c(k)$ even if he is rationed and unable to trade. Only agents who choose not to participate avoid this cost.

Let $\mathcal{E} = \{S, T, N, u, v, c\}$ denote the economy with incomplete information.

2.1 Equilibrium

The buyers and sellers who want to trade contract θ are randomly matched. An agent does not know the type of agent he will be matched with, but he does know the probability of being matched with any particular type of agent. The equilibrium matching probabilities are described by a *probability assessment* $\mu : \Theta \times K \to \mathbf{R}_+$, where $\mu(\theta, t)$ denotes the probability that a seller choosing contract θ will exchange the contract with a type-t buyer and $\mu(\theta, s)$ denotes the probability that a buyer choosing contract θ will exchange the contract with a type-s seller.

The number of buyers and sellers who want to trade θ may not be equal. In that case, the market clears through rationing. A buyer's probability of trading θ is $\sum_s \mu(\theta, s)$. Likewise, a seller's probability of trading θ is $\sum_t \mu(\theta, t)$. Thus, the set of admissible probability assessments is

$$M = \left\{ \mu : \Theta \times K \to \mathbf{R}_+ | \sum_s \mu(\theta, s) \leq 1, \sum_t \mu(\theta, t) \leq 1 \right\}.$$

The probability of being rationed (unable to trade) is $1 - \sum_s \mu(\theta, s) \geq 0$ for buyers and $1 - \sum_t \mu(\theta, t) \geq 0$ for sellers. Note that the equilibrium probability assessment μ is common for all agents. Thus, all sellers have identical beliefs about the trading possibilities open to them and all buyers have identical beliefs about trading possibilities open to them.

It is important to distinguish contracts which are demanded by a positive measure of agents from those which are not demanded by anyone. Matching probabilities are well defined in markets for contracts that are traded in equilibrium, but for non-traded contracts the matching probabilities can be more or less arbitrary. Since Θ is supposed to represent the set of all possible contracts, most of the contracts in Θ will not be actively traded in equilibrium. This means that the equilibrium probability assessment μ is to a large extent arbitrary and this arbitrariness can give rise to a large set of equilibria, as we shall see.

For any attainable alloction f, let $\lambda_f(\theta)$ measure the long side of the market for contract θ, that is,

$$\lambda_f(\theta) = \max \left\{ \sum_s f(\theta, s), \sum_t f(\theta, t) \right\}.$$

The market for contract θ is called *active* if $\lambda_f(\theta) > 0$. Otherwise, the market for θ is called *inactive*. In active markets, beliefs are determined by rational expectations and the random matching process. Since the random matching process treats all buyers and all sellers symmetrically, the probability assessment $\mu(\theta, \cdot)$ must be the same for all agents if the market for θ is active. If the market is inactive, however, the agents' beliefs are more or less arbitrary (i.e., not determined by the matching technology) and here the assumption of common beliefs represents a mild symmetry condition.

The matching rules treat all agents on the same side of the market identically and maximize the probability of trade. A probability assessment is consistent with an allocation if it reflects the actual matching probabilities determined by the allocation in each active market. Formally, the probability assessment μ is *consistent* with the allocation f if

$$\lambda_f(\theta)\mu(\theta, k) = f(\theta, k),$$

for any θ and any k. If the market for θ is active, then $\lambda_f(\theta) > 0$ and the probability $\mu(\theta, k)$ is uniquely determined by the allocation f. If $\lambda_f(\theta) = 0$, the consistency condition is automatically satisfied and does not place any restrictions on the equilibrium probability assessment.

Now we are ready to define an equilibrium. Intuitively, an equilibrium requires each agent to choose the contract that maximizes his expected utility, taking as given the probability assessment, that is, the probability of trading each contract and the distribution of types with whom he may be matched. The probability assessment is determined jointly by the choices of all the agents. Formally, an *equilibrium* consists of an attainable allocation f and a consistent probability assessment μ such that, for every type s and any contract θ, a positive measure $f(\theta, s) > 0$ of agents choose θ only if it is optimal

$$\sum_t u(\theta, s, t)\mu(\theta, t) = u^*(s) = \max_\theta \left\{ \sum_t u(\theta, s, t)\mu(\theta, t) \right\} \geq c(s),$$

and for every type t and any contract θ, a positive measure $f(\theta, t) > 0$ of agents choose θ only if it is optimal

$$\sum_s v(\theta, s, t)\mu(\theta, s) = v^*(t) = \max_\theta \left\{ \sum_s v(\theta, s, t)\mu(\theta, s) \right\} \geq c(t).$$

The significance of the participation cost is simply to ensure that the equilibrium payoff from trading is positive if agents choose to participate. That is, $u^*(s) > 0$ if $f(\theta, s) > 0$ for some θ and $v^*(t) > 0$ if $f(\theta, t) > 0$ for some θ.

2.2 *Perfection and stability*

As has already been noted, the probability assessment $\mu(\theta, \cdot)$ is more or less arbitrary when the market for θ is inactive. The problem this poses for the theory is that there may be many different equilibrium allocations supported by different beliefs about trading probabilities in inactive markets. Some of these equilibria are of little interest because they depend on implausible beliefs about the trading possibilities in inactive markets. For example, if we assume that $\mu(\theta_0, k) = 0$ for some fixed but arbitrary θ_0 and any k, then it is easy to see that no one will attempt to trade the contract θ_0 in equilibrium. Then $f(\theta_0, k) = 0$ for every k, and the probability assessment $\mu(\theta_0, \cdot)$ will be consistent with the allocation $f(\theta_0, \cdot)$. In this way we can "close" the market for any contract θ_0 without violating the equilibrium conditions. By closing markets, we can generate a large number of different equilibrium allocations; but these equilibria are not very interesting.

To rule out such trivial equilibria, it is usually assumed that markets must be orderly, which means that at most one side of the market can be rationed in equilibrium (Hahn and Negishi, 1962; Dreze, 1975; Hahn, 1978). The probability assessment μ is *orderly* if, for every θ,

$$\max \left\{ \sum_t \mu(\theta, t), \sum_s \mu(\theta, s) \right\} = 1.$$

This restriction rules out some equilibria, but it does not eliminate the multiplicity of equilibria caused by arbitrary beliefs in inactive markets. To rule out these equilibria, a further restriction of beliefs in inactive markets is required. There are various strategies for refining an equilibrium concept. One is based on the idea of the 'trembling hand' introduced by Selten (1975). Here it is the 'invisible hand' of the market that 'trembles'. The essential idea is to perturb the economy so that all markets are active, find an equilibrium for the perturbed economy, and then let the perturbation become vanishingly small. The limit of this sequence of perturbed equilibria will be an equilibrium of the unperturbed economy.

Formally, a *perturbation* is an attainable allocation g such that $\lambda_g(\theta) > 0$ for all θ. For any perturbation g, denote the set of attainable allocations of the perturbed economy by $F(g)$, where

$$F(g) = \{f \in F | f \geq g\}.$$

The perturbed economy is denoted by $\mathcal{E}(g) = \{S, T, N, u, v, c, g\}$. Note that the parameters are the same as in the original economy $\mathcal{E}$; only the set of attainable allocations $F(g)$ has changed.

Define an equilibrium for the perturbed economy $\mathcal{E}(g)$ to be an attainable allocation $f \in F(g)$ and a consistent probability assessment μ such that, for each type s,

$$f(\cdot, s) \in \arg\max_{F(g)} \sum_{\theta} f(\theta, s) \left\{ \sum_{t} u(\theta, s, t)\mu(\theta, t) - c(s) \right\}$$

and, for each type t,

$$f(\cdot, t) \in \arg\max_{F(g)} \sum_{\theta} f(\theta, t) \left\{ \sum_{s} v(\theta, s, t)\mu(\theta, s) - c(t) \right\}.$$

A *perfect equilibrium* is the limit (f^0, μ^0) of a sequence of equilibria $\{(f^n, \mu^n)\}$ where, for each n, (f^n, μ^n) is an equilibrium of the perturbed economy $\mathcal{E}(n^{-1} \cdot g)$. Standard arguments suffice to show that there exists a perfect equilibrium for any economy $\mathcal{E}$.

In the perturbed economy, all markets are active, so the equilibrium probability assessment μ is uniquely determined by the equilibrium allocation $f \in F(g)$. A perfect equilibrium, being the limit of perturbed equilibria, has a probability assessment that is the limit of probability assessments generated by attainable allocations.

Note that an equilibrium of a perturbed economy is orderly by construction. For any θ, $\lambda_f(\theta) > 0$ so the consistency condition implies that $\mu(\theta, k) = f(\theta, k)/\lambda_f(\theta)$ and

$$\begin{aligned} \max\left\{ \sum_t \mu(\theta, t), \sum_s \mu(\theta, s) \right\} &= \max\left\{ \frac{\sum_t f(\theta, t)}{\lambda_f(\theta)}, \frac{\sum_s f(\theta, s)}{\lambda_f(\theta)} \right\} \\ &= \frac{\max\left\{\sum_t f(\theta, t), \sum_s f(\theta, s)\right\}}{\lambda_f(\theta)} \\ &= \frac{\lambda_f(\theta)}{\lambda_f(\theta)} = 1. \end{aligned}$$

A perfect equilibrium, being the limit of a sequence of orderly equilibria, is also orderly.

Although perfection imposes some restrictions on the equilibrium beliefs, they are pretty mild. In order to restrict beliefs further we have to use a stronger refinement. This refinement is related to the notion of strategic stability introduced by Kohlberg and Mertens (1986). A perfect equilibrium (f, μ) is robust to a single perturbation g in the sense that, for any $\varepsilon > 0$ and each n sufficiently large, there is an equilibrium of the perturbed economy $\mathcal{E}(n^{-1} \cdot g)$ which is ε-close to (f, μ). The Kohlberg-Mertens notion of stability requires this kind of robustness in the face of all possible perturbations. It may not be possible to find a single equilibrium that has such a property, so we consider sets of equilibria. Call S a *stable set* if it is a minimal set of equilibria with the property that:

for any perturbation g and any $\varepsilon > 0$ there exists a number $n_0 > 0$ such that for any $n > n_0$ there exists an equilibrium of the perturbed economy $\mathcal{E}(n^{-1} \cdot g)$ that is ε-close to S.

One reason why we need to consider sets of equilibria is that by choosing different perturbations we generate different probability assessments in the inactive markets. So sequences of equilibria corresponding to different perturbations may have different limiting probability assessments. However, this may be the only difference between the equilibria belonging to the stable set. In that case, there is a unique allocation that satisfies the stability criterion. If all the equilibria in S have the same allocation f then f is called a *unique stable allocation*. When there is no risk of ambiguity we refer to f as a *stable allocation* for short.

The existence of a unique stable allocation is established for generic economies with incomplete information in Gale (1992).

2.3 Stable beliefs

When an economy is perturbed, the probability assessments are forced to change. If the probability assessments change in a way that makes some unused contract θ more attractive, it may cause a large number of agents to deviate to θ, in which case the original equilibrium is deemed to have been non-robust or unstable. On the other hand, if the number of agents deviating to θ is small and if the probability assessment changes in such a way that no one strictly prefers θ to the equilibrium payoff, then a small perturbation has led to a small change in the equilibrium and the equilibrium is considered robust. In other words, if an allocation is stable then any perturbation can be stabilized by the endogenous re-allocation of a small number of agents. This principle plays a crucial role in what follows. The next step is to characterize exactly what this means for beliefs in a stable equilibrium.

Let f be a stable allocation and let g be a fixed but arbitrary perturbation. There is a sequence of equilibria $\{(f^n, \mu^n)\}$, where (f^n, μ^n) is an equilibrium of the perturbed economy $\mathcal{E}(n^{-1} \cdot g)$ for each n and

$$\lim_{n \to \infty} (f^n, \mu^n) = (f, \mu) \in S.$$

As was noted above, the probability assessment μ may depend on the particular sequence $\{(f^n, \mu^n)\}$. Let $u^*(k)$ denote the equilibrium payoff of type k in the limiting equilibrium (f, μ). Suppose that for some fixed but arbitrary contract θ_1, $u^*(s) > \sum_t \mu(\theta_1, t) u(\theta_1, s, t)$. Since the payoff functions are continuous, $\max_\theta \sum_t \mu^n(\theta, t) u(\theta, s, t) > \sum_t \mu^n(\theta_1, t) u(\theta_1, s, t)$, for all n sufficiently large and this implies that no agent of type s voluntarily chooses contract θ_1, i.e., $f^n(\theta_1, s) = n^{-1} \cdot g(\theta_1, s)$, for all n sufficiently large. A symmetric conclusion holds for buyers. Thus, we have established that, for any contract θ,

$$\left[u^*(s) > \sum_t \mu(\theta, t) u(\theta, s, t)\right] \Longrightarrow \left[\mu^n(\theta, s) = \frac{n^{-1} g(\theta, s)}{\lambda^n(\theta)}\right], \forall s$$

$$\left[v^*(t) > \sum_s \mu(\theta, s) v(\theta, s, t)\right] \Longrightarrow \left[\mu^n(\theta, t) = \frac{n^{-1} g(\theta, t)}{\lambda^n(\theta)}\right], \forall t,$$

for all n sufficiently large. Taking limits as $n \to \infty$ immediately proves the following result.

Theorem 1 *Suppose that f is a unique stable allocation and g is a fixed but arbitrary perturbation. Then there exists a probability assessment μ such that (f, μ) is an equilibrium and, for any contract θ and for some constant $\gamma(\theta) \geq 0$,*

$$\left[u^*(s) > \sum_t \mu(\theta, t)u(\theta, s, t)\right] \Longrightarrow [\mu(\theta, s) = \gamma(\theta)g(\theta, s)],$$

for any seller type s and

$$\left[v^*(t) > \sum_s \mu(\theta, s)v(\theta, s, t)\right] \Longrightarrow [\mu(\theta, t) = \gamma(\theta)g(\theta, t)],$$

for any buyer type t.

To sum up, a perturbation changes the equilibrium probability assessment, but exactly how it changes depends on the equilibrium responses of the agents. So the relationship between the equilibrium probability assessment and the perturbation in the market for a contract θ depends on whether agents find it optimal to choose that contract in equilibrium. In particular, if no one finds it optimal to choose θ in equilibrium, then $f^n(\theta, \cdot) = n^{-1}g(\theta, \cdot)$ and the probability assessment $\mu^n(\theta, \cdot)$ is determined by the perturbation $n^{-1}g(\theta, \cdot)$. On the other hand, if the perturbation $g(\theta, \cdot)$ by itself would have made θ attractive to some types (would give them a payoff higher than their equilibrium payoff) then in equilibrium a small number of agents must be endogenously re-allocated to θ in order to make θ less attractive and prevent a large deviation by other agents. So one way to show that an allocation is not stable is to find a perturbation g that cannot be stabilized by a small re-allocation of agents.

2.4 General contract spaces

Even in finite games, the definition of a perturbation requires some care (see Kohlberg and Mertens, 1986). There is no comparable development of the theory for infinite games or economies with an infinite number of contracts. Similarly, the existence of equilibrium in an economy with a finite number of contracts is a straightforward matter (it uses a standard fixed point argument) but the existence of a unique stable allocation is more difficult (see Gale, 1992). Again, the theory has not been developed to deal with an infinite number of contracts.

The assumption of a finite number of contracts is a convenient simplification, but for some purposes it is more convenient to have a continuum of contracts. In particular, when it comes to characterizing the degree of separation in equilibrium it is nice to be able to consider neighboring contracts. The theory can be extended from a finite to an infinite set of contracts by taking limits (this was the approach taken in Gale, 1992), but for simplicity an alternative approach is adopted here. I take as a definition of stability the characterization of stable beliefs derived in Theorem 1. From now on it is assumed that Θ is an open subset of some finite-dimensional Euclidean space endowed with the usual topology. The payoff functions $u(\cdot, s, t)$ and $v(\cdot, s, t)$ are assumed to be continuously differentiable functions of θ on Θ, for every pair (s, t).

When the contract space is infinite, an allocation is represented by a measure. To keep things simple (and it really does not make much difference to the analysis to follow), we assume that the number of contracts traded in equilibrium is finite. In that case, we can continue to define an allocation as a function $f : \Theta \times K \to \mathbf{R}_+$ with finite support, where $f(\theta, k)$ is the number of agents of type k that choose $\theta \in \Theta$. The allocation f is attainable if

$$\sum_{\theta} f(\theta, k) \leq N(k), \forall k.$$

An orderly probability assessment is a measurable function $\mu : \Theta \times K \to \mathbf{R}_+$ such that

$$\max \left\{ \sum_t \mu(\theta, t), \sum_s \mu(\theta, s) \right\} = 1, \forall \theta.$$

The probability assessment μ is consistent with an attainable allocation f if

$$\lambda_f(\theta)\mu(\theta, k) = f(\theta, k)$$

for every θ, where $\lambda_f(\theta) = 0$ for all but a finite number of contracts θ.

An attainable allocation f is said to be *stable* if it satisfies the condition that

for any attainable allocation g, there exists an orderly probability assessment μ consistent with f such that for any contract θ and some constant $\gamma(\theta) \geq 0$,

$$\left[u^*(s) > \sum_t \mu(\theta, t) u(\theta, s, t) \right] \Longrightarrow [\mu(\theta, s) = \gamma(\theta) g(\theta, s)] \quad (1)$$

for any seller type s and

$$\left[v^*(t) > \sum_s \mu(\theta, s) v(\theta, s, t) \right] \Longrightarrow [\mu(\theta, t) = \gamma(\theta) g(\theta, t)] \quad (2)$$

for any buyer type t, where $u^*(s)$ and $v^*(t)$ are the payoffs of types s and t respectively in the allocation f.

3 Separation with one-sided uncertainty

Paralleling the familiar signaling models in the literature, we first consider the special case of an economy in which there is a single type of buyer $T = \{t_0\}$. This is a special case of one-sided uncertainty (OSU).[1] We can suppress the reference to the buyers' type and write the payoff functions as $u(\theta, s)$ and $v(\theta, s)$. Normalize the number of buyers to 1. Then the economy $\mathcal{E} = \{S, N, \Theta, u(\cdot), v(\cdot), c(\cdot)\}$ is defined by the set of seller types S, the distribution of seller types $N(s)$, the set of contracts Θ, the payoff functions $u(\theta, s)$ and $v(\theta, s)$, and the participation cost function $c(k)$.

The result presented in this section gives conditions that are sufficient to rule out pooling of different types of sellers at a single contract if the allocation is stable. We are only interested in contracts that are actually traded in equilibrium, so there is no loss of generality in focusing attention on a contract θ_0 such that

$$\sum_s f(\theta_0, s) > 0 \text{ and } f(\theta_0, t_0) > 0.$$

Let S_0 denote the set of seller types that find θ_0 optimal, that is,

$$S_0 = \{s \in S | \mu(\theta_0, t_0) u(\theta_0, s) = u^*(s)\}.$$

Then it is assumed that there is a single type s_0 in S_0 that is most preferred by the buyers, that is,

$$v(\theta_0, s_0) > v(\theta_0, s), \forall s \in S_0, s \neq s_0.$$

This condition merely rules out ties and avoids some complications that do not seem to be important in the general analysis. The main condition in the theorem is that there exists a contract θ_1 that is arbitrarily close to θ_0 and that is preferred by type s_0 and not by any types $s \in S_0, s \neq s_0$:

$$(S1)\ u(\theta_1, s_0) > u(\theta_0, s_0)$$
$$(S2)\ u(\theta_1, s) < u(\theta_0, s), \forall s \in S_0, s \neq s_0.$$

This condition is much weaker than the famous single-crossing property, although it is implied by the latter. It requires only that we can find some dimension of the contract and some direction in that dimension that is preferred by type s_0 and only

[1] An economy is characterized by OSU (one-sided adverse selection) if the agents on one side of the market, say the sellers, are indifferent about being matched with different types of buyers. If there is only one type of buyer, this condition is automatically satisfied, but it would also be satisfied when there are heterogeneous types of buyers, as long as sellers do not care which type of buyer they are matched with. Formally, OSU requires $u(\theta, s, t) = u(\theta, s, t')$ for every θ, s, t and t'. Since sellers do not care about the buyers' types, they do not face an adverse selection problem and this simplifies the analysis considerably. The existence of heterogeneous types of buyers may still be important in equilibrium, however. Because different buyer types have different preferences over contracts and types of sellers, they will typically choose different contracts in equilibrium. Gale (1992) shows how this can lead to positive assortative matching in equilibrium. This phenomenon, which cannot arise when all buyers are identical, illustrates one way in which the model encompasses a richer set of equilibrium possibilities than the classical signaling game, even when they are superficially similar.

by that type. Beyond this, the argument is much the same as in the standard analysis, except that it works through perturbations of the game and has to take account of the possiblity of rationing, which does not appear in models with market-clearing prices.

Under these conditions, it is impossible for two types of sellers to pool at a contract θ_0. The formal proof is left until Section 5. A heuristic proof follows. First, consider a perturbation of the game that assigns only type s_0 to contracts that are not used in the stable allocation f. If no type finds θ_1 weakly optimal, then buyers must believe that only type s_0 will exchange contract θ_1. By assumption, sellers of type s_0 prefer θ_1 to θ_0 and, for θ_1 close to θ_0, buyers will always prefer to trade θ_1 with s_0 rather than trade θ_0 with a mixture of s_0 and less desirable types. Since markets are orderly, some agents can trade θ_1 with probability one and this contradicts the definition of stability. So at least one type of seller other than s_0 must find θ_1 weakly optimal, but then condition $(S1)$ implies that s_0 strictly prefers θ_1 to θ_0, again contradicting the stability condition.

This proves the following result.

Theorem 2 *Let f be a stable allocation and let θ_0 be a contract in Θ that is traded in equilibrium: $f(\theta_0, s) > 0$ for some s and $f(\theta_0, t_0) > 0$. Suppose (i) that there is a unique best type $s_0 \in S_0 = \{s : \mu(\theta_0, t_0)u(\theta_0, s) = u^*(s) \geq c(s)\}$, and (ii) that for any $\varepsilon > 0$ there is a contract θ_1 that is ε-close to θ_0 and satisfies conditions $(S1)$ and $(S2)$. Then $f(\theta_0, s) = 0$ for any $s \neq s_0$.*

While Theorem 2 provides conditions under which at most one type chooses θ_0, it will often be the case that θ_0 will be optimal for more than one type. In other words, the self-selection constraint is binding in equilibrium.

3.1 Discussion

The model analyzed in this section is a special case of the model described in Gale (1992). In particular, Gale (1992) allows for private information on one side of the market and heterogeneous agents on both sides. However, the assumptions used in Gale (1992) to prove separation are much stronger than the assumptions of Theorem 2. In particular, in order to prove that all private information is revealed in a stable equilibrium, Gale (1992) assumes that preferences satisfy the Spence-Mirrlees or single-crossing condition.

The separation conditions $(S1\text{-}S2)$ say that, for any contract θ_0, we can find a nearby contract θ_1 that is better for s_0 and worse for every type $s \in S_0$, $s \neq s_0$. Since S_0 is determined endogenously, conditions $(S1\text{-}S2)$ are not an assumption about primitives. However, it is easy to find a condition on primitives that is sufficient for conditions $(S1\text{-}S2)$. Say that $s < s_0$ if $v(\theta_0, s) < v(\theta_0, s_0)$. Then assume that for any s_0 and all $s < s_0$ there exists a contract θ_1 arbitrarily close to θ_0 such that

$$u(\theta_1, s_0) > u(\theta_0, s_0)$$

and

$$u(\theta_1, s) < u(\theta_0, s), \forall s < s_0.$$

This condition can be interpreted as a condition on the 'richness' of the contract set Θ. More precisely, it is a joint condition on the rank of the Jacobian matrix

$$\left[\frac{\partial u(\theta_0, s)}{\partial \theta}\right]$$

and the dimension of the contract space Θ. If the dimension of Θ is greater than $|S|$ and the Jacobian matrix has full rank, then the conditions ($S1$-$S2$) will be satisfied. We can thus see the conditions ($S1$-$S2$) as a combination of a regularity assumption and an assumption on the relative dimensions of Θ and S. This discussion is formalized in the following result.

Corollary 3 *Let f be a stable allocation and let θ_0 be a contract in Θ that is traded in equilibrium: $f(\theta_0, s) > 0$ for some s and $f(\theta_0, t_0) > 0$. Suppose (i) that there is a unique best type $s_0 \in S_0 = \{s : \mu(\theta_0, t_0)u(\theta_0, s) = u^*(s) \geq c(s)\}$, and (ii) the dimension of Θ is greater than $|S|$ and the Jacobian matrix has full rank. Then $f(\theta_0, s) = 0$ for any $s \neq s_0$.*

Note that it is not enough to assume that Θ is 'big' because some dimensions of Θ may not be payoff relevant. That is why the regularity (full rank) condition has to be added.

4 Separation with two-sided uncertainty

Returning to the 'general' model of an economy with two-sided uncertainty (TSU), it is interesting to see how the conditions for separation change. Theorem 2 shows that conditions on preferences over contracts are sufficient for equilibrium separation of types. More precisely, conditions on preferences over contracts ensure that s_0 is more likely to choose θ_1 than any other type of seller and this ensures that the pooling equilibrium is destabilized when a small measure of type s_0 are assigned to θ_1.

In an economy with TSU, things are more complicated. For all agents, the payoff to a contract $\theta_1 \neq \theta_0$ depends on both the contract θ_1 and the distribution of types $\mu(\theta_1, \cdot)$ associated with θ_1. Looking at it from the point of view of sellers, one cannot say which type of seller will be attracted to θ_1 until one knows the distribution of buyers that are expected to trade θ_1. Similarly, one cannot say which type of buyer will be attracted to θ_1 until one knows the distribution of sellers that are expected to trade θ_1. Even if sellers of type s_0 prefer θ_1 to θ_0, other things being equal, the probability assessment $\mu(\theta_1, t)$ can make θ_0 more attractive than θ_1. Furthermore, Theorem 1 tells us that a reasonable probability assessment $\mu(\theta_1, t)$ depends on which types of buyers find it optimal to choose θ_1. For this reason, one cannot determine what is a reasonable probability assessment $\mu(\theta, \cdot)$ by looking at one side of the market only. To determine what is a reasonable belief one must look at both sides of the market simultaneously.

4.1 The inclusion principle

To generalize the separation condition used in the case of OSU, we need to be able to make the following kind of statement: if s_0 chooses a contract θ_0 then there exists a nearby contract θ_1 such that whenever a less desirable type $s < s_0$ finds θ_1 weakly optimal, s_0 will find θ_1 strictly optimal. In other words,

$$\left[\sum_t \mu(\theta_1,t)u(\theta_1,s,t) \geq \sum_t \mu(\theta_0,t)u(\theta_0,s,t)\right]$$
$$\Longrightarrow \left[\sum_t \mu(\theta_1,t)u(\theta_0,s_0,t) > \sum_t \mu(\theta_0,t)u(\theta_0,s_0,t)\right].$$

If this is true for all contracts θ_1 arbitrarily close to θ_0 then, given the continuity of the payoff functions, we can take the limit as $\theta_1 \to \theta_0$ to get

$$\left[\sum_t [m(t) - \mu(\theta_0,t)]u(\theta_0,s,t) \geq 0\right] \tag{3}$$
$$\Longrightarrow \left[\sum_t [m(t) - \mu(\theta_0,t)]u(\theta_0,s_0,t) \geq 0\right],$$

where $m(t)$ represents the limit of the probability assessments $\mu(\theta_1,t)$ as $\theta_1 \to \theta_0$. Since we have little information about the probability assessments that might obtain in equilibrium, if we want the implication (3) to hold in equilibrium we will have to assume that it holds for all possible probability assessments. That is, for any vector of weights $\lambda = (\lambda(t))$ we must assume that

$$\left[\sum_t \lambda(t)u(\theta_0,s,t) \geq 0\right] \Longrightarrow \left[\sum_t \lambda(t)u(\theta_0,s_0,t) \geq 0\right]. \tag{4}$$

The implication (4) is generally valid if and only if $u(\theta_0,s,\cdot)$ is a positive scalar multiple of $u(\theta_0,s_0,\cdot)$, that is,

$$u(\theta_0,s,t) = \alpha(\theta_0,s)u(\theta_0,s_0,t), \forall t,$$

for any $s < s_0$ and for some constant $\alpha(\theta_0,s) > 0$.

If this relationship holds for every contract θ and every type s (how else can we guarantee that it will hold in equilibrium as required?), then the payoff functions can be written in the form

$$u(\theta,s,t) = a(\theta,s)b(\theta,t), \forall(\theta,s,t) \in \Theta \times S \times T.$$

In other words, the preferences are separable and each type of seller has identical marginal preferences over different types of buyers. This is a strong but natural condition that is sufficient for extending the argument used in Section 3 to the case of TSU. Later, we consider the extent to which this condition might be necessary as well.

4.2 Separation with separable preferences

Assume that all the agents on one side of the market have similar preferences over the types of agents on the other side and vice versa, that is,

$$u(\theta, s, t) = a(\theta, s)b(\theta, t)$$
$$v(\theta, s, t) = c(\theta, t)d(\theta, s).$$

Let f be a stable allocation and θ_0 a contract that is traded in f. Let $S_0 = \{s | \sum_t \mu(\theta_0, t)u(\theta_0, s, t) = u^*(s) \geq c(s)\}$ be the set of seller types for which θ_0 is an optimal choice. We assume that there is a unique best type $s_0 \in S_0$. Then $s \in S_0$ and $s \neq s_0$ imply that $d(\theta_0, s) < d(\theta_0, s_0)$. We shall also assume that $d(\theta_0, s_0) \neq d(\theta_0, s)$ for any $s \notin S_0$. The separation condition is the following: for any $\varepsilon > 0$ there is a contract θ_1 that is ε-close to θ_0 and satisfies

$$(S3)\ \frac{a(\theta_1, s_0)}{a(\theta_0, s_0)} > \frac{a(\theta_1, s)}{a(\theta_0, s)}, \forall t, \forall s \ni d(\theta_0, s) < d(\theta_0, s_0).$$

Note that the separation condition $(S3)$ restricts the preferences of all the types s that are less attractive than s_0 to buyers, and not just the types $s \in S_0, s \neq s_0$. A similar condition is required for the other side of the market:

$$(S4)\ \frac{c(\theta_1, s_0)}{c(\theta_0, s_0)} > \frac{c(\theta_1, s)}{c(\theta_0, s)}, \forall s, \forall t \ni b(\theta_0, s) < b(\theta_0, s_0).$$

Note that ($S3$-$S4$) does not require that θ_1 is preferred to θ_0 by types s_0 and t_0 other things being equal. This would be an implausible condition given the nature of trade. Rather, it requires that types s_0 and t_0 dislike θ_1 relatively less than types $s < s_0$ and $t < t_0$, respectively.

With these assumptions we can show that a stable allocation must be separating.

Theorem 4 *Suppose that the payoff functions have the additively separable form*

$$u(\theta, s, t) = a(\theta, s)b(\theta, t)$$
$$v(\theta, s, t) = c(\theta, t)d(\theta, s).$$

Let f be a stable allocation and θ_0 a contract belonging to the interior of Θ that is traded in equilibrium. Let S_0 (resp. T_0) denote the set of seller types (resp. buyer types) that find θ_0 optimal in equilibrium. Let s_0 (resp. t_0) denote the best type in S_0 (resp. T_0). Assume that $d(\theta_0, s) \neq d(\theta_0, s_0)$ if $s \neq s_0$ (resp. $b(\theta_0, t) \neq b(\theta_0, t_0)$ if $t \neq t_0$) and that for any $\varepsilon > 0$ we can find a contract θ_1 that is ε-close to θ_0 and satisfies ($S3$-$S4$). Then $f(\theta_0, s) = 0$ for $s \in S_0, s \neq s_0$ and $f(\theta_0, t) = 0$ for $t \in T_0, t \neq t_0$.

Proof. See Section 5. □

Again, we can see condition ($S3$-$S4$) is implied by the joint assumption that the space of contracts Θ is sufficiently rich (has a high dimension) and that the payoff functions are regular (the Jacobian matrix has full rank).

Corollary 5 *Suppose that the payoff functions have the additively separable form and let f be a stable allocation and θ_0 a contract belonging to the interior of Θ that is traded in equilibrium. Let s_0 (resp. t_0) denote the best type in S_0 (resp. T_0), the types of sellers (resp. buyers) that find θ_0 optimal in equilibrium. Assume that $d(\theta_0, s) \neq d(\theta_0, s_0)$ if $s \neq s_0$ (resp. $b(\theta_0, t) \neq b(\theta_0, t_0)$ if $t \neq t_0$). If Θ has a sufficiently high dimension and the payoff functions are regular (the Jacobian matrix has full rank) then $f(\theta_0, s) = 0$ for $s \in S_0, s \neq s_0$ and $f(\theta_0, t) = 0$ for $t \in T_0, t \neq t_0$.*

Separability is a natural assumption to make in this context in order to extend the argument used in the OSU case. Nonetheless, separability is restrictive so it is important to ask how far one can relax this assumption and still guarantee separation of types in a stable allocation.

4.3 Separation without separability

Without separability, there is no hope of applying the general line of argument used above, but there is additional structure that might be used to argue that private information will be fully revealed even if preferences are not separable. To illustrate the problems and possibilities of this approach, consider the case where there are two types on each side of the market, $S = \{s_1, s_2\}$ and $T = \{t_1, t_2\}$.

To simplify, the types are ranked in order of attractiveness to the other side of the market. Types s_2 and t_2 are the 'good' types and s_1 and t_1 are the 'bad' types. This means that, for any contract θ, and for any seller type s

$$u(\theta, s, t_1) < u(\theta, s, t_2),$$

and for any buyer type t

$$v(\theta, s_1, t) < v(\theta, s_2, t).$$

Suppose that f is a stable allocation and let θ_0 be a traded contract. Let S_0 and T_0 denote the types of sellers and buyers, respectively, for whom θ_0 is an optimal choice. If S_0 and T_0 are singletons, there is nothing to prove; if one of them is a singleton, the argument used in Section 3 will suffice to establish separation. So, without loss of generality, we can assume that $S_0 = \{s_1, s_2\}$ and $T_0 = \{t_1, t_2\}$.

Suppose then, contrary to what we want to prove, that $f(\theta_0, s_1) > 0$ and $f(\theta_0, t_1) > 0$. Let θ_1 be an arbitrary contract that is very close to θ_0. Consider a perturbation g such that $g(\theta_1, s_2) = g(\theta_1, t_2) > 0$ and $g(\theta_1, s_1) = g(\theta_1, t_1) = 0$. Let (f, μ) be the equilibrium satisfying (1-2) relative to the perturbation g.

Markets are orderly, so at most one side of the market is rationed. Suppose that the buyers are constrained in the market for the contract θ_1 (the other case is exactly symmetrical). Then the sellers are unconstrained and their probability of trade is $\sum_t \mu(\theta_1, t) = 1$.

For any θ_1 sufficiently close to θ_0, the continuity of the payoff functions and the assumption that type t_2 is better (more desirable) than type t_1 implies that

$$\begin{aligned} u(\theta_1, s_2, t_2) &> \sum_t \mu(\theta_0, t) u(\theta_0, s_2, t) \\ &= u^*(s_2). \end{aligned}$$

So the optimality conditions require that $\mu(\theta_1, t_2) < 1$ and $\mu(\theta_1, t_1) > 0$. Given the construction of the perturbation g and the stability condition (2) this can only be true if

$$v^*(t_1) = \sum_s \mu(\theta_1, s) v(\theta_1, s, t_1). \tag{5}$$

There are two cases to consider.

Case 1. First, suppose that θ_1 is not optimal for sellers of type s_1. Then (1) and the construction of g imply that $\mu(\theta_1, s_1) = 0$. Then the optimality condition (5) reduces to

$$\sum_s \mu(\theta_0, s) v(\theta_0, s, t_1) = \mu(\theta_1, s_2) v(\theta_1, s_2, t_1).$$

and the optimality condition for buyers of type t_2 can be written

$$\sum_s \mu(\theta_0, s) v(\theta_0, s, t_2) \geq \mu(\theta_1, s_2) v(\theta_1, s_2, t_2).$$

Dividing these conditions by $v(\theta_1, s_2, t_1)$ and $v(\theta_1, s_2, t_2)$ respectively and letting θ_1 converge to θ_0 gives

$$\mu(\theta_0, s_1) \frac{v(\theta_0, s_1, t_1)}{v(\theta_0, s_2, t_1)} + \mu(\theta_0, s_2) = m(s_2)$$

and

$$\mu(\theta_0, s_1) \frac{v(\theta_0, s_1, t_2)}{v(\theta_0, s_2, t_2)} + \mu(\theta_0, s_2) \geq m(s_2),$$

where $m(s)$ denotes the limiting value of $\mu(\theta_1, s)$ as $\theta_1 \to \theta_0$. These conditions are mutually inconsistent if the relative variation in payoffs for type t_1 is greater than for type t_2:

$$\frac{v(\theta, s_1, t_2)}{v(\theta, s_2, t_2)} > \frac{v(\theta, s_1, t_1)}{v(\theta, s_2, t_1)}, \forall \theta. \tag{6}$$

Thus, condition (6) is sufficient to rule out pooling in this case.

Case 2. The second case assumes that θ_1 is optimal for sellers of type s_1. Then

$$\sum_t \mu(\theta_1, t) u(\theta_1, s_1, t) = \sum_t \mu(\theta_0, t) u(\theta_0, s_1, t). \tag{7}$$

We need to show that condition (7) implies that sellers of type s_2 will strictly prefer the contract θ_1. The following two conditions will be shown to be sufficient. The first condition is that type s_2 is more sensitive than type s_1 to the type of buyer he is matched with:

$$u(\theta, s_2, t_2) - u(\theta, s_2, t_1) > u(\theta, s_1, t_2) - u(\theta, s_1, t_1) \tag{8}$$

for any contract θ. The second condition is that type s_1 has a stronger relative preference for θ_0 over θ_1 than does type s_2:

$$0 < u(\theta_0, s_1, t) - u(\theta_1, s_1, t) > u(\theta_0, s_2, t) - u(\theta_1, s_2, t), \forall t. \tag{9}$$

The condition (7) can be rewritten as

$$\sum_t \left[\mu(\theta_1, t) - \mu(\theta_0, t)\right] u(\theta_1, s_1, t) = \sum_t \mu(\theta_0, t) \left[u(\theta_0, s_1, t) - u(\theta_1, s_1, t)\right]. \tag{10}$$

The right hand side of (10) is positive (condition (9) says that s_1 prefers θ_0 to θ_1) so the left hand side must be positive and this implies that the probability distribution $\mu(\theta_1, t)$ puts more weight on type t_2 relative to $\mu(\theta_0, t)$:

$$\mu(\theta_1, t_2) - \mu(\theta_0, t_2) > 0 > \mu(\theta_1, t_1) - \mu(\theta_0, t_1).$$

Then condition (8) implies that

$$\sum_t \left[\mu(\theta_1, t) - \mu(\theta_0, t)\right] u(\theta_1, s_2, t) > \sum_t \left[\mu(\theta_1, t) - \mu(\theta_0, t)\right] u(\theta_1, s_1, t)$$

and condition (9) implies that

$$\sum_t \mu(\theta_0, t) \left[u(\theta_0, s_1, t) - u(\theta_1, s_1, t)\right] > \sum_t \mu(\theta_0, t) \left[u(\theta_0, s_2, t) - u(\theta_1, s_2, t)\right].$$

Combining these conditions with (10) gives

$$\begin{aligned}
\sum_t \left[\mu(\theta_1, t) - \mu(\theta_0, t)\right] u(\theta_1, s_2, t) &> \sum_t \left[\mu(\theta_1, t) - \mu(\theta_0, t)\right] u(\theta_1, s_1, t) \\
&= \sum_t \mu(\theta_0, t) \left[u(\theta_0, s_1, t) - u(\theta_1, s_1, t)\right] \\
&> \sum_t \mu(\theta_0, t) \left[u(\theta_0, s_2, t) - u(\theta_1, s_2, t)\right],
\end{aligned}$$

which proves the θ_1 is strictly preferred to θ_0 by type s_2, contradicting the optimality conditions.

A symmetric argument applies for the case in which the sellers are constrained in the market for θ_1, but note that in order to deal with both cases we need to find a single contract θ_1 that satisfies the condition (6) for the buyers and the analogue for sellers, and that satisfies conditions (8) and (9) for sellers and their analogues for the buyers. These are not innocuous conditions but they are not extremely restrictive either. They are, however, much more restrictive than the conditions required in Section 3.

Theorem 6 *Suppose that there are two types of buyers and sellers, $S = \{s_1, s_2\}$ and $T = \{t_1, t_2\}$, and that the types can be ranked as follows:*

$$u(\theta, s, t_1) < u(\theta, s, t_2), \forall s,$$

$$v(\theta, s_1, t) < v(\theta, s_2, t), \forall t.$$

Let f be a stable allocation and θ_0 a contract belonging to the interior of Θ that is traded in equilibrium. Assume that for any contract θ

$$\frac{v(\theta, s_1, t_2)}{v(\theta, s_2, t_2)} > \frac{v(\theta, s_1, t_1)}{v(\theta, s_2, t_1)},$$
$$\frac{u(\theta, s_2, t_1)}{u(\theta, s_2, t_2)} > \frac{u(\theta, s_1, t_1)}{u(\theta, s_1, t_2)},$$

and that

$$u(\theta, s_2, t_2) - u(\theta, s_2, t_1) > u(\theta, s_1, t_2) - u(\theta, s_1, t_1)$$
$$v(\theta, s_2, t_2) - v(\theta, s_1, t_2) > v(\theta, s_2, t_1) - v(\theta, s_1, t_1).$$

For any $\varepsilon > 0$ suppose that there is a contract θ_1 that is ε-close to θ_0 and that

$$0 < u(\theta_0, s_1, t) - u(\theta_1, s_1, t) > u(\theta_0, s_2, t) - u(\theta_1, s_2, t), \forall t,$$
$$0 < v(\theta_0, s, t_1) - v(\theta_1, s, t_1) > v(\theta_0, s, t_2) - v(\theta_1, s, t_2), \forall s.$$

Then if θ_0 is optimal for types s_2 and t_2, it is not traded by types s_1 and t_1, that is, $f(\theta_0, s_1) = 0 = f(\theta_0, t_1)$.

4.4 Discussion

For the case of OSU, Theorem 2 shows that conditions on preferences over contracts ($S1$-$S2$) are sufficient for separation of types in a stable allocation. With the assumption of separability, it is possible to extend this result to the case of TSU. Theorem 4 shows that separation of types in a stable allocation is implied by the conditions ($S3$-$S4$), which only refer to agents' preferences over contracts. Furthermore, these conditions on preferences over contracts can be interpreted as requiring that the contract space be "sufficiently rich".

Separability is a natural condition to impose if we want to extend the separation argument from OSU to TSU. At the same time it is restrictive and so it would be nice to relax it. The 2×2 example suggests that sufficient conditions for full revelation of private information will have to be much stronger under TSU without separability than under OSU or under TSU with separability. Essentially, we have seen that with OSU a stable allocation will be fully revealing as long as the contract space is sufficiently rich. With TSU, a lot more structure has to be placed on the preferences of the agents to ensure that a stable allocation is separating. In particular, we need to put restrictions on the *intensity* of agents' preferences over the different types of agents on the other side of the market.

The conditions obtained so far are sufficient but not necessary. However, the restrictiveness of the sufficient conditions does suggest that full revelation may be less likely under TSU without separability than in the other cases.

5 Proofs

Proof of Theorem 2. Let θ_1 be a contract arbitrarily close to θ_0 that satisfies conditions $(S1)$ and $(S2)$. Let g denote a perturbation such that for any choice of θ_1 close to θ_0

$$g(\theta_1, s) = \begin{cases} \delta > 0 & s = s_0 \\ 0 & s \neq s_0 \end{cases}.$$

From the definition of a stable allocation, there exists an equilibrium (f, μ) whose probability assessment satisfies conditions (1-2).

I claim that $\mu(\theta_1, s) = 0$ for any $s \neq s_0$. To prove this, we start by noting that the equilibrium condition for s_0 implies that

$$\mu(\theta_1, t_0)u(\theta_1, s_0) \leq \mu(\theta_0, t_0)u(\theta_0, s_0). \tag{11}$$

From conditions ($S1$-$S2$), we know that $u(\theta_1, s_0) > u(\theta_0, s_0) > 0$, so inequality (11) implies that

$$\mu(\theta_1, t_0) < \mu(\theta_0, t_0). \tag{12}$$

This fact is enough to show that θ_1 is not optimal for any $s \notin S_0$. To see this, we consider two cases. If $\mu(\theta_1, t_0)u(\theta_1, s) < 0$, the conclusion is obvious so suppose that $0 \leq \mu(\theta_1, t_0)u(\theta_1, s) \leq \mu(\theta_0, t_0)u(\theta_1, s)$. If θ_0 is not optimal for s, then either some other contract is strictly preferred or no trade is strictly preferred:

$$\mu(\theta_0, t_0)u(\theta_0, s) - c(s) < \max\{u^*(s) - c(s), 0\}. \tag{13}$$

Then (12) and (13) imply that

$$\mu(\theta_1, t_0)u(\theta_1, s) - c(s) < \max\{u^*(s) - c(s), 0\}$$

because $u(\theta_1, s)$ is approximately equal to $u(\theta_0, s)$ for θ_1 arbitrarily close to θ_0. In other words, θ_1 is not an optimal choice for $s \notin S_0$, as claimed.

Now consider the types $s \in S_0, s \neq s_0$. To show that θ_1 is not an optimal choice for s there are three cases that need to be considered. If $\mu(\theta_1, t_0) = 0$ there is nothing to prove since $u^*(s) > 0$. If $\mu(\theta_1, t_0) > 0$ and $u(\theta_1, s) < 0$ then again there is nothing to prove since $\mu(\theta_1, t_0)u(\theta_1, s) < 0$ violates individual rationality. So we are left with the case where $\mu(\theta_1, t_0) > 0$ and $u(\theta_1, s) \geq 0$. Then $\mu(\theta_1, t_0) < \mu(\theta_0, t_0)$ and $u(\theta_1, s) < u(\theta_0, s)$ imply $\mu(\theta_1, t_0)u(\theta_1, s) < \mu(\theta_0, t_0)u(\theta_0, s)$ as required.

We have shown that θ_1 is not an optimal choice for any $s \neq s_0$ if θ_1 is chosen close enough to θ_0. Then $(S1)$ and the definition of g imply that $\mu(\theta_1, s) = 0$ for every $s \neq s_0$, as claimed.

Orderliness and $\mu(\theta_1, t_0) < 1$ imply that buyers are not rationed at θ_1, that is, $\mu(\theta_1, s_0) = \sum_s \mu(\theta_1, s) = 1$. The optimality condition for type t_0 requires

$$\begin{aligned} \mu(\theta_1, s_0)v(\theta_1, s_0) &= \sum_s \mu(\theta_1, s)v(\theta_1, s) \\ &\leq \sum_s \mu(\theta_0, s)v(\theta_0, s). \end{aligned}$$

For θ_1 sufficiently close to θ_0, we have $v(\theta_1, s_0) \approx v(\theta_0, s_0) > v(\theta_0, s)$ for every $s \in S_0, s \neq s_0$, so the equilibrium condition will only be satisfied if $\mu(\theta_0, s) = 0$ for every $s \in S_0, s \neq s_0$. Since θ_0 is traded in equilibrium, this implies that $f(\theta_0, s) = 0$ for all $s \neq s_0$. □

Proof of Theorem 4. Let θ_1 be a contract arbitrarily close to θ_0 that satisfies $(S3\text{-}S4)$. We want to show that θ_1 is not optimal for any type s such that $d(\theta_0, s) < d(\theta_0, s_0)$. The optimality condition

$$\sum_t \mu(\theta_1, t) u(\theta_1, s_0, t) \leq \sum_t \mu(\theta_0, t) u(\theta_0, s_0, t)$$

can be rewritten as

$$\frac{a(\theta_1, s_0)}{a(\theta_0, s_0)} \sum_t \mu(\theta_1, t) b(\theta_1, t) \leq \sum_t \mu(\theta_0, t) b(\theta_0, t). \tag{14}$$

If $\sum_t \mu(\theta_1, t) b(\theta_1, t) = 0$ then θ_1 is clearly not optimal for any type s so without loss of generality we can assume that $\sum_t \mu(\theta_1, t) b(\theta_1, t) > 0$. Then (14) and the separation condition imply that

$$\frac{a(\theta_1, s)}{a(\theta_0, s)} \sum_t \mu(\theta_1, t) b(\theta_1, t) < \sum_t \mu(\theta_0, t) b(\theta_0, t),$$

for s such that $d(\theta_0, s) < d(\theta_0, s_0)$ and this last inequality can be reorganized to give

$$\sum_t \mu(\theta_1, t) u(\theta_1, s, t) < \sum_t \mu(\theta_0, t) u(\theta_0, s, t)$$

for s such that $d(\theta_0, s) < d(\theta_0, s_0)$. This proves that θ_1 is not optimal for $s < s_0$.

A similar argument applies to the other side of the market.

To complete the proof, consider a perturbation g such that

$$g(\theta_1, k) = \begin{cases} \delta > 0 & k = s_0, t_0 \\ 0 & k \neq s_0, t_0 \end{cases}.$$

Since f is stable there exists a consistent probability assessment μ satisfying the conditions (1-2), which tells us that $\mu(\theta_1, s) = 0$ if $d(\theta_0, s) < d(\theta_0, s_0)$ and $\mu(\theta_1, t) = 0$ if $b(\theta_0, t) < b(\theta_0, t_0)$. The orderly markets condition requires that either $\sum_t \mu(\theta_1, t) = 1$ or $\sum_s \mu(\theta_1, s) = 1$. Then for some θ_1 sufficiently close to θ_0, some types will strictly prefer θ_1 to θ_0, contradicting the equilibrium conditions, unless $f(\theta_0, s) = 0$ for $s \neq s_0$ and $f(\theta_0, t) = 0$ for $t \neq t_0$. This proves the desired result. □

References

Banks, J., Sobel, J.: Equilibrium selection in signalling games. Econometrica **55**, 647–661 (1987)
Bester, H.: Screening vs rationing in credit markets with imperfect information. American Economic Review **75**, 850–855 (1985)
Boyd, J., Prescott, E.: Financial intermediary coalitions. Journal of Economic Theory **38**, 211–232 (1986)
Cho, I.K., Sobel, J.: Strategic stability and uniqueness in signaling games. Journal of Economic Theory **50**, 381–414 (1990)
Cho, I.K., Kreps, D.: Signaling games and stable equilibria. Quarterly Journal of Economics **102**, 179–221 (1987)
Dreze, J.: Existence of an equilibrium under price rigidity and quantity rationing. International Economic Review **16**, 301–320 (1975)
Gale, D.: Incomplete mechanisms and efficient allocation in labour markets. Review of Economic Studies **58**, 823–851 (1991)
Gale, D.: A Walrasian theory of markets with adverse selection. Review of Economic Studies **59**, 229–255 (1992)
Gale, D.: Equilibria and pareto optima of markets with adverse selection. Economic Theory **7**, 207–235 (1996)
Grossman, S., Perry, M.: perfect sequential equilibrium. Journal of Economic Theory **39**, 97–119 (1986)
Hahn, F.: On non-Walrasian equilibria. The Review of Economic Studies **45**, 1–18 (1978)
Hahn, F., Negishi, T.: A non-tatonnement model of adjustment. Econometrica **30**, 463–469 (1962)
Harris, M., Townsend, R.: Resource allocation under asymmetric information. Econometrica **49**, 33–64 (1982)
Hellwig, M.: Some recent developments in the theory of competition in markets with adverse selection. European Economic Review Papers and Proceedings **31**, 319–325 (1987)
Jerez, M.B.: General equilibrium with asymmetric information: a dual approach. UCLA (1999) (unpublished)
Kohlberg, E., Mertens, J.-F.: On the strategic stability of equilibria. Econometrica **54**, 1003–1038 (1986)
Koutsougeras, L., Yannelis, N.: Incentive compatibility and information superiority of the core of an economy with differential information. Economic Theory **43**, 195–216 (1993)
Krasa, S., Yannelis, N.: The value allocation of an economy with differential information. Econometrica **62**, 881–900 (1994)
Mailath, G.: An abstract two-period game with simultaneous signalling—existence of separating equilibria. Journal of Economic Theory **46**, 413–424 (1988)
Mailath, G., Okuno-Fujiwara, M., Postlewaite, A.: Belief-based refinements in signalling games. Journal of Economic Theory **60**, 241–276 (1993)
Myerson, R.: Sustainable matching plans with adverse selection. Games and Economic Behavior **9**, 35–65 (1995)
Prescott, E., Townsend, R.: Pareto optima and competitive equilibria with adverse selection and moral hazard. Econometrica **52**, 21–45 (1984)
Prescott, E., Townsend, R.: General competitive analysis in an economy with private information. International Economic Review **25**, 1–20 (1984)
Riley, J.: Informational equilibrium. Econometrica **47**, 331–359 (1979)
Rothschild, M., Stiglitz, J.: Equilibrium in competitive insurance markets: an essay on the economics of imperfect information. Quarterly Journal of Economics **90**, 629–650 (1976)
Selten, R.: Reexamination of the perfectness concept for equilibrium points in extensive games. International Journal of Game Theory **4**, 25–55 (1975)
Spence, M.: Job market signaling. Quarterly Journal of Economics **87**, 355–374 (1973)
Stiglitz, J., Weiss, A.: Credit rationing in markets with imperfect information. American Economic Review **71**, 393–410 (1981)
Wilson, C.: A model of insurance markets with incomplete information. Journal of Economic Theory **16**, 167–207 (1977)
Yannelis, N.: The core of an economy with differential information. Economic Theory **1**, 183–198 (1991)

Non-myopic learning in differential information economies: the core★

Konstantinos Serfes

Department of Economics, SUNY at Stony Brook, Stony Brook, NY 11794-4384, USA
(e-mail: kserfes@datalab2.sbs.sunysb.edu)

Received: May 25, 1999; revised version: August 9, 1999

Summary. We study the process of learning in a differential information economy, with a continuum of states of nature that follow a Markov process. The economy extends over an infinite number of periods and we assume that the agents behave non-myopically, i.e., they discount the future. We adopt a new equilibrium concept, the non-myopic core. A realized agreement in each period generates information that changes the underlying structure in the economy. The results we obtain serve as an extension to the results in Koutsougeras and Yannelis (1999) in a setting where agents behave non-myopically. In particular, we examine the following two questions: 1) If we have a sequence of allocations that are in an approximate non-myopic core (we allow for bounded rationality), is it possible to find a subsequence that converges to a non-myopic core allocation in a limit full information economy? 2) Given a non-myopic core allocation in a limit full information economy can we find a sequence of approximate non-myopic core allocations that converges to that allocation?

Keywords and Phrases: Non-myopic learning, Differential information, Core.

JEL Classification Numbers: D82, D50, D83.

1 Introduction

In this paper, we address the issue of learning in a differential information economy i.e., an economy with a finite number of agents, where each agent is characterized by a state dependent utility function, a state dependent initial endowment, a private information set (which is a partition of an exogeneously given probability measure

★ This is part of my Ph.D. thesis at the University of Illinois at Champaign-Urbana under the supervision of Professors N.C. Yannelis and W.J. Shafer to whom I am heavily indebted. I would also like to thank Professor D. Bernhardt for helpful comments and suggestions.

space) and a prior. The equilibrium concept we employ is the non-myopic core which is an extension of the private core (see Yannelis, 1991) to dynamic economies with non-myopic agents. Our economy extends over an infinite number of periods and agents discount the future. Consequently, the utility functions depend not only on current consumption, but also on future consumption. Agents are risk averse and hence they want to smooth their consumption. In each period, they agree upon a contract that specifies the terms of exchange, contingent on the states of nature. This agreement is based on each agent's private information and has the property that there does not exist a coalition of agents who can redistribute their initial endowments using their private information and make everybody in the coalition better off. A realized agreement in each period generates information that changes the underlying information structure in the economy.

We are studying the exchange of goods and information that takes place at the interim stage i.e., after the agents have observed the events that contain the realized state of nature. To be more precise, all contracts are negotiated at the beginning of the history of the economy and from then on all actions are determined by the already chosen acts. There is no need to revise any strategies, because the choice of the strategies has already taken account of the structure of information in the future i.e., what information will be available at each date. The process through which learning occurs is the following: At the end of each period the agents observe the non-myopic core equilibrium outcome plus the endowments of the current period and they refine their information partitions. The link between today and the future is the information that each agent possesses. So, agents by deciding upon the trade that will take place today, affect their information partitions tomorrow, which in turn affects the future expected utility. Learning itself is not the goal of the agents, but rather a result of actions by agents who are concerned with the expected utility.

It becomes apparent from the above discussion that the information agents possess restricts their consumption and trade choices. A question that naturally arises, and the one we address is: Can the agents through the process of exchange reach a non-myopic core equilibrium allocation that is in a limit full information economy? (i.e., in an economy where everything that could be learned has been learned.)

This work draws upon the results obtained by Koutsougeras and Yannelis (1999). They addressed the issue of learning in a pure exchange economy with differential information using the private core (Yannelis, 1991) as an equilibrium concept by assuming that the agents behave myopically.[1] In their model agents only care about current consumption and their utility does not depend on future allocations at all.

There is a substantial literature that deals with the issue of non-myopic learning in dynamic games, a small subset of which are the papers by: Kalai and Lehrer (1993), Nyarko (1998) and Serfes and Yannelis (1998). However, they put the problem in a different setting than we do. In particular, the first two papers consider an infinitely repeated game where agents have subjective beliefs about their opponents' strategies. They prove convergence of the actual play to Nash equilibrium (Kalai and Lehrer), or convergence of beliefs to subjective Nash equilibria

[1] Henceforth, we will be calling the equilibrium concept in Koutsougeras and Yannelis, myopic core.

(Nyarko). Serfes and Yannelis (1998) address the same questions that are addressed in this paper in an infinitely repeated game setting by employing the Bayesian Nash as the equilibrium concept.

What we add to the existing literature and in particular to Koutsougeras and Yannelis (1999), is the study of the learning problem when agents behave non-myopically and the states of nature follow a Markov process. To do this, we introduce a new equilibrium concept, the non-myopic core. The result is that we may get allocations and learning processes that may differ, depending on the equilibrium concept i.e., myopic versus non-myopic core. Our equilibrium concept is more general, since as the discount factor goes to zero, our model reduces to that of Koutsougeras and Yannelis (1999).

The paper contains the following results: We prove the non-emptiness of the set of non-myopic core allocations. Next, we define the concept of a limit full information economy and ask the following: If we have a sequence of allocations that are in an approximate non-myopic core (allowing for bounded rationality), is it possible to find a subsequence that converges to a non-myopic core allocation in a limit full information economy? And given a non-myopic core allocation in a limit full information economy can we find a sequence of approximate non-myopic core allocations that converges to that allocation?

The rest of the paper is organized as follows. In Section 2, we have collected the results that we are going to use in the sequel. In Section 3, we present the myopic core. In Section 4, we outline our model and prove the existence theorem. In Section 5, we describe the learning process, present an example and state the main Theorems. Finally, in Section 6 we prove our main learning theorems.

2 Mathematical preliminaries

If X and Y are sets, the *graph* of the set-valued function (or correspondence), $\phi : X \to 2^Y$ is denoted by

$$G_\phi = \{(x,y) \in X \times Y : y \in \phi(x)\}.$$

Let $(\Omega, \mathcal{F}, \mu)$ be a complete, finite measure space, and X be a separable Banach space. The set-valued function $\phi : \Omega \to 2^X$ is said to have a *measurable graph* if $G_\phi \in \mathcal{F} \otimes \beta(X)$, where $\beta(X)$ denotes the Borel σ-algebra on X and $\otimes$ denotes the product σ-algebra. The set-valued function $\phi : \Omega \to 2^X$ is said to be *lower measurable* or just *measurable* if for every open subset V of X, the set

$$\{\omega \in \Omega : \phi(\omega) \cap V \neq \emptyset\}$$

is an element of $\mathcal{F}$.

Let $(\Omega, \mathcal{F}, \mu)$ be a finite measure space and X be a Banach space. Following Diestel-Uhl (1977) the function $f : \Omega \to X$ is called *simple* if there exist $x_1, x_2, ..., x_n$ in X and $\alpha_1, \alpha_2, ..., \alpha_n$ in $\mathcal{F}$ such that $\sum_{i=1}^{n} x_i \chi_{\alpha_i}$ where $\chi_{\alpha_i}(\omega) = 1$ if $\omega \in \alpha_i$ and $\chi_{\alpha_i}(\omega) = 0$ if $\omega \notin \alpha_i$. A function $f : \Omega \to X$ is said to be μ-*measurable* if there exists a sequence of simple functions $f_n : \Omega \to X$ such that $\lim_{n\to\infty} \|f_n(\omega) - f(\omega)\| = 0$ for almost all $\omega \in \Omega$. A μ-*measurable* function

$f : \Omega \to X$ is said to be *Bochner integrable* if there exists a sequence of simple functions $\{f_n : n = 1, 2, ...\}$ such that

$$\lim_{n\to\infty} \int_\Omega \|f_n(\omega) - f(\omega)\| d\mu(\omega) = 0.$$

In this case we define for each $E \in \mathcal{F}$ the integral to be

$$\int_E f(\omega)d\mu(\omega) = \lim_{n\to\infty} \int_E f_n(\omega)d\mu(\omega).$$

It can be shown (see Diestel-Uhl, 1977, Theorem 2, p.45) that if $f : \Omega \to X$ is a μ-measurable function then, f is Bochner integrable if and only if $\int_\Omega \|f(\omega)\|d\mu(\omega) < \infty$.

For $1 \leq p < \infty$, we denote by $L_p(\mu, X)$ the space of equivalence classes of X-valued Bochner integrable functions $x : \Omega \to X$ normed by

$$\|x\|_p = (\int_\Omega \|x(\omega)\|^p d\mu(\omega))^{\frac{1}{p}}.$$

It is a standard result that normed by the functional $\|\cdot\|_p$ above, $L_p(\mu, X)$ becomes a Banach space (see Diestel-Uhl, 1977, p.50).

Let $X : \Omega \to 2^Y$, be a correspondence, where Y is a Banach space. Also let $u : \Omega \times Y \to R$ be a real-valued function. Ω can be decomposed into an atomless part Ω_1 and a countable union of atoms Ω_2. A result due to Balder and Yannelis (1993), [Theorem 2.8] says that if: 1. a.e. in Ω_1, $X(\omega)$ is convex and closed, 2. $u(\omega, \cdot)$ is concave and upper semicontinuous on $X(\omega)$, 3. $u(\omega, \cdot)$ is integrably bounded, 4. for all $\omega \in \Omega_2$, $X(\omega)$ is weakly closed, and 5. $u(\omega, \cdot)$ is weakly upper semicontinuous on $X(\omega)$ then,

$$U(x) = \int_\Omega u(\omega, x(\omega))d\mu(\omega)$$

is weakly upper semicontinuous on the weakly closed set $L_X = \{y \in L_1(\mu, Y) : y(\omega) \in X(\omega)$ and y is $\mathcal{F} -$ measurable$\}$.

Another result due to Balder and Yannelis (1993), [Theorem 2.1] tells us that if $X(\omega)$ is convex and closed a.e. in Ω_1 and weakly closed a.e. in Ω_2, then L_X is weakly closed.

Now we present some basic results on Banach lattices (see Aliprantis-Burkinshaw, 1985). Recall that a *Banach lattice* is a Banach space L equipped with an order relation $\geq$ (i.e., $\geq$ is reflexive, antisymmetric, and transitive relation) satisfying:

(i) $x \geq y$ implies $x + z \geq y + z$ for every z in L,
(ii) $x \geq y$ implies $\lambda x \geq \lambda y$ for all $\lambda \geq 0$,
(iii) for all x, y in L there exists a supremum (least upper bound) $x \vee y$ and an infimum (gretest lower bound) $x \wedge y$,
(iv) $|x| \geq |y|$ implies $\|x\| \geq \|y\|$ for all x, y in L.

As usual $x^+ = x \vee 0$, $x^- = (-x) \vee 0$ and $|x| = x \vee (-x) = x^+ + x^-$; we call x^+, x^- the *positive* and *negative* parts of x, respectively and $|x|$ the *absolute value* of x. The symbol $\|\cdot\|$ denotes the norm on L. If x, y are elements of the Banach lattice L, then we define the order interval $[x, y]$ as follows:

$$[x, y] = \{z \in L : x \leq z \leq y\}.$$

Note that $[x, y]$ is norm closed and convex (hence weakly closed). A Banach lattice L is said to have an *order continuous norm* if, $x_\alpha \downarrow 0$ [2] in L implies $\|x_\alpha\| \downarrow 0$. A very useful result that will play an important role in the sequel is that if L is a Banach lattice then the fact that L has order continuous norm is equivalent to weak compactness of the order interval $[x, y] = \{z \in L : x \leq z \leq y\}$ for every x, y in L [see for instance Aliprantis-Brown-Burkinshaw (1990), Theorem 2.3.8, p.104 or Lindenstrauss-Tzafriri (1979), p.28].

We note that Cartwright (1974) has shown that if X is a Banach lattice with order continuous norm (or equivalently X has weakly compact order intervals) then $L_1(\mu, X)$, has weakly compact order intervals, as well.

We close this section by defining the notion of a martingale and stating the martingale convergence theorem. Let I be a directed set and let $\{\mathcal{F}_i : i \in I\}$ be a monotone increasing net of sub σ-fields of $\mathcal{F}$ (i.e., $\mathcal{F}_{i_1} \subseteq \mathcal{F}_{i_2}$ for $i_1 \leq i_2, i_1, i_2$ in I). A net $\{x_i : i \in I\}$ in $L_1(\mu, X)$ is *a martingale* if

$$E(x_i|\mathcal{F}_{i_1}) = x_{i_1}, \forall i \geq i_1.$$

We will denote the above martingale by $\{x_i, \mathcal{F}_i\}_{i \in I}$. The proof of the following *martingale convergence theorem* can be found in Diestel-Uhl (1977, p.126). A martingale $\{x_i, \mathcal{F}_i\}_{i\in I}$ in $L_1(\mu, X)$ converges in the $L_1(\mu, X)$-norm if and only if there exists x in $L_1(\mu, X)$ such that $E(x|\mathcal{F}_i) = x_i$ for all $i \in I$. Finally, recall (see for instance Diestel–Uhl, 1977, p.129) that if the martingale $\{x_i, \mathcal{F}_i\}_{i\in I}$ converges in the $L_1(\mu, X)$-norm to $x \in L_1(\mu, X)$, it also converges almost everywhere, i.e., $\lim_{i\to\infty} x_i = x$ almost everywhere.

3 The Yannelis core

The definition of the core of an exchange economy with differential information is given as follows (see also Yannelis, 1991).

Let Y, which denotes the commodity space,[3] be a separable Banach lattice with an order continuous norm and Y_+ be its positive cone. Let $(\Omega, \mathcal{F}, \mu)$ be a probability space. An exchange economy with differential information,

$$\mathcal{E} = \{(X_i, u_i, e_i, \mathcal{F}_i, q_i) : i = 1, 2, ..., n\}$$

is a set of quintuples $(X_i, u_i, e_i, \mathcal{F}_i, q_i)$ where,
(1) $X_i : \Omega \to 2^{Y_+}$ is the *random consumption set* of agent i.

[2] $x_\alpha \downarrow 0$, means that x_α is a decreasing net with $\inf_\alpha x_\alpha = 0$.

[3] It is important to note that even if we assume that our commodity space is $Y = R^l$, the space $L_p(\mu, R^l), 1 \leq p \leq \infty$ is still infinite dimensional (in view of the continuum of states). Hence, even with one good we still need to work with an infinite dimensional space.

(2) $u_i : \Omega \times Y_+ \to R$ is the *random utility function* of agent i.
(3) $\mathcal{F}_i$ is a sub-σ-algebra of $(\Omega, \mathcal{F})$ which denotes the private information of agent i.
(4) $e_i : \Omega \to Y_+$ is the *random initial endowment* of agent i, $e_i(\cdot)$ is $\mathcal{F}_i$-measurable, Bochner integrable and $e_i(\omega) \in X_i(\omega)$ for all i, $\mu - a.e.$.
(5) $q_i : \Omega \to R_{++}$ is the *prior* of agent i, (i.e., q_i is the Radon-Nikodym derivative having the property that $\int_{t\in\Omega} q_i(t)d\mu(t) = 1$).

Denote by L_{X_i}, the *set of all Bochner integrable and $\mathcal{F}_i$-measurable selections from the consumption set* X_i of agent i, i.e.,

$$L_{X_i} = \{x_i \in L_1(\mu, Y_+) : x_i : \Omega \to Y_+ \text{is } \mathcal{F}_i - \text{measurable} \\ \text{and } x_i(\omega) \in X_i(\omega), \mu - a.e.\}.$$

For each i, $(i = 1, 2, \ldots, n)$ denote by $E_i(\omega)$ the event in $\mathcal{F}_i$ containing the realized state of nature $\omega \in \Omega$ and suppose that $\int_{t\in E_i(\omega)} q_i(t)d\mu(t) > 0$. Given $E_i(\omega)$ in $\mathcal{F}_i$ define the *interim expected utility* of agent i, $V_i : \Omega \times L_{X_i} \to R$ by,

$$V_i(\omega, x_i) = \int_{k\in E_i(\omega)} u_i(k, x_i(k))q_i(k|E_i(\omega))d\mu(k)$$

where

$$q_i(k|E_i(\omega)) = \begin{cases} 0 & \text{if } k \notin E_i(\omega) \\ \frac{q_i(k)}{\int_{k\in E_i(\omega)} q_i(k)d\mu(k)} & \text{if } k \in E_i(\omega). \end{cases}$$

Below we give the definitions of the core and the ϵ-core of the above economy.

Definition 3.1 (Yannelis, 1991). We say that $x = (x_1, x_2, ..., x_n) \in \Pi_{i=1}^n L_{X_i}$ is a core allocation for $\mathcal{E}$ if,

i) $\sum_{i=1}^n x_i = \sum_{i=1}^n e_i$ and,

ii) it is not true that there exists $S \subset \{1, 2, ..., n\}$ and $y \in \Pi_{i\in S} L_{X_i}$, such that $\sum_{i\in S} y_i = \sum_{i\in S} e_i$, and $V_i(\omega, y_i) > V_i(\omega, x_i), \forall i \in S$ and for almost all ω.

Definition 3.2 (Yannelis, 1991). An allocation $x \in \Pi_{i=1}^n L_{X_i}$, is said to be an ϵ**-core allocation** for $\mathcal{E}$ if in addition to $i)$ above it satisfies

ii′) it is not true that there exists $S \subset \{1, 2, ..., n\}$ and $y \in \Pi_{i\in S} L_{X_i}$ such that $\sum_{i\in S} y_i = \sum_{i\in S} e_i$, and $V_i(\omega, y_i) > V_i(\omega, x_i) + \epsilon, \forall i \in S$ and for almost all ω.

Theorem 3.1 (Yannelis, 1991). *Suppose that an exchange economy with differential information satisfies for each agent i the following assumptions,*
(a.3.1) $X_i : \Omega \to 2^{Y_+}$ *is a convex, closed, non-empty valued and* $\mathcal{F}$*-measurable correspondence.*
(a.3.2) for each $\omega \in \Omega$, $u_i(\omega, \cdot)$ *is continuous and integrably bounded and,*
(a.3.3) for each $\omega \in \Omega$, $u_i(\omega, \cdot)$ *is concave.*

Then a private core allocation exists in $\mathcal{E}$.

4 The non-myopic core

Let T be a countable set denoting the time horizon. Let Y be a Banach lattice with an order continuous norm and $(\Omega, \mathcal{F})$ be a measurable space with initial probability measure λ_0 and *transition function* Q.[4] The set Ω contains the states of nature which follow a *Markov* process overtime. Let $(\Omega^\infty, \mathcal{F}^\infty, \mu^\infty(\lambda_0, \cdot))$ be an infinite product probability measure space. The interpretation is that any sequence of *shocks* will lie in this space and μ^∞ gives the probability of that sequence occurring. Each state in Ω^∞ determines the entire history of all aspects of the economy that are beyond the control of any of the agents (see Savage, 1974, Ch. 2, for a detailed discussion of this concept).

Now let $x_{it} : \Omega^\infty \to Y_+$ be a vector-valued function that denotes the allocation of agent i in period t contingent on the history of realizations up to that period. We denote by $\bar{x}_i = (x_{i1}, \dots, x_{it}, \dots)$ an infinite sequence of such vector-valued functions for agent i. By $\bar{x}$ we denote such a sequence for all agents. Hence, $\bar{x}$ can be viewed as a stochastic process on $(\Omega^\infty, \mathcal{F}^\infty, \mu^\infty(\lambda_0, \cdot))$. Also the endowments $e_{it} : \Omega^\infty \to Y_+, t = 1, 2, \dots$ define a stochastic process on the same space. All contracts are negotiated at the beginning of the history of the economy, and from then on all actions are determined by already chosen strategies. Such strategies may, of course, take account of new information as it becomes available. An exchange economy with differential information is actually a sequence of economies

$$\{\mathcal{E}^t : t \in T\}$$

where for each t,

$$\mathcal{E}^t = \{(X_i, u_i, e_{it}, \mathcal{F}_{it}, q_i) : i = 1, 2, \dots, n\}$$

is a set of quintuples $(X_i, u_i, e_{it}, \mathcal{F}_{it}, q_i)$ where,
(1) $X_i : \Omega^\infty \to 2^{Y_+}$, is a *random consumption correspondence of agent* i.
(2) $u_i : \Omega^\infty \times Y_+ \to R$, is a *state dependent utility function* of agent i.
(3) $\mathcal{F}_{it}$ is a sub-σ-algebra of $(\Omega^\infty, \mathcal{F}^\infty)$ which denotes the private information of agent i in period t.
(4) $e_{it} : \Omega^\infty \to Y_+$ is the *random initial endowment* of agent i in period t, $e_{it}(\cdot)$ is $\mathcal{F}_{it}$-measurable, Bochner integrable and $e_{it}(\omega^\infty) \in X_i(\omega^\infty)$ for all i, $\mu^\infty - a.e.$.
(5) $q_i : \Omega^\infty \to R_{++}$ is the *prior* of agent i, (i.e., q_i is the Radon–Nikodym derivative having the property that $\int_{k \in \Omega^\infty} q_i(k) d\mu^\infty(k) = 1$).

As in R.J. Aumann (1987), we assume that the economy is common knowledge.

Denote by $L_{X_{it}}$, the set of all *Bochner integrable and* $\mathcal{F}_{it}$*-measurable selections* from the consumption set X_i of agent i, in period t i.e.,

$$L_{X_{it}} = \{x_{it} \in L_1(\mu^\infty, Y_+) : x_{it} : \Omega^\infty \to Y_+ \text{ is } \mathcal{F}_{it} - \text{measurable and}$$
$$x_{it}(\omega^\infty) \in X_i(\omega^\infty), \mu^\infty - a.e.\}.$$

Thus, $\bar{x}_i = (x_{i1}, \dots, x_{it}, \dots)$ is an element of $L_{\bar{X}_i} = L_{X_{i1}} \times \dots \times L_{X_{it}} \times \dots$.

[4] A transition function is a function $Q : \Omega \times \mathcal{F} \to [0, 1]$ such that,
a. for each $\omega \in \Omega, Q(\omega, \cdot)$ is a probability measure on $(\Omega, \mathcal{F})$; and
b. for each $A \in \mathcal{F}, Q(\cdot, A)$ is a $\mathcal{F}$-measurable function.

For each i, $(i = 1, 2, \dots, n)$ and $t \in T$, denote by $E_{it}(\omega^\infty)$ the event in $\mathcal{F}_{it}$ containing the realized state of nature $\omega^\infty \in \Omega^\infty$ and suppose that $\int_{k \in E_{it}(\omega^\infty)} q_i(k) d\mu(k) > 0$ for all $t \in T$.

For each i, $(i = 1, 2, \dots, n)$ and ω^∞ define the *total discounted interim expected utility* of agent i, $\bar{V}_i : \Omega^\infty \times L_{\bar{X}_i} \to R$ by

$$\bar{V}_i(\omega^\infty, \bar{x}_i) = \sum_{t=0}^{\infty} \delta^t \int_{k \in E_{it}(\omega^\infty)} u_i(k, x_{it}(k)) q_i(k | E_{it}(\omega^\infty)) d\mu(k) \tag{4.1}$$

where $\delta \in [0, 1)$ is the discount factor and $q_i(k | E_{it}(\omega^\infty))$ was defined in Section 3.

We are now ready to define the first central notion of the paper.

Definition 4.1. We say that $\bar{x} = (\bar{x}_1, \bar{x}_2, \dots, \bar{x}_n) \in \Pi_{i=1}^n L_{\bar{X}_i}$ is a *non-myopic core allocation* for the economy $\{\mathcal{E}^t : t \in T\}$ if,

(i) $\sum_{i=1}^n x_{it} = \sum_{i=1}^n e_{it}$, for all $t \in T$ and,

(ii) it is not true that there exist $S \subset \{1, 2, ..., n\}$ and $(\bar{y}_i)_{i \in S} \in \Pi_{i \in S} L_{\bar{X}_i}$ such that $\sum_{i \in S} y_{it} = \sum_{i \in S} e_{it}$, for all $t \in T$ and $\bar{V}_i(\omega^\infty, \bar{y}_i) > \bar{V}_i(\omega^\infty, \bar{x}_i), \forall i \in S$ and for almost all ω^∞.[5]

Definition 4.2. We say that $\bar{x} = (\bar{x}_1, \bar{x}_2, \dots, \bar{x}_n) \in \Pi_{i=1}^n L_{\bar{X}_i}$ is an *approximate or ϵ-non-myopic core allocation* for the economy $\{\mathcal{E}^t : t \in T\}$ if in addition to (i) above it satisfies,

(ii') it is not true that there exist $S \subset \{1, 2, ..., n\}$ and $(\bar{y}_i)_{i \in S} \in \Pi_{i \in S} L_{\bar{X}_i}$ such that $\sum_{i \in S} y_{it} = \sum_{i \in S} e_{it}$, for all $t \in T$ and $\bar{V}_i(\omega^\infty, \bar{y}_i) > \bar{V}_i(\omega^\infty, \bar{x}_i) + \epsilon, \forall i \in S$ and for almost all ω^∞.

We are now ready to state our first main result:

Theorem 4.1. *Let $\{\mathcal{E}^t : t \in T\}$ be an exchange economy with differential information as defined above which satisfies the following assumptions, for each i, $(i = 1, 2, \dots, n)$,*

(a.4.1) $X_i : \Omega^\infty \to 2^{Y_+}$ is convex, closed, non-empty valued, and $\mathcal{F}^\infty$-measurable correspondence.

(a.4.2) for each ω^∞, u_i is upper semicontinuous on $X_i(\omega^\infty)$ and integrably bounded.

(a.4.3) for each ω^∞, u_i is concave.

Then the set of non-myopic core allocations for $\{\mathcal{E}^t : t \in T\}$ is a non-empty subset of $\Pi_{i=1}^n L_{\bar{X}_i}$.

Lemma 4.1. *Under assumptions (a.4.1)–(a.4.3), the total discounted interim expected utility $\bar{V}_i$ (4.1) is weakly upper semicontinuous for each i and for each ω^∞.*

Proof. By assumption, the utility function u_i is upper semicontinuous. Then, by Theorem 2.8 in Balder and Yannelis (1993) (see also Section 2),

$$\int_{k \in E_{it}(\omega^\infty)} u_i(k, x_{it}(k)) q_i(k | E_{it}(\omega^\infty)) d\mu(k)$$

[5] What we mean is "for μ^∞ almost all $\omega^\infty \in \Omega^\infty$," but for convenience from now on we simply write "for almost all ω^∞."

is weakly upper semicontinuous.

By another application of the same Theorem and since T consists of a countable union of atoms, $\bar{V}_i$ is weakly upper semicontinuous as well. □

Next we present the proof of our Theorem.

Proof of Theorem 4.1. The set $L_{\bar{X}_i}$ is convex since each $L_{X_{it}}$ is convex. It is also weakly closed, since again each $L_{X_{it}}$ is weakly closed (see Thm. 2.1 in Balder and Yannelis (1993)). Now let's define an n-person game $\hat{V}$ by

$$\hat{V}(S) = \{x \in R^n : \text{ there exists an allocation } \bar{y} \in \mathcal{A}_S \text{ such that}$$
$$x_i \leq \bar{V}_i(\omega^\infty, \bar{y}_i), \forall i \in S \text{ and for almost all } \omega^\infty\}$$

where $\mathcal{A}_S$ is defined as

$$\mathcal{A}_S = \{\bar{y} \in \Pi_{i\in S} L_{\bar{X}_i} : \sum_{i\in S} y_{it} = \sum_{i\in S} e_{it}, \text{ for all } t \in T\}.$$

Notice that $\mathcal{A}_S$ is weakly compact because $\bar{y}_i \in [0, \sum e_{i1}] \times [0, \sum e_{i2}] \times \ldots, \forall i \in S$, the order intervals $[0, \sum_{i=1}^n e_{it}], \forall t \in T$ are weakly compact (by Cartwright's Theorem) and $\Pi_{i\in S} L_{\bar{X}_i}$ is weakly closed. It is also nonempty since $e_{it} \in L_{X_{it}}$, for all $t \in T$.

The n-person game satisfies the properties of Scarf's Theorem (Scarf, 1967). Notice that the comprehensiveness follows immediately. The fact that $\hat{V}$ is bounded from above follows from the fact that $\forall i \in N$, $\bar{V}_i$ is a weakly upper semicontinuous real-valued function (Lemma 4.1) on the non-empty, weakly compact set $\mathcal{A}_S$.

We need to show that $\hat{V}(S)$ is closed. To this end, let a sequence $(x_1^k, ..., x_n^k)$ of some $\hat{V}(S)$ satisfy $(x_1^k, ..., x_n^k) \to (x_1, .., x_n)$ in R^n. We must show that $(x_1, .., x_n)$ belongs to $\hat{V}(S)$. For each k pick an allocation $(\bar{y}_1^k, \ldots, \bar{y}_n^k)$ satisfying $x_i^k \leq \bar{V}_i(\omega^\infty, \bar{y}_i^k), \forall i \in S$ and for almost all ω^∞, and $\sum_{i\in S} y_{it}^k = \sum_{i\in S} e_{it}$, for all $t \in T$. Since $y_{it}^k \in [0, e_t]$ (where $e_t = \sum_{i=1}^n e_{it}$, for all $t \in T$) holds for all i and all k and $[0, e_t]$ is weakly compact, we can assume by passing to an appropriate subsequence that $\bar{y}_i^k \to y_i$ weakly for all i. Clearly, $(y_1, ..., y_n)$ is an allocation and $\sum_{i\in S} y_{it} = \sum_{i\in S} e_{it}$, for all $t \in T$. Since $\bar{V}_i$ is weakly upper semicontinuous it follows that

$$x_i = \limsup_k x_i^k \leq \limsup_k \bar{V}_i(\omega^\infty, \bar{y}_i^k) \leq \bar{V}_i(\omega^\infty, y_i)$$

for all $i \in S$ and for almost all ω^∞. Therefore, $(x_1, ..., x_n) \in \hat{V}(S)$ and so each $\hat{V}(S)$ is closed. Hence the market game $(\hat{V}, N)$ is balanced and has therefore a non-empty core (Scarf's Theorem). Standard arguments now can be applied (see for instance Aliprantis, Brown and Burkinshaw, 1990, pp.48–49) to show that non-emptiness of the core of the game $(\hat{V}, N)$ implies non-emptiness of the core of the economy $\{\mathcal{E}^t : t \in T\}$. □

Also an *approximate or* ϵ*-private non-myopic core allocation* exists since the set of all non-myopic core allocations, denoted by $C(\{\mathcal{E}^t : t \in T\})$, is a subset of the set of all ϵ-private non-myopic core allocations denoted by $C_\epsilon(\{\mathcal{E}^t : t \in T\})$.

Next we turn to the question of learning.

5 Convergence and approximation theorems for the non-myopic private core and ϵ-non-myopic private core

5.1 The process of learning

Let T be any countably infinite set denoting the time horizon. We are going to study the learning process described by Koutsougeras and Yannelis (1999), by using the non-myopic core as the equilibrium concept of our economy. There are two advantages of using the non-myopic core. First, it is a general concept and one can recover all the fundamental results of Koutsougeras and Yannelis (1999) by simply letting the discount factor go to zero. Second, and most important, the agents in our framework look into the future which may capture allocations and learning processes that cannot be captured by the myopic core. We also make a further generalization by allowing the states of nature to follow a Markov process.

The economy extends over an infinite number of periods. Since the agents are risk-averse, they want to smooth their consumption. Therefore, in each period they agree upon a contract which specifies the terms of the exchange contingent upon the realized state of nature.

Hence, each agent's private information in each period is generated by his/her endowment in current and all past periods, his/her utility function and the equilibrium allocations in previous periods i.e.,

$$\mathcal{F}_{it} = \sigma(\{e_{it'}, t' = 1, \dots, t\}, u_i, \{x_{t'}, t' = 1, \dots, t-1\}).$$

In this scenario, the private information of agent i in period $t+1$ will be $\mathcal{F}_{it}$ together with the information that the endowment, the utility function and the private core allocations generate i.e.,

$$\mathcal{F}_{it+1} = \mathcal{F}_{it} \vee \sigma(e_{it+1}, x_t).$$

Clearly, in period $t+2$ the private information set of agent i will be, $\mathcal{F}_{it+2} = \mathcal{F}_{it+1} \vee \sigma(e_{it+2}, x_{t+1})$ and so on. Consequently, for each agent i and each time period, we have that

$$\mathcal{F}_{it} \subseteq \mathcal{F}_{it+1} \subseteq \mathcal{F}_{it+2} \subseteq \dots \quad .$$

The above expression represents a learning process for agent i and it generates a sequence of differential information economies i.e., $\{\mathcal{E}^t : t \in T\}$.

Next we define a *limit full information economy,*

$$\mathcal{E}^\infty = \{(X_i, u_i, e_{i\infty}, \bar{\mathcal{F}}_i, q_i) : i = 1, 2, ..., n\}$$

to be the set of quintuples $(X_i, u_i, e_{i\infty}, \bar{\mathcal{F}}_i, q_i)$ where, $\bar{\mathcal{F}}_i = \vee_{t=0}^{\infty} \mathcal{F}_{it}$ is the pooled information of agent i over the entire time horizon, X_i, u_i, q_i have been defined previously and $e_{i\infty}$ denotes an endowment function in a limit full information economy which is $\bar{\mathcal{F}}_i$-measurable.

Denote by $C(\mathcal{E}^\infty)$ and $C_\epsilon(\mathcal{E}^\infty)$ the *set of all limit full information non-myopic core allocation and the limit full information non-myopic ϵ-core allocation* respectively for the economy $\mathcal{E}^\infty$.

Throughout our analysis we will assume that a private information economy $\{\mathcal{E}^t : t \in T\}$ as well as a limit full information economy $\mathcal{E}^\infty$, satisfy the assumptions (a.4.1), (a.4.2) and (a.4.3) and therefore, $C(\{\mathcal{E}^t : t \in T\}) \neq \emptyset$ and $C(\mathcal{E}^\infty) \neq \emptyset$. Since, $C(\{\mathcal{E}^t : t \in T\}) \subset C_\epsilon(\{\mathcal{E}^t : t \in T\})$ the latter set is non-empty as well.

It is apparent that the information structure of the economy largely determines the resulting allocation. The example we present next illustrates the above argument as well as how the learning takes place in our economy.

5.2 Example

Consider the following two person economy ($I = \{1, 2\}$) with two commodities i, j, ($X = R^2_+$) and four different states ($\Omega = \{a, b, c, d\}$). To simplify the example, the economy extends to only two periods t and $t+1$. To be consistent with our notation in the previous section the state space in each period is $Z = \{\omega_1, \omega_2\}$ and $\Omega = Z \times Z$. Hence, $a = \omega_1\omega_1, b = \omega_1\omega_2, \ldots$. The idea as Debreu (1960) puts it is the following: Nature makes a choice (state) from a number of possibilities (states). These possibilities are states at time $t+1$ (in our example). Once a state is given, atmospheric conditions, technological knowledge, natural disasters, ... are determined for the entire period under consideration. At time t economic agents have some information about the state at $t+1$ which will occur. This knowledge in our economy comes from observing the endowments. Additional knowledge in each period is acquired by the allocation in the previous period. This information can be described by a partition of the set of states at $t+1$ into sets called events at t.

Each state occurs with probability $\frac{1}{4}$. The random initial (period t) endowment and private information sets are given by

$$e_{1t} = ((10,0), (10,0), (10,0), (10,0)), \mathcal{F}_{1t} = \{\{a, b, c, d\}\}$$

$$e_{2t} = ((0,10), (0,10), (0,0), (0,0)), \mathcal{F}_{2t} = \{\{a, b\}, \{c, d\}\}.$$

Note that the initial endowment of each agent is measurable with respect to his/her partition. The utility function of both agents is given by

$$u(\omega, x) = \sqrt{x_i} + \sqrt{x_j}, \text{ for all } \omega.$$

The agents before they observe any event they will agree on the following contract. In period t there will be no trade due to measurability constraints. Note that the net trade must be measurable with respect to each agent's private information at period t. Since agent 1 has trivial information the net trade must be constant across all states. But agent 2 has nothing to give at states c and d. Thus, the net trade must be zero. This implies that the allocation in period t is

$$x_{1t} = ((10,0), (10,0), (10,0), (10,0))$$

$$x_{2t} = ((0,10), (0,10), (0,0), (0,0)).$$

The information that this allocation generates is

$$\sigma(x_t) = \{\{a,b\},\{c,d\}\}.$$

Hence, the agents in the second period $(t+1)$ will possess the following information (we assume that the endowments are the same)

$$\mathcal{F}_{1t+1} = \mathcal{F}_{1t} \vee \sigma(x_t) = \{\{a,b\},\{c,d\}\}$$

$$\mathcal{F}_{2t+1} = \mathcal{F}_{2t} \vee \sigma(x_t) = \{\{a,b\},\{c,d\}\}$$

and the allocation in that period will be

$$x_{1t+1} = ((5,5),(5,5),(10,0),(10,0))$$

$$x_{2t} = ((5,5),(5,5),(0,0),(0,0)).$$

Notice that the allocation is measurable with respect to each agent's information and that both agents became better off. Therefore, the agreed upon contract is

$$x = (x_t, x_{t+1})$$

as described above.

Below we state and prove the main theorems of this section.

5.3 Learning theorems

We assume that the sequence of endowments satisfies the following condition: There exists $\sum_{i\in S} e_{i\infty} \in L_1(\mu, Y)$ such that for all $S \subset N$

$$E[\sum_{i\in S} e_{i\infty} | \wedge_{i\in S} \mathcal{F}_{it}] = \sum_{i\in S} e_{it}, \forall t \in T.$$

Theorem 5.3.1. *Let $\{\mathcal{E}^t : t \in T\}$ be a sequence of private information economies satisfying the following assumption:*

$\forall S \subset N$, where N is the set of agents, $\{\sum_{i\in S} e_{it}, \wedge_{i\in S}\mathcal{F}_{it}\}_{t\in T}$ is a martingale.

If the sequence $\{x_t : t \in T\}$ belongs to $C_\epsilon(\{\mathcal{E}^t : t \in T\})$, then we can extract a subsequence $\{x_{t_m} : m = 1,2,...,\}$ from the sequence x_t which converges weakly to $x^* \in C(\mathcal{E}^\infty)$.

Theorem 5.3.2. *Let $\{\mathcal{E}^t : t \in T\}$ be a sequence of private information economies satisfying the following assumptions:*

(i) $\forall i$, $\{e_{it}, \mathcal{F}_{it}\}_{t\in T}$ is a martingale,

(ii) $\{\sum_{i=1}^n e_{it}, \wedge_{i=1}^n \mathcal{F}_{it}\}_{t\in T}$ is a martingale.

Let x^* be a limit full information non-myopic core allocation for the economy $\mathcal{E}^\infty$, i.e., $x^* \in C(\mathcal{E}^\infty)$. Then, there exists a $t' \in T$ big enough and a sequence of allocations $\{x_t : t \in T\}$ such that $\{x_t\}_{t\geq t'} \in C_\epsilon(\{\mathcal{E}^t : t \geq t'\})$ and $\{x_t\}_{t\in T}$ converges in the L^1-norm to x^*.

An immediate conclusion of Theorem 1 is the following result.

Corollary 5.3.1. *Let $\{\mathcal{E}^t : t \in T\}$ be a sequence of private information economies satisfying assumption the of Theorem 5.3.1. If the sequence $\{x_t : t \in T\}$ belongs to $C(\{\mathcal{E}^t :, t \in T\})$, then we can extract a subsequence from the sequence x_t which converges weakly to $x^* \in C(\mathcal{E}^\infty)$.*

Discussion: In both Theorems, the aggregate endowments in the economy obey a "stability" property given by the Martingale assumption. For a more detailed discussion about the assumptions see Koutsougeras and Yannelis (1999). Theorem 5.3.1 states that non-myopic and non-fully-rational agents can, by repetition, reach an equilibrium allocation in an economy that everything that could be learned has been learned. Theorem 5.3.2 states the converse. That is, given an equilibrium allocation in such an economy, non-myopic and non-fully-rational agents will find the way through trading to reach that allocation. This may be viewed as a kind of "stability" property of the non-myopic core.

Remark 1. When the discount factor δ goes to zero, the above two Theorems reduce to the ones in Koutsougeras and Yannelis (1999) i.e., Theorems 3.3.1 and 3.3.2.

Remark 2. For the above two Theorems we want the total discounted interim expected utility (4.1) to be weakly continuous and not just weakly upper semicontinuous. If in addition we assume that for all i, $\mathcal{F}_{it}$ is a partition, and the utility function u_i is weakly continuous, then $\bar{V}_i$ is weakly continuous [for more details see Yannelis (1991), Claim 4.1 and Balder and Yannelis (1993), Corollary 2.9].

6 Proofs of the theorems

6.1 Proof of Theorem 5.3.1

For each i, let $\bar{L}_{X_i}$ be the set of all Bochner integrable and $\bar{\mathcal{F}}_i$-measurable selections from the consumption correspondence X_i i.e.,

$$\bar{L}_{X_i} = \{x_{i\infty} \in L_1(\mu^\infty, Y_+) : x_{i\infty} : \Omega^\infty \to Y_+ \text{ is } \bar{\mathcal{F}}_i - \text{measurable and}$$
$$x_{i\infty}(\omega^\infty) \in X_i(\omega^\infty), \mu^\infty - a.e.\}.$$

An allocation in a limit full information economy belongs in the above set i.e., $x_i^* \in \bar{L}_{X_i}$. Let $e_{i\infty} \in \bar{L}_{X_i}$,[6] be the endowments for agent i in a limit full information economy. Note that for each t, each feasible consumption $x_{it} \in L_{X_{it}}$, lies in the order interval $[0, \sum_{i=1}^n e_{it}] \subset \sum_{i=1}^n L_{X_{it}}$. By the *Cartwright Theorem*, $[0, e_t]$ (and any order interval) (where $e_t = \sum_{i=1}^n e_{it}$), is weakly compact. Finally, $\bar{V}_i(\omega^\infty, \cdot)$ is *weakly continuous* for each i and for each ω^∞.

Let $\bar{x} = \{x_t : t \in T\}$ be in $C_\epsilon(\{\mathcal{E}^t : t \in T\})$. Obviously, $x_{it} \in [0, \sum_{i=1}^n e_{it}]$, for all i and $t \in T$. Since e_t is a martingale, it converges to e_∞ in the $L_1(\mu, Y)$-norm. Moreover, $L_1(\mu, Y)$ is a Banach lattice. By a standard result [e.g. Aliprantis-Burkinshaw (1985)] we can extract a subsequence (for convenience we still denote it by e_t) and find a positive element z in $L_1(\mu, Y)$ such that $|e_t - e_\infty| < \frac{1}{2^k} z$, (where the superscript k is the index of the subsequence). Hence, we can conclude

[6] Notice, that since we are working with partitions, the allocations are essentially in l^1.

that the subsequence e_t is order bounded above by an element say v in $L_1(\mu, Y)$ and below by 0 i.e., e_t belongs to the order interval $[0, v]$ in $L_1(\mu, Y)$. Therefore, a subsequence of the allocation $\bar{x} = \{x_t : t \in T\}$ belongs to the order interval $[0, v]$. By the *Eberlein-Smulian Theorem* we can extract a further subsequence (still denoted by $\{x_t\}$) which converges weakly to $x^* \in [0, v]^n$ (where $[0, v]^n$ is the n-fold product of $[0, v]$). Notice that in what follows we are dealing with the subsequence of the subsequence of the original allocation $\bar{x}$ and endowments e_t for which we kept the same indices.

We need to show that x^* is in $C(\mathcal{E}^\infty)$. Note that for each $t \in T, \sum_{i=1}^n x_{it} = \sum_{i=1}^n e_{it}$, $\{x_t\}_{t\in T}$ converges weakly to x^* and $\{e_t\}_{t\in T}$ converges weakly to e_∞ (since by the Martingale Convergence Theorem e_{it} converges in the L^1-norm to $e_{i\infty}$ and hence weakly). So, we conclude that $\sum_{i=1}^n x_i^* = \sum_{i=1}^n e_{i\infty}$. Thus, $x_i^* \in [0, e_\infty] \subset \sum_{i=1}^n \bar{L}_{X_i}$ (where $e_\infty = \sum_{i=1}^n e_{i\infty}$), and therefore each x_i^* is $\bar{\mathcal{F}}_i$-measurable.

Hence, all it remains to be shown is that

there is no coalition S and $y_\infty \in \Pi_{i\in S}\bar{L}_{X_i}$ such that $\sum_{i\in S} y_{i\infty} = \sum_{i\in S} e_{i\infty}$, and $\bar{V}_i(\omega^\infty, y_{i\infty}) > \bar{V}_i(\omega^\infty, x_i^*), \forall i \in S$ and for almost all ω^∞.

Suppose by way of contradiction that this is not true. Then, there exists a coalition S and $y_\infty \in \Pi_{i\in S}\bar{L}_{X_i}$ such that $\sum_{i\in S} y_{i\infty} = \sum_{i\in S} e_{i\infty}$, and $\bar{V}_i(\omega^\infty, y_{i\infty}) > \bar{V}_i(\omega^\infty, x_i^*), \forall i \in S$ and for almost all ω^∞.

For each $i \in S$ and each $t \in T$, let $y_{it} = E[y_{i\infty} | \wedge_{i\in S} \mathcal{F}_{it}]$. Notice that

$$E[y_{i\infty}|\wedge_{i\in S}\mathcal{F}_{it}] = E[E[y_{i\infty}|\wedge_{i\in S}\mathcal{F}_{it'}]|\wedge_{i\in S}\mathcal{F}_{it}] = E[y_{it'}|\wedge_{i\in S}\mathcal{F}_{it}], \text{ for } t' \geq t.$$

Hence, $\{y_{it}, \wedge_{i\in S}\mathcal{F}_{it}\}_{t\in T}$ is a martingale and

$$\begin{aligned}\sum_{i\in S} y_{it} &= \sum_{i\in S} E[y_{i\infty}| \wedge_{i\in S} \mathcal{F}_{it}] = E\left[\sum_{i\in S} y_{i\infty}| \wedge_{i\in S} \mathcal{F}_{it}\right] \\ &= E\left[\sum_{i\in S} e_{i\infty}| \wedge_{i\in S} \mathcal{F}_{it}\right] = \sum_{i\in S} e_{it}.\end{aligned}$$

This is true for all $t \in T$ and hence $\{y_{it}\}_{t\in T}$ is feasible for the coalition S.

By virtue of the Martingale Convergence Theorem, $\{y_{it}\}_{t\in T}$ converges to $y_{i\infty}$ in the L_1-norm and therefore weakly. Since $\{x_t\}_{t\in T}$ also converges weakly to x^* and $\bar{V}_i$ is weakly continuous we may choose $t' \in T$ so that

$$|\bar{V}_i(\omega^\infty, y_{i\infty}) - \bar{V}_i(\omega^\infty, \{y_{it}\}_{t\geq t'})| < \frac{\delta - \epsilon}{2} \quad \text{and}$$
$$|\bar{V}_i(\omega^\infty, \{x_{it}\}_{t\geq t'}) - \bar{V}_i(\omega^\infty, x_i^*)| < \frac{\delta - \epsilon}{2}$$

where $\delta = \bar{V}_i(\omega^\infty, y_i^\infty) - \bar{V}_i(\omega^\infty, x_i^*) > \epsilon$. Thus,

$$\begin{aligned}&|\bar{V}_i(\omega^\infty, y_{i\infty}) - \bar{V}_i(\omega^\infty, \{y_{it}\}_{t\geq t'}) + \bar{V}_i(\omega^\infty, \{x_{it}\}_{t\geq t'}) - \bar{V}_i(\omega^\infty, x_i^*)| \\ &\leq |\bar{V}_i(\omega^\infty, y_{i\infty}) - \bar{V}_i(\omega^\infty, \{y_{it}\}_{t\geq t'})| + |\bar{V}_i(\omega^\infty, \{x_{it}\}_{t\geq t'}) - \bar{V}_i(\omega^\infty, x_i^*)| \\ &< \frac{\delta - \epsilon}{2} + \frac{\delta - \epsilon}{2} = \delta - \epsilon.\end{aligned}$$

Therefore, $\bar{V}_i(\omega^\infty, y_{i\infty}) - \bar{V}_i(\omega^\infty, \{y_{it}\}_{t\geq t'}) + \bar{V}_i(\omega^\infty, \{x_{it}\}_{t\geq t'}) - \bar{V}_i(\omega^\infty, x_i^*) < \delta - \epsilon \iff -\bar{V}_i(\omega^\infty, \{y_{it}\}_{t\geq t'}) + \bar{V}_i(\omega^\infty, \{x_{it}\}_{t\geq t'}) < -\epsilon$ or $\epsilon + \bar{V}_i(\omega^\infty, \{x_{it}\}_{t\geq t'}) < \bar{V}_i(\omega^\infty, \{y_{it}\}_{t\geq t'}), \forall i \in S$ and for almost all ω^∞.

So, the allocation y_t is feasible and $\bar{V}_i(\omega^\infty, \{y_{it}\}_{t\geq t'}) > \bar{V}_i(\omega^\infty, \{x_{it}\}_{t\geq t'}) + \epsilon, \forall i \in S$ and for almost all ω^∞, a contradiction to the fact that $\bar{x} \in C_\epsilon(\{\mathcal{E}^t : t \in T\})$. □

6.2 *Proof of Theorem 5.3.2*

Let x_∞ be an element of $C(\mathcal{E}^\infty)$. Consider the allocation $x_t = E[x_\infty | \wedge_{i=1}^n \mathcal{F}_{it}]$ and notice that for $r \geq t$

$$x_t = E[x_\infty | \wedge_{i=1}^n \mathcal{F}_{it}] = E[E[x_\infty | \wedge_{i=1}^n \mathcal{F}_{ir}] | \wedge_{i=1}^n \mathcal{F}_{it}] = E[x_r | \wedge_{i=1}^n \mathcal{F}_{it}].$$

Hence $\{x_t, \wedge_{i=1}^n \mathcal{F}_{it}\}_{t\in T}$ is a martingale and by virtue of the Martingale Convergence Theorem $\{x_t\}_{t\in T}$ converges in the L^1-norm to x_∞. By the definition of the conditional expectation we know that for each i and $t \in T$, x_{it} is $\mathcal{F}_{it}$-measurable. We must show that there exists a t' big enough such that the sequence $\{x_t : t \geq t'\}$ lies in $C_\epsilon(\{\mathcal{E}^t : t \geq t'\})$. We first show that $\{x_t\}_{t\in T}$ is feasible for the grand coalition. Note that, for all $t \in T$

$$\sum_{i=1}^n x_{it} = \sum_{i=1}^n E[x_{i\infty} | \wedge_{i=1}^n \mathcal{F}_{it}] = E[\sum_{i=1}^n x_{i\infty} | \wedge_{i=1}^n \mathcal{F}_{it}] = E[\sum_{i=1}^n e_{i\infty} | \wedge_{i=1}^n \mathcal{F}_{it}]$$
$$= \sum_{i=1}^n e_{it},$$

and we can conclude that $\{x_t\}_{t\in T}$ is feasible. We now show that there exists a t' such that the allocation $\{x_t\}_{t\geq t'}$ cannot be ϵ-blocked by any coalition i.e.,

there do not exist coalition S and allocation $\{y_t\}_{t\geq t'} \in \Pi_{i\in S} L_{\bar{X}_i}$ such that $\sum_{i\in S} y_{it} = \sum_{i\in S} e_{it}$, for all $t \geq t'$ and $\bar{V}_i(\omega^\infty, \{y_{it}\}_{t\geq t'}) > \bar{V}_i(\omega^\infty, \{x_{it}\}_{t\geq t'}) + \epsilon, \forall i \in S$ and for almost all ω^∞.

Suppose by way of contradiction that the above statement is false. Then, there exists a coalition S and a sequence $\{y_t\}_{t\in T}$, $y_t \in \Pi_{i\in S} L_{X_{it}} \subset \Pi_{i\in S} L_{\bar{X}_i}$ having the property that $\sum_{i\in S} y_{it} = \sum_{i\in S} e_{it}$, for all $t \in T$ and $\bar{V}_i(\omega^\infty, \{y_{it}\}_{t\geq t'}) > \bar{V}_i(\omega^\infty, \{x_{it}\}_{t\geq t'}) + \epsilon, \forall i \in S$ for almost all ω^∞ and for all t'.

By adopting an argument similar to the one in the previous proof, $\{y_t\}_{t\in T}$ lies in the order interval $[0, v]^{|S|}$ which is weakly compact (recall Cartwright's Theorem). Hence, by the weak compactness of $[0, v]^{|S|}$ we can find a further subsequence $\{y_{t_m}\}$ that converges weakly to $y_\infty \in [0, v]^{|S|}$. For this subsequence we have

$$\sum_{i\in S} y_{it_m} = \sum_{i\in S} e_{it_m}, \text{ for all } t_m \in T.$$

Since e_{it_m} converges to $e_{i\infty}$ in the L^1-norm and hence weakly and y_{it_m} converges to $y_{i\infty}$ weakly we have that

$$\sum_{i\in S} y_{i\infty} = \sum_{i\in S} e_{i\infty}.$$

By our assumption, we also have that

$$\bar{V}_i(\omega^\infty, \{y_{it_m}\}_{t_m \geq t'}) > \bar{V}_i(\omega^\infty, \{x_{it_m}\}_{t_m \geq t'}) + \epsilon$$

$\forall i \in S$, for almost all ω^∞ and for all t'. By the weak continuity of $\bar{V}_i(\omega^\infty, \cdot)$, $\bar{V}_i(\omega^\infty, y_{i\infty}) \geq \bar{V}_i(\omega^\infty, x_{i\infty}) + \epsilon, \forall i \in S$ and for almost all ω^∞. Hence, $\bar{V}_i(\omega^\infty, y_{i\infty}) > \bar{V}_i(\omega^\infty, x_{i\infty}), \forall i \in S$ and for almost all ω^∞ and consequently the coalition S qualifies to block x_∞ a contradiction to the fact that $x_\infty \in C(\mathcal{E}^\infty)$. □

References

Aliprantis, C.D., Burkinshaw, O.: Positive operators. New York London: Academic Press 1985

Aliprantis, C.D., Brown, D.J., Burkinshaw, O.: Existence and optimality of competitive equilibria. Berlin Heidelberg New York: Springer 1990

Aumann, R.J.: Correlated equilibrium as an expression of Bayesian rationality. Econometrica **55**, 1–18 (1987)

Balder, E.J., Yannelis, N.C.: On the continuity of the expected utility. Economic Theory **3**, 625–643 (1993)

Cartwright, D.: The order completeness of some vector-valued functions. Bulletin of the Australian Mathematical Society **11**, 57–61 (1974)

Debreu, G.: Economics under uncertainty. In Mathematical economics: Twenty Papers of G. Debreu. Cambridge: Cambridge University Press 1960

Diestel, J., Uhl, J.J.: Vector measures. Mathematical surveys 15. Providence, RI: American Mathematical Society 1977

Kalai, E., Lehrer, E.: Rational learning leads to Nash equilibrium. Econometrica **61**, 1019–1045 (1993)

Koutsougeras, L., Yannelis, N.C.: Convergence and approximation results for non-cooperative Bayesian games: learning theorems. Economic Theory **4**, 843–857 (1994)

Koutsougeras, L., Yannelis, N.C.: Bounded rational learning in differential information economies: Core and value. Journal of Mathematical Economics **31**, 373–391 (1999)

Nyarko, Y.: Bayesian learning and convergence to Nash equilibria without common priors. Economic Theory **3**, 643–655 (1998)

Savage, L.J.: Foundations of statistics. New York: Dover 1974

Scarf, H.: The core of an n-person game. Econometrica **35**, 50–69 (1967)

Serfes, K., Yannelis, N.C.: Learning in Bayesian games by boundedly rational players II: Non-myopia. Macroeconomic Dynamics **2**, 141–155 (1998)

Yannelis, N.C.: The core of an economy with differential information. Economic Theory **1**, 183–198 (1991)

PART 5

VALUE ALLOCATIONS AND THE BARGAINING SET

Cooperative games with incomplete information*

R.B. Myerson

Kellogg Graduate School of Management; Northwestern University, Evanston, L 60201, USA

Summary. A bargaining solution concept which generalizes the Nash bargaining solution and the Shapley NTU value is defined for cooperative games with incomplete information. These bargaining solutions are efficient and equitable when interpersonal comparisons are made in terms of certain virtual utility scales. A player's *virtual utility* differs from his real utility by exaggerating the difference from the preferences of false types that jeopardize his true type. In any incentive-efficient mechanism, the players always maximize their total virtual utility *ex post*. Conditionally transferable virtual utility is the strongest possible transferability assumption for games with incomplete information.

1 Introduetion

In a cooperative game the players must bargain to select an outcome which is efficient for them. Each player wants to demand the outcome that is best for himself, so the players inust moderate their demands to reach a feasible agreement. In general, the amount that a player can realistically demand in such bargaining will depend on his power in the game situation. Here, power means the ability to alternatively help or hurt other players at will, and to defend oneself against the threats of others. A solution concept in cooperative game theory is an attempt to systematically predict which outcomes on the Pareto frontier would be selected by the players, in any cooperative games in such a way that each player's payoff is commensurate with his power. This paper will develop a general solution concept for games with incomplete information.

The Nash [1950, 1953] bargaining solution, defined for two-Person bargaining problems, and the Shapley [1953] value, defined for n-person games with trans-

* Research for this paper was supported by the Kellogg Center for Advanced Study in Managerial Economics and Decision Sciences, and by a research fellowship from I.B.M.

ferable utility, are the most conceptually elegant and appealing solution theories in cooperative game theory. Each is derived as the unique fair allocation rule satisfying a set of compelling and (seemingly) weak axioms. Harsanyi [1963] showed that these two important solution concepts are special cases of a more general solution concept, called a *nontransferable-utility value* or *NTU value*, that is defined for all complete-information cooperative games, with any number of players, with or without transferable utility. Shapley [1969] developed a simplified version of the NTU value. These solution concepts may be viewed as generalizations and extensions of the equal-gains principle (that any two players should gain equally from cooperating with each other) to games with more than two players and without obvious symmetries or interpersonally-comparable utility scales.

Harsanyi/Selten [1972] developed a generalized Nash solution for games with incomplete information, a modified version of which was presented by Myerson [1979]. However, this solution concept had serious theoretical drawbacks, and no n-person generalization value could be found. Myerson [1983] analyzed the problem of inscrutable selection of a mechanism by a player who has all of the bargaining ability but also needs to conceal his preferences and private information. This work led to a new generalization of the Nash bargaining, solution for two-player games with incomplete information where both players have equal bargaining ability. This new generalized Nash solution was derived from axioms in Myerson [1982].

In this paper, we will construct a bargaining solution concept that will extend the solution concept of Myerson [1982] and the NTU value of Shapley [1969] to general cooperative games with incomplete information, using the Bayesian formulation of Harsanyi [1967–68]. Our bargaining solution will not be derived from axioms here. Its justification will be that it generalizes and unifies the three basic axiomatically derived theories of Nash [1950], Shapley [1953], and Myerson [1982].

It is reasonable to ask why we should be interested in finding unified cooperative solution concepts of such great generality. One goal is to have a common framework within which to analyze and compare a wide variety of games: Another goal is to use generalizability as a test of solution concepts themselves. That is if there are two solution concepts which appear equally plausible for a limited class of games, but only one is naturally generalizable to a much broader class, then that is evidence in favor of the conceptual significance of the generalizable concept. In this sense, perhaps the bargaining solution concept in this paper should be viewed as a further justification of the Shapley value and the Nash bargaining solution.

But the most important gain from developing a unified solution concept for general cooperative games with incomplete information maybe that it forces us to systematically survey the basic logical issues involved in cooperation under uncertainty. In this paper, we will be developing conceptual structures and perspectives which may prove to have significance beyond the specific solution concept to which they are applied in this paper. In particular, the ideas of *virtual utility* and *maximal linear extensions*; developed in Sections 3 and 4 respectively, might also be applied to develop alternative solution concepts for cooperative games with incomplete information. Also, the interpretation of the rational-threats criterion developed in Section 6 may also help justify the Shapley NTU value against the recent criticism of Roth [1980] and Shafer [1980].

In Section 2, the general structure of cooperative games with incomplete informations formalized: Incentive-efficient mechanisms for, such games satisfy a parametric linear programming problem, which is characterized in Section 3. Virtual utility is defined so that the Lagrangian function for this parametric optimization problem can be expressed as the expected sum of the players' virtual utility. Thus, in an efficient agreement subject to incentive constraints, it may appear *ex post* that the players have maximized their virtual utilities, rather than their real utilities. This suggests the following *virtual utility hypothesis*: that when incentive constraints (necessary for players to trust each other) are binding in bargaining situation, players may act as if they want to maximize their virtual utilities, rather than their real utilities.

The concept of transferable utility was extremely important in the first development of cooperative game theory. However, for games with incomplete information, linear activities like side payments can serve a signalling purpose as well as a transfer purpose, which makes matters more complicated. In Section 4, it is shown that, for a game with incomplete information, the most transferability that can be allowed, without totally replacing the efficient frontier, is transferability of virtual utility conditionally on the state of information in the game.

In Section 5, the ideas of Sections 3 and 4 are applied to construct the general solution concept. With complete information, a Shapley NTU value is an allocation for which there exist nonnegative weighting factors for all players' utility scales such that the allocation would be both equitable (as evaluated by the Shapley value) and efficient if interpersonal comparisons and transfers could be made in terms of these weighted utility scales. For games with incomplete information, a bargaining solution is an incentive-compatible mechanism for which there exist virtual utility scales such that the mechanism would be both equitable and efficient if interpersonal comparisons and transfers could be made in terms of these virtual utility scales. The main results of this paper are the existence and individual rationality of these general bargaining solutions.

The rational-threat criterion used in our solution concept is reconsidered in Section 6. We show that the rational-threat criterion may be most appropriate in games where the coalitions can commit themselves to threats in advance, when they anticipate only a small probability of actually carrying out the threats. In such a situation, a single coalition's threat against its complement does not need to be either equitable or incentive compatible. Instead, it should be evaluated as part of a plan of threat and agreement that must be equitable and incentive-compatible overall.

Section 7 contains the longer proofs.

2 Basic definitions

Let $N = \{1, 2, ..., n\}$ denote the set of players, and CL denote the set of possible conditions or nonempty subsets of N, so that

$$CL = \{S | S \subseteq N, S \neq \emptyset\}.$$

For any coalition S, we let D_S denote the set of collective actions or decisions feasible for the members of S if they cooperate with each other. For example, in a market game, D_S might be the set of possible trades among the members of S. For any two disjoint coalitions R and S, we assume that

$$D_R \times D_S \subseteq D_{R \cup S}.$$

That is, $R \cup S$ can implement any decisions: feasible for R and S separately, if $R \cap S = \emptyset$.

For any player i in N, we let T_i denote the set of possible *types* for player i, where each type t_i in T_i is a complete description of i's private information about preferences, endowments, and any other factors relevant to the players. For any coalition S, we let

$$T_S = \mathop{\mathbf{X}}_{i \in S} T_i,$$

so any t_S in T_S denotes a possible combination of types for the members of S. For mathematical simplicity, we will assume that all T_i and D_S are nonempty finite sets. The decision spaces and type spaces for the grand coalition N will play a major role here, so we may drop the subscript N for these sets; that is,

$$D = D_N, \quad T = T_N.$$

For any d in D and t in T, we let $u_i(d, t)$ denote the payoff to player i, measured in some von Neumann-Morgenstern utility scale, if t is the combination of types for the players and d represents the decisions made by the players.

Throughout this paper, whenever t, t_S, and t_i appear in the same formula, then t_i denotes the i-th componentof the vector t in T, and $t_S = (t_j)_{j \in S}$. We also use the notation $N - i - N \backslash \{i\}$, and we may write $t = (t_{N-i}, t_i)$. Similarly, $t_{N-i}, s_i)$ is the vector of types differing from t, in that the i-th component is changed to s_i.

For any t in T, we let $p_i, (t_{N-i}|t_i)$ denote the conditional probability that t_{N-1} is the combination of types for players other than i, as would be assessed by player i if t_i were his type. We will assume that these probabilities are *consistent* in the sense of Harsanyi [1967–68]. That is, there exists some probability distribution p over T such that

$$p_i(t_{N-i}|t_i) = p(t)/p^i(t_i) \qquad \forall i \in N, \quad \forall t \in T \tag{2.1}$$

where

$$p^i(s_i) = \sum_{s_{N-i} \in T_{N-i}} p(s) \qquad \forall i \in N, \quad \forall s_i \in T_i. \tag{2.2}$$

We will also assume that no types-vector has zero probability, so

$$p(t) > 0, \quad \forall t \in T. \tag{2.3}$$

(These consistency and positivity assumptions (2.1)–(2.3) will be needed only to simplify the interpretation of our results. The solution concept developed in this paper will satisfy the *probability-invanance axiom* described in Myerson [1982], and so it can be extended using this axiom to games without consistent positive probability distributions.)

Thus a *cooperative game with incomplete information* is defined by these structures:

$$\Gamma = ((D_S)_{S\in CL}, (T_i, u_i)_{i\in N}, p).$$

We assume that this structure Γ is common knowledge among the players when they play the game, plus each player also knows in own true type. We may refer to a vector of the players' types as a *state* of the game.

Any coalition S, if it were to form, could plan to determine its collective decision randomly as a function of its members' information. We let M_S devote the set of all functions from T_S into the set of probability distributions over D_S. That is, $\mu_S \in M_S$ iff

$$\mu_S(d_S|t_S) \geq 0 \text{ and } \sum_{c_S\in D_S} \mu_S(c_S|t_S) = 1 \qquad \forall d_S \in D_S, \quad \forall t_S \in T_S. \tag{2.5}$$

Any such μ_S in M_S may be referred to as a *mechanism* for coalition S.

If $R \cap S = \emptyset$, then we can embed $M_R \times M_S$ in $M_{R\cup S}$ in the obvious way. That is, if $\mu_R \in M_R$ and $\mu_S \in M_S$, then (μ_R, μ_S) in $M_{R\cup S}$ is defined by

$$(\mu_R, \mu_S)(d_R, d_S|t_R, t_S) = \mu_R(d_R|t_R) \cdot \mu_S(d_S|t_S)$$
$$\text{if } (d_R, d_S) \in D_R \times D_S \subseteq D_{R\cup S},$$

and

$$(\mu_R, \mu_S)(d_{R\cup S}|t_R, t_S) = 0 \text{ if } d_{R\cup S} \notin D_R \times D_S.$$

We shall assume that, in the cooperative game, only the mechanism chosen by the grand coalition N will actually be implemented. As a threat during bargaining, each coalition S may commit itself to some mechanism μ_S in M_S, to be carried out if the other players refuse to cooperate with the members of S. Such threats will be significant only to the extent that they may influence the mechanism μ_N chosen by the grand coalition. In the rest of this section and in Sections 3 and 4, we will only consider mechanisms in M_N, to develop the theory of efficient mechansm for the grand coalition. In Section 5 we will reconsider the threats of all coalitions and construct our bargaining solution.

We let $U_i^*(\mu_N, s_i|t_i)$ denote the expected utility for player i from the mechanism μ_N in M_N,, if i's true type is t_i but he reports type s_i, while all other players are expected to report their types truthfully. That is

$$U_i^*(\mu_N, s_i|t_i) =$$
$$= \sum_{t_{N-i}\in T_{N-i}} p_i(t_{N-i}|t_i) \sum_{d\in D} \mu_N(d|t_{N-i}, s_i) u_i(d, t). \tag{2.6}$$

We let

$$U_i(\mu_N|t_i) = U_i^*(\mu_N, t_i|t_i)$$
$$= \sum_{t_{N-i}\in T_{N-i}} p_i(t_{N-i}|t_i) \sum_{d\in D} \mu_N(d|t) u_i(d, t). \tag{2.7}$$

That is, $U_i(\mu_N|t_i)$ is the expected utility for player i from the mechanism μ_N, if i's true type is t_i and all players are expected to report their types truthfully in implementing μ_N.

We shall assume that each player's type is not observable by other players, so that the types are unverifiable. Thus, if a player had some incentive to lie about his type when the grand coalition N implements its mechanism μ_N, then he would do so. A mechanism is *incentive compatible* [or, more correctly, *Bayesian incentive compatible* in the sense of d'Aspremont/Gerard-Varet [1979]] iff

$$U_i(\mu_N|t_i) \geq U_i^*(\mu_N, s_i|t_i) \qquad \forall i \in N, \quad \forall t_i \in T_i, \quad \forall s_i \in T_i. \tag{2.8}$$

That is, μ_N is incentive compatible iff it would be a Bayesian Nash equilibrium for all players to plan to report their types honestly in the mechanism μ_N, assuming that they are asked to report their types simultaneously and confidentially. Thus, with unverifiable types, the players must choose an incentive-compatible mechanism if honest reporting of types is to be induced. It has been argued elsewhere [See Myerson, 1979, for example] that any Bayesian equilibrium of possibly dishonest reporting strategies in any mechanism can be simulated by an equivalent incentive-compatible mechanism with honest reporting. So without loss of generality, we may assume that the mechanism selected by the grand coalition N must be incentive compatible.

In some games, it may be possible for some types to costlessly prove that other types are false.[1] For example, if a person can play the piano, then he can prove that he is not a non-pianist simply by playing a few bars. On the other hand, the nonpianist cannot prove that he is not really a pianist unless he is given the proper incentives. If player i, when s_i is his true type, could costlessly prove that he is not type t_i, then we should drop the corresponding constraint (saying that t_i must not be tempted to report s_i) in (2.8). With this modification, our analysis in this paper can be extended to cover the case of verifiable or semi-verifiable types. Henceforth in this paper we will consider only the case of unverifiable types.

3 The primal and dual problems and virtual utility

A mechanism μ_N in M_N *is incentive-efficient* iff it is incentive compatible and there does not exist any other incentive-compatible mechanism $\hat{\mu}_N$ such that

$$U_i(\hat{\mu}_n|t_i) \geq U_i(\mu_N|t_i), \qquad \forall i \in N, \quad \forall t_i \in T_i, \tag{3.1}$$

with $U_j(\hat{\mu}_N|t_j) > U_j(\mu_N|t_j)$ for at least one type t_j of some player j. If the players can bargain effectively, then they should be able to ultimately agree on some incentive-efficient mechanism. Otherwise, it would be common knowledge that all players could agree to a change to some other mechanism $\hat{\mu}_N$ satisfying (3.1). See Holmström/Myerson [1983] for an analysis of this and other concepts of efficiency for games with incomplete information.

[1] I am indebted to Paul Milgrom for pointing out this issue.

Let Λ be the following simplex in $\mathbf{X}_{i\in N}\mathbf{R}^{T_i}$,

$$\Lambda = \left\{ \lambda \in \mathop{\mathbf{X}}_{i\in N} \mathbf{R}^{T_i} | \lambda_i(t_i) \geq 0, \quad \forall i \in N, \forall t_i \in T_i, \sum_{j\in N} \sum_{s_j\in T_j} \lambda_j(s_j) = n \right\}. \tag{3.2}$$

Let Λ° denote the relative interior of Λ,

$$\Lambda^\circ = \{\lambda \in \Lambda | \lambda_i(t_i) > 0, \quad \forall i \in N, \quad \forall t_i \in T_i\}. \tag{3.3}$$

Since we are assuming that D and T are finite sets, the set of all incentive-compatible mechanisms is a closed convex polyhedron in M_N, defined by the linear inequalities (2.5) and (2.8). (Notice that $U_i(\mu_N|t_i)$ and $U_i^*(\mu_N, s_i|t_i)$ are both linear functions μ_N.) Thus, by the supporting hyperplane theorem, μ_N, is incentive-efficient iff there exists some λ in Λ° such that μ_N is an optimal solution to the-problem

$$\operatorname*{maximize}_{\mu_N\in M_N} \sum_{i\in N} \sum_{t_i\in T_i} \lambda_i(t_i) U_i(\mu_N|t_i) \tag{3.4}$$

subject to the incentive constraints (2.8).

We shall refer to this optimization problem (3.4) as the *primal problem* for λ.

Given λ, the primal (3.4) is a linear programming problem. Let us now formulate its dual. We shall generally let $\alpha_i(s_i|t_i)$ denote the dual variable (or shadow price) corresponding to the incentive constraint (2.8) that asserts that player i should not be tempted to claim to be type s_i if his true type is t_i. We let

$$\begin{aligned} \underset{\sim}{A} = \Big\{ &\alpha \in \mathop{\mathbf{X}}_{i\in N} \mathbf{R}^{T_i\times T_i} | \alpha_i(s_i|t_i) \geq 0, \alpha_i(t_i|t_i) = 0 \\ &\forall i \in N, \quad \forall t_i \in T_i, \quad \forall s_i \in T_i \Big\}. \end{aligned} \tag{3.5}$$

That is, $\underset{\sim}{A}$ is the set of all possible vectors of dual variables for the incentive constraints. ((2.8) holds trivially when $s_i = t_i$, so the shadow price $\alpha_i(t_i|t_i)$ will be zero.)

We now come to an important definition. Given any λ and Λ and α in $\underset{\sim}{A}$, let

$$\begin{aligned} \mathrm{v}_i(d, t, \lambda, \alpha) = \Big(&\Big(\lambda_i(t_i) + \sum_{s_i\in T_i} \alpha_i(s_i|t_i)\Big) p_i(t_{N-i}|t_i) u_i(d, t) \\ &- \sum_{s_i\in T_i} \alpha_i(t_i|s_i) p_i(t_{N-i}|s_i) u_i(d, (t_{N-i}, s_i))\Big)\Big/p(t) \end{aligned} \tag{3.6}$$

for any i in N, d in D, and t in T. We shall refer to $v_i(d, t, \lambda, \alpha)$ as player i's *virtual utility* for decision d in state t, with respect to λ and α.

If we multiply the incentive constraints by their dual variables and add them into the primal objective function, then we get the following *Lagrangian function:*

$$\sum_{i \in N} \sum_{t_i \in T_i} \lambda_i(t_i) U_i(\mu_N|t_i) + \sum_{i \in N} \sum_{t_i \in T_i} \sum_{s_i \in T_i} \alpha_i(t_i|s_i)(U_i(\mu_n|t_i) - U_i^*(\mu_N, s_i|t_i)) = \sum_{t \in T} p(t) \sum_{d \in D} \mu_N(d|t) \sum_{i \in N} \mathrm{v}_i(d, t, \lambda, \alpha). \tag{3.7}$$

The equality in (3.7) follows by straightforward manipulation from the Definitions (2.6), (2.7) and (3.6). So the Lagrangian function for the primal problem is just the expected sum of the players' virtual utilities.

By standard Lagrangian analysis, an incentive-compatible mechanism μ_N will be an optimal solution of the primal problem for λ if and only if there is some α in $\underset{\sim}{A}$ such that

$$\alpha_i(s_i|t_i)(U_i(\mu_N|t_i) - U_i^*(\mu_N, s_i|t_i)) = 0, \quad \forall i \in N, \quad \forall t_i \in T_i, \forall s_i \in T_i,$$

and μ_N maximizes the Lagrangian function subject only to the probability constraints (2.3). Obviously, this Lagrangian function is maximized by putting all probability weight, in each $\mu_N(\cdot|t)$ distribution, on the decisions that maximize the sum of the players' virtual utilities. That is, μ_N maximizes the Lagrangian function over all mechanisms in M_N if and only if

$$\sum_{d \in D} \mu_N(d|t) \sum_{t \in T} \mathrm{v}_i(d, t, \lambda, \alpha) = \underset{d \in D}{\text{maximum}} \sum_{i \in N} \mathrm{v}_i(d, t, \lambda, \alpha), \quad \forall t \in T. \tag{3.8}$$

The appropriate vector a for use in this Lagrangian analysis is the vector that solves the dual of (3.4). This *dual problem* for λ can be written

$$\underset{\alpha \in \underset{\sim}{A}}{\text{minimize}} \sum_{t \in T} p(t) \cdot \underset{d \in D}{\text{maximum}} \sum_{i \in N} \mathrm{v}_i(d, t, \lambda, \alpha). \tag{3.9}$$

Each $\mathrm{v}_i(d, t, \lambda, \alpha)$ is linear in α, so this dual problem is indeed a linear programming problem.

The virtual utility functions will play an important role in our theory of bargaining, so it is worthwhile to try to develop some intuitive understanding of them. So let us assume that μ_N is an incentive-efficient mechanism. Let λ in Λ° and α in $\underset{\sim}{A}$ be such that μ_N solves the primal for λ and α solves the dual for λ. We say that one type s_i *jeopardizes* another type t_i of player i, in the incentive-efficient mechanism μ_N iff the constraint that says s_i should not gain by claiming to be t_i is binding (that is, $U_i(\mu_N|s_i) = U_i^*(\mu_N, t_i|s_i)$) and its shadow price $\alpha_i(t_i|s_i)$ is positive. Then player i's virtual utility when he is of type t_i differs from his real utility in a way that exaggerates the difference from the types that jeopardize t_i.

That is, equation (3.6) defines i's virtual utility for d at t as a positive multiple of his real Utility for d at t, minus a multiple of what his utility for d would be if his type were changed to a type that jeopardizes t_i. To see this more clearly, notice that (3.6) may be rewritten as

$$p^i(t_i)\mathrm{v}_t(d,t,\lambda,\alpha) = (\lambda_i(t_i) + \sum_{s_i} \alpha_i(s_i|t_i))u_i(d,t)$$
$$-\sum_{s_i} \alpha_i(t_i|s_i)u_i(d,(t_{N-i},s_i))(p_i(t_{N-i}|s_i)/p_i(t_{N-i}|t_i)),$$

where the probability-correction ratio in the last term vanishes to one if the players' types are stochastically-independent.

For type t_i of player i, the expected virtual utility from the mechanism μ_N (honestly implemented) is

$$\sum_{t_{N-i}\in T_{N-i}} \sum_{d\in D} p_i(t_{N-i}|t_i)\mu_N(d|t)\mathrm{v}_i(d,t,\lambda,\alpha) \tag{3.10}$$
$$= ((\lambda_i(t_i) + \sum_{s_i} \alpha_i(s_i|t_i))U_i(\mu_N|t_i) - \sum_{s_i} \alpha_i(t_i|s_i)U_i^*(\mu_N, t_i|s_i))/p^i(t_i).$$

If μ_N solves the primal for λ and a solves the dual for λ, then by complementary slackness this formula can be further simplified to

$$((\lambda_i(t_i) + \sum_{s_i} \alpha_i(s_i|t_i))U_i(\mu_N|t_i) - \sum_{s_i} \alpha_i(t_i|s_i)U(\mu_N|s_i))/p^i(t_i). \tag{3.11}$$

Let us consider now an application of the virtual utility concept. An incentive-efficient mechanism need not be efficient *ex post*, after the players learn each other's type. That is because, in order to satisfy incentive constraints, it may be necessary to accept a positive probability of an outcome that is bad for both players. For example, in union-management negotiations, if the management is of the "type" that can only afford to pay lower wages, then it might have to accept a positive probability of a strike before it can get a reduction in the wage rate. The strike is needed to prove to the workers that management is not of the type with high ability to pay. But it may be difficult to understand how the players can commit themselves to implement a strike of any duration, since management's low type is revealed as soon as the strike begins, and then both sides would prefer to settle at a low wage.

By (3.8), an incentive-efficient mechanism always maximizes the sum of the players' virtual utilities (with respect to the appropriate λ and α) in every state t. Thus an incentive-efficient mechanism would appear efficient *ex post* if the players' payoffs were measured in virtual utility, instead of real utility. Instead of saying that the incentive constraints (2.8) force the players to accept *ex post* inefficiency, we may say that the incentive constraints force each player to transform his effective preferences from his real to his virtual utility, function, to exaggerate the difference between his true type and the false types that jeopardize it. This idea, that players in bargaining may act as if they want to maximize their virtual utilities instead of their actual utilities, may be referred to as the *virtual-utility hypothesis*.

In Section 5, we will extend this virtual-utility hypothesis by assuming that the players also make interpersonal equity comparisons in terms of virtual utility, to compute their fair payoffs or *warranted claims*. But first, we consider generalizations of the classical tansferable-utility assumption for games with incomplete information.

4 Transferable utility and linear activities

The assumption of *transferable utility* has played an important role in the development of cooperative game theory. Of course, bounded utility transfers can be accomodated within the model described in Section 2 (by interpreting, the decisions in each D_S as including specifications of how much utility should be transferred between each pair of players in S), so there is very little loss of generality in restricting ourselves to this model. Nevertheless, to understand cooperation under uncertainty, it is useful to see how the assumption of tansferable utility extends to games with incomplete information.

Transfer of utility between players is just a special kind of *linear activity* which could be permitted in a game. In general, a linear activity can be represented by a function $f : T \to \mathbf{R}^n$, such that $f_i(t)$ is the utility gained by player i in state t if the players do one unit of activity f. Let F be any finite set of such activities, so that F is a subset of $\mathbf{R}^{N \times T}$. Given Γ as in (2.1), the game Γ *extended by* F refers to the game in which the grand coalition can also use any linear combination of activities in F as a function of the players type-reports. (More generally, we could introduce a set of feasible linear activities F_S for each coalition S, with $F_S \subseteq F_R$ if $S \subseteq R$, but we will only be concerned with the grand coalition N in this section.)

In the extended game with linear activities, the set of mechansm for N becomes $M_N \times \mathbf{R}^{F \times T}$. That is, a mechanism is a pair (μ_N, e) in $M_N \times \mathbf{R}^{F \times T}$ where $e(f|t)$ is interpreted as the level of activity f to be performed if the players report their vector of types as t. Notice that we allow that $e(f|t)$ maybe positive or negative.

The expected utility gained by a player i from linear activities in the mechanism (μ_N, e), given that i's type is t_i, is

$$G_i(e|t_i) = \sum_{t_{N-i} \in T_{N-i}} p_i(t_{N-i}|t_i) \sum_{f \in F} f_i(t) e(f|t) \tag{4.1}$$

if all players are honest, and is

$$G_i^*(e, s_i|t_i) = \sum_{t_{N-i}} p_i(t_{N-i}|t_i) \sum_{f} f_i(t) e(f|t_{N-i}, s_i) \tag{4.2}$$

if all players are honest except for i who reports s_i. The mechanism (μ_N, e) is *incentive-compatible* iff

$$U_i(\mu_N|t_i) + G_i(e|t_i) \geq U_i^*(\mu_N, s_i|t_i) + G_i^*(e, s_i|t_i)$$
$$\forall i \in N, \quad \forall t_i \in T_i, \quad \forall s_i \in T. \tag{4.3}$$

With linear activities, the *extended primal problem* for λ is defined to be

$$\underset{\mu_N,e}{\text{maximize}} \sum_{i\in N}\sum_{t_i\in T_i} \lambda_i(t_i)(U_i(\mu_N|t_i) + G_i(e|t_i)) \tag{4.4}$$
$$\text{subject to } \mu_N \in M_N \text{ and (4.3).}$$

This extended primal problem differs from the original primal problem (3.4) only in that it has more variables, in the vector e. Thus, the *extended dual problem* for λ differs from the original dual problem (4.9) in that it has more constraints, one new constraint for each variable $e(f|t)$, as follows:

$$\sum_{i\in N}((\lambda_i(t_i) + \sum_{s_i\in T_i} \alpha_i(s_i|t_i))p_i(t_{N-i}|t_i)f_i(t)$$
$$- \sum_{s_i\in T_i} \alpha_i(t_i|s_i)p_i(t_{N-i}|s_i)f_i(_{N-i}, s_i)) = 0, \qquad \forall f \in F, \quad \forall t \in T. \tag{4.5}$$

(Notice that (4.5) is a linear constraint on α.)

The assumption of *transferable utility* means that, for any two players j and k, F includes an activity of transfering one unit of utility from j to k. We may denote this activity by f^{jk}, where

$$f_j^{jk}(t) = +1, \quad f_k^{jk}(t) = -1, \quad f_i^{jk}(t) = 0 \text{ if } i \notin \{j,k\}, \quad \forall t \in T.$$

Let us suppose that the players' types are stochastically independent random variables, so that

$$p(t) = \mathop{\Pi}_{i\in N} p^i(t_i), \quad \forall t \in T.$$

Then (4.5) for $f = f^{jk}$ becomes

$$(\lambda_j(t_j) + \sum_{s_j\in T_j} \alpha_j(s_j|t_j) - \sum_{s_j\in T_j} \alpha_j(t_j|s_j)) \mathop{\Pi}_{i\in N-j} p^i(t_i)$$
$$= (\lambda_k(t_k) + \sum_{s_k\in T_k} \alpha_k(s_k|t_k) - \sum_{s_k\in T_k} \alpha_k(t_k|s_k)) \sum_{j\in N-k} p^i(t_i), \quad \forall t \in T.$$

Dividing both sides of this equation by $p(t)$ gives us

$$\left(\lambda_j(t_j) + \sum_{s_j} \alpha_j(s_j|t_i) - \sum_{s_j} \alpha_j(t_j|s_j)\right) / p^j(t_j) \tag{4.6}$$
$$= (\lambda_k(t_k) + \sum_{s_k} \alpha_k(s_k|t_k) - \sum_{s_k} \alpha_k(t_k|s_k))/p^k(t_k), \quad \forall t_j \in T_j, \quad \forall t_k \in T_k.$$

Thus, if the players' types are independent and if utility is transferable between all players then, for any λ and Λ, a satisfies the dual constraint (4.5) iff

$$\lambda_i(t_i) + \sum_{s_i\in T_i} \alpha_i(s_i|t_i) - \sum_{s_i\in T_i} \alpha_i(t_i|s_i) = p^i(t_i), \quad \forall i \in N, \quad \forall t_i \in T_i. \tag{4.7}$$

(The constant ratio in (4.6) must be 1, because the at $\lambda_i(t_i)$ sum to n, for λ in Λ. Recall (3.2).)

Equation (4.7) can be very helpful for solving applied problems. However, it also illustrates why transferable utility is less useful as an assumption for games with incomplete information than it was for games with complete information. With complete information each player has only one type, so that (4.7) becomes simply $\lambda_i(t_i)$; that is, all players must be given equal weight in the primal problem or else the dual is infeasible and the primal has no finite optimum. Thus, transferable utility with complete information implies that all efficient mechanisms solve the same primal problem, and the Pareto-efficient frontier is a hyperplane. With incomplete information, (4.7) implies that

$$\sum_{t_i \in T_i} \lambda_i(t_i) = 1, \quad \forall i \in N,$$

but this still leaves $\sum_{i \in n}(|T_i| - 1)$ degrees of freedom in choosing λ, and the dual problems are generally nontrivial to solve. Under incomplete information; the incentive-efficient frontier in $\mathbf{X}_{j \in N}\mathbf{R}^{T_i}$ is generally *not* a hyperplane for games with transferable utility.

To get a conceptual simplification comparable to that offered by transferable utility under complete information, we must introduce a larger class of linear activities. So let us re-examine (4.5), but think of it now as a constraint on f for some given λ and α. The expression in brackets in (4.5) is just $p(t)$ times i's virtual utility for one unit of activity f in state t. Thus, (4.5) asserts that f must transfer virtual utility between the players in each state.

Thus, instead of transferable utility, let us consider the assumption of *conditionally transferable virtual utility*. Given any two players j and k and given any type-vector in T, let g^{jks} denote the activity that transfers one unit of virtual utility (with respect to λ and α) from player j to player k conditionally on s being the true vector of types, and that transfers zero units of virtual utility otherwise. That is, g^{jks} satisfies the following equations, for every i in N and t in T:

$$\begin{aligned} &\Big(\Big(\lambda_i(t_i) + \sum_{r_i \in T_i} \alpha_i(r_i|t_i)\Big)p_i(t_{N-i}|t_i)g_i^{jks}(t) \\ &\quad - \sum_{r_i \in T_i} \alpha_i(t_i|r_i)p_i(t_{N-i}|r_i)g_i^{jks}(t_{N-i}, r_i)\Big)/p(t) \\ &= \begin{cases} -1 \text{ if } i = j \text{ and } t = s \\ +1 \text{ if } i = k \text{ and } t = s \\ 0 \text{ if } i \notin \{j,k\} \text{ or } t \neq s. \end{cases} \end{aligned} \tag{4.8}$$

If $\lambda \in \Lambda^\circ$, $\alpha \in \underset{\sim}{A}$, and all $p_i(t_{N-i}|t_i) > 0$, then (4.8) has a unique solution g^{jks}, and this vector satisfies:

$$\begin{aligned} &g_i^{jks}(t) = 0 \text{ if } i \notin \{j,k\} \text{ or } t_{N-i} \neq s_{N-i} \\ &g_j^{jks}(s_{N-j}, t_j) \leq 0 \quad \forall t_j \in T_j \\ &g_k^{jks}(s_{N-k}, t_k) \geq 0 \quad \forall t_k \in T_k. \end{aligned}$$

(These properties follow from (4.8) using Lemma 1 in Sect. 7). Notice that, although g^{jks} gives no *virtual utility* to player k except in state s, g^{jks} may in fact give him positive amounts of *real utility* in states where his own type differs from s_k.

Given any λ in Λ° and α in $\underset{\sim}{A}$, we let $\bar{F}^{\lambda\alpha}$ denote the set of all such g^{jks} generated by λ and α. That is:

$$\bar{F}^{\lambda\alpha} = \{g^{jks} | j \in N, , k \in N, s \in T, \text{ and (4.8) is satisfed with } \lambda \text{ and } \alpha\}.$$

If Γ is extended by $\bar{F}^{\lambda\alpha}$ then we may say that virtual utility with respect to λ and α is conditionally transferable. (Here "conditionally transferable" refers to the fact that the transfers can be conditioned on the players' true types, rather than just their reported types. So conditional transferability is a stronger property than simple transferability).

The following theorem states that, if we try to extend a game in such a way as to preserve at least one of its incentive-efficient mechanisms, then the maximal extension is to allow conditionally-transferable virtual utility.

Theorem 1. Let Γ be as in (2.1) and let μ_N be an incentive-efficient mechanism for the grand coalition in Γ. Let F be any set of linear activities. Then $(\mu_N, \underset{\sim}{0})$ is incentive-efficient in Γ extended by F if and only if there exists some λ in Λ° and α in $\underset{\sim}{A}$ such that μ_N is an optimal solution of the primal problem for λ, α is an optimal solution of the dual problem for λ, and F is contained in the linear span of $\bar{F}^{\lambda\alpha}$ (here $(\mu_N, \underset{\sim}{0})$ is just μ_N without using any activities in F).

Proof. $(\mu_N, \underset{\sim}{0})$ is incentive-efficient in Γ extended by F if and only if there is some λ in Λ° such that $(\mu_N, \underset{\sim}{0})$ is optimal in the extended primal for λ. But this holds if and only if there is some α in $\underset{\sim}{A}$ such that α is feasible in the extended dual for λ and the value of the primal objective function at μ_N, equals the value of the dual objective function at α. This in turn holds if and only if μ_N is optimal in the (unextended) primal for λ, α is optimal in the (unextended) dual for λ, and α is feasible in the extended dual. But the linear span of $\bar{F}^{\lambda\alpha}$ is just the set of all activities f that satisfy (4.5) for λ and α. (To check this, observe that all activities in $\bar{F}^{\lambda\alpha}$ satisfy (4.5); there are $(n-1)|T|$ linearly independent vectors g^{jks} in $\bar{F}^{\lambda\alpha}$; and the set of vectors satisfying (4.5) has $(n-1)|T|$ dimensions.) So α is feasible in the extended dual for λ if and only if F is contained in the linear span of $\bar{F}^{\lambda\alpha}$. □

To understand Theorem 1, it is helpful to recognize that linear activities in games with incomplete information can be used for *signalling*, that is, for helping to satisfy incentive compatibility, as well as for transferring utility. For example, if real utility (instead of virtual utility) were conditionally transferable, then a player could perfectly signal his type by agreeing to transfer large amounts of utility to other players conditionally on his type being anything other than what he reports. In general, any linear activity that affects different types of a player differently may be used for signalling, to help prove that the player is not of the type that loses more from the activity. The activity g^{jks}, which transfers virtual utility from j to k

conditionally on state s, can affect the real utility payoffs of j or k in states other than s; so its potential for signalling purposes is less than that of an activity that transfers real utility from j to k conditionally on state s.

5 The general bargaining solution

The construction of Shapley's NTU value for games with complete information may be sketched as follows. First, select any outcome on the Pareto-efficient frontier for the grand coalition. Now extend the game by a maximal collection of linear activities such that the selected outcome is still on the efficient frontier of the extended game. These linear activities can be characterized as transfers of weighted utility between players, where each player's "weighted utility" payoff is some constant λ_i, times his original utility. In the extended game, let each coalition choose a threat; let the worth of each coalition be the total weighted utility that would be earned by its members if it and its complement both carried out their threats; and let the grand coalition N act so as to give each player weighted utility equal to his Shapley-value allocation computed from these coalitional worths. If this hypothetical behavior in the extended game, when each coalition chooses its threat optimally for its members, turns out to give the players payoffs equal to what they were getting in the originally selected outcome (which was feasible in the original game), then we say that that outcome is a Shapley NTU value for the original game. That is, the Shapley NTU value is defined as a Shapley value for an extended game with transfers, that is also feasible in the original game without transfers.

In Section 3 we saw that, when players face binding incentive constraints, they may appear to act according to the preferences of their virtual utility functions. In Section 4 we saw that, with incomplete information, the maximal linear extension (without completely replacing the efficient frontier) is to let virtual utility (w.r.t. some λ and α) be conditionally transferable in every state. Thus, to follow the logic of the Shapley NTU value, we should let coalitional worths and Shapley values be computed in terms of virtual utilities. This key insight, to look at the game with transferable *virtual utility* rather than *weighted utility*, was not evident to this author until after eight years of search; but with it we can readily construct a bargaining solution which generalizes the Shapley-NTU value and has satisfactory mathematical properties, including individual rationality and existence.

In our model of bargaining, every coalition makes a threat against the complementary coalition, and then these threats form the basis for computing the warranted claims of each player. We let

$$M = \mathop{\mathbf{X}}_{S \in CL} M_S$$

denote the set of possible combinations of mechanisms that the coalitions might select as threats. That is, any vector $\mu = (\mu_S)_{S \in CL}$ in M includes a specification of the mechanism μ_S that each coalition $S \subset N$ threatens to use in the case that its complement $N \backslash S$ refuses to cooperate with it.

For any coalition S, we let $W_S(\mu, t, \lambda, \alpha)$ denote the sum of the virtual utilities (with respect to λ and α) that the members of S would expect in state t, if S and

$N\backslash S$ carried out their threats. That is, if $S \neq N$,

$$W_S(\mu, t, \lambda, \alpha) = \tag{5.1}$$
$$= \sum_{d_S \in D_S} \sum_{d_{N\backslash S} \in D_{N\backslash S}} \mu_S(d_S|t_S)\mu_{N\backslash S}(d_{N\backslash S}(d_{N\backslash S}|t_{N\backslash S})$$
$$\times \sum_{i \in S} \mathrm{v}_i((d_S, d_{N\backslash S}), t, \lambda, \alpha).$$

In the case of $S = N$, there is no complementary coalition to threaten, so (5.1) simply reduces to:

$$W_N(\mu, t, \lambda, \alpha) = \sum_{d \in D} \mu_N(d|t) \sum_{i \in N} \mathrm{v}_i(d, t, \lambda, \alpha).$$

We let $W(\mu, t, \lambda, \alpha) = (W_S(\mu, t, \lambda, \alpha))_{S \in CL}$ denote the characteristic function game with these coalitional worths. Its Shapley value for player i is:

$$\phi_i(W(\mu, t, \lambda, \alpha)) = \sum_{\substack{S \in CL \\ S \supseteq \{i\}}} \frac{(|S|-1)!(n-|S|)!}{n!} \cdot$$
$$\cdot (W_S(\mu, t, \lambda, \alpha) - W_{N\backslash S}(\mu, t, \lambda, \alpha)). \tag{5.2}$$

(This formula is equivalent to the more familiar formula with $S - i$ replacing $N\backslash S$. We let $W_\Phi = 0$.) Thus, if the coalitions make threats μ in the game with conditionally transferable virtual utility, then the Shapley value gives type t_i of player i on expected virtualutility payoff equal to:

$$\sum_{t_{N-i} \in T_{N-i}} p_i(t_{N-i}|t_i)\phi_i(W(\mu, t, \lambda, \alpha)).$$

We want to know what allocation of real utility corresponds to this allocation of virtual utility. By (3.11) we know that, if each type s_i of player i gets expected (real) utility $\omega_i(s_i)$ from an incentive-compatible mechanism which maximizes the sum of the virtual utilities, then the corresponding virtual utility expected by type t_i is:

$$((\lambda_i(t_i) + \sum_{s_i} \alpha_i(s_i|t_i))\omega_i(t_i) - \sum_{s_i} \alpha_i(t_i|s_i)\omega_i(s_i))/p^i(t_i).$$

Equating these two formulas (and multiplying through by $p^i(t_i)$) we see that the allocation of real expected utilities corresponding to the Shapley value allocation of virtual utilities should satisfy:

$$(\lambda_i(t_i) + \sum_{s_i \in T_i} \alpha_i(s_i|t_i))\omega_i(t_i) - \sum_{s_i \in T_i} (t_i|s_i)\omega_i(s_i)$$
$$= \sum_{t_{N-i} \in T_{N-i}} p(t)\phi_i(W(\mu, t, \lambda, \alpha)), \qquad \forall i \in N, \quad \forall t_i \in T_i. \tag{5.3}$$

A vector ω in $\mathbf{X}_{i\in N}\mathbf{R}^{T_i}$ which satisfies (5.3) is said to be *warranted* by λ, α, and μ; $\omega_i(s_i)$ is then the *warranted claim* of type s_i. Thus, the warranted claims are real utility payoffs corresponding to an allocation which would give each type of each player his expected Shapley value, if the players made interpersonal equity comparisons in terms of their virtual utility scales.

For any λ in Λ° and α in $\underset{\sim}{A}$, equations (5.3) have a unique solution in ω, by Lemma 1 in Section 7. Furthermore, these solutions are monotone increasing (weakly) in the right-hand sides. That is, inereasing the right-hand side of (5.3) for any type of player i weakly increases the warranted claims of all types of player i. Thus, to maximize i's warranted claim in any type, player i wants to maximize his expected virtual allocation from the Shapley value in all his types.

The threat μ_S affects the Shapley value allocation only through the difference $W_S - W_{N\backslash S}$, which all members of S want to maximize. Thus we say that μ in M is a vector of *rational threats* with respect to λ and α if

$$\sum_{t\in T} p(t)(W_S(\mu,t,\lambda,\alpha) - W_{N\backslash S}(\mu,t,\lambda,\alpha)) = \tag{5.4}$$

$$= \max_{\nu_S\in M_S} \sum_{t\in T} p(t)(W_S(\mu_S,\nu_S),t,\lambda,\alpha) - W_{N\backslash S}((\mu_S,\nu_S)t,\lambda,\alpha)), \quad \forall S\in CL.$$

(Here (μ_{-s},ν_S) is the vector where ν_s replaces μ_S in μ.) Notice that (5.4) really depends only on μ_S and $\mu_{N\backslash S}$, so the two complementary coalitions are involved in a two-person zero-sum game when they choosing their rational threats. We do not require that rational threats to be incentive compatible; we only require that μ_S must be in M_S, satisfying the probability constraints (2.5). (The set of incentive-compatible mechanisms for coalition S could depend discontinuously on the mechanism chosen by $N\backslash S$. So the threat-selection game between S and $N\backslash S$ would be a pseudogame and would not necessarily have any equilibrium, if we required that each threat be incentive-compatible given the other.)

Condition (5.4) includes the case of $S = N$, using $W_\phi = 0$. Thus, if μ is a vector of rational threats with respect to λ and α then

$$W_N(\mu,t,\lambda,\alpha) = \max_{d\in D} \sum_{i\in N} \mathrm{v}_i(d,t,\lambda,\alpha).$$

That is, μ_N maximizes the Lagrangian function (3.7).

The essential idea in defining our general bargaining solution is that if the warranted claims for a set of rational threats can actually be achieved by an incentive-compatible mechanism, then this mechanism may be called a bargaining solution for the game. Some care is needed in formulating this idea precisely, to permit an existence theorem to be proven. The problem is that the warrant equations (5.3) are only known to be solvable if all $\lambda_i(t_i)$ are strictly positive, so that $\lambda \in \Lambda^\circ$. But the Kakutani [1941] fixed point theorem cannot be applied to the interior of a simplex. We solve this dilemma by allowing that some of our positive $\lambda_i(t_i)$ weights may be infinitesimal. In standard analysis, this is done by considering a sequence of vectors in Λ^0, some of whose components may converge to zero.

(This dilemma also arises in the case of complete information, where the resolution proposed by Shapley [1969] is not quite satisfactory. For Shapley's definition, if there is a dummy in a game then any feasible allocation will be an NTU value, with all nondummies having $\lambda_i = 0$. The definition developed below refines Shapley's definition in a way that rules out such perverse solutions without losing existence.)

We say that $\bar{\mu}_N$ is a *bargaining solution* (or an NTU *value*) for Γ iff $\bar{\mu}_N$ is an incentive-efficient mechanism and there exists a sequence $\{(\lambda^k, \alpha^k, \mu^k, \omega^k)\}_{k=1}^{\infty}$ such that

$$\alpha^k \in \underset{\sim}{A}, \mu^k \in M, \text{ and } \lambda^k \in \Lambda^0 \text{ (so all } \lambda_i^k(t_i) > 0, \forall k; \tag{5.5}$$

$$\mu_k \text{ is a vector of rational threats for } \lambda^k \text{ and } \alpha^k, \text{` } \forall k; \tag{5.6}$$

$$\omega^k \text{ is warranted by } \lambda^k, \alpha^k, \text{ and } \mu_k, \quad \forall k; \tag{5.7}$$

$$\limsup_{k\to\infty} \omega_i^k(t_i) \le U_i(\bar{\mu}_N|t_i), \quad \forall i \in N, \quad \forall t_i \in T_i. \tag{5.8}$$

That is, a bargaining solution is an incentive-efficient mechanism such that there is a vector of warranted claims, supported by positive utility-weights and rational threats, in which no type's warranted claim exceeds the utility that it expects from the mechanism by more than an arbitrarily small amount.

We can now state our main existence and individual-rationality theorems.

Theorem 2. There exists at least one bargaining solution $\bar{\mu}_N$ for Γ.

Theorem 3. If $\bar{\mu}_N$ is a bargaining solution then

$$U_i(\bar{\mu}_N|t_i) \ge \underset{\mu_{N-i}\in M_{N-i}}{\text{minimum}} \ \underset{\mu_i \in M_{\{i\}}}{\text{maximum}} \, U_i(\mu_{N-i}, \mu_i|t) \quad \forall i \in N, \quad \forall t_i \in T_i.$$

Proofs are deferred to Section 7.

For any positive number δ, (5.5)–(5.8) imply that, for all sufficiently large k, $\omega^k(t_i) \le U_i(\bar{\mu}_N|t_i) + \delta$ for every i and t_i, and so

$$\begin{aligned} &\sum_{t\in T} p(t) \max_{d\in D} \sum_{i\in N} \mathrm{v}_i(d, t, \lambda^k, \alpha^k) \\ &= \sum_{t\in T} p(t) W_N(\mu^k, t, \lambda^k, \alpha^k) \\ &= \sum_{t\in T} p(t) \sum_{i\in N} \phi_i(W(\mu^k, t, \lambda^k, \alpha^k)) \\ &= \sum_{t\in T} \sum_{i\in N} \lambda_i^k(t_i) \omega_i^k(t_i) \\ &\le \sum_{i\in N} \sum_{t_i\in T_i} \lambda_i^k(t_i) U_i(\bar{\mu}_N|t_i) + n\delta. \end{aligned} \tag{5.9}$$

(Here the first equality holds because μ_N^k is a rational strategy for N with respect to λ^k and α^k. The second equality is the Pareto-optirnality of the Shapley value. The third equality follows from summing the warrant equations (5.3) over i and t_i. The final inequality follows from (5.8) and the fact that the $\lambda_i(t_i)$ sum to n, since $\lambda \in \Lambda^0$.) But $\bar{\mu}_N$ is incentive compatible; so by duality, if $\delta > 0$ then, for all sufficiently large $k, \bar{\mu}_N$ and α^k are respectively within $n\delta$ of the optimum in the primal and dual problems for λ^k.

The following theorem follows from (5.9), and lists some convenient necessary conditions for a bargaining solution. Notice that these conditions seem to be well. determined, in the sense that (5.10)–(5.13) can determine $\bar{\mu}_N, \alpha, \mu$, and ω, and (5.14) has one equation for each component in λ. This suggests a conjecture that the set of bargaining solutions might be generically finite.

Theorem 4. If $\bar{\mu}_N$ is a bargaining solution for Γ then there exist $(\lambda, \alpha, \mu, \omega)$ such that

$$\bar{\mu}_N \text{ is an optimal solution of the primal problem for } \lambda; \tag{5.10}$$

$$\alpha \text{ is an optimal solution of the dual problem for } \lambda; \tag{5.11}$$

$$\mu \text{ is a vector of rational threats for } \lambda \text{ and } \alpha, \text{ and } \mu_N = \bar{\mu}_N; \tag{5.12}$$

$$\omega \text{ is warranted by } \lambda, \alpha, \text{ and } \mu; \tag{5.13}$$

$$\lambda_i(t_i) \geq 0, \omega_i(t_i) \leq U_i(\bar{\mu}_N|t_i), \text{ and} \tag{5.14}$$
$$\lambda_i(t_i)\omega_i(t_i) = \lambda_i(t_i)U_i(\bar{\mu}_N|t_i), \quad \forall i \in N, \quad \forall t_i \in T_i;$$

$$(\lambda, \alpha) \neq (\underset{\sim}{0}, \underset{\sim}{0}). \tag{5.15}$$

(That is, λ may be any vector in the nonnegative orthant $\mathbf{X}_{i\in N}\mathbf{R}_+^{T_i}$, not necessarily in the simplex Λ. But, to avoid trivial solutions, λ and α cannot both be zero vectors.) See Section 7 for the proof of this theorem.

6 Interpretation of the rational-threat criterion

Our rational-threat criterion (5.4) postulates that each coalition should seek to maximize the expected difference between the total virtual utility that its members would earn and the total virtual utility that the complementary coalition would earn, if both carried out their threats. The rational threats for coalitions other than N are not required to satisfy any equity or incentive-compatibility constraints. These aspects of our rational-threat criterion deserve some interpretive discussion.

In any bargaining situation, a coalition's threat normally has both defensive and offensive objectives. The defensive objective is to show that the coalition could maintain high payoffs for its members if the complementary coalition refused to cooperate. The offensive objective is to show that the complementary coalition's members would be hurt by such a breakdown in cooperation. Obviously, a threat that is strong both defensively and offensively would be the ideal; but the best defensive threat will generally not be the best for offensive purposes. Thus, a coalition may have to make some tradeoff between these two objectives. As observed by Harsanyi [1963], the Shapley value implicitly defines such a tradeoff, since it only depends on the difference $W_S - W_{N\setminus S}$. The defensive and offensive objectives are combined with this tradeoff in our rational-threats criterion.

This interpretation of the rational-threat criterion relies on our identifying W_S as the natural defensive objective function for coalition S. Once this identification is made, then the natural offensive objective function is $-W_{N\backslash S}$ (the opposite of the complement's defensive objective), and $W_S - W_{N\backslash S}$ is a natural combination of defensive and offensive objectives. But, in what sense is $W_S(\mu, t, \alpha, \lambda)$ an appropriate measure of the defensive strength of coalition S?

We can best understand the purely defensive aspect of threats by studying games in which each coalition can only influence its own members' payoffs, so that there are no offensive possibilities to consider. In the terminology of Shapley/Shubik [1973], these are games with orthogonal coalitions. That is, a game Γ has orthogonal coalitions iff,

$$u_i((d_S, d_{N\backslash S}), t) = u_i((d_S, \hat{d}_{N\backslash S}), t) \tag{6.1}$$
$$\forall S \in CL,\ \forall i \in S,\ \forall d_S \in D_S,\ \forall d_{N\backslash S} \in D_{N\backslash S},\quad \forall \hat{d}_{N\backslash S} \in D_{N\backslash S},\quad \forall t \in T,$$

so that the threat of coalition $N\backslash S$ cannot affect the payoffs to members of S. Market games of pure exchange are examples of games with orthogonal coalitions.

In a game with orthogonal coalitions, suppose $i \in S$. Then, we can let $u_i(d_S, t)$ denote the utility payoff for i in state t if d_S in D_S is carried out. That is, $u_i(d_S, t) = u_i((d_S, d_{N\backslash S}), t)$ for any $d_{N\backslash S}$ in $D_{N\backslash S}$. (Recall $D_S \times D_{N\backslash S} \subseteq D$.) We similarly define $v_i(d_S, t, \lambda, \alpha)$ as $v_i((d_S, d_{N\backslash S}), t, \lambda, \alpha)$ for any $d_{N\backslash S}$. Then for any μ_S in M_S, the obvious generalizations of (2.6), (2.7) and (5.1) are:

$$U_i^*(\mu_S, r_i|t_i) = \sum_{t_{N-1}} p_i(t_{N-i}|t_i) \sum_{d_S} \mu_S(d_S|t_{S-i}, r_i) u_i(d_S, t)$$
$$U_i(\mu_S|t_i) = U_i^*(\mu_S, t_i|t_i)$$
$$W_S(\mu_S, t, \lambda, \alpha) = \sum_{d_S} \mu_S(d_S|t_S) \sum_{j\in S} \mathrm{v}_j(d_S, t, \lambda, \alpha).$$

Let us now consider what threats would be defensively optimal for a given player, say player 1, in a game with orthogonal coalitions. To be specific, suppose that player 1 is acting as a coordinator or leader for all the coalitions to, which he can belong. Suppose that, to maintain his leadership, player 1 must use a threat-plan that offers each type t_i of each player i at least some minimal expected utility $w_i(t_i)$. For any coalition $S \supseteq \{1\}$, let q_S denote the probability that S will be the coalition forming under player 1's leadership. Ordinarily, q_S would depend on how much player 1 offers the other players, but for simplicity let us suppose that q_N will be some fixed number close to one and all other q_S will be small positive numbers, for any threat-plan that gives all players at least their $w_i(t_i)$ payoffs. (Then q_S for $S \neq N$ may be thought of as a "trembling-hand" probability of the coalition S forming instead of N.)

In such a situation, if player 1's type is t_1, then he wants to choose his threat-plan $(\mu_S)_{S\subseteq\{1\}}$ in $\mathbf{X}_{S\supseteq\{1\}} M_S$ so as to maximize his expected utility $\sum_{S\supseteq\{1\}} q_S U_1(\mu_S|t_i)$ subject to the *minimum payoff* constraints

$$\sum_{S\supseteq\{1,i\}} q_S U_i(\mu_S|t_i) \geq \left(\sum_{S\supseteq\{1,i\}} q_S\right) w_i(t_i), \quad \forall i \in N-1, \quad \forall t_i \in T_i, \tag{6.2}$$

and the *incentive-compatibility* constraints

$$\sum_{S\supseteq\{1,i\}} q_S U_i(\mu_S|t_i) \geq \sum_{S\supseteq\{1,i\}} q_S U_i^*(\mu_S, r_i|t_i), \quad \forall i\in N,\ \forall t_i\in T_i,\ \forall r_i\in T_i. \tag{6.3}$$

This constraint (6.3) asserts that no player i should have any incentive to lie about this type when agreeing to follow player 1 in a coalition. We assume that 1 can negotiate separately with each other player i; so that i agrees without knowing which coalition $S\supseteq\{1,i\}$ will actually form.

To conceal his own type, player 1 must use a threat-plan which achieves some balance between the objectives of his various types. [See Myerson, 1983 for detailed discussion of this issue.] At the very least, however, player 1 should choose a threatplan such that there is no other threat-plan satisfying (6.2) and (6.3) that gives higher expected utility to all types t_1 in T_1 . For any such undominated threat-plan there must exist some vector λ such that $(\mu_S)_{S\supseteq\{1\}}$ maximizes

$$\sum_{t_1\in T_1} \lambda_1(t_1) \sum_{S\supseteq\{1\}} q_S U_1(\mu_S|t_1) \tag{6.4}$$

subject to the constraints (2.6) and (6.3).

So optimal defensive threats for player 1 should maximize (6.4) over $(\mu_S)_{S\supseteq\{1\}}$, subject to (6.2) and (6.3). The Lagrangian for this problem can be written as follows

$$\begin{aligned} &\sum_{t_1\in T_1} \lambda_1(t_1) \sum_{S\supseteq\{1\}} q_S U_1(\mu_s|t_1) \\ &+ \sum_{i\in N-1}\sum_{t_i\in T_i} \lambda_i(t_i) \sum_{S\supseteq\{1,i\}} q_S\big(U_i(\mu_S|t_i) - w_i(t_i)\big) \\ &+ \sum_{i\in N}\sum_{t_i\in T_i}\sum_{r_i\in T_i} \alpha_i(r_i|t_i) \sum_{S\supseteq\{1,i\}} q_S\big(U_i(\mu_S|t_i) - U_i^*(\mu_S, r_i|t_i)\big) \\ &= \sum_{S\supseteq\{1\}} q_S \left(\sum_{t\in T} p(t) W_S(\mu_S, t, \lambda, \alpha) - \sum_{i\in S-1}\sum_{t_i\in T_i} \lambda_i(t_i) w_i(t_i)\right). \end{aligned} \tag{6.5}$$

(This equality follows straightforward from the definitions of U_i, W_S, and virtual utility).

Thus, by (6.5), in any plan of optimal defensive threats for player 1, there must exist some λ and a such that every coalition is choosing a threat that maximizes the expected sum of its members' virtual utilities with respect to λ and α. The maximum value of 1's weighted objective function (6.4) is equal to the expected sum of these virtual utilities for the coalition forming around player 1, minus terms in (6.5) that do not depend on the threat-plans. This is exactly the result that we wanted, since it shows that the sum of virtual utilities can be a valid measure of the defensive strength of a coalition.

In particular, suppose that q_N is almost one, and all other q_S are only infinitesimal probabilities. Then constraints (6.2) and (6.3) require that μ_N must be (almost) incentive compatible and equitable overall. Thus, our rational-threat criterion can be justified in situations where the coalitions commit themselves to their threats and gather type-reports from their members *before* the realized coalition structure is determined, provided that all players believe that the probability of the grand coalition forming is close to one. In such situations, the value of a threat for a coalition $S (S \neq N)$ depends on what it can contribute to the required utility and incentive-compatibility of the overall plan. With appropriate shadow prices λ and α, expected virtual utility W_S measures this contribution. For $S \neq N$, the threat μ_S does not need to be either equitable or incentive compatible itself, because the members of S do not expect to carry out this threat when they agree to make it part of their threat-plans.

7 Proofs

First, we cite a basic lemma.

Lemma 1. Given any player i, α in $\underset{\sim}{A}$, λ in Λ°, and h_i in $\mathbf{R}^{T_i}$, there is a unique vector W_i in $\mathbf{R}^{T_i}$ satisfying

$$(\lambda_i(t_i) + \sum_{s_i} \alpha_i(S_i|t_i))w_i(t_i) - \sum_{s_i} \alpha_i(t_i|s_i)w_i(s_i) = h_i(t_i), \quad \forall t_i \in T_i. \tag{7.1}$$

Furthermore, the solution w_i to these linear equations is increasing in the vector h_i. (That is, if $h_i'(t_i) \geq h_i(t_i) \forall t_i$, and w_i' solves (7.1) for h_i' instead of h_i, then $w_i'(t_i) \geq w_i(t_i) \forall t_i$.)

This result is proven as "Lemma 1" in Myerson [1983].

Lemma 2. Suppose that μ is a vector of rational threats with respect to λ and α, and w is the vector of warranted claims for λ, α, and μ, where $\lambda \in \Lambda^0$ and $\alpha \in \underset{\sim}{A}$. Then

$$w_i(t_i) \geq \underset{\nu_{N-i} \in M_{N-i}}{\text{minimum}} \ \underset{\nu_i \in M\{i\}}{\text{maximum}} \, U_i(\upsilon_{N-i}, \text{v}_i|t_i).$$

for any player i and type t_i.

Proof. Let i be any fixed player. For any coalition $S \subseteq \{i\}$, let μ_i^S be a mechanism in $M_{\{i\}}$ such that

$$\mu_i^S \in \underset{\nu_i \in M_{\{i\}}}{\text{argmax}} \, U_i(\mu_{S-i}, \nu_i, \mu_{N \backslash S}|t_i)$$

for every t_i and T_i. That is, $\mu_i^S(d_i|t_i) > 0$ only if d_i would be a best response for player i if this type were t_i and the coalitions $S - i$ and $N \backslash S$ were expected to independently implement their threats from μ. Then let $\hat{\mu}$ in M be defined so that

$$\hat{\mu}_S = (\mu_{S-i}, \mu_i^S) \text{ if } i \in S$$
$$\hat{\mu}_S = \mu_S \text{ if } i \notin S.$$

For the threat-vector $\hat{\mu}$, no coalition changes its threat when player i joins it. So $W_S(\hat{\mu}, t, \lambda, \alpha)$ and $W_{S-i}(\hat{\mu}, t, \lambda, \alpha)$ differ only by the addition of i's expected virtual utility in state t when μ_{S-i}, μ_i^S, and $\mu_{N\backslash S}$ are carried out. Thus, for any t_i,

$$\sum_{t_{N-i}} p(t)(W_S(\hat{\mu}, t, \lambda, \alpha) - W_{S-i}(\hat{\mu}, t, \lambda, \alpha))$$
$$= (\lambda_i(t_i) + \sum_{r_i} \alpha_i(r_i|t_i))U_i(\mu_{S-i}, \mu_i^S, \mu_{N\backslash S}|t_i)$$
$$-\sum_{r_i} \alpha_i(t_i|r_i)U_i^*((\mu_{S-i}, \mu_i^S, \mu_{N\backslash S}), t_i|r_i).$$

Let

$$\eta_i(t_i) = \sum_{S \subseteq \{i\}} \frac{(|S > -1)!(n-|S|)!}{n!} U_i(\mu_{S-i}, \mu_i^S, \mu_{N\backslash S}|t_i)$$
$$\eta_i^*(t_i|r_i) = \sum_{S \supseteq \{i\}} \frac{(|S|-1)!(n-|S|)!}{n!} U_i^*((\mu_{S-i}, \mu_i^S, \mu_{N\backslash S}), t_i|r_i).$$

Since $(\mu_i^S \cdot |r_i)$ is the best response for type r_i against $(\mu_{S-i}, \mu_{N\backslash S})$, for each S, it follows that $\eta_i(r_i) \geq \eta_i^*(t_i|r_i)$.

Consider now the following chain of inequalities

$$(\lambda_i(t_i) + \sum_{r_i} \alpha_i(r_i|t_i)\eta_i(t_i) - \sum_{r_i} \alpha_i(t_i|r_i)\eta_i(r_i)$$
$$\leq (\lambda_i(t_i) + \sum_{r_i} \alpha_i(r_i|t_i))\eta_i(t_i) - \sum_{r_i} \alpha_i(t_i|r_i)\eta_i^*(t_i|r_i)$$
$$= \sum_{t_{N-i}} p(t) \sum_{S \supseteq \{i\}} \frac{(|S|-1)!(n-|S|)!}{n!} (W_S((\hat{\mu}, t, \lambda, \alpha) - W_{S-i}(\hat{\mu}, t, \lambda, \alpha))$$
$$= \sum_{t_{N-i}} p(t) \sum_{S \supseteq \{i\}} \frac{|S|-1)!(n-|S|)!}{n!} (W_S((\mu_{-S}, \hat{\mu}_S), t, \lambda, \alpha) -$$
$$-W_{N\backslash S}((\mu_{-S}, \hat{\mu}_S), t, \lambda, \alpha))$$
$$\leq \sum_{t_{N-i}} p(t) \sum_{S \supseteq \{i\}} \frac{(|S|-1)!(n-|S|)!}{n!} (W_S(\mu, t, \lambda, \alpha) - W_{N\backslash S}(\mu, t, \lambda, \alpha))$$
$$= (\lambda_i(t_i) + \sum_{r_i} \alpha_i(r_i|t_i))w_i(t_i) - \sum_{r_i} \alpha_i(t_i|r_i)w_i(r_i).$$

In this chain, the fourth line holds because $W_S(\hat{\mu}, t, \lambda, \alpha)$ and $W_{N\backslash S}(\hat{\mu} t, \lambda, \alpha)$ depend only on $\hat{\mu}_S$ and $\hat{\mu}_{N\backslash S} = \mu_{N\backslash S}$. Then the next inequality uses the fact that μ is a vector of rational threats, and the last equality uses the fast that w is the vector warranted claims.

Since the above chain of inequalities holds for all t_i, Lemma 1 implies that $w_i(t_i) \geq \eta_i(t_i)$ for all t_i. But $\eta_i(t_i)$ is an average of best-response payoffs for type t_i against a variety of mechanisms for $N - i$, and so $\eta_i(t_i)$ is not smaller than the right-band side of the inequality in Lemma 2. □

Proof of Theorem 2. We begin with some definitions. For any k larger than $\sum_{i=1}^{n} |T_i|$, let

$$\Lambda^k = \{\lambda \in \Lambda | \lambda_i(t_i) \geq 1/k. \qquad \forall i, \quad \forall t_i\}.$$

There exists a compact convex set $\underset{\sim}{A}^*$ such that $\underset{\sim}{A}^* \subseteq \underset{\sim}{A}$ and, for each λ in Λ there is some α in $\underset{\sim}{A}^*$ such that α is an optimal solution of the dual for λ. The proof of this fact is given in the proof of Theorem 6 in Myerson [1983].

Let

$$B = \underset{i,d,t}{\text{maximum}} |u_i(d,t)| + 1$$

and let

$$X = \left\{ w \in \underset{i \in N}{\mathbf{X}} \mathbf{R}^{T_i} | -B \leq w_i(t_i) \leq B, \quad \forall i, \quad \forall t_i \right\}.$$

For each k greater than $\sum_{i \in N} |T_i|$, we define a correspondence $Z^k : M \times \underset{\sim}{A}^* \times X \times \Lambda^k \Rightarrow M \times \underset{\sim}{A}^* \times X \times \Lambda^k$ so that $(\hat{\mu}, \hat{\alpha}, \hat{w}, \hat{\lambda}) \in Z^k(\mu, \alpha, w, \lambda)$ iff the following conditions are satisfied

$$\hat{\mu}_N \text{ is an optimal solution of the primal problem for } \lambda; \tag{7.2}$$

$$\hat{\mu}_S \in \underset{v_S \in M_S}{\text{argmax}} \sum_{r \in T} p(t)(W_S((\mu_{-S}, \nu_S)t, \lambda, \alpha) - W_{N \backslash S}((\mu_{-S}, v_S), \lambda, \alpha)). \tag{7.3}$$

$$\hat{\alpha} \text{ is an optimal solution of the dual for } \lambda; \tag{7.4}$$

$$\hat{w}_i(t_i) = \max\{-B, \min\{B, \tilde{w}_i(t_i)\}\}, \forall i, \forall t_i, \text{ where } \tilde{w} \text{ is the vector of claims warranted by } \lambda, \alpha, \text{ and } \mu; \tag{7.5}$$

$$\lambda_i(t_i) = 1/k \text{ for every } t_i \text{ such that } w_i(t_i) - U_i(\mu_N|t_i) < \underset{j,s_j}{\text{maximum}}(w_j(s_j) - U_j(\mu_N|s_j)). \tag{7.6}$$

By the Kakutani fixed-point theorem, for each k there exists some $(\mu^k, \alpha^k, w^k, \lambda^k)$ such that

$$(\mu^k, \alpha^k, w^k, \lambda^k) \in Z^k(\mu^k, \alpha^k, w^k, \lambda^k). \tag{7.7}$$

Since this sequence of fixed points is in a compact domain, there exists a convergent sequence, converging to some $(\bar{\mu}, \bar{\alpha}, \bar{w}, \bar{\lambda}$ in $M \times \underset{\sim}{A}^* \times X \times \Lambda$. We will show that $\hat{\mu}_N$ is a bargaining solution.

By the fixed-point condition, each μ^k is a vector of rational threats for λ^k and α^k.

Let $\bar{w}^k$ be the vector of claims warranted by λ^k, α^k, and μ^k. By Lemma 2, since $\tilde{w}^k$ is a vector of warranted claims supported by rational threats, $\tilde{w}_i^k(t_i) \geq -B$ for every i and t_i. Thus $w_i^k(t_i)$ can differ from $\tilde{w}_i^k(t_i)$ only if $\tilde{w}_i^k(t_i) > B$, in which case $w_i^k(t_i) \leq \tilde{w}_i^k(t_i)$.

By summing the warrant equations, we get

$$\sum_{i\in N}\sum_{t_i\in T_i} \lambda_i^k(t_i)\tilde{w}_i^k(t_i) = \sum_{t_i\in T_i} \lambda_i^k(t_i)U_i(\mu_N^k|t_i).$$

For any i and t_i, if $\tilde{w}_i^k(t_i) < U_i(\mu_N^k|t_i)$ then, by (7.6) and (7.7), $\lambda^k(t_i) = 1/k$; thus

$$\text{if } \liminf_{k\to\infty} \tilde{w}_i^k(t_i) < U_i(\bar{\mu}_N|t_i) \text{ then } \lim_{k\to\infty} \lambda_i^k(t_i) = 0 \tag{7.8}$$

Now, suppose that there were some j and r_j such that

$$\limsup_{k\to\infty} \tilde{w}_j^k(r_j) > U_j(\bar{\mu}_N|r_j) = \lim_{k\to\infty} U_j(\mu_N^k|r_j). \tag{7.9}$$

Then (7.8) could be strengthened to:

$$\text{if } \liminf_{k\to\infty} \tilde{w}_i^k(t_i) \le U_i(\bar{\mu}_N|t_i) \text{ then } \lim_{k\to\infty} \lambda_i^k(t_i) = 0.$$

Since each λ^k is in the simplex Λ, we could find some j and r_j, satisfying (7.9) such that $\bar{\lambda}_j(r_j) > 0$. But then we would get

$$\begin{aligned} 0 &< \limsup_{k\to\infty} \lambda_j^k(r_j)(\tilde{w}_j^k(r_j) - U_j(\mu_N^k|r_j)) \\ &= \limsup_{k\to\infty} \sum_{(i,t_j)\neq(j,r_j)} \lambda_i^k(t_i)(U_i(\mu_N^k|t_i) - \tilde{w}_i^k(t_i)) \le 0, \end{aligned}$$

using (7.8) (and the fact that $\tilde{w}_i^k(t_i)$ does not diverge to $-\infty$ as $k \to \infty$, since it is bounded below by $-B$) to get the last inequality. But $0 < 0$ is impossible, so no $j, r_j)$ pair satisfying (7.9) can exist. That is, for every i and t_i,

$$\limsup_{k\to\infty} \tilde{w}_i^k(t_i) \le U_i(\bar{\mu}_N|t_i).$$

So $\{(\lambda^k, \alpha^k, \mu^k, \tilde{w}^k)\}_{k=1}^{\infty}$ form a sequence verifying (5.5)–(5.8) for $\bar{\mu}_N$. □

Proof of Theorem 3. Theorem 3 follows immediately from Lemma 2 and the definition of a bargaining solution. □

Proof of Theorem 4. Given the bargaining solution $\bar{\mu}$, let $\{(\lambda^k, \alpha^k, \mu^k, w^k)\}_{k=1}^{\infty}$ satisfy (5.5)–(5.8). Let $\hat{\lambda}^k$ and $\hat{\alpha}^k$ be defined by

$$\begin{aligned} \hat{\lambda}_i^k(t_i) &= \lambda_i^k(t_i)/(|\lambda^k| + |\alpha^k|) \\ \hat{\alpha}_i^k(r_i|t_i) &= \alpha_i^k(r_i|t_i)/(|\lambda^k| + |\alpha^k|) \end{aligned}$$

where

$$|\lambda^k| + |\alpha^k| = \sum_{i\in N}\sum_{t_i\in T_i} (\lambda_i^k(t_i) + \sum_{r_i\in T_i} \alpha_i^k(r_i|t_i)) \ge n.$$

So for each k, $(\hat{\lambda}^k, \hat{\alpha}^k)$ lies in a unit simplex. By the linear homogeneity of all formulas concerned, μ^k is a vector of rational threats for $\hat{\lambda}^k$ and $\hat{\alpha}^k$, as well as for

λ^k and α^k, and w^k is warranted by $\hat{\lambda}^k, \hat{\alpha}^k$, and μ^k. By Lemma 2 and condition (5.8), each w^k must lie within the compact set X defined in the proof of Theorem 2, and each μ^k is in the compact, set M. So there must exist a subsequence $\{(\hat{\lambda}^k, \hat{\alpha}^k, \mu^k, w^k)\}_k$ that is convergent to some limit $(\lambda, \alpha, \mu, w)$.

The vectors λ and α cannot both be zero, because (λ, α) has a summation-norm of one. By continuity of the rational-threat and warranted-claim conditions, μ is a vector of rational threats for λ and α, and w is warranted by λ, α, and μ. By (5.8), $w_i(t_i) \leq U_i(\bar{\mu}_N|t_i)$ for every i and t_i. From (5.9) (dividing through by $|\lambda^k| + |\alpha^k|$, and letting $\delta \to 0$ as $k \to \infty$, we get

$$\sum_{t \in T} p(t) \max_{d \in D} \sum_{i \in N} \mathrm{v}_i(d, t, \lambda, \alpha) \leq \sum_i \sum_{t_i} \lambda_i(t_i) U_i(\bar{\mu}_N|t_i),$$

and so by duality $\bar{\mu}_N$ and α are optimal solutions of the primal and dual for λ, respectively. Duality also implies that the above inequality must be an equality, which gives us the complementary slackness conditions in (5.14).

Thus we have all of the conditions in Theorem 4, except that letting $\mu_N = \lim_{k \to \infty} \mu_N^k$ does *not* imply $\mu_N = \bar{\mu}_N$. However, since $\bar{\mu}_N$ is an optimal solution of the primal for λ, it must also maximize the sum of the virtual utilities in every state. Thus, if we redefine μ_N as being equal to $\bar{\mu}_N$, we do not change $W_N(\mu, t, \lambda, \alpha)$ for any t, and so w is still warranted by λ, α, μ. □

References

D'Aspremont, C., Gerard-Varet, L.-A.: Incentives and incomplete information. Journal of Public Economics **11**, 25–45 (1979)

Harsanyi, J.C.: Simplified bargaining model for the n-person cooperative game. International Economic Review **4**, 194–220 (1963)

Harsanyi, J.C.: Games with incomplete information played by "Bayesian" players. Management Science **14**, 159–189, 320–334, 486–502 (1967–1968)

Harsanyi, J.C., Selten, R.: A generalized Nash bargaining solution for two-person games with incomplete information. Management Science **18**, P80–P106 (1972)

Holmström, B., Myerson, R.B.: Efficient and durable decision rules with incomplete information. Econometrica **51**, 1799–1819 (1983)

Kakutani, S.: A generalization of Brouwer's fixed point theorem. Duke Journal of Mathematics **8**, 457–459 (1941)

Myerson, R.B.: Incentive compatibility and the bargaining problem. Econometrica **47**, 61–73 (1979)

Myerson, R.B.: Mechanism design by an informed principal. Econometrica **51**, 1767–1797 (1983)

Myerson, R.B.: Two-person bargaining problems with incomplete information. Mimeo, Northwestern University: Evanston 1982. Subsequently published in Econometria **52**, 461–487 (1984)

Nash, J.F.: The bargaining problem. Econometrica **18**, 155–162 (1950)

Nash, J.F.: Two-person cooperative games. Econometrica **21**, 128–140 (1953)

Roth, A.E.: Values for games without sidepayments: some difficulties with current concepts. Econometrica **48**, 457–465 (1980)

Shafer, W.J.: On the existence and interpretation of value allocation. Econometrica **48**, 467–476 (1980)

Shapley, L.S.: A value for n-person games. In: Kuhn, H.W., Tucker, A.W. (eds.) Contributions to the theory of games, II, pp. 307–317. Princeton 1953

Shapley, L.S.: Utility comparison and the theory of games. In: La decision, pp. 251–263. Paris 1969. Reprinted in: Roth, A.E. (ed.) The Shapley Value: essays in honor of Lloyd S. Shapley, chapter 19, pp. 3307–319. Cambridge U Press 1988

Shapley, L.S., Shubik, M.: Game theory in economics, Chap. 6: Characteristic function, core, and stable set. Santa Monica 1973. And published in: Shubik, M. (ed.) Game theory in the social sciences: concepts and solution, chapter 6, pp. 127–178. Mit Press 1982

The value allocation of an economy with differential information*

Stefan Krasa and Nicholas C. Yannelis

Department of Economics, University of Illinois at Urbana-Champaign, Champaign IL 61820, USA

Summary. We analyze the Shapley value allocation of an economy with differential information. Since the intent of the Shapley value is to measure the sum of the expected marginal contributions made by an agent to any coalition that he/she belongs to, the value allocation of an economy with differential information provides an interesting way to measure the information advantage of an agent. This feature of the Shapley value allocation is not necessarily shared by the rational expectation equilibrium. Thus, we analyze the informational structure of an economy with differential information from a different and new viewpoint.

In particular we address the following questions: How do coalitions of agents share their private information? How can one measure the information advantage or superiority of an agent? Is each agent's private information verifiable by other members of a coalition? Do coalitions of agents pool their private information? Do agents have an incentive to report their true private information? What is the correct concept of a value allocation in an economy with differential information? Do value allocations exist in an economy with differential information? We provide answers to each of these questions.

Keywords and Phrases: Shapley value, Differential information economies, Coalitional incentive compatibility.

JEL Classification Numbers: C71, D10, D82.

* We wish to thank David Ballard, Leonidas Koutsougeras, Wayne Shafer, Anne Villamil and three referees for useful comments, discussions and suggestions. We are also very indebted to Andreu Mas-Colell whose thoughtful suggestions inspired Section 4.2. As usual we are responsible for any remaining shortcomings.

1 Introduction

The concept of a (cardinal) value allocation was first introduced by Shapley (1969). Roughly speaking, it is defined as a feasible allocation which yields to each agent in the economy a "utility level" which is equal to the sum of the agent's expected marginal contributions to all coalitions that he/she is a member of. Because the Shapley value measures the sum of an agent's expected marginal contributions to coalitions, we argue in this paper that it provides an interesting way to measure the "worth" of an agent's information advantage in an economy with differential information. This feature of the Shapley value allocation is not necessarily shared by the traditional rational expectation equilibrium. Thus, we analyze the informational structure of an economy with differential information from a different viewpoint. In particular, we address the following questions: How can one measure the information advantage or superiority of an agent? How do coalitions of agents share their private information? Is each agent's private information verifiable by all other members of a coalition? Do agents have an incentive to report their private information truthfully? What is the correct concept of a value allocation in the presence differential information? Do value allocations exist in an economy with differential information? We provide answers to each of these questions.

Similar to Radner (1968), we consider a two-period economy where each agent i is characterized by a utility function, a random second period endowment, a prior belief about the distribution of all agents' second period endowments, and private information about the actual endowment realizations after uncertainty is resolved in the second period. Although the value allocation is a cooperative solutions concept, there is a non-cooperative aspect inherent in a private information economy framework. In particular, a coalition of agents might agree on how to share their own private information. Nonetheless, some agents within the coalition may have an incentive to misreport their true private information. For example, consider an economy in which agents are exposed to idiosyncratic endowment shocks (cf., Example 1). Although a coalition of agents might agree ex-ante on insuring its members if they report a low endowment realization, once the state of nature is realized, some members of this coalition may have an incentive to agree on misreporting their actual endowments to the other agents. In particular, even if they did not receive a low endowment realization, they may report differently in order to obtain the insurance payment. As long as the coalition includes all agents who would be able to verify these reports using their own private information, agents of the complementary coalition cannot detect that they are being cheated (although they might suspect it). In order to alleviate this problem, we consider two incentive compatible ways of information sharing which we refer to as weak and strong coalitional incentive compatibility, respectively. The central idea of both notions is that it should not be profitable for coalitions of agents to agree on misreporting their information in such a way that none of the remaining agents are able to detect the misreports. We then provide conditions under which strong coalitional incentive compatibility implies that no actual exchange of information is taking place.

Formally, this means that each agent's net-trade must be measurable with respect to his/her own information.[1]

We define three alternative notions of a value allocation. In the weak and the strong value allocation information sharing within coalitions must fulfill weak and strong coalitional incentive compatibility, respectively. In the private value allocation each agent's net-trade must be measurable with respect to his/her information. This corresponds to the (private) core notion of Yannelis (1991). We first show that in the private value allocation information asymmetries matter. That is, an agent with superior private information can make a Pareto improvement to the economy, and he/she will be rewarded for it. This can happen, even if the agent has zero initial endowment. It also indicates that agents can benefit from the information of other agents even if they do not acquire that information for themselves. Moreover, we show that the private value allocation fulfills strong coalitional incentive compatibility. It should be noted that in the weak value allocation agents with superior information are rewarded as well. In this case agents not only benefit from the information of other agents, but they exchange information as well. It seems to us, however, that the private value is the most useful of these concepts, since it is technically very tractable and also has properties which are very similar to those of the weak and the strong value, i.e., agents are rewarded for superior information. In contrast, both the weak and the strong value allocation are technically not as tractable.

Before we proceed we would like to mention two early seminal papers in cooperative game theory with incomplete information which differ from our paper. The first is by Wilson (1978) who analyzed the core of an economy with differential information. The second is by Myerson (1984) who extended the Nash bargaining solution and the NTU value to an incomplete information setting. It should be noted that Myerson analyzes a mechanism design problem whereas our approach considers an exchange economy with differential information.

The paper proceeds as follows: Section 2 contains the description of the model (i.e., the exchange economy with differential information). In Section 3 we introduce the alternative notions of value allocation with differential information. The interpretation of the private value is discussed in Section 4. Section 5 discusses other concepts of a value allocation. Finally, Section 6 contains some concluding thoughts.

2 The economy with differential information

Consider an exchange economy which extends over two time periods $t = 0, 1$ where consumption takes place at $t = 1$. Let $\mathbb{R}^l_+$ denote the commodity space.[2] At $t = 0$ there is uncertainty over the state of nature described by a probability space $(\Omega, \mathcal{F}, \mu)$. Let $I = \{1, \ldots, n\}$ denote the set of all agents. At $t = 0$ agents agree on net-trades which may be contingent on the state of nature at $t = 1$. However,

[1] This essentially means that an agent cannot make his consumption contingent on two states between which he/she cannot distinguish.

[2] $\mathbb{R}^l_+$ denotes the positive cone of $\mathbb{R}^l$.

they have differential information with respect to the true state of nature that will be realized. This is modeled as follows: At $t = 1$ agents do not necessarily know which state $\omega \in \Omega$ has actually occurred. They know their own endowment realization, and every agent i may also have some additional information about the state. This information is described by a measurable partition $\mathcal{F}_i$ of Ω.[3] Hence if $\bar{\omega}$ is the true state of the economy at $t = 1$ agent i observes the event $E_i(\bar{\omega})$ in the partition $\mathcal{F}_i$ which contains $\bar{\omega}$. Since agents always observe their own endowment realization, we can assume without loss of generality that agent i's endowment is measurable with respect to $\mathcal{F}_i$.[4]

In summary, an *exchange economy with differential information* is given by $\mathcal{E} = \{(X_i, u_i, e_i, \mathcal{F}_i, \mu) \colon i = 1, \ldots, n\}$ where

(1) $X_i = \mathbb{R}^l_+$ is the *consumption set* of agent i;
(2) $u_i \colon \mathbb{R}^l_+ \to \mathbb{R}$ is the *utility function* of agent i;[5]
(3) $\mathcal{F}_i$ is a partition of Ω denoting the *private information* of agent i;
(4) $e_i \colon \Omega \to \mathbb{R}^l_+$ is the *initial endowment* of agent i, where each e_i is $\mathcal{F}_i$-measurable and $e_i(\omega) \in X_i$ μ-a.e.;
(5) μ is a probability measure on Ω denoting the *common prior* of each agent.

Finally, the *expected utility* of agent i is given by $\int_\Omega u_i(x_i(\omega))\, d\mu(\omega)$.[6] Throughout the paper, we assume that the utility function u_i of each agent i is monotone, continuous, concave and integrably bounded.

3 Value allocations in economies with differential information

3.1 Coalitional incentive compatibility

When agents have differential information, arbitrary allocations are not generally viable. In particular, arbitrary allocations might not be incentive compatible in the

[3] A measurable partition of Ω is a collection of sets A_j, $j \in J$, with the following properties: (a) J is finite or countable; (b) $A_j \in \mathcal{A}$ for every j, i.e., the sets are measurable; (c) $\bigcup_{j \in J} A_j = \Omega$; (d) $A_j \cap A_k = \emptyset$ for all $j \neq k$.

[4] By (slight) abuse of notation we identify the partition with the σ-algebra generated by the partition.

[5] All the results of the paper remain valid if we assume the utility function is random, i.e., u_i is a real valued function defined on $\Omega \times \mathbb{R}^l_+$.

[6] Bayesian updating of priors can be introduced as follows: Let $q_i : \Omega \to \mathbb{R}_{++}$ be a Radon-Nikodym derivative (density function) denoting the prior of agent i. For each $i = 1, \ldots, n$, denote by $E_i(\omega)$ the event in $\mathcal{F}_i$ containing the realized state of nature $\omega \in \Omega$ and suppose that $\int_{t \in E_i(\omega)} q_i(t)\, d\mu(t) > 0$. Given $E_i(\omega) \in \mathcal{F}_i$, define the *conditional expected utility* of agent i as follows:

$$\int_{t \in E_i(\omega)} u_i(t, x_i(t)) q_i(t|E_i(\omega))\, d\mu(t),$$

where

$$q_i(t|E_i(\omega)) = \begin{cases} 0 & \text{if } t \notin E_i(\omega) \\ \dfrac{q_i(t)}{\int_{t \in E_i(\omega)} q_i(t)\, d\mu(t)} & \text{if } t \in E_i(\omega). \end{cases}$$

As in Yannelis (1991) all results of the paper remain valid under this conditional expected utility formulation, but we choose not to do so for the simplicity of the exposition.

sense that groups of agents may misreport their information without other agents noticing it, and hence achieve different payoffs ex post. To obviate this problem, we introduce in Definitions 1 and 2 below two alternative notions of incentive compatibility.

The cooperative solution concept which we apply to economies with differential information captures the idea that groups of agents come together, exchange information, or trade with each other given their informational constraints. However, this exchange of information has a non-cooperative aspect. In particular, although agents might agree to exchange information, they might not have an incentive to reveal their information truthfully. We take this into account by considering two alternative definitions of coalitional incentive compatibility.

Definition 1. *Let $x\colon \Omega \to \prod_{i=1}^{n} X_i$ be a feasible allocation.*[7] *Then x fulfills strong coalitional incentive compatibility if and only if the following does not hold: There exists a coalition $S \subset I$ and states a, b, $a \neq b$ which members of $I \setminus S$ are unable to distinguish (i.e., a and b are in the same event of the partition for every agent not in S) and such that after an appropriate endowment redistribution members of S are strictly better off by announcing b whenever a has actually occurred. Formally, strong coalitional incentive compatibility implies that there does not exist a coalition S, states a, b with $b \in E_i(a)$ for every $i \notin S$ and a net-trade vector z^i, $i \in S$ such that $\sum_{i \in S} z^i = 0$, $e^i(a) + x^i(b) - e^i(b) + z^i \in \mathbb{R}^l_+$ for every $i \in S$ and*

$$u^i(e^i(a) + (x^i(b) - e^i(b)) + z^i) > u^i(x^i(a)), \textit{for every } i \in S. \qquad (C1)$$

Strong coalitional incentive compatibility models the idea that it is impossible for any coalition to cheat the complementary coalition by misreporting the state and making side-payments to each other which agents who are not members of this coalition cannot observe.[8] The restriction $e^i(a) + x^i(b) - e^i(b) + z^i \in \mathbb{R}^l_+$ for every $i \in S$ is obviously required in order to ensure that the consumption of the agents is still in the commodity space.

When side-payments are observable we get the following weaker notion of incentive compatibility. (Set $z_i = 0$ for every $i \in S$ in (C1) to get weak coalitional incentive compatibility.) Formally, we have the following definitions.

Definition 2. *Let $x\colon \Omega \to \prod_{i=1}^{n} X_i$ be a feasible allocation. Then x fulfills weak coalitional incentive compatibility if and only if the following does not hold: There exists a coalition $S \subset I$ and states a, b, $a \neq b$ which members of $I \setminus S$ are unable to distinguish (i.e., a and b are in the same event of the partition for every agent not in S) and such that members of S are strictly better off by announcing b whenever a has actually occurred. Formally, $b \in E_i(a)$ for every $i \notin S$, $e^i(a) + x^i(b) - e^i(b) \in \mathbb{R}^l_+$ for every $i \in S$ and*

$$u^i(e^i(a) + (x^i(b) - e^i(b)) > u^i(x^i(a)), \textit{for every } i \in S. \qquad (C2)$$

[7] An allocation $x\colon \Omega \to \prod_{i \in I} X_i$ is said to be *feasible* if $\sum_{i \in I} x_i = \sum_{i \in I} e_i$.

[8] If we want incentive compatibility to hold almost everywhere then it must be required that states in which agents misreport occur with probability zero.

Weak incentive compatibility is weaker in the sense that allocations which fulfill strong coalitional incentive compatibility must also fulfill weak coalitional incentive compatibility. It is easy to find examples where the reverse is not true. In what follows we discuss both notions. It turns out that in certain cases strong coalitional incentive compatibility is analytically more tractable since it corresponds to individual measurability of the net-trade of each agent as proved in Lemma 1 below. We now introduce some notation. In particular, let $U^w(S)$ and $U^s(S)$ denote the utility allocations which coalition S can attain, and which fulfill weak and strong coalitional incentive compatibility, respectively. That is

$U^w(S) = \{(w_1, \dots, w_{|S|}) \in \mathbb{R}^{|S|}$: there exists an allocation x_i, $i \in S$ such that $\sum_{i \in S} x_i = \sum_{i \in S} e_i$, where x_i, $i \in S$ fulfills weak coalitional incentive compatibility, and where $w_i \leq \int u_i(x_i)\, d\mu$ for every $i \in S\}$.

$U^s(S) = \{(w_1, \dots, w_{|S|}) \in \mathbb{R}^{|S|}$: there exists an allocation x_i, $i \in S$ such that $\sum_{i \in S} x_i = \sum_{i \in S} e_i$, where x_i, $i \in S$ fulfills strong coalitional incentive compatibility, and where $w_i \leq \int u_i(x_i)\, d\mu$ for every $i \in S\}$.

For both notions of incentive compatibility we will analyze the corresponding concept of a value allocation. The definitions are introduced in Section 3.3.

3.2 Private measurability of allocations

We now continue by characterizing the set $U^s(S)$. It turns out that $U^s(S)$ corresponds to the attainable utility allocations which fulfill individual measurability of the net-trade of each agent for an economy with one commodity per state.

Let $U^p(S) = \{(w_1, \dots, w_n)$: there exist net-trades z_i such that $\sum_{i \in S} z_i = 0$, where z_i is $\mathcal{F}_i$-measurable , and $w_i \leq \int u_i(e_i + z_i)\, d\mu\}$.

Lemma 1. *Assume that there is one commodity per state; then* $U^p(S) = U^s(S)$.

Proof. We first show that $U^p(S) \subset U^s(S)$. Let $w \in U^p(S)$. Then there exist an allocation x_i, $i \in S$ such that for every agent $i \in I$ the net-trades $y_i = x_i - e_i$ are $\mathcal{F}_i$-measurable, $\sum_{i \in S} y_i = 0$, and $w_i \leq \int u_i(x_i)\, d\mu$. Now assume by way of contradiction that strong coalitional incentive compatibility does not hold. Then there exist a coalition $T \subset S$, and two states a, b which members of $S \setminus T$ cannot distinguish, such that members of T are strictly better off by redistributing $z_i, i \in T$ (that is $\sum_{i \in T} z_i = 0$) and by reporting b whenever a has actually occurred. Since net trades are individually measurable it follows that $\sum_{i \in S \setminus T} y_i(a) = \sum_{i \in S \setminus T} y_i(b)$. Thus, since the feasibility constraint holds with equality we get $\sum_{i \in T} y_i(a) = \sum_{i \in T} y_i(b)$. Since there is one commodity per state and preferences are monotonic it follows that $y_i(b) + z_i > y_i(a)$ for every $i \in T$. We therefore get

$$\sum_{i \in T} y_i(b) = \sum_{i \in T} (y_i(b) + z_i) > \sum_{i \in T} y_i(a) = \sum_{i \in T} y_i(b),$$

which provides the contradiction.

It remains to prove that $U^s(S) \subset U^p(S)$ for every S. Suppose by way of contradiction that the net-trade of one agent, say agent j is not $\mathcal{F}_j$-measurable. Hence, there exist two states a and b which agent j cannot distinguish and for which

$z^j(a) \neq z^j(b)$. Without loss of generality assume that $z^j(a) > z^j(b)$. Then in state b the coalition $T = S \setminus \{j\}$ can announce state a and agent j is unable to verify it and they can redistribute their excess income and make all members of T better off. This provides the contradiction to strong coalitional incentive compatibility. Hence $U^s(S) \subset U^p(S)$. This concludes the proof.

Lemma 1 implies that when there is one commodity per state of nature the sets $U^s(S)$ and $U^p(S)$ coincide for every S. The fact that strong coalitional incentive compatibility—i.e., the assumption that coalitions of agents can trade with each other in such a way that agents of the complementary coalition cannot observe that such trades are taking place—implies individual measurability of net-trades, means that no exchange of information takes place. That is, after the state of nature is realized agents still only know the event of their own information partition which contains the true state. Nevertheless, agents can benefit from each other's information as we show in Examples 1 and 2 below. In contrast, we show in Example 4 that agents may exchange information *and* benefit from each others information if we require only weak coalitional incentive compatibility. Which of the two incentive compatibility assumptions is more appropriate will clearly depend on the specific economic question which we want to analyze.

In the case of more than one commodity, Lemma 1 does not apply, and we can therefore define a third concept of information sharing. In particular, we assume that all net-trades are individually measurable, i.e., no actual exchange of information takes place. As mentioned earlier agents can nevertheless benefit from each others information. More importantly, however, private information sharing is analytically more tractable than weak or strong coalitional incentive compatibility. At the same time the results are qualitatively not very different. This seems to indicate that private information sharing is the most useful concept to analyze economies with private information.

3.3 Value allocations

The strategy in this section is to derive a game with transferable utility from the economy with differential information, $\mathcal{E}$, in which each agent's utility is weighted by a factor λ_i which allows interpersonal utility comparisons. In the value allocation itself no side-payments are necessary. This claim is justified by appealing to the *principle of irrelevant alternatives:* "If restriction of the feasible set, by eliminating side-payments, does not eliminate some solution point, then that point remains a solution" (Shapley (1969)).[9] We thus get a game with side-payments as follows:

Definition 3. *A game with side-payments $\Gamma = (I, V)$ consists of a finite set of agents $I = \{1, \ldots, n\}$ and a superadditive, real valued function V defined on 2^I such that $V(\emptyset) = 0$. Each $S \subset I$ is called a coalition and $V(S)$ is the "worth" of the coalition S.*

[9] See Emmons and Scafuri (1985, p. 60) or Shafer (1980, p. 468) for further discussion.

The Shapley value of the game Γ, (Shapley (1953)) is a rule which assigns to each agent i a "payoff" Sh_i given by the formula[10]

$$\mathrm{Sh}_i(V) = \sum_{\substack{S \subset I \\ S \supset \{i\}}} \frac{(|S|-1)!(|I|-|S|)!}{|I|!} [V(S) - V(S \setminus \{i\})].$$

The Shapley value has the property that $\sum_{i \in I} \mathrm{Sh}_i(V) = V(I)$, i.e., the Shapley value is Pareto optimal. Moreover, it is individually rational, i.e., $\mathrm{Sh}_i(V) \geq V(\{i\})$ for all $i \in I$.

We now start by defining the private value allocation. For each economy with differential information $\mathcal{E}$ and each set of weights $\{\lambda_i : i = 1, \ldots, n\}$, we can now associate a game with side-payments (I, V_λ^p), (we also refer to this as a "transferable utility" (TU) game) according to the rule:

For $S \subset I$ let

$$V_\lambda^p(S) = \max_{x_i} \sum_{i \in S} \lambda_i \int u_i(x_i(\omega))\, d\mu(\omega) \tag{3.1}$$

subject to

(i) $\sum_{i \in S} x_i(\omega) = \sum_{i \in S} e_i(\omega)$, μ-a.e.

(ii) the net trades $x_i - e_i$ are $\mathcal{F}_i$-measurable for every $i \in S$.

We now define the private value allocation for economies with differential information.

Definition 4. *An allocation $x \colon \Omega \to \prod_{i=1}^n X_i$ is said to be a private value allocation of the economy with differential information $\mathcal{E}$ if the following holds:*

(i) The net trade $e_i - x_i$, is $\mathcal{F}_i$ measurable for all $i \in I$.
(ii) $\sum_{i=1}^n x_i(\omega) = \sum_{i=1}^n e_i(\omega)$, μ-a.e.
(iii) There exist $\lambda_i \geq 0$, for every $i = 1, \ldots, n$ which are not all equal to zero, with $\lambda_i \int u_i(x_i(\omega))\, d\mu(\omega) = \mathrm{Sh}_i(V_\lambda^p)$ for every i, where $\mathrm{Sh}_i(V_\lambda^p)$ is the Shapley value of agent i derived from the game (I, V_λ^p), defined in (3.1).

Condition (i) says that net-trades must be measurable with respect to private information; (ii) is a market clearing condition; and (iii) says that the expected utility of each agent multiplied by his/her weight λ_i must be equal to his/her Shapley value derived from the TU game (I, V_λ^p).

Alternatively, we can also define the weak and the strong value allocation. In order to define the weak value allocation, first replace (ii) in (3.1) by (ii') $x_i, i \in S$ fulfills weak coalitional incentive compatibility.

We thus derive a TU-game V_λ^w. In order to define the weak value allocation, just substitute V_λ^w for V_λ^p in Definition 4 and replace (i) in Definition 4 by (i') $x_i, i \in I$ fulfills weak coalitional incentive compatibility.

[10] The Shapley value is the sum of the expected marginal contributions an agent can make to each coalitions of which he/she is a member (see Shapley (1953)).

Similarly, we can define a strong value allocation by replacing replace (ii) in (3.1) by (ii") x_i, $i \in S$ fulfills strong coalitional incentive compatibility.
This defines the TU-game V_λ^s. We now define the strong value allocation in the obvious way. However, by Lemma 1 the strong and the private value allocation coincide if there is one commodity per state. We therefore only discuss the properties of the private and the weak value allocations in the examples in the subsequent sections.

3.4 Existence and incentive compatibility of the private value allocation

A private value allocation always exists for a differential information economy. This follows from the existence results in Emmons and Scafuri (1985) or Shapley (1969). Specifically, these papers show that if agents' utility functions are concave, and continuous; and if the consumption set of each agent is bounded from below, closed and convex; then a value allocation exists.

In order to apply their existence results, we first define the consumption space to be the set of all functions $x_i\colon \Omega \to \mathbb{R}_+^l$ which are $\mathcal{F}_i$-measurable. One can then show that the expected utility is (weakly) continuous on this consumption set. Thus, the above results can be applied (for details see Krasa and Yannelis (1991)).

We next show that the private value fulfills strong coalitional incentive compatibility. In the case of one commodity per state this follows from Lemma 1. In the case of more than one commodity, we add to Definition 1 the requirement that agents in coalition S must agree on the state a that they misreport, i.e., equation (C1) must hold for every $s \in \bigwedge_{i\in S} E_i(a)$.[11] In other words the event in which the agents in coalition S cheat must be common knowledge to its members.[12]

Lemma 2. *The private value allocation fulfills strong coalitional incentive compatibility.*

Proof. Let x_i, $i \in I$ be a private value allocation. Assume by way of contradiction that the allocation does not fulfill strong coalitional incentive compatibility. Let $T \subset I$ be a coalition for which it is optimal to report state b whenever state a has actually occurred. Members in $I \setminus T$ cannot distinguish state a from state b. Since net-trades are individually measurable it follows that $x_i(a) = e_i(a) + (x_i(b) - e_i(b))$ for every $i \in I \setminus T$. Let z_i, $i \in T$ denote the redistribution within coalition T. Then we define a new allocation x_i^*, $i \in I$ as follows: For every $i \in T$ let

$$x_i^*(s) = \begin{cases} e_i(s) + (x_i(b) - e_i(b)) + z_i & \text{for every } s \in \bigwedge_{i\in T} E_i(a); \\ x_i(s) & \text{otherwise.} \end{cases}$$

Let $x_i^* = x_i$ for every $i \in I \setminus T$. Then x_i^* is a feasible allocation and all net-trades $x_i - e_i$ are $\mathcal{F}_i$-measurable. Furthermore, members of T are strictly better off with the new allocation than with the original allocation. Members of $I \setminus T$ are indifferent.

[11] $\bigwedge_{i\in S} E_i(a)$ is the event in $\bigwedge_{i\in S} \mathcal{F}_i$ which contains a.

[12] For related incentive compatibility results for the core of an economy with differential information see Koutsougeras and Yannelis (1993).

This, however, is a contradiction to the Pareto efficiency of the value allocation. Thus, the value allocation must fulfill strong coalitional incentive compatibility.

4 Interpretation of the value allocation

4.1 The private value allocation

In this section we discuss the properties of the private value allocation. We illustrate these properties by way of simple examples. In the first example, we consider an economy in which two agents, denoted by I and J are subject to independent endowment shocks in the second period. Specifically, each agent's endowment can be either high or low. Neither of the agents, however, is able to verify whether or not the other agent's endowment is high or low. We assume that there is a third agent, denote by K, in the economy whose superior information potentially allows verification of the endowment shocks. Example 1 addresses the question to what extend the superior information of agent K makes an insurance arrangement between agents I and J possible. Moreover, it is also shown that agent K is rewarded for his/her superior information.

Example 1. Consider an economy with three agents denoted by I, J, and K, and four states of nature a, b, c, and d. Assume there is only one consumption commodity in each state. The random endowment of the agents are given by $(4, 4, 0, 0)$ for I; $(4, 0, 4, 0)$ for J; and $(1, 1, 1, 1)$ for K. Agent K has an information set $\mathcal{F}_K$. We consider the case where $\mathcal{F}_K$ is trivial – i.e., $\mathcal{F}_K = \{\{a, b, c, d\}\}$ – and the case where $\mathcal{F}_K$ corresponds to full information – i.e., $\mathcal{F}_K = \{\{a\}, \{b\}, \{c\}, \{d\}\}$. Further, assume that agent I cannot distinguish state a from b, and state c from d; and finally, agent J cannot distinguish a from c, and b from d, i.e., $\mathcal{F}_I = \{\{a, b\}, \{c, d\}\}$ and $\mathcal{F}_J = \{\{a, c\}, \{b, d\}\}$. Let all agents have the same von Neumann-Morgenstern utility function $\sqrt{x}$, and assume that each state occurs with the same probability.

Consider first the case where agent K has full information. We now derive the set of all attainable utility allocations for the private value allocation. For the one-agent coalitions we get: $U^p(\{i\}) = \{w_i \colon w_i \leq 1\}$, for $i = I, J, K$. Further, all trades between agents I and J must be state independent.[13] However, since each agent's consumption must be non negative it follows that no trade between agents I and J is possible. Thus, $U^p(\{I, J\}) = \{(w_I, w_J) \colon w_I \leq 1$, and $w_J \leq 1\}$. This, does not apply to the other two agent coalitions. $U^p(\{I, K\}) = U^p(\{J, K\}) = \{(w_1, w_2) \colon w_1 \leq (1/2)\sqrt{4 + t_1} + (1/2)\sqrt{t_2}, w_2 \leq (1/2)\sqrt{1 - t_1} + (1/2)\sqrt{1 - t_2}$, such that $-1 \leq t_1 \leq 1$ and $0 \leq t_2 \leq 1$, for $i = 1, 2\}$. Similarly, $U^p(\{I, J, K\}) = \{(w_1, w_2, w_3) \colon$ there exist state independent net-trades z_i, $i = I, J, K$ where z_i is $\mathcal{F}_i$-measurable for $i = J, K$, where $\sum_{i=I,J,K} z_i = 0$, and $w_i \leq \sum_{s=a,b,c,d} (1/4)\sqrt{e_i + z_i}$, for $i = I, J, K\}$. In view of Section 3.4, a strong value exists. Further, it is

[13] This follows immediately since the net-trade z between both agents must be measurable with respect to the information of each agent. However, only state independent net-trades are measurable with respect to both agents' information.

obvious that $V_\lambda^p(\{I,J,K\}) \geq V_\lambda^p(\{I,J\}) + V_\lambda^p(\{K\})$ for all $\lambda > 0$.[14] Moreover, $V_\lambda^p(\{i,K\}) > V_\lambda^p(\{i\}) + V_\lambda^p(\{K\})$, $i = I, J$ for all $\lambda > 0$.[15] Hence, $\mathrm{Sh}_K(V_\lambda) > \int u_K(e_K)\, d\mu$, i.e., agent K must get a higher utility than he/she derives from the initial endowment.[16]

Next consider the case where $\mathcal{F}_K$ is trivial. Now only constant net-trades are possible within all coalitions. Since the private value allocation is individually rational all agents must consume their initial endowment.

Example 1 shows that agent K's information clearly matters. It shows that when agent K has full information and I, J do not, he/she can use this information advantage to act as an intermediary to allow trade between agents I and J that would not otherwise be possible. This becomes clear when looking at $U^p(\{I,J,K,\})$ and $U^p(\{I,J\})$. Without agent K, agents I and J cannot make any beneficial trades. This changes when agent K enters the coalition. Now all trades basically go through agent K since $U^p(\{I,J,K,\})$ is essentially the union of $U^p(\{I,K\})$ and $U^p(\{J,K\})$. It is important to note that agents do not exchange their private information. Nonetheless, using the superior information of agent K, trade made everybody in the economy better off. Moreover, agent K is compensated for his/her intermediation service by getting a strictly higher utility than in the case where he/she is less well informed and is unable to facilitate trade. However, it is essential for agent K to have a strictly positive endowment in every state. If agent K's endowment is for example 0, then he/she is not able to trade with agents I and J, i.e., to increase their consumption in the low-income state and decrease it in the high-income state since this would require agent K to hold a positive initial endowment in state d. The private value allocation in such a case assigns to every agent the initial endowment. This changes immediately if we consider endowments which are not independent.[17]

We now consider the case where endowments are not independent in the context of another simple example. This modification of the previous example demonstrates

---

[14] This inequality is true in general since the game derived from the differential information economy is superadditive. In particular, the sum of the utilities of the agents must be at least as great when utility is jointly maximized than when utility is only maximized for agents I and J, and agent K consumes his/her initial endowment.

[15] This follows immediately from the first order condition. Note that $V_\lambda^p(\{i,K\})$ can be found by solving

$$\max_{t_i, t_i'} \frac{\lambda_i}{2}\sqrt{4+t_i'} + \frac{\lambda_i}{2}\sqrt{t_i} + \frac{\lambda_K}{2}\sqrt{1-t_i'} + \frac{\lambda_K}{2}\sqrt{1-t_i},$$

where $i = I, J$. The first order conditions imply that it is optimal to choose the transfer t_i to be strictly positive. Thus the agents are strictly better off by trading than by consuming their initial endowment which implies that the strict inequality holds.

[16] The Shapley value of agent K is a weighted sum of K's marginal contribution $V(S) - V(S\backslash\{K\})$ to all coalition S of which agent K is a member of. By the above arguments we have $V(S) - V(S\setminus \{K\}) \geq V_\lambda^p(\{K\})$ for all coalitions S which contain K. Moreover, the strict inequality holds if $S = \{I,K\}$ and if $S = \{I,J\}$. Thus, agent K must receive a higher utility in the value allocation than $V_\lambda^p(\{K\})$.

[17] Let L and H denote the endowment realizations in the low and in the high states respectively. Then independence of endowments means that $P(\{e_I = L\}) = P(\{e_I = L \mid e_J = L\}) = P(\{e_I = L \mid e_J = H\})$, and similar for the high state and for the other agents (where P denotes the probability).

that the information superiority of an agent can matter, even though this agent may have no initial endowment at all. As long as this agent's private information is useful to the rest of the economy, he/she can always trade his/her superior information for actual goods.

Example 2. Consider an economy as in Example 1 but assume that the endowments of agent I and J are given by $(4,4,0,4)$ and $(4,0,4,4)$, respectively. Assume that $\mathcal{F}_I = \{\{a,b,d\},\{c\}\}$, $\mathcal{F}_J = \{\{a,c,d\},\{b\}\}$, and that agent K has full information. We also assume that agent K has a zero endowment in all states. The derivation of $U^p(S)$ is similar to the derivation in Example 1. We therefore omit any specification of the set of attainable utility allocations and show directly that $\mathrm{Sh}_K(V^p_\lambda) > 0$ for every $\lambda \geq 0$.[18]

We now show that $V^p_\lambda(\{I,J,K\}) > V^p_\lambda(\{I,J\})$. Let t_i be the net-trade of agent $i = I, J$ in the low-income state, and let t'_i denote the net-trade of agent $i = I, J$ in the high-income state.

We first consider the case where $\lambda_K > 0$. The first order conditions immediately imply that it is never optimal to choose $t_i = t'_i$ for $i = I, J$.[19] This, however, means that the weighted sum of the expected utilities will always be lower if we restrict agents I, J, and K to state-independent net trades. However, state-independent trades are the only ones which are incentive compatible in the two agent coalition $\{I,J\}$. Thus, $V^p_\lambda(\{I,J,K\}) > V^p_\lambda(\{I,J\})$.

We now consider the case where $\lambda_K = 0$. The first order conditions imply that it is never optimal to choose $t_i = t'_i = 0$.[20] In addition, agents I and J cannot receive

---

[18] By $\lambda \geq 0$ we mean that some, but not all of the weights can be zero. In particular, we must consider the case where $\lambda_K > 0$.

[19] The agents solve

$$\max_{t_i,t'_i} \sum_{i=I,J} \left[\frac{3\lambda_i}{4}\sqrt{4+t'_i} + \frac{\lambda_i}{4}\sqrt{t_i}\right] + \frac{\lambda_K}{2}\sqrt{-t'_I - t'_J} + \frac{\lambda_K}{4}\sqrt{-t'_I - t_J} + \frac{\lambda_K}{4}\sqrt{-t_I - t'_J}.$$

Without loss of generality assume that $\lambda_I > 0$. Then the first order conditions with respect to t_I and t'_I are

$$\frac{3\lambda_I}{\sqrt{4+t'_I}} = \frac{2\lambda_K}{\sqrt{-t'_I - t'_J}} + \frac{\lambda_K}{\sqrt{-t'_I - t_J}}$$

$$\frac{\lambda_I}{\sqrt{t_I}} = \frac{\lambda_K}{\sqrt{-t_I - t'_J}}.$$

If $t'_I = t_I$ and $t'_J = t_J$ then the first order conditions yield $\sqrt{4+t_I} = \sqrt{t_I} = 1$, which is clearly impossible. Thus, there does not exist a solution if $t'_I = t_I$ and $t'_J = t_J$.

[20] The argument is similar as the one above. Just consider the first order conditions of

$$\max_{t_I,t'_I} \frac{3\lambda_I}{4}\sqrt{4+t'_I} + \frac{\lambda_I}{4}\sqrt{t_I} + \frac{3\lambda_J}{4}\sqrt{4-t_I} + \frac{\lambda_J}{4}\sqrt{-t'_I},$$

and choose $t_I = t'_I = 0$. We can substitute the constraints $t_I + t'_J \leq 0$ and $t'_I + t_J \leq 0$ in the maximization problem, since they must obviously hold with equality.

a negative net-transfer in all states since this would imply negative consumption. Thus, it can never be the case that $t_i = t'_i \neq 0$.

Since there cannot be state-independent net-transfers in the value allocation, it follows that $V^p_\lambda(\{I, J, K\}) - V^p_\lambda(\{I, J\}) > 0$ for all weights λ which are candidates for utility comparison weights. Consequently agent K must have a positive Shapley value,[21] and he/she must get positive consumption in the value allocation. Moreover, it is also the case that the value allocation assigns agent K a strictly positive consumption in all states. This follows since the value allocation must be a solution to the Pareto problem in which each agent i's utility is multiplied with weight λ_i. Since $\lambda_K > 0$, the first order conditions stated in footnote 19 immediately implies the result, i.e., that agent K's consumption is strictly positive in all states.

Now compare this result with the case where agent K has no information (i.e., $\mathcal{F}_k = \{\{a, b, c\}\}$). Then his/her Shapley value is 0, and the initial allocation is the only equilibrium. This demonstrates that information superiority matters. Finally, note that any notion of a rational expectation equilibrium in this economy give zero consumption to agent K since his/her budget set is zero.

When agent K has useful private information the value allocation assigns positive consumption to agent K if the endowments of agents I and J are not independent, but assigns zero consumption if the endowments are independent. This occurs for the following reason: In both cases agents I and J attempt to insure against low-income realizations. Because of differential information, however, they need agent K as an intermediary to execute the correct trades. This arrangement works even if agent K has a zero-endowment as long as only one of the agents has a low endowment realization, because the claim of this particular agent can then be covered by the agent who has the high endowment-realization. This is the essence of Example 2. If both agents have low endowment realizations at the same time (which can occur if endowments are independent as in Example 1) then they both want a positive net-transfer. Agent K cannot fulfill his/her payment obligations because his/her endowment is zero, and K claims insolvency. However, this claim is problematic because agents I and J cannot verify whether agent K is in fact insolvent. Thus, the problem is to find an incentive compatible way to let agent K announce bonafide insolvencies. Clearly this is possible if agents I and J are able to observe state d.[22]

An alternative way to permit bonafide insolvencies by agent K is to weaken the incentive compatibility requirements. We do this in Example 4

[21] This follows immediately, since $(1/3)\left(V^p_\lambda(\{I, J, K\}) - V^p_\lambda(\{I, J\})\right)$ is one of the summands in the formula for the Shapley value. None of the summands is negative since the game is superadditive (see footnote 16).

[22] This can be done simply by assuming in Example 1 that the information partitions of agents I and J are given by $\{\{a, b\}, \{c\}, \{d\}\}$ and $\{\{a, c\}, \{b\}, \{d\}\}$, respectively. The proof that agent K's Shapley value (and hence consumption) is strictly positive is along the same lines as Example 2.

4.2 Risk sharing versus informational effects: The Shafer example

Readers familiar with the Roth (1980) and Shafer (1980) examples (as well as the debate on the value allocation Aumann (1985, 1987), Roth (1983), Scafuri and Yannelis (1984), Yannelis (1983)) will notice that our Examples 2 and 3 have a similar flavor in the sense that an agent with a zero endowment ends up with positive consumption. In the Roth and Shafer examples this effect may be attributed to risk aversion: the agent with a zero initial endowment is less risk averse. However, this is clearly *not* the case in our setting since all agents have *identical* utility functions. Also notice that in our Example 2 agent K gets zero consumption in the value allocation if he/she has no information and this is also the *only* Shapley value allocation. However, the agent gets positive consumption if he/she has full information. In fact, when agents I and J implicitly "use" agent K's information, this leads to a Pareto improvement for the whole economy. Thus, it is solely informational effects that drive our Examples rather than risk sharing. To illustrate this point we introduce differential information in the Shafer example (1980, Example 2, pp. 471–472).

Example 3. Assume there are three agents denoted by I, J, K, two possible states of nature a, b, and one commodity per state. The endowments of agents I and J are given by $(4, 0)$ and $(0, 4)$, respectively. Of course, I and J have full information. That is $\mathcal{F}_I = \mathcal{F}_J = \{\{a\}, \{b\}\}$. They have the same utility function given by $W^i(x_a, x_b) = \left((1/2)\sqrt{x_a} + (1/2)\sqrt{x_b}\right)^2$ for $i = I, J$. Agent K's endowment is $(0, 0)$ and he/she is risk neutral, i.e., the utility function is given by $W^K(x_a, x_b) = (1/2)x_a + (1/2)x_b$. Shafer's example corresponds in our differential information framework to the complete information case, i.e., $\mathcal{F}_I = \mathcal{F}_J = \mathcal{F}_K$. It can be shown[23] that for $\lambda_I = \lambda_J = \lambda_K = 1$ there exists a value allocation which gives positive consumption to the agent with zero initial endowment. In particular, in the value allocation, agents I and J receive $(11/6, 11/6)$ and agent K receives $(2/6, 2/6)$.

Now consider the case where agent K has trivial information, i.e., $\mathcal{F}_K = \{\{a, b\}\}$ and agents I and J have full information, i.e., $\mathcal{F}_I = \mathcal{F}_J = \{\{a\}, \{b\}\}$. We will show that the private value allocation assigns zero consumption to agent K (despite the fact that agent K is less risk averse). Assume by way of contradiction that there exists a private value allocation which assigns positive consumption to agent K. In such a value allocation K must also have a positive weight λ_K. However, agent K cannot enter into any trades with agent I or with agent J separately, since those trades would have to be state-independent and would therefore assign negative consumption to one of the agents in each state (so agents I and J both prefer their initial endowment in a two agent coalition with K). Thus, $V_\lambda(\{I, K\}) = V_\lambda(\{J, K\}) = V_\lambda(\{I\}) = V_\lambda(\{J\})$, where the first and the third equality follow from symmetry. Furthermore, $V_\lambda(\{K\}) = 0$, since K has a zero initial endowment. Thus, K's Shapley value is given by

$$\mathrm{Sh}_K(V_\lambda) = \frac{1}{3}\Big(V_\lambda(\{I, J, K\}) - V_\lambda(\{I, J\})\Big). \tag{4.1}$$

[23] For explicit computations see Yannelis (1983, pp. 291–292).

Let (x^I, x^J, x^K) be the private value allocation. Then since no side-payments are necessary in the value allocation we must have

$$V_\lambda(\{I, J, K\}) = \sum_{i=I,J,K} \lambda_i W^i(x^i). \tag{4.2}$$

Furthermore, since $x^I + x^J \leq e^I + e^J = (4, 4)$ it follows that

$$V_\lambda(\{I, J\}) \geq \lambda_I W^I(x^I) + \lambda_J W^J(x^J). \tag{4.3}$$

(4.1), (4.2) and (4.3) imply

$$\mathrm{Sh}_K(V_\lambda) \leq \frac{1}{3}\lambda_K W^K(x^K) < \lambda_K W^K(x^K).$$

However, this means that agent K gets strictly more than his/her Shapley value, a contradiction to the fact that (x^I, x^J, x^K) is assumed to be a value allocation. Thus, the *only* value allocation which exists in this example assigns zero consumption and a weight of zero to agent K.

The introduction of differential information in the Shafer example enables us to draw the following two conclusions.

(a) It resolves the problem noted by Roth and Shafer that a "dummy player" ends up with positive consumption.
(b) More importantly, the example indicates that there is an essential difference between the risk aversion effect which drives the Roth-Shafer examples and the informational asymmetries which drive our results.

It is important to note that in *all* our examples all agents have the *same* utility function and therefore the same risk attitude. Nonetheless, in situations where an agent with zero consumption ended up with positive consumption this was due purely to the information superiority of the agent. Furthermore, if we consider the economy as a transferable utility game (i.e., we fix the weights λ and allow side-payments in equilibrium), our Examples 1 and 2 show that the agent with superior information gets strictly positive consumption. This is independent of the agent's risk aversion and holds for *any choice* of λ, i.e., even if the agent with a zero endowment has zero weight. The agent with a zero endowment still receives a strictly positive Shapley value due to his/her superior information.

5 Other value concepts

5.1 The weak value allocation

In all of our examples, the private and the strong value coincide. In contrast, weak incentive compatibility increases in general the set of attainable utility allocations, thus resulting in a different value allocation. Similar to the private value allocation, the weak value allocation rewards agents with superior information. On the other hand, however, in the weak value allocation agents do not only benefit from each others' information but they can also exchange information. Consider the following example.

Example 4. Consider the economy of Example 1, except assume that agent K's endowment is given by $(0, 0, 0, 0)$ and that the agent has full information. We now analyze the weak coalitional incentive compatible value.

It is clear that the $U^p(S) = U^w(S)$ for the coalitions $S = \{I\}$, $S = \{J\}$ and $S = \{I, J\}$. Further, for the coalitions $S = \{I, K\}$ and $S = \{J, K\}$ we can derive $U^w(S)$ by a similar procedure as $U^p(S)$, taking into account that agent K has zero endowment. The attainable utility allocations differ in an interesting way when we consider the grand coalition. We show that $U^w(\{I, J, K\})$ corresponds to the attainable utility allocations under full information:

Consider an allocation (x_I, x_J, x_K) which is Pareto optimal under full information. Let $x_i(s)$ denote the consumption of agent i in state s. We now show that this allocations fulfills weak coalitional incentive compatibility. Clearly, $x_i(b) = x_i(c)$ for $i = I, J, K$, since the aggregate endowment in states b and c coincides. Further, $x_K(a) \geq x_K(b) \geq x_K(d)$ since the aggregate endowment in state a is higher than the aggregate endowment in state b, and since the aggregate endowment in state b is higher than the aggregate endowment in state d. Note that agent K cannot misreport if state d occurs because one of the other agents will disagree.[24] The same is true if state b or state c occurs. Finally, agent K has no incentive to misreport in state a since this is the state where he/she gets the highest net-transfer.

The remainder of the argument is similar to that in Example 2. We must show that agent K must get a strictly positive consumption of the good in each state of nature in the value allocation.[25] This, however, follows immediately since the above computations imply that $V_\lambda^w(\{I, J, K\}) - V_\lambda^w(\{I, J\}) > 0$ for every λ and hence $\mathrm{Sh}_K(V_\lambda^w) > 0$.[26] Since $\mathrm{Sh}_K(V_\lambda^w) = \lambda_k \int u_K(x_K)\, d\mu$, where x_K is the consumption assigned to agent K in the value allocation, x_K must be strictly positive.

The weak, and the private value allocations explicitly take into account an agent's information superiority in an economy with differential information. Example 1 shows that the information of agent K matters (as does the relative lack of information of all other agents) and Examples 2 and 4 show that agent K can be an "intermediary" between agents I and J even without having a positive endowment (agent K simply announces the true state of nature and as compensation gets a positive net-transfer for this service). However, in the weak value allocation exchange of information takes place. Consider the above example. Agent I's and agent J's net-trades assigned in the weak value allocation are not individually measurable.

---

[24] Agent K can only announce either states b or c when attempting to misreport the occurrence of d. Hence either agent I or J must agree. Without loss of generality assume it is agent I. This, however, is impossible since I would have to pay a net-transfer corresponding to a high income state. Such a transfer is always strictly higher than the net-transfer in a low income state.

[25] Note that the sets $U^w(S)$ are compact and convex, and that superadditivity is fulfilled. Hence the conditions for the existence of a value allocation hold (cf., Shapley (1969) or Emmons and Scafuri (1985)). Hence a value allocation exists.

[26] The inequality follows since the grand coalition can attain all unconstrained Pareto efficient allocations. None of them can be attained via state-independent net-transfers.

Specifically, the agents' net-trades are different in the four states. Thus, once the state of nature is realized, there is no differential information ex-post. In this sense, agents I and J have "received information" from agent K.

The role of agent K in Examples 2 and 4 is also interesting in connection with the literature on financial intermediation. For example, Boyd and Prescott (1986) argue that coalitional structures, i.e., cooperative games with differential information, are important for understanding financial intermediation. As our examples indicate, the Shapley value can be an interesting tool for such an analysis and might provide an alternative to the core which Boyd and Prescott use. It is important to point out, that we do not need to assume the existence of a "state verification technology" at the outset as it is standard in this literature. Rather, verification evolves endogenously in our model. Whenever there is "doubt" about the state of the economy, agent I and agent J can turn to agent K who then announces the true state. Further, agent K is compensated for this service by a positive net-transfer. This net-transfer to agent K can be interpreted as a "cost of state verification" paid by agent I and agent J, and the magnitude of this cost is determined endogenously. For further work on intermediation and the relationship between intermediation and media of exchange in a differential information economy with pairwise trade see Huggett and Krasa (1993).

5.2 The Coarse and the Fine Value

We now define the Coarse and the Fine Value Allocation.

(a) A coalition of agents S always pools all the information of its members. Formally, let $\mathcal{F}_i$ be the information partition of agent i, then the *pooled* information $\bigvee_{i\in S}\mathcal{F}_i$ is given by the union of the information partitions of the individual agents.[27]

(b) A coalition of agents makes their trades contingent solely on common knowledge information. This can be interpreted as assuming that trades within a coalition must be verifiable by all members of the coalition. Formally, *common knowledge* information is given by $\bigwedge_{i\in S}\mathcal{F}_i$ which is the intersection of the information partitions $\mathcal{F}_i$ of all agents.[28]

Wilson (1978) refers to the core with information sharing as in (a) as the *fine core,* and to the core with information sharing as in (b) as the *coarse core.* In this section we want to discuss briefly the analogous definitions for the Shapley value, and show that they are problematic.[29] By exchanging constraint (ii) in (3.1) by "$x_i - e_i$ is $\bigvee_{i\in S}\mathcal{F}_i$-measurable for every $i \in S$" and (i) in Definition 6 by "$x_i - e_i$ is $\bigvee_{i\in I}\mathcal{F}_i$-measurable for every $i \in I$" we get the definition of the "fine value." Similarly, we can define a "coarse value."

---

[27] Strictly speaking, we must take the coarsest partition which contains the information partition of every agent. Thus, two states a and b are indistinguishable for the coalition S if and only if for all $A_i \in \mathcal{F}_i$ it is the case that $\bigcap_{i\in S} A_i \neq \emptyset$ implies $\{a,b\} \subset \bigcap_{i\in S} A_i$.

[28] Thus, two states a and b are indistinguishable with respect to the common knowledge information of coalition S, if and only if there exists an agent $i \in S$ and a set $A_i \in \mathcal{F}_i$ such that $\{a,b\} \subset A_i$.

[29] For subsequent work on the coarse and the fine core see Koutsougeras and Yannelis (1993).

The first obvious problem of the fine value is that allocations will in general not be incentive compatible. Consider, for example two agents I, J with the same endowments and the same utility functions as I and J in Example 1. Since agents pool their information, there is full information in coalition $S = \{I, J\}$. Consequently arbitrary trades can be achieved. For example, agent I can promise agent J the net-trade $(-1, -1, 1, 1)$. However, since J's information partition is $\mathcal{F}_J = \{\{a, c\}, \{b, d\}\}$, agent I always has the incentive to announce state a. Such trades are ruled out by weak as well as strong coalitional incentive compatibility as well as by the private value allocation. However, they are admissible in the fine value.

The second problem of the fine value is that information asymmetries are not taken into account. In particular, consider again the economy of Example 1. Since the coalition $S = \{I, J\}$ always has full information, the information of K is irrelevant. In contrast to the value concepts discussed above, agent K is not needed since there are no informational problems between I and J any more. Similarly, the value allocation will not change if we change the information of I and J.[30] It is relatively easy to see that this observation is true in general. Thus, the fine value does not seem to be a useful concept.

On the other hand, although the coarse value takes information asymmetries into account, it has some rather strange features. For example, add to an arbitrary economy an agent who has zero initial endowment and no information. Then the common knowledge information of the grand coalition immediately becomes trivial, and hence only trivial net-trades are possible. Thus, an agent who should be irrelevant in the economic allocation process, influences the outcome in the coarse value in a major way. The reason is that "bad" information of one agent poses a negative externality on all other agents in the economy. Thus, rather than measuring the marginal contributions of an agent, the coarse value measures this negative externality of an agents on all other coalitions. This seems to us to be very much against the general idea of what the Shapley value is supposed to describe.

For a more thorough discussion of these two concepts see Krasa and Yannelis (1991). There the existence of the fine value as well as examples of non-existence of the coarse value are presented. Independently of our work, Allen (1991) has also examined the existence and the non-existence of the coarse, the private and the fine value.

6 Concluding remarks

In this paper we study several value allocation concepts in an economy with differential information. We show that the private value allocation provides an interesting way to measure the "worth" of an agents's information advantage. In particular, the Shapley value provides an explicit way not only to measure the information superiority of an agent, but also to reward the agent for making a Pareto improvement for the economy as a whole by using his/her informational advantage. Moreover, this

30 In this particular example this means that the agent has more information, since the initial endowment of an agent must always be measurable with respect to his/her information.

concept ensures truthful revelation of information within a coalition because incentive compatibility is inherent in it. Furthermore, our examples suggests that the value allocation may be suitable for analyzing problems of financial intermediation.

It should be noted that our different value allocation concepts do not provide a dynamic procedure which explains how the final equilibrium outcome is reached. However, we know from the work of Winter (1992) that the Shapley value of a TU-game can be rationalized as a solution to a non-cooperative game in extensive form.[31] His results require convex games and risk neutrality which are stronger conditions than those adapted in our modeling. It would be of interest to see if one can provide non-cooperative foundations of our results, and explain the dynamics of reaching equilibrium outcomes. This seems to be an important open question.

Finally, it is well known that cardinal value allocations characterize Walrasian equilibrium allocations in large economies. However, in a differential information economy framework it is not only unknown to us whether such a result can be obtained, but it is not even clear what should be the correct definition of a Walrasian equilibrium in a differential information economy.

References

Allen, B.: Market games with asymmetric information: the value. Mimeo, University of Pennsylvania (1991)

Aumann, R.J.: On the non-transferable utility value: a comment on the Roth-Shafer examples. Econometrica **53**, 667–677 (1985)

Aumann, R.J.: Value, symmetry and equal treatment: a comment on Scafuri-Yannelis. Econometrica **55**, 1461–1464 (1987)

Boyd, J., Prescott, E.C.: Financial intermediary coalitions. Journal of Economic Theory **38**, 211–232 (1986)

Emmons, D., Scafuri, A.J.: Value allocations – An exposition. In: Aliprantis, C.D., et al. (eds.) Advances in equilibrium theory. Berlin, Heidelberg, New York: Springer 1985

Gul, F.: Bargaining foundations of the Shapley value. Econometrica **57**, 81–95 (1989)

Huggett, M., Krasa, S.: Media of exchange in economies with differential information. Mimeo, University of Illinois (1993)

Koutsougeras, L., Yannelis, N.C.: Incentive compatibility and information superiority of the core of an economy with differential information. Economic Theory **3**, 195–216 (1993)

Krasa, S., Yannelis, N.C.: The value allocation of an economy with differential information. Mimeo, University of Illinois (1991)

Myerson, R.B.: Cooperative games with incomplete information. International Journal of Game Theory **13**, 69–96 (1984)

Radner, R.: Competitive equilibrium under uncertainty. Econometrica **36**, 31–58 (1968)

Roth, A.: Values of games without side payments: some difficulties with current concepts. Econometrica **48**, 457–465 (1980)

Roth, A.: Some further thoughts on the NTU value and related matters: a rejoinder to Aumann. Mimeo, University of Pittsburgh (1983) (a shortened version appeared in: Econometrica **55**, 981–984 (1986)

Scafuri, A.J., Yannelis, N.C.: Non-symmetric cardinal value allocations. Econometrica **52**, 1365–1368 (1984)

Shafer, W.J.: On the existence and interpretation of value allocation. Econometrica **48**, 467–474 (1980)

[31] Gul (1989) has also provided a non-cooperative (bargaining) foundation for the Shapley value of a TU-game. However, his model is based on bilateral bargaining rather than multilateral, contrary to Winter's (1992) model. The restriction to bilateral bargaining appears to make his results inapplicable to our framework.

Shapley, L.A.: A value of n-person games. In: Kuhn, H.W., Tucker, A.W. (eds.) Contributions to the theory of games, Vol. II, pp. 307–317. Princeton University Press 1953

Shapley, L.A.: Utility comparisons and the theory of games. In: Guilbaud, (ed.) La decision. Paris, France: Edition du CNRS, 1969

Winter, E.: The demand commitment bargaining and snowballing cooperation. Economic Theory (1992) (forthcoming)

Wilson, R.: Information, efficiency, and the core of an economy. Econometrica **46**, 807–816 (1978)

Yannelis, N.C.: Existence and fairness of value allocation without convex preferences. Journal of Economic Theory **31**, 282–292 (1983)

Yannelis, N.C.: The core of an economy with differential information. Economic Theory **1**, 183–198 (1991)

Existence and properties of a value allocation for an economy with differential information*

Stefan Krasa and Nicholas C. Yannelis

Department of Economics, University of Illinois at Urbana-Champaign, Champaign, IL 61820, USA

Summary. We prove the existence of a private value allocation for an economy with differential information where the commodity space may be infinite dimensional, and there is a continuum of states. We also discuss the existence, non-existence, and properties of two alternative value allocation concepts.

1 Introduction

In a companion paper Krasa and Yannelis (1994) introduced the concept of a private value allocation for an economy with differential information. This concept was presented as an alternative to the rational expectation equilibrium notion. In particular, we demonstrated that the private value allocation is coalitionally incentive compatible and it takes into account the informational advantage or superiority of an individual. Moreover, we showed by means of examples that the private value allocation yields sensible and reasonable outcomes in situations where any rational expectation equilibrium notion fails to do so.

In view of the fact that the private value allocation seems to be a successful alternative of the rational expectation equilibrium it is important to know conditions that guarantee its existence. Although in Krasa and Yannelis it was mentioned that the private value allocation may exist under fairly mild assumptions, no existence proof was provided. It is the main purpose of the present paper to present sufficient conditions for the existence of a private value allocation. Moreover, we examine the existence and interpretation of two alternative value allocation concepts, i.e., the coarse and the fine value allocation.

It should be noted that for the deterministic value allocation (either ordinal or cardinal value) several general existence results are available in the literature,

* We wish to thank two anonymous referees for helpful comments. As always we are responsible for any remaining errors. This research was supported by the Campus Research Board of the University of Illinois.

e.g., Shapley (1969), Shafer (1980), Yannelis (1983), Emmons and Scafuri (1985). However, none of these results can be applied to differential information economies directly. In particular, the presence of a continuum of states even with a finite dimensional commodity space necessitates the use of functional analytic and measure theoretic methods.

The paper proceeds as follows. In Section 2 the economy with differential information is introduced. Section 3 defines the private value allocation. Section 4 is focused on the existence proof, and finally Section 5 discusses the coarse and the fine value allocations.

2 The model

Let Y denote the commodity space. In our existence result in Section 4, Y may be infinite dimensional. Hence infinitely many commodities are permissible.[1] We consider an exchange economy which extends over two time periods $t = 0, 1$ where consumption takes place in $t = 1$. At $t = 0$ there is uncertainty over the state of nature described by a complete probability space $(\Omega, \mathcal{F}, \mu)$. Let $I = \{1, \ldots, n\}$ denote the set of all agents. In $t = 0$ agents will agree on net-trades which may be contingent on the state of nature in $t = 1$. However, agents are differentially informed with respect to the true state of nature. Specifically, we assume that at $t = 1$ agents do not necessarily know which state $\omega \in \Omega$ has actually occurred. They know their own endowment realization and every agent i might have some additional information about the state described by a σ-algebra $\mathcal{F}_i$ with $\mathcal{F}_i \subset \mathcal{F}$. Although all our results we will be proved for arbitrary information σ-algebras, it easier to understand the interpretation of the information σ-algebras by considering partitions. Thus, assume for example that $\mathcal{F}_i$ is generated by a countable partition A_k, $k \in \mathbb{N}$. Let $\bar{\omega}$ be the true state of the economy in $t = 1$. Then agent i observes the event A_k which contains $\bar{\omega}$. However, he/she does not know which state $\omega \in A_k$ has actually occurred.

By assumption agents can always observe their own endowment realization. Thus, we can assume without loss of generality that agent i's initial endowment e_i is measurable with respect to $\mathcal{F}_i$.

In summary, an *exchange economy with differential information* is given by $\mathcal{E} = \{(X_i, u_i, e_i, \mathcal{F}_i, \mu)\colon\ i = 1, \ldots, n\}$ where

(1) $X_i\colon \Omega \to 2^{Y_+}$ is the *consumption set* of agent i;
(2) $u_i\colon Y_+ \to \mathbb{R}_+$ is the *utility function* of agent i;[2]
(3) $\mathcal{F}_i$ is a σ-algebra with $\mathcal{F}_i \subset \mathcal{F}$ denoting the *private information* of agent i;
(4) $e_i\colon \Omega \to Y_+$ is the *initial endowment* of agent i, where each e_i is $\mathcal{F}_i$-measurable, (Bochner) integrable[3] and $e_i(\omega) \in X_i(\omega)$ μ-a.e.;

[1] Y can be any Banach lattice with an order continuous norm. Y_+ denotes the positive cone of Y (see Section 4 for the appropriate definitions).

[2] One may also assume that the utility function is random, i.e., u_i is a real valued function defined on $\Omega \times Y$. All the results of the paper will remain valid.

[3] See Section 4 for a definition of the Bochner integral. If $X = \mathbb{R}^n$ this is the standard Lebesgue integral.

(5) μ is a probability measure on Ω denoting the *common prior* of each agent.

The *expected utility* of agent i is given by

$$\int_{\Omega} u_i(x_i(\omega))\, d\mu(\omega).^4$$

Throughout the paper, we assume that the utility function u_i of each agent i is (weakly) continuous, concave and bounded.

3 The private value allocation

We now describe the notion of a private value allocation. This notion is the analog of the private core of Yannelis (1991) and was introduced in Krasa and Yannelis (1994). The main idea for this concept is that each agent i's trades are restricted to those which are measurable with respect to their private information. Since the Shapley value measures the marginal contribution of each agent to any coalition the agent is member off, the assumption that an agent's trade within the coalition must be measurable with respect to the agent's information implies that information asymmetries matter. Since the main focus of this paper is to provide a general existence result, the interested reader is referred to Krasa and Yannelis (1994) for a further discussion of the properties of the private value allocation.

As in the definition of the standard value allocation concept, we must first derive a transferable utility (TU-) game in which each agent's utility is weighted by a factor λ_i which allows interpersonal utility comparisons. In the value allocation itself no side-payments are necessary.[5] A game with side-payments is then defined as follows.

Definition 1 *A game with side-payments $\Gamma = (I, V)$ consists of a finite set of agents $I = \{1, \ldots, n\}$ and a superadditive, real valued function V defined on 2^I such that $V(\emptyset) = 0$. Each $S \subset I$ is called a coalition and $V(S)$ is the "worth" of the coalition S.*

The Shapley value of the game Γ, [Shapley (1953)] is a rule which assigns to each agent i a "payoff" Sh_i given by the formula[6]

$$\mathrm{Sh}_i(V) = \sum_{\substack{S \subset I \\ S \supset \{i\}}} \frac{(|S|-1)!(|I|-|S|)!}{|I|!} [V(S) - V(S \setminus \{i\})].$$

The Shapley value has the property that $\sum_{i \in I} \mathrm{Sh}_i(V) = V(I)$, i.e., the Shapley value is Pareto efficient.

[4] Different priors and updating of priors could be introduced as in Krasa and Yannelis (1994, footnote 7).

[5] See Emmons and Scafuri (1985, p. 60) or Shafer (1980, p. 468) for further discussion.

[6] The Shapley value measures is the sum of the expected marginal contributions an agent can make to all the coalitions that he/she is a member of [see Shapley (1953)].

We now define for each economy with differential information $\mathcal{E}$ and for each set of weights $\{\lambda_i : i = 1, \ldots, n\}$ the associated game with side-payments (I, V_λ^p) [we also refer to this as a "transferable utility" (TU) game] as follows:

For every coalition $S \subset I$ let

$$V_\lambda^p(S) = \max_{x_i} \sum_{i \in S} \lambda_i \int u_i(x_i(\omega))\, d\mu(\omega) \tag{3.1}$$

subject to

(i) $\sum_{i \in S} x_i(\omega) = \sum_{i \in S} e_i(\omega)$, μ-a.e.

(ii) $x_i - e_i$ is $\mathcal{F}_i$-measurable for every $i \in S$.

We are now ready to define the private value allocation.

Definition 2 *An allocation $x\colon \Omega \to \prod_{i=1}^n Y_i$ with $x_i(\omega) \in X_i(\omega)$ μ-a.e. for all i is said to be a private value allocation of the economy with differential information $\mathcal{E}$ if the following holds:*

(i) Each net-trade $x_i - e_i$ is $\mathcal{F}_i$-measurable.
(ii) $\sum_{i=1}^n x_i(\omega) = \sum_{i=1}^n e_i(\omega)$, μ-a.e.
(iii) There exist $\lambda_i \geq 0$, for every $i = 1, \ldots, n$ which are not all equal to zero, with $\lambda_i \int u_i(x_i(\omega))\, d\mu(\omega) = \mathrm{Sh}_i(V_\lambda^p)$ for all i, where $\mathrm{Sh}_i(V_\lambda^p)$ is the Shapley value of agent i derived from the game (I, V_λ^p), defined in (3.1).

Condition (i) requires individual measurability of net-trades, i.e., net-trades can only be contingent on each agent's individual information $\mathcal{F}_i$. (ii) is the market clearing condition. (iii) says that the expected utility of each agent multiplied with his/her weight λ_i must be equal to his/her Shapley value derived from the TU game (I, V_λ^p).

An immediate consequence of Definition 2 is that $\mathrm{Sh}_i(V_\lambda^p) \geq \lambda_i \int u_i(e_i)\, d\mu$ for every i, i.e., the value allocation is individually rational. This follows immediately from the fact that the game (V_λ^p, I) is superadditive for all weights λ. Similarly, efficiency of the Shapley value for games with side payments immediately implies that the value allocation is constrained Pareto efficient.

We now state our main existence result.

Theorem 1. *Let $\mathcal{E} = \{(X_i, u_i, e_i, \mathcal{F}_i, \mu)\colon i = 1.\ldots, n\}$ be a finite exchange economy with differential information satisfying the following assumptions for each agent:*

(A1) The commodity space Y_+ is the positive cone of a Banach lattice Y with an order continuous norm.
(A2) $X_i\colon \Omega \to 2^{Y_+}$ is a convex, closed, non-empty valued correspondence.
(A3) $u_i\colon Y \to \mathbb{R}_+$ is weakly continuous, bounded and concave.

Then a private value allocation exists in $\mathcal{E}$.

The technical conditions in (A1) are explained in the following section. Note that (A1) is automatically fulfilled for all finite dimensional spaces. Other basic examples of Banach lattices with an order continuous norm are the Lebesgue spaces $L^p, 1 \leq p < \infty$ of $\mathbb{R}^n$ valued functions.

4 The existence proof

The goal of this section is to provide a general existence result. In particular we show that private value allocations exists in a setting where there is an infinite number of commodities and an infinite number of states of nature. Before proving our existence result, we outline some mathematical preliminaries.

4.1 Mathematical preliminaries

Let $(\Omega, \mathcal{F}, \mu)$ be a probability space, and X be a Banach space. We denote by $L^1_X(\mu)$ the space of all equivalence classes of X-valued Bochner integrable[7] functions $f\colon \Omega \to X$ normed by

$$\|f\| = \left(\int \|f(\omega)\|^p \, d\mu(\omega) \right)^{\frac{1}{p}} .$$

It is a standard result that when normed by the functional $\|1\|\cdot$ above, $L^1_X(\mu)$ becomes a Banach space [see Diestel and Uhl (1977), p. 50].

We now collect some basic results on Banach lattices [for an excellent treatment see Aliprantis and Burkinshaw (1985)]. Recall that a Banach space X is a *Banach lattice* if there exists an ordering $\geq$ on X with the following properties:

(i) $x \geq y$ implies $x + z \geq y + z$ for every $z \in X$;
(ii) $x \geq y$ implies $\lambda x \geq \lambda y$ for every scalar $\lambda \geq 0$;
(iii) for all $x, y \in X$ there exists a supremum (denoted by $x \vee y$) and an infimum (denoted by $x \wedge y$).
(iv) $|x| \geq |y|$ implies $\|x\| \geq \|y\|$ for all $x, y \in X$.

As usual, $x^+ = x \vee 0$, $x^- = (-x) \vee 0$, and $|x| = x^+ + x^-$, we call x^+ and x^- the positive and negative parts of x, respectively and $|x|$ the absolute value of x. For $x, y \in X$ we define the *order interval* $[x, y]$ as follows:

$$[x, y] = \{z \in X : x \leq z \leq y\}.$$

Note that $[x, y]$ is convex and norm closed, hence weakly closed (recall Mazur's Theorem). A Banach lattice L is said to have an order continuous norm if $x_\alpha \downarrow 0$[8] in L implies $\|x_\alpha\| \downarrow 0$. A very useful result which will play an important role is that if X is a Banach lattice then the fact that X has an order continuous norm is equivalent to the weak compactness of order intervals [see for example Aliprantis and Burkinshaw (1985)].

We finally note that Cartwright (1974) has shown that if X is a Banach lattice with order continuous norm (or equivalently X has weakly compact order intervals) then $L^1_X(\mu)$ has weakly compact order intervals, as well. Cartwright's Theorem will play a crucial role in our existence proof.

[7] The Bochner integral of a function f can be obtained by approximating f by a sequence of simple functions f_n (for a definition of simple functions see footnote 15). This is the same construction which is used to define the Lebesgue integral for $X = \mathbb{R}^n$.

[8] $x_\alpha \downarrow 0$, means that x_α is a decreasing net with $\inf_\alpha x_\alpha = 0$.

4.2 Proof of Theorem 1

Let $\{(X_i, u_i, e_i)\colon i = 1, 2, \ldots, n\}$ be an exchange economy where

(a) $X_i \subset \mathbb{R}^l_+$ is the *consumption set* of agent i;
(b) $u_i\colon X_i \to \mathbb{R}_+$ is the *utility function* of agent i;
(c) $e_i \in X_i$ is the *initial endowment* of agent i.

Given an economy $\{(X_i, u_i, e_i)\colon i = 1, 2, \ldots, n\}$ and a set of weights $\{\lambda_i\colon i = 1, \ldots, n\}$, where $\lambda_i \geq 0$ for every i and $\sum_{i=1}^n \lambda_i = 1$, define the game

$$V_\lambda(S) = \max_{x_i \in X_i} \sum_{i \in S} \lambda_i u_i(x_i), \text{ subject to } \sum_{i \in S} x_i = \sum_{i \in S} e_i.$$

Denote by $\mathrm{Sh}_i(V_\lambda)$ the Shapley value of agent i. The allocation

$$x = (x_1, \ldots, x_n) \in \prod_{i=1}^n X_i$$

is said to be a λ-*transfer value allocation* or a *cardinal value allocation* for the economy $\{(X_i, u_i, e_i)\colon i = 1, \ldots, n\}$ if

(i) $\sum_{i=1}^n x_i = \sum_{i=1}^n e_i$, and
(ii) there exist $\{\lambda_i \geq 0\colon i = 1, \ldots, n\}$ with $\sum_{i=1}^n \lambda_i = 1$ such that $\lambda_i u_i(x_i) = \mathrm{Sh}_i(V_\lambda)$ for each i.

Emmons and Scafuri (1985) or Shapley (1969) show that if u_i is concave, and continuous; and if X_i is bounded from below, closed and convex; then a cardinal value allocation exists for the economy $\{(X_i, u_i, e_i)\colon, i = 1, \ldots, n\}$. Let L_{X_i} denote the set of all functions $x_i\colon \Omega \to Y$ which are $\mathcal{F}_i$-measurable, Bochner integrable, and for which $x_i(\omega) \in X_i(\omega)$, μ-a.e. Define $W_i\colon L_{X_i} \to \mathbb{R}$ by $W_i(x) = \int u_i(x_i(\omega))\, d\mu(\omega)$.

Following Bewley's (1972) argument, we will prove our Theorem by considering its trace in finite dimensions and appealing to the Emmons and Scafuri (1985) existence result. We first need to prove some simple facts:

(F1) L_{X_i} is non-empty.
(F2) L_{X_i} is convex, norm closed and bounded from below.
(F3) W_i is weakly upper semi-continuous on L_{X_i}.
(F4) W_i is concave on L_{X_i}.

Fact (F1) follows immediately. Since by assumption e_i is $\mathcal{F}_i$-measurable and Bochner integrable, we can conclude that $e_i \in L_{X_i}$. Fact (F2) follows directly from assumption (A2). Fact (F3) is proved in Balder and Yannelis (1993, Theorem 2.8) and (F4) follows directly from the concavity of u_i.

Now consider the economy $\bar{\mathcal{E}} = \{(L_{X_i}, W_i, e_i)\colon i = 1, \ldots, n\}$, where L_{X_i} denotes the consumption set of agent i, where W_i is the utility function of agent i, and where $e_i \in L_{X_i}$ denotes the initial endowment of agent i. Note that the existence of a value allocation in $\bar{\mathcal{E}}$ implies the existence of a value allocation for the original economy $\mathcal{E}$.

Let $\mathcal{A}$ be the set of all finite dimensional subspaces of $L^1_Y(\mu)$ containing the initial endowments. For each $\alpha \in \mathcal{A}$, let $L^\alpha_{X_i} = L_{X_i} \cap \alpha$ be the consumption set of agent i and $W^\alpha_i \colon L^\alpha_{X_i} \to \mathbb{R}$ be the utility function of agent i. Note that W^α_i is continuous, since it is the expected utility over a finite dimensional space.[9] For each $\alpha \in \mathcal{A}$, we have an economy $\bar{\mathcal{E}}^\alpha$ with a finite dimensional consumption space. Further, for each $\alpha \in \mathcal{A}$, the economy $\bar{\mathcal{E}}^\alpha$ fulfills the assumptions of Emmons and Scafuri (1985). Thus, there exists a value allocation, i.e., there exist $x^\alpha \in \prod_{i=1}^n L^\alpha_{X_i}$ such that

(i) $\sum_{i=1}^n x^\alpha_i = \sum_{i=1}^n e_i$;
(ii) there exist $\lambda^\alpha_i \geq 0$ with $\sum_{i=1}^n \lambda^\alpha_i = 1$, such that $\lambda^\alpha_i W^\alpha_i(x^\alpha_i) = \mathrm{Sh}_i(V_{\lambda^\alpha W^\alpha})$ for every i, where $\mathrm{Sh}_i(V_{\lambda^\alpha W^\alpha})$ is the Shapley value of agent i derived from the game $(I, V_{\lambda^\alpha W^\alpha})$.[10]

By (i) we have that

$$0 \leq \sum_{i=1}^n x^\alpha_i = \sum_{i=1}^n e_i = e.$$

Hence each x^α_i lies in the order interval $[0, e]$ in $\sum_{i=1}^n L_{X_i} \subset L^1_Y(\mu)$, which is weakly compact by Cartwright's Theorem [see Cartwright (1974) or Section 4.1].

Order the set $\mathcal{A}$ by inclusion. Then $\{(x^\alpha_1, \dots, x^\alpha_n, \lambda^\alpha_1, \dots, \lambda^\alpha_n) \colon \alpha \in \mathcal{A}\}$ is a net in $K = \prod_{i=1}^n [0, e] \times \Delta$, where Δ denotes the $(n-1)$-dimensional simplex. Since K is compact we can therefore assume without loss of generality that the net converges to a point $(x_1, \dots, x_n, \lambda_1, \dots, \lambda_n)$.[11] To complete the proof we must show that this limit is a value allocation of our original economy $\mathcal{E}$, i.e., that conditions (i) and (ii) in Definition 2 hold. (i) follows immediately since by Mazur's Theorem L_{X_i} is weakly closed and hence contains the limit points of the net x^α_i, $\alpha \in \mathcal{A}$.

We now prove (ii). We first show that $\lim_\alpha V_{\lambda^\alpha W^\alpha}(S) = V_{\lambda W}(S)$. Note that weak upper-semicontinuity implies that

$$\begin{aligned}
\limsup_\alpha V_{\lambda^\alpha W^\alpha}(S) &= \limsup_\alpha \sum_{i \in S} \lambda^\alpha_i W^\alpha_i(x^\alpha_i) \\
&= \sum_{i \in S} \limsup_\alpha \lambda^\alpha_i W_i(x^\alpha_i) \\
&\leq \sum_{i \in S} \lambda_i W_i(\lim_\alpha x^\alpha_i) \\
&= \sum_{i \in S} \lambda_i W_i(x_i) = V_{\lambda W}(S).
\end{aligned}$$

[9] This follows automatically from the Lebesgue dominated convergence theorem since each u_i is by assumption continuous and bounded.

[10] Clearly, the game $(I, V_{\lambda^\alpha W^\alpha})$ is defined as follows: For every coalition $S \subset I$ let

$$V_{\lambda^\alpha W^\alpha}(S) = \max_{x_i \in L^\alpha_{X_i}} \sum_{i \in S} \lambda^\alpha_i W^\alpha_i(x_i) \text{ subject to } \sum_{i \in S} x_i = \sum_{i \in S} e_i.$$

[11] More precisely, there exists a subnet which converges. For simplicity, we choose again $\mathcal{A}$ as the index set.

Hence,

$$\limsup_{\alpha} V_{\lambda^\alpha W^\alpha}(S) \leq V_{\lambda W}(S). \tag{4.1}$$

Now choose x_i^* such that $V_{\lambda W}(S) = \sum_{i \in S} \lambda_i W_i(x_i^*)$.[12] Then for every $\varepsilon > 0$ there exist $\bar{\alpha}$ such that

$$V_{\lambda W}(S) = \sum_{i \in S} \lambda_i W_i(x_i^*) \leq \sum_{i \in S} \lambda_i^\alpha W_i(x_i^*) + \varepsilon, \tag{4.2}$$

for every $\alpha > \bar{\alpha}$. Now choose $\bar{\beta} > \bar{\alpha}$ such that $\bar{\beta}$ contains the space spanned by $\{x_1^*, \dots, x_n^*\}$. Then

$$\sum_{i \in S} \lambda_i^\alpha W_i(x_i^*) \leq \max_{x_i \in L_{X_i} \cap \alpha} \sum_{i \in S} \lambda_i^\alpha W_i(x_i) = V_{\lambda^\alpha W^\alpha}(S), \tag{4.3}$$

for all $\alpha > \bar{\beta}$ since $x_i^* \in L_{X_i} \cap \alpha$. Thus, (4.2) and (4.3) imply

$$\liminf_{\alpha} V_{\lambda^\alpha W^\alpha}(S) \geq V_{\lambda W}(S). \tag{4.4}$$

Hence, (4.1) and (4.4.) imply that $\lim_\alpha V_{\lambda^\alpha W^\alpha}(S) = V_{\lambda W}(S)$. It follows from the continuity of the Shapley value that $\lim_\alpha \mathrm{Sh}_i(V_{\lambda^\alpha W^\alpha}) = \mathrm{Sh}_i(V_{\lambda W})$. By taking the lim sup on both sides of the equation $\lambda_i^\alpha W_i^\alpha(x_i^\alpha) = \mathrm{Sh}_i(V_{\lambda^\alpha W^\alpha})$ and by using the weak upper semicontinuity of W_i we can conclude that

$$\lambda_i W_i(x_i) \geq \limsup_{\alpha} \lambda_i^\alpha W_i(x_i^\alpha) = \limsup_{\alpha} \mathrm{Sh}_i(V_{\lambda^\alpha W^\alpha}) = \mathrm{Sh}_i(V_{\lambda W}). \tag{4.5}$$

However, the Pareto efficiency of the Shapley value implies that the equality must hold in (4.5). Thus, condition (ii) for a Shapley value allocation holds. This completes the proof of the theorem.

5 The coarse and the fine value allocation

5.1 Definitions

In this section we introduce two alternative notions of a value allocation for an economy with differential information. The difference stems from the measurability restriction on the type of allocations that are allowed. Both notions are the analogs of the coarse and the fine cores of Wilson (1978). We begin by defining these concepts. First, note that for arbitrary σ-algebras $\mathcal{F}_i$, $i = 1, \dots, n$, *common knowledge* information is given by $\bigwedge_{i=1}^n \mathcal{F}_i$ which is the intersection of all σ-algebras $\mathcal{F}_i$, $i = 1, \dots, n$. In contrast $\bigvee_{i=1}^n \mathcal{F}_i$ which is the σ-algebra generated by the union of the σ-algebras $\mathcal{F}_i$, $i = 1, \dots, n$ is the *pooled* information.

For each economy with differential information $\mathcal{E}$ and each set of weights $\{\lambda_i : i = 1, \dots, n\}$, we associate a game with side-payments (I, V_λ^c), [we also refer to this as a "transferable utility" (TU-) game] according to the rule:

[12] Note that x_i^* exists since each W_i is weakly upper semicontinuous, and since the x_i in the optimization problem in footnote 5 are restricted to the weakly compact order interval $\left[0, \sum_{i \in S} e_i\right]$ because of feasibility.

For every coalition $S \subset I$ let

$$V_\lambda^c(S) = \max_{x_i} \sum_{i \in S} \lambda_i \int u_i(x_i(\omega))\, d\mu(\omega) \tag{5.1}$$

subject to

(i) $\sum_{i \in S} x_i(\omega) = \sum_{i \in S} e_i(\omega)$, μ-a.e.

(ii) $x_i - e_i$ is $\bigwedge_{i \in S} \mathcal{F}_i$-measurable for every $i \in S$.

The *coarse value allocation* can now be defined by replacing condition (i) in Definition 2 by:

Each net-trade $x_i - e_i$ is $\bigwedge_{i=1}^n \mathcal{F}_i$-measurable; and by replacing V^p by V^c in condition (iii).

Thus, in contrast to the private value we now require that net-trades within a coalition can only be based on the common knowledge information of an agent.

The second concept that we introduce in this section is the fine value. For each economy with differential information $\mathcal{E}$ and each set of weights $\{\lambda_i : i = 1, \ldots, n\}$, we associate a game with side-payments (I, V_λ^f) according to the rule:

$$V_\lambda^f(S) = \max_{x_i} \sum_{i \in S} \lambda_i \int u_i(x_i(\omega))\, d\mu(\omega) \tag{5.2}$$

subject to

(i) $\sum_{i \in S} x_i(\omega) = \sum_{i \in S} e_i(\omega)$, μ-a.e.

(ii) $x_i - e_i$ is $\bigvee_{i \in S} \mathcal{F}_i$-measurable for every $i \in S$.

The *fine value allocation* can now be defined by replacing condition (i) in Definition 2 by:

Each net-trade $x_i - e_i$ is $\bigvee_{i=1}^n \mathcal{F}_i$-measurable; and by replacing V^p by V^f in condition (iii).

Thus, in contrast to the private value and the coarse value we now require that net-trades within a coalition can only be based on the pooled information of a coalition.

5.2 The fine value allocation

One can easily check that the existence of a fine value allocation follows immediately from the existence result for the private value allocation. However, it does not take information asymmetries into account as the Theorem below indicates.

Theorem 2. *Let $\mathcal{E}$ be a differential information economy for which the conditions of Theorem 1 hold. Let $\tilde{\mathcal{E}}$ denote the economy with complete information, i.e., where we replace each $\mathcal{F}_i$ by $\mathcal{F}$. Then for every fine value allocation x^i, $i \in I$ of $\mathcal{E}$ there* *exists a fine value allocation* $\tilde{x}^i$, $i \in I$ *for the economy* $\tilde{\mathcal{E}}$ *such that weights* λ_i *and the agents' expected utilities in both economies are the same.*

In order to prove the Theorem we need the following result.

Lemma 1. *Let* $\mathcal{G}_i$ *be the* σ*-algebra generated by* e_i *and let* $\mathcal{G} = \bigvee_{i \in S} \mathcal{G}_i$. *Then for every allocation* x^i, $i \in S$ *which is feasible for coalition* S *there exists an allocation* $\tilde{x}^i$, $i \in S$ *which is feasible for coalition* S *such that*

$$\int u_i(\tilde{x}_i(\omega))\, d\mu(\omega) \geq \int u_i(x_i(\omega))\, d\mu(\omega), \tag{5.3}$$

for every $i \in S$ *and such that each* $\tilde{x}_i$ *is* $\mathcal{G}$*-measurable.*

Proof. We prove Lemma 1 in two steps. In Step 1 we assume that $\mathcal{G}$ can be represented by a partition A_k, $k \in \mathbb{N}$ of Ω. In Step 2 we prove the general case.

Step 1. Without loss of generality we can assume that $\mu(A_k) > 0$ for every $k \in \mathbb{N}$ since we can modify x_i, $i \in S$ arbitrarily on sets of measure 0. Let $e_S = \sum_{i \in S} e_i$. Then e_S is $\mathcal{G}$-measurable and therefore constant on each set A_k. For each agent $i \in S$ and for each $k \in \mathbb{N}$ define

$$c_i^k = \frac{1}{\mu(A_k)} \int_{A_k} x_i(\omega)\, d\mu(\omega). \tag{5.4}$$

Let $\tilde{x}_i = \sum_{k \in \mathbb{N}} c_i^k \mathbf{1}_{A_k}$, where $\mathbf{1}_{A_k}$ denotes the characteristic function of the set A_k.[13] Note that Jensen's inequality[14] and (5.4) implies that

$$u_i(c_i^k) \geq \frac{1}{\mu(A_k)} \int_{A_k} u_i(x_i^k)\, d\mu(\omega). \tag{5.5}$$

Now (5.3) follows by first multiplying both sides of (5.5) with $\mu(A_k)$ and then summing both sides with respect to $k \in \mathbb{N}$.

We must now prove that x_i, $i \in S$ is feasible for the coalition S. This follows immediately by summing both sides of (5.4) with respect to $i \in S$. Thus, for any $\bar{\omega} \in A_k$ we get

$$\begin{aligned}\sum_{i \in S} \tilde{x}_i(\bar{\omega}) = \textstyle\sum_{i \in S} c_i^k \\ &= \frac{1}{\mu(A_k)} \sum_{i \in S} \int_{A_k} x_i(\omega)\, d\mu(\omega) \\ &= \frac{1}{\mu(A_k)} \int_{A_k} \sum_{i \in S} x_i(\omega)\, d\mu(\omega) \\ &= \frac{1}{\mu(A_k)} \int_{A_k} e_S(\omega)\, d\mu(\omega) = \frac{1}{\mu(A_k)} e_S(\bar{\omega})\mu(A_k) = e_S(\bar{\omega}).\end{aligned}$$

[13] That is, $\mathbf{1}_{A_k}(\omega) = 1$ if and only if $\omega \in A_k$ and $\mathbf{1}_{A_k}(\omega) = 0$, otherwise.

[14] If E denotes the expected value and if u is concave and bounded then Jensen's inequality implies $u(E(f)) \geq E(u(f))$, for any integrable function f. In our case, $E(f) = (1/\mu(A_k)) \int_{A_k} f(\omega)\, d\mu(\omega)$.

Note that we can take e_S out of the integrals since e_S is constant on A_k. Hence $\tilde{x}_i$, $i \in S$ is feasible. This concludes the proof of Step 1.

Step 2. We now use an approximation argument to prove the result for arbitrary $\mathcal{G}$. Note that by the definition of the Bochner integral there exists a sequence of simple functions[15] e_i^n such that e_i^n converges in the norm to e_i as for each $i \in S$ and for which $e_i^n \leq e_i$. Similarly, we can also find a sequence of functions x_i^n, $n \in \mathbb{N}$ such that x_i^n converges in the norm to x_i and $\sum_{i \in S} x_i^n = \sum_{i \in S} e_i^n$ for each $n \in \mathbb{N}$.

Let $\mathcal{G}_i^n$ be the σ-algebra generated by e_i^n and define $\mathcal{G}^n = \bigvee_{i \in S} \mathcal{G}_i^n$. Moreover, we can assume that $\bigvee_{n \in \mathbb{N}} \mathcal{G}^n \subset \mathcal{G}$. Step 1 implies that for each $n \in \mathbb{N}$ there exist an allocation $\tilde{x}_i^n$, $i \in S$ which is feasible for coalition S, which is $\mathcal{G}^n$-measurable and for which

$$\int u_i(\tilde{x}_i^n(\omega))\, d\mu(\omega) \geq \int u_i(x_i^n(\omega))\, d\mu(\omega) \tag{5.6}$$

holds. Since $e_i^n \leq e_i$ it follows that each $\tilde{x}_i^n$ is an element of the order interval $[0, e_S]$ (recall that $\tilde{x}_i^n$, $i \in S$ is feasible for the coalition S). Since by Cartwright's Theorem the order interval $[0, e_S]$ is weakly compact, we can assume without loss of generality that each $\tilde{x}_i^n$ converges weakly to $\tilde{x}_i$. Since the set of all Bochner integrable and $\mathcal{G}$-measurable functions is weakly closed (recall Mazur's Theorem) it follows that each $\tilde{x}_i$ is $\mathcal{G}$-measurable. Moreover, the allocation $\tilde{x}_i$, $i \in S$ is feasible for coalition S. It now remains to take the limit on both sides of (5.6). Since u is bounded and continuous, the expected utility is norm continuous by virtue of the Lebesgue Dominated Convergence Theorem. Thus, the right-hand side of (5.6) converges to the right-hand side of (5.3). Since the expected utility is weakly upper semicontinuous[16] the lim sup of the left-hand side of (5.6) is less or equal to the left-hand side of (5.3). This proves Lemma 1.

It should be clear that in Step 1 the function $\tilde{x}_i$ is the conditional expectation of x_i given $\mathcal{G}$ and that we use in essence Jensen's inequality for conditional expectations. The approximation argument in Step 2 can be used to show the existence of a conditional expectation for Bochner integrals and to extend a version of Jensen's inequality. This would provide an alternative proof of Theorem 4 (Chapter 5) of Diestel and Uhl (1977).

We are now ready to prove Theorem 2.

Proof of Theorem 2. In order to prove the Theorem we first show the following.

Claim. Let x_i, $i \in S$ be a solution to

$$\max_{x_i} \sum_{i \in S} \lambda_i \int u_i(x_i(\omega))\, d\mu(\omega) \tag{5.7}$$
$$\text{subject to}$$
$$\sum_{i \in S} x_i(\omega) = \sum_{i \in S} e_i(\omega),\ \mu\text{-a.e.}$$

[15] A function f is simple if and only if there exists a countable partition A_i, $i \in \mathbb{N}$ of Ω and $b_i \in Y$, $k \in \mathbb{N}$ such that $f = \sum_{i=1}^{\infty} b_k \mathbf{1}_{A_k}$ (recall that Y is the Banach space).

[16] See Balder and Yannelis (1993, Theorem 2.8).

Then there exist an alternative allocation $\tilde{x}_i, \in S$ which is feasible for coalition S, which is $\mathcal{G}$-measurable, and which gives the same expected utility to the agents as allocation x^i, $i \in S$.

In order to prove this claim note that Lemma 1 implies the existence of an allocation $\tilde{x}_i$, $i \in S$ which is feasible for coalition S, which is $\mathcal{G}$-measurable and for which (5.3) holds. Since the original allocation x_i, $i \in S$ is a solution to (5.7) the equality must hold in (5.3) for all agents $i \in S$. This, proves the claim.

The Theorem now follows immediately from the claim. First note that each $\mathcal{G}$-measurable function is automatically $\bigvee_{i \in S} \mathcal{F}_i$ measurable. This follows since by assumption (iv) in Section 2 each e_i is $\mathcal{F}_i$-measurable. Hence, the information σ-algebras do not matter in the optimization problems. As a consequence, the induced TU-game is independent of the information σ-algebras. Moreover, if x_i, $i \in I$ is a fine value allocation for $\mathcal{E}$ then x_i, $i \in I$ must solve (5.7) for $S = I$. Thus, we can use again the claim to replace x_i, $i \in I$ by a $\mathcal{G}$-measurable allocation $\tilde{x}_i$, $i \in I$ without affecting the agents' expected utilities. Thus, $\tilde{x}_i$, $i \in I$ is a fine value allocation. This concludes the proof of the Theorem.

Theorem 2 demonstrates that the fine value allocation does not take into account the information superiority of an agent. Intuitively, this is the case because the only information which is relevant to a coalition S are the endowment realizations of all members. However, this information is contained in $\bigvee_{i \in S} \mathcal{G}_i$ and hence in $\bigvee_{i \in S} \mathcal{F}_i$. As a consequence, the fine value allocation does not appear to be a useful concept for measuring informational asymmetries. This together with the fact that the fine value allocation need not be coalitionally incentive compatible [see Krasa and Yannelis (1994)] make this concept less attractive than the private value allocation which takes information asymmetries into account and which is also coalitionally incentive compatible.

5.3 The coarse value allocation

The TU-game (V_λ^c, I) derived from a differential information economy need not be superadditive,[17] i.e., there can exist coalitions S, T, with $S \cap T = \emptyset$ and $V_\lambda^c(S) + V_\lambda^c(T) > V_\lambda^c(S \cup T)$. On the one hand this causes problems with the existence of a coarse value allocation. On the other hand, and (most importantly) this indicates that the coarse value allocation does not measure the marginal contributions of an agent to the coalitions he/she is member off. In particular, consider an agent i whose information is relatively coarse. Assume that i joins a coalition S. Then trades within $S \cup \{i\}$ must be measurable with respect to the common knowledge information $\bigwedge_{i \in S \cup \{i\}} \mathcal{F}_i$. As a consequence, the trading opportunities of the members of S decrease, and members of S become worse off if agent i joins. Thus, $V_\lambda^c(S \cup \{i\}) - V_\lambda^c(S) < 0$, which indicate that we measure the disutility agent i imposes

[17] Allen (1991) has made a related observation for the TU case which also applies to our framework. She has also proved existence results using different argument than ours. Also, Myerson (1984) has proved existence results for the Nash bargaining solution with incomplete information. However, all of these results are of different nature and do not cover our differential information economy framework.

on members of S rather than agent i's contribution to S. This is clearly not in the spirit of the Shapley value.

In order to illustrate this point, consider the following example.

Example 1. Consider and economy with three agents denoted by I, J and K and two states $\Omega = \{a, b\}$. Each state occurs with the same probability. There is one commodity in each state. All agents have the same von Neumann-Morgenstern utility function $u(x) = \sqrt{x}$. The endowments are given by $e^I = (9, 0)$, $e^J = (0, 9)$ and $e^K = (0, 0)$. Assume that I and J have full information, and that agent K has only trivial information (i.e., $\mathcal{F}_I = \mathcal{F}_J = \{\{a\}, \{b\}\}$, and $\mathcal{F}_K = \{\{a, b\}\}$.

In this economy, the coalition $\{I, J\}$ can achieve perfect risk sharing. However, whenever agent K joins, only state independent net-trades are possible. This leads to Shapley values of agents I, J which are so high that they cannot be compensated any more by state-independent net-trades within the grand coalition. Similarly, the Shapley value of agent K is negative for all weights λ_i, $i = I, J, K$.

In order to see this, note that the marginal contribution $V_\lambda^c(S \cup \{K\}) - V_\lambda^c(S) \leq 0$ for all coalitions S. Moreover, if $S = \{I, J\}$ the strict inequality holds, since the agents can trade before K joins. However, in the grand coalition no trade is feasible.[18] Since the Shapley value of agent K is the weighted sum of these marginal contributions, we can conclude that $\mathrm{Sh}_K(V_\lambda^c) < 0$. However, since expected utility is strictly positive it therefore follows that there cannot exist an allocation x_i, $i \in I$ such that $\mathrm{Sh}_i(V_\lambda^c) = \lambda_i W_i(x_i)$, for all agents i.

In the above example, we choose agent K's endowment to be zero in order to simplify the presentation. The continuity of the Shapley value immediately implies that $\mathrm{Sh}_K(V_\lambda^c) < 0$ also holds if we perturb the endowment slightly. Hence, the non-existence result is robust.

References

Aliprantis, C.D., Burkinshaw, O.: Positive operators. New York London: Academic Press 1985

Allen, B.: Market games with asymmetric information: the value. Mimeo, University of Pennsylvania (1991)

Balder, E., Yannelis, N.C.: On the continuity of expected utility. Economic Theory **3**, 625–643 (1993)

Bewley, T.: Existence of equilibria in economies with infinitely many commodities. Journal of Economic Theory **4**, 541–540 (1972)

Cartwright, D.: The order completeness of some spaces of vector valued functions. Bulletin of the Australian Mathematical Society **11**, 57–61 (1974)

Diestel, J., Uhl, J.J.: Vector measures. Mathematical Surveys, vol. 15. American Mathematical Society, Providence, Rhode Island (1977)

Emmons, D., Scafuri, A.J.: Value allocations – an exposition. In: Aliprantis, C.D., et al. (eds.) Advances in equilibrium theory. Berlin Heidelberg New York: Springer 1985

Krasa, S., Yannelis, N.C.: The value allocation of an economy with differential information. Econometrica **62**, 881–900 (1994)

Myerson, R.B.: Cooperative games with incomplete information. International Journal of Game Theory **13**, 69–96 (1984)

Shafer, W.J.: On the existence and interpretation of value allocation. Econometrica **48**, 467–474 (1980)

[18] Because of the measurability restriction, all net-trades must be not state contingent. Thus, unless the net trades are zero, the consumption of some agents would become negative.

Shapley, L.A.: A value of n-person games. In: Kuhn, H.W., Tucker, A.W. (eds.) Contributions to the theory of games, vol. II, pp. 307–317. Princeton University Press 1953

Shapley, L.A.: Utility comparisons and the theory of games. In: Guilbaud (ed.) La decision. Paris, France: Edition du CNRS 1969

Wilson, R.: Information, efficiency, and the core of an economy. Econometrica **46**, 807–816 (1978)

Yannelis, N.C.: Existence and fairness of value allocation without convex preferences. Journal of Economic Theory **31**, 282–292 (1983)

Yannelis, N.C.: The core of an economy with differential information. Economic Theory **1**, 183–198 (1991)

The bargaining set of a large economy with differential information*

Ezra Einy[1], Diego Moreno[2], and Benyamin Shitovitz[3]

[1] Department of Economics, Ben Gurion University of the Negev, Beer Sheva, 84105 ISRAEL (e-mail: einy@bgumail.bgu.ac.il)
[2] Departamento de Economía, Universidad Carlos III de Madrid, 28903 Getafe, SPAIN (e-mail: dmoreno@eco.uc3m.es)
[3] Department of Economics, University of Haifa, Haifa, 31905 ISRAEL (e-mail: binya@econ.haifa.ac.il)

Received: February 15, 1999; revised version: August 9, 1999

Summary. We study the Mas-Colell bargaining set of an exchange economy with differential information and a continuum of traders. We established the equivalence of the private bargaining set and the set of Radner competitive equilibrium allocations. As for the weak fine bargaining set, we show that it contains the set of competitive equilibrium allocations of an associated symmetric information economy in which each trader has the "joint information" of all the traders in the original economy, but unlike the weak fine core and the set of fine value allocations, it may also contain allocations which are not competitive in the associated economy.

Keywords and Phrases: General equilibrium in large exchange economies with differential information, Weak fine bargaining set, Core, Value.

JEL Classification Numbers: D50, D82, C70.

1 Introduction

There is a large literature studying the cooperative foundations of competitive equilibria. In this literature the core is introduced as a solution concept that, without appealing to an specific institutional framework, identifies the allocations that may result from a multilateral bargaining situation in which traders discuss alternative

* We thank the associate editor and a referee for helpful comments. This work was done while Einy and Shitovitz visited the Department of Economics of the Universidad Carlos III de Madrid. Einy acknowledges the financial support of the Universidad Carlos III de Madrid. Moreno acknowledges the support of the Spanish Ministry of Education (DGES), grant PB97-0091. Shitovitz acknowledges the support of the Spanish Ministry of Education, grant SAB98-0059.

Correspondence to: D. Moreno

mutually beneficial trades. A difficulty of the core as a solution concept is that when a coalition threats to break an agreement it does not take into account how other coalitions may react to this threat; i.e., coalitional objections to a proposed allocation are not required to be robust to possible counterobjections. To address this issue, Aumann and Maschler (1964) introduce the notion of bargaining set for cooperative games with finitely many players. In the definition of the bargaining set, coalitional objections to a proposed agrement that admit counterobjections are disregarded; that is, when demanding improvements, coalitions must take account of the reactions of other coalitions–for a discussion of this issue (see Maschler, 1976, 1992). Mas-Colell (1989) introduces a new notion of bargaining set and shows that in a complete information exchange economy with a continuum of traders it coincides with the set of competitive allocations. [The equivalence of the core, the set of value allocations and the set of competitive equilibrium allocations in this context was established by Aumann (1964, 1975).]

Radner (1968, 1982) introduces a model of exchange economy with differential information in which every trader is characterized by a state dependent utility function, a random initial endowment, an information partition, and a prior belief. In this framework, traders arrange contingent contracts for trading commodities before they obtain any information about the realized state of nature. Radner (1968) extends the notion of Arrow-Debreu competitive equilibrium to this model. In the definition of competitive equilibrium (in the sense of Radner), the information of an agent places a restriction on his feasible trades (i.e., his budget set): better information allows for more contingent trades (i.e., enlarges the agent's budget set). Thus, a Radner competitive equilibrium rewards the information advantage of a trader.

In this paper we study the relation of the Mas-Colell bargaining set and the set of competitive allocations of an economy with differential information and a continuum of traders. Our aim is not only to determine whether there are equivalence results similar to those found for complete information economies, but also to explore whether the bargaining set discriminates between traders with differential information.

In the context of exchange economies with differential information and finitely many traders, Yannelis (1991) introduces the concept of private core, and proves that it is non-empty. Krasa and Yannelis (1994) introduce the notion of private value allocation, and discuss examples where the private value rewards the information advantage of a trader. In this approach, the traders of a coalition use only their private information (i.e., there is no information exchange). Einy, Moreno and Shitovitz (1998) show that in a Radner type economy with a continuum of traders the private core coincides with the set of Radner competitive equilibrium allocations, and Einy and Shitovitz (1998) establish the analogous result for the set of private value allocations. Thus, as pointed out by Koutsougeras and Yannelis (1993) and Krasa and Yannelis (1994), the private core and private value reward the information advantage of a trader.

Our findings in the present paper confirm these results: we introduce the notion of Mas-Colell private bargaining set, and we show that in a Radner type economy with a continuum of traders this set coincides with the set of Radner competitive allocations. Our proof that the Mas-Colell private bargaining set coincides with the set of Radner competitive equilibrium allocations is along the lines of the proof of Mas-Colell (1989), although the details of some of the arguments require more involved constructions because we must deal with the measurability restrictions imposed by the traders differential information, and also with the possibility that competitive prices may not be strictly positive.

An interesting question is whether the information advantage of a trader is rewarded when we account for the possibility that traders in a coalition may communicate and share some of their information. These possibilities are captured by the notion of fine core due to Wilson (1978). Einy, Moreno and Shitovitz (1998) show that the set of (weak) fine core allocations of a Radner type economy with a continuum of traders coincides with set of competitive equilibrium allocations of an associated symmetric information economy in which each trader has the "joint information" of all the traders in the original economy. Einy, Moreno and Shitovitz (1999) establish an analogous result for fine value allocations.

These results suggest that when the possibility of sharing information is introduced the information advantage of a trader is worthless. Interestingly, this is not the case when we use the weak fine bargaining set as the solution concept: we find that in a Radner type economy with a continuum of traders the weak fine bargaining set contains the competitive allocations of the associated symmetric information economy, albeit it may also contain other allocations where the traders with an information advantage are more favorably treated. Thus, in contrast with the weak fine core and the set of weak fine value allocations, the weak fine bargaining set may reward the information advantage of a trader.

2 The model

We consider a Radner-type exchange economy $\mathcal{E}$ with differential information (e.g., Radner (1968, 1982)).

The space of traders is a measure space (T, Σ, μ), where T is a set (the set of *traders*), Σ is a σ-field of subsets of T (the set of *coalitions*), and μ is a non-atomic measure on Σ. The commodity space is $\Re^l_+$. The space of states of nature is a finite set Ω. The economy extends over two time periods, $\tau = 0, 1$. Consumption takes place at $\tau = 1$. At $\tau = 0$ there is uncertainty over the state of nature; in this period traders arrange contracts that may be contingent on the realized state of nature at $\tau = 1$. At $\tau = 1$ traders do not necessarily know which state of nature $\omega \in \Omega$ actually occurred, although they know their own endowments, and may also have some additional information about the state of nature. We do not assume, however, that traders know their own utility function.

The information of a trader $t \in T$ is described by a partition Π_t of Ω. We denote by $\mathcal{F}_t$ the field generated by Π_t. If ω_0 is the true state of nature, at $\tau = 1$ trader t observes the member of Π_t which contains ω_0. Every trader $t \in T$ has a probability distribution q_t on Ω which represents his *prior beliefs*. The preferences of a trader $t \in T$ are represented by a *state dependent utility function,* $u_t : \Omega \times \Re^l_+ \to \Re$ such that for every $(t, x) \in \Omega \times \Re^l_+$, the mapping $(t, x) \to u_t(\omega, x)$ is $\Sigma \times \mathcal{B}$ measurable, where ω is a fixed member of Ω, and $\mathcal{B}$ is the σ-field of Borel subsets of $\Re^l_+$. If x is a random bundle (i.e., a function from Ω to $\Re^l_+$) we denote by $h_t(x)$ the *expected utility* of trader $t \in T$ from x. That is

$$h_t(x) = \sum_{\omega \in \Omega} q_t(\omega) u_t(\omega, x(\omega)).$$

An *assignment* is a function $\mathbf{x} : T \times \Omega \to \Re^l_+$ such that for every $\omega \in \Omega$ the function $\mathbf{x}(\cdot, \omega)$ is μ-integrable on T. There is a fixed *initial assignment* $\mathbf{e}$; $\mathbf{e}(t, \omega)$ represents the *initial endowment* of trader $t \in T$ in the state of nature $\omega \in \Omega$. We assume that $\mathbf{e}(t, \omega)$ is in $\Re^l_{++}$ for every $(t, \omega) \in T \times \Omega$, and for every $t \in T$ the function $\mathbf{e}(t, \cdot)$ is $\mathcal{F}_t$-measurable.

In the rest of the paper, an economy $\mathcal{E}$ is an atomless economy with differential information as described above. Also we use the following notation: For two vectors $x = (x_1, \ldots, x_l)$ and $y = (y_1, \ldots, y_l)$ in $\Re^l$ we write $x \geq y$ when $x_k \geq y_k$ for all $1 \leq k \leq l$, $x > y$ when $x \geq y$ and $x \neq y$, and $x \gg y$ when $x_k > y_k$ for all $1 \leq k \leq l$.

Let $\mathcal{E}$ be an economy. A *private allocation* is an assignment $\mathbf{x}$ such that

(2.1) for almost all $t \in T$ the function $\mathbf{x}(t, \cdot)$ is $\mathcal{F}_t$-measurable, and

(2.2) $\int_T \mathbf{x}(t, \omega) d\mu \leq \int_T \mathbf{e}(t, \omega) d\mu$ for all $\omega \in \Omega$.

A *price system* is a non-zero function $p : \Omega \to \Re^l_+$. Let $t \in T$. Write M_t for the set of all $\mathcal{F}_t$-measurable functions from Ω to $\Re^l_+$. For a price system p, define the *budget set* of t by

$$B(p, t) = \left\{ x \mid x \in M_t \text{ and } \sum_{\omega \in \Omega} p(\omega) \cdot x(\omega) \leq \sum_{\omega \in \Omega} p(\omega) \cdot \mathbf{e}(t, \omega) \right\}.$$

A *competitive equilibrium* (in the sense of Radner) is a pair $(p, \mathbf{x})$ where p is a price system and $\mathbf{x}$ is private allocation such that

(2.3) for almost all $t \in T$ the function $\mathbf{x}(t, \cdot)$ maximizes h_t on $B(p, t)$, and

(2.4) $\sum_{\omega \in \Omega} p(\omega) \cdot \int_T \mathbf{x}(t, \omega) d\mu = \sum_{\omega \in \Omega} p(\omega) \cdot \int_T \mathbf{e}(t, \omega) d\mu$.

A *competitive allocation* is a private allocation $\mathbf{x}$ for which there exists a price system p such that $(p, \mathbf{x})$ is a competitive equilibrium.

Note that since Ω is a finite set there is a finite family $\{\Pi_i\}_{i=1}^n$ of partitions of Ω such that for all $t \in T$ there is $1 \leq i \leq n$ with $\Pi_t = \Pi_i$. We assume that for all $1 \leq i \leq n$, the set $T_i = \{t \in T \mid \Pi_t = \Pi_i\}$ is measurable, and $\mu(T_i) > 0$. For all $1 \leq i \leq n$ we denote by $\mathcal{F}_i$ the field generated by Π_i.

Throughout the paper we assume that for all $t \in T$ and $\omega \in \Omega$ the function $u_t(\omega, \cdot)$ is strictly increasing and continuous on $\Re^l_+$. (A function $u : \Re^l_+ \to \Re$ is *strictly increasing* if for all $x, y \in \Re^l_+$, $x > y$ implies $u(x) > u(y)$.)

Einy, Moreno and Shitovitz (1998) have shown that if the utility functions of the traders are continuous and strictly increasing, and if every commodity is present in the market (i.e., $\int_T \mathbf{e}(t, \omega) d\mu \gg 0$ for all $\omega \in \Omega$), then a competitive equilibrium (in the sense of Radner) exists when the economy is irreducible (see Theorem A in Einy, Moreno and Shitovitz, 1998). Since in our model the initial endowments of the traders are in $\Re^l_{++}$, the economies we consider here are irreducible (see Proposition 3.1 in Einy, Moreno and Shitovitz, 1998), and therefore always have a competitive equilibrium.

3 The private bargaining set

In this section we introduce the notion of (Mas-Colell) private bargaining set, and show that it coincides with the set of (Radner) competitive allocations. We begin by extending to our model the definition of private core due to Yannelis (1991).

Let $\mathcal{E}$ be an economy, and let $\mathbf{x}$ be a private allocation. A *private objection to* $\mathbf{x}$ is a pair $(S, \mathbf{y})$ such that

(3.1) $\mu(S) > 0$,

(3.2) $\mathbf{y}(t, \cdot)$ is $\mathcal{F}_t$-measurable for almost all $t \in S$,

(3.3) $\int_S \mathbf{y}(t, \omega) d\mu \le \int_S \mathbf{e}(t, \omega)$ for all $\omega \in \Omega$,

(3.4) $h_t(\mathbf{y}(t, \cdot)) \ge h_t(\mathbf{x}(t, \cdot))$ for almost all $t \in S$, and

(3.5) $\mu(\{t \in S \mid h_t(\mathbf{y}(t, \cdot)) > h_t(\mathbf{x}(t, \cdot))\}) > 0$.

An assignment $\mathbf{x}$ is a *private core allocation of* $\mathcal{E}$ if it has no private objection. The *private core of* $\mathcal{E}$ is the set of all private core allocations of $\mathcal{E}$.

In defining the core, usually the inequalities (3.4) are strict, and (3.5) is omitted. Since in our framework the utility functions of the traders are continuous and strictly increasing in every state of nature, these alternative definitions of the core are equivalent.

Let $\mathcal{E}$ be an economy, let $\mathbf{x}$ be a private allocation and let $(S, \mathbf{y})$ be a private objection to $\mathbf{x}$. A *private counterobjection* to $(S, \mathbf{y})$ is a pair $(Q, \mathbf{z})$ such that

(3.6) $\mu(Q) > 0$,

(3.7) $\mathbf{z}(t, \cdot)$ is $\mathcal{F}_t$-measurable for almost all $t \in Q$,

(3.8) $\int_Q \mathbf{z}(t, \omega) d\mu \le \int_Q \mathbf{e}(t, \omega)$ for all $\omega \in \Omega$,

(3.9) $h_t(\mathbf{z}(t, \cdot)) > h_t(\mathbf{y}(t, \cdot))$ for almost all $t \in Q \cap S$, and

(3.10) $h_t(\mathbf{z}(t, \cdot)) > h_t(\mathbf{x}(t, \cdot))$ for almost all $t \in Q \backslash S$.

A private objection to $\mathbf{x}$, $(S, \mathbf{y})$, is *justified* if it has no private counterobjection. The *(Mas-Colell) private bargaining set* is the set of private allocations which have no justified private objection. Note that the private core of an economy $\mathcal{E}$ is a subset of the private bargaining set of $\mathcal{E}$.

Theorem A. *The private bargaining set of an economy $\mathcal{E}$ coincides with the set of Radner competitive allocations of $\mathcal{E}$.*

Einy, Moreno and Shitovitz (1998) have established that the set of Radner competitive equilibrium allocations of an economy $\mathcal{E}$ as defined here coincides with the private core of $\mathcal{E}$. Since the private core is a subset of the private bargaining set, in order to prove Theorem A it suffices to show that every private bargaining set allocation of $\mathcal{E}$ is a competitive allocation of $\mathcal{E}$. Our proof of this result is along the lines of the proof of Theorem 1 in Mas-Colell (1989), although the details of some of the arguments require more involved constructions because we must deal with the measurability restrictions imposed by the traders differential information, and also with the possibility that competitive prices may not be strictly positive. (In spite of the fact that traders utility functions are strictly increasing, in an economy with differential information we cannot guarantee that competitive prices are strictly positive.) In establishing this result, the notion of competitive objection will be useful.

A private objection $(S, \mathbf{y})$ to $\mathbf{x}$ is a *competitive objection* if there is a price system p such that for almost all $t \in T$

(3.11) if $t \in S$ and $z \in M_t$ satisfies $h_t(z) \geq h_t(\mathbf{y}(t,\cdot))$, then $\sum_{\omega\in\Omega} p(\omega)\cdot z(\omega) \geq \sum_{\omega\in\Omega} p(\omega)\cdot \mathbf{e}(t,\omega)$, and

(3.12) if $t \in T\backslash S$ and $z \in M_t$ satisfies $h_t(z) \geq h_t(\mathbf{x}(t,\cdot))$, then $\sum_{\omega\in\Omega} p(\omega)\cdot z(\omega) \geq \sum_{\omega\in\Omega} p(\omega)\cdot \mathbf{e}(t,\omega)$.

Theorem A is a consequence of the following two lemmata.

Lemma 3.1. *Every competitive objection $(S, \mathbf{y})$ to a private allocation $\mathbf{x}$ is justified.*

Proof. Let $(S, \mathbf{y})$ be competitive objection to an allocation $\mathbf{x}$, and let p be the price system associated with $(S, \mathbf{y})$. Assume contrary to our claim that there is a private counterobjection $(Q, \mathbf{z})$ to $(S, \mathbf{y})$. Then $h_t(\mathbf{z}(t,\cdot)) > h_t(\mathbf{y}(t,\cdot))$ for almost all $t \in Q\cap S$, and $h_t(\mathbf{z}(t,\cdot)) > h_t(\mathbf{x}(t,\cdot))$ for almost all $t \in Q\backslash S$. Since for all $t \in T$ and all $\omega \in \Omega$, $u_t(\omega,\cdot)$ is strictly increasing and $\mathbf{e}(t,\omega) \gg 0$, and since $(S, \mathbf{y})$ is a competitive objection, for almost all $t \in Q$ we have

$$\sum_{\omega\in\Omega} p(\omega)\cdot \mathbf{z}(t,\omega) > \sum_{\omega\in\Omega} p(\omega)\cdot \mathbf{e}(t,\omega).$$

This contradicts that for all $\omega \in \Omega$, $\int_Q \mathbf{z}(t,\omega)d\mu \leq \int_Q \mathbf{e}(t,\omega)d\mu$. □

Lemma 3.2. *If $\mathbf{x}$ is not a competitive allocation, then there is a competitive objection to $\mathbf{x}$.*

Proof. Throughout the proof we assume without loss of generality that $\mu(T) = 1$. Assume that $\mathbf{x}$ is not a competitive allocation. We construct a competitive objection to $\mathbf{x}$. Define

$$P = \bigcap_{i=1}^{n} \left\{ p \in (\Re^l_+)^\Omega \mid \sum_{\omega\in\Omega}\sum_{j=1}^{l} p_j(\omega) = 1 \text{ and } \sum_{\omega\in A} p(\omega) \gg 0, \text{ for all } A \in \Pi_i \right\}.$$

Then P is a non-empty convex subset of $(\Re^l_+)^\Omega$. Now for $p \in P$ and $t \in T$, the budget set $B(p,t)$ is a compact subset of M_t. Therefore the function h_t attains a maximum on $B(p,t)$. For all $p \in P$ and all $t \in T$ let

$$D(p,t) = \{x \in M_t \mid x \text{ maximizes } h_t \text{ on } B(p,t)\},$$

and

$$F(p,t)=\begin{cases} D(p,t) & \text{if } h_t(D(p,t))>h_t(\mathbf{x}(t,\cdot))\\ D(p,t)\cup\{\mathbf{e}(t,\cdot)\} & \text{if } h_t(D(p,t))=h_t(\mathbf{x}(t,\cdot))\\ \{\mathbf{e}(t,\cdot)\} & \text{if } h_t(D(p,t))<h_t(\mathbf{x}(t,\cdot)).\end{cases}$$

Let

$$\alpha=1+\sum_{\omega\in\Omega}\sum_{j=1}^{l}\int_T e_j(t,\omega)d\mu,$$

and let

$$K=\left\{x\in(\Re^l_+)^\Omega \mid \sum_{\omega\in\Omega}\sum_{j=1}^{l}x_j(\omega)\le\alpha\right\},$$

and

$$\hat{K}=\left\{x\in K \mid \sum_{\omega\in\Omega}\sum_{j=1}^{l}x_j(\omega)=\alpha\right\}.$$

Note that K is a non-empty compact convex subset of $(\Re^l_+)^\Omega$. Write $\bar{P}$ for the closure of P, and define a correspondence $\phi:\bar{P}\times T\twoheadrightarrow 2^K$ by

$$\phi(p,t)=\begin{cases} F(p,t)\cap K & \text{if } p\in P \text{ and } D(p,t)\cap K\neq\emptyset\\ \hat{K}\cap\{\lambda d\mid d\in F(p,t),\ \lambda\ge 0\} & \text{if } p\in P \text{ and } D(p,t)\cap K=\emptyset\\ B(p,t)\cap\hat{K} & \text{if } p\in\bar{P}\backslash P.\end{cases}$$

For every $p\in\bar{P}$ define

$$\psi(p)=\int_T\phi(p,t)d\mu-\int_T\mathbf{e}(t,\cdot)d\mu.$$

Then for every $p\in\bar{P}$, $\psi(p)$ is a non-empty convex subset of the compact convex set K. The proof that ψ is also upper semicontinuous on $\bar{P}$ is standard. From the definition of ψ it is clear that for all $p\in\bar{P}$ we have $p\cdot\psi(p)\le 0$. Therefore by (1) in Section 5.6 of Debreu (1959), there exists $p^*\in\bar{P}$ and $z^*\in\psi(p^*)$ such that $z^*\le 0$. We show that $p^*\notin\bar{P}\backslash P$. Suppose $p^*\in\bar{P}\backslash P$; then $z^*\in(\int_T(B(p^*,t)\cap\hat{K})d\mu-\int_T\mathbf{e}(t,\cdot)d\mu)$. Therefore

$$\sum_{\omega\in\Omega}\sum_{j=1}^{l}z^*_j(\omega)=\alpha-\sum_{\omega\in\Omega}\sum_{j=1}^{l}\int_T e_j(t,\omega)d\mu=1,$$

which contradicts $z^*\le 0$. Thus $p^*\notin\bar{P}\backslash P$. As $z^*\le 0$, we have

$$z^*\in\int_T(F(p^*,t)\cap K)d\mu-\int_T\mathbf{e}(t,\cdot)d\mu.$$

Hence there exists an integrable function f on T such that $f(t) \in F(p^*, t)$ for all $t \in T$ and $z^* = \int_T f(t)d\mu - \int_T \mathbf{e}(t, \cdot)d\mu$. Write

$$S = \{t \in T \mid f(t) \in D(p^*, t)\},$$

and

$$C(p^*) = \{t \in T \mid h_t(D(p^*, t)) > h_t(\mathbf{x}(t, \cdot))\}.$$

Since $\mathbf{x}$ is not a competitive allocation, we have $\mu(C(p^*)) > 0$. As $C(p^*) \subset S$, $\mu(S) > 0$.

Now for all $(t, \omega) \in T \times \Omega$ let

$$\mathbf{y}(t, \omega) = (f(t))(\omega).$$

We show that $(S, \mathbf{y})$ is a competitive objection to $\mathbf{x}$. As noted above, $\mu(S) > 0$. Since $z^* \leq 0$ and $f(t) = \mathbf{e}(t, \cdot)$ for $t \in T \backslash S$, we have

$$\int_S \mathbf{y}(t, \omega)d\mu \leq \int_S \mathbf{e}(t, \omega)d\mu,$$

for all $\omega \in \Omega$. By the definition of $\mathbf{y}$ we have $h_t(\mathbf{y}(t, \cdot)) \geq h_t(\mathbf{x}(t, \cdot))$ for all $t \in S$, and $h_t(\mathbf{y}(t, \cdot)) > h_t(\mathbf{x}(t, \cdot))$ for all $t \in C(p^*)$. If $t \in S$ and $z \in M_t$ satisfies $h_t(z) \geq h_t(\mathbf{y}(t, \cdot))$, then $h_t(D(p^*, t)) \leq h_t(z)$. Therefore

$$\sum_{\omega \in \Omega} p^*(\omega) \cdot z(\omega) \geq \sum_{\omega \in \Omega} p^*(\omega) \cdot \mathbf{e}(t, \omega).$$

Let $t \in T \backslash S$. Then $h_t(D(p^*, t)) \leq h_t(\mathbf{x}(t, \cdot))$. Therefore if $z \in M_t$ satisfies $h_t(z) \geq h_t(\mathbf{x}(t, \cdot))$, then $h_t(z) \geq h_t(D(p^*, t))$, and thus

$$\sum_{\omega \in \Omega} p^*(\omega) \cdot z(\omega) \geq \sum_{\omega \in \Omega} p^*(\omega) \cdot \mathbf{e}(t, \omega).$$

This completes the proof that $(S, \mathbf{y})$ is a competitive objection to $\mathbf{x}$. □

4 The weak fine bargaining set

In this section we introduce the notion of weak fine bargaining set and study its relation with the set of competitive allocations.

Let $\mathcal{E}$ be an economy, and let $S \in \Sigma$. Define

$$I(S) = \{i \mid 1 \leq i \leq n \text{ and } \mu(S \cap T_i) > 0\}.$$

where n and T_i are defined in Section 2. A *weak fine allocation* is an assignment $\mathbf{x}$ such that

(4.1) For all $t \in T$, $\mathbf{x}(t, \cdot)$ is $\bigvee_{i=1}^n \mathcal{F}_i$-measurable, and

(4.2) $\int_T \mathbf{x}(t, \omega)d\mu \leq \int_S \mathbf{e}(t, \omega)$ for all $\omega \in \Omega$.

Let $\mathbf{x}$ be a weak fine allocation. A *weak fine objection to* $\mathbf{x}$ is a pair $(S, \mathbf{y})$ such that

(4.3) $\mu(S) > 0$,

(4.4) $\mathbf{y}(t, \cdot)$ is $\bigvee_{i \in I(S)} \mathcal{F}_i$-measurable for all $t \in S$,

(4.5) $\int_S \mathbf{y}(t, \omega) d\mu \leq \int_S \mathbf{e}(t, \omega)$ for all $\omega \in \Omega$,

(4.6) $h_t(\mathbf{y}(t, \cdot)) \geq h_t(\mathbf{x}(t, \cdot))$ for almost all $t \in S$, and

(4.7) $\mu(\{t \in S \mid h_t(\mathbf{y}(t, \cdot)) > h_t(\mathbf{x}(t, \cdot))\}) > 0$.

A *weak fine core allocation of* $\mathcal{E}$ is a weak fine allocation $\mathbf{x}$ which has no weak fine objection. The *weak fine core* of $\mathcal{E}$ is the set of all weak fine core allocations of $\mathcal{E}$.

The weak fine core was introduced in Yannelis (1991), Allen (1991), and Koutsougeras and Yannelis (1993). In order to define the weak fine bargaining set we need to introduce the definition of weak fine counterobjection.

Let $\mathcal{E}$ be an economy, let $\mathbf{x}$ be a weak fine allocation, and let $(S, \mathbf{y})$ be a weak fine objection to $\mathbf{x}$. A *weak fine counterobjection* to $(S, \mathbf{y})$ is a pair $(Q, \mathbf{z})$ such that

(4.8) $\mu(Q) > 0$,

(4.9) $\mathbf{z}(t, \cdot)$ is $\bigvee_{i \in I(Q)} \mathcal{F}_i$-measurable for almost all $t \in Q$,

(4.10) $\int_Q \mathbf{z}(t, \omega) d\mu \leq \int_Q \mathbf{e}(t, \omega)$ for all $\omega \in \Omega$,

(4.11) $h_t(\mathbf{z}(t, \cdot)) > h_t(\mathbf{y}(t, \cdot))$ for almost all $t \in Q \cap S$, and

(4.12) $h_t(\mathbf{z}(t, \cdot)) > h_t(\mathbf{x}(t, \cdot))$ for almost all $t \in Q \backslash S$.

A weak fine objection $(S, \mathbf{y})$ to a weak fine allocation $\mathbf{x}$ is *justified* if it has not weak fine counterobjection. The *(Mas-Colell) weak fine bargaining set* is the set of weak fine allocations which have no justified weak fine objection.

Let $\mathcal{E}$ be an economy. Denote by $\mathcal{E}^*$ the economy obtained from $\mathcal{E}$ by giving to each trader in $\mathcal{E}$ the joint information of all the traders in $\mathcal{E}$, i.e., for all $t \in T$, $\mathcal{F}_t^* = \bigvee_{i=1}^n \mathcal{F}_i$, and leaving the rest of his characteristics unchanged. Note that in $\mathcal{E}^*$ all traders have the same information (i.e., $\mathcal{E}^*$ is an economy with symmetric information). For each $t \in T$, we denote by M_t^* the set of all $\mathcal{F}_t^*$-measurable functions from Ω to $\Re_+^l$. Let $p : \Omega \to \Re_+^l$ be a price system. In the economy $\mathcal{E}^*$ the budget set of $t \in T$ with respect to p is

$$B^*(p, t) = \left\{ x \mid x \in M_t^*, \ \sum_{\omega \in G} p(\omega) \cdot x(\omega) \leq \sum_{\omega \in G} p(\omega) \cdot \mathbf{e}(t, \omega) \right\}.$$

A competitive equilibrium of $\mathcal{E}^*$ (in the sense of Radner) is now defined as in Section 2.

Proposition 4.1. *Every competitive allocation of* $\mathcal{E}^*$ *is in the weak fine bargaining set of* $\mathcal{E}$.

Proof. It is easy to see that an economy $\mathcal{E}$ as defined in Section 2 satisfies the assumptions of Theorem C in Einy, Moreno and Shitovitz (1998), which establishes that the set of competitive allocations of $\mathcal{E}^*$ coincides with the weak fine core of $\mathcal{E}$. Since the weak fine core is a subset of the weak fine bargaining set, Proposition 4.1 readily follows from this result. □

As the following example shows, the analog of Theorem C in Einy, Moreno and Shitovitz (1998) for the weak fine bargaining set does not hold: there are allocations in the weak fine bargaining set that are not competitive allocations of $\mathcal{E}^*$. For the analysis of the example we need some notation and a lemma which is interesting on its own.

If $\mathcal{E}$ is an economy and S is a coalition with $\mu(S) > 0$, we denote by $\mathcal{E}_S$ the restriction of $\mathcal{E}$ to S; that is, $\mathcal{E}_S$ is an economy for which the space of traders is (S, Σ_S, μ_S), where $\Sigma_S = \{Q \mid Q \in \Sigma, Q \subset S\}$, and μ_S is the restriction of μ to Σ_S.

Lemma 4.2. *Let $\mathcal{E}$ be an economy. Assume that $(S, \mathbf{y})$ is a justified weak fine objection to a weak fine allocation $\mathbf{x}$ in $\mathcal{E}$. Then the restriction of $\mathbf{y}$ to $S \times \Omega$ is a competitive allocation of $\mathcal{E}_S^*$.*

Proof. Assume by way of contradiction that the restriction $\hat{\mathbf{y}}$ of $\mathbf{y}$ to $S \times \Omega$ is not competitive in $\mathcal{E}_S^*$. Then by Theorem C of Einy, Moreno and Shitovitz (1998), $\hat{\mathbf{y}}$ is not in the weak fine core of $\mathcal{E}_S$. Therefore $\hat{\mathbf{y}}$ has a weak fine objection $(Q, \hat{\mathbf{z}})$ in $\mathcal{E}_S$. Let $\mathbf{z}$ be an extension of $\hat{\mathbf{z}}$ to an assignment in $\mathcal{E}$. As $Q \subset S$, $(Q, \mathbf{z})$ is a weak fine counterobjection to $(S, \mathbf{y})$ in $\mathcal{E}$. But this contradicts our assumption that $(S, \mathbf{y})$ is a weak fine justified objection to $\mathbf{x}$ in $\mathcal{E}$. □

Example 4.3. Consider an economy $\mathcal{E}$ in which the commodity space is $\Re_{++}$, and the set of traders is $([0,3], \mathcal{B}, \mu)$, where $\mathcal{B}$ is the σ-field of Borel subsets of $[0,3]$ and μ is the Lebesgue measure. The space of states of nature is $\Omega = \{\omega_1, \omega_2\}$. All traders have the same utility function, given for $(\omega, x) \in \Omega \times \Re_{++}$ by

$$u(\omega, x) = \ln x.$$

The initial assignment is $\mathbf{e}(t, \omega) = 2$, for all $(t, \omega) \in T \times \Omega$. Let $T_1 = [0,1]$, $T_2 = (1,2]$, and $T_3 = (2,3]$. The information partition of a trader $t \in T_1 \cup T_2$ is $\Pi_1 = \Pi_2 = \{\Omega\}$, and that of the traders t in T_3 is $\Pi_3 = \{\{\omega_1\}, \{\omega_2\}\}$. The priors of the traders in T_1, T_2, and T_3 are, respectively, $q_1 = (\frac{1}{4}, \frac{3}{4})$, $q_2 = (\frac{3}{4}, \frac{1}{4})$, and $q_3 = (\frac{1}{2}, \frac{1}{2})$. We construct a weak fine bargaining set allocation of $\mathcal{E}$ which is not competitive in $\mathcal{E}^*$.

Define an assignment $\mathbf{x} : T \times \Omega \to \Re_{++}$ by

$$\mathbf{x}(t, \omega_1) = \begin{cases} 1 & t \in T_1 \\ 2.55 & t \in T_2 \\ 2.45 & t \in T_3, \end{cases}$$

and

$$\mathbf{x}(t, \omega_2) = \begin{cases} 2.55 & t \in T_1 \\ 1 & t \in T_2 \\ 2.45 & t \in T_3. \end{cases}$$

Then $\mathbf{x}$ is a weak fine allocation in $\mathcal{E}$. We show that $\mathbf{x}$ is in the weak fine bargaining set of $\mathcal{E}$, but it is not competitive in $\mathcal{E}^*$. Assume, by way of contradiction, that $(S, \mathbf{y})$

is a justified weak fine objection to $\mathbf{x}$ in $\mathcal{E}$. Then by Lemma 4.2 the restriction $\hat{\mathbf{y}}$ of $\mathbf{y}$ to $S \times \Omega$ is a competitive allocation in $\mathcal{E}_S^*$. Now if $\mu(S \cap T_3) = 0$, then we must have that $\hat{\mathbf{y}}(t, \omega) = \mathbf{e}(t, \omega)$ for all $(t, \omega) \in S \times \Omega$. As $h_t(\mathbf{e}(t, \cdot)) < h_t(\mathbf{x}(t, \cdot))$ for all $t \in T_1 \cup T_2$, this leads to a contradiction. Assume that $\mu(S \cap T_3) > 0$. Let p be a price system such that $(p, \hat{\mathbf{y}})$ is a competitive equilibrium of $\mathcal{E}_S^*$. Then $p(\omega_1) > 0$ and $p(\omega_2) > 0$. Without loss of generality assume that $p(\omega_2) = 1$, and denote $r = p(\omega_1)$. Then the first order conditions for utility maximization imply that for almost all $t \in S$,

$$\hat{\mathbf{y}}(t, \omega_1) = \begin{cases} \frac{1}{2}(1 + \frac{1}{r}) & t \in T_1 \\ \frac{3}{2}(1 + \frac{1}{r}) & t \in T_2 \\ 1 + \frac{1}{r} & t \in T_3, \end{cases}$$

and

$$\hat{\mathbf{y}}(t, \omega_2) = \begin{cases} \frac{3}{2}(1 + r) & t \in T_1 \\ \frac{1}{2}(1 + r) & t \in T_2 \\ 1 + r & t \in T_3. \end{cases}$$

Since $\mu(S) > 0$ and $(p, \hat{\mathbf{y}})$ is a competitive equilibrium of $\mathcal{E}_S^*$, we have

$$\left(1 + \frac{1}{r}\right)(1 + r) \leq \frac{16}{3} < (2.45)^2 .$$

Therefore for almost all $t \in S \cap T_3$ we have

$$h_t(\mathbf{y}(t, \cdot)) = \frac{1}{2} \ln \left[\left(1 + \frac{1}{r}\right)(1 + r)\right] < \frac{1}{2} \ln (2.45)^2 = h_t(\mathbf{x}(t, \cdot)).$$

As $\mu(S \cap T_3) > 0$, this contradicts the assumption that $(S, \mathbf{y})$ is a weak fine objection to $\mathbf{x}$.

The above argument shows that if, in particular, $\mathbf{z}$ is a competitive allocation of $\mathcal{E}^*$, then for almost all $t \in T_3$

$$h_t(\mathbf{z}(t, \cdot)) < h_t(\mathbf{x}(t, \cdot)).$$

Therefore $\mathbf{x}$ is not a competitive allocation of $\mathcal{E}^*$. Note that the last inequality implies that the informed traders (i.e., the traders in T_3) are better off in $\mathbf{x}$ than in any competitive allocation of $\mathcal{E}^*$.

References

Al Allen, B.: Market games with asymmetric information and non-transferable utility: representations results, and the core. Mimeo, University of Pennsylvania (1991)

Au1 Aumann, R.J.: Markets with a continuum of traders. Econometrica **32**, 39-50 (1964)

Au2 Aumann, R.J.: Values of markets with a continuum of traders. Econometrica **43**, 611–646 (1975)

AM Aumann, R.J. and Maschler, M.: The bargaining set for cooperative games. In: Dresher, M., Shapley, L.S., Tucker, A.W. (eds.) Advances in game theory, Annals of Mathematical Studies, no. **52**, pp. 443-476. Princeton, NJ: Princeton University Press 1964

D Debreu, G.: Theory of value, New Haven: Yale University Press 1959

EMS1 Einy, E., Moreno, D., Shitovitz, B.: Existence and equivalence of competitive and core allocations in large exchange economies with differential information. University Carlos III de Madrid, WP 98-71 (1998)

EMS2 Einy, E., Moreno, D., Shitovitz, B.: Fine value allocations in large exchange economies with differential information. Mimeo (1999)

ES Einy, E., and Shitovitz, B.: Equivalence of value and competitive allocations in large exchange economies with differential information. Mimeo (1998)

KY Koutsougeras, L.C., and Yannelis, N.C.: Incentive compatibility and information superiority of the core of an economy with differential information. Economic Theory **3**, 195-216 (1993)

KrY Krasa, S., and Yannelis, N.C.: The value allocation of an economy with differential information. Econometrica **62**, 881-900 (1994)

M1 Maschler, M.: An advantage of the bargaining set over the core. Journal of Economic Theory **13**, 184-194 (1976)

M2 Maschler, M.: The bargaining set, kernel, and nucleolus. In: Aumann, R.J., Hart, S. (eds.) Handbook of game theory, vol. 1, pp. 591-667. Amsterdam: Elsevier 1992

MA Mas-Colell, A.: An equivalence theorem for a bargaining set. Journal of Mathematical Economics **18**, 129-139 (1989)

R1 Radner, R.: Competitive equilibrium under uncertainty. Econometrica **36**, 31-58 (1968)

R2 Radner, R.: Equilibrium under uncertainty. In: Arrow, K.J., Intriligator, M.D. (eds.) Handbook of Mathematical Economics, vol. II. Amsterdam: North Holland 1982

W Wilson, R.: Information, efficiency, and the core of an economy. Econometrica **40**, 807-816 (1978)

Y Yannelis, N.C.: The core of an economy with differential information. Economic Theory **1**, 183-198 (1991)

Coalition structure values in differential information economies: Is unity a strength?

Stefan Krasa[1], Akram Temimi[2], and Nicholas C. Yannelis[1]

[1] Department of Economics, University of Illinois at Urbana-Champaign, 1206 South Sixth Street, Champaign IL 61820, USA
[2] Department of Economics, Finance and Legal Studies, University of Alabama, Tuscaloosa, AL 35487.
Temimi acknowledges financial support from the Culverhouse College of Commerce and Business Administration at the University of Alabama, USA

Summary. The coalition structure (CS) value, introduced by Owen [9] and Hart and Kurz [5], generalizes the Shapley value to social situations where coalitions form for the purpose of bargaining. This paper introduces the CS value to economies with differential information. We show that the private CS values exists and is Bayesian incentive compatible. Moreover, we construct examples that go against the intuitive viewpoint that "unity is strength." In particular, we consider a three person economy in which two agents bargain as a unit against the third agent. We show that bargaining as a unit is advantageous if and only if information is complete. This result sheds new light on bargaining under differential information.

1 Introduction

Recent work on differential information economies has indicated that cooperative solution concepts, such as the core, the Shapley value, and the bargaining set provide successful alternatives to the rational expectations equilibrium (see Allen and Yannelis [1] and the references therein). In particular, as first shown in Krasa and Yannelis [7], the Shapley value is sensitive to information asymmetries and rewards agents with superior information, features that are not shared by the traditional rational expectations equilibrium.

In this paper, we pursue this line of research further, going beyond Krasa and Yannelis [7,8], by introducing differential information into the coalition structure (CS) value concept of Owen [9] and Hart and Kurz [5]. One of the main properties of the CS value is that if agents are allowed to form coalitions to bargain as a unit, they may do so to strengthen their situation, that is, to increase their payoffs. The intent of the coalition structure value is to take into account situations where groups of players organize themselves for the purpose of bargaining with the rest of the players. Examples include political parties, unions, and cartels. In particular, Hart and Kurz [5] construct a three agent economy, in which the first two agents obtain higher payoffs whenever they bargain as a unit against the third agent, compared to the standard Shapley value that allows arbitrary coalition formation.

Given the appealing features of the CS value, it is of interest to know how this concept behaves in differential information economies. To this end we introduce differential information into the CS value concept. In particular, we consider two new equilibrium concepts: The private CS value and the Bayesian incentive compatible (BIC) CS value. In general, the two concepts differ, but we show that (i) private CS values are always incentive compatible and that (ii) the set of attainable utilities of the transferable utility (TU) game derived from an economy with differential information is always a subset of the set of BIC attainable utilities. It turns out that the private CS value always exists, but the BIC value need not exist because the set of all BIC allocations need not be a convex set.

What we found most surprising, however, is that the intuitive statement that "unity is strength" ceases to be true for the CS value once differential information is introduced. In particular, we construct a three person economy in which two agents bargain as a unit against the third agent. We show that bargaining as a unit is advantageous if and only if information is complete.

The paper proceeds as follows. In Section 2 we introduce the economy with differential information. In Section 3 we show two alternative ways how a TU game can be derived from an economy with differential information, and a comparison is provided. In Section 4 we introduce the main concepts of this paper, the coalition structure values of a differential information economy. Section 5 shows that coalitional bargaining may not be advantageous if informational asymmetries matter. All proofs are in the Appendix.

2 The economy with differential information

We consider an exchange economy that extends over two time periods, $t = 0, 1$, where consumption takes place in $t = 1$. At $t = 0$ there is uncertainty over the state of nature described by a probability space $(\Omega, \mathcal{F}, \mu)$. Agents are indexed by $i \in I = \{1, \ldots, n\}$.

In each state ω there are ℓ goods. The commodity space is therefore $\mathbb{R}^\ell_+$. Each agent i's endowment is given by $e_i \colon \Omega \to \mathbb{R}^\ell_+$.

At $t = 0$ agents will agree on net-trades that may be contingent on the state of nature at $t = 1$. However, agents are differentially informed with respect to the true state of nature. Specifically, we assume that at $t = 1$ agents do not necessarily know which state $\omega \in \Omega$ has actually occurred. They know their own endowment realization, and every agent i might have some additional information about the state described by a σ-algebra $\mathcal{F}_i$ with $\mathcal{F}_i \subset \mathcal{F}$. We assume that $\mathcal{F}_i$ is generated by a countable partition of Ω. With a slight abuse of notation we will write $\mathcal{F}_i$ both for agent i's σ-algebra and for agent i's partition. By assumption, agents can always observe their own endowment realization, i.e., e_i is $\mathcal{F}_i$-measurable for each agent $i \in I$.

In summary, an *exchange economy with differential information* is given by

$$\mathcal{E} = \{(X_i, u_i, \mathcal{F}_i, e_i, \mu) : i = 1, \ldots, n\}$$

where:

1. $X_i(\omega) = \mathbb{R}^\ell_+$, for all $\omega \in \Omega$, is agent i's *consumption set*;
2. $u_i : \Omega \times \mathbb{R}^\ell_+ \to \mathbb{R}$ is agent i's *utility function*;
3. $\mathcal{F}_i$ is a measurable countable partition of Ω, denoting the *private information* of agent i;
4. $e_i : \Omega \to \mathbb{R}^\ell_+$ is agent i's *initial endowment*, where each e_i is $\mathcal{F}_i$-measurable and integrable;
5. μ is the probability measure on Ω denoting the *common prior* of all agents.

Agent i's *ex-ante expected utility* of consuming $x_i : \Omega \to \mathbb{R}^\ell_+$ is

$$\int_\Omega u_i(\omega, x_i(\omega))d\mu(\omega).$$

Let $\omega^\star$ be the true state of the economy in $t = 1$. Then at the *interim*, agent i observes the event $E_i(\omega^\star)$ in the partition $\mathcal{F}_i$ which contains $\omega^\star$. Agent i's updated prior is then given by $\mu(\cdot|E_i(\omega^\star))$.[1]

An allocation will be denoted by $(x_i)_{i\in I}$. An allocation is *feasible* if $\sum_{i\in I} x_i(\omega) = \sum_{i\in I} e_i(\omega)$, μ-a.e.

Throughout the paper, we assume for each $\omega \in \Omega$ that the utility function $u_i(\omega, \cdot)$ of each agent i is continuous and concave.

3 The TU game

As in the definition of the standard value allocation concept, we must first derive a transferable utility (TU) game in which each agent's utility is weighted by a factor λ_i, $(i = 1, \ldots, n)$, which allows interpersonal utility comparisons. In the value allocation itself no side payments are necessary. A game with side-payments is then defined as follows.

Definition 1 *A game with side-payments (I, V) consists of a finite set of agents $I = \{1, \ldots, n\}$ and a superadditive, real valued function V defined on 2^I such that $V(\emptyset) = 0$. Each $S \subset I$ is called a coalition and $V(S)$ is the "worth" of coalition S.*

We now define for each economy with differential information, $\mathcal{E}$, and for each set of weights, $\{\lambda_i : i = 1, \ldots, n\}$, the associated game with side-payments. Clearly, each coalition of agents S can obtain only those allocations that are feasible for the coalition, i.e., $\sum_{i\in S} x_i(\omega) = \sum_{i\in S} e_i(\omega)$, μ-a.e. However, trade among agents is also restricted because of private information. We consider two alternative ways to specify the set of allocations that a coalition S can obtain. The first way is to consider trades, which are private information measurable. That is, agents can

[1] Thus, $\mu(A|E_i(\omega^\star)) = \mu(A \cap E_i(\omega^\star))/\mu(E_i(\omega^\star))$ for any $A \in \mathcal{F}$, where $\mu(E_i(\omega^\star)) > 0$.

only make net trades based on their own information. The second way allows for trades, which are Bayesian incentive compatible. The relationship between these two alternative scenarios is shown in Theorem 1 below. More formally,

1. In the first specification, we assume that each coalition of agents, S, can achieve any feasible allocation $(x_i)_{i\in S}$ that fulfills private measurability, i.e, x_i is $\mathcal{F}_i$-measurable for all agents i in S.
2. Alternatively, we assume that each coalition can obtain any feasible allocation $(x_i)_{i\in S}$ that is Bayesian incentive compatible (BIC) for coalition S and $\bigvee_{i\in S}\mathcal{F}_i$-measurable.

Before deriving the TU game for the differential information economy, we define Bayesian incentive compatibility (see Glycopantis et al. [4] for a related definition).

Definition 2 *An allocation $x = (x_i)_{i\in S}$ is Bayesian incentive compatible (BIC) for coalition S if and only if there does not exist an agent $j \in S$ and states $\omega^\star, \omega' \in \Omega$ such that*

$$\begin{aligned}&\int_{Z_S(\omega^\star)} u_j(\omega, e_j(\omega) + x_j(\omega') - e_j(\omega'))\, d\mu(\omega \mid E_j(\omega^\star))\\&\quad > \int_{Z_S(\omega^\star)} u_j(\omega, x_j(\omega))\, d\mu(\omega \mid E_j(\omega^\star))\end{aligned} \tag{1}$$

where $Z_S(\omega^\star) = \bigcap_{i\in S} E_i(\omega^\star)$, $\omega' \in \bigcap_{i\in S\setminus\{j\}} E_i(\omega^\star)$ and $e_j(\omega^\star) + x_j(\omega') - e_j(\omega') \in \mathbb{R}^\ell_+$.

Inequality (1) states that agents cannot improve by misreporting their information. Definition 2 also requires that agent j's false report cannot always be detected by other agents in the coalition. In particular, in order for the false report report ω' not to be detected when $\omega^\star$ is the true state, $\omega' \in \bigcap_{i\in S\setminus\{j\}} E_i(\omega^\star)$ must hold. Moreover, note that if (1) holds then $\mu\big(Z_S(\omega^\star) \bigm| E_j(\omega^\star)\big) > 0$. Thus, if agent j can improve by a misrepresenting his information, then his false report should not be detectable with certainty (i.e., the set of states $\omega \neq \omega'$ that are consistent with ω' has positive measure).[2]

The reader might wonder why we do not allow misreports that can be detected by other agents with certainty. In particular, we interpret the example in Section 5 as a model in which agents sign contracts ex-ante to insure each other against low endowment realizations. If agent i receives a high realization he must make a payment to agents who received a lower realization. Now assume that agent j makes a report which is inconsistent with those of the other agents. If such a report leads to no trade, then agent j can keep his endowment, effectively reneging on his ex-ante agreement to insure the other agents. Therefore, in order to make insurance contracts enforceable, we must assume that agents cannot make reports that are inconsistent.

[2] In order to see what consistency of reports means in a simple example, assume that $\Omega = \{a, b, c\}$, that there are two agents $I = \{1, 2\}$ and $\mathcal{F}_1 = \{\{a, b\}, \{c\}\}$, $\mathcal{F}_2 = \{\{a, c\}, \{b\}\}$. Assume for example that agent 1 reports c and agent 2 reports b. Then these reports are not consistent because $\{c\} \cap \{b\} = \emptyset$. In contrast, the report $\{a, b\}$ by agent 1 and $\{b\}$ by agent 2 is consistent.

We are now ready to define the two alternative versions of the TU game. First, for every coalition $S \subset I$ let

$$\begin{aligned} V_\lambda^p(S) = \max_{(x_i)_{i\in S}} & \sum_{i\in S} \lambda_i \int u_i(x_i(\omega))d\mu(\omega), \\ \text{s.t. (i)} & \sum_{i\in S} x_i(\omega) = \sum_{i\in S} e_i(\omega), \mu\text{-a.e.} \\ & \text{(ii) } x_i \text{ is } \mathcal{F}_i\text{-measurable}, \forall i \in S. \end{aligned} \tag{2}$$

Second, we define

$$\begin{aligned} V_\lambda^{BIC}(S) = \max_{(x_i)_{i\in S}} & \sum_{i\in S} \lambda_i \int u_i(x_i(\omega))d\mu(\omega), \\ \text{s.t. (i)} & \sum_{i\in S} x_i(\omega) = \sum_{i\in S} e_i(\omega), \mu\text{-a.e.} \\ & \text{(ii) } x_i \text{ is } \bigvee_{i\in S} \mathcal{F}_i\text{-measurable}, \forall i \in S. \\ & \text{(iii) } (x_i)_{i\in S} \text{ is BIC for coalition } S. \end{aligned} \tag{3}$$

Note that (ii) in (3) is needed to ensure that agents can obtain allocation $(x_i)_{i\in S}$, given the available information.

We characterize the relationship between the two TU games in Theorem 1 below. The proof is in the Appendix.

Theorem 1 *Assume that $\lambda_i > 0$ for all $i \in I$. Then*

1. *$V_\lambda^p(S) \leq V_\lambda^{BIC}(S)$ for all $S \subset I$.*
2. *Consider a feasible allocation $(x_i)_{i\in S}$ for coalition S that is measurable with respect to private information, i.e., x_i is $\mathcal{F}_i$-measurable for all $i \in S$. Then $(x_i)_{i\in S}$ is BIC for coalition S.*

Clearly, the private and the BIC value allocation differ when $V_\lambda^p(I) < V_\lambda^{BIC}(I)$. An example where this is the case can be found in Krasa [6], p. 164.

4 The coalition structure (CS) value allocation

The standard Shapley value of the game (I, V) (Shapley [10]), is a rule that assigns to each agent i a payoff Sh_i, given by the formula:

$$\text{Sh}_i = \sum_{S\subset I, S\supset\{i\}} \frac{(|S|-1)!(|I|-|S|)!}{|I|!} \left[V(S) - V(S\setminus\{i\})\right]. \tag{4}$$

Following the treatment in Hart and Kurz [5], we first generalize the Shapley value formula to account for coalition formation (see also Owen [9]).

Let $\mathcal{B} = \{B_1, \ldots, B_m\}$ be a partition of the set of agents I. We refer to $\mathcal{B}$ as a *coalition structure*. As in the standard Shapley value, we wish to measure each agent's expected contribution to a coalition that he/she is a member of. However, this expected contribution should be compatible with the coalition structure $\mathcal{B}$.

Consider all possible ways to order the set of coalitions in $\mathcal{B}$ and the agents within each coalition. We say that a complete linear order on I is *consistent* with $\mathcal{B}$ if, for all $k = 1, \ldots, m$ and all i, j in B_k, all elements of I between i and j also belong to B_k. There are $m!b_1! \ldots b_m!$ such consistent orderings. Assume that each ordering is equally likely. The CS value of the game (I, V) is a rule which assigns to each agent i the expected marginal contribution to every coalition that agent i is a member of, respecting the coalition structure $\mathcal{B}$. That is, agent i's expected marginal contribution is given by

$$\phi_i(V, \mathcal{B}) = E[V(P^i \cup \{i\}) - V(P^i)], \tag{5}$$

where the expectation is over all random orders on I that are consistent with $\mathcal{B}$, and P^i denotes the set of random predecessors of player i.

For example, if no agent forms a coalition with another agent, then the coalition structure $\mathcal{B} = \big\{\{1\}, \ldots, \{n\}\big\}$. In this case the CS Shapley value (5) value coincides with the standard Shapley value (4).

As another example, assume that $I = \{1, 2, 3\}$, and $\mathcal{B} = \big\{\{1, 2\}, \{3\}\big\}$. Then there are the following consistent orderings

$$1 \prec 2 \prec 3 \qquad 2 \prec 1 \prec 3 \qquad 3 \prec 1 \prec 2 \qquad 3 \prec 2 \prec 1$$

Therefore,

$$\begin{aligned}
\phi_3(V, \mathcal{B}) &= \frac{1}{4}\Big(V(\{1,2,3\}) - V(\{1,2\})\Big) + \frac{1}{4}\Big(V(\{2,1,3\}) - V(\{2,1\})\Big) \\
&\quad + \frac{1}{4}\Big(V(\{3\}) - V(\emptyset)\Big) + \frac{1}{4}\Big(V(\{3\}) - V(\emptyset)\Big) \\
&= \frac{1}{2}\Big(V(\{1,2,3\}) - V(\{1,2\})\Big) + \frac{1}{2}V(\{3\}).
\end{aligned} \tag{6}$$

Agent 3's CS Shapley value is therefore the same as the standard value of a game where we treat agents 1 and 2 as a single player. The insight that coalitions of agents in the CS value can be treated as single agents in the standard Shapley value is true in general (see Corollary 2.4 in Hart and Kurz [5]). The CS Shapley value therefore measures agents' expected contributions after coalitions have formed for the purpose of bargaining. It is also important to note that the CS Shapley value is Pareto efficient, i.e., $\sum_{i \in I} \pi_i(V, \mathcal{B}) = V(I)$. Quoting Hart and Kurz [5], "the efficiency of the CS value is an essential feature. It differs from the Aumann and Dreze [2]) approach, where each coalition $B_k \in B$ gets only its worth (i.e $V(B_k)$). The idea is that coalitions form not in order to get their worth, but to be in a better position when bargaining with the others on how to divide the maximal amount available."

We now describe the notion of a CS value allocation for an economy with private information. These concepts generalize those of Shapley value allocation introduced in Krasa and Yannelis [7] (see also Einy and Shitovitz [3]).

In (2) and (3) of Section 3 we have introduced two alternative ways to assign a TU game to an economy with differential information. In (2) agents' trades are required to be measurable with respect to private information. In (3) we replace private measurability with incentive compatibility. As in Krasa and Yannelis [7] we therefore get two alternative definitions of a Shapley value allocation. Note that if $\mathcal{B}$ is replaced by the fine partition, then the following two concepts coincide with those of [7].

Definition 3 *An allocation $(x_i)_{i\in I}$ is a* private CS value allocation *of the economy with differential information, $\mathcal{E}$, if the following holds:*

(i) x_i *is $\mathcal{F}_i$-measurable for all $i \in I$.*
(ii) $\sum_{i=1}^n x_i(\omega) = \sum_{i=1}^n e_i(\omega), \mu$*−a.e.,*
(iii) There exist $\lambda_i \geq 0$, for every $i = 1, \ldots, n$, which are not all equal to zero, with $\lambda_i \int u_i(x_i(\omega))d\mu(\omega) = \phi_i(V_\lambda^P, \mathcal{B})$, where $\phi_i(V_\lambda^P, \mathcal{B})$ is the CS Shapley value of agent i derived from the game (I, V_λ^P), defined in (2).

Definition 4 *An allocation $(x_i)_{i\in I}$ is a* Bayesian incentive compatible (BIC) CS value allocation *of the economy with differential information, $\mathcal{E}$, if the following holds:*

(i) x_i *is $\bigvee_{i\in I}\mathcal{F}_i$-measurable for all $i \in I$, and $(x_i)_{i\in I}$ is BIC for coalition I.*
(ii) $\sum_{i=1}^n x_i(\omega) = \sum_{i=1}^n e_i(\omega), \mu$*−a.e.,*
(iii) There exist $\lambda_i \geq 0$, for every $i = 1, \ldots, n$, which are not all equal to zero, with $\lambda_i \int u_i(x_i(\omega))d\mu(\omega) = \phi_i(V_\lambda^{BIC}, \mathcal{B})$, where $\phi_i(V_\lambda^{BIC}, \mathcal{B})$ is the CS Shapley value of agent i derived from the game (I, V_λ^{BIC}), defined in (3).

Note that the existence proof of Shapley [10] can be easily modified to cover the coalition structure value. Given this result one can adapt the existence proof in Krasa and Yannelis [8] to show that a private coalition structure value exists. Theorem 1 then immediately implies that the private CS value allocation is Bayesian incentive compatible. We now state this result formally.

Theorem 2 *The private CS value allocation of a differential information economy exists and is BIC for the grand coalition I.*

In contrast to the private CS value allocation, the BIC CS value allocation need not always exist because the set of feasible incentive compatible allocations need not be convex. In the economy of Section 5 the two concepts coincide. Hence, our conclusions about the advantages or disadvantages of coalitional bargaining hold for both concepts.

5 Is unity a strength?

We now consider an economy with three agents $i = 1, 2, 3$. In this economy agents 1 and 2 seek insurance against their uncertain endowment realization. Agents 1 and 2 can insure each other, but they can also seek insurance from agent 3 who is risk neutral.

First consider the case of complete information, i.e., once the state is realized it becomes public information. As one would expect, agents 1 and 2 are better off if they collectively bargain against agent 3. That is, if information is complete the CS Shapley value for agents 1 and 2 is strictly higher than the standard Shapley value. However, once asymmetric information is introduced the result is completely reversed. Now agents 1 and 2 are strictly better off if they do not form a coalition. In this sense, "unity is strength" does not hold when information is asymmetric.

Assume there are four states $\omega \in \Omega = \{a, b, c, d\}$ that are equally likely. There is one consumption good in each state. The agents' utility functions are given by

$$u_1(x,\omega) = \begin{cases} x & \text{if } \omega = a, b, d \\ 2\sqrt{x} & \text{if } \omega = c; \end{cases} \qquad u_2(x,\omega) = \begin{cases} x & \text{if } \omega = a, c, d \\ 2\sqrt{x} & \text{if } \omega = b; \end{cases}$$

$$u_3(x,\omega) = x, \ \forall \omega.$$

The agents' endowments are

$$e_1(\omega) = \begin{cases} 2 & \text{if } \omega = a, b \\ 0 & \text{if } \omega = c, d; \end{cases} \qquad e_2(\omega) = \begin{cases} 2 & \text{if } \omega = a, c \\ 0 & \text{if } \omega = b, d; \end{cases}$$

$$e_3(\omega) = \begin{cases} \frac{1}{4} & \text{if } \omega = a, b, c \\ 10 & \text{if } \omega = d. \end{cases}$$

Agents 1 and 2 are risk averse in states c and b, respectively. The agents therefore seek insurance against the low endowment realization in these states. Note that the role of state d is to ensure that the weights λ_i in the TU game are all equal.

In order to derive the value allocation, we must find the utility weights λ_i, $i = 1, 2, 3$. However, because agents' utilities are quasilinear all weights λ_i must be equal, i.e., $\lambda_1 = \lambda_2 = \lambda_3 = 1$. This is shown formally in Lemma 1 in the Appendix. Therefore, the economy corresponds to a game with transferable utility and we can write $V(S)$ instead of $V_\lambda(S)$ to denote the payoff of coalition S.

Complete Information Assume that all agents learn the true state ω after it is realized. For any coalition S, let $V(S)$ be the maximum attainable utility, i.e.,

$$V(S) = \max_{\{x_i(\omega) | i \in S, \omega \in \Omega\}} \sum_{i \in S} \sum_{\omega \in \Omega} \frac{1}{4} u_i(\omega, x_i(\omega))$$

$$\text{s.t.} \sum_{i \in S} x_i(\omega) = \sum_{i \in S} e_i(\omega), \ \forall \omega \in \Omega.$$

For example, assume that agent 1 belongs to S. Then it follows immediately that it is optimal to choose $x_1(c) = 1$. The same is true for agent 2 and state b.

The TU game is therefore given by

$$V(\{1\}) = V(\{2\}) = 1, \ V(\{3\}) = \frac{43}{16}$$

$$V(\{1,2\}) = \frac{5}{2}, \ V(\{1,3\}) = V(\{2,3\}) = \frac{31}{8}$$

$$V(\{1,2,3\}) = \frac{83}{16}$$

The resulting Shapley values are

$$\mathrm{Sh}_1 = \mathrm{Sh}_2 = \frac{39}{32}, \quad \mathrm{Sh}_3 = \frac{11}{4}.$$

A Shapley value allocation is given by

$$x_1 = x_2 = \left(1, 1, 1, \frac{7}{8}\right);\, x_3 = \left(\frac{9}{4}, \frac{1}{4}, \frac{1}{4}, \frac{33}{3}\right). \tag{7}$$

Now assume agents 1 and 2 form a coalition. Then by Corollary 2.4 in Hart and Kurz [5] agent 3's Shapley value can be computed from the Shapley value of the game where $\{1, 2\}$ is treated as a single player. Moreover, because of symmetry the Shapley values of agents 1 and 2 must be exactly one half of the Shapley value of the "single player" $\{1, 2\}$. Then

$$\phi_1(\mathcal{B}) = \phi_2(\mathcal{B}) = \frac{5}{4}, \quad \phi_3(\mathcal{B}) = \frac{43}{16}$$

A CS Shapley value allocation is given by

$$x_1 = x_2 = (1, 1, 1, 1);\, x_3 = \left(\frac{9}{4}, \frac{1}{4}, \frac{1}{4}, 8\right). \tag{8}$$

Therefore $\phi_3(\mathcal{B}) < \mathrm{Sh}_3$ and $\phi_i(\mathcal{B}) > \mathrm{Sh}_i$, for $i = 1, 2$, i.e., agents 1 and 2 are strictly better off when they bargain together. Hence, "unity is strength" in this case.

Differential Information Assume that agent 1 cannot distinguish states a and b, and that agent 2 cannot distinguish states a and c. Agent 3 has complete information.[3] Again, the economy corresponds to a TU game. The payoffs in the TU game are the same as above except that

$$V^p(\{1, 2\}) = V^{BIC}(\{1, 2\}) = 2$$

We first show that all coalitions $S \neq \{1, 2\}$ have the same payoff under differential information as under complete information. This is obvious for all single agent coalitions. Next, note that (7) indicates an allocation $(x_i)_{i \in I}$, for which each x_i is $\mathcal{F}_i$-measurable, and which yields the payoff $V(I)$. Theorem 1 implies that $(x_i)_{i \in I}$ is BIC for coalition I. Therefore, $V^p(I) = V^{BIC}(I) = V(I)$. Next, consider coalition $S = \{1, 3\}$. The payoff $V(\{1, 3\})$ can be obtained by choosing $x_1 = (1, 1, \frac{1}{4}, 0)$ and $x_3 = (\frac{5}{4}, \frac{5}{4}, 0, 10)$. This allocation for coalition S is measurable with respect to each agent's private information and therefore also BIC. The argument for $S = \{2, 3\}$ is similar.

Finally, consider coalition $S = \{1, 2\}$. The only feasible allocation $(x_i)_{i \in S}$ that is measurable with respect to private information consists of the agents' endowments. In particular feasibility implies $x_1 - e_1 = x_2 - e_2$. Private measurability implies that $x_1 - e_1$ is $\mathcal{F}_1$-measurable and that $x_2 - e_2$ is $\mathcal{F}_2$-measurable. Therefore each $x_i - e_i$ must be $\mathcal{F}_1 \wedge \mathcal{F}_2$-measurable. Since $\mathcal{F}_1 \wedge \mathcal{F}_2 = \{\{a, b, c\}, \{d\}\}$

[3] Formally, $\mathcal{F}_1 = \{\{a, b\}, \{c\}, \{d\}\}$, $\mathcal{F}_2 = \{\{a, c\}, \{b\}, \{d\}\}$, and $\mathcal{F}_3 = \{\{a\}, \{b\}, \{c\}, \{d\}\}$.

the only feasible trades are trivial. It is also not possible to improve upon autarky via BIC net trades. In order to improve upon autarky, either $x_1(c)$ or $x_2(b)$ must be strictly positive. Let $\omega^\star = a$ and assume that agent 1 reports $\omega' = c$. Then $Z_S(\omega^\star) = \{a\}$. Therefore, incentive compatibility implies $x_1(a) \geq 2 + x_1(c)$. Similarly, $x_2(a) \geq 2 + x_2(b)$ must hold. Therefore, $x_1(c) + x_2(b) \leq 0$ which implies $x_1(c) = x_2(b) = 0$.

The Shapley values are therefore given by

$$\mathrm{Sh}_1 = \mathrm{Sh}_2 = \frac{109}{96}, \quad \mathrm{Sh}_3 = \frac{35}{12}$$

A private or BIC Shapley value allocation is given by

$$x_1 = x_2 = \left(1, 1, 1, \frac{13}{24}\right); \; x_3 = \left(\frac{9}{4}, \frac{1}{4}, \frac{1}{4}, \frac{107}{12}\right).$$

Now assume that agents 1 and 2 form a coalition. Then,

$$\phi_1(\mathcal{B}) = \phi_2(\mathcal{B}) = \frac{9}{8}, \quad \phi_3(\mathcal{B}) = \frac{47}{16}.$$

A BIC or a private CS Shapley value allocation is given by

$$x_1 = x_2 = \left(1, 1, 1, \frac{1}{2}\right); \; x_3 = \left(\frac{9}{4}, \frac{1}{4}, \frac{1}{4}, 9\right).$$

Therefore $\phi_3(\mathcal{B}) > \mathrm{Sh}_3$ and $\phi_i(\mathcal{B}) < \mathrm{Sh}_i$, for $i = 1, 2$. As a consequence, when information is incomplete agents 1 and 2 are strictly worse off it they bargain as a coalition. Therefore, in this case unity among coalition members is not a strength.

5.1 Interpretation

Although there is no explicit time structure in our model, equation (6), and more generally Corollary 2.4 in Hart and Kurz [5], show that we can interpret the CS value as follows.

First, agents 1 and 2 decide to form a coalition. They sign an agreement that determines how any surplus will be split among them. After the agreement is signed, coalition $\{1, 2\}$ bargains with agent 3. Because agents 1 and 2 can provide insurance to each other without the help of an outside agent, agent 3's service is not essential. Coalition $\{1, 2\}$ is therefore in the strongest possible bargaining position, and agent 3 will receive only a small percentage of the surplus. In contrast, when agent 1 or 2 bargains separately with agent 3, then agent 3 can provide a significant service by insuring that agent. As a consequence, agent 3 is able to extract a higher percentage of the surplus when agents 1 and 2 bargain separately.

As another example, consider a coalition of firms that bargains with an outside firm for a service. If the coalition is sufficiently diverse, it is likely that the outsider's service could be replaced by that of a coalition member, and is therefore not that essential for the coalition. As a consequence, the outside firm will find itself in a weak bargaining position. Unity among coalition members is therefore a strength.

What goes wrong with this intuition when information is asymmetric? In our example, agent 3 has information that is essential for agents 1 and 2. The value of agent 3's information increases with the number of agents that need agent 3's information. Formally, $V(\{1,2,3\}) - V(\{3\}) > V(\{1,3\}) - V(\{3\}) = V(\{2,3\}) - V(\{3\})$. In other words, the information becomes more important the larger the coalition of agents that depends on it. Therefore, ceteris paribus, agent 3 will extract a high percentage of the surplus if he only bargains with large coalitions.

Consider again the example of the coalition of firms, but now assume that the outside firm has information that is crucial to the coalition. Then the outside firm will be able to sell its information for a higher price (i.e., a higher percentage of the surplus) the larger the coalition. Intuitively, the outside firm can hold the coalition "hostage" to its information. "Unity is a strength" may therefore cease to be true when informational asymmetries matter.

6 Appendix

Proof of Theorem 1. Because statement 2 implies statement 1, it is sufficient to prove 2.

Assume by contradiction that $(x_i)_{i\in S}$ is feasible for coalition S, but not BIC for coalition S.

Then there exists an agent $j \in S$, an $\omega^\star, \omega'$ such that

$$\begin{aligned}\int_{Z_S(\omega^\star)} u_j\Big(\omega, e_j(\omega) - \big(x_j(\omega') - e_j(\omega')\big)\Big)\, d\mu\big(\omega \mid E_j(\omega^\star)\big) \\ > \int_{Z_S(\omega^\star)} u_j\big(\omega, x_j(\omega)\big)\, d\mu\big(\omega \mid E_j(\omega^\star)\big)\end{aligned} \tag{9}$$

where $\omega' \in \bigcap_{i\in S\setminus\{j\}} E_i(\omega^\star)$ and $Z_S(\omega^\star) = \bigcap_{i\in S} E_i(\omega^\star)$. For the fixed $\omega^\star \in \Omega$, define $y_i\colon \Omega \to \mathbb{R}^\ell_+$ for each agent i by

$$y_i(\omega) = \begin{cases} e_i(\omega) + x_i(\omega') - e_i(\omega') & \text{for } \omega \in Z_S(\omega^\star); \\ x_i(\omega) & \text{for } \omega \notin Z_S(\omega^\star). \end{cases} \tag{10}$$

It can be easily checked that $\sum_{i\in S} y_i(\omega) = \sum_{i\in S} e_i(\omega)$, μ-a.e.

Consider an agent $i \in S \setminus \{j\}$. Let $\omega \in Z_S(\omega^\star)$. Then $\omega' \in \bigcap_{k\in S\setminus\{j\}} E_k(\omega^\star) \subset E_i(\omega^\star)$ and $Z_S(\omega^\star) \subset E_i(\omega^\star)$ imply that agent i cannot distinguish ω from ω'. Therefore, $\mathcal{F}_i$-measurability of x_i implies that $x_i(\omega) = x_i(\omega')$ for all $\omega \in Z_S(\omega^\star)$. Thus, $y_i = x_i$ for all $i \in S \setminus \{j\}$. This and feasibility imply

$$y_j = \sum_{i\in S} e_i - \sum_{i\in I\setminus\{j\}} y_i = \sum_{i\in S} e_i - \sum_{i\in I\setminus\{j\}} x_i = x_j. \tag{11}$$

Equation (11) immediately implies

$$\int_{Z_S(\omega^\star)} u_j\Big(\omega, y_j(\omega)\Big)\, d\mu\big(\omega \mid E_j(\omega^\star)\big) = \int_{Z_S(\omega^\star)} u_j\big(\omega, x_j(\omega)\big)\, d\mu\big(\omega \mid E_j(\omega^\star)\big) \tag{12}$$

However, in view of (10), the left-hand side of (12) is equal to (9). Therefore (12) contradicts (9). Consequently, $(x_i)_{i \in S}$ is BIC for coalition S. □

Lemma 1 *Consider the exchange economies of Section 5. Then all utility weights are identical in all coalition structure and in all standard Shapley value allocations both with complete and incomplete information.*

Proof. Assume by way of contradiction that the utility weights differ. In the case of complete information, the agents' consumption in the Shapley value allocation must solve

$$\max_{\{x_i(\omega) | i \in I, \omega \in \Omega\}} \sum_{i \in I} \sum_{\omega \in \Omega} \frac{1}{4} \lambda_i u_i(\omega, x_i(\omega)) \tag{13}$$

$$\text{s.t. (i)} \sum_{i \in I} x_i(\omega) = \sum_{i \in I} e_i(\omega), \ \forall \omega \in \Omega;$$

With differential information we must add one of the following constraints.

$$\text{(iia) } x_i \text{ is BIC } \forall i \in I,$$

$$\text{(iib) } x_i \text{ is } \mathcal{F}_i\text{-measurable } \forall i \in I,$$

It follows immediately, that (iia) is always slack, and can therefore be omitted. In particular, because agent 3 has complete information, any misreport by agent 1 or agent 2 is immediately detected, i.e., the sets $\bigcap_{j \in I \setminus \{i\}} E_j(\omega^\star)$ are singletons for $i = 1, 2$, which implies $\omega^\star = \omega'$. Moreover, because $\mathcal{F}_1 \vee \mathcal{F}_2 = \mathcal{F}_3$, agent 3 can also not misreport his information, i.e., $E_1(\omega^\star) \cap E_2(\omega^\star)$ only contains $\omega^\star$.

We now prove by way of contradiction that all weights λ_i are identical. It should be noted that is does not matter in the argument whether or not (iib) is imposed.

First, assume that $\lambda_1 < \lambda_3$. Then agent 1's consumption in states a, b, and d must be zero. Therefore, agent 1's expected utility is $\mu(\{c\}) 2\sqrt{x_3(c)}$. Because of feasibility $x_3(c) \leq \frac{9}{4}$. Individual rationality is therefore violated for agent 1. Thus, $\lambda_1 \geq \lambda_3$. A similar argument shows that $\lambda_2 \geq \lambda_3$.

Now assume that $\lambda_3 < \lambda_1$ or $\lambda_3 < \lambda_2$. Then $x_3(d) = 0$. However, because of feasibility, $x_3(a) \leq 4$ and $x_3(b), x_3(c) \leq 2$. Therefore individual rationality would be violated for agent 3. This implies $\lambda_1 = \lambda_2 = \lambda_3$. □

References

1. Allen, B., Yannelis, N.C.: Differential information economies: Introduction. Economic Theory **18**, 263–273 (2001)
2. Aumann, R.J., Dreze, J.: Cooperative games with coalition structures. International Journal of Game Theory **3**, 214–237 (1974)
3. Einy, E., Shitovitz, B.: Private value allocations in large economies with differential information. Games and Economic Behavior **34**, 287–311 (2001)
4. Glycopantis, D., Muir, A., Yannelis, N.C.: An extensive form interpretation of the private core. Economic Theory **18**, 293–319 (2001)
5. Hart, S., Kurz, M.: Endogenous formation of coalitions. Econometrica **51**, 1047–1064 (1983)
6. Krasa, S.: Unimprovable allocations in economies with differential information. Journal of Economic Theory **87**, 144–168 (1999)

7. Krasa, S., Yannelis, N.C.: The value allocation of an economy with differential information. Econometrica **62**, 881–900 (1994)
8. Krasa, S., Yannelis, N.C.: Existence and properties of a value allocation for an economy with differential information. Journal of Mathematical Economics **25**, 165–179 (1996)
9. Owen, G.: Values of games with a priori unions. In: Essays in Mathematical Economics and Game Theory. Springer, Berlin Heidelberg New York 1977
10. Shapley, L.S.: A value for n-person games. In: Contributions to the theory of games. Princeton University Press, 1953

PART 6

IMPLEMENTATION

Coalitional Bayesian Nash implementation in differential information economies*

Guangsug Hahn[1] **and Nicholas C. Yannelis**[2]

[1] Korea Economic Research Institute, Seoul, 150-756, KOREA (e-mail: hahn@keri.org)
[2] Department of Economics, University of Illinois at Urbana-Champaign, Champaign, IL 61820, USA (e-mail: nyanneli@uiuc.edu)

Received: January 12, 1998; revised version: March 30, 2000

Summary. A mechanism *coalitionally implements* a social choice set if any outcome of the social choice set can be achieved as a coalitional Bayesian Nash equilibrium of a mechanism and vice versa. We say that a social choice set is *coalitionally implementable* if there is a mechanism which coalitionally implements it. Our main theorem proves that a social choice set is coalitionally implementable if and only if it is interim individually rational, interim efficient, coalitional Bayesian incentive compatible, and satisfies a coalitional Bayesian monotonicity condition as well as a closure condition. As an application of our main result, we show that the private core and the private Shapley value of an economy with differential information are coalitionally implementable.

Keywords and Phrases: Implementation, Differential information, Cooperative games, Incentive compatibility, Interim private core, Interim private value.

JEL Classification Numbers: C71, D51, D78, D82.

1 Introduction

An economy with differential information consists of a finite set of agents, each of whom is characterized by a random utility function, a random initial endowment, a private information set, and a prior (a precise definition can be found in Section 2.1).

The traditional notion which has been adopted in the literature to analyze trade in a differential information economy is the (Walrasian) rational expectations equilibrium. One of the criticisms of the above notion is that it does not provide a

* We wish to thank the referees for several useful comments. Special thanks to D. Glycopantis for pointing out a calculation error in Example 5.1.
Correspondence to: N. C. Yannelis

mechanism which describes how the equilibrium prices reflect information asymmetries in the economy. To this end we adopt the private core (Yannelis, 1991) and the private value (Krasa and Yannelis, 1994) in order to analyze the trading procedure in a differential information economy. The private core and the private value are not fully cooperative in a differential information economy framework, because within a coalition agents make redistributions of their initial endowments based on their own private information (without necessarily sharing it). Hence, despite the fact that coalitions of agents get together and make redistributions (the cooperative aspect of the concepts), there is a noncooperative element in that agents in the coalition bargain using their differential information. This noncooperative feature of the private core and the private value results in allocations which are always coalitionally incentive compatible.[1] Moreover, these concepts provide sensible and reasonable outcomes in situations where the traditional rational expectations equilibrium fails to do so [for examples of this effect, see Koutsougeras and Yannelis (1993, pp. 206-207) and Krasa and Yannelis (1994, pp. 890-892) as well as the Example 3.1 in Section 3 of this paper].[2]

The outcomes generated by the private core or value are of interest because they resemble contracts, and contracts are a common means by which agents execute trade. In particular, in a contract it is common for agents to make an agreement ex ante (or interim), which is executed ex-post (for example insurance contracts). The allocation rules we consider, the private core and the private value, have the following properties that we believe are desirable. First, information asymmetries matter and agents benefit from superior information. Second, optimal contracts (i. e., private core or private value allocations) always exist, which is not the case for the rational expectations equilibrium. This matches the observation that contracts are more common than competitive markets in situations where differential information makes trade difficult.[3] In view of these attractive features that the private core and the private value possess, it is important to know whether or not they are implementable, *i.e.*, can a game be constructed whose equilibrium outcomes coincide with the private core or the private value? This knowledge will enable us not only to understand better the outcomes that these allocation rules generate but also to distinguish and compare them from the traditional (Walrasian) rational expectations equilibrium.

Our implementation results indicate that indeed information asymmetries matter and the stringent informational conditions needed for the Bayesian Nash implementation of the Walrasian expectations equilibrium (see for example Blume and Easley, 1990; Palfrey and Srivastava, 1987; Postlewaite and Schmeidler, 1986) are not needed. In particular, Palfrey and Srivastava (1987) have shown that the core (their core notion is different than the one adopted in this paper) of an economy with differential information may not be implementable as a Bayesian Nash equi-

[1] See Section 2.4 for a precise definition.

[2] See also Ichiishi and Radner (1999) and Ichiishi and Sertel (1998) for related core notions.

[3] These points are made formally in Example 3.1 of Section 3 where we refer the reader for further discussion.

librium.[4] However, despite the negative result of Palfrey and Srivastava (1987), we demonstrate that indeed our private core notion is implementable as a coalitional (strong) Bayesian Nash equilibrium, *i.e.*, we can construct a game (mechanism) whose coalitional Bayesian Nash equilibrium outcomes coincide with the private core. By focusing on the coalitional implementation of a social choice set, we reconsider the problem of implementation in differential information economies studied in a series of papers by Blume and Easley (1987), Jackson (1991), Palfrey and Srivastava (1986, 1987, 1989), and Postlewaite and Schmeidler (1986, 1987). To be precise, we say that a mechanism *coalitionally implements* a social choice set if any outcome of the social choice set can be achieved as a coalitional Bayesian Nash equilibrium of a mechanism, and vice versa. We say that a social choice set is *coalitionally implementable* if there is a mechanism which coalitionally implements it.

The main purpose of this paper is to show that a social choice set is coalitionally implementable if and only if it is interim individually rational, interim efficient, coalitional Bayesian incentive compatible, and satisfies a coalitional Bayesian monotonicity condition as well as a closure condition. As an application of this result, we show that the private core and the private Shapley value of an economy with differential information are coalitionally implementable. In doing so, we build on the incomplete information monotonicity condition of Jackson (1991), Palfrey and Srivastava (1987, 1989), and Postlewaite and Schmeidler (1986), and introduce new concepts. We define a coalitional form of monotonicity which is appropriate for our model.

Finally, it should be mentioned that we not only examine the problem of coalitional Bayesian implementation for the first time and provide characterization results, but we also make several other advances. First, we are able to address the incentive compatibility issue in a coalitional way. This is of great importance because individually incentive compatible contracts may not be necessarily coalitional incentive compatible Hence, if one considers multilateral contracts, the individually incentive compatibility may not be sufficient to guarantee that the contract may be viable. Secondly, we implement the Shapley value without restricting trade to be bilateral (e. g., Gul, 1989) or to the transferable utility case (e. g., Winter, 1994). Hence, we also contribute to the literature of finding ways to rationalize the Shapley value. Thirdly, we offer a new construction of a mechanism which takes into account coalitional deviations, is not wasteful, and is feasible.

The paper is organized as follows: In Section 2, we describe the model and characterize the coalitional implementation. In Sections 3 and 4, we show that the private core and the private value are coalitionally implementable. Concrete examples are presented in Section 5. Finally, some concluding remarks are collected in Section 6.

---

[4] That is, one cannot construct a game whose set of Bayesian Nash equilibria coincides with their core notion.

2 Coalitional implementation

We begin with some notation and definitions. $|A|$ denotes the number of elements in the set A. If A is a set, we denote by χ_A the characteristic function having the property that $\chi_A(\omega)$ is one if $\omega \in A$ and it is zero otherwise. $\setminus$ denotes the set theoretic subtraction.

2.1 Differential information economies

Below we define the notion of an economy with differential information. Let $(\Omega, \mathcal{F}, \mu)$ be a probability measure space denoting the states of the world, $\boldsymbol{R}^\ell$ be an Euclidean space denoting the commodity space and $I = \{1, 2, \ldots, N\}$ be a finite set of agents. For simplicity, we assume that $\Omega = \{\omega_1, \omega_2, \ldots, \omega_n\}$ is a finite set of states.[5] An *economy with differential information* is described by $\mathcal{E} = \{(X_i, u_i, \mathcal{F}_i, \mu, e_i) : i \in I\}$, where

(1) $X_i = \boldsymbol{R}^\ell_+$ is the *consumption set* of agent $i \in I$,
(2) $u_i : \Omega \times \boldsymbol{R}^\ell_+ \to \boldsymbol{R}$ is the *state-dependent utility function* of agent $i \in I$,
(3) $\mathcal{F}_i$ is a finite measurable partition of Ω denoting the *private information* of agent $i \in I$,
(4) μ is a probability measure on Ω denoting the *common prior* of each agent,
(5) $e_i : \Omega \to \boldsymbol{R}^\ell_{++}$ is an $\mathcal{F}_i$-measurable function[6] denoting the *state-dependent initial endowment* of agent $i \in I$.

We assume that the structure of the differential information economy is common knowledge among all agents. We call a set of states an event. An event E_i, which is an element of the partition $\mathcal{F}_i$, is the largest set of states that agent i cannot distinguish. Let $E_i(\omega)$ denote the event of $\mathcal{F}_i$ which contains $\omega \in \Omega$. This means that when the true state ω occurs, agent i knows only that the event $E_i(\omega)$ occurs. Assume that $\mu(\omega) > 0$ for every $\omega \in \Omega$.

Let $L_\mathcal{F}$ be the set of $\mathcal{F}$-measurable functions which maps Ω to $\boldsymbol{R}^\ell_+$, L_{X_i} the set of $\mathcal{F}_i$-measurable functions which maps Ω to $\boldsymbol{R}^\ell_+$, and L_i the set of $\mathcal{F}_i$-measurable functions which maps Ω to $\boldsymbol{R}^\ell$. The *conditional expected utility function* of agent i is a function $V_i : \Omega \times L_\mathcal{F} \to \boldsymbol{R}$ defined by[7]

$$V_i(\omega, x_i) = \frac{1}{\mu(E_i(\omega))} \sum_{\omega' \in E_i(\omega)} u_i(\omega', x_i(\omega'))\mu(\omega').$$

An element $x = (x_i)_{i \in I} \in L_X := \prod_{i \in I} L_{X_i}$ is called an *allocation*. The *set of feasible allocations* is given by $\boldsymbol{A} = \{x \in L_X : \sum_{i \in I} x_i = \sum_{i \in I} e_i\}$. For each i, an element $z_i \in L_i$ with $z_i = x_i - e_i$ is a *net trade* of agent i. The *set of feasible net trades* is given by $\boldsymbol{Z} = \{z \in L : \sum_{i \in I} z_i = 0\}$ where $L = \prod_{i \in I} L_i$. Let $\hat{Z} = \{\hat{z} \in \prod_{i \in I} Y_i : \sum_{i \in I} \hat{z}_i = 0\}$, where $Y_i = \boldsymbol{R}^\ell$ for every $i \in I$. Notice that the initial endowment vector denoted by $e = (e_i)_{i \in I}$ is an element of L_X.

---

[5] One would allow for infinitely many states and infinitely many commodities. We refer the reader to Hahn and Yannelis (1995) for the details.

[6] A function $f : \Omega \to \boldsymbol{R}$ is $\mathcal{F}_i$*-measurable* if $f(\omega) = f(\omega')$ for every $\omega, \omega' \in E_i \in \mathcal{F}_i$.

[7] One could allow agents to have different priors.

2.2 Coalitional implementation

A *social choice set* Γ is a subset of $\boldsymbol{A}$. A *mechanism* for an economy with differential information $\mathcal{E}$ is a pair (M, f) where $M = \prod_{i \in I} M_i$ is the set of messages and $f : M \to \hat{Z}$ is an outcome function. If $M = F$ with $F = \prod_{i \in I} \mathcal{F}_i$, the mechanism (F, f) is a *direct revelation mechanism*. A *strategy* for agent i is a function $\sigma_i : \mathcal{F}_i \to M_i$. We use the following notation: $\sigma = (\sigma_i)_{i \in I}$, $\sigma(E(\omega)) = (\sigma_i(E_i(\omega)))_{i \in I}$, $f(\sigma)(\omega) = f(\sigma(E(\omega)))$, $E = (E_i)_{i \in I}$. For $S \subset I$, $\sigma_S = (\sigma_i)_{i \in S}$, $\sigma_{-S} = (\sigma_i)_{i \notin S}, \sigma_S(E_S(\omega)) = (\sigma_i(E_i(\omega)))_{i \in S}, \sigma_{-S}(E_{-S}(\omega)) = (\sigma_i(E_i(\omega)))_{i \notin S}, E_S = (E_i)_{i \in S}$.

Definition 2.2.1. *A strategy vector σ is a* Bayesian Nash equilibrium (BNE) *for the mechanism (M, f) if for every $i \in I$, for every $\omega \in \Omega$, and for every strategy $\sigma_i' : \mathcal{F}_i \to M_i$,*

$$V_i(\omega, e_i + f_i(\sigma)) \geq V_i(\omega, e_i + f_i(\sigma_i', \sigma_{-i})).$$

When agents are allowed to form coalitions, one may define a stronger equilibrium concept.

Definition 2.2.2. *A strategy vector σ is a* coalitional Bayesian Nash equilibrium (CBNE) *for the mechanism (M, f) if it is not true that there exists a state $\omega \in \Omega$, a coalition $S \subset I$, and a strategy $\sigma_S' : \prod_{i \in S} \mathcal{F}_i \to \prod_{i \in S} M_i$ such that*

$$V_i(\omega, e_i + f_i(\sigma_S', \sigma_{-S})) > V_i(\omega, e_i + f_i(\sigma)), \; \forall i \in S.$$

In this paper, we consider full implementation which requires that the set of equilibrium outcomes of the mechanism exactly coincide with the given social choice set. This does not allow the existence of any undesirable equilibrium outcome in the mechanism.

Definition 2.2.3. *A mechanism (M, f)* coalitionally implements (c-implements) *a social choice set Γ if*

(1) For any $x \in \Gamma$, there exists a coalitional Bayesian Nash equilibrium σ for (M, f) such that $\mathrm{e} + f(\sigma) = x$,
(2) If σ is a coalitional Bayesian Nash equilibrium for (M, f), then $\mathrm{e} + f(\sigma) \in \Gamma$.

A social choice set Γ is *coalitionally implementable (c-implementable)* if there is a mechanism (M, f) which c-implements Γ. Given a mechanism (M, f), we assume that for every $i \in I$ and every strategy vector σ, there is a strategy σ_i' for agent i such that $f_i(\sigma_i', \sigma_{-i}) = 0$. That is, we restrict attention to such mechanisms.

2.3 Interim efficiency and interim individual rationality

A social choice set Γ is *interim efficient (IE)* if for every $x \in \Gamma$, there is no $x' \in \boldsymbol{A}$ such that for some $\omega \in \Omega$, $V_i(\omega, x_i') > V_i(\omega, x_i)$ for every $i \in I$[8] Since for each $\omega \in \Omega$, $u_i(\omega, \cdot)$ is monotone and continuous, so is $V_i(\omega, \cdot)$. Therefore, one can easily show that the above definition of interim efficiency coincides with that of a stronger interim efficiency, *i.e.*, a social choice set Γ is interim efficient if for every $x \in \Gamma$, there is no $x' \in \boldsymbol{A}$ such that for some $\omega \in \Omega$, $V_i(\omega, x_i') \geq V_i(\omega, x_i)$ for every $i \in I$ with strict inequality for some $i \in I$.

Proposition 2.3.1. *If a social choice set Γ is c-implementable by a mechanism (M, f) and $f : M \to \hat{Z}$ is onto, then it is IE.*

Proof. Suppose, by way of contradiction, that (M, f) c-implements Γ but Γ is not IE. Then, there exists an allocation $x = e + z \in \Gamma$ such that for some state $\omega \in \Omega$ and for some allocation $x' = e + z' \in \boldsymbol{A}$,

$$V_i(\omega, e_i + z_i') > V_i(\omega, e_i + z_i), \forall i \in I.$$

Since Γ is c-implementable, we have a CBNE σ such that $f(\sigma) = z$. Because f is onto, there is a strategy profile σ' such that $f(\sigma'(\omega)) = z'(\omega) \in \hat{Z}$ for every $\omega \in \Omega$. Hence, for every $i \in I$,

$$V_i(\omega, e_i + f_i(\sigma')) > V_i(\omega, e_i + f_i(\sigma)),$$

a contradiction to the fact that σ is a CBNE. □

A social choice set Γ is *interim individually rational (IIR)* if for every $x \in \Gamma$ and for every $\omega \in \Omega$, $V_i(\omega, x_i) \geq V_i(\omega, e_i)$ holds for every $i \in I$. One can easily show that interim individual rationality is a necessary condition for coalitional implementation. The following result is the counterpart of that by Hurwicz et al. (1984, Proposition, p. 14).

Proposition 2.3.2. *If a social choice set Γ is c-implementable, then it is IIR.*

Proof. Suppose (M, f) c-implements Γ but Γ is not IR. Then there is $x = e+z \in \Gamma$ such that there exists $\omega \in \Omega$ and $i \in I$ such that $V_i(\omega, x_i) < V_i(\omega, e_i)$. Since Γ is c-implementable, we have a CBNE σ such that $f(\sigma) = z$. Since we assume that for every i and every σ, there exists σ_i' such that $f(\sigma_i', \sigma_{-i}) = 0$, we have $V_i(\omega, e_i + f(\sigma_i', \sigma_{-i})) > V_i(\omega, e_i + f(\sigma))$, a contradiction. □

2.4 Coalitional Bayesian incentive compatibility

When agents have differential information, arbitrary allocations are not generally viable. In particular, arbitrary allocations might not be incentive compatible in the

[8] Note that our interim efficiency notion is different than usual one (e. g., Holmström-Myerson, 1983) in that we use a weaker quantifier "for some $\omega \in \Omega$." Due to the private information measurability, our notion is not so strong as it seems and cannot be directly compared with the usual one. For comparisons of different efficiency notions in differential information economies, see Hahn and Yannelis (1997).

sense that groups of agents may misreport their information without other agents noticing it, and hence achieve different payoffs. We will show that a social choice set must satisfy an incentive compatibility criterion in order to be coalitionally implementable.

An allocation $x = e + z \in \boldsymbol{A}$ is *coalitionally Bayesian incentive compatible* if it is not true that there exists a coalition S and states ω^*, ω' ($\omega^* \neq \omega'$) with $\omega' \in \bigcap_{i \notin S} E_i(\omega^*)$ such that

$$\frac{1}{\mu(E_i(\omega^*))} \sum_{\omega \in E_i(\omega^*)} u_i(\omega, e_i(\omega) + z_i(\omega'))\mu(\omega)$$
$$> \frac{1}{\mu(E_i(\omega^*))} \sum_{\omega \in E_i(\omega^*)} u_i(\omega, e_i(\omega) + z_i(\omega))\mu(\omega)$$

for every $i \in S$. In essence, this concept assures that no coalition S can make redistributions among themselves in states that the complementary coalition cannot distinguish, and become better off. In other words, if state ω^* occurs and the agents in the coalition $I \setminus S$ cannot distinguish between the state ω^* and ω', it must be the case that the agents of coalition S cannot become better off by announcing ω' instead of the actually occurred ω^*. The measurability implies that $\omega' \notin E_i(\omega^*)$ for every agent i in the coalition S.

As in Palfrey and Srivastava (1989), a *deception* for agent i is a function $\alpha_i : \mathcal{F}_i \to \mathcal{F}_i$. Let $\alpha_i^* : \mathcal{F}_i \to \mathcal{F}_i$ be the truth-telling strategy for agent i. A deception vector $\alpha = (\alpha_i)_{i \in I}$ is *compatible* with F if $\alpha(\omega) := \bigcap_{i \in I} \alpha_i(E_i(\omega)) \neq \emptyset$ for every $\omega \in \Omega$. In a direct revelation mechanism, a deception is a strategy such that for every $\omega \in \Omega$, agent i reports $\alpha_i(E_i(\omega))$ instead of $E_i(\omega)$. Notice that when $\sigma_i : \mathcal{F}_i \to M_i$ is a strategy and α_i is a deception of agent i, their composition $\sigma_i \circ \alpha_i : \mathcal{F}_i \to M_i$ is also a strategy of agent i.

We use the following notation: $E^S(\omega) = \bigcap_{i \in S} E_i(\omega)$, $E^{-S}(\omega) = \bigcap_{i \notin S} E_i(\omega)$, $\alpha_S(\omega) = E^S_\alpha(\omega) = \bigcap_{i \in S} \alpha_i(E_i(\omega))$, $\alpha_{-S}(\omega) = E^{-S}_\alpha(\omega) = \bigcap_{i \notin S} \alpha_i(E_i(\omega))$, $(\sigma \circ \alpha)_S = (\sigma_i \circ \alpha_i)_{i \in S}$.[9] Let $z \in \boldsymbol{Z}$ be a feasible net trade. If $\alpha(\omega) \neq \emptyset$, let $z \circ \alpha(\omega) = z(\alpha(\omega)) = z(\omega')$ for all $\omega' \in \alpha(\omega)$, otherwise let $(z \circ \alpha)(\omega) = 0$. Note that $(z \circ \alpha)_i = z_i \circ \alpha$ and $(z \circ \alpha^*)(\omega) = z(\omega)$. Recall from Lemma 1 of Palfrey and Srivastava (1989, p. 120) that for every $i \in I$, if $\omega' \in E_i(\omega)$, then $\alpha(\omega') \subset E_i(\alpha(\omega))$ for every $i \in I$, where $E_i(\alpha(\omega))$ is the event that contains $\alpha(\omega)$. In view of this Lemma, we immediately conclude that if $z \in \boldsymbol{Z}$, then $z \circ \alpha \in \boldsymbol{Z}$ for every deception α.

Using the notion of deception, we can define coalitional Bayesian incentive compatibility as follows.

---

[9] For example, consider the following information structure:

$$\mathcal{F}_1 = \{\{\omega_1, \omega_2\}, \{\omega_3\}\}, \mathcal{F}_2 = \{\{\omega_1, \omega_3\}, \{\omega_2\}\}, \mathcal{F}_3 = \{\{\omega_1\}, \{\omega_2\}, \{\omega_3\}\}$$

Let us define a deception α as follows: for every ω, $\alpha_i(E_i(\omega)) = E_i(\omega_1), \forall i = 1, 2$ and $\alpha_3(E_3(\omega)) = E_3(\omega)$. Then for the coalition $S = \{1, 3\}$, $\alpha^*_S(\omega_3) = E^S(\omega_3) = \{\omega_3\}$, $\alpha^*_{-S}(\omega_3) = E^{-S}(\omega_3) = \{\omega_1, \omega_3\}$, $\alpha_S(\omega_3) = E^S_\alpha(\omega_3) = \{\omega_1\}$.

Definition 2.4.1. *A social choice set Γ is said to be* coalitionally Bayesian incentive compatible (CBIC) *if for every $x = e + z \in \Gamma$, it is not true that there exists a state $\omega \in \Omega$, a coalition $S \subset I$, and a deception $\alpha_S : \prod_{i \in S} \mathcal{F}_i \to \prod_{i \in S} \mathcal{F}_i$ such that for every $i \in S$,*

$$V_i(\omega, e_i + [z \circ (\alpha_S, \alpha^*_{-S})]_i) > V_i(\omega, x_i),$$

where $e + z \circ (\alpha_S, \alpha^*_{-S}) \in \boldsymbol{A}$.

This notion of incentive compatibility states that it is not possible for any coalition S to become better off by announcing a false event. Observe that if S is a singleton, then the CBIC condition is reduced to standard Bayesian incentive compatibility. It is straightforward to show (see Theorem 2.4.1 below) that the coalitional Bayesian incentive compatibility is a necessary condition for coalitional implementation, *i.e.*, if a social choice set is implementable as a coalitional Bayesian Nash equilibrium, then it is coalitionally Bayesian incentive compatible. Note that this is the counterpart of the standard Bayesian Nash implementation results (see, for example, Jackson, 1991; Palfrey and Srivastava, 1989; Postlewaite and Schmeidler, 1986), *i.e.*, if a social choice set is implementable as a Bayesian Nash equilibrium, then it is Bayesian incentive compatible.

Theorem 2.4.1. *If a social choice set Γ is c-implementable, then it is CBIC.*

Proof. Let (M, f) c-implement Γ and $x = e + z \in \Gamma$. Then there is a CBNE σ^* with $f(\sigma^*) = z$. Now suppose that x is not CBIC, then there exists a state $\omega \in \Omega$, a coalition $S \subset I$, and a deception $\alpha_S : \prod_{i \in S} \mathcal{F}_i \to \prod_{i \in S} \mathcal{F}_i$ such that for every $i \in S$,

$$V_i(\omega, e_i + [z \circ (\alpha_S, \alpha^*_{-S})]_i) > V_i(\omega, x_i),$$

with $e + [z \circ (\alpha_S, \alpha^*_{-S})] \in \boldsymbol{A}$, which is equivalent to

$$V_i(\omega, e_i + f_i((\sigma^* \circ \alpha)_S, \sigma^*_{-S})) > V_i(\omega, e_i + f_i(\sigma^*)),$$

a contradiction to the fact that σ^* is a CBNE for the mechanism (M, f). Hence, Γ is CBIC. □

2.5 Coalitional Bayesian monotonicity

In the literature of Nash implementation with complete information, Maskin (1977) first recognized that a monotonicity condition is necessary. The Maskin-type monotonicity condition states the following: Denote the ex post preference of agent i at the state ω by $\succeq_i(\omega)$. If the outcome x is in a social choice set $\Gamma(\omega)$ and $x \notin \Gamma(\omega')$ where $\omega' \neq \omega$, then there exist an agent i and an outcome x' such that $x \succeq_i(\omega)\, x'$ but $x' \succ_i(\omega')\, x$ (see also Saijo, 1988; Williams, 1986). In an incomplete information setting, Palfrey and Srivastava (1987, 1989) and Postlewaite and Schmeidler (1986) introduced a Bayesian monotonicity condition, which is an extension of that of Maskin (1977). Below we introduce a coalitional form of Bayesian monotonicity.

Definition 2.5.1. *A social choice set Γ satisfies* Bayesian monotonicity (BM) *if for every $x = e + z \in \Gamma$, whenever $e + z \circ \alpha \in \boldsymbol{A} \setminus \Gamma$ for α compatible with*

F, there exists a state $\omega \in \Omega$, an agent $i \in I$, and a net trade $z' \in \boldsymbol{Z}$ such that $e + z' \circ \alpha \in \boldsymbol{A}$, $e + z' \circ (\alpha_i, \alpha^*_{-i}) \in \boldsymbol{A}$,

(1) $V_i(\omega, e_i + (z' \circ \alpha)_i) > V_i(\omega, e_i + (z \circ \alpha)_i)$, and
(2) $V_i(\omega', e_i + z_i) \geq V_i(\omega', e_i + [z' \circ (\alpha_i, \alpha^*_{-i})]_i)$, $\forall \omega' \in \Omega$.

In our context, Palfrey and Srivastava (1989) require instead of (2) above that:

(2′) $V_i(\omega', e_i + z_i) \geq V_i(\omega', e_i + [z' \circ (\alpha^{\omega}_i, \alpha^*_{-i})]_i)$, $\forall \omega' \in E^{-i}_{\alpha}(\omega)$,

where $\alpha^{\omega}_i(E_i) = \alpha_i(E_i(\omega))$ for every $E_i \in \mathcal{F}_i$. Our definition of Bayesian monotonicity is not directly comparable with those of Palfrey and Srivastava (1989) and Jackson (1991) because of the private information measurability. But if we impose the private information measurability on their notions, our definition is the same with that of Jackson (1991) since his conditions must hold for all deceptions (including incompatible ones) but incompatible deceptions here make the first condition violated. However, our definition is stronger than that of Palfrey and Srivastava (1989) since they require the second condition to hold only for the states which the other agents collectively report and for the restricted deceptions. Below we introduce a coalitional form of the above definition.

Definition 2.5.2. *A social choice set Γ satisfies* coalitional Bayesian monotonicity (CBM) *if for every $x = e + z \in \Gamma$, whenever $e + z \circ \alpha \in \boldsymbol{A} \setminus \Gamma$ for α compatible with F, there exists a state $\omega \in \Omega$, a coalition $S \subset I$, and a net trade $z' \in \boldsymbol{Z}$ such that $e + z' \circ \alpha \in \boldsymbol{A}$, $e + z' \circ (\alpha_S, \alpha^*_{-S}) \in \boldsymbol{A}$,*

(1) $\forall i \in S$, $V_i(\omega, e_i + (z' \circ \alpha)_i) > V_i(\omega, e_i + (z \circ \alpha)_i)$, and
(2) $\exists i \in S$, $V_i(\omega', e_i + z_i) \geq V_i(\omega', e_i + [z' \circ (\alpha_S, \alpha^*_{-S})]_i)$, $\forall \omega' \in \Omega$.

Note that if S is a singleton, the coalitional Bayesian monotonicity is equivalent to the Bayesian monotonicity. Since $\{i\}$ is a coalition, the Bayesian monotonicity implies the coalitional Bayesian monotonicity but not vice versa. This means that if an allocation is eliminated by the Bayesian Nash equilibrium criterion, then it must be excluded by the coalitional Bayesian Nash equilibrium criterion. The following theorem and its proof shed light on the implications of coalitional Bayesian monotonicity to the coalitional implementation. It is the coalitional counterpart of a result in Palfrey and Srivastava (1989, Theorem 2, p. 124).

Theorem 2.5.1. *If a social choice set Γ is c-implementable, then it satisfies the CBM condition.*

Proof. Let (M, f) c-implement Γ and $x = e+z \in \Gamma$. Then there exists a CBNE σ of (M, f) with $f(\sigma) = z$. Assume that for some α compatible with F, $e+z\circ\alpha \in \boldsymbol{A}\setminus\Gamma$. Note that $f(\sigma \circ \alpha) = z \circ \alpha$. Since Γ is c-implementable and $e + f(\sigma \circ \alpha) = e+z\circ\alpha \in \boldsymbol{A}\setminus\Gamma$, the strategy vector $\sigma\circ\alpha$ is not a CBNE. Therefore, there exists a state $\omega \in \Omega$, a coalition $S \subset I$, and a strategy vector $\sigma'_S \in \prod_{i \in S} \mathcal{F}_i \to \prod_{I \in S} M_i$ such that for every $i \in S$, $V_i(\omega, e_i + f_i(\sigma'_S, (\sigma\circ\alpha)_{-S})) > V_i(\omega, e_i + f_i(\sigma\circ\alpha))$.[10]

[10] Since it does not matter which message agent $i \in S$ sends at $\omega' \notin E_i(\omega)$, without loss of generality, we can choose σ'_S such that $\sigma'_i(E_i(\omega')) = \sigma'_i(E_i(\omega))$ at every $\omega' \in \Omega$ for all $i \in S$.

Now for every $i \in S$, define $\overline{\sigma}_i$ by $\overline{\sigma}_i(E_i) = \sigma'_i(E_i(\omega))$ for every $E_i \in \mathcal{F}_i$ and let $z' = f(\overline{\sigma}_S, \sigma_{-S})$. Then since $z' \circ \alpha = f(\sigma'_S, (\sigma \circ \alpha)_{-S})$, it follows that $V_i(\omega, e_i + (z' \circ \alpha)_i) > V_i(\omega, e_i + (z \circ \alpha)_i)$ for every $i \in S$. Note that $z' \circ (\alpha_S, \alpha^*_{-S}) = f(\sigma'_S, \sigma_{-S})$. Since σ is a CBNE, the coalition S with the strategy σ'_S cannot improve upon σ. That is, there exists some $i \in S$ such that $V_i(\omega', e_i + z_i) = V_i(\omega', e_i + f_i(\sigma)) \geq V_i(\omega', e_i + f_i(\sigma'_S, \sigma_{-S})) = V_i(\omega', e_i + [z' \circ (\alpha_S, \alpha^*_{-S})]_i)$ for every $\omega' \in \Omega$. Hence Γ satisfies the CBM condition. □

2.6 Closure

Denote by $\bigwedge_{i \in I} \mathcal{F}_i$ the *finest common coarsening* of $\{\mathcal{F}_i : i \in I\}$, *i.e.*, the finest partition of Ω which is coarser than $\mathcal{F}_i$ for every $i \in I$. An event E is said to be *common knowledge at* ω if $(\bigwedge_{i \in I} \mathcal{F}_i)(\omega) \subset E$ where $(\bigwedge_{i \in I} \mathcal{F}_i)(\omega)$ is the event of $\bigwedge_{i \in I} \mathcal{F}_i$ containing ω. Notice that $(\bigwedge_{i \in I} \mathcal{F}_i)(\omega)$ itself is common knowledge at ω. We also call $\bigwedge_{i \in I} \mathcal{F}_i$ the *common knowledge partition* of Ω.

Following Postlewaite and Schmeidler (1986), we define z^* to be the *common knowledge concatenation* of $\{z^k \in L : k = 1, \ldots, m\}$ if $z^*(\omega) = \sum_{k=1}^m z^k(\omega)\chi_{E^k}(\omega)$ where $\{E^k : k = 1, \ldots, m\}$ is the common knowledge partition of Ω. Let $\{z^k \in L : k = 1, \ldots, m\}$ be a collection of net trades such that $e + z^k \in \Gamma$. If the common knowledge concatenation z^* of $\{z^k : k = 1, \ldots, m\}$ has the property that $e + z^* \in \Gamma$, then Γ is said to satisfy *closure (C)*. It turns out that a c-implementable social choice set Γ satisfies the closure condition as Lemma below indicates.

Lemma 2.6.1. *If a social choice set Γ is c-implementable, then it satisfies the condition C.*

Proof. Suppose that (M, f) c-implements Γ. Let $\{E^k : k = 1, \ldots, m\}$ be the common knowledge partition and $e + z^k \in \Gamma$ for $k = 1, \ldots, m$. Define $z^* = \sum_{k=1}^m z^k \cdot \chi_{E^k}$. We must show that $e + z^* \in \Gamma$. Let σ^k be a CBNE such that $f(\sigma^k) = z^k$. Then the strategy vector σ defined by $\sigma(E(\omega)) = \sum_{k=1}^m \sigma^k(E(\omega))\chi_{E^k}(\omega)$ is also a CBNE. For otherwise there exists a state $\omega \in \Omega$, a coalition $S \subset I$, and $\sigma'_S : \prod_{i \in S} \mathcal{F}_i \to \prod_{i \in S} M_i$ such that for every $i \in S$,

$$V_i(\omega, e_i + f_i(\sigma'_S, \sigma_{-S})) > V_i(\omega, e_i + f_i(\sigma)),$$

which is equivalent to

$$V_i(\omega, e_i + f_i(\sigma'_S, \sigma^k_{-S})) > V_i(\omega, e_i + f_i(\sigma^k)),$$

where $\omega \in E^k$ for some k. Then σ^k is not a CBNE, a contradiction. Furthermore, $f(\sigma) = z^*$. Since Γ is c-implementable, $e + z^* \in \Gamma$. □

2.7 Sufficient conditions for coalitional implementation

In this section, we will show that interim individual rationality, interim efficiency, coalitional Bayesian incentive compatibility, coalitional Bayesian monotonicity, and closure are sufficient conditions for coalitional implementation. As in the previous literature, the proof is constructive. It is an extension of the constructions in Postlewaite and Schmeidler (1986) and Palfrey and Srivastava (1989), which allows us to consider deviations by coalitions. It should be noted that, as in Hurwicz et al. (1984), our mechanism is not wasteful and also maintains the feasibility of the outcomes out of equilibrium.

Before stating the main theorem, it is worth mentioning the case where there is only one good in the economy. If there is only one good, the measurability of allocations implies that the set of interim efficient allocations is equivalent to the set of feasible allocations. In this case, the initial endowment is the unique interim efficient and interim individually rational allocation (see footnote 15) and it is clearly c-implementable. It is enough to consider the mechanism (M, f) where $f(\sigma) = 0$ for every strategy profile σ. Hence, in the theorem below it is assumed that there is more than one good.

Theorem 2.7.1. *Assume that $N \geq 3$. If a social choice set Γ is IIR, IE, CBIC, and satisfies CBM and C, then it is c-implementable.*

Proof. Consider the message space of agent i, $M_i = \{m_i = (E_i, z^i, n_i) \in \mathcal{F}_i \times \boldsymbol{Z} \times \boldsymbol{N}_0 : e + z^i \in \Gamma\}$ for every i, where $\boldsymbol{N}_0 = \{0, 1, 2, 3, \dots\}$. Thus every agent i reports his/her private information event E_i, net trade profile $z^i = (z^i_j)_{j \in I}$ of the economy, and a nonnegative integer n_i. In principle, we can divide the message space M into two main groups. One is a region where the reported private information events have nonempty intersection. In this region, the mechanism designer cannot tell whether someone is lying about his/her private information event. This region consists of M^0, M^1, M^4, and M^5 (see below for the definitions of the regions and outcome function). In the region M^6 where the reported information events have empty intersection, some agent reports a non-zero integer. In M^6, the mechanism designer knows that some one is lying about his/her private information event. The outcome function in this region assigns no trade. The remaining regions are M^2 and M^3. When all agents report the integer zero and every agent except agent 1 reports the same net trade configuration, the message belongs to the region M^2. The mechanism makes agent 1 give away his/her reported endowments to the other agents who will equally share them with each other. In the region M^3, where all agents report the integer 0 but the message does not belong to either M^0 or M^2, agent 1 takes the reported endowments of all the other agents.

For the former regions (*i.e.*, M^0, M^1, M^4, M^5), more explanation is needed. First of all, in the region M^0, every agent agrees on the net trade profile of the economy and the integer zero, but reports his/her own private information event. In this case, the outcome function assigns the agreed net trade at the consented states. In the region $M^1(S)$, agents in the coalition S unanimously report the net trade profile of the economy and a nonzero integer, but they report their own private information events. However, the agents in the complementary coalition use strategies in the same fashion as in M^0. The outcome function assigns the net trade suggested by the coalition S at the agreed states in $M^{11}(S)$ where some agent in the coalition S does not prefer his/her proposed net trade to the net trade proposed by the complementary coalition at their agreed states. The outcome function assigns no trade in $M^{12}(S)$. In this case, every agent in the coalition S prefers the net trade of the coalition at some agreed state of the complementary coalition. In the region $M^4(S)$, agents in the coalition S use strategies without unanimity. In the complementary coalition, agents send messages in the same fashion as in M^0. In this region, the outcome is determined by the "integer game", *i.e.*, the agent who has the lowest index among the agents reporting the highest integer receives the reported endowments of all the other agents. Finally, the region M^5 collects all the messages which are not in M^0, M^1, M^2, M^3, M^4 and M^6. In particular, there is no agent who reports the integer zero. As in M^4, the outcome is determined by the integer game, but there is no tie-breaker of choosing the agent with the lowest index so that the winners (the agents reporting the highest number) evenly share the sum of the endowments of the losers.

We now formalize the above discussion. Let S be a nonempty proper coalition of I and let us write $m_i = (m_i^1, m_i^2, m_i^3)$ for each $i \in I$. For every $i \in I$, define $z_i[E_S](\omega) = z_i(\omega')$ where $\omega' \in E^S \cap E^{-S}(\omega)$. Define the sets:

$$\begin{aligned}
M^0 &= \{m \in M : m_i = (E_i, z, 0), \forall i \in I; \bigcap_{i \in I} E_i \neq \emptyset\}, \\
M^1(S) &= \{m \in M \setminus M^0 : m_i = (E_i, z', n), n \neq 0, \forall i \in S; m_i = (E_i, z, 0), \\
&\qquad \forall i \notin S; \bigcap_{i \in I} E_i \neq \emptyset\}, \\
M^{11}(S) &= \{m \in M^1(S) : \exists i \in S, V_i(\omega, e_i + z_i) \geq V_i(\omega, e_i + z_i'[E_S]), \\
&\qquad \forall \omega \in E^{-S}\}, \\
M^{12}(S) &= M^1(S) \setminus M^{11}(S), \\
M^1 &= \bigcup_S M^1(S), \\
M^2 &= \{m \in M : m_1 = (E_1, z', 0); m_i = (E_i, z, 0), \forall i \neq 1\}, \\
M^3 &= \{m \in M \setminus (M^0 \cup M^2) : m_i^3 = 0, \forall i \in I\}, \\
M^4(S) &= \{m \in M \setminus \bigcup_{k=0}^{3} M^k : m_i = (E_i, z, 0), \forall i \notin S; \bigcap_{i \in I} E_i \neq \emptyset\}, \\
M^4 &= \bigcup_S M^4(S), \\
M^5 &= \{m \in M \setminus \bigcup_{k=0}^{4} M^k : \bigcap_{i \in I} m_i^1 \neq \emptyset\}, \\
M^6 &= \{m \in M \setminus M^3 : \bigcap_{i \in I} m_i^1 = \emptyset\}.
\end{aligned}$$

Define the outcome function $f : M \to \hat{Z}$ as follows: For every $i \in I$,

$$f_i(m) = \begin{cases} z_i(\omega), & \omega \in \bigcap_{i\in I} m_i^1 \quad \text{if } m \in M^0, \\ z_i'(\omega), & \omega \in \bigcap_{i\in I} m_i^1 \quad \text{if } m \in M^{11}(S) \text{ for some } S, \\ 0 & \text{if } m \in M^{12}(S) \text{ for some } S, \\ e_1(\omega)/(N-1), & \omega \in m_1^1 \quad \text{if } m \in M^2, i \neq 1, \\ -e_1(\omega), & \omega \in m_1^1 \quad \text{if } m \in M^2, i = 1, \\ \sum_{j\neq i} e_j(\omega_j), & \omega_j \in m_j^1 \quad \text{if } m \in M^3, i = 1, \\ -e_i(\omega_i), & \omega_i \in m_i^1 \quad \text{if } m \in M^3, i \neq 1, \\ \sum_{j\neq i} e_j(\omega), & \omega \in \bigcap_{i\in I} m_i^1 \quad \text{if } m \in M^4, i = \min\{k : k \in K\}, \\ -e_i(\omega), & \omega \in \bigcap_{i\in I} m_i^1 \quad \text{if } m \in M^4, i \neq \min\{k : k \in K\}, \\ \frac{1}{|K|}\sum_{j\notin K} e_j(\omega), & \omega \in \bigcap_{i\in I} m_i^1 \quad \text{if } m \in M^5, i \in K, \\ -e_i(\omega), & \omega \in \bigcap_{i\in I} m_i^1 \quad \text{if } m \in M^5, i \notin K, \\ 0 & \text{if } m \in M^6, \end{cases}$$

where $K = \{k \in I : m_k^3 = \max_{i\in I} m_i^3\}$.

Since the mechanism is common knowledge, every agent knows that his/her (implicitly) reported endowment can be confiscated. Therefore, it is must be the case that when each agent reports his/her private information (and implicitly reports his/her initial endowment), he/she cannot overreport his/her initial endowment. That is, for every $i \in I$ and every α_i, $e_i(\omega') \leq e_i(\omega)$ for every $\omega' \in m_i^1 = \alpha_i(E_i(\omega))$ when ω occurs. Since $e_i(\omega) + z_i^j(\omega') \geq e_i(\omega') + z_i^j(\omega') \geq 0$ for $\omega' \in \bigcap_{i\in I} m_i^1$ with $m \in M^0 \cup M^1(S)$ for some S and for every $i, j \in I$ when state ω occurs, it follows that the allocations induced by the mechanism are always positive, *i.e.*, $e_i(\omega) + f_i(m) \geq 0$ for every i, every ω and every m. Furthermore, $\sum_{i\in i} f_i(m) = 0$ for every $m \in M$. Hence the mechanism is feasible.

By Lemma 2.7.2 below, for every $x = e + z \in \Gamma$, we have a CBNE σ for (M, f) such that $f(\sigma) = z$. By Lemma 3.7.4 below, we conclude that for every CBNE strategy σ for (M, f), $e + f(\sigma) \in \Gamma$. Hence Γ is c-implementable. □

Lemma 2.7.2 below establishes that the mechanism of Theorem 2.7.1 satisfies the first requirement for coalitional implementation (condition (1) of Definition 2.2.3). Lemma 2.7.4 below shows that the mechanism of Theorem 2.7.1 satisfies the second requirement for coalitional implementation (condition (2) of Definition 2.2.3).

Lemma 2.7.2. *For every $x = e + z \in \Gamma$, let σ be such that $\sigma_i(E_i(\omega)) = (E_i(\omega), z, 0)$ for all i and for all ω. Then σ is a CBNE for the mechanism (M, f) and $f(\sigma) = z$.*

Proof. See Appendix. □

For the proof of Lemma 2.7.4 we need a result (Lemma 2.7.3) which guarantees that no CBNE message lies outside of the region M^0. Indeed, if a message lies inside the region M^0, there is no profitable coalitional deviation. Moreover, if a message lies outside of the region M^0, there is always a profitable coalitional deviation.

Lemma 2.7.3. *If σ is a CBNE for (M, f), then $\sigma(E(\omega)) \in M^0$ for all $\omega \in \Omega$.*

Proof. See Appendix. □

Although all agents do not truthfully report their private information events, the equilibrium still belongs to the social choice set as long as they agree on the net trade configuration and the integer zero.

Lemma 2.7.4. *If σ is a CBNE for (M, f), then $e + f(\sigma) \in \Gamma$.*

Proof. See Appendix. □

Remark 2.7.1. If there is only one agent in the economy, the initial endowment is the unique feasible allocation, which is trivially c-implementable. Assume that $N = 2$ and that the initial endowment is not interim efficient. If a social choice set Γ is IR, IE, CBIC, and satisfies CBM and C, then it is c-implementable. The proof of Theorem 2.7.1 can be modified as follows. The mechanism f assigns the same net trade as before except on M^2, where f assigns no trade. Thus, there is no profitable deviation from M^0 to M^2 and there is a profitable deviation from M^2 to M^0, since the initial endowment is not Pareto optimal. Note that $M^3 = \emptyset$. The other arguments continue to hold.

Remark 2.7.2. It should be noted that for the complete information case Dutta and Sen (1991) provided a strong Nash implementation theorem, which is different than ours. In particular, they do not require individual rationality. In economic environments, they identified only the sufficiency conditions for strong Nash implementation. One can substitute the strong Maskin monotonicity condition with the individual rationality condition and the coalitional monotonicity condition (which is weaker than the strong Maskin monotonicity) to get a full charaterization result for strong Nash implementation.

3 Coalitional implementation of the private core

The core notion defined below (see also Yannelis, 1991) does not necessarily allow agents in a coalition to share their private information. In fact, allowing agents in a coalition to use either their common knowledge information or their pooled information, one may face serious problems as Example 3.1 will indicate (see also Koutsougeras and Yannelis, 1993, Section 5). More importantly, however, we will show in Section 5 that a core notion which allows for pooling of information may not be c-implementable.

Definition 3.1. *An allocation $x \in \boldsymbol{A}$ is an* (ex ante) private core allocation *of the economy with differential information $\mathcal{E}$ if it is not true that there exists a coalition $S \subset I$ and $(y_i)_{i \in S} \in \prod_{i \in S} L_{X_i}$ such that*

(1) $\sum_{i \in S} y_i = \sum_{i \in S} e_i$, and
(2) for every $i \in S$,

$$\sum_{\omega \in \Omega} u_i(\omega, y_i(\omega))\mu(\omega) > \sum_{\omega \in \Omega} u_i(\omega, x_i(\omega))\mu(\omega).$$

The (ex ante) private core *is the set of all ex ante private core allocations for $\mathcal{E}$.*

Definition 3.2. *An allocation $x \in \boldsymbol{A}$ is an* (interim) private core allocation *of the economy with differential information $\mathcal{E}$ if it is not true that there exists a state $\omega \in \Omega$, a coalition $S \subset I$, and $(y_i)_{i \in S} \in \prod_{i \in S} L_{X_i}$ such that*

(1) $\sum_{i \in S} y_i = \sum_{i \in S} e_i$, *and*
(2) for every $i \in S$, $V_i(\omega, y_i) > V_i(\omega, x_i)$.

The (interim) private core *is the set of all interim private core allocations for $\mathcal{E}$ and it is denoted by $\boldsymbol{C}(\mathcal{E})$.*

The only difference between the two above concepts is that agents in the ex ante private core use their ex ante expected utility functions and in the interim private core, their interim expected utility functions. The example below will illustrate that despite the fact that they have the same properties, i. e., they are coalitional incentive compatible and they take into account the informational superiority of an individual, they may be different.

Example 3.1. Consider an economy with differential information with three agents, one good, and three states (*i.e.*, $\Omega = \{\omega_1, \omega_2, \omega_3\}$) with equal probability (*i.e.*, $\mu(\{\omega\}) = 1/3$ for every $\omega \in \Omega$), where utility functions, initial endowment, and private information sets are given as follows:

$$\begin{aligned} u_1(\omega, x) &= \sqrt{x},\ e_1 = (10, 10, 0),\ \mathcal{F}_1 = \{\{\omega_1, \omega_2\}, \{\omega_3\}\}, \\ u_2(\omega, x) &= \sqrt{x},\ e_2 = (10, 0, 10),\ \mathcal{F}_2 = \{\{\omega_1, \omega_3\}, \{\omega_2\}\}, \\ u_3(\omega, x) &= \sqrt{x},\ e_3 = (0, 0, 0),\quad \mathcal{F}_3 = \{\{\omega_1\}, \{\omega_2\}, \{\omega_3\}\}. \end{aligned}$$

It can be shown that the allocation $x^* = (x_1, x_2, x_3)$ is in the ex ante private core where

$$x_1^* = (8, 8, 2),\ x_2^* = (8, 2, 8),\ x_3^* = (4, 0, 0). \tag{3.1}$$

In the above example, agents 1 and 2 cannot undertake any risk sharing among themselves (the trades between agents 1 and 2 are state independent and these trades do not make them better off) without agent 3. Since agent 3 has superior information, she acts as an intermediary who executes the correct trades (makes a Pareto improvement) and as a consequence gets rewarded for this service.

It should be noted that the allocation (3.1) in the ex ante private core is entirely different than that of any traditional rational expectations equilibrium (REE). Indeed, in any REE, agent 3 gets zero because his/her budget set is zero in each state. However, in any ex ante private core allocation, agent 3 gets positive consumption[11] in state ω_1. It follows that agent 3 plays the role of an intermediary who makes a Pareto improvement for the economy as a whole and he/she gets rewarded for this. It is important to note that if the private information of agent

[11] This can be proved as follows: Suppose not and let x be an ex ante private core allocation such that agent 3's consumption at state ω_1 is zero. Since $x \in \boldsymbol{A}$, $x_3 = (0, 0, 0)$ and $x_1 - e_1 = -(x_2 - e_2)$. Since $x_i - e_i$ is $\mathcal{F}_i$-measurable for $i = 1, 2$, $x_i - e_i$ is $(\mathcal{F}_1 \wedge \mathcal{F}_2)$-measurable for $i = 1, 2$. Note that $\mathcal{F}_1 \wedge \mathcal{F}_2 = \{\{\omega_1, \omega_2, \omega_3\}\}$. Therefore, $x_1 - e_1 = (c, c, c)$ and $x_2 - e_2 = (-c, -c, -c)$ for some $c \in \boldsymbol{R}$. If $c < 0$, agent 1 blocks x since $x_1 < e_1$. If $c > 0$, agent 2 blocks x for the same reason. Thus $x = e$. However, the grand coalition with the allocation given in the above Example blocks $x = e$, a contradiction.

3 is $\mathcal{F}'_3 = \{\{\omega_1, \omega_2, \omega_3\}\}$, then agent 3 gets $(0, 0, 0)$ and in this case the initial endowment is the unique ex ante private core allocation. Hence, contrary to the REE,[12] changes in the private information of an agent have effects on the resulting ex ante private core.[13]

Suppose now that agents 1 and 2 pool their information to obtain the allocation:

$$x_1^* = x_2^* = (10, 5, 5),\ x_3^* = (0, 0, 0).$$

However, such a contract may not be viable because the above allocation is not incentive compatible. Simply notice that agent 1 becomes better off by reporting state ω_3 if state ω_1 occurs and agent 2 cannot distinguish ω_1 from ω_3. Using the same reasoning, one can easily see that agent 2 has an incentive to report ω_2 whenever state ω_1 occurs and agent 1 cannot detect his because he/she is not able to distinguish ω_1 from ω_2. Hence, pooling of information violates coalitional incentive compatibility (see also Example 5.1 in Section 5).

Finally, notice that the initial endowment is the *unique* interim private core allocation[14] and it is not in the ex ante private core since the initial endowment is blocked by the grand coalition with the allocation x^* given by (3.1).

In the next example, we have an interim private core allocation which is not the initial endowment. Hence, the example below indicates that an interim private core allocation exists.

Example 3.2. Consider an economy with differential information with two agents, two goods (*i.e.*, x^1, x^2), and three equally probable states, where utility functions, random initial endowments, and private information structures are given as follows:

$$u_1(\omega, x^1, x^2) = \sqrt{x^1 x^2},\ \forall\omega\quad e_1 = ((3,1),(3,1),(5,3)),\ \mathcal{F}_1 = \{\{\omega_1, \omega_2\}, \{\omega_3\}\},$$
$$u_2(\omega, x^1, x^2) = \sqrt{x^1 x^2},\ \forall\omega,\ e_2 = ((1,3),(3,5),(3,5)),\ \mathcal{F}_2 = \{\{\omega_1\}, \{\omega_2, \omega_3\}\}.$$

The allocation

$$x^* = (x_1^*, x_2^*) = (((2,2),(2,2),(4,4)),((2,2),(4,4),(4,4)))$$

is the unique interim private core allocation which is different from the initial endowment.

[12] Notice that by changing the private information of agent 3 from $\mathcal{F}_3 = \{\{\omega_1\}, \{\omega_2\}, \{\omega_3\}\}$ to $\mathcal{F}'_3 = \{\{\omega_1, \omega_2, \omega_3\}\}$, the REE does not affect the consumption of agent 3, *i.e.*, he/she always gets zero since his/her budget set is zero in every state.

[13] Similar examples can be constructed for the interim private core, the ex ante private value, and the interim private value (see also Krasa and Yannelis, 1994, Section 4).

[14] In an economy with one good per state, *i.e.*, the interim private core is the initial endowment. First, notice that the initial endowment is in the interim private core. Otherwise, there exists a state ω, a coalition S, and $(y_i)_{i \in S}$ such that $\sum_{i \in S} y_i = \sum_{i \in S} e_i$ and $V_i(\omega, y_i) > V_i(\omega, e_i)$ for every $i \in S$. Since there is only one good, by monotonicity and measurability, we have $y_i(\omega) > e_i(\omega)$ for every $i \in S$, a contradiction. If there is another interim private core allocation $x \neq e$, the feasibility implies that there is an agent $i \in I$ such that $e_i(\omega) > x_i(\omega)$ for some $\omega \in \Omega$. Since there is only one good, $V_i(\omega, e_i) > V_i(\omega, x_i)$ by measurability and monotonicity. This implies that this agent is a blocking coalition against x at ω, a contradiction.

In order to show that the interim private core is c-implementable, we will need some Lemmata.

Lemma 3.1. *The interim private core $\boldsymbol{C}(\mathcal{E})$ is IIR and IE.*

Proof. It is immediate from the definition. □

Lemma 3.2. *The interim private core $\boldsymbol{C}(\mathcal{E})$ satisfies the CBM condition.*

Proof. Let $x = e + z \in \boldsymbol{C}(\mathcal{E})$ and $e + z \circ \alpha \in \boldsymbol{A} \setminus \boldsymbol{C}(\mathcal{E})$. We must show that there exists a state $\omega^* \in \Omega$, a coalition $S \subset I$, and a net trade $z' \in \boldsymbol{Z}$ such that $e + z' \circ \alpha \in \boldsymbol{A}$, $e + z' \circ (\alpha_S, \alpha^*_{-S}) \in \boldsymbol{A}$,

(a) for every $i \in S$, $V_i(\omega^*, e_i + (z' \circ \alpha)_i) > V_i(\omega^*, e_i + (z \circ \alpha)_i)$, and
(b) for some $i \in S$, $V_i(\omega', e_i + z_i) \geq V_i(\omega', e_i + [z' \circ (\alpha_S, \alpha^*_{-S})]_i)$ for all $\omega' \in \Omega$.

Since $e + z \circ \alpha \notin \boldsymbol{C}(\mathcal{E})$, there exists a state $\omega^* \in \Omega$, a coalition $S \subset I$, and $z^* \in \boldsymbol{Z}$ such that $\sum_{i \in S} z_i^* = 0$ and for every $i \in S$,

$$V_i(\omega^*, e_i + z_i^*) > V_i(\omega^*, e_i + (z \circ \alpha)_i). \tag{3.2}$$

Now define $z' = (z_i')_{i \in I} \in \boldsymbol{Z}$ by

$$z_i'(\omega') = \begin{cases} z_i^*(\omega^*) & \text{if } \omega' \in (\bigwedge_{i \in I} \mathcal{F}_i)(\alpha(\omega^*)), \\ 0 & \text{otherwise.} \end{cases}$$

Then $e + z' \circ \alpha \in \boldsymbol{A}$ and $(z' \circ \alpha)_i(\omega^*) = z_i^*(\omega^*)$ for every $i \in S$. Thus, it follows from (3.2) that for every $i \in S$,

$$V_i(\omega^*, e_i + (z' \circ \alpha)_i) > V_i(\omega^*, e_i + (z \circ \alpha)_i).$$

Thus, condition (a) holds.

Also note that for every $\omega' \in \Omega$,

$$[z' \circ (\alpha_S, \alpha^*_{-S})](\omega') = \begin{cases} z^*(\omega^*), & \text{if } E^S_\alpha(\omega') \subset (\bigwedge_{i \in I} \mathcal{F}_i)(\alpha(\omega^*)), \\ 0, & \text{otherwise,} \end{cases}$$

which implies that $e + z' \circ (\alpha_S, \alpha^*_{-S}) \in \boldsymbol{A}$. Since $e + z \in \boldsymbol{C}(\mathcal{E})$, it must be the case that for some $i \in S$,

$$V_i(\omega', e_i + z_i) \geq V_i(\omega', e_i + [z' \circ (\alpha_S, \alpha^*_{-S})]_i), \ \forall \omega' \in \Omega.^{15} \tag{3.3}$$

Hence condition (b) holds, and this completes the proof of the Lemma. □

Lemma 3.3. *The interim private core $\boldsymbol{C}(\mathcal{E})$ is CBIC.*

Proof. Let $x = e + z \in \boldsymbol{C}(\mathcal{E})$ and suppose that x is not CBIC. Then there exists a state $\omega \in \Omega$, a coalition S, and a deception $\alpha_S : \prod_{i \in S} \mathcal{F}_i \to \prod_{i \in S} \mathcal{F}_i$ such that $(\alpha_S, \alpha^*_{-S})$ is compatible with F and for every $i \in S$,

$$V_i(\omega, e_i + [z \circ (\alpha_S, \alpha^*_{-S})]_i) > V_i(\omega, x_i),$$

[15] For if (3.3) does not hold, then there is a state $\omega \in \Omega$ such that $V_i(\omega, e_i + [z' \circ (\alpha_S, \alpha^*_{-S})]_i) > V_i(\omega, e_i + z_i)$ for every $i \in I$, a contradiction to the fact that $e + z \in \boldsymbol{C}(\mathcal{E})$.

where $e + z \circ (\alpha_S, \alpha^*_{-S}) \in \boldsymbol{A}$. Since for every $\omega' \in E^S_\alpha(\omega) \bigcap E^{-S}(\omega)$ it holds that $z_i(\omega') = z_i(\omega)$, *i.e.*, $[(z \circ (\alpha_S, \alpha^*_{-S})]_i(\omega) = z_i(\omega)$ for every $i \notin S$, it must be the case that for every $i \notin S$,

$$V_i(\omega, e_i + [z \circ (\alpha_S, \alpha^*_{-S})]_i) = V_i(\omega, x_i). \tag{3.4}$$

Since $V_i(\omega, \cdot)$ is continuous for every $i \in I$, there exists an $\varepsilon > 0$ such that for every $i \in S$,

$$V_i(\omega, e_i + [z \circ (\alpha_S, \alpha^*_{-S})]_i - \varepsilon \cdot \mathbf{1}) > V_i(\omega, x_i). \tag{3.5}$$

Now define $x' = (x'_i)_{i \in I}$ by

$$x'_i = \begin{cases} e_i + [z \circ (\alpha_S, \alpha^*_{-S})]_i - \varepsilon \cdot \mathbf{1} & \text{if } i \in S, \\ e_i + [z \circ (\alpha_S, \alpha^*_{-S})]_i + \frac{|S|}{|I \setminus S|}\, \varepsilon \cdot \mathbf{1} & \text{if } i \notin S. \end{cases}$$

Note that x'_i is $\mathcal{F}_i$-measurable and x' is a feasible allocation since $e + z \circ (\alpha_S, \alpha^*_{-S}) \in \boldsymbol{A}$. However, (3.5) implies that $V_i(\omega, x'_i) > V_i(\omega, x_i)$ for every $i \in S$. Because $V_i(\omega, \cdot)$ is monotone for every $i \in I$, (3.4) implies that $V_i(\omega, x'_i) > V_i(\omega, x_i)$ for every $i \notin S$, a contradiction to the fact that $x \in \boldsymbol{C}(\mathcal{E})$. □

Lemma 3.4. *The interim private core* $\boldsymbol{C}(\mathcal{E})$ *satisfies the condition C.*

Proof. Let $\{E^k : k = 1, \ldots, m\}$ be the common knowledge partition and $e + z^k \in \boldsymbol{C}(\mathcal{E})$ for $k = 1, \ldots, m$. Define $z^* = \sum_{i=1}^m z^k \cdot \chi_{E^k}$. Suppose, by way of contradiction, that $e + z^* \notin \boldsymbol{C}(\mathcal{E})$. Then there exists a state ω, a coalition $S \subset I$, and $x_S \in \prod_{i \in S} L_{X_i}$ such that $\sum_{i \in S} x_i = \sum_{i \in S} e_i$ and for every $i \in S$,

$$V_i(\omega, x_i) > V_i(\omega, e_i + z^*_i),$$

which is equivalent to

$$V_i(\omega, x_i) > V_i(\omega, e_i + z^k_i),$$

where $\omega \in E^k$ for some k. Then $e + z^k$ is not an interim private core allocation, a contradiction. □

Theorem 3.5. *If* $N \geq 3$*, the interim private core* $\boldsymbol{C}(\mathcal{E})$ *is c-implementable.*

Proof. By Lemmata 3.1 - 3.4, the interim private core $\boldsymbol{C}(\mathcal{E})$ is IIR, IE, CBIC, and satisfies CBM and C. Thus, by virtue of Theorem 2.7.1, we can conclude that the interim private core is c-implementable. □

Note that when there are two agents in the economy and the initial endowment is not interim efficient, the interim private core $\boldsymbol{C}(\mathcal{E})$ is implementable.

4 Coalitional implementation of the private value

As is the case with the private core notions defined above, the private value (see also Krasa and Yannelis, 1994) does not necessarily allow agents to share their private information. The problems that arise whenever coalitions of agents either pool their information or use their common knowledge information are discussed in Krasa and Yannelis (1994, 1996).

We introduce an interim version of a private value allocation. For each economy with differential information $\mathcal{E}$, for each state $\omega \in \Omega$, and for each set of weights $\{\lambda_i(\omega) : i \in I\}$, we can now associate a TU-game $G = (I, W)$ according to the following rule: For each $\omega \in \Omega$ and each $S \subset I$, let

$$W(\omega, S) = \max \left\{ \sum_{i \in S} \lambda_i(\omega) V_i(\omega, x_i) : \sum_{i \in S} x_i = \sum_{i \in S} e_i; x_i \in L_{X_i} \right\}. \quad (4.1)$$

The *interim Shapley value* of the TU-game $G = (I, W)$ is a rule which assigns to each agent i a payoff $\Psi_i(\omega, W)$ at each state ω, which is given by:

$$\Psi_i(\omega, W) = \sum_{S \subset I, S \ni i} \frac{(|S| - 1)!(N - |S|)!}{N!} [W(\omega, S) - W(\omega, S \setminus \{i\})].$$

Note that the interim Shapley value is individually rational and Pareto optimal, *i.e.*, $\Psi(\omega, W) \geq W(\omega, \{i\})$ for every $\omega \in \Omega$ and for every $i \in I$, and $\sum_{i \in I} \Psi(\omega, W) = W(\omega, I)$ for every ω.

Definition 4.1. *An allocation $x \in \boldsymbol{A}$ is an* (interim) private value allocation *of the economy with differential information $\mathcal{E}$ if for every $\omega \in \Omega$, there exist $\lambda(\omega) = (\lambda_i(\omega))_{i \in I} \in \boldsymbol{R}_+^N \setminus \{0\}$ such that for each $i \in I$,*

$$\lambda_i(\omega) V_i(\omega, x_i) = \Psi_i(\omega, W),$$

where $\Psi_i(\omega, W)$ is the interim Shapley value derived from the TU-game $G = (I, W)$ defined by (4.1). The interim private value *is the set of all interim private value allocations for $\mathcal{E}$ and it is denoted by $\boldsymbol{V}(\mathcal{E})$.*

Theorem 4.1. *If $N \geq 3$ and $\lambda \gg 0$, then the interim value $\boldsymbol{V}(\mathcal{E})$ is c-implementable.*

Proof. Since the interim private value $\boldsymbol{V}(\mathcal{E})$ is IIR, IE, CBIC, and satisfies CBM and C (see Hahn and Yannelis, 1995, for the details), Theorem 2.7.1 implies that the interim private value is c-implementable. □

Similarly with the interim private core, if there are two agents in the economy and the initial endowment is not interim efficient, the interim value $\boldsymbol{V}(\mathcal{E})$ with $\lambda \gg 0$ is c-implementable.

5 Examples of non-c-implementation

According to Palfreya and Srivastava (1987), the rational expectations equilibrium (REE) social choice set is Bayesian Nash implementable but neither the interim efficient social choice set nor the interim core is Bayesian Nash implementable because neither one satisfies the Bayesian monotonicity condition. Note that Palfrey and Srivastava define an interim efficiency notion without information sharing at all and the initial endowment of each agent e_i is not $\mathcal{F}_i$-measurable. Hence their core notion is quite different than ours and one can easily show that it is not c-implementable. Also, one can construct examples to show that the interim private core is not Bayesian Nash implementable. We show below that the interim fine core (which allows for information pooling within a coalition) is not c-implementable.

Definition 5.1. *A feasible allocation x with x_i being $\bigvee_{i \in I} \mathcal{F}_i$-measurable for every $i \in I$ is an* interim fine core allocation *of the economy with differential information $\mathcal{E}$ if it is not true that there exist a state $\omega \in \Omega$, a coalition $S \subset I$, and $(y_i)_{i \in S}$ such that y_i is $\bigvee_{i \in S} \mathcal{F}_i$-measurable for every $i \in S$, $\sum_{i \in S} y_i = \sum_{i \in S} e_i$, and $V_i(\omega, y_i) > V_i(\omega, x_i)$ for every $i \in S$.*

The above core concept is different from the (interim) private core in that agents in a coalition now are allowed to pool their information instead of making redistributions based on their individual private information only (as the private core requires). This notion is analogous to fine core notion of Wilson (1978) (see also Srivastava, 1984a, b; Yannelis, 1991). The interim fine core and the interim fine value need not be c-implementable because they violate the CBIC condition as the following example indicates.

Example 5.1. Consider an economy with differential information with three agents, two goods (*i.e.*, x^1, x^2), and three equally probable states, where utility functions, random initial endowments, and private information sets are given as follows:

$$\begin{aligned}
u_1(\omega, x^1, x^2) &= \sqrt{x^1 x^2},\ e_1 = ((7,1),(7,1),(4,1)),\quad \mathcal{F}_1 = \{\{\omega_1, \omega_2\}, \{\omega_3\}\},\\
u_2(\omega, x^1, x^2) &= \sqrt{x^1 x^2},\ e_2 = ((7,1),(7,1),(4,1)),\quad \mathcal{F}_2 = \{\{\omega_1, \omega_2\}, \{\omega_3\}\},\\
u_3(\omega, x^1, x^2) &= \sqrt{x^1 x^2},\ e_3 = ((1,10),(1,7),(1,7)),\ \mathcal{F}_3 = \{\{\omega_1\}, \{\omega_2, \omega_3\}\}.
\end{aligned}$$

The allocation $x^* = (x_1^*, x_2^*, x_3^*)$ with

$$\begin{aligned}
x_1^* &= ((33/8,\ 33/10), (13/3,\ 13/5), (5/2,\ 5/2))\\
x_2^* &= ((33/8,\ 33/10), (13/3,\ 13/5), (5/2,\ 5/2))\\
x_3^* &= ((54/8,\ 54/10), (19/3,\ 19/5), (4, 4))
\end{aligned}$$

is an interim fine core allocation. But it is not CBIC. To see this, suppose that ω_2 is realized and let $z^* = x^* - e$. Consider the coalition $S = \{1, 2\}$ and the deception $\alpha_i(E_i(\omega)) = \{\omega_3\}$ for every $\omega \in \Omega$ and for $i \in S$. Since

$$V_i(\omega_2, e_i + [z^* \circ (\alpha_S, \alpha^*_{-S})]_i) > V_i(\omega_2, x_i^*)$$

for $i \in S$, it follows that x^* is not CBIC. Therefore, the interim fine core is not c-implementable.[16]

We will show that the fine core allocation in Example 5.1 is also a fully revealing REE allocation[17] which in turn violates the CBIC condition and therefore it is not c-implementable: Consider the same economy and the same allocation x^* as in Example 5.1. The price-allocation pair $(p^*, x^*) = (p^*, x_1^*, x_2^*, x_3^*)$ with $p^* = ((4/5, 1), (3/5, 1), (1, 1))$ constitutes a fully revealing rational expectations equilibrium. However, since the allocation x^* is not CBIC as it is shown in Example 5.1, we can conclude that the set of REE allocations is not c-implementable. It should be noted that one cannot c-implement the ex ante private core because such allocation rules are not necessarily interim individually rational.

Finally, one may wonder as to whether or not an extension of the coalition-proof Nash equilibrium concept to differential information economies can be adopted here instead of the coalitional Bayesian Nash equilibrium concept. The answer is no because such a concept will yield outcomes which are not necessarily interim efficient.

6 Conclusions

We introduced the idea of coalitional Bayesian implementation with the main objective to examine solution concepts that the standard Bayesian Nash implementation literature does not cover. In particular, we presented necessary and sufficient conditions for the coalitional implementation of a social choice set, and as a consequence of this, we showed that the private core and private value are indeed coalitionally implementable.

It is important to note that our c-implementation results do not rule out information asymmetries, *i.e.*, an agent who has superior information that is useful to the rest of the economy will be rewarded. This is in sharp contrast with the Bayesian Nash implementation results of the rational expectations equilibrium, where the stringent informational assumptions rule out information asymmetries.

It should be emphasized that the idea of coalitional Bayesian implementation is quite natural for resource allocation concepts because the outcomes that the game generates are always interim efficient, contrary to the standard Bayesian Nash implementation. Moreover, the assumptions needed for the c-implementation of our solution concepts are quite attractive from a normative viewpoint.

Finally our results support the conjectures and findings of Wilson (1978), Srivastava (1984a, b), Yannelis (1991), Krasa and Yannelis (1994, 1996), and Koutsougeras and Yannelis (1993). Specifically, these authors note that some information sharing (*e. g.*, pooling of information) may not be incentive compatible and may

[16] A similar example can be constructed for the interim private value allocation (see Hahn and Yannelis (1995) for the details).

[17] When prices and allocations are $(\bigvee_{i \in I} \mathcal{F}_i)$-measurable, one can define the notion of a $(\bigvee_{i \in I} \mathcal{F}_i)$-revealing REE and it can be easily checked (the proof is similar to that of Debreu-Scarf) that the set of $(\bigvee_{i \in I} \mathcal{F}_i)$-revealing REE allocations is contained in the fine core. A related result has been proved by Srivastava (1984b).

also rule out the information superiority of an individual. Our examples in Section 6 indicate that indeed such core and value notions which allow for pooling information need not be c-implementable (as is the case for the REE). In that sense, our main c-implementation theorem is especially useful, because it not only delineates a set of reasonable and mild conditions which are necessary and sufficient for c-implementation, but it also enables us to conclude that the private core and private value do provide a successful alternative to the (Walrasian) rational expectation equilibrium.

Appendix

Proof of Lemma 2.7.2. First notice that $\sigma(E(\omega)) \in M^0$ for every $\omega \in \Omega$. Observe that $f(\sigma) = z$ by the definition of f. Consider an arbitrary state $\omega \in \Omega$. Let S be a coalition deviating from the strategy vector σ and denote the deviating strategy of S by σ'_S with $\sigma'_i(E_i(\omega)) \neq \sigma_i(E_i(\omega))$ and $\sigma'_i(E_i(\omega')) = \sigma_i(E_i(\omega')), \forall \omega' \notin E_i(\omega)$ for every $i \in S$. Let $\sigma' = (\sigma'_S, \sigma_{-S})$.

First, consider a proper coalition S. Then $\sigma'(E(\omega)) \in [\bigcup_{k=0}^{4} M^k] \cup M^6$. Notice that it is impossible that $\sigma'(E(\omega)) \in M^5$ because no agent reports the integer zero in M^5. If $\sigma'(E(\omega)) \in M^0$, CBIC of z implies that the coalition S cannot misreport to become better off. If $\sigma'(E(\omega)) \in M^{11}(S)$, the definition of f and the property of V_i on this region implies that there exists at least one agent in the coalition S who cannot become better off by deviating. If $\sigma'(E(\omega)) \in M^{12}(S)$, or $\sigma'(E(\omega)) \in M^6$, the new outcome is no trade and IIR of $e+z$ implies that no agent in the coalition S can become better off. If $\sigma'(E(\omega)) \in M^2$, where only agent 1 deviates, it is clear that agent 1 becomes worse off. Suppose that $\sigma'(E(\omega)) \in M^3$, then by the definition of f, at least one agent is worse off since every agent (except agent 1) transfers his/her reported endowment to agent 1. If $\sigma'(E(\omega)) \in M^4(S)$ with $|S| \geq 2$, then IIR of z and the monotonicity of preferences would imply that an agent in the coalition S who is not the winner of the "integer game" would be worse off.

Let S be a grand coalition, *i.e.*, $S = I$, $\sigma'(E(\omega)) \in \bigcup_{k=0}^{6} M^k$. The interim efficiency of $e + z$ implies that every agent in the grand coalition cannot become better off.

Since no coalitional deviation from σ is profitable, we conclude that σ is a CBNE and $f(\sigma) = z$. □

In the argument below, we set $\sigma_i(E_i(\omega)) = (\sigma_i^1(E_i(\omega)), \sigma_i^2(E_i(\omega)), \sigma_i^3(E_i(\omega)))$.

Proof of Lemma 2.7.3. Suppose, by way of contradiction, that $\sigma(E(\omega)) \notin M^0$ for some ω. Let us define $\tilde{\sigma}_S$ to be a deviation from σ_S by the coalition S as follows: For every $i \in S$,

$$\tilde{\sigma}_i(E_i(\omega)) = (\sigma_i^1(E_i(\omega)), \sigma_i^2(E_i(\omega)), n^*),$$
$$\tilde{\sigma}_i(E_i(\omega')) = \sigma_i(E_i(\omega')), \forall \omega' \notin E_i(\omega),$$

where $n^* = 1 + \max\{\sigma_i^3(E_i(\omega)) : i \in I\}$. Let $\tilde{\sigma} = (\tilde{\sigma}_S, \sigma_{-S})$. Then there are the following cases to consider:

(1) Suppose that $\sigma(E(\omega)) \in M^{11}(S)$ for some S. Let $\sigma_i(E_i(\omega)) = (E_i, z', n)$ for every $i \in S$. Observe that $0 \leq e_i + z'_i \leq \sum_{i \in I} e_i$ for every $i \in I$. Consider any $\omega' \in \bigcap_{i \in I} \sigma_i^1(E_i(\omega))$. Then $z'_i(\omega') < \sum_{j \neq i} e_j(\omega')$ for every $i \in I$. For, otherwise there is an agent k in I such that $z'_k(\omega') = \sum_{i \neq k} e_i(\omega')$. By feasibility, $\sum_{i \neq k} z'_i(\omega') = -\sum_{i \neq k} e_i(\omega)$. Since $z'_i(\omega') + e_i(\omega') \geq 0$ for every $i \neq k$, it follows that $z'_i(\omega') = -e_i(\omega') \ll 0$ for every $i \neq k$, a contradiction to the IIR of z'. Hence, some agent i in $I \setminus S$ will deviate to $M^4 \cup M^5$ using the strategy $\tilde{\sigma}_i$ to win the integer game and become better off.

(2) Suppose that $\sigma(E(\omega)) \in M^{12}(S)$ for some S. Note that $\sum_{j \neq i} e_j(\omega') \gg 0$ for every $i \in I$ and every $\omega' \in \bigcap_{j \in I} \sigma_j^1(E_j(\omega))$. Some agent i in $I \setminus S$ will deviate to $M^4 \cup M^5$ by using strategy $\tilde{\sigma}_i$. For he/she who gets $-e_i(\omega')$ at σ would become the winner of the integer game, obtain $\sum_{j \neq i} e_j(\omega')$, and become better off, since the new message $\tilde{\sigma}(E(\omega))$ would belong to $M^4 \cup M^5$.

(3) Suppose that $\sigma(E(\omega)) \in M^2$. If $\bigcap_{i \in I} E_i \neq \emptyset$, then an agent $i \in I \setminus \{1\}$ will deviate using the strategy $\tilde{\sigma}_i$. Since the new message $\tilde{\sigma}(E(\omega))$ belongs to M^4, he/she wins the integer game and becomes better off. If $\bigcap_{i \in I} E_i = \emptyset$, agent 1 will deviate to M^6 using the strategy $\tilde{\sigma}_1$ and get no trade.

(4) Suppose that $\sigma(E(\omega)) = (E_i, z^i, 0)_{i \in I} \in M^3$. We first consider the case where $\bigcap_{i \in I} E_i \neq \emptyset$. An agent $i \in I \setminus \{1\}$ will deviate using the strategy $\tilde{\sigma}_i$ to become better off. Since the new message $\tilde{\sigma}(E(\omega))$ lies in M^4 and he/she wins the integer game, he/she becomes better off. If $\bigcap_{i \in I} E_i = \emptyset$, an agent $i \in I \setminus \{1\}$ will deviate using the strategy $\tilde{\sigma}_i$ to become better off. Since the new message $\tilde{\sigma}(E(\omega))$ lies in M^6 and he/she obtains no trade, he/she becomes better off.

(5) If $\sigma(E(\omega)) \in M^4 \cup M^5$, an agent i who is one of the losers in the integer game will use the strategy $\tilde{\sigma}_i$ and become better off. Since the new message $\tilde{\sigma}(E(\omega))$ lies in $M^4 \cup M^5$, agent i becomes the winner of the integer game and gets $\sum_{j \neq i} e_j(\omega') \gg -e_i(\omega')$ for every $\omega' \in \bigcap_{j \in I} \sigma_j^1(E_j(\omega))$.

(6) Suppose that $\sigma(E(\omega)) = (E_i, z^i, n_i)_{i \in I} \in M^6$. Fix any agent k such that $\sigma_k(E_k(\omega)) = (E_k, z^k, n_k)$ with $n_k \neq 0$. Then the coalition $S = I \setminus \{k\}$ will deviate by using the strategy $\bar{\sigma}_S$ such that $\bar{\sigma}_i(E_i(\omega)) = (\bar{E}_i, \bar{z}^i, n_k + 1)$ and $\bar{\sigma}_i(E_i(\omega')) = \sigma_i(E_i(\omega')), \forall \omega' \notin E_i(\omega)$ for every $i \in S$, where $E_k \cap [\bigcap_{i \neq k} \bar{E}_i] \neq \emptyset$ and $\bar{z}^i \neq \bar{z}^j$ for some $i, j \in I \setminus \{k\}$. Since the new message $\bar{\sigma}(E(\omega))$ with $\bar{\sigma} = (\bar{\sigma}_S, \sigma_{-S})$ lies in M^5, the agent k, who is not the winner of the integer game, gives to the coalition S his/her reported endowment $e_k(\omega') \gg 0$ with $\omega' \in E_k \cap [\bigcap_{i \neq k} \bar{E}_i]$, which all agents in the coalition S evenly share. Therefore, every agent in the coalition S becomes better off.

From (1) through (6), it follows that σ is not a CBNE for (M, f), a contradiction.
□

Proof of Lemma 2.7.4. By Lemma 2.7.3, $\sigma(E(\omega)) \in M^0$ for all ω. Since Γ satisfies C, we get $\sigma_i^2(E_i(\omega)) = z^*$ with $e + z^* \in \Gamma$ for every $i \in I$ and for every $\omega \in \Omega$. Define $\alpha_i(E_i(\omega)) = \sigma_i^1(E_i(\omega))$ for every $i \in I$. Then it follows from the definition of the mechanism that $f(\sigma) = z^* \circ \alpha$. We have to show that $e + f(\sigma) = e + z^* \circ \alpha \in \Gamma$. Suppose, by way of contradiction, that $e + z^* \circ \alpha \in \mathbf{A} \setminus \Gamma$.

By CBM, there exists a state $\omega^* \in \Omega$, a coalition $S \subset I$, and $z' \in \boldsymbol{Z}$ such that $e + z' \circ \alpha \in \boldsymbol{A}, e + z' \circ (\alpha_S, \alpha^*_{-S}) \in \boldsymbol{A}$,

(1) $\forall i \in S,\ V_i(\omega^*, e_i + (z' \circ \alpha)_i) > V_i(\omega^*, e_i + (z^* \circ \alpha)_i)$, and
(2) $\exists i \in S,\ V_i(\omega', e_i + z_i^*) \geq V_i(\omega', e_i + [z' \circ (\alpha_S, \alpha^*_{-S})]_i),\ \forall\, \omega' \in \Omega.$

If they use the strategy σ'_S with $\sigma'_i(E_i(\omega^*)) = (\alpha_i(E_i(\omega^*)), z', 1)$ and $\sigma'_i(E_i(\omega)) = \sigma_i(E_i(\omega)), \forall \omega \notin E_i(\omega^*)$ for every $i \in S$, by (2) they move from M^0 to $M^{11}(S)$. Furthermore, since $f(\sigma'_S, \sigma_{-S}) = z' \circ \alpha$, (1) implies that every agent $i \in S$ becomes better off at ω^*, *i.e.*, $V_i(\omega^*, e_i + f_i(\sigma'_S, \sigma_{-S})) > V_i(\omega^*, e_i + f_i(\sigma))$, a contradiction to the fact that σ is a CBNE for (M, f). □

References

Berliant, M.: On income taxation and the core. Journal of Economic Theory **56**, 121–141 (1992)

Blume, L., Easley, D.: Implementation of Walrasian expectations equilibria. Journal of Economic Theory **51**, 207–227 (1990)

Dutta, B., Sen, A.: Implementation under strong equilibrium: a complete characterization. Journal of Mathematical Economics **20**, 49–67 (1991)

Gul, F.: Bargaining foundations of the Shapley value. Econometrica **57**, 81–95 (1989)

Hahn, G., Yannelis, N.C.: Coalitional Bayesian Nash implementation in differential information economies. Working Paper 95–0135, University of Illinois at Urbana-Champaign (1995)

Hahn, G., Yannelis, N.C.: Efficiency and incentive compatibility in differential information economies. Economic Theory **10**, 383–411 (1997)

Holmstöm, B., Myerson, R: Efficient and durable decision rules with incomplete information. Econometrica **49**, 1477–1500 (1983)

Hong, L.: Feasible Bayesian implementation with state dependent feasible sets. Journal of Economic Theory **80**, 201–221 (1998)

Hurwicz, L., Maskin, E., Postlewaite, A.: Feasible implementation of social choice correspondences by Nash equilibria. Mimeo (1984)

Ichiishi, T., Radner, R.: A profit-center game with incomplete information. Review of Economic Design **4**, 307–343 (1999)

Ichiishi, T., Sertel, M.R.: Cooperative interim contract and re-contract: Chandler's M-form firm. Economic Theory **11**, 523–543 (1998)

Jackson, M.O.: Bayesian implementation. Econometrica **59**, 461–477 (1991)

Koutsougeras, L., Yannelis, N.C.: Incentive compatibility and information superiority of the core of an economy with differential information. Economic Theory **3**, 195–216 (1993)

Krasa, S., Yannelis, N.C.: The value allocation of an economy with differential information. Econometrica **62**, 881–900 (1994)

Krasa, S., Yannelis, N.C.: Existence and properties of a value allocation for an economy with differential information. Journal of Mathematical Economics **25**, 165–179 (1996)

Maskin, E.: Nash equilibrium and welfare optimality. Mimeo, MIT (1977)

Maskin, E.: Implementation and strong Nash-equilibrium. In: Laffont, J.-J. (ed.) Aggregation and revelation of preferences. Amsterdam: North-Holland 1979

Myerson, R.: Incentive compatibility and the bargaining problem. Econometrica **47**, 61–74 (1979)

Palfrey, T., Srivastava, S.: Private information in large economies. Journal of Economic Theory **39**, 34–58 (1986)

Palfrey, T., Srivastava, S.: On Bayesian implementable allocations. Review of Economic Studies **54**, 193–208 (1987)

Palfrey, T., Srivastava, S.: Implementation with incomplete information in exchange economies. Econometrica **57**, 115–134 (1989)

Postlewaite, A., Schmeidler, D.: Implementation in differential information economies. Journal of Economic Theory **39**, 14–33 (1986)

Postlewaite, A., Schmeidler, D.: Differential information and strategic behavior in economic environments: a general equilibrium approach. In: Groves, T., Radner, R., Reiter, S. (eds.) Information, incentives, and economic mechanisms: essays in honor of Leonid Hurwicz. London: Basil Blackwell 1987

Saijo, T.: Strategy space reduction in Maskin's theorem: sufficient conditions for Nash implementation. Econometrica **56**, 693–700 (1988)

Shapley, L.A.: A value of n-person games. In: Kuhn, H.W., Tucker, A.W.(eds.) Contributions to the theory of games, Vol. II, pp. 307–317. Princeton: Princeton University Press 1953

Srivastava, S.: A limit theorem on the core with differential information. Economics Letters **134**, 111–116 (1984a)

Srivastava, S.: Rational expectations equilibria and the core. Working Paper 1382–83, Carnegie-Mellon University (1984b)

Williams, S.R.: Realization and Nash implementation: two aspects of mechanism design. Econometrica **54**, 139–151 (1986)

Wilson, R.: Information, efficiency, and the core of an economy. Econometrica **46**, 807–816 (1978)

Winter, E.: The demand commitment bargaining and snowballing cooperation. Economic Theory **4**, 255–273 (1994)

Yannelis, N.C.: The core of an economy with differential information. Economic Theory **1**, 183–198 (1991)

An extensive form interpretation of the private core*

Dionysius Glycopantis[1], **Allan Muir**[2], **and Nicholas C. Yannelis**[3]

[1] Department of Economics, City University, Northampton Square, London EC1V 0HB, UK (e-mail: d.glycopantis@city.ac.uk)
[2] Department of Mathematics, City University, Northampton Square, London EC1V 0HB, UK (e-mail: a.muir@city.ac.uk)
[3] Department of Economics, University of Illinois at Urbana-Champaign, Champaign, IL 61820, USA (e-mail: nyanneli@uiuc.edu)

Received: July 21, 2000; revised version: January 16, 2001

Summary. The private core of an economy with differential information, (Yannelis (1991)), is the set of all state-wise feasible and private information measurable allocations which cannot be dominated, in terms of ex ante expected utility functions, by state-wise feasible and private information measurable net trades of any coalition. It is coalitionally Bayesian incentive compatible and also takes into account the information superiority of an individual. We provide a noncooperative extensive form interpretation of the private core for three person games. We construct game trees which indicate the sequence of decisions and the information sets, and explain the rules for calculating ex ante expected payoffs. In the spirit of the Nash programme, the private core is thus shown to be supported by the perfect Bayesian equilibrium of a noncooperative game. The discussion contributes not only to the development of ideas but also to the understanding of the dynamics of how coalitionally incentive compatible contracts can be realized.

Keywords and Phrases: Differential information economy, Private core, Weak fine core, Coalitional Bayesian incentive compatibility, Game trees, Perfect Bayesian equilibrium, Contracts, Nash Programme.

JEL Classification Numbers: 020, 226.

---

* We would like to thank the referees of this journal for their invaluable comments. In particular one of the referees made extensive suggestions which improved the final version of the paper.

1 Introduction

An economy with differential information consists of a finite set of agents each of which is characterized by a random utility function, a random consumption set, random initial endowments, a private information set and a prior probability distribution. The private core of a differential information economy (see Yannelis (1991)) is the set of all state-wise feasible and private information measurable allocations which cannot be dominated, in terms of expected utility, by any coalition's state-wise feasible and private information measurable net trades.

The private core is not susceptible to the criticism of the traditional rational expectations equilibrium (REE). In particular, the REE does not provide an explanation as to how prices reflect the information asymmetries in the economy. On the contrary the private core not only takes into account the information asymmetries but also rewards agents with "superior" information as shown in Example 3.1 in Section 3. Furthermore it is coalitionally Bayesian incentive compatible (see Koutsougeras and Yannelis, 1993). Hence the private core can be used to explain how incentive compatible contracts are written.

The main purpose of this paper is to provide a noncooperative, extensive form interpretation of the private core. Generally speaking we investigate whether or not cooperative core concepts, i.e. the private core and the weak fine core, defined below, can be supported as a perfect Bayesian equilibrium.

This investigation falls in the area of the Nash programme, which is a research agenda originated by Nash (1953) and emphasized by Binmore (1980a,b). The idea is to provide support and justification of cooperative solutions to economic problems through noncooperative formulations. More generally the issue is the relation between dynamic and static considerations. Our approach provides a dynamic interpretation of the static private core notion. Consequently it helps to explain the dynamics of how incentive compatible contracts are realized.

In our analysis, in order to provide support for the private core, we introduce game trees. They show the prior probability with which nature chooses and make explicit the sequential moves, i.e., which player makes announcements or moves first. They also take into account the private information sets of each player as well as the measurability of decisions.

Given the above structure of the game tree, we specify rules, i.e., the terms of a contract, which imply specific redistributions of the random initial endowments in different events. The rules are a statement as to the consequences of actions by the players under all possible states of nature. Having specified the rules, we obtain the payoffs in terms of quantities and then we are looking for an appropriate refinement of Nash equilibrium for games with imperfect or differential information.

We require an equilibrium concept which adopts a probabilistic approach with respect to the nodes of an information set and reduces to subgame perfect equilibrium in case the information sets are singletons. Such a concept is the Kreps and Wilson (1982) sequential equilibrium and its variants which are either weaker versions or refinements. We adopt here the perfect Bayesian equilibrium, described in Tirole (1988), where also a comparison is made with other, similar type ideas.

A *perfect Bayesian equilibrium* consists of a set of players' optimal behavioral strategies and, consistent with these, a set of beliefs which attach a probability distribution to the nodes of each information set. Consistency requires that the decision from an information set is optimal given the particular player's beliefs about the nodes of this set and the strategies from all other sets, and that beliefs are formed from updating using the available information. If the optimal play of the game enters an information set then updating of beliefs must be Bayesian. Otherwise appropriate beliefs are assigned arbitrarily to the nodes of the set.

The term "implementation" is used below in the sense of realization of an allocation and not in the sense of implementation theory or mechanism design which requires the introduction of a planner. Recent work in this area is by Trockel (2000) which contributes to the Nash programme and casts the implementation discussion in its context.

The main results in this paper are the following. Despite the fact that "pooled" information core allocations, (i.e., the weak fine core), exist under mild assumptions, we construct a game tree, with reasonable rules for calculating payoffs, which shows that a redistribution of this nature cannot be supported as a perfect Bayesian equilibrium. Indeed, such contracts (allocations) need not be Bayesian incentive compatible which suggests a difficulty in implementing them.

On the other hand, we construct a three player example which indicates that the private core, which is Bayesian incentive compatible, can be supported as a perfect Bayesian equilibrium. The above results not only provide a first step into the noncooperative extensive form interpretation of the core of economies with differential information, but also enable us to understand how coalitionally Bayesian incentive compatible contracts are realized.

Finally we provide a generalization of the private core existence result of Yannelis (1991) by relaxing the continuity assumption of the random utility functions. This enables us to include private information sets which not only can be measurable partitions of the exogenously given probability measure space, but can also be sub-σ-algebras.

To the best of our knowledge the present paper is the first attempt to provide a noncooperative foundation for core concepts in economies with differential information.

We note that the complete information works of Lagunoff (1994), Perry and Reny (1994), Serrano (1995) and Serrano and Vohra (1997), which discuss a non-cooperative approach to the core, do not apply to the differentiable economy framework that we are considering here.

The paper is organized as follows. Section 2 contains the definition of the differential information economy. Section 3 contains the core concepts employed in this paper as well as a new core existence result. Section 4 discusses ideas of incentive compatibility on the basis of which core allocations can be classified. Section 5 discusses the non-implementation of the weak fine core and Section 6 the implementation of the private core in extensive form games. Section 7 offers brief concluding remarks. Appendix I proves, under certain conditions, the existence of a private core allocation and Appendix II derives the private core allocations in an explicit example which is used in the text.

2 Differential information economy

Although we shall be concerned with a special model we repeat briefly, for completeness, the notation used and the definition of the private core in a general case. We define below the notion of a finite-agent economy with differential information. Let $(\Omega, \mathcal{F}, \mu)$ be a complete probability measure space and Y be a separable *Banach lattice*[1] with an order continuous norm denoting the commodity space. The positive cone of Y is denoted by Y_+.

A differential information economy $\mathcal{E}$ is a set $\{((\Omega, \mathcal{F}, \mu), X_i, \mathcal{F}_i, u_i, e_i) : i = 1, \ldots, n\}$ where

1. $X_i : \Omega \rightarrow 2^{Y_+}$ is the set-valued function giving the *random consumption set* of Agent (Player) i, who is denoted also by Pi,
2. $\mathcal{F}_i$ is a partition (or sub-σ-algebra) of $\mathcal{F}$, denoting the *private information*[2] of Pi,
3. $u_i : \Omega \times Y_+ \rightarrow \mathbb{R}$ is the *random utility* function of Pi,
4. $e_i : \Omega \rightarrow Y_+$ is the *random initial endowment* of Pi, where $e_i(\cdot)$ is $\mathcal{F}_i$-measurable and *Bochner integrable*[3], and $e_i(\omega) \in X_i(\omega)$ μ-a.e., and
5. μ denotes the common *prior* of all agents.

The *ex ante expected utility* of Pi is given by

$$v_i(x_i) = \int_{\Omega} u_i(\omega, x_i(\omega)) d\mu(\omega). \tag{1}$$

Denote by $E_i(\omega)$ the event in the partition $\mathcal{F}_i$ of Agent i which contains the realized state of nature, $\omega \in \Omega$. The *interim expected utility* function of Agent i is given by

$$v_i(\omega, x_i) = \frac{1}{\mu(E_i(\omega))} \int_{\omega' \in E_i(\omega)} u_i(\omega', x_i(\omega')) d\mu(\omega'), \tag{2}$$

where $\mu(E_i(\omega))$ is assumed to be positive.

Despite the fact that the differential information economy is static, we can provide a two-period interpretation as follows. In the first period agents make contracts in the ex ante stage. In the interim stage, i.e., after they have received a signal[4] as to what is the event containing the realized state of nature, one considers the incentive compatibility of the contract.

3 The private core and the weak fine core

First we define the notion of the private core (Yannelis (1991)). We begin with some notation. Denote by $L_1(\mu, Y)$ the space of all equivalence classes of Bochner integrable functions.

[1] See Appendix I.

[2] Following Aumann (1987) we assume that the players' information partitions are common knowledge.

[3] See Appendix I.

[4] A signal to a player is a function from states of nature to the possible observations specific to the player, which induces on Ω a sub-σ-algebra of $\mathcal{F}$.

L_{X_i} is the set of all Bochner integrable and $\mathcal{F}_i$-measurable selections from the random consumption set of Agent i, i.e.,

$$L_{X_i}=\{x_i \in L_1(\mu, Y) : x_i : \Omega \to Y \text{ is } \mathcal{F}_i\text{-measurable and } x_i(\omega)\in X_i(\omega)\mu\text{-}a.e.\}$$

and let $L_X = \prod_{i=1}^{n} L_{X_i}$.

Also let

$$\bar{L}_{X_i} = \{x_i \in L_1(\mu, Y) : \ x_i(\omega) \in X_i(\omega) \ \mu\text{-}a.e.\}$$

and let $\bar{L}_X = \prod_{i=1}^{n} \bar{L}_{X_i}$.

An element $x = (x_1, \ldots, x_n) \in \bar{L}_X$ will be called an *allocation*. For any subset of players S, an element $(y_i)_{i\in S} \in \prod_{i\in S} \bar{L}_{X_i}$ will also be called an allocation, although strictly speaking it is an allocation to S.

Definition 3.1. *An allocation $x \in L_X$ is said to be a* private core allocation *if*

(i) $\sum_{i=1}^{n} x_i = \sum_{i=1}^{n} e_i$ *and*
(ii) *there do not exist coalition S and allocation $(y_i)_{i\in S} \in \prod_{i\in S} L_{X_i}$ such that*

$\sum_{i\in S} y_i = \sum_{i\in S} e_i$ *and* $v_i(y_i) > v_i(x_i)$ *for all* $i \in S$.

Hence, a private core allocation is feasible, reflects the private information of each agent, i.e., each $x_i(\cdot)$ is $\mathcal{F}_i$-measurable, and has the property that no coalition of agents can redistribute their initial endowments, based on their own private information, and make each of its members better off. It is important to notice that since initial endowments are private information measurable, net trades $x_i(\cdot)-e_i(\cdot)$ are also $\mathcal{F}_i$-measurable.

Observe that despite the fact that a coalition of agents get together they do not necessarily share their own information. On the contrary, the redistributions of the initial endowments are based only on their own private information. This is quite important because the resulting private core allocation has desirable properties, i.e., it is coalitionally incentive compatible, as we shall see below, and takes into account the information superiority of an individual.[5]

Although several private core existence results can be found in the literature, as for example in Yannelis (1991), Allen (1991), Koutsougeras and Yannelis (1993), Balder and Yannelis (1994), Page (1997) and Lefebvre (2001), among others, the proof of the theorem below appears to be the shortest, simplest and quite general. It improves on the original one of Yannelis (1991).

---

[5] See Koutsougeras and Yannelis (1993) and Example 3.1 below. Notice that in Definition 3.1 the ex ante expected utility function is used. The (interim) private core is also defined similarly by replacing (ii) in Definition 3.1 by
(ii) there do not exist coalition S and allocation $(y_i)_{i\in S} \in \prod_{i\in S} L_{X_i}$ such that $\sum_{i\in S} y_i = \sum_{i\in S} e_i$ and $v_i(\omega, y_i) > v_i(\omega, x_i)$ for all $i \in S$ and μ-a.e.
Both private cores (ex ante and interim) exist and also have similar qualitative properties (see Hahn and Yannelis, 2000).

Theorem 3.1: *Let $\mathcal{E} = \{((\Omega, \mathcal{F}, \mu), X_i, \mathcal{F}_i, u_i, e_i) : i = 1, \ldots, n\}$ be a differential information economy satisfying for each i the following assumption:*
u_i is concave, upper semicontinuous (u.s.c.) and integrably bounded.
Then a private core allocation exists in $\mathcal{E}$.

Proof. See Appendix I.

The theorem in Yannelis (1991) is generalized in the following way. The utility functions need not be weakly continuous, but only u.s.c. in the norm topology. However in the presence of concavity they become weakly u.s.c. (Balder - Yannelis (1993)). The latter enables us to generalize the private information sets $\mathcal{F}_i$ of each agent from partitions to a sub-σ-algebra. Furthermore, we do not need to assume that the dual of Y has the Radon - Nikodym property. In the examples below the σ-algebras will be generated from partitions.

The example below illustrates the private core.

Example 3.1 Consider the following three agents economy, $I = \{1, 2, 3\}$ with one commodity, i.e. $X_i = \mathbb{R}_+$ for each i, and three states of nature $\Omega = \{a, b, c\}$.

The agents are characterized by their initial endowments, their private information and their utility functions. We assume that the structure is

$$\begin{aligned} e_1 &= (5, 5, 0), & \mathcal{F}_1 &= \{\{a, b\}, \{c\}\}; \\ e_2 &= (5, 0, 5), & \mathcal{F}_2 &= \{\{a, c\}, \{b\}\}; \\ e_3 &= (0, 0, 0), & \mathcal{F}_3 &= \{\{a\}, \{b\}, \{c\}\}. \end{aligned}$$

Notice that the initial endowment of each agent is $\mathcal{F}_i$-measurable. It is also assumed that $u_i(\omega, x_i(\omega)) = x_i^{\frac{1}{2}}$, which is a typical strictly concave and monotone function in x_i, and that each state of nature occurs with the same probability, i.e. $\mu(\{\omega\}) = \frac{1}{3}$, for $\omega \in \Omega$. For convenience, in the discussion below expected utilities are multiplied by 3.

It can be shown[6] that a private core allocation of this economy is $x_1 = (4, 4, 1)$, $x_2 = (4, 1, 4)$ and $x_3 = (2, 0, 0)$. Clearly this allocation is feasible and $\mathcal{F}_i$-measurable. It is important to observe that in spite of the fact that Agent 3 has zero initial endowments, her superior information allows him to make a Pareto improvement for the economy as a whole and clearly he was rewarded for doing so. In other words, Agent 3 traded her superior information for actual consumption in state a. In return Agent 3 provided insurance to Agent 1 in state c and to Agent 2 in state b. Notice that if the private information set of Agent 3 is the trivial partition, i.e., $\mathcal{F}_3^{'} = \{a, b, c\}$, then no trade takes place and clearly in this case she gets zero utility. Thus the private core is sensitive to information asymmetries.

Contrary to the private core any rational expectation Walrasian equilibium notion will always give zero to Agent 3 since her budget set is zero in each state. This is so irrespective of whether her private information is the full information partition $\mathcal{F}_3 = \{\{a\}, \{b\}, \{c\}\}$ or the trivial partition $\mathcal{F}_3^{'} = \{a, b, c\}$. Hence the rational expectations equilibrium does not take into account the informational superiority of an agent.

Next we define another core concept, the weak fine core (see Yannelis, 1991, p. 188; Koutsougeras and Yannelis, 1993). This concept is a refinement of the fine

[6] See Appendix II.

core of Wilson (1978). Recall that the fine core notion of Wilson as well as the fine core in Yannelis, and Koutsougeras and Yannelis may be empty in well behaved economies. It is exactly for this reason that we are working with a different concept.

Definition 3.2. *An allocation* $x = (x_1, \ldots, x_n) \in \bar{L}_X$ *is said to be a* weak fine core *allocation if*

(i) *each* $x_i(\cdot)$ *is* $\bigvee_{i=1}^{n} \mathcal{F}_i$*-measurable* [7]
(ii) $\sum_{i=1}^{n} x_i(\omega) = \sum_{i=1}^{n} e_i(\omega)$ *μ-a.e. and*
(iii) *there do not exist coalition* S *and allocation* $(y_i)_{i\in S} \in \prod_{i\in S} \bar{L}_{X_i}$ *such that* $y_i(\cdot) - e_i(\cdot)$ *is* $\bigvee_{i\in S} \mathcal{F}_i$*-measurable for all* $i \in S$, $\sum_{i\in S} y_i(\omega) = \sum_{i\in S} e_i(\omega)$ *μ-a.e., and* $v_i(y_i) > v_i(x_i)$ *for all* $i \in S$.

Notice that now in the weak fine core, coalitions of agents are allowed to pool their own information. Identical assumptions with those in Theorem 3.1 and a similar argument shows that a weak fine core allocation exists in $\mathcal{E}$. The example below illustrates this concept.

Example 3.2 Consider Example 3.1 without Agent 3. Then if Agents 1 and 2 pool their own information a possible allocation is $x_1 = x_2 = (5, 2.5, 2.5)$. Notice that this allocation is $\bigvee_{i=1}^{2} \mathcal{F}_i$-measurable and cannot be dominated by any coalition of agents using their pooled information. Hence it is a weak fine core allocation.

4 Incentive compatibility

A careful examination of Example 3.1 indicates that the private core allocation is incentive compatible in the sense that no coalition of agents has an incentive to misreport the realized state of nature and become better off. The argument which supports this conclusion is as follows. Agent 3 can presumably lie to Agents 1 and 2 if the realized state of nature is a since Agent 1 cannot distinguish state a from state b and Agent 2 state a from state c. However, Agent 3 has no incentive to do so since only in state a does she get positive consumption. Hence, Agent 3 who would potentially cheat in state a has no incentive to do so.

We could consider the example in more detail. We ask the question whether the coalition $S = \{1, 3\}$ can cheat P2. This is not possible because P3 would become worse off. For suppose that the state of nature is a but S reports c. Then $u_1(e_1(a) + x_1(c) - e_1(c)) = u_1(5+1-0) > u_1(x_1(b)) = u_1(x_1(a)) = u_1(4)$ but $u_3(e_3(a) + x_3(c) - e_3(c)) = u_3(0) < u_3(x_3(a)) = u_3(2)$. Similarly the coalition $S = \{2, 3\}$ cannot form, and the coalitions $S = \{1, 2\}$, $S = \{1\}$ and $S = \{2\}$ cannot misreport to P3.

Generalizing we have a coalition S and the complementary set which we denote by $I \setminus S$. The members of S will be denoted by i and the members of $I \setminus S$ by

---

[7] $\bigvee_{i=1}^{n} \mathcal{F}_i$ denotes the smallest σ-algebra containing each $\mathcal{F}_i$.

j. Suppose that the realized state of nature is ω^*. A member $i \in S$ sees $E_i(\omega^*)$. Obviously not all $E_i(\omega^*)$ need be the same since different i's have different information sets. However they all know from their information that the actual state of nature could be ω^*.

Consider now a state of nature ω' with the following property. For all $j \in I \setminus S$ we have $\omega' \in E_j(\omega^*)$ and for at least one $i \in S$ we have $\omega' \notin E_i(\omega^*)$ (otherwise ω' would be indistinguishable from ω^* for all players so in effect could be considered as the same element of Ω). Now the coalition S decides that each member i will announce that she has seen her own set $E_i(\omega')$ which, of course, definitely contains a lie. On the other hand we have that $\omega' \in \bigcap_{j \notin S} E_j(\omega^*)$ (we also denote $j \in I \setminus S$ by $j \notin S$).

Now the idea is that if all members of $I \setminus S$ believe the statements of the members of S then each $i \in S$ expects to gain. For *coalitional Bayesian incentive compatibility* (CBIC) of an allocation we require that this is not possible.

A formal definition of the notion of CBIC[8] is:

Definition 4.1. *An allocation $x = (x_1, \ldots, x_n) \in \bar{L}_X$ with $\sum_{i=1}^n x_i = \sum_{i=1}^n e_i$ is said to be CBIC if it is not true that there exist coalition S and states ω^*, ω', with ω^* different than ω', and $\omega' \in \bigcap_{j \notin S} E_j(\omega^*)$ such that*

$$\frac{1}{\mu(Z_i(\omega^*))} \int_{\omega \in Z_i(\omega^*)} u_i(\omega, e_i(\omega) + x_i(\omega') - e_i(\omega'))d\mu(\omega)$$
$$> \frac{1}{\mu(Z_i(\omega^*))} \int_{\omega \in Z_i(\omega^*)} u_i(\omega, x_i(\omega))d\mu(\omega) \qquad (3)$$

for all $i \in S$, where $Z_i(\omega^*) = E_i(\omega^*) \cap (\bigcap_{j \notin S} E_j(\omega^*))$ and $\mu(Z_i(\omega^*))$ is assumed to be positive.

The integrals above can be evaluated since, due to the common knowledge assumption of Section 2, each player knows all the information sets of the other players and therefore can calculate the relevant intersection $Z_i(\omega^*)$.

This definition implies that no coalition of agents has an incentive to misreport the realized state of nature to the complementary set, despite the fact that the latter cannot distinguish the actual state from the misreported one. They do not expect that, by misreporting, each member of the coalition could become better off. If, for example, the realized state of nature is ω^* and for all $j \notin S$, $\omega' \in E_j(\omega^*)$, while for at least $i \in S$ it is true that $\omega' \notin E_i(\omega^*)$, it must be the case that the agents in S have no incentive to report state ω'. I.e., they do not expect that it is possible to become better off if they are believed, by adding to their initial endowment the net trade in state ω'. If $S = \{i\}$ the above definition reduces to *individual Bayesian incentive compatibility* (IBIC).

It has been shown in Koutsougeras - Yannelis (1993) that if the utility functions are monotone and continuous then private core allocations are *always* CBIC. On

[8] See also Krasa and Yannelis (1994), Hahn and Yannelis (2001) for other CBIC concepts.

the other hand the weak fine core allocations are not always CBIC, as the above Example 3.2 with proposed redistribution $x_1 = x_2 = (5, 2.5, 2.5)$ shows.

Indeed, if Agent 1 observes $\{a, b\}$, she has an incentive to report c, as Agent 2 cannot distinguish between a and c. Agent 1 stands to gain if she is believed, which is a possibility as a might be the true state and Agent 2 believes the statement that it is c. In this case Agent 1 keeps the 5 units of the initial endowments in state a, and also gets an additional 2.5 units from Agent 2. In terms of the Definition 4.1, the fact that $u_1(e_1(a) + x_1(c) - e_1(c)) = u_1(5 + 2.5 - 0) > u_1(5) = u_1(x_1(a))$ implies that the proposed allocation is not CBIC. Similarly Agent 2 has an incentive to report b when he observes $\{a, c\}$.

Now in employing game trees in the analysis, as it is done below, we will adopt the definition of IBIC. The equilibrium concept employed will be that of perfect Bayesian equilibrium the application of which is explained below.

A core allocation will be IBIC if there is a profile of optimal behavioral strategies and equilibrium paths along which no player misreports the state of nature he has observed. This allows for the possibility, as we shall see later, that such strategies could imply that players have an incentive to lie from information sets which are not visited by an optimal play. The definition of a play of the game is a directed path from the initial to a terminal node.

The issue is whether core allocations can be obtained as perfect Bayesian equilibria. That is whether the cooperative core solutions can also be supported through an appropriate noncooperative solution concept. The analysis in Sections 5 and 6 below shows that the private core which is CBIC can be supported by a perfect Bayesian equilibrium while for the weak fine core, which may not be CBIC, we find that a reasonable extensive form game does not support it.

5 Non-implementation of the weak fine core in an extensive game

In this section we investigate, by considering sequential decisions, whether in Example 3.2, a particular contract between P1 and P2, with a distribution which is Pareto superior to the initial allocation, will be signed or not.

In particular we consider the weak fine core allocation (5, 2.5, 2.5) in Example 3.2. As we saw in the previous section this is not CBIC which suggests a difficulty in implementing it by means of a contract. We construct a game tree and employ reasonable rules for describing the outcomes of combinations of states of nature and actions of the players. We find that although the Pareto superior allocation (5, 2.5, 2.5) is possible, the optimal strategies of the players imply no trade because of lack of IBIC. Hence there is no advantage in signing such a contract.

One of the issues that has been considered is whether, in order to implement the allocation $(5, 2.5, 2.5)$, the information of P1 and P2 can be pooled into $\mathcal{F}_1 \bigvee \mathcal{F}_2 = \{\{a\}, \{b\}, \{c\}\}$ through the two agents informing each other. The proposed allocation (5, 2.5, 2.5) is measurable with respect to $\mathcal{F}_1 \bigvee \mathcal{F}_2$ and it is a Pareto improvement over the initial endowments.

When the agents form their coalition, they do so in order to sign a contract. The contract depends on their realization that together they could know the state of nature. If each player announces truthfully what he sees, the state of nature would then be common knowledge. Having written the contract, another issue then arises. That is whether the players have an incentive to lie about what they have seen in the interim state. It is this second stage that the game tree is analysing. The game is played before the state is revealed and as the extensive form indicates, in the interim stage each player has an incentive to lie. Therefore the pooling of information does not take place because of lack of incentive compatibility.

We discuss the possible realization of the allocation $(5, 2.5, 2.5)$ through the analysis of a specific sequence of decisions and information sets shown in the game tree in Figure 1. The players are given choices to tell the truth or to lie, i.e., we model the idea that agents truly inform each other about what states of nature they observe, or deliberately aim to mislead their opponent. The issue is what type of behaviour is optimal and therefore whether a proposed contract will be signed or not.

Figures 1 and 2 show that the allocation $(5, 2.5, 2.5)$ will be rejected by the players. It is not IBIC and the proposed contract will not be signed. Notice that vectors at the terminal nodes of a game tree will refer to payoffs of the players, in terms of allocations. The first element will be the payoff to P1, etc.

The explanation of Figure 1 is as follows. Nature chooses states a, b or c with equal probabilities. This choice is flashed on a screen which both players can see. P1 cannot distinguish between a and b, and P2 between a and c. This accounts for the information sets I_1, I_2 and I_2' with more than one node. A player to which such an information set belongs cannot distinguish between these nodes and therefore his decisions are common to all of them. A behavioral strategy of a player is an assignment of a probability distribution per information set that belongs to him over the choices available from that set. This is irrespective of whether a particular play of the game will imply that all these choices will have an effect on the payoffs. Indistinguishable nodes imply the $\mathcal{F}_i$-measurability of decisions.

P1 moves first and has two choices. That is he can either play $A_1 = \{a, b\}$ or $c_1 = \{c\}$, i.e., he can say "I have seen $\{a, b\}$ being unable to distinguish between the two", or "I have seen c". Obviously one of these declarations will be true and the other a lie. Following a choice by P1 then P2 is to respond saying that the signal he has seen on the screen is $A_2 = \{a, c\}$ or that it is $b_2 = \{b\}$. One of these statements is of course a lie.

Strictly speaking the notation for choices should vary with the information set but, for simplicity, we do not modify it, as there is no danger of confusion here. Finally notice that the structure of the game tree is such that when P2 is to act he knows exactly what P1 has chosen before him. This is an assumption about the relation between decisions. In general, in forming game trees the sequence of events and the information of the agents must be specified explicitly.

Next, given the sequence of decisions of the players, shown on the tree, we specify the rules for calculating the payoffs, i.e. we specify the terms of the contract. This is a statement of what to do under all possible states of nature and declarations by the players.

The *rules* are:

(i) If the declarations by the two players are incompatible, that is (c_1, b_2) then at least one of the players is lying and, moreover, the opponent of a lying player detects that lie. This is the case when state c occurs and agent 1 reports state c and agent 2 state b. In state a both agents can lie and the lie cannot be detected by either agent (however, the agents are in the events $\{a, b\}$ and $\{a, c\}$, respectively and they get five units of the initial endowments). Therefore, whenever the declarations are incompatible, no trade takes place and the players retain their initial endowments.

(ii) If the declarations are (A_1, A_2) then even if one of the players is lying, this cannot be detected by his opponent who believes that state a has occured and both players have received endowment 5. Hence no trade takes place.

(iii) If the declarations are (A_1, b_2) then a lie can be beneficial and undetected, and P1 is trapped and must hand over half of his endowment to P2. Obviously if his endowment is zero then he has nothing to give.

(iv) If the declarations are (c_1, A_2) then again a lie can be beneficial and undetected. P2 is now trapped and must hand over half of his endowment to P1. Obviously if his endowment is zero then he has nothing to give.

The calculations of payoffs do not require the revelation of the actual state of nature. Optimal decisions from an information set will be denoted by a heavy line. If either decision is optimal then both will be shown with a heavy line. We could assume that a player does not lie if he cannot get a higher payoff by doing so.

Assuming that each player chooses optimally from the information sets which belong to him, the game in Figure 1 folds back to the one in Figure 2. This is achieved by considering the optimal decisions of P2 and applying backward induction. Inspection of Figure 1 reveals that from the information set I_2 he can play b_2 with probability 1. (A heavy line A_2 indicates that this choice also would not affect the analysis). This accounts for the payoff $(2.5, 7.5)$ and the first payoff $(0, 5)$ from left to right in Figure 2. Similarly we undo all other information sets of P2 and we arrive at Figure 2. Inspection of this figure reveals also the optimal strategies of P1.

Summarizing, the optimal behavioral strategy for P1 is to play c_1 from I_1, i.e to lie, and from the singleton to play any mixture of options, and we have chosen $(A_1, \frac{1}{2}; c_1, \frac{1}{2})$. This is the meaning of $\frac{1}{2}$ on the branches from the singleton. Optimal behavioral strategy of P2 is to play b_2 with probability 1 from both I_2 and $I_2^{'}$, i.e. to lie, and from the singletons he can either tell the truth or lie, or spin a wheel to decide what to do.

Finally we point out that in Figures 1 and 2 the fractions next to the nodes in the information sets correspond to beliefs of the agents obtained, wherever possible, through Bayesian updating. I.e., they are consistent with the choice of a state by nature and the optimal strategies of the players. Hence strategies and beliefs satisfy the conditions of a perfect Bayesian equilibrium. This is a concept employed in analyzing games with information sets with more than one node. As explained above, it requires that given the beliefs, the strategies are optimal, and given the strategies, the beliefs are, wherever possible, obtainable through Bayesian updating.

These probabilities are calculated as follows. We give labels to the nodes of the information sets: From left to right, in I_1, we denote them by j_1 and j_2, in I_2 by n_1

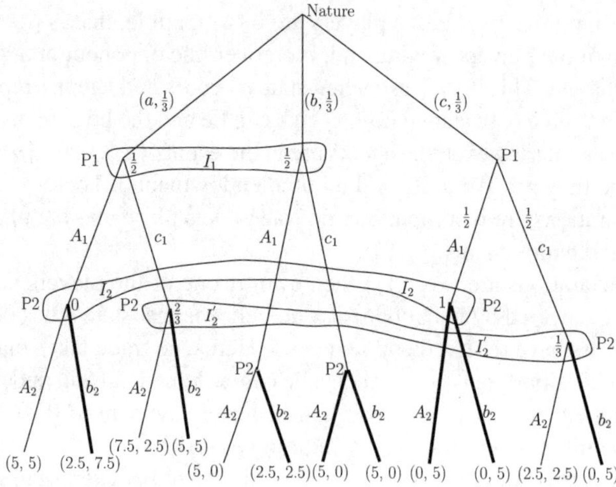

Figure 1

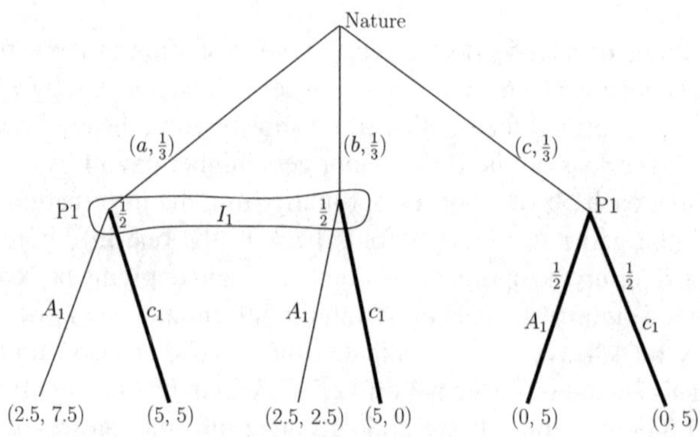

Figure 2

and n_2 and in I_2' by n_3 and n_4. The probabilities attached to the nodes in I_1 follow from the fact that the probability with which nature chooses state a is the same as the one with which it chooses state b. Given the choices by nature, the strategies of the players described above and using the Bayesian formula for updating beliefs we also calculate the conditional probabilities

$$\begin{aligned} Pr(n_1/A_1) &= \frac{Pr(A_1/n_1) \times Pr(n_1)}{Pr(A_1/n_1) \times Pr(n_1) + Pr(A_1/n_2) \times Pr(n_2)} \\ &= \frac{1 \times 0}{1 \times 0 + 1 \times \frac{1}{3} \times \frac{1}{2}} = 0 \end{aligned} \tag{4}$$

and

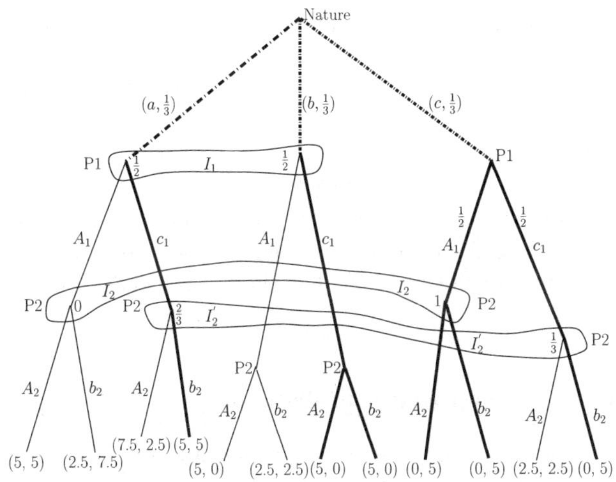

Figure 3

$$Pr(n_3/c_1) = \frac{Pr(c_1/n_3) \times Pr(n_3)}{Pr(c_1/n_3) \times Pr(n_3) + Pr(c_1/n_4) \times Pr(n_4)}$$
$$= \frac{1 \times \frac{1}{3}}{1 \times \frac{1}{3} + 1 \times \frac{1}{2} \times \frac{1}{3}} = \frac{2}{3}\,. \tag{5}$$

Obviously from the above we obtain $Pr(n_2/A_1) = 1$ and $Pr(n_4/c_1) = \frac{1}{3}$.

Therefore the perfect Bayesian equilibrium obtained above confirms the initial endowments and the decisions to lie imply that the contract $(5, 2.5, 2.5)$ cannot be realized and the players will not sign.

In Figure 3 we indicate, through heavy lines, plays of the game which are the outcome of the choices by nature and the optimal behavioral strategies by the players. The interrupted heavy lines at the beginning of the tree signify that nature does not take an optimal decision, as it has no payoff function, but simply chooses among three alternatives, with equal probabilities. From each such choice the play of the game continues through the optimal decisions by the agents to a specific terminal node. The directed path (a, c_1, b_2) with payoffs (5, 5) occurs with probability $\frac{1}{3}$. The paths (b, c_1, A_2) and (b, c_1, b_2) lead to payoffs (5, 0) and occur with probability $\frac{1}{3}(1-q)$ and $\frac{1}{3}q$, respectively. The values $(1-q)$ and q denote the probabilities with which P2 decides to choose between A_2 and b_2 from the singleton node at the end of (b, c_1). Of course no matter what q is selected this does not affect the payoffs. The paths (c, A_1, b_2) (c, c_1, b_2) lead to payoffs (0, 5) and occur, each, with probability $\frac{1}{3} \times \frac{1}{2}$, as, by assumption, from the singleton node at the end of (c), P1 chooses between A_1 and c_1 with probability $\frac{1}{2}$. This of course is not significant because any other probabilities attached to A_1 and c_1 would not affect the payoffs.

Summarizing, we note that the implied equilibrium paths are as follows. If nature chooses a or b, P1 responds by playing c_1, i.e. he lies. Then P2 lies from

I_2' and from the singleton node at the end of (b, c_1) he can tell the truth or lie. The players end up with their initial endowments. If nature chooses c, P1 can tell the truth, or even lie, but P2 will play b_2, i.e. he will lie. Again the players end up with their initial endowments. It follows that for all choices by nature, at least one of the players tells a lie on the optimal play. The players by lying avoid the possibility of having to make a payment to their opponent.

We have constructed an extensive form game and employed reasonable rules for calculating payoffs and shown that the proposed allocation (5, 2.5, 2.5) will not be realized. The same conclusion would have been reached if P2 were assumed to move first.

6 The implementation of private core allocations

Next we investigate the role of P3 in the implementation, or realization, of private core allocations in Example 3.1 of Section 3. We have seen that such core allocations are CBIC, which is a desirable property of the cooperative solution. We shall now show how they can be supported as perfect Bayesian equilibrium of a noncooperative game. This falls into the agenda of the Nash programme.

We use as an example the private core allocation

$$\begin{pmatrix} 4\,4\,1 \\ 4\,1\,4 \\ 2\,0\,0 \end{pmatrix}.$$

The ith line refers to Player i and the columns from left to right to states a, b and c.

When P3 enters the scene he is characterized by $e_3 = (0, 0, 0)$ with $\mathcal{F}_3 = \{\{a\}, \{b\}, \{c\}\}$. P3 announces his observation and this implies that, if he is believed, P1 and P2 will now be able to figure out all states of nature. We shall show how the payoffs of the matrix above will be realized from the optimal decisions of the players in a sequential game.

P1 and P2 see on a screen the announced state but P1 cannot distinguish between states a and b and P2 between a and c. P3 sees the correct state and moves first. However he can either announce exactly what he saw or he can lie. Obviously he can lie in two ways. Following the announcement of P3 it is the turn of P1 to act. When he comes to decide he has his information from the screen and also he knows the strategy that P3 played. Then it is the turn of P2 to act. When he comes to decide he has his information from the screen and he also knows what P3 and P1 played before him. Both P1 and P2, when it is their turn to act, can either tell the truth about what they saw on the screen or they can lie.

We must distinguish between the announcements of the players designed to maximize their expected returns, and the true state of nature. The former, with the players' temptations to lie, cannot be used to determine the true state which is needed for the purpose of making payoffs, which include any imposition of penalties for lying. P3 has a special status but he should also take into account that in the end the lie will be detected and this can affect his payoff. The terms of the contract, which we propose to examine below, take this into account.

The *rules* of calculating payoffs, i.e. the terms of the contract, are as follows: If P3 tells the truth we implement the redistribution in the matrix above which is proposed for this particular choice of nature.
If P3 lies then we look into the strategies of P1 and P2 and decide as follows:

(i) If the declaration of P1 and P2 are incompatible we go to the initial endowments and each player keeps his.
(ii) If the declarations are compatible we expect the players to honour their commitments for the state in the overlap, using the endowments of the true state, provided these are positive. If a player's endowment is zero then no transfer from that agent takes place as he has nothing to give.

We are looking for a perfect Bayesian equilibrium, i.e. a set of optimal behavioral strategies consistent with a set of beliefs. The beliefs are indicated by the probabilities attached to the nodes of the information sets in Figure 4 with arbitrary r, s, q, p and t between 0 and 1. Given these beliefs optimal decisions of P2 are indicated with heavy lines and the tree in Figure 4 folds up to the one in Figure 5. In this, optimal decisions of P1 are indicated with heavy lines. Figure 5 then folds up into Figure 6 which shows with heavy lines optimal decisions of P3.

In summary, an optimal behavioral strategy for P3 is to tell the truth, i.e. to play, with probability 1, a from a, b from b and c from c. An optimal behavioral strategy for P1 is to play A_1 from both I_1^1 and I_1^2, i.e. to tell the truth, and to play c_1 from I_1^3, i.e. to lie. From the singletons he plays c_1, i.e. he tells the truth. Finally optimal behavioral strategy for P2 is to play b_2 from the singletons, i.e. to tell the truth, to play A_2 from I_2^1 and I_2^6, i.e. to tell the truth, and to play b_2 from I_2^2, I_2^3, I_2^4 and I_2^5, i.e. to lie. Each player is rational and reaches the conclusion that P3 has no incentive to lie, before any revelation of the actual state of nature.

It is possible to check that the beliefs indicated next to the nodes are consistent with these strategies. Hence optimal behavioral strategies and beliefs form a perfect Bayesian equilibrium. We note that the implied equilibrium paths are as follows. If nature chooses a, P3 follows with a, P1 responds with A_1 and P2 declares A_2, and the payoffs are (4, 4, 2). If nature chooses b, P3 follows with b, P1 responds with A_1 and P2 declares b_2, and the payoffs are now (4, 1, 0). Finally if nature chooses c, P3 plays c, P1 follows with c_1 and P2 responds with A_2, and the payoffs are (1, 4, 0).

Along the optimal paths nobody has an incentive to misrepresent the realized state of nature and hence the private core allocation is incentive compatible. On the other hand the explicit considerations through a game tree show clearly that even optimal behavioral strategies, which of course are fully rational, can imply that players might have an incentive to lie from certain information sets, which though are not visited by the optimal play of the game. For example, P1, although he knows that nature has chosen a or b, has an incentive to declare c_1 from I_1^3, trying to take advantage of a possible lie by P3. Similarly P2, although he knows that nature has chosen a or c, has an incentive to declare b_2 from I_2^2, I_2^3, I_2^4 and I_2^5, trying to take advantage of possible lies by the other players. For example, the right hand side node of I_2^3 is reached by both P3 and P1 lying. Incentive compatibility has now

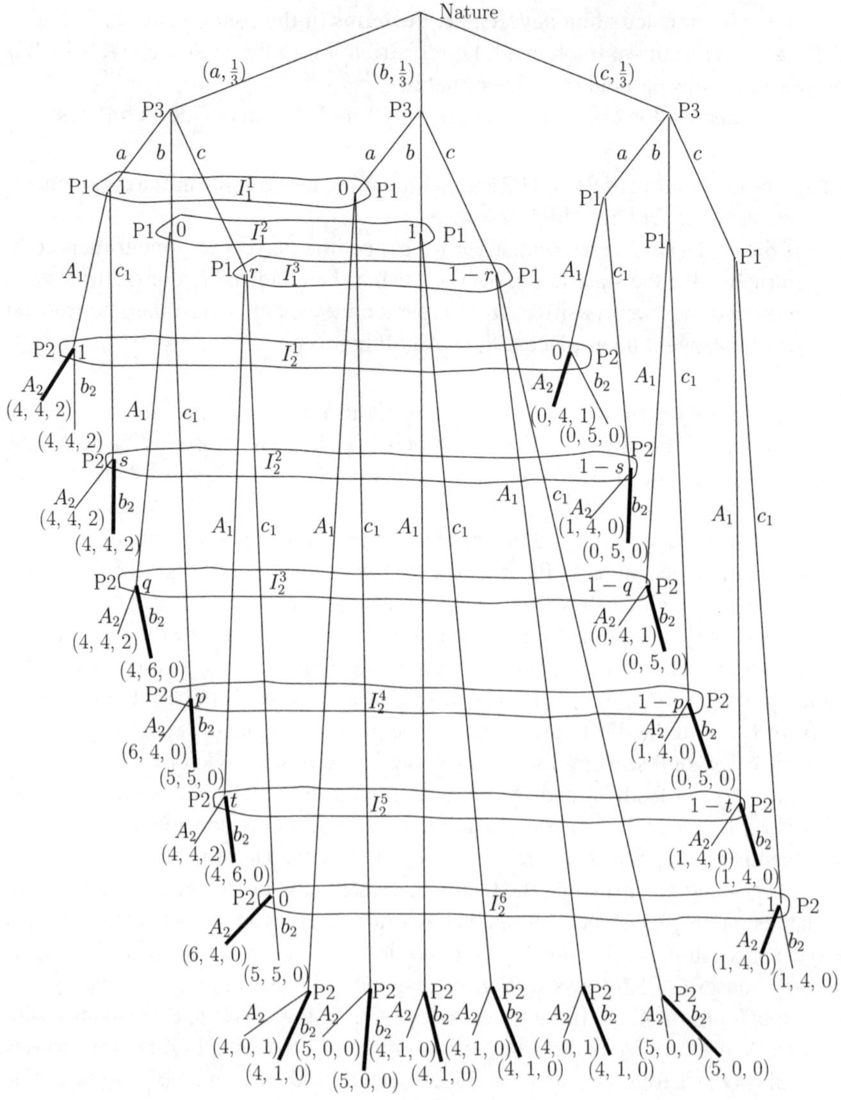

Figure 4

been defined to allow that the optimal behavioral strategies can contain lies, while there must be an optimal play which does not.

In Figure 7 we indicate through heavy lines the equilibrium paths obtained above. Again, the interrupted heavy lines at the beginning of the tree signify that nature does not take an optimal decision, as it has no payoff function, but simply chooses among three alternatives, with equal probabilities. The directed paths (a, a, A_1, A_2) with payoffs (4, 4, 2), (b, b, A_1, b_2) with payoffs (4, 1, 0) and (c, c, c_1, A_2) with payoffs (1, 4, 0) occur, each, with probability $\frac{1}{3}$. It is clear that

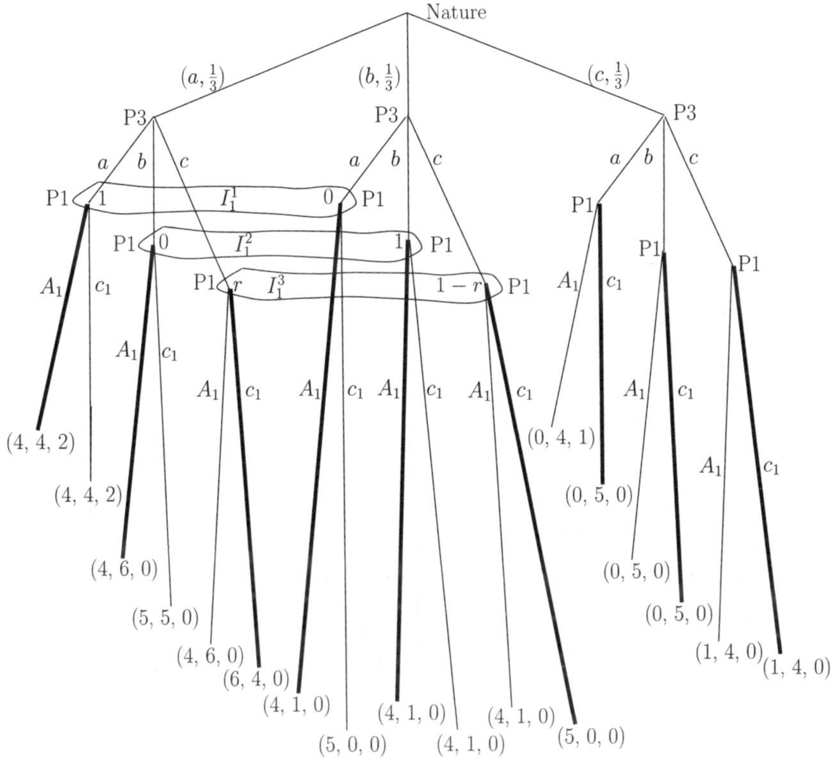

Figure 5

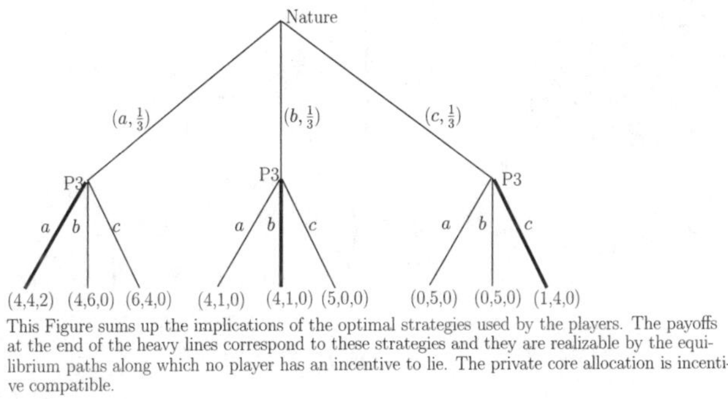

Figure 6 This figure sums up the implications of the optimal strategies used by the players. The payoffs at the end of the heavy lines correspond to these strategies and they are realizable by the equilibrium paths along which no player has an incentive to lie. The private core allocation is incentive compatible

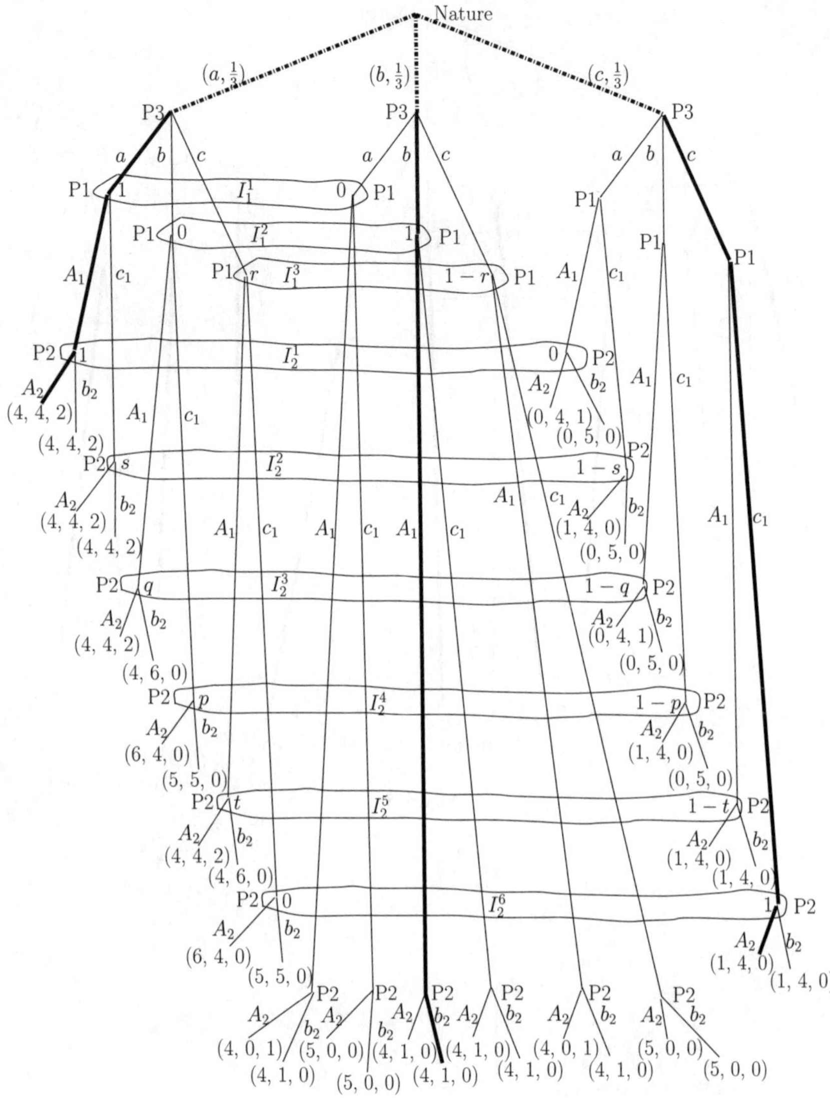

Figure 7

nobody lies on the optimal paths and that the proposed reallocation is incentive compatible and hence it will be realized.

Off the equilibrium strategies even P3 has considered the possibility of lying. For example when nature chooses b he would consider playing a, hoping that P1 will respond with A_1 and P2 with A_2. However such a move is dismissed because he knows that the other players are rational.

Analogous conclusions as above would have been reached if, following the announcement of P3, it was assumed that P2 moves first.

7 Concluding remarks

We consider the area of incomplete and differential information and how it is modeled important for the development of economic theory. Efforts are being made in breaking new ground using formulations which are promising but rather difficult. It is hoped that the use of game trees in the analysis helps in the development of ideas in that it makes them more discussable.

Our discussion in Section 5 suggests that core notions which may not be CBIC, i.e., the weak fine core, cannot easily be supported as a perfect Bayesian equilibrium. On the other hand, as we saw in Section 6, the private core which is CBIC can be supported as a perfect Bayesian equilibrium. The discussion above provides a noncooperative interpretation or foundation of the private core while making, through the game tree, the individual decisions transparent. In this way a better and possibly deeper understanding of how CBIC contracts are formed is obtained.

The positive result for the private core is not a general theorem but rather a 3-agent differential information economy example. However we believe that graph theory techniques may be adopted to construct a general result. We have not attempted this since it would complicate the technical analysis while it is not certain that it would advance our economic insights or knowledge very much. At the moment we leave this as an open question.

Appendix I: Proof of Theorem 3.1

Before we engage in the proof of Theorem 3.1, we will need some definitions. Let $(\Omega, \mathcal{F}, \mu)$ be a finite measure space, and X be a Banach space. Following Diestel and Uhl (1977), the function $f : \Omega \to X$ is called *simple* if there exist $x_1, x_2, \dots, x_n$ in X and $A_1, A_2, \dots, A_n$ in $\mathcal{F}$ such that $f = \sum_{i=1}^{n} x_i \mathcal{X}_{A_i}$ where $\mathcal{X}_{A_i}$ denotes the indicator function. A function $f : \Omega \to X$ is said to be *μ-measurable* if there exists a sequence of simple functions $f_n : \Omega \to X$ such that $\lim_{n\to\infty} \| f_n(\omega) - f(\omega) \| = 0$ for almost all $\omega \in \Omega$. A μ-measurable function $f : \Omega \to X$ is *Bochner integrable* if there exists a sequence of simple functions $\{f_n : n = 1, 2, ...\}$ such that

$$\lim_{n\to\infty} \int_{\Omega} \| f_n(\omega) - f(\omega) \| d\mu(\omega) = 0. \tag{I.1}$$

In this case, for each $A \in \mathcal{F}$, we define the integral to be

$$\int_A f(\omega) d\mu(\omega) = \lim_{n\to\infty} \int_A f_n(\omega) d\mu(\omega). \tag{I.2}$$

The integral is of course independent of the approximating sequence of simple functions.[9]

[9] Let $\Phi = \{f_n : n = 1, \dots, n\}$ be a sequence of simple functions from Ω to X for which $\lim \int f_n(\omega) d\mu(\omega)$ exists with respect to the norm topology and take this limit to define a quantity

It can be shown (see Diestel and Uhl, 1977), Theorem 2, pp. 45) that if $f : \Omega \to X$ is a μ-measurable function, then f is Bochner integrable if and only if $\int_\Omega \| f(\omega) \| d\mu < \infty$. It is important to note that the *Dominated Convergence Theorem* holds for Bochner integrable functions. In particular, if $\{f_n : \Omega \to X : n = 1, 2, ...\}$ is a sequence of Bochner integrable functions such that $\lim_{n\to\infty} f_n(\omega) = f(\omega)$ μ-a.e., and $\| f_n(\omega) \| \leq g(\omega)$ μ-a.e., where $g : \Omega \to R$ is an integrable function, then f is Bochner integrable and $\lim \int_\Omega \| f_n(\omega) - f(\omega) \| d\mu(\omega) = 0$, (see Diestel and Uhl, 1977), Theorem 3, pp. 45).

Denote by $L_p(\mu, X)$ with $1 \leq p < \infty$ the space of equivalence classes of X-valued Bochner integrable functions $x : \Omega \to X$ normed by

$$\| x \|_p = \left(\int_\Omega \| x(\omega) \|^p d\mu(\omega)\right)^{\frac{1}{p}} < \infty. \tag{I.6}$$

It is a standard result that normed by the functional $\| \,.\, \|_p$ above, $L_p(\mu, X)$ becomes a Banach space (see Diestel and Uhl, 1977, p. 50). It is also well-known that $L_q(\mu, X^*)$ is the dual of $L_p(\mu, X)$, where $1 \leq p < \infty$ and $1/p + 1/q = 1$, and the value $w \cdot x$ of $x \in L_p(\mu, X)$ at $w \in L_q(\mu, X^*)$ is defined by

---

$I(\Phi)$. We can certainly use linearity, particularly in the form

$$I(\Phi - \mathcal{G}) = I(\Phi) - I(\mathcal{G}), \tag{I.3}$$

for *any* two such sequences.

We have defined that a μ-measurable function is Bochner integrable if there exists a sequence Φ for which

$$\lim_{n\to\infty} \int_\Omega \|f_n(\omega) - f(\omega)\| d\mu(\omega) = 0. \tag{I.4}$$

Now we argue that if two sequences $\Phi, \mathcal{G}$ both satisfy this, for some given f, then $I(\Phi) = I(\mathcal{G})$. This will establish that the value obtained only depends upon f and so can be used to define its integral. We proceed as follows.

$$\begin{aligned}
\|I(\Phi) - I(\mathcal{G})\| &= \|I(\Phi - \mathcal{G})\| \quad \text{from (I.3)} \\
&= \| \lim_{n\to\infty} \int_\Omega [f_n(\omega) - g_n(\omega)] d\mu(\omega)\| \quad \text{from (I.4)} \\
&= \lim_{n\to\infty} \| \int_\Omega [f_n(\omega) - g_n(\omega)] d\mu(\omega)\| \\
&\leq \lim_{n\to\infty} \int_\Omega \|f_n(\omega) - g_n(\omega)\| d\mu(\omega) \quad \text{see Note * below} \qquad \text{(I.5)} \\
&= \lim_{n\to\infty} \int_\Omega \|[f_n(\omega) - f(\omega)] - [g_n(\omega) - f(\omega)]\| d\mu(\omega) \\
&\leq \lim_{n\to\infty} \int_\Omega \|f_n(\omega) - f(\omega)\| d\mu(\omega) + \int_\Omega \|g_n(\omega) - f(\omega)\| d\mu(\omega) \\
&= 0.
\end{aligned}$$

Then $\|I(\Phi) - I(\mathcal{G})\| = 0$ implies $I(\Phi) = I(\mathcal{G})$.

Note * : This inequality can be used since it only involves the finite sum employed in the definition of the integral of a simple function.

$$w \cdot x = \int_{\Omega} [w(\omega) \cdot x(\omega)] d\mu(\omega). \tag{I.7}$$

Recall that $\sigma(L_p(\mu, X), L_q(\mu, X^*))$ is defined as the weakest topology on $L_p(\mu, X)$ for which a net $\{x^\lambda : \lambda \in \Lambda\}$ converges to x if and only if $\{w \cdot x^\lambda\} \to w \cdot x$ for all $w \in L_q(\mu, X^*)$. We call this topology as *weak topology* and the convergence as *weak convergence*. A function $f : X \to R$ is *weakly upper semicontinuous* if $\limsup f(x^\lambda) \leq f(x)$, *weakly lower semicontinuous* if $\liminf f(x^\lambda) \geq f(x)$, and weakly continuous if it is both weakly upper semicontinuous and weakly lower semicontinuous, whenever $\{x^\lambda\} \to x$ weakly.

We now define a Banach lattice (see Aliprantis and Burkinshaw, 1985). A Banach space X is a Banach lattice if there is an ordering $\geq$ on X with the following properties:

(i) $x \geq y$ implies $x + z \geq y + z$ for every $z \in X$,
(ii) $x \geq y$ implies $\lambda x \geq \lambda y$ for $\lambda \in \mathbb{R}_+$,
(iii) for all $x, y \in X$, there exist a supremum $x \vee y$ and an infimum $x \wedge y$,
(iv) $| x | \geq | y |$ implies $\| x \| \geq \| y \|$ for every $x, y \in X$.

If X is a Banach lattice[10] then for any $x, y \in X$, define the *order interval* $[x, y] = \{z \in X : x \leq z \leq y\}$. Note that $[x, y]$ is convex and norm closed, hence weakly closed (Mazur's Theorem). Cartwright (1974) has shown that if X is a Banach lattice with an *order continuous norm*[11] (or equivalently has weakly compact order intervals), then $L_p(\mu, X)$ with $1 \leq p < \infty$ has weakly compact order intervals as well. With the above preliminaries out of the way we can proceed with the proof.

Proof of Theorem 3.1. For each $i = 1, 2, \ldots, n$ let L_{X_i} be the set of all Bochner integrable and $\mathcal{F}_i$-measurable selections from the consumption set correspondence $X_i : \Omega \to 2^{Y_+}$ of Player i, i.e.

$$L_{X_i} = \{x_i \in L_1(\mu, Y) : \; x_i(\cdot) \text{ is } \mathcal{F}_i\text{-measurable and } x_i(\omega) \in X_i(\omega) \; \mu\text{-}a.e.\}. \tag{I.8}$$

This means that for each agent we select from her consumption correspondence an element per ω and form a function. We require this function to be in $L_1(\mu, Y)$, and measurable with respect to the agent's information partition.

Since by assumption each $e_i : \Omega \to Y$ is $\mathcal{F}_i$-measurable and Bochner integrable, it follows that $e_i \in L_{X_i}$ for all i. Therefore each L_{X_i} is non-empty and so is $L_X = \prod_{i=1}^{n} L_{X_i}$.

For each i, define the correspondence $P_i : L_{X_i} \to 2^{L_{X_i}}$ by

[10] An example of a Banach lattice is $\mathbb{R}^n$ with the usual vector partial ordering, the sum of moduli as norm, and absolute value of an element, the vector of absolute values of its coordinates.

[11] $\{x^\lambda\} \downarrow 0$ means that $\{x^\lambda : \lambda \in \Lambda\}$ is a decreasing net with $\inf x^\lambda = 0$. A Banach lattice X is said to have an *order continuous norm* if $\{x^\lambda\} \downarrow 0$ in X implies $\| x^\lambda \| \downarrow 0$. If X is a Banach lattice, X has an order continuous norm if and only if any order interval is weakly compact.

$$P_i(x_i) = \{y_i \in L_{X_i} : v_i(y_i) > v_i(x_i)\}. \tag{I.9}$$

Since for each i, and each fixed $\omega \in \Omega$, $u_i(\omega, \cdot)$ is concave, upper semicontinuous (u.s.c.) and integrably bounded, by Theorem 2.8 in Balder and Yannelis, $v_i(\cdot)$ is weakly-u.s.c. Hence, the set

$$P_i^{-1}(y_i) = \{x_i \in L_{X_i} : y_i \in P_i(x)\} = \{x_i \in L_{X_i} : v_i(y_i) > v_i(x_i)\} \tag{I.10}$$

is weakly open in L_{X_i}. Notice that since for any fixed $\omega \in \Omega$, $u_i(\omega, \cdot)$ is concave the set $P_i(x_i)$ for all $x_i \in L_{X_i}$ is convex and also $x_i \notin P_i(x_i)$ for all $x_i \in L_{X_i}$. Hence the correspondence $P_i : L_{X_i} \to 2^{L_{X_i}}$ is convex valued and irreflexive.

We now have an infinite dimensional commodity space economy

$$\bar{\mathcal{E}} = \{(L_{X_i}, P_i, e_i) : i = 1, 2, \ldots, n\} \tag{I.11}$$

where

(a) L_{X_i} denotes the consumption set of Pi,
(b) $P_i : L_{X_i} \to 2^{L_{X_i}}$ is the preference correspondence of Pi, and
(c) $e_i \in L_{X_i}$, is the initial endowments of Pi.

In the new economy that has been constructed, a good is also characterized by the state of nature, and $v_i(x_i)$, on which the preference correspondence is based, can be thought of as a utility, rather than an expected utility, function. It is as if uncertainty and information partitions have vanished from the scene. However they are present since L_{X_i}, the consumption set of Agent i, takes into account the information partition $\mathcal{F}_i$.

We show that a core allocation exists in $\bar{\mathcal{E}}$, i.e., there exists $x^* \in L_X$ satisfying the following two conditions:

(1) $\sum_{i=1}^{n} x_i^* = \sum_{i=1}^{n} e_i$, and
(2) there do not exist coalition S and allocation $(y_i)_{i \in S} \in \prod_{i \in S} L_{X_i}$ such that $\sum_{i \in S} y_i = \sum_{i \in S} e_i$, $y_i \in P_i(x_i^*)$ for all $i \in S$.

It can easily be checked that the existence of a core allocation in $\bar{\mathcal{E}}$ implies the existence of a private core allocation in the original economy $\mathcal{E} = \{((\Omega, \mathcal{F}, \mu), X_i, \mathcal{F}_i, u_i, e_i) : i = 1, \ldots, n\}$.

Let $\mathcal{A}$ be the set of all finite dimensional subspaces of $L_1(\mu, Y)$ containing the initial endowments. For each $\alpha \in \mathcal{A}$ define $L_{X_i}^{\alpha} = L_{X_i} \cap \alpha$ and $P_i^{\alpha} : L_{X_i}^{\alpha} \to 2^{L_{X_i}^{\alpha}}$ by $P_i^{\alpha}(x_i) = P_i(x_i) \cap \alpha$. We have constructed an economy $\bar{\mathcal{E}}^{\alpha} = \{(L_{X_i}^{\alpha}, P_i^{\alpha}, e_i) : i = 1, 2, \ldots, n\}$ in a finite dimensional commodity space where

(1′) $L_{X_i}^{\alpha}$ is the consumption set of Pi,
(2′) $P_i^{\alpha} : L_{X_i}^{\alpha} \to 2^{L_{X_i}^{\alpha}}$ is the preference correspondence of Pi,
(3′) $e_i \in L_{X_i}^{\alpha}$ is the initial endowment of Pi.

The economy constructed is finite dimensional in that each consumption set can be spanned by a finite number of vectors. For every such, finite dimensional, α-economy one can prove the existence of a core allocation. This implies in the limit, as the number of dimensions tends to infinity, the existence of a core allocation for $\bar{\mathcal{E}}$, which has been approximated through the net of economies.

It can easily be checked that for each $\alpha \in \mathcal{A}$, $\bar{\mathcal{E}}^\alpha$ satisfies all the assumptions of Florenzano's (1989) core existence theorem and therefore there exists $x^\alpha \in \prod_{i=1}^{n} L^\alpha_{X_i} = L^\alpha_X$ such that

(4′) $\sum_{i=1}^{n} x_i^\alpha = \sum_{i=1}^{n} e_i$

(5′) and it is not true that there exist $S \subset \{1, 2, \ldots, n\}$ and $(y_i)_{i \in S} \in \prod_{i \in S} L^\alpha_{X_i}$ such that $\sum_{i \in S} y_i = \sum_{i \in S} e_i$ and $y_i \in P_i^\alpha(x_i^\alpha)$ for all $i \in S$.

From (4′) it follows that for each $\alpha \in \mathcal{A}$ we have that every $x_i^\alpha \in [0, \sum_{i=1}^{n} e_i]$. Since by assumption Y is a Banach lattice with an order continuous norm by the Cartwright theorem so is $L_1(\mu, Y)$ and therefore we can conclude that the order interval $[0, \sum_{i=1}^{n} e_i]$ in $\sum_{i=1}^{n} L_{X_i}$ is weakly compact.

Direct the set $\mathcal{A}$ by inclusion so that $\{(x_1^\alpha, x_2^\alpha, \ldots, x_n^\alpha) : \alpha \in \mathcal{A}\}$ forms a net in $\prod_{i=1}^{n} L_{X_i}$. Since each x_i^α lies in $[0, \sum_{i=1}^{n} e_i]$ which is weakly compact we can extract a subnet

$$\{(x_1^{\alpha(m)}, x_2^{\alpha(m)}, \ldots, x_n^{\alpha(m)}) : m \in M\},$$

(where M is directed by " $\geq$"), from the net $\{(x_1^\alpha x_2^\alpha, \ldots, x_n^\alpha) : \alpha \in \mathcal{A}\}$ which converges weakly to some vector $(x_1, x_2, \ldots, x_n)$ in $[0, \sum_{i=1}^{n} e_i]$.

We will show that $(x_1, x_2, \ldots, x_n)$ is a core allocation for the economy $\bar{\mathcal{E}}$. Notice that since for each $m \in M$, $\sum_{i=1}^{n} x_i^{\alpha(m)} = \sum_{i=1}^{n} e_i$ and $x_i^{\alpha(m)}$ converges weakly to $x_i \in L_X$ we have that $\sum_{i=1}^{n} x_i = \sum_{i=1}^{n} e_i$, i.e., $(x_1, x_2, \ldots, x_n)$ is a feasible allocation.

In order to complete the proof we must show that:

($\star$) It is not true that there exists $S \subset \{1, 2, \ldots, n\}$ and $(y_i)_{i \in S} \in \prod_{i \in S} L_{X_i}$ such that $\sum_{i \in S} y_i = \sum_{i \in S} e_i$ and $y_i \in P_i(x_i)$ for all $i \in S$.

Suppose that ($\star$) is not true, then there exist coalition S and $(y_i)_{i \in S} \in \prod_{i \in S} L_{X_i}$ such that $\sum_{i \in S} y_i = \sum_{i \in S} e_i$ and $y_i \in P_i(x_i)$ for all $i \in S$. Since $x_i^{\alpha(m)}$ converges weakly to x_i and P_i has weakly open lower sections, there exists $m_0 \in M$ such that $y_i \in P_i(x_i^{\alpha(m)})$ for all $m \geq m_0$, and for all $i \in S$. Choose $m_1 \geq m_0$ so that if $m \geq m_1$, $y_i \in L_{X_i}^{\alpha(m)}$ for all $i \in S$. Then $y_i \in P_i^{\alpha(m)}(x_i^{\alpha(m)})$, for all $m \geq m_1$, and for all $i \in S$, a contradiction to (5$'$), which means that ($\star$) holds.

Finally, the fact that $\bar{\mathcal{E}}$ has been derived from the original economy $\mathcal{E}$ by integrating over the states of nature implies that a core allocation in the former is also a private core[12] allocation in the latter, and this completes the proof of Theorem 3.1.

Appendix II: The private core allocations of Example 3.1

In this section we show that the redistribution

$$\begin{pmatrix} 4\,4\,1 \\ 4\,1\,4 \\ 2\,0\,0 \end{pmatrix}.$$

where again the ith line refers to Player i and the columns from left to right to states a, b and c, is a private core allocation.

An $\mathcal{F}_i$-measurable redistribution of the endowments of the three agents above is given by

$$\begin{pmatrix} 5-\varepsilon & 5-\varepsilon & \delta_1 \\ 5-\delta & \varepsilon_1 & 5-\delta \\ \varepsilon+\delta & \varepsilon_2 & \delta_2 \end{pmatrix}$$

with $\varepsilon = \varepsilon_1 + \varepsilon_2$ and $\delta = \delta_1 + \delta_2$.

We show below that the private core allocations are obtained from

Problem

Maximize $\mathcal{U}_3 = (\varepsilon+\delta)^{\frac{1}{2}} + \varepsilon_2^{\frac{1}{2}} + \delta_2^{\frac{1}{2}}$

Subject to

$$\begin{aligned} &\mathcal{U}_1 = 2(5-\varepsilon)^{\frac{1}{2}} + \delta_1^{\frac{1}{2}} \geq \alpha_1^{\frac{1}{2}} \\ &\mathcal{U}_2 = 2(5-\delta)^{\frac{1}{2}} + \varepsilon_1^{\frac{1}{2}} \geq \alpha_2^{\frac{1}{2}}, \\ &\varepsilon_1 + \varepsilon_2 = \varepsilon \leq 5 \text{ and } \delta_1 + \delta_2 = \delta \leq 5, \\ &\varepsilon_i, \delta_i \geq 0 \end{aligned}$$

for Pareto optimality, and $\alpha_1^{\frac{1}{2}}, \alpha_2^{\frac{1}{2}} \geq 2(5^{\frac{1}{2}}) = 20^{\frac{1}{2}}$ for individual rationality. We shall not give characterizations of optimality through Lagrange or Kuhn-Tucker conditions because the utility functions, although continuous on their domains of definition, are not differentiable at the origin.

The solution to the problem exists because of the compactness of the feasible set, which follows from the fact that the values of all variables are bounded between 0 and 5 and the set defined by the utility constraints is closed, and the maximum is unique due to the concavity of the functions. Pareto optimality of the solution follows from the fact that there is no possible improvement to the values of all

[12] This can easily be shown by contradiction. I.e. one picks a core allocation x in the economy $\bar{\mathcal{E}}$ and supposes that it is not a core allocation in $\mathcal{E}$ and reaches a contradiction.

three utility functions, because if there were then we could increase the value of $\mathcal{U}_3$ without violating the constraints. Individual rationality follows from the fact that the initial endowments of the players imply utility $2(5^{\frac{1}{2}}) = 20^{\frac{1}{2}}$. Finally it is not possible for any pair of traders to redistribute their initial endowments and become better off, while retaining measurability. Hence the solution to the Problem is in the core.

Next we note that the solution to the problem above always satisfies the utility constraints of P1 and P2 with equality. For suppose, say, the first constraint was satisfied with an inequality. Then it would be possible to increase ε and ε_2, without disturbing measurability, and thus increase $\mathcal{U}_3$.

The question arises whether there exist core allocations which cannot be captured as solutions to a problem of the above type. Consider any allocation in the core and formulate the above problem with $\mathcal{U}_1, \mathcal{U}_2$ taking the corresponding values. From the fact that it is maximized, we should get for $\mathcal{U}_3$ at least the value of the proposed allocation, and if we actually obtain a higher one then it must be for a different allocation for at least one of the utilities, say $\mathcal{U}_1$. Now through concavity we can improve the proposed values of $\mathcal{U}_3$ and $\mathcal{U}_1$ and then through a redistribution from $\mathcal{U}_1$ to $\mathcal{U}_2$ and $\mathcal{U}_3$, by a small increase in ε and ε_1, we can improve all utilities in relation to the proposed allocation, which therefore was not Pareto optimal.

We shall now discuss properties of the private core allocations. We shall call symmetric allocations those with $\varepsilon = \delta$ and $\varepsilon_i = \delta_i$. First we consider the case when $\alpha_1 = \alpha_2 = \alpha$. This condition implies that the solution is symmetric. For otherwise it would not be unique. A further restriction on the symmetric solutions is when $\varepsilon_1 = \varepsilon$. In order to investigate this we look at the function $y = 2(5 - \varepsilon)^{\frac{1}{2}} + \varepsilon^{\frac{1}{2}}$. It is routine to show that it is a strictly concave function attaining its maximum value 5 at $\varepsilon = 1$. We are interested in the values of ε for which $y \geq 20^{\frac{1}{2}}$.

Suppose now that the common value of α is equal to 25 which is the maximum possible such value, since $\alpha^{\frac{1}{2}} = 2(5-\varepsilon)^{\frac{1}{2}} + \varepsilon_1^{\frac{1}{2}} \leq y = 2(5-\varepsilon)^{\frac{1}{2}} + \varepsilon^{\frac{1}{2}} \leq (25)^{\frac{1}{2}}$. Then we must have $\varepsilon_1 = \varepsilon = 1$, for otherwise the constraint will not be satisfied. The implied value for $\mathcal{U}_3$ is $2^{\frac{1}{2}}$ and this confirms that the redistribution at the beginning of this appendix is a private core allocation.

Next let the common admissible value of α be less than 25. We investigate whether it is now possible that the solution implies $\varepsilon_1 = \varepsilon$. In such a case the structure of the function y above would mean that there are two such values of ε, one smaller and one greater than 1. But then by strict concavity of the functions we could obtain feasible ε which would satisfy the constraint and increase the value of the objective function. It follows that although the solution is symmetric we do not have $\varepsilon_1 = \varepsilon$ which would have implied the corner solution $\varepsilon_2 = 0$.

Finally we look at the case where $\alpha_1 \neq \alpha_2$. Obviously the solution cannot be symmetric. The question arises whether we should have $\varepsilon_1 = \varepsilon$ and $\delta_1 = \delta$. On a (δ, ε) plane we consider the iso-level curves $2(5 - \varepsilon)^{\frac{1}{2}} + \delta^{\frac{1}{2}} = 5$ and $2(5-\delta)^{\frac{1}{2}} + \varepsilon^{\frac{1}{2}} = 5$. In this plane the first one is a concave and the second a convex function. Their unique common point is (1, 1). Now consider a slightly lower in value iso-level curve of the second type while the one of the first type stays the same. The two curves cross at a point with $\varepsilon, \delta > 1$ and $\varepsilon < \delta$. However there is no obvious reason why in the solution of the maximization problem we should have both $\varepsilon_1 = \varepsilon$ and $\delta_1 = \delta$.

References

Aliprantis, C. D., Burkinshaw, O.: Positive operators. New York London: Academic Press 1985

Allen., B.: Market games with asymmetric information and non-transferable utility: representation results and the core. Mimeo (1991)

Aumann. R. J.: Correlated equilibria as an expression of Bayesian rationality. Econometrica **55**, 1–18 (1987)

Balder, E. J., Yannelis, N. C. : On the continuity of expected utility. Economic Theory **3**, 625–643 (1993)

Binmore, K.: Nash bargaining theory I, ICERD Discussion Paper No 9, London School of Economics, pp. 1–40 (1980a). Published in: Binmore, K., Dasgupta, P. (eds.) The economics of bargaining, pp. 27–46. Oxford: Blackwell 1987

Binmore, K.: Nash bargaining theory II, ICERD Discussion Paper No 14, London School of Economics, pp. 1–28 (1980b). Published in: Binmore, K., Dasgupta, P. (eds.) The economics of Bargaining, pp. 61–76. Oxford: Blackwell 1987

Cartwright, D.: The order completeness of some spaces of vector valued functions. Bull. Aust. Math. Soc. **11**, 57–61 (1974)

Diestel, J., Uhl, J. J.: Vector measures. In: Mathematical surveys, vol. 15. Providence, RI: American Mathematical Society 1997

Florenzano, M.: On the nonemptiness of the core of a coalitional production economy without ordered preferences. Journal of Mathematical Analysis and Applications **141**(2), 207–219 (1989)

Hahn, G., Yannelis, N.C.: Coalitional Bayesian Nash implementation in differential information economies. Economic Theory (this issue) (2001)

Koutsougeras, L., Yannelis, N.C.: Incentive compatibility and information superiority of the core of an economy with differential information. Economic Theory **3**, 195–216 (1993)

Krasa, S., Yannelis, N. C.: The value allocation of an economy with differential information. Econometrica **62**, 881–900 (1994)

Kreps, M.D., Wilson, R.: Sequential equilibrium. Econometrica **50**, 889–904 (1982)

Lagunoff, R.: A simple noncooperative core story. Games and Economic Behaviour **7**, 54–61 (1994)

Lefebvre, I.: An alternative proof of the non-emptiness of the private core. Economic Theory (this issue) (2001)

Nash, J.F.: Two-person cooperative games. Econometrica **21**, 128–40 (1953)

Page, F.: Market games with differential information and infinite dimensional commodity spaces: the core. Economic Theory **9**, 151–159 (1997)

Perry, M., Reny, P. J.: A non-cooperative view of coalition formation and the core. Econometrica **62**, 795–817 (1994)

Serrano, R.: A market to implement the core. Journal of Economic Theory **67**, 285–294 (1995)

Serrano, R., Vohra, R.: Non-cooperative implementation of the core. Social Choice and Welfare **14**, 513–525 (1997)

Tirole, J.: The theory of industrial organization. Cambridge, MA London: MIT Press 1988

Trockel, W.: Implementations of the Nash solution based on its Walrasian characterization (revised version of former IMW working Paper No. 306). Economic Theory **16**, 277–294 (2000)

Wilson, R.: Information, efficiency, and the core of an economy. Econometrica **46**, 807–816 (1978)

Yannelis, N. C.: The core of an economy with differential information. Economic Theory **1**, 183–198 (1991)

On extensive form implementation of contracts in differential information economies*

Dionysius Glycopantis[1], **Allan Muir**[2], **and Nicholas C. Yannelis**[3]

[1] Department of Economics, City University, Northampton Square, London EC1V 0HB, UK (e-mail: d.glycopantis@city.ac.uk)
[2] Department of Mathematics, City University, Northampton Square, London EC1V 0HB, UK (e-mail: a.muir@city.ac.uk)
[3] Department of Economics, University of Illinois at Urbana-Champaign, IL 61820, USA (e-mail: nyanneli@uiuc.edu)

Received: November 19, 2001; revised version: April 17, 2002

Summary. In the context of differential information economies, with and without free disposal, we consider the concepts of Radner equilibrium, rational expectations equilibrium, private core, weak fine core and weak fine value. We look into the possible implementation of these concepts as perfect Bayesian or sequential equilibria of noncooperative dynamic formulations. We construct relevant game trees which indicate the sequence of decisions and the information sets, and explain the rules for calculating ex ante expected payoffs. The possibility of implementing an allocation is related to whether or not it is incentive compatible. Implementation through an exogenous third party or an endogenous intermediary is also considered.

Keywords and Phrases: Differential information economy, Private core, Radner equilibrium, Rational expectations equilibrium, Weak fine core, Weak fine value, Free disposal, Coalitional Bayesian incentive compatibility, Game trees, Perfect Bayesian equilibrium, Sequential equilibrium, Contracts.

JEL Classification Numbers: 020, 226.

1 Introduction

An economy with differential information consists of a finite set of agents each of which is characterized by a random utility function, a random consumption set,

* This paper comes out of a visit by Nicholas Yannelis to City University, London, in December 2000. We are grateful to Dr A. Hadjiprocopis for his invaluable help with the implementation of Latex in a Unix environment. We also thank Leon Koutsougeras and a referee for several, helpful comments.
Correspondence to: N. C. Yannelis

random initial endowments, a private information set defined on the states of nature, and a prior probability distribution on these states. For such an economy there are a number of cooperative and non-cooperative equilibrium concepts.

We have the noncooperative concepts of the generalized Walrasian equilibrium ideas of Radner equilibrium and rational expectations equilibrium (REE) defined in Radner (1968), Allen (1981) and Einy, Moreno, and Shitovitz (2000, 2001).[1] We also have the cooperative concepts of the private core (Yannelis, 1991), of the weak fine core, defined in Yannelis (1991) and Koutsougeras and Yannelis (1993), and that of the weak fine value (Krasa and Yannelis, 1994). The last two concepts allow the agents to pool their information.[2]

In a comparison of the equilibrium concepts we note that contrary to the private core any rational expectations Walrasian equilibium notion will always give zero quantities to an agent whose initial endowments are zero in each state. This is so irrespective of whether his private information is the full partition or the trivial partition of the states of nature. Hence the Radner as well as the REE do not register the informational superiority of an agent.

In Glycopantis, Muir, and Yannelis (2001) we provided a noncooperative interpretation of the private core for a three persons economy without free disposal. We constructed game trees which indicate the sequence of decisions and the information of the agents, and explained the rules for calculating ex ante, expected payoffs, through the reallocation of initial endowments. We showed that the private core can be given a dynamic interpretation as a perfect Bayesian equilibrium (PBE) of a noncooperative extensive form game.

The term implementation is used in the sense of realization of an allocation and not in the formal sense of implementation theory or mechanism design. Implementation or support of an allocation is sought through the PBE concept, described in Tirole (1988), which is a variant of the Kreps-Wilson (1982) idea of sequential equilibrium.

A PBE consists of a set of players' optimal behavioral strategies, and consistent with these, a set of beliefs which attach a probability distribution to the nodes of each information set. Consistency requires that the decision from an information set is optimal given the particular player's beliefs about the nodes of this set and the strategies from all other sets, and that beliefs are formed from updating, using the available information. If the optimal play of the game enters an information set then updating of beliefs must be Bayesian. Otherwise appropriate beliefs are assigned arbitrarily to the nodes of the set. This equilibrium concept is further looked at in Appendix I.

Our main observation in Glycopantis, Muir, and Yannelis (2001) was that Bayesian incentive compatible concepts, like the private core, can be implemented as a PBE of a noncooperative, extensive form game. Moreover we provided a counter example which demonstrates that core concepts which are not necessarily Bayesian incentive compatible, as for example the weak fine core, cannot be supported, under reasonable rules, in a dynamic framework. In the present paper we

---

[1] Kurz (1994) has provided the alternative idea of rational belief equilibria.

[2] See also Allen and Yannelis (2001) for additional references.

examine further the issue of extensive form implementation and obtain additional results.

Firstly, we consider cooperative and noncooperative solution concepts *with* and *without* free disposal. To our surprise, as it was not intuitively obvious, we found that solution concepts which are Bayesian incentive compatible without free disposal, do not retain this property under free disposal. In particular, not only free disposal *destroys* incentive compatibility but a problem also appears in verifying that an agent has actually destroyed part of his initial endowment.

Secondly, we provide examples which demonstrate that with free disposal cooperative and noocooperative solution concepts are not implementable as a PBE. However implementation becomes possible by introducing a third party, such as a court which has perfect knowledge in order to be able to penalize the lying agents.

Thirdly, for the purpose of implementation of the (non-free) disposal private core, we follow an alternative approach. We consider the (non-free) disposal private core example of the one-good, three-agent economy discussed in Glycopantis, Muir, and Yannelis (2001). The introduction of a third party results in the implementation of the private core allocation as a PBE. We show here that it can also be implemented as a sequential equilibrium (Kreps and Wilson, 1982).

Finally we provide a full characterization of our Bayesian incentive compatibility concept in the case of one good per state.

The analysis suggests that if an allocation is not incentive compatible, i.e. the agents do not find that it is in accordance with their interests, then there is a difficulty in implementing it in a dynamic framework. On the other hand incentive compatible allocations are implementable through contracts with reasonable conditions. We note that the implementation analysis is independent of the equilibrium notion. It applies to contracts in general which can be analysed by a similar tree structure.

Parts of the investigation fall into the area of the Nash programme the purpose of which has been, as explained in Glycopantis, Muir, and Yannelis (2001), to provide support and justification of cooperative solutions through noncooperative formulations. On the other hand we extend here the investigation into more general areas by discussing explicitly the possible implementation of noncooperative concepts such as Radner equilibrium and REE. It appears that in general the issue is the relation between dynamic and static considerations, not necessarily between cooperative and noncooperative formulations.

The paper is organized as follows. Section 2 defines a differential information exchange economy. Section 3 contains the equilibrium concepts discussed in this paper. Section 4 describes ideas of incentive compatibility. Section 5 discusses the non-implementation of free disposal private core allocations and Section 6 the implementation of private core and Radner equilibria through the courts. Section 7 discusses the implementation of non-free disposal private core allocations through an endogenous intermediary. Section 8 offers concluding remarks. Appendix I contains further remarks on PBE.

2 Differential information economy

We define the notion of a finite-agent economy with differential information, confining ourselves to the case where the set of states of nature, Ω, is finite and there is a finite number of goods, l, per state. $\mathcal{F}$ is a σ-algebra on Ω, I is a set of n players and $\mathbb{R}^l_+$ will denote the positive orthant of $\mathbb{R}^l$.

A differential information exchange economy $\mathcal{E}$ is a set $\{((\Omega,\mathcal{F}), X_i, \mathcal{F}_i, u_i, e_i, q_i) : i = 1,\dots,n\}$ where

1. $X_i : \Omega \to 2^{\mathbb{R}^l_+}$ is the set-valued function giving the *random consumption set* of Agent (Player) i, who is denoted also by Pi;
2. $\mathcal{F}_i$ is a partition of Ω, denoting the *private information*[3] of Pi;
3. $u_i : \Omega \times \mathbb{R}^l_+ \to \mathbb{R}$ is the *random utility* function of Pi;
4. $e_i : \Omega \to \mathbb{R}^l_+$ is the *random initial endowment* of Pi, assumed to be constant on elements of $\mathcal{F}_i$, with $e_i(\omega) \in X_i(\omega)$ for all $\omega \in \Omega$;
5. q_i is an $\mathcal{F}$-measurable probability function on Ω giving the *prior* of Pi. It is assumed that on all elements of $\mathcal{F}_i$ the aggregate q_i is positive. If a common prior is assumed it will be denoted by μ.

We will refer to a function with domain Ω, constant on elements of $\mathcal{F}_i$, as $\mathcal{F}_i$*-measurable*, although, strictly speaking, measurability is with respect to the σ-algebra generated by the partition. We can think of such a function as delivering information to Pi which does not permit discrimination between the states of nature belonging to any element of $\mathcal{F}_i$.

In the first period agents make contracts in the ex ante stage. In the interim stage, i.e., after they have received a signal[4] as to what is the event containing the realized state of nature, one considers the incentive compatibility of the contract.

For any $x_i : \Omega \to \mathbb{R}^l_+$, the *ex ante expected utility* of Pi is given by

$$v_i(x_i) = \sum_{\omega \in \Omega} u_i(\omega, x_i(\omega)) q_i(\omega). \tag{1}$$

Denote by $E_i(\omega)$ the element in the partition $\mathcal{F}_i$ which contains the realized state of nature, $\omega \in \Omega$. It is assumed that $q_i\big(E_i(\omega)\big) > 0$ for all $\omega \in \Omega$. The *interim expected utility* function of Pi is given by

$$v_i(\omega, x_i) = \sum_{\omega' \in \Omega} u_i(\omega', x_i(\omega')) q_i\big(\omega' | E_i(\omega)\big), \tag{2}$$

where

$$q_i\big(\omega' | E_i(\omega)\big) = \begin{cases} 0 & \text{for } \omega' \notin E_i(\omega) \\ \frac{q_i(\omega')}{q_i\left(E_i(\omega)\right)} & \text{for } \omega' \in E_i(\omega). \end{cases}$$

[3] Following Aumann (1987) we assume that the players' information partitions are common knowledge. Sometimes $\mathcal{F}_i$ will denote the σ-algebra generated by the partition, in which case $\mathcal{F}_i \subseteq \mathcal{F}$, as it will be clear from the context.

[4] A *signal* to Pi is an $\mathcal{F}_i$-measurable function from Ω to the set of the possible distinct observations specific to the player; that is, it induces the partition $\mathcal{F}_i$, and so gives the finest discrimination of states of nature directly available Pi.

3 Private core, weak fine core, Radner equilibrium, REE and weak fine value

We define here the various equilibrium concepts in this paper, distinguishing between the free disposal and the non-free disposal case. A comparison is also made between these concepts. All definitions are in the context of the exchange economy $\mathcal{E}$ in Section 2.

We begin with some notation. Denote by $L_1(q_i, \mathbb{R}^l)$ the space of all equivalence classes, with respect to q_i, of $\mathcal{F}$-measurable functions $f_i : \Omega \to \mathbb{R}^l$.

L_{X_i} is the set of all $\mathcal{F}_i$-measurable selections from the random consumption set of Agent i, i.e.,

$$L_{X_i} = \big\{x_i \in L_1(q_i, \mathbb{R}^l) : x_i : \Omega \to \mathbb{R}^l \text{ is } \mathcal{F}_i\text{-measurable and } x_i(\omega) \in X_i(\omega)\ q_i\text{-}a.e.\big\}$$

and let $L_X = \prod_{i=1}^{n} L_{X_i}$.

Also let

$$\bar{L}_{X_i} = \Big\{x_i \in L_1(q_i, \mathbb{R}^l) : x_i(\omega) \in X_i(\omega)\ q_i\text{-}a.e.\Big\}$$

and let $\bar{L}_X = \prod_{i=1}^{n} \bar{L}_{X_i}$.

An element $x = (x_1, \ldots, x_n) \in \bar{L}_X$ will be called an *allocation*. For any subset of players S, an element $(y_i)_{i \in S} \in \prod_{i \in S} \bar{L}_{X_i}$ will also be called an allocation, although strictly speaking it is an allocation to S.

We note that the above notation is employed also for purposes of comparisons with the analysis in Glycopantis, Muir, and Yannelis (2001). In case there is only one good, i.e. $l = 1$, we shall use the notation $L^1_{X_i}$, $\bar{L}^1_{X_i}$ etc. When a common prior is also assumed $L_1(q_i, \mathbb{R}^l)$ will be replaced by $L_1(\mu, \mathbb{R}^l)$.

First we define the notion of the (ex ante) private core[5] (Yannelis, 1991).

Definition 3.1. An allocation $x \in L_X$ is said to be a *private core allocation* if

(i) $\sum_{i=1}^{n} x_i = \sum_{i=1}^{n} e_i$ and
(ii) there do not exist coalition S and allocation $(y_i)_{i \in S} \in \prod_{i \in S} L_{X_i}$ such that $\sum_{i \in S} y_i = \sum_{i \in S} e_i$ and $v_i(y_i) > v_i(x_i)$ for all $i \in S$.

Notice that the definition above does not allow for free disposal. If the feasibility condition (i) is replaced by (i)′ $\sum_{i=1}^{n} x_i \leq \sum_{i=1}^{n} e_i$ then *free disposal* is allowed.

Example 3.1. Consider the following three agents economy, $I = \{1, 2, 3\}$ with one commodity, i.e. $X_i = \mathbb{R}_+$ for each i, and three states of nature $\Omega = \{a, b, c\}$.

We assume that the initial endowments and information partitions of the agents are given by

[5] The private core can also be defined as an interim concept. See Yannelis (1991) and Glycopantis, Muir, and Yannelis (2001).

$$\begin{aligned} e_1 &= (5,5,0), & \mathcal{F}_1 &= \{\{a,b\},\{c\}\}; \\ e_2 &= (5,0,5), & \mathcal{F}_2 &= \{\{a,c\},\{b\}\}; \\ e_3 &= (0,0,0), & \mathcal{F}_3 &= \{\{a\},\{b\},\{c\}\}. \end{aligned}$$

It is also assumed that $u_i(\omega, x_i(\omega)) = x_i^{\frac{1}{2}}$, which is a typical strictly concave and monotone function in x_i, and that every player expects that each state of nature occurs with the same probability, i.e. $\mu(\{\omega\}) = \frac{1}{3}$, for $\omega \in \Omega$. For convenience, in the discussion below expected utilities are multiplied by 3.

It was shown in Appendix II of Glycopantis, Muir, and Yannelis (2001) that, without free disposal, a private core allocation of this economy is $x_1 = (4,4,1)$, $x_2 = (4,1,4)$ and $x_3 = (2,0,0)$. It is important to observe that in spite of the fact that Agent 3 has zero initial endowments, his superior information allows him to make a Pareto improvement for the economy as a whole and he was rewarded for doing so. In other words, Agent 3 traded his superior information for actual consumption in state a. In return Agent 3 provided insurance to Agent 1 in state c and to Agent 2 in state b. Notice that if the private information set of Agent 3 is the trivial partition, i.e., $\mathcal{F}_3' = \{a,b,c\}$, then no-trade takes place and clearly in this case he gets zero utility. Thus the private core is sensitive to information asymmetries.

Next we define another core concept, the weak fine core (Yannelis, 1991; Koutsougeras and Yannelis, 1993). This is a refinement of the fine core concept of Wilson (1978). Recall that the fine core notion of Wilson as well as the fine core in Koutsougeras and Yannelis may be empty in well behaved economies. It is exactly for this reason that we are working with a different concept.

Definition 3.2. An allocation $x = (x_1, \ldots, x_n) \in \bar{L}_X$ is said to be a *weak fine core allocation* if

(i) each $x_i(\cdot)$ is $\bigvee_{i=1}^{n} \mathcal{F}_i$-measurable [6]
(ii) $\sum_{i=1}^n x_i = \sum_{i=1}^n e_i$ and
(iii) there do not exist coalition S and allocation $(y_i)_{i \in S} \in \prod_{i \in S} \bar{L}_{X_i}$ such that $y_i(\cdot) - e_i(\cdot)$ is $\bigvee_{i \in S} \mathcal{F}_i$-measurable for all $i \in S$, $\sum_{i \in S} y_i = \sum_{i \in S} e_i$ and $v_i(y_i) > v_i(x_i)$ for all $i \in S$.

Existence of private core and weak fine core allocations is discussed in Glycopantis, Muir, and Yannelis (2001). The weak fine core is also an ex ante concept. As with the private core the feasibility condition can be relaxed to (ii)′ $\sum_{i=1}^n x_i \le \sum_{i=1}^n e_i$. Notice however that now coalitions of agents are allowed to pool their own information and all alocations will exhaust the resource. The example below illustrates this concept.

Example 3.2. Consider the Example 3.1 without Agent 3. Then if Agents 1 and 2 pool their own information a possible allocation is $x_1 = x_2 = (5, 2.5, 2.5)$. Notice

[6] $\bigvee_{i=1}^{n} \mathcal{F}_i$ denotes the smallest σ-algebra containing each $\mathcal{F}_i$.

that this allocation is $\bigvee_{i=1}^{2} \mathcal{F}_i$-measurable and cannot be dominated by any coalition of agents using their pooled information. Hence it is a weak fine core allocation.[7]

Next we shall define a Walrasian equilibrium notion in the sense of Radner. In order to do so, we need the following. A *price system* is an $\mathcal{F}$-measurable, non-zero function $p : \Omega \to \mathbb{R}_+^l$ and the *budget set* of Agent i is given by

$$B_i(p) = \Big\{ x_i : \ x_i : \Omega \to \mathbb{R}^l \text{ is } \mathcal{F}_i\text{-measurable } \ x_i(\omega) \in X_i(\omega) \\ \text{and} \sum_{\omega \in \Omega} p(\omega) x_i(\omega) \leq \sum_{\omega \in \Omega} p(\omega) e_i(\omega) \Big\}.$$

Notice that the budget constraint is across states of nature.

Definition 3.3. A pair (p, x), where p is a price system and $x = (x_1, \ldots, x_n) \in L_X$ is an allocation, is a *Radner equilibrium* if

(i) for all i the consumption function maximizes v_i on B_i
(ii) $\sum_{i=1}^n x_i \leq \sum_{i=1}^n e_i$ (free disposal), and
(iii) $\sum\limits_{\omega \in \Omega} p(\omega) \sum_{i=1}^n x_i(\omega) = \sum\limits_{\omega \in \Omega} p(\omega) \sum_{i=1}^n e_i(\omega)$.

Radner equilibrium is an ex ante concept. We assume free disposal, for otherwise it is well known that a Radner equilibrium with non-negative prices might not exist. This can be seen through straightforward calculations in Example 3.1.

Next we turn our attention to the notion of REE. We shall need the following. Let $\sigma(p)$ be the smallest sub-σ-algebra of $\mathcal{F}$ for which $p : \Omega \to \mathbb{R}_+^l$ is measurable and let $\mathcal{G}_i = \sigma(p) \vee \mathcal{F}_i$ denote the smallest σ-algebra containing both $\sigma(p)$ and $\mathcal{F}_i$. We shall also condition the expected utility of the agents on $\mathcal{G}_i$ which produces a random variable.

Definition 3.4. A pair (p, x), where p is a price system and $x = (x_1, \ldots, x_n) \in \bar{L}_X$ is an allocation, is a *rational expectations equilibrium* (REE) if

(i) for all i the consumption function $x_i(\omega)$ is $\mathcal{G}_i$-measurable.
(ii) for all i and for all ω the consumption function maximizes

$$v_i(x_i|\mathcal{G}_i)(\omega) = \sum_{\omega' \in E_i^{\mathcal{G}_i}(\omega)} u_i(\omega', x_i(\omega')) \frac{q_i(\omega')}{q_i\big(E_i^{\mathcal{G}_i}(\omega)\big)}, \tag{3}$$

(where $E_i^{\mathcal{G}_i}(\omega)$ is the event in $\mathcal{G}_i$ which contains ω and $q_i(E_i^{\mathcal{G}_i}(\omega)) > 0$) subject to

$$p(\omega) x_i(\omega) \leq p(\omega) e_i(\omega)$$

i.e. the budget set at state ω, and
(iii) $\sum_{i=1}^n x_i(\omega) = \sum_{i=1}^n e_i(\omega)$ for all ω.

[7] See Koutsougeras and Yannelis (1993).

This is an interim concept because we condition expectations on information received from prices as well. In the definition, free disposal can easily be introduced. The idea of conditioning on the σ-algebra, $v_i(x_i|\mathcal{G}_i)(\omega)$, is rather well known.

REE can be classified as (i) *fully revealing* if the price function reveals to each agent all states of nature, (ii) *partially revealing* if the price function reveals some but not all states of nature and (iii) *non-revealing* if it does not disclose any particular state of nature.

Finally we define the concept of *weak fine value allocation* (see Krasa and Yannelis, 1994). As in the definition of the standard value allocation concept, we must first define a transferable utility (TU) game in which each agent's utility is weighted by a factor λ_i $(i = 1, \ldots, n)$, which allows interpersonal comparisons. In the value allocation itself no side payments are necessary.[8] A game with side payments is then defined as follows.

Definition 3.5. A game with side payments $\Gamma = (I, V)$ consist of a finite set of agents $I = \{1, \ldots, n\}$ and a superadditive, real valued function V defined on 2^I such that $V(\emptyset) = 0$. Each $S \subset I$ is called a coalition and $V(S)$ is the 'worth' of the coalition S.

The Shapley value of the game Γ (Shapley, 1953) is a rule that assigns to each Agent i a 'payoff', Sh_i, given by the formula[9]

$$Sh_i(V) = \sum_{\substack{S \subseteq I \\ S \supseteq \{i\}}} \frac{(|S| - 1)!(|I| - |S|)!}{|I|!} [V(S) - V(S \setminus \{i\})]. \tag{4}$$

The Shapley value has the property that $\sum_{i \in I} Sh_i(V) = V(I)$, i.e. it is Pareto efficient.

We now define for each economy with differential information, $\mathcal{E}$, and a common prior, and for each set of weights, $\lambda_i : i = 1, \ldots, n$, the associated game with side payments (I, V_λ) (we also refer to this as a 'transferable utility' (TU) game) as follows:

For every coalition $S \subset I$, let

$$V_\lambda(S) = \max_x \sum_{i \in S} \lambda_i \sum_{\omega \in \Omega} u_i(\omega, x_i(\omega))\mu(\omega) \tag{5}$$

subject to

(i) $\sum_{i \in S} x_i(\omega) = \sum_{i \in S} e_i(\omega)$, μ–a.e.,
(ii) $x_i - e_i$ is $\bigvee_{i \in S} \mathcal{F}_i$–measurable.

We are now ready to define the weak fine value allocation.

---

[8] See Emmons and Scafuri (1985, p. 60) for further discussion.

[9] The Shapley value measure is the sum of the expected marginal contributions an agent can make to all the coalitions of which he/she is a member (see Shapley, 1953).

Definition 3.6. An allocation $x = (x_1, \ldots, x_n) \in \bar{L}_X$ is said to be a *weak fine value allocation* of the differential information economy, $\mathcal{E}$, if the following conditions hold

(i) Each net trade $x_i - e_i$ is $\bigvee_{i=1}^{n} \mathcal{F}_i$-measurable,

(ii) $\sum_{i=1}^{n} x_i = \sum_{i=1}^{n} e_i$ and

(iii) There exist $\lambda_i \geq 0$, for every $i = 1, \ldots, n$, which are not all equal to zero, with $\sum\limits_{\omega \in \Omega} \lambda_i u_i(\omega, x_i(\omega))\mu(\omega) = Sh_i(V_\lambda)$ for all i, where $Sh_i(V_\lambda)$ is the Shapley value of Agent i derived from the game (I, V_λ), defined in (5) above.

Condition (i) requires the pooled information measurability of net trades, i.e. net trades are measurable with respect to the "join". Condition (ii) is the market clearing condition and (iii) says that the expected utility of each agent multiplied by his/her weight, λ_i, must be equal to his/her Shapley value derived from the TU game (I, V_λ).

An immediate consequence of Definition 3.6 is that

$$Sh_i(V_\lambda) \geq \lambda_i \sum_{\omega \in \Omega} u_i(\omega, e_i(\omega))\mu(\omega)$$

for every i, i.e. the value allocation is individually rational. This follows immediately from the fact that the game (V_λ, I) is superadditive for all weights λ. Similarly, efficiency of the Shapley value for games with side payments immediately implies that the value allocation is weak-fine Pareto efficient.

On the basis of the definitions and the analysis of Example 3.1 of an exhange economy with 3 agents and of Example 3.2 with 2 agents we make comparisons between the various equilibrium notions. The calculations of all, cooperative and noncooperative, equilibrium allocations are straightforward.

Contrary to the private core any rational expectation Walrasian equilibium notion, such as Radner equilibrium or REE, will always give zero to an agent who has no initial endowments. For example, in the 3-agent economy of Example 3.1, Agent 3 receives no consumption since his budget set is zero in each state. This is so irrespective of whether his private information is the full information partition $\mathcal{F}_3 = \{\{a\}, \{b\}, \{c\}\}$ or the trivial partition $\mathcal{F}_3^{'} = \{a, b, c\}$. Hence the Walrasian, competitive equilibrium ideas do not take into account the informational superiority of an agent.

The set of Radner equilibrium allocations, with and without free disposal, are a subset of the corresponding private core allocations. Of course it is possible that a Radner equilibrium allocation might not exist. In the two-agent economy of Example 3.2, assuming non-free disposal the unique private core is the initial endowments allocation while no Radner equilibrium exists. On the other hand, assuming free disposal, for the same example, the REE coincides with the initial endowments allocation which does not belong to the private core. It follows that *the REE allocations need not be in the private core.*

We also have that *a REE need not be a Radner equilibrium.* In Example 3.2, without free disposal no Radner equilibrium with non-negative prices exists but REE does. It is unique and it implies no-trade.

As for *the comparison between private and weak fine core allocations* the two sets could intersect but there is no definite relation. Indeed the measurability requirement of the private core allocations separates the two concepts. In Example 3.2 the allocation (5, 2.5, 2.5) to Agent 1 and (5, 2.5, 2.5) to Agent 2, as well as (6, 3, 3) and (4, 2, 2) belong to the weak fine core but not to the private core. There are many weak fine core allocations which do not satisfy the measurability condition.

For $n = 2$ one can easily verify that the weak fine value belongs to the weak fine core. However it is known (see, for example, Scafuri and Yannelis, 1984) that for $n \geq 3$ *a value allocation may not be a core allocation, and therefore may not be a Radner equilibrium.*

Also, in Example 3.1 *a private core allocation is not necessarily in the weak fine core.* Indeed the division (4, 4, 1), (4, 1, 4) and (2, 0, 0), to Agents 1, 2 and 3 respectively, is a private core but not a weak fine core allocation. The first two agents can get together, pool their information and do better. They can realize the weak fine core allocation, (5, 2.5, 2.5), (5, 2.5, 2.5) and (0, 0, 0) which does not belong to the private core.

Finally notice that even with free disposal no allocation which does not distribute the total resource could be in the weak fine core. The three agents can get together, distribute the surplus and increase their utility.

In the next section we shall discuss whether core and Walrasian type allocations have certain desirable properties from the point of view of incentive compatibility. Following this, we shall turn our attention in later sections to the implementation of such allocations.

4 Incentive compatibility

The basic idea is that an allocation is incentive compatible if no coalition can misreport the realized state of nature to the complementary set of agents and become better off.

Let us suppose we have a coalition S, with members denoted by i, and the complementary set $I \setminus S$ with members j. Let the realized state of nature be ω^*. A member $i \in S$ sees $E_i(\omega^*)$. Obviously not all $E_i(\omega^*)$ need be the same, however all Agents i know that the actual state of nature could be ω^*.

Consider now a state of nature $\omega^{'}$ with the following property. For all $j \in I \setminus S$ we have $\omega^{'} \in E_j(\omega^*)$ and for at least one $i \in S$ we have $\omega^{'} \notin E_i(\omega^*)$ (otherwise $\omega^{'}$ would be indistinguishable from ω^* for all players and, by redefining utilities appropriately, could be considered as the same element of Ω). Now the coalition S decides that each member i will announce that she has seen her own set $E_i(\omega^{'})$ which, of course, definitely contains a lie. On the other hand we have that $\omega^{'} \in \bigcap_{j \notin S} E_j(\omega^*)$, (we also denote $j \in I \setminus S$ by $j \notin S$).

Now the idea is that if all members of $I \setminus S$ believe the statements of the members of S then each $i \in S$ expects to gain. For *coalitional Bayesian incentive compatibility* (CBIC) of an allocation we require that this is not possible. This is the incentive compatibility condition used in Glycopantis, Muir, and Yannelis (2001) where we gave a formal definition.

We showed there that in the three-agent economy without free disposal the private core allocation $x_1 = (4,4,1)$, $x_2 = (4,1,4)$ and $x_3 = (2,0,0)$ is incentive compatible. This follows from the fact that Agent 3 who would potentially cheat in state a has no incentive to do so. It has been shown in Koutsougeras and Yannelis (1993) that if the utility functions are monotone and continuous then private core allocations are *always* CBIC.

On the other hand the weak fine core allocations are not always incentive compatible, as the proposed redistribution $x_1 = x_2 = (5, 2.5, 2.5)$ in the two-agent economy shows. Indeed, if Agent 1 observes $\{a, b\}$, he has an incentive to report c and Agent 2 has an incentive to report b when he observes $\{a, c\}$.

CBIC coincides in the case of a two-agent economy with *Individually Bayesian Incentive Compatibility* (IBIC) which corresponds to the case in which S is a singleton.

The concept of *Transfer Coalitionally Bayesian Incentive Compatible* (TCBIC) allocations, used in this paper,[10] allows for transfers between the members of a coalition, and is therefore a strengthening of the concept of Coalitionally Bayesian Incentive Compatibility (CBIC).

Definition 4.1. An allocation $x = (x_1, \ldots, x_n) \in \bar{L}_X$, with or without free disposal, is said to be *Transfer Coalitionally Bayesian Incentive Compatible* (TCBIC) if it is not true that there exists a coalition S, states ω^* and ω', with ω^* different from ω' and $\omega' \in \bigcap_{i \notin S} E_i(\omega^*)$ and a random net-trade vector, z, among the members of S,

$$(z_i)_{i \in S}, \sum_S z_i = 0$$

such that for all $i \in S$ there exists $\bar{E}_i(\omega^*) \subseteq Z_i(\omega^*) = E_i(\omega^*) \cap (\bigcap_{j \notin S} E_j(\omega^*))$, for which

$$\begin{aligned} &\sum_{\omega \in \bar{E}_i(\omega^*)} u_i(\omega, e_i(\omega) + x_i(\omega') - e_i(\omega') + z_i) q_i\big(\omega | \bar{E}_i(\omega^*)\big) \\ &> \sum_{\omega \in \bar{E}_i(\omega^*)} u_i(\omega, x_i(\omega)) q_i\big(\omega | \bar{E}_i(\omega^*)\big). \end{aligned} \tag{6}$$

Notice that the z_i' s above are not necessarily measurable. The definition is cast in terms of all possible z_i' s. It follows that $e_i(\omega) + x_i(\omega') - e_i(\omega') + z_i(\omega) \in X_i(\omega)$ is not necessarily measurable. The definition means that no coalition can form with the possibility that by misreporting a state, every member will become better off if the announcement is believed by the members of the complementary set.

Returning to Definition 4.1, one then can define CBIC to correspond to $z_i = 0$ and then IBIC to the case when S is a singleton. Thus we have (not IBCI) $\Rightarrow$ (not CBIC) $\Rightarrow$ (not TCBIC). It follows that TCBIC $\Rightarrow$ CBIC $\Rightarrow$ IBIC.

We now provide a *characterization* of TCBIC:

[10] See Krasa and Yannelis (1994) and Hahn and Yannelis (1997) for related concepts.

Proposition 4.1. *Let $\mathcal{E}$ be a one-good differential information economy as described above, and suppose each agent's utility function, $u_i = u_i(\omega, x_i(\omega))$ is monotone in the elements of the vector of goods x_i, that $u_i(., x_i)$ is $\mathcal{F}_i$-measurable in the first argument, and that an element $x = (x_1, \ldots, x_n) \in \bar{L}^1_X$ is a feasible allocation in the sense that $\sum_{i=1}^n x_i(\omega) = \sum_{i=1}^n e_i(\omega)\ \forall \omega$. Consider the following conditions:*

(i) $x \in L^1_X = \prod_{i=1}^n L^1_{X_i}$ *and*
(ii) x *is TCBIC.*

Then (i) is equivalent to (ii).

Proof. First we show that (i) implies (ii) by showing that (i) and the negation of (ii) lead to a contradiction.

Let $x \in L_X$ and suppose that it is not TCBIC. Then, varying the notation for states to emphasize that Definition 4.1 does not hold, there exists a coalition S, states a and b, with $a \neq b$ and $b \in \bigcap_{i \notin S} E_i(a)$ and a net-trade vector, z, among the members of S,

$$(z_i)_{i \in S}, \quad \sum_S z_i = 0$$

such that for all $i \in S$ there exists $\bar{E}_i(a) \subseteq Z_i(a) = E_i(a) \cap (\bigcap_{j \notin S} E_j(a))$, for which

$$\sum_{c \in \bar{E}_i(\alpha)} u_i(c, e_i(c) + x_i(b) - e_i(b) + z_i) q_i\big(c|\bar{E}_i(a)\big) \qquad (7)$$

$$> \sum_{c \in \bar{E}_i(a)} u_i(c, x_i(c)) q_i\big(c|\bar{E}_i(a)\big).$$

For $c \in \bar{E}_i(a), e_i(c) = e_i(a)$ since e_i is $\mathcal{F}_i$-measurable, so

$$e_i(c) + x_i(b) - e_i(b) + z_i = e_i(a) + x_i(b) - e_i(b) + z_i$$

and hence also

$$u_i\big(c, e_i(c) + x_i(b) - e_i(b) + z_i\big) = u_i\big(a, e_i(a) + x_i(b) - e_i(b) + z_i\big),$$

by the assumed $\mathcal{F}_i$-measurability of u_i.

Since, by (i), $x_i(c) = x_i(a)$ for $c \in \bar{E}_i(a)$, we similarly have $u_i\big(c, x_i(c)\big) = u_i\big(a, x_i(a)\big)$. Thus in equation (7) the common utility terms can be lifted outside the summations giving

$$u_i\big(a, e_i(a) + x_i(b) - e_i(b) + z_i\big) > u_i\big(a, x_i(a)\big)$$

and hence $e_i(a) + x_i(b) - e_i(b) + z_i > x_i(a)$, by monotonicity of u_i.

Consequently,

$$\sum_{i \in S} \big(x_i(b) - e_i(b)\big) > \sum_{i \in S} \big(x_i(a) - e_i(a)\big). \qquad (8)$$

On the other hand for $i \notin S$ we have $x_i(b) - e_i(b) = x_i(a) - e_i(a)$ from which we obtain

$$\sum_{i \notin S} \big(x_i(b) - e_i(b)\big) = \sum_{i \notin S} \big(x_i(a) - e_i(a)\big). \tag{9}$$

Taking equations (8),(9) together we have

$$\sum_{i \in I} \big(x_i(b) - e_i(b)\big) > \sum_{i \in I} \big(x_i(a) - e_i(a)\big), \tag{10}$$

which is a contradiction since both sides are equal to zero, by feasibility.[11]

We now show that (ii) implies (i). For suppose not. Then there exists some Agent j and states a, b with $b \in E_j(a)$ such that $x_j(a) \neq x_j(b)$. Without loss of generality, we may assume that $x_j(a) > x_j(b)$. Since $e_j(.)$ is $\mathcal{F}_j$-measurable $e_j(b) = e_j(a)$ and therefore

$$x_j(a) - e_j(a) > x_j(b) - e_j(b). \tag{11}$$

Let $S = I \backslash \{j\}$. From the feasibility of x and (11) it follows that

$$\begin{aligned}\sum_{i \in S} \big(x_i(a) - e_i(a)\big) &= -\big(x_j(a) - e_j(a)\big) < -\big(x_j(b) - e_j(b)\big) \\ &= \sum_{i \in S} \big(x_i(b) - e_i(b)\big).\end{aligned} \tag{12}$$

From (12) we have that

$$\delta = \sum_{i \in S} \big(e_i(a) + x_i(b) - e_i(b) - x_i(a)\big) > 0. \tag{13}$$

For each $i \in S$, let

$$z_i = x_i(a) - e_i(a) - x_i(b) + e_i(b) + \frac{\delta}{n-1}$$

so that $\sum_{i \in S} z_i = 0$ and

$$e_i(a) + x_i(b) - e_i(b) + z_i > x_i(a).$$

By monotonicity of u_i, we can conclude that

$$u_i(a, e_i(a) + x_i(b) - e_i(b) + z_i) > u_i(a, x_i(a)), \tag{14}$$

for all $i \in S$, a contradiction to the fact that x is TCBIC as the role of $\bar{E}_i$ in the definition can be played by $\{a\}$.

Finally note that a particular case of $\mathcal{F}_i$-measurability of u_i is when it is independent of ω. This completes the proof of Proposition 4.1. □

[11] Koutsougeras and Yannelis (1993) and Krasa and Yannelis (1994) show that (i) implies (ii) for any number of goods, but for ex post utility functions. This means that the contract is made ex ante and after the state of nature is realized we see that we have incentive compatibility. Hahn and Yannelis (1997) show that (i) implies (ii) for any number of goods and for interim utility functions. Notice that since the non-free disposal Radner equilibrium is a subset of the non-free disposal ex ante private core, it follows from Hahn and Yannelis that the non-free disposal Radner equilibrium is TCBIC.

In the lemma that follows we refer to CBIC, as TCBIC does not make much sense since z_i is not available. CBIC is obtained when all z_i's are set equal to zero.

Lemma 4.1. *Under the conditions of the Proposition, if there are only two agents then (ii) x is CBIC, which is the same as IBIC, implies (i).*

Proof. For suppose not. Then lack of $\mathcal{F}_i$-measurability of the allocations implies that there exist Agent j and states a, b, where $b \in E_j(a)$, such that $x_j(b) < x_j(a)$ and therefore

$$x_j(b) - e_j(b) < x_j(a) - e_j(a). \tag{15}$$

Feasibility implies

$$x_i(b) - e_i(b) + x_j(b) - e_j(b) = x_i(a) - e_i(a) + x_j(a) - e_j(a) \tag{16}$$

from which we obtain

$$x_i(b) - e_i(b) > x_i(a) - e_i(a). \tag{17}$$

By monotonicity and the one-good per state assumption it follows that,

$$u_i(a, e_i(a) + x_i(b) - e_i(b)) > u_i(a, x_i(a)). \tag{18}$$

This implies that we have

$$u_i(a, e_i(c) + x_i(b) - e_i(b)) > u_i(a, x_i(c)) \tag{19}$$

which contradicts the assumption that x is CBIC. This completes the proof of the lemma. □

The above results characterize TCBIC and CBIC in terms of private individual measurability, i.e. $\mathcal{F}_i$-measurability, of allocations. These results will enable us to conclude whether or not, in case of non-free disposal, any of the solution concepts, i.e. Radner equilibrium, REE, private core, weak fine core and weak fine value will be TCBIC whenever feasible allocations are $\mathcal{F}_i$-measurable.

It follows from the lemma that the redistribution shown in the matrix below, which is a weak fine core allocation of Example 3.2, where the ith line refers to Player i and the columns from left to right to states a, b and c,

$$\begin{pmatrix} 5 & 2.5 & 2.5 \\ 5 & 2.5 & 2.5 \end{pmatrix}$$

is not CBIC as it is not $\mathcal{F}_i$-measurable. Thus, *a weak fine core allocation may not be* CBIC.

On the other hand the proposition implies that, in Example 3.2, the no-trade allocation

$$\begin{pmatrix} 5 & 5 & 0 \\ 5 & 0 & 5 \end{pmatrix}$$

is incentive compatible. This is a non-free disposal REE, and a private core allocation.

We note that the Proposition 4.1 refers to non-free disposal. As a matter of fact Proposition 4.1 is not true if we assume free disposal. Indeed if free disposal is allowed $\mathcal{F}_i$-measurability PBE *does not* imply incentive compatibility.

In the case with *free disposal, private core and Radner equilibrium need not be incentive compatible.* In order to see this we notice that in Example 3.2 the (free disposal) Radner equilibrium is $x_1 = (4,4,1)$ and $x_2 = (4,1,4)$. The above allocation is clearly $\mathcal{F}_i$-measurable and it can easily be checked that it belongs to the (free disposal) private core. However it is not TBIC since if state a occurs Agent 1 has an incentive to report state c and gain.

Now in employing game trees in the analysis, as it is done below, we will adopt the definition of IBIC. The equilibrium concept employed will be that of PBE. The definition of a play of the game is a directed path from the initial to a terminal node.

In terms of the game trees, a core allocation will be IBIC if there is a profile of optimal behavioral strategies and equilibrium paths along which no player misreports the state of nature he has observed. This allows for the possibility, as we shall see later, that such strategies could imply that players have an incentive to lie from information sets which are not visited by an optimal play.

In view of the analysis in terms of game trees we comment again on the general idea of CBIC. First we look at it once more, in a similar manner to the one in the beginning of Section 4.

Suppose the true state of nature is $\bar{\omega}$. Any coalition can only see that the state lies in $\bigcap\limits_{i\in S} E_i(\bar{\omega})$ when they pool their observations. If they decide to lie they must first guess at what is the true state and they will do so at some $\omega^* \in \bigcap\limits_{i\in S} E_i(\bar{\omega})$. Then of course we have $\bigcap\limits_{i\in S} E_i(\bar{\omega}) = \bigcap\limits_{i\in S} E_i(\omega^*)$. Having decided on ω^* as a possible true state, they now pick some $\omega^{'} \in \bigcap\limits_{j\notin S} E_j(\omega^*)$ and (assuming the system is not CBIC) they hope, by announcing (each of them) that they have seen $E_i(\omega^{'})$ to secure better payoffs.

This is all contingent on their being believed by $I \setminus S$. This, in turn, depends on their having been correct in their guessing that $\omega^* = \bar{\omega}$, in which case they might be believed. If $\omega^* \neq \bar{\omega}$, i.e they guess wrongly, then since $\bigcap\limits_{j\notin S} E_j(\omega^*) \neq \bigcap\limits_{j\notin S} E_j(\bar{\omega})$ they may be detected in their lie, since possibly $\omega^{'} \notin \bigcap\limits_{j\notin S} E_j(\bar{\omega})$.

This is why the definition of CBIC can only be about possible existence of situations where a lie might be beneficial. It is not concerned with what happens if the lie is detected. On the other hand the extensive form forces us to consider that alternative. It requires statements concerning earlier decisions by other players to lie or tell the truth and what payoffs will occur whenever a lie is detected, through observations or incompatibility of declarations. Only in this fuller description can players really make a decision whether to risk a lie, since only then can they balance the gains from not being caught against a definitely declared payoff if they are.

The issue is whether cooperative and noncooperative static solutions can be obtained as perfect Bayesian or sequential equilibria. That is whether such allocations can also be supported through an appropriate noncooperative solution concept. The analysis below shows that CBIC allocations can be supported by a PBE while lack of incentive compatibility implies non-support, in the sense that the two agents, left on their own, do not sign the contract. It is also shown how implementation of allocations becomes possible through the introduction in the analysis of an exogenous third party or an endogenous intermediary.

5 Non-implementation of free disposal private core and Radner equilibria, and of weak fine core allocations

The main point here is that lack of IBIC implies that the two agents based on their information cannot sign a proposed contract because both of them have an incentive to cheat the other one and benefit. Indeed PBE leads to no-trade. This so irrespective of whether in state a the contract specifies that they both get 5 or 4.

Note that to impose free disposal in state a causes certain problems, because the question arises as to who will check that the agents have actually thrown away 1 unit. In general, free disposal is not always a very satisfactory assumption in differential information economies with monotone preferences.

We shall investigate the possible implementation of the allocation

$$\begin{pmatrix} 4 & 4 & 1 \\ 4 & 1 & 4 \end{pmatrix}$$

in Example 3.2, contained in a contract between P1 and P2 when no third party is present. For the case with free disposal, this is both a private core and a Radner equilibrium allocation.

This allocation is not IBIC because, as we explained in the previous section, if Agent 1 observes $\{a, b\}$, he has an incentive to report c and Agent 2 has an incentive to report b when he observes $\{a, c\}$.

We construct a game tree and employ reasonable rules for describing the outcomes of combinations of states of nature and actions of the players. In fact we look at the contract

$$\begin{pmatrix} 5 & 4 & 1 \\ 5 & 1 & 4 \end{pmatrix}$$

in which the agents get as much per state as under the private core allocation above. The latter can be obtained by invoking free disposal in state a.

The investigation is through the analysis of a specific sequence of decisions and information sets shown in the game tree in Figure 1. Notice that vectors at the terminal nodes of a game tree will refer to payoffs of the players in terms of quantities. The first element will be the payoff to P1, etc.

The players are given strategies to tell the truth or to lie, i.e., we model the idea that agents truly inform each other about what states of nature they observe, or deliberately aim to mislead their opponent. The issue is what type of behavior is

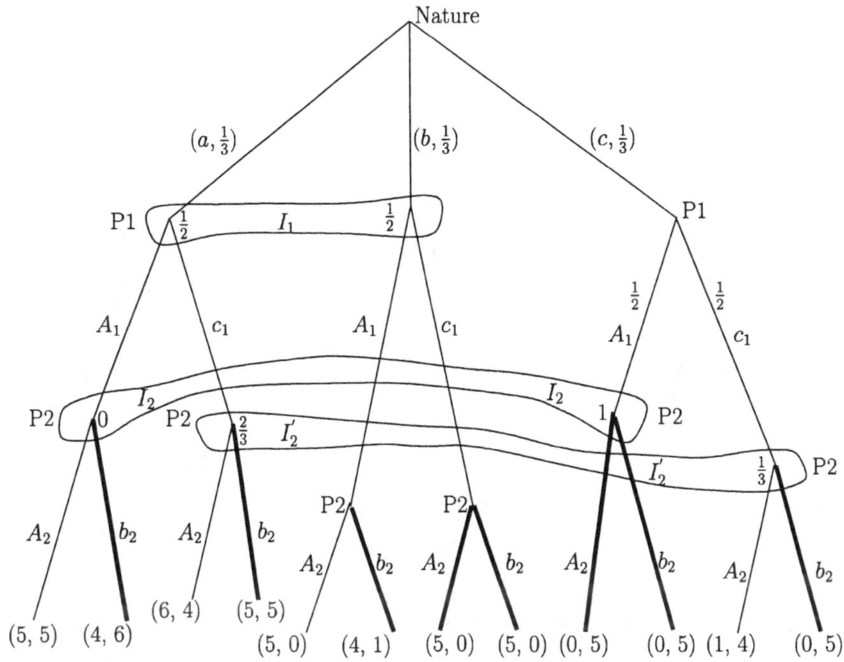

Figure 1

optimal and therefore whether a proposed contract will be signed or not. We find that the optimal strategies of the players imply no-trade.

Figures 1 and 2 show that the allocation $(5, 4, 1)$ and $(5, 1, 4)$ will be rejected by the players. They prefer to stay with their initial endowments and will not sign the proposed contract as it offers to them no advantage.

In Figure 1, nature chooses states a, b or c with equal probabilities. This choice is flashed on a screen which both players can see. P1 cannot distinguish between a and b, and P2 between a and c . This accounts for the information sets I_1, I_2 and I_2' which have more than one node. A player to which such an information set belongs cannot distinguish between these nodes and therefore his decisions are common to all of them. A behavioral strategy of a player is to declare which choices he would make, with what probability, from each of his information sets. Indistinguishable nodes imply the $\mathcal{F}_i$-measurability of decisions.

P1 moves first and he can either play $A_1 = \{a, b\}$ or $c_1 = \{c\}$, i.e., he can say "I have seen $\{a, b\}$ or "I have seen c". Of course only one of these declarations will be true. Then P2 is to respond saying that the signal he has seen on the screen is $A_2 = \{a, c\}$ or that it is $b_2 = \{b\}$. Obviously only one of these statements is true.

Strictly speaking the notation for choices should vary with the information set but there is no danger of confusion here. Finally notice that the structure of the game tree is such that when P2 is to act he knows exactly what P1 has chosen.

Next we specify the *rules* for calculating the payoffs, i.e. the terms of the contract:

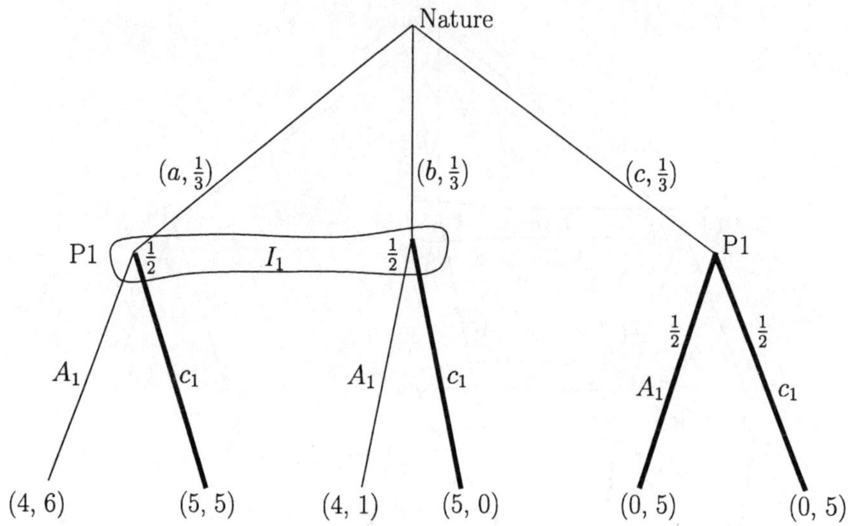

Figure 2

(i) If the declarations by the two players are incompatible, that is (c_1, b_2) then no-trade takes place and the players retain their initial endowments. That is the case when either state c, or state b occurs and Agent 1 reports state c and Agent 2 state b. In state a both agents can lie and the lie cannot be detected by either of them. They are in the events $\{a, b\}$ and $\{a, c\}$ respectively, they get 5 units of the initial endowments and again they are not willing to cooperate. Therefore whenever the declarations are incompatible, no trade takes place and the players retain their initial endowments.

(ii) If the declarations are (A_1, A_2) then even if one of the players is lying, this cannot be detected by his opponent who believes that state a has occured and both players have received endowment 5. Hence no-trade takes place.

(iii) If the declarations are (A_1, b_2) then a lie can be beneficial and undetected. P1 is trapped and must hand over one unit of his endowment to P2. Obviously if his initial endowment is zero then he has nothing to give.

(iv) If the declarations are (c_1, A_2) then again a lie can be beneficial and undetected. P2 is now trapped and must hand over one unit of his endowment to P1. Obviously if his initial endowment is zero then he has nothing to give.

The calculations of payoffs do not require the revelation of the actual state of nature. Optimal decisions will be denoted by a heavy line. We could assume that a player does not lie if he cannot get a higher payoff by doing so.

Assuming that each player chooses optimally from his information sets, the game in Figure 1 folds back to the one in Figure 2. Inspection of Figure 1 reveals that from the information set I_2 agent P2 can play b_2 with probability 1. (A heavy line A_2 indicates that this choice also would not affect the analysis). This accounts for the payoff $(4, 6)$ and the first payoff $(0, 5)$ from left to right in Figure 2. Similarly by considering the optimal decisions from all other information sets of P2 we arrive at Figure 2. Analyzing this figure we obtain the optimal strategies of P1.

In conclusion, the optimal behavioral strategy for P1 is to play c_1 with probability 1 from I_1, i.e to lie, and from the singleton to play any probability mixture of options, and we have chosen $(A_1, \frac{1}{2}; c_1, \frac{1}{2})$. The optimal strategy of P2 is to play b_2 from both I_2 and $I_2^{'}$, i.e. to lie, and from the second singleton he can either tell the truth or lie, or spin a wheel, divided in proportions corresponding to A_1 and c_1, to decide what to choose.

In Figures 1 and 2, the fractions next to the nodes in the information sets correspond to beliefs of the agents obtained, wherever possible, through Bayesian updating. I.e., they are consistent with the choice of a state by nature and the optimal behavioral strategies of the players. This means that strategies and beliefs satisfy the conditions of a PBE.

These probabilities are calculated as follows. From left to right, we denote the nodes in I_1 by j_1 and j_2, in I_2 by n_1 and n_2 and in $I_2^{'}$ by n_3 and n_4. Given the choices by nature, the strategies of the players described above and using the Bayesian formula for updating beliefs we can calculate, for example, the conditional probabilities

$$\begin{aligned} Pr(n_1/A_1) &= \frac{Pr(A_1/n_1) \times Pr(n_1)}{Pr(A_1/n_1) \times Pr(n_1) + Pr(A_1/n_2) \times Pr(n_2)} \\ &= \frac{1 \times 0}{1 \times 0 + 1 \times \frac{1}{3} \times \frac{1}{2}} = 0 \end{aligned} \tag{20}$$

and

$$\begin{aligned} Pr(n_3/c_1) &= \frac{Pr(c_1/n_3) \times Pr(n_3)}{Pr(c_1/n_3) \times Pr(n_3) + Pr(c_1/n_4) \times Pr(n_4)} \\ &= \frac{1 \times \frac{1}{3}}{1 \times \frac{1}{3} + 1 \times \frac{1}{2} \times \frac{1}{3}} = \frac{2}{3}. \end{aligned} \tag{21}$$

In Figure 3 we indicate, through heavy lines, plays of the game which are the outcome of the choices by nature and the optimal behavioral strategies by the players. The interrupted heavy lines signify that nature does not take an optimal decision but simply chooses among three alternatives, with equal probabilities. The directed path (a, c_1, b_2) with payoffs (5, 5) occurs with probability $\frac{1}{3}$. The paths (b, c_1, A_2) and (b, c_1, b_2) lead to payoffs (5, 0) and occur with probability $\frac{1}{3}(1-q)$ and $\frac{1}{3}q$, respectively. The values $(1-q)$ and q denote the probabilities with which P2 chooses between A_2 and b_2 from the singleton node at the end of (b, c_1). The paths (c, A_1, b_2) (c, c_1, b_2) lead to payoffs (0, 5) and occur, each, with probability $\frac{1}{3} \times \frac{1}{2}$.

For all choices by nature, at least one of the players tells a lie on the optimal play. The players by lying avoid the possibility of having to make a payment to their opponent and stay with their initial endowments. The PBE obtained above confirms the initial endowments. The decisions to lie imply that the players will not sign the contract $(5, 4, 1)$ and $(5, 1, 4)$.

We have constructed an extensive form game and employed reasonable rules for calculating payoffs and shown that the proposed allocation (5, 4, 1) and (5, 1, 4) will not be realized. A similar conclusion would have been reached if we

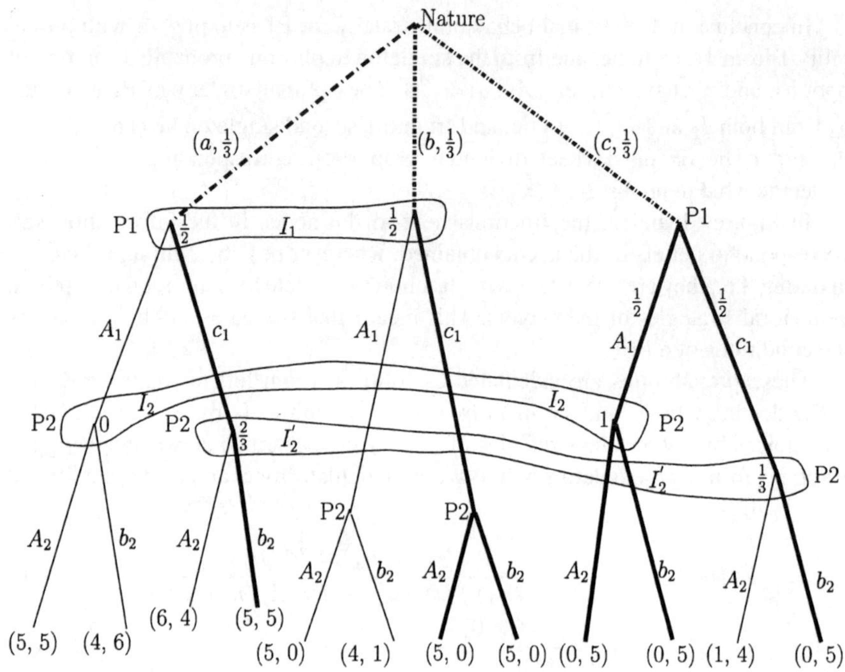

Figure 3

investigated the allocation (4, 4, 1) and (4, 1, 4) which would have been brought about by considering free disposal.

Finally suppose we were to modify (iii) and (iv) of the *rules* and adopt those in Section 5 of Glycopantis, Muir, and Yannelis (2001):

(iii) If the declarations are (A_1, b_2) then a lie can be beneficial and undetected, and P1 is trapped and must hand over half of his endowment to P2. Obviously if his endowment is zero then he has nothing to give.
(iv) If the declarations are (c_1, A_2) then again a lie can be beneficial and undetected. P2 is now trapped and must hand over half of his endowment to P1. Obviously if his endowment is zero then he has nothing to give.

The new rules would imply, starting from left to right, the following changes in the payoffs in Figure 1. The second vector would now be (2.5, 7.5), the third vector (7.5, 2.5), the sixth vector (2.5, 2.5) and the eleventh vector (2.5, 2.5). The analysis in Glycopantis, Muir, and Yannelis (2001) shows that the weak fine core allocation in which both agents receive (5, 2.5, 2.5) cannot be implemented as a PBE. Again this allocation is not IBIC.

Since we have two agents, the weak fine value belongs to the weak fine core. We can also check through routine calculations that the non-implementable allocation $x_1 = x_2 = (5, 2.5, 2.5)$ belongs to the weak fine value, with the two agents receiving equal weights.

Finally we note that, in the context of Figure 1, the perfect Bayesian equilibrium implements the initial endowments allocation

$$\begin{pmatrix} 5 & 5 & 0 \\ 5 & 0 & 5 \end{pmatrix}.$$

In the case of non-free disposal, no-trade coincides with the REE and it is implementable. However as it is shown in Glycopantis, Muir, and Yannelis (2002) a REE is not in general implementable.

6 Implementation of private core and Radner equilibria through the courts; implementation of weak fine core

We shall show here how the free disposal private core and also Radner equilibrium allocation

$$\begin{pmatrix} 4 & 4 & 1 \\ 4 & 1 & 4 \end{pmatrix}$$

of Example 3.2 can be implemented as a PBE by invoking an exogenous third party, which can be interpreted as a court which imposes penalties when the agents lie.

We shall assume that the agents do not hear the choice announced by the other player or that they do not pay much attention to each other because the court will verify the true state of nature.

It should be noted that now if the two players see the events (A_1, A_2) the exogenous agent will not allow them to misreport the state of nature by imposing a penalty for lying. Therefore the contract will be enforced exogenously.

The analysis is through the figures below. Figure 4 contains the information sets of the two agents, P1 and P2, their sequential decisions and the payoffs in terms of quantities. Each agent can choose either to tell the truth about the information set he is in, or to lie.

Nature chooses states a, b and c with equal probabilities. P1 acts first and cannot distinguish between a and b. When P2 is to act he has two kinds of ignorance. Not only he cannot distinguish between a and c but also he does not know what P1 has chosen before him. This is an assumption about the relation between decisions. The one unit that the courts take from a lying agent can be considered to cover the costs of the court.

Next given the sequence of decisions of the two players, shown on the tree, we specify the rules for calculating payoffs in terms of quantities, i.e we specify the terms of the contract. They will, of course include the penalties that the court would impose to the agents for lying.

The *rules* are:

(i) If a player lies about his observation, then he is penalized by 1 unit of the good. If both players lie then they are both penalized. For example if the declarations are (c_1, b_2) and state a occurs both are penalized. If they choose (c_1, A_2) and state a occurs then the first player is penalized. If a player lies and the other agent has a positive endowment then the court keeps the quantity

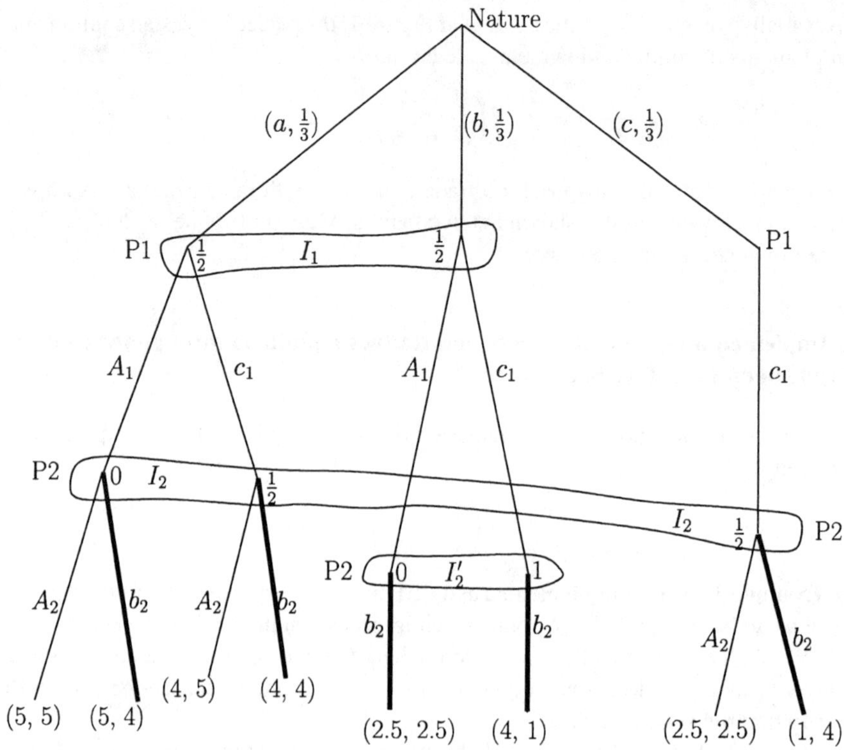

Figure 4

substracted for itself. However, if the other agent has no endowment, then the court transfers to him the one unit subtracted from the one who lied.

(ii) If the declarations of the two agents are consistent, that is (A_1, A_2) and state a occurs, (A_1, b_2) and state b occurs, (c_1, A_2) and state c occurs, then they divide equally the total endowments in the economy.

One explanation of the size of the payoffs is that if the agents decide to share, they do so voluntarily. On the other hand the court feel that they can punish them for lying but not to the extent of forcing them to share their endowments.

Assuming that each player chooses optimally, given his stated beliefs, from the information sets which belong to him, P2 chooses to play b_2 with probability 1 from both I_2 and I_2' and the game in Figure 4 folds back to the one in Figure 5. The choice of b_2 is justified as follows. We ignore for the moment the specific conditional probabilities attached to the nodes of I_2. On the other hand, starting from left to right, the sum of the probabilities of the first two nodes must be equal to $\frac{1}{2}$, and this implies that strategy b_2 overtakes, in utility terms, strategy A_2, as $\frac{1}{2}5^{\frac{1}{2}} + \frac{1}{2}2.5^{\frac{1}{2}} < 4^{\frac{1}{2}}$. It follows that P2 chooses to play the behavioral strategy b_2 with probability 1. Now inspection of Figure 5 implies that P1 will choose c_1 from I_1. The conditional probabilities on the nodes of I_1 follow from the fact that nature

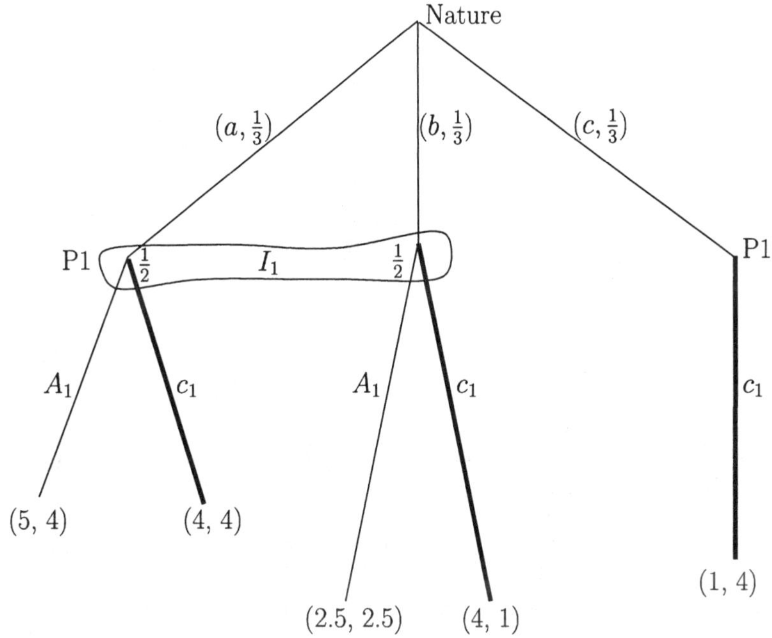

Figure 5

chooses with equal probabilities and the optimal choice of c_1 with probability 1 follows again from the fact that $\frac{1}{2}5^{\frac{1}{2}} + \frac{1}{2}2.5^{\frac{1}{2}} < 4^{\frac{1}{2}}$.

Figure 6 indicates, through heavy lines, plays of the game which are the outcome of choices by nature and the optimal strategies of the players. The fractions next to the nodes of the information sets are obtained through Bayesian updating. I.e. they are consistent with the choice of a state by nature and the optimal behavioral strategies of the players. We have thus obtained a PBE and the above argument implies that it is unique.

The free disposal private core allocation that we are concerned with is implemented, always, by at least one of the agents lying. The reason is that they make the same move from all the nodes of an information set and the rules of the game imply that they are not eager to share their endowments. They prefer to suffer the penalty of the court.

Finally notice the following. Suppose that the penalties are changed as follows. The court is extremely severe when an agent lies while the other agent has no endowment. It takes all the endowment from the one who is lying and transfers it to the other player. Everything else stays the same. Then the game is summarized in a modified Figure 4. Numbering the end points from left to right, the 2nd vector will be replaced by (5, 0), the 3rd by (0, 5), the 4th by (0, 0), the 6th by $(0, 5)$ and the 8th one by $(5, 0)$.

The analysis of the game implies now that P2 will play A_2 from I_2 and P1 will play A_1 from I_1. Therefore invoking an exogenous agent implies that the PBE will

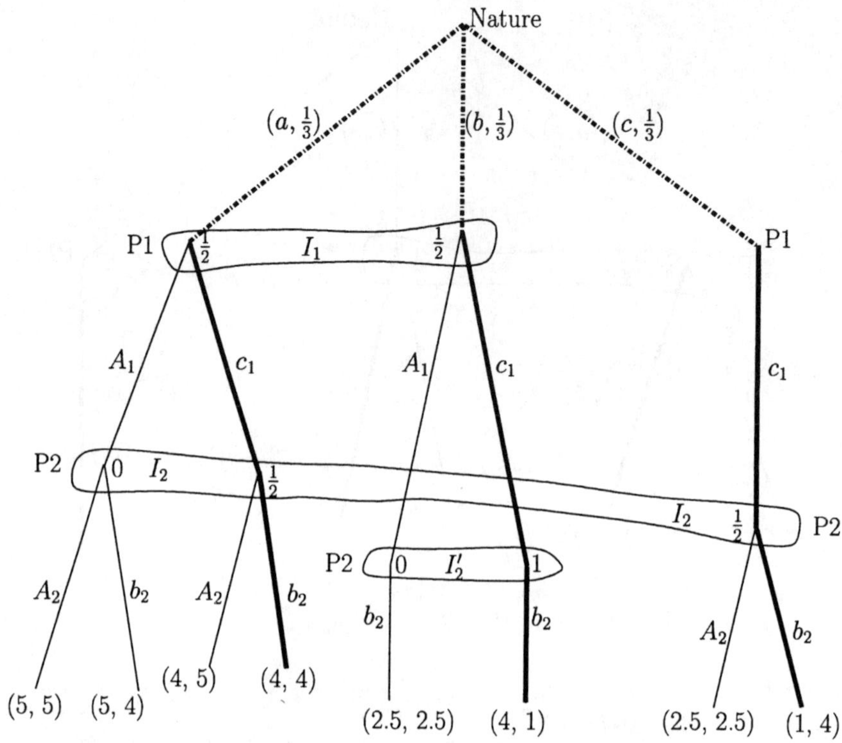

Figure 6

now implement the weak fine core allocation

$$\begin{pmatrix} 5 & 2.5 & 2.5 \\ 5 & 2.5 & 2.5 \end{pmatrix}.$$

7 Implementation of non-free disposal private core through an endogenous intermediary

Here we draw upon the discussion in Glycopantis, Muir, and Yannelis (2001) but we add the analysis that the optimal paths obtained are also part of a sequential equilibrium. Hence we obtain a stronger conclusion, in the sense that we implement the private core allocation as a sequential equilibrium, which requires more conditions than PBE.

In the case we consider now there is no court and the agents in order to decide must listen to the choices of the other players before them. The third agent, P3, is endogenous and we investigate his role in the implementation, or realization, of private core allocations.

Private core without free disposal seems to be the most satisfactory concept. The third agent who plays the role of the intermediary implements the contract and gets rewarded in state a. We shall consider the private core allocation, of Example 3.1,

$$\begin{pmatrix} 4 & 4 & 1 \\ 4 & 1 & 4 \\ 2 & 0 & 0 \end{pmatrix}.$$

We know that such core allocations are CBIC and we shall show now how they can be supported as a perfect Bayesian equilibrium of a noncooperative game.

P1 cannot distinguish between states a and b and P2 between a and c. P3 sees on the screen the correct state and moves first. He can either announce exactly what he saw or he can lie. Obviously he can lie in two ways. When P1 comes to decide he has his information from the screen and also he knows what P3 has played. When P2 comes to decide he has his information from the screen and he also knows what P3 and P1 played before him. Both P1 and P2 can either tell the truth about the information they received from the screen or they can lie.

We must distinguish between the announcements of the players and the true state of nature. The former, with the players' temptations to lie, cannot be used to determine the true state which is needed for the purpose of making payoffs. P3 has a special status but he must also take into account that eventually the lie will be detected and this can affect his payoff.

The *rules* of calculating payoffs, i.e. the terms of the contract, are as follows:

If P3 tells the truth we implement the redistribution in the matrix above which is proposed for this particular choice of nature.

If P3 lies then we look into the strategies of P1 and P2 and decide as follows:

(i) If the declaration of P1 and P2 are incompatible we go to the initial endowments and each player keeps his.

(ii) If the declarations are compatible we expect the players to honour their commitments for the state in the overlap, using the endowments of the true state, provided these are positive. If a player's endowment is zero then no transfer from that agent takes place as he has nothing to give.

The extensive form game is shown in Figure 7, in which the heavy lines can be ignored in the first instance. We are looking for a PBE, i.e. a set of optimal behavioral strategies consistent with a set of beliefs. The beliefs are indicated by the probabilities attached to the nodes of the information sets, with arbitrary r, s, q, p and t between 0 and 1. The folding up of the game tree through optimal decisions by P2, then by P1 and subsequently by P3 is explained in Glycopantis, Muir, and Yannelis (2001).

In Figure 7 we indicate through heavy lines the equilibrium paths. The interrupted heavy lines at the beginning of the tree signify that nature does not take an optimal decision but simply chooses among three alternatives, with equal probabilities. The directed paths (a, a, A_1, A_2) with payoffs (4, 4, 2), (b, b, A_1, b_2) with payoffs $(4, 1, 0)$ and (c, c, c_1, A_2) with payoffs $(1, 4, 0)$ occur, each, with probability $\frac{1}{3}$. It is clear that nobody lies on the optimal paths and that the proposed reallocation is incentive compatible and hence it will be realized.

Along the optimal paths nobody has an incentive to misrepresent the realized state of nature and hence the private core allocation is incentive compatible. However even optimal strategies can imply that players might have an incentive to lie

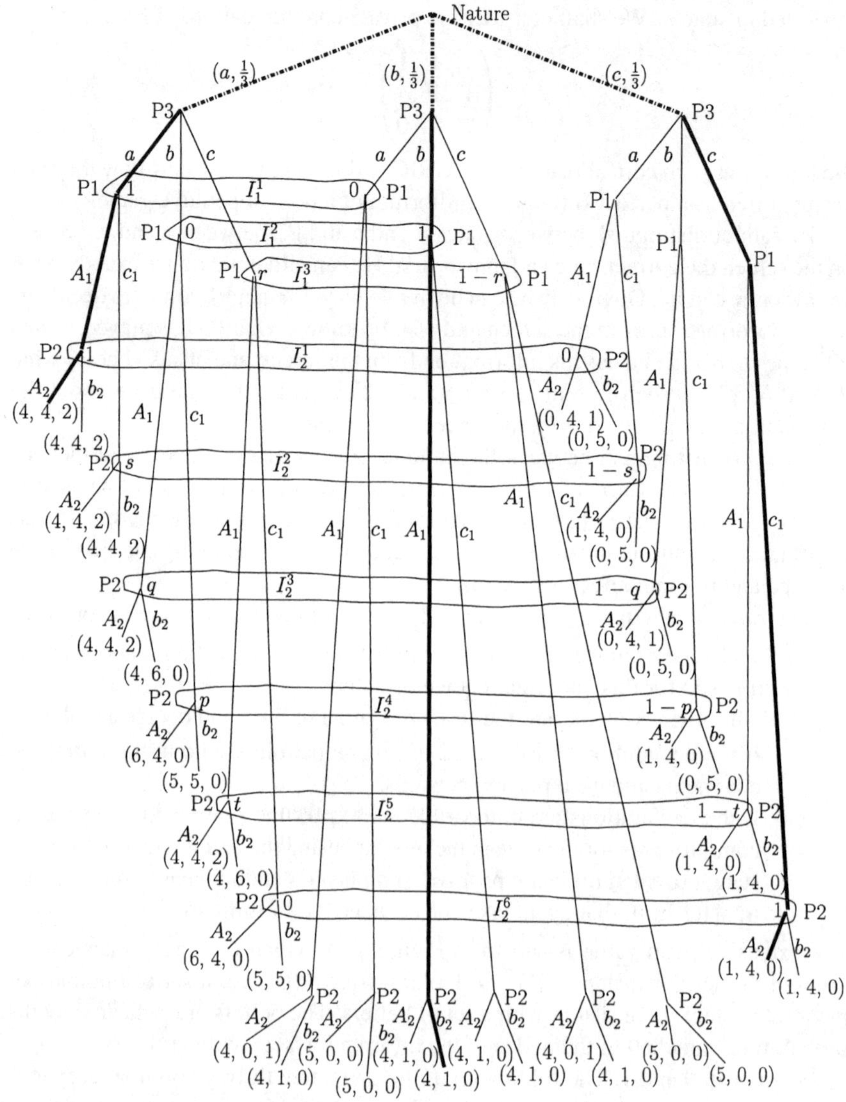

Figure 7

from information sets which are not visited by the optimal play of the game. For example, P1, although he knows that nature has chosen a or b, has an incentive to declare c_1 from I_1^3, trying to take advantage of a possible lie by P3. Similarly P2, although he knows that nature has chosen a or c, has an incentive to declare b_2 from I_2^2, I_2^3, I_2^4 and I_2^5, trying to take advantage of possible lies by the other players. Incentive compatibility has now been defined to allow that the optimal strategies can contain lies, while there must be an optimal play which does not.

We also note that the same payoffs, i.e. (4, 4, 2), (4, 1, 0) and (1, 4, 0), can be confirmed as a PBE for all possible orders of the players.

Next we turn our attention to obtaining a sequential equilibrium. This adds further conditions to those of a PBE. Now, it is also required that the optimal behavioral strategies and the beliefs consistent with these are the limit of a sequence consisting of completely stochastic behavioral strategies, that is all choices are played with positive probability, and the implied beliefs. Throughout the sequence it is only required that beliefs are consistent with the strategies. The latter are not expected to be optimal.

We discuss how the PBE shown in Figure 7 can also be obtained as a sequential equilibrium in the sense of Kreps and Wilson (1982). Therefore, we are looking for a sequence of positive probabilities attached to all the choices from each information set and beliefs consistent with these such that their limits are the results given in Figure 7.

First we specify the positive probabilities, i.e. the completely stochastic strategies, with which the players choose the available actions. The sequence is obtained through $\{n = 2, 3, \dots\}$.

In the first instance we consider the singletons from left to right belonging to P3. At the first one the positive probabilities attached to the various actions are given by $(a, 1-\frac{2}{n};\ b, \frac{1}{n};\ c, \frac{1}{n})$, at the second one by $(a, \frac{1}{n};\ b, 1-\frac{2}{n};\ c, \frac{1}{n})$ and at the third one by $(a, \frac{1}{n};\ b, \frac{1}{n};\ c, 1-\frac{2}{n})$.

Then we come to the probabilities with which P1 chooses his actions from the various information sets belonging to him. From I_1^1 and I_1^2 the choices and the probabilities attached to these are $(A_1, 1-\frac{1}{n};\ c_1, \frac{1}{n})$, and from I_1^3, as well as from all the singletons, they are $(A_1, \frac{1}{n};\ c_1, 1-\frac{1}{n})$.

With respect to P2 choices and probabilities are given as follows. From I_2^1 and I_2^6 they are $(A_2, 1-\frac{1}{n};\ b_2, \frac{1}{n})$ and from I_2^2, I_2^3, I_2^4 and I_2^5 they are $(A_2, \frac{1}{n};\ b_2, 1-\frac{1}{n})$. With respect to the singletons belonging to P2 we have for all of them $(A_2, \frac{1}{n};\ b_2, 1-\frac{1}{n})$.

Beliefs are indicated by the probabilities attached to the nodes of the information sets. Below by the left (right) probability we mean the consistent with the above behavioral strategies belief that the player attaches to being at the left (right) corner node of an information set. We also give the limit of these beliefs as n tends to ∞.

In I_1^1 the left probability is $\dfrac{1-\frac{2}{n}}{1-\frac{1}{n}}$ and the right probability is $\dfrac{\frac{1}{n}}{1-\frac{1}{n}}$. The limit is $(1, 0)$.

In I_1^2 the left probability is $\dfrac{\frac{1}{n}}{1-\frac{1}{n}}$ and the right probability is $\dfrac{1-\frac{2}{n}}{1-\frac{1}{n}}$. The limit is $(0, 1)$.

In I_1^3 the left probability is $\dfrac{1}{2}$ and the right probability is $\dfrac{1}{2}$. The limit is $(\frac{1}{2}, \frac{1}{2})$.

In I_2^1 the left probability is $\dfrac{(1-\frac{1}{n})(1-\frac{2}{n})}{(1-\frac{2}{n})(1-\frac{1}{n})+(\frac{1}{n})^2}$ and the right probability is $\dfrac{(\frac{1}{n})^2}{(1-\frac{1}{n})(1-\frac{2}{n})+(\frac{1}{n})^2}$. The limit is $(1, 0)$.

In I_2^2 the left probability is $\dfrac{(1-\frac{2}{n})\frac{1}{n}}{(1-\frac{2}{n})\frac{1}{n}+(1-\frac{1}{n})\frac{1}{n}}$ and the right probability is $\dfrac{(1-\frac{1}{n})\frac{1}{n}}{(1-\frac{2}{n})\frac{1}{n}+(1-\frac{1}{n})(\frac{1}{n})}$. The limit is $(\dfrac{1}{2},\dfrac{1}{2})$.

In I_2^3 the left probability is $\dfrac{(1-\frac{1}{n})\frac{1}{n}}{(1-\frac{1}{n})\frac{1}{n}+(\frac{1}{n})^2}$ and the right probability is $\dfrac{(\frac{1}{n})^2}{(1-\frac{1}{n})\frac{1}{n}+(\frac{1}{n})^2}$. The limit is $(1,0)$.

In I_2^4 the left probability is $\dfrac{(\frac{1}{n})^2}{(1-\frac{1}{n})\frac{1}{n}+(\frac{1}{n})^2}$ and the right probability is $\dfrac{(1-\frac{1}{n})\frac{1}{n}}{(1-\frac{1}{n})\frac{1}{n}+(\frac{1}{n})^2}$. The limit is $(0,1)$.

In I_2^5 the left probability is $\dfrac{(\frac{1}{n})^2}{(1-\frac{2}{n})\frac{1}{n}+(\frac{1}{n})^2}$ and the right probability is $\dfrac{(1-\frac{2}{n})\frac{1}{n}}{(1-\frac{2}{n})\frac{1}{n}+(\frac{1}{n})^2}$. The limit is $(0,1)$.

In I_2^6 the left probability is $\dfrac{(1-\frac{1}{n})\frac{1}{n}}{(1-\frac{1}{n})\frac{1}{n}+(1-\frac{1}{n})(1-\frac{2}{n})}$ and the right probability is $\dfrac{(1-\frac{1}{n})(1-\frac{2}{n})}{(1-\frac{1}{n})\frac{1}{n}+(1-\frac{1}{n})(1-\frac{2}{n})}$. The limit is $(0,1)$.

The belief attached to each singleton is that it has been reached with probability 1.

The limits of the sequence of strategies and beliefs confirm a particular Bayesian equilibrium as a sequential one. In an analogous manner, sequential equilibria can also be obtained for the models analyzed in the previous sections.

8 Concluding remarks

As we have already emphasized in Glycopantis, Muir, and Yannelis (2001), we consider the area of incomplete and differential information and its modelling important for the development of economic theory. We believe that the introduction of game trees, which gives a dynamic dimension to the analysis, helps in the development of ideas.

The discussion in that paper is in the context of one-good examples without free disposal. The conclusion was that core notions which may not be CBIC, such as the weak fine core, cannot easily be supported as a PBE. On the other hand, in the presence of an agent with superior information, the private core which is CBIC can be supported as a PBE. The discussion provided a noncooperative interpretation or foundation of the private core while making, through the game tree, the individual decisions transparent. In this way a better understanding of how incentive compatible contracts are formed is obtained.

In the present paper we investigate, in a one-good, two-agent economy, with and without free disposal, the implementation of private core, of Radner equilibrium, of weak fine core and weak fine values allocations. We obtain, through the construction of a tree with reasonable rules, that free disposal private core allocations, to which also the Radner equilibrium belongs, are not implementable. A brief comparison of the idea of CBIC in the static presentation with the case when the analysis is in terms of game trees is made.

It is surprising that free disposal destroys incentive compatibility and creates problems for implementation. Implementation in this case can be achieved by invoking an exogenous third party which can be thought of as a court that penalizes lying agents. It is of course possible that rational agents, once they realize that they can be cheated, might decide not to trade rather than rely on a third party which has to prove that he has perfect knowledge and can execute the correct trades. Notice that the third, exogenous party, in this case the court, plays the role of the mechanism designer in the relevant implementation literature (see Hahn and Yannelis, 2001, and the references there).

Similarly, implementation of a private core allocation becomes possible through the introduction of an endogenous third party with zero endowments but with superior information. In this case the third party is part of the model, i.e. an agent whose superior information allows him to play the role of an intermediary. The analysis overlaps with the one in Glycopantis, Muir, and Yannelis (2001). On the other hand we show here that implementation can also be achieved through a sequential equilibrium. It should be noted that the endogenous third agent is rewarded for his superior information by receiving consumption in a particular state, in spite of the fact that he has zero initial endowments in each state. However, both Radner equilibria and REE would not recognize a special role to such an agent. These Walrasian type notions would award to him zero consumption in all states of nature.

In summary, the analysis here considers the relation between, cooperative and noncooperative, static equilibrium concepts and noncooperative, game theoretic dynamic processes in the form of game trees. We have examined the possible support and implementation as perfect Bayesian equilibria of the cooperative concepts of the private core and the weak fine core, and the noncooperative generalized, Walrasian type equilibrium notions of Radner equilibrium and REE. In effect what we are doing is to look directly into the Bayesian incentive compatibility of the corresponding allocations, as if they were contracts, and then consider their implementability.

Appendix I: A note on PBE

In this note we look briefly at equilibrium notions when sequential decisions are taken by the players, i.e. in the context of game trees. For strategies we shall employ the idea of a *behavioral strategy* for a player being an assignment to each of his information sets of a probability distribution over the options available from that set. For a game of perfect recall, Kuhn (1953) shows that analysis of the game in terms of behavioral strategies is equivalent to that in terms of, the more familiar, mixed strategies. In any case, behavioral strategies are more natural to employ with an extensive form game. Sometimes we shall refer to them simply as *strategies*.

Consider an extensive form game and a given profile of behavioral strategies

$$s = \{s_i : i \in I\}$$

where I is the set of players.

When s is used each node of the tree is reached with probability obtained by producting the option probabilities given by s along the path leading to that node. In particular, there is a probability distribution over the set of terminal nodes so the expected payoff E_i to each player Pi may be expressed in terms of option probabilities from each information set.

Consider any single information set J owned by Pi, with corresponding option probabilities $(1-\pi_J, \pi_J)$, where for simplicity of notation we assume binary choice. The dependence of E_i on π_J is determined only by the paths which pass through J. Taking any one of these paths, on the assumption that the game is of perfect recall, the term it contributes to E_i will only involve π_J once in the corresponding product of probabilities. Thus, on summing over all such paths, the dependence of E_i on π_J is seen to be linear, with coefficients depending on the remaining components of s.

This allows the formation of a reaction function expressing π_J in terms of the remaining option probabilities, by optimizing π_J while holding the other probabilities constant; hence the Nash equilibria are obtained, as usual, as simultaneous solutions of all these functional relations. We are here adopting an *agent* form for a player, where optimization with respect to each of his decisions is done independently from all the others. A solution is guaranteed by the usual proof of existence for Nash equilibria.

For example, consider the tree in Figure 4, denoting the option probabilities from I_1, I_2 by $(1-\alpha, \alpha), (1-\beta, \beta)$ respectively. The payoff functions are then (apart from the factor $\frac{1}{3}$ expressing the probability of Nature's choice, and leaving out terms not involving α which come from paths not passing through I_1, I_2

$$\begin{aligned} E_1 &= 5(1-\alpha)(1-\beta) + 5(1-\alpha)\beta \\ &\quad +4\alpha(1-\beta) + 4\alpha\beta + 2.5(1-\alpha) + 4\alpha + \dots \\ &= 7.5 + 0.5\alpha + \dots ; \\ E_2 &= 5(1-\alpha)(1-\beta) + 4(1-\alpha)\beta \\ &\quad +5\alpha(1-\beta) + 4\alpha\beta + 2.5(1-\beta) + 4\beta + \dots \\ &= 7.5 + 0.5\beta + \dots . \end{aligned}$$

Since the coefficient of α in E_1 is positive, the optimal choice of α, i.e. the reaction function of Agent 1 is 1. Similarly for β in E_2 we obtain the value 1, and this is the reaction function of Agent 2.

Note that in any such calculation, only the coefficient of each π_J is important for the optimization – the rest of E_i is irrelevant. We may similarly treat the 21 option probabilities in Figure 7, obtaining 21 conditions which they must satisfy. These are quite complex and there are, probably, many solutions but it may be checked that the one given satisfies all conditions.

When an equilibrium profile is used, it is possible that some nodes are visited with zero probability. This means that the restriction of the strategy profile to subsequent nodes has no effect on the expected payoffs, so may be chosen arbitrarily. To eliminate this redundancy in the set of Nash equilibria, a refinement of the equilibrium concept to that of perfect equilibrium, was introduced for games of perfect information – that is, games in which each information set is a singleton. This requires an equilibrium strategy also to be a Nash equilibrium for *all* sub-games of the given game. In other words, the strategy profile should be a Nash equilibrium for the game which might be started from any node of the given tree, not just the nodes actually visited in the full game.

Any attempt to extend this notion to general games encounters the problem that sub-trees might start from nodes which are not in singleton information sets. In such a case, the player who must move first cannot know for certain at which node he is located within that set. He can only proceed if he adopts *beliefs* about where he might be, in the form of a probability distribution over the nodes of the information set. Moreover, these beliefs must be common knowledge, for the other players to be able to respond appropriately, so the desired extension of the equilibrium concept must take into account both strategies and beliefs of the players. The game will be played from any information set as if the belief probabilities had been realised by an act of nature.

We need, therefore, to consider pairs (s, μ), consisting of a behavioral strategy profile s and a belief profile

$$\mu = \{\mu_J : J \in \mathcal{J}\}.$$

Here, $\mathcal{J}$ denotes the set of information sets and μ_J is a probability distribution over the nodes of information set J, expressing the beliefs of the player who might be required to play from that set. Given the belief profile, we then require that the strategy profile give a perfect equilibrium, in the sense of being optimal for each player starting from every information set. But we need also to consider the source of the beliefs.

Given any behavioral strategy profile s denote the probability of reaching any node a, using s, by $\nu(a)$. Consider first an information set, J, not all of whose nodes are visited with zero probability when using s. We may calculate the conditional probability of being at a node $a \in J$ given that it is in J by

$$\nu(a|J) = \frac{\nu(\{a\} \cap J)}{\nu(J)} = \frac{\nu(a)}{\nu(J)}$$

since $a \in J \Rightarrow \{a\} \cap J = \{a\}$. Thus the belief probabilities $\mu_J(a) = \nu(a|J)$ for J are just the relative probabilities of reaching the nodes of J.

For example, returning to Figure 4 and employing the only Nash solution $\alpha = \beta = 1$ noted above, the probabilities of reaching the nodes of I_2 are $0, \frac{1}{3}, \frac{1}{3}$ which relativises, given the condition that we reach I, to $0, \frac{1}{2}, \frac{1}{2}$ as stated.

Thus for a *PBE*, the behavioral strategy-belief profile pair (s, μ) should satisfy two conditions:

(i) For the given belief profile μ, the strategy profile s should be a perfect equilibrium, as defined above;

(ii) For the given strategy profile s, the belief profile μ should be calculated at each information set for which $\nu(I) \neq 0$ by the formula above.

Justifications of the concept of perfect equilibrium in games of perfect information will argue that the players need to have good strategies to employ, even were something to go wrong with the intended play so that the game accidentally enters sub-trees which ought not to be accessed. One way to argue this is through the notion of a trembling hand which makes errors, so possibly choosing the wrong move. Employing this same idea in the context of perfect Bayesian equilibria, we can allow small perturbations in the strategies, such that all information sets are visited with non-zero probability. Then the relation determining beliefs from strategies is well posed and we may consider only beliefs which arise as limiting cases of such perturbations. This more restrictive definition of equilibrium is called a *sequential equilibrium.*

References

Allen, B.: Generic existence of completely revealing equilibria with uncertainty, when prices convey information. Econometrica **49**, 1173–1199 (1986)

Allen, B., Yannelis, N. C.: Differential information economies. Economic Theory **18**, 263–273 (2001)

Aumann, R. J.: Correlated equilibria as an expression of Bayesian rationality. Econometrica **55**, 1–18 (1987)

Emmons, D., Scafuri, A. J.: Value allocations – An exposition. In: Aliprantis, C. D., Burkinshaw, O., Rothman, N. J. (eds.) Advances in economic theory. Lecture notes in economics and mathematical systems, pp. 55–78. Berlin Heidelberg New York: Springer 1985

Einy, E., Moreno, D., Shitovitz, B.: Rational expectations equilibria and the ex post core of an economy with asymmetric information. Journal of Mathematical Economics **34**, 527–535 (2000)

Einy, E., Moreno, D., Shitovitz, B.: Competitive and core allocations in large economies with differential information. Economic Theory **34**, 321–332 (2001)

Glycopantis, D., Muir, A., Yannelis, N. C.: An extensive form interpretation of the private core. Economic Theory **18**, 293–319 (2001)

Glycopantis, D., Muir, A., Yannelis, N. C.: On the extensive form implementation of REE. Mimeo (2002)

Hahn, G., Yannelis, N. C.: Efficiency and incentive compatibility in differential information economies. Economic Theory **10**, 383–411 (1997)

Hahn, G., Yannelis, N. C.: Coalitional Bayesian Nash implementation in differential information economies. Economic Theory **18**, 485–509 (2001)

Koutsougeras, L., Yannelis, N. C.: Incentive compatibility and information superiority of the core of an economy with differential information. Economic Theory **3**, 195–216 (1993)

Krasa, S., Yannelis, N. C.: The value allocation of an economy with differential information. Econometrica **62**, 881–900 (1994)

Kreps, M. D., Wilson, R.: Sequential equilibrium. Econometrica **50**, 889–904 (1982)

Kuhn, H. W.: Extensive games and the problem of information. In: Kuhn, H. W., Tucker, A. W. (eds.) Contributions to the theory of games, Vol. II. Annals of mathematical studies, Vol. 28, pp. 193–216. Princeton, NJ: Princeton University Press 1953

Kurz, M.: On rational belief equilibria. Economic Theory **4**, 859–876 (1994)

Radner, R.: Competitive equilibrium under uncertainty. Econometrica **36**, 31–58 (1968)

Scafuri, A. J., Yannelis, N. C.: Non-symmetric cardinal value allocations. Econometrica **52**, 1365–1368 (1984)

Shapley, L. S.: A value for n-person games. In: Kuhn, H. W., Tucker, A. W. (eds.) Contributions to the theory of games, Vol. II. Annals of mathematical studies, Vol. 28, pp. 307–317. Princeton, NJ: Princeton University Press 1953

Tirole, J.: The theory of industrial organization. Cambridge, MA London: The MIT Press 1988

Wilson, R.: Information, efficiency, and the core of an economy. Econometrica **46**, 807–816 (1978)

Yannelis, N. C.: The core of an economy with differential information. Economic Theory **1**, 183–198 (1991)